# TAURINE 3

Cellular and Regulatory Mechanisms

# ADVANCES IN EXPERIMENTAL MEDICINE AND BIOLOGY



Recent Volumes in this Series

Volume 433
RECENT ADVANCES IN PROSTAGLANDIN, THROMBOXANE, AND LEUKOTRIENE RESEARCH
Edited by Helmut Sinzinger, Bengt Samuelsson, John R. Vane, Rodolfo Paoletti, Peter Ramwell, and Patrick Y-K Wong

Volume 434
PROCESS-INDUCED CHEMICAL CHANGES IN FOOD
Edited by Fereidoon Shahidi, Chi-Tang Ho, and Nguyen van Chuyen

Volume 435
GLYCOIMMUNOLOGY 2
Edited by John S. Axford

Volume 436
ASPARTIC PROTEINASES: Retroviral and Cellular Enzymes
Edited by Michael N. G. James

Volume 437
DRUGS OF ABUSE, IMMUNOMODULATION, AND AIDS
Edited by Herman Friedman, John J. Madden, and Thomas W. Klein

Volume 438
LACRIMAL GLAND, TEAR FILM, AND DRY EYE SYNDROMES 2: Basic Science and Clinical Relevance
Edited by David A. Sullivan, Darlene A. Dartt, and Michele A. Meneray

Volume 439
FLAVONOIDS IN THE LIVING SYSTEM
Edited by John Manthey and Béla Buslig

Volume 440
CORONAVIRUSES AND ARTERIVIRUSES
Edited by Luis Enjuanes, Stuart G. Siddell, and Willy Spaan

Volume 441
SKELETAL MUSCLE METABOLISM IN EXERCISE AND DIABETES
Edited by Erik A. Richter, Henrik Galbo, Bengt Saltin, and Bente Kiens

Volume 442
TAURINE 3: Cellular and Regulatory Mechanisms
Edited by Stephen Schaffer, John B. Lombardini, and Ryan J. Huxtable


A Continuation Order Plan is available for this series. A continuation order will bring delivery of each new volume immediately upon publication. Volumes are billed only upon actual shipment. For further information please contact the publisher.

# TAURINE 3

## Cellular and Regulatory Mechanisms

Edited by

Stephen Schaffer

University of South Alabama School of Medicine
Mobile, Alabama

John B. Lombardini

Texas Tech University Health Sciences Center
Lubbock, Texas

and

Ryan J. Huxtable

University of Arizona College of Medicine
Tucson, Arizona

PLENUM PRESS • NEW YORK AND LONDON

Library of Congress Cataloging in Publication Data

Taurine 3: cellular and regulatory mechanisms / edited by Stephen Schaffer, John B. Lombardini, and Ryan J. Huxtable.
p. cm.—(Advances in experimental medicine and biology; 442)
"Proceedings of the International Taurine Symposium '97: Cellular and Regulatory Mechanisms, held July 15–19, 1997, in Tucson, Arizona"—T.p. verso.
Includes bibliographical references and index.
ISBN 0-306-45914-0
1. Taurine—Physiological effect—Congresses. I. Schaffer, Stephen. II. Lombardini, J. Barry. III. Huxtable, Ryan J. IV. International Taurine Symposium '97: Cellular and Regulatory Mechanisms: 1997: Tucson, Ariz. V. Series.
QP801.T3T373 1998
612'.0157—dc21 98-7494
CIP

Proceedings of the International Taurine Symposium '97: Cellular and Regulatory Mechanisms, held July 15–19, 1997, in Tucson, Arizona

ISBN 0-306-45914-0


A Division of Plenum Publishing Corporation
233 Spring Street, New York, N.Y. 10013

http://www.plenum.com

10 9 8 7 6 5 4 3 2 1



Printed in the United States of America

## CONTRIBUTORS

J. Alberti
Departments of Immunology and Developmental Biochemistry
NY State Institute for Basic Research in Developmental Disabilites
Staten Island, NY 10314

F.J. Alonso
Servicio de Neurobiología
Depto. de Investigación
Hospital Ramón y Cajal
Crta Colmenar Km 9, Spain
28034 Madrid, Spain

F. Amicarelli
Dipartimento di Biologia di Base ed Applicata
Università de L'Aquila
L'Aquila, Italy

J. Azuma
Department of Clinical Evaluation of Medicines and Therapeutics
Faculty of Pharmaceutical Science
Osaka University
Osaka, Japan

M. Azuma
Department of Clinical Evaluation of Medicines and Therapeutics
Faculty of Pharmaceutical Science
Osaka University
Osaka, Japan

A. Baba
Department of Pharmacology
Faculty of Pharmaceutical Science
Osaka University
Osaka, Japan

C. Ballard-Croft
Department of Pharmacology
University of South Alabama School of Medicine
Mobile, AL 36688

D. Baker
Department of Pharmacodynamics
College of Pharmacy
JHMHC Box 100487
University of Florida
Gainesville, FL 32610

L. Bianchi
Dipartimento di Farmacologia Preclinica e Clinica "M. Aiazzi Mancini"
Università degli Studi di Firenze
Vle G.G. Morgagni 65
5014 Florence, Italy

M. Bitoun
INSERUM U 433
Faculté de Médecine Laennec
F-69372 Lyon, France

J.P. Bolam
Istituto di Scienze Farmacologiche
Università degli Studi di Siena
Via E.S. Piccolomini 170
53100 Siena, Italy

J. Bouchier-Hayes
Royal College of Surgeons in Ireland
Department of Surgery
Beaumont Hospital
Dublin 9, Ireland

K. Brendel
Department of Pharmacology
University of Arizona
College of Medicine
Tucson, AZ 85724

A.M. Budreau
Department of Pediatrics
University of Tennessee and the Crippled
Children's Foundation Research
Center at Le Bonheur
Children's Medical Center
Memphis, TN 38103

J. Bustamante
Depto. Fisiologia
Facultad de Medicina
Universidad Complutense
28040 Madrid, Spain

C. Camerino
Department of Pharmacology
College of Medicine
University of Arizona
Tucson, AZ 85724

A. Cantafora
Instituto Superiore di Sanitá
Laboratorio di Metabolismo e
Biochimica Patologica
Rome, Italy

K.J. Chang
Department of Food and Nutrition
Inha University
Inchon, 402-751, Korea

Y.S. Chang
Department of Nuclear Medicine
College of Medicine
Seoul National University
Seoul, 110-744, Korea

X.-C. Chen
Institute of Nutrition and Food Hygiene
CAPM
29 Nan Wei Road
Beijing, 100050, China

R.W. Chesney
Department of Pediatrics
University of Tennessee and the Crippled
Children's Foundation Research
Center at Le Bonheur
Children's Medical Center
Memphis, TN 38103

K.-H. Cho
Department of Food Science and Nutrition
Dankook University
Seoul, 140-714, Korea

M. Christensen
Departments of Pediatrics and Clinical
Pharmacy
The Pediatric Pharmacology Research Unit
and Crippled Children's Foundation
Research Center at Le Bonheur
Children's Medical Center
The University of Tennessee College of
Pharmacy
Memphis, TN 38103

J.-K. Chung
Department of Nuclear Medicine
College of Medicine
Seoul National University
Seoul, 110-744, Korea

M.A. Colivicchi
Dipartimento di Farmacologia Preclinica e
Clinica "M. Aiazzi Mancini"
Università degli Studi di Firenze
Vle G.G. Morgagni 65
50134 Florence, Italy

D. Conte-Camerino
Unit of Pharmacology
Department of Pharmacobiology
Faculty of Pharmacy
University of Bari
Via Orabona 4
70125 Bari, Italy

M. Costa
Dipartimento di Scienze Biochimiche
Centro di Biologia Molecolare del C.N.R.
Universitá di Roma "La Spienza
Rome, Italy

S. Cubillos
Laboratorio de Neuroquímica
Centro de Biofísica y Bioquímica
Instituto Venezolano de Investigaciones Científicas
Apdo. 21827
Caracas 1020-A, Venezuela

W.E. Davies
Department of Pharmacology
Medical School
University of Birmingham
Birmingham, England

R. Dawson, Jr.
Department of Pharmacodynamics College of Pharmacy
University of Florida
JHMHC Box 100487
Gainesville, FL 32610

A. De Luca
Unit of Pharmacology
Department of Pharmacobiology
Faculty of Pharmacy
University of Bari
Via Orabona 4, 70125
Bari, Italy

L. Della Corte
Dipartimento di Farmacologia Preclinica e
Clinica "M. Aiazzi Mancini"
Università degli Studi di Firenze
Vle G.B. Morgagni 65
50134 Florence, Italy

N. Del Olmo
Servicio de Neurobiología
Depto. de Investigación
Hospital Ramón y Cajal
28034 Madrid, Spain

S.K. Dhillon
Hearing Services Centre
City Hospital Birmingham
Birmingham, England

R. Dumas
New York State Institute for Basic Research and Developmental Disabilities
1050 Forest Hill Road
Staten Island, NY 10314

S. Dupre
Dipartimento di Scienze Biochimiche Centro di Biologia Molecolare del C.N.R.
Universitá di Roma "La Spienza"
Rome, Italy

A. El Idrissi
New York State Institute for Basic Research and Developmental Disabilities
1050 Forest Hill Road
Staten Island, NY 10314

B. Eppler
Department of Pharmacodynamics College of Pharmacy
University of Florida
JHMHC Box 100487
Gainesville, FL 23610

J.A. Evers
Department of Emergency Medicine
Wright State University
Dayton, OH 45401

M.D. Faiman
Department of Pharmacology and Toxicology
Univeristy of Kansas
Lawrence, KS 66045

M. Frosini
Istituto di Scienze Farmacologiche
Università di Siena
Siena, Italy

F. Fusi
Istituto di Scienze Farmacologiche
Università di Siena
Siena, Italy

F. Galeffi
Dipartimento di Farmacologia Preclinica e Clinica "M. Aiazzi Mancini"
Università degli Studi di Firenze
Vle G.B. Morgagni 65
50134 Florence, Italy

E. Giné
Depto. Bioquìmica
Facultad de Medicina
Universidad Complutense
28040 Madrid, Spain

C. Giovannini
Instituto Superiore di Sanitá
Laboratorio di Metabolismo e Biochimica Patologica
Rome, Italy

P. Green
Department of Pharmacodynamics
University of Florida
Gainesville, FL 32610

X. Han
Department of Pediatrics
University of Tennessee and the Crippled Children's Foundation Research Center at Le Bonheur Children's Medical Center
Memphis, TN 38103

J.E. Haroutounian
Buniatian Institute of Biochemistry
National Academy of Sciences of Armenia
Sevag St. 5/1
Yerevan 14, Armenia

C. Harris
CSI/IBR Center for Developmental Neuroscience and Developmental Disabilities
2800 Victory Blvd.
Staten Island, NY 10314

S.J. Hasal
Department of Pharmacology
University of Arizona
College of Medicine
Tucson, AZ 85724

T. Hashimoto
Department of Pharmacology
Kyoto Prefectural University of Medicine
Kawaramachi-Hirokoji
Kamikyo-ku
Kyoto 602, Japan

R.A. Helms
Departments of Pediatrics and Clinical Pharmacy
The Pediatric Pharmacology Research Unit and Crippled Children's Foundation Research Center at Le Bonheur Children's Medical Center
The University of Tennessee College of Pharmacy
Memphis, TN 38103

H. Hern
Department of Pharmacodynamics College of Pharmacy
University of Florida
JHMHC Box 100487
Gainesville, FL 32610

P.C. Hopkins
Department of Pharmacology
Medical School
University of Birmingham
Birmingham, England

M. Hu
Department of Pharmacodynamics College of Pharmacy
University of Florida
JHMHC Box 100487
Gainesville, FL 32610

R.J. Huxtable
Department of Pharmacology
University of Arizona
College of Medicine
Tucson, AZ 85724

H. Imaki
Department of Developmental Biochemistry
Institute for Basic Research in Developmental Disabilities
1050 Forest Hill Road
Staten Island, NY 10314

W. Isselhard
Institute for Experimental Medicine
University of Cologne
Robert-Koch-Str. 10
50931 Cologne, Germany

E. Jaffé
Laboratorio de Neuroquímica
Centro de Biofísica y Bioquímica
Instituto Venezolano de Investigaciones
Científicas
Apdo. 21827
Caracas 1020-A, Venezuela

J.M. Jeong
Department of Nuclear Medicine
College of Medicine
Seoul National University
Seoul, 110-744, Korea

J. Jia
Institute of Nutrition and Food Hygiene
Chinese Academy of Preventive Medicine
Beijing 100050, China

L.R. Jones
Indiana University School of Medicine
Indianapolis, IN 46202

H.K. Kang
Department of Pharmacology
Seoul National University
College of Medicine
Chongno-gu
Seoul, 110-799, Korea

B.K. Kim
Department of Microbial Chemistry
College of Pharmacy
Seoul National University
Seoul, 151-742, Korea

B.-K. Kim
Department of Microbial Chemistry
College of Pharmacy
Seoul National University
Seoul, 110-744, Korea

C. Kim
Indiana University School of Medicine
Herman B Wells Center for
Pediatric Research
1044 W. Walnut
Indianapolis, IN 46202

E.-S. Kim
Department of Food Science and Nutrition
Dankook University
Seoul, 140-714, Korea

H. Kim
Department of Food and Nutrition
Seoul National University
Seoul, 151-742, Korea

H.W. Kim
Department of Life Science
The Seoul City University of Seoul
Seoul, 130-743, Korea

J.-S. Kim
Department of Food Science and Nutrition
Dankook University
Seoul, 140-714, Korea

W.B. Kim
Research Laboratories of Dong-A
Pharmaceutcal Co., Ltd.
Yongin-shi
Kyunggi-do 449-900, Korea

J. Ko
Research Laboratories of Dong-A
Pharmaceutcal Co., Ltd.
Yongin-shi
Kyunggi-do 449-900, Korea

H. Kodama
Department of Chemistry
Kochi Medical School
Okocho, Nankoku
Kochi 783, Japan

C.-S. Koh
Department of Nuclear Medicine
College of Medicine
Seoul National University
Seoul, 110-744, Korea

E. Kriegesmann
Institute for Experimental Medicine
University of Cologne
Robert-Koch-Str. 10
D- 50924 Cologne, Germany

K. Kuriyama
Department of Pharmacology
Kyoto Prefectural University of Medicine
Kawaramachi-Hirokoji
Kamikyo-ku
Kyoto 602, Japan

I.H. Lambert
The August Krogh Institute
Biochemical Department
University of Copenhagen
Universitetsparken 13, DK-2100
Copenhagen Ø, Denmark

C.A. Lau-Cam
St. John's University
College of Pharmacy and Allied Health Professions
8000 Utopia Parkway
Jamaica, NY 11439

R.O. Law
Department of Cell Physiology and Pharmacology
University of Leicester
Leicester Le 1 9HN, England

E.J. Lee
Department of Microbial Chemistry
College of Pharmacy
Seoul National University
Seoul 151-742, Korea

K. Lee
Department of Food and Nutrition
Yonsei University
134 Shinchon-dong, Sudaemun-ku
Seoul, 120-749, Korea

K.-H. Lee
Department of Food Science and Nutrition
Dankook University
Seoul, 140-714, Korea

M.C. Lee
Department of Nuclear Medicine College of Medicine
Seoul National University
Seoul, 110-744, Korea

Y.J. Lee
Department of Nuclear Medicine College of Medicine
Seoul National University
Seoul, 110-744, Korea

C. Licht
Department of Pediatrics
University of Cologne
Joseph-Stelzmann-Str. 9
D - 50924 Cologne, Germany

E. Lim
Department of Food and Nutrition
Seoul National University
Seoul, 151-742, Korea

L. Lima
Laboratorio de Neuroquímica
Centro de Biofísica y Bioquímica
Instituto Venezolano de Investigaciones Científicas
Apdo. 21827
Caracas 1020-A, Venezuela

Y. Lin
Institute of Nutrition and Food Hygiene
Chinese Academy of Preventive Medicine
Beijing 100050, China

D.-S. Liu
Institute of Nutrition and Food Hygiene
CAPM
29 Nan Wei Road
Beijing 100050, China

M.V.T. Lobo
Servicio de Neurobiología
Depto. de Investigación
Hospital Ramón y Cajal
Crta Colmenar Km 9
28034 Madrid, Spain

J.B. Lombardini
Departments of Pharmacology,
Ophthalmology and Visual Sciences
Texas Tech University
Health Sciences Center
Lubbock, TX 79430

A. Macone
Dipartimento di Scienze Biochimiche
Centro di Biologia Molecolare del C.N.R.
Universitá di Roma "La Spienza
Rome, Italy

R. Martin del Rìo
Servicio de Neurobiología
Depto. de Investigación
Hospital Ramón y Cajal
Crta Colmenar Km 9
28034 Madrid, Spain

R. Masella
Instituto Superiore di Sanitá
Laboratorio di Metabolismo e
Biochimica Patologica
Rome, Italy

J. Messing
Department of Developmental Biochemistry
Institute for Basic Research in
Developmental Disabilities
1050 Forest Hill Road
Staten Island, NY 10314

D.V. Michalk
Department of Pediatrics
University of Cologne
Joseph-Stelzmann-Str. 9
D - 50924 Cologne, Germany

J.D. Militante
Departments of Pharmacology
Texas Tech University
Health Sciences Center
Lubbock, TX 79430

T. Minor
Institute for Experimental Medicine
University of Cologne
Robert-Koch-Str. 10
50931 Cologne, Germany

H. Mochizuki
School of Food and Nutritional Sciences
The University of Shizuoka
52-1 Yada
Shizuoka 422, Japan

D. Modesti
Istituto Superiore de Sanità
Laboratorio di Metabolismo e Biochimica
Patologica
Rome, Italy

J.R. Moorman
University of Virginia
Charlottesville, VA

J. Moràn
Institute of Cell Physiology
National University of Mexico
Mexico City, Mexico

N. Munoz
Department of Physiology and Biophysics
Wright State University
Dayton, OH 45401

F. Niessen
Institute for Experimental Medicine
University of Cologne
Robert-Koch-Str. 10
50931 Cologne, Germany

H. Oda
Department of Applied Biological Sciences
Nagoya University
Nagoya 464-01, Japan

S.S. Oja
Tampere Brain Research Center
University of Tampere Medical School
Box 607
FIN-33101 Tampere, Finland

J.E. Olson
Department of Emergency Medicine
Department of Physiology and Biophysics
Wright State University
Dayton, OH 45401

J. Owens
Department of Pharmacology and Physiology
Oklahoma State Univeristy
Tulsa, OK 74107

M. Palmi
Istituto di Scienze Farmacologiche
Università di Siena
Siena, Italy

Z.-L. Pan
Institute of Nutrition and Food Hygiene
CAPM
29 Nan Wei Road
Beijing 100050, China

E. Park
Departments of Immunology and Developmental Biochemistry
NY State Institute for Basic Research in Developmental Disabilities
Staten Island, NY 10314

S. Park
Department of Food and Nutrition
Seoul National University
Seoul, 151-742, Korea

T. Park
Department of Food and Nutrition
Yonsei University
134 Shinchon-dong, Sudaemun-Ku
Seoul, 120-749, Korea

H. Pasantes-Morales
Institute of Cell Physiology
National University of Mexico
Mexico City, Mexico

A.M. Petrosian
Buniatian Institute of Biochemistry
National Academy of Sciences of Armenia
Sevag St. 5/1
Yerevan 14, Armenia

S. Pierno
Unit of Pharmacology
Department of Pharmacobiology
Faculty of Pharmacy
University of Bari
Via Orabona 4
70125 Bari, Italy

G. Pitari
Dipartimento di Biologia di Base ed Applicata
Università de L'Aquila
L'Aquila, Italy

R.W. Putnam
Department of Physiology and Biophysics
Wright State University
Dayton, OH 45401

O. Quesada
Institute of Cell Physiology
National University of Mexico
Mexico City, Mexico

M.R. Quinn
Developmental Biochemistry
NY State Institute for Basic Research in Developmental Disabilities
Staten Island, NY 10314

A. Rabe
New York State Institute for Basic Research and Developmental Disabilities
1050 Forest Hill Road
Staten Island, NY 10314

M.R. Rao
Department of Cardiovascular Pharmacology
Nanjing Medical University
Nanjing, P.R. China

H.P. Redmond
Royal College of Surgeons in Ireland
Department of Surgery
Beaumont Hospital
Dublin 9, Ireland

I. Reymond
INSERUM U 433
Faculté de Médecine Laennec
F-69372 Lyon, France

S.J. Rose
Department of Pediatrics
Heartlands Hospital
Birmingham, England

P. Rossi
Dipartimento di Biologia di Base ed
Applicata
Università de L'Aquila
L'Aquila, Italy

P. Saransaari
Tampere Brain Research Center
University of Tampere Medical School
Box 607
FIN-33101 Tampere, Finland

H. Satoh
Department of Pharmacology
Division of Molecular and Cellular Biology
Nara Medical University
Kashihara, Nara 634, Japan

S.W. Schaffer
Department of Pharmacology
University of South Alabama
School of Medicine
Mobile, AL 36688

J.V. Schloss
Department of Medicinal Chemistry
University of Kansas
Lawrence, KS 66045

G. Schuller-Levis
Departments of Immunology and
Developmental Biochemistry
NY State Institute for Basic Research in
Developmental Disabilities
Staten Island, NY 10314

A. Sergeant
INSERUM U 412
Ecole Normale Supérieure
F-69634 Lyon, France

G.P. Sgaragli
Istituto di Scienze Farmacologiche
Università di Siena
Siena, Italy

D. Shih
Department of Pharmacodynamics
College of Pharmacy
University of Florida
JHMHC Box 100487
Gainesville, FL 32610

M.J. Shim
Department of Life Science
The Seoul City University of Seoul
Seoul 130-743, Korea

J.M. Solìs
Servicio de Neurobiología
Depto. de Investigación
Hospital Ramón y Cajal
Crta Colmenar Km 9, Spain
28034 Madrid, Spain

M. Son
Research Laboratories of Dong-A
Pharmaceutcal Co., Ltd.
Yongin-shi
Kyunggi-do 449-900, Korea

A. Spirito
Dipartimento di Scienze Biochimiche
Centro di Biologia Molecolare del C.N.R.
Universitá di Roma "La Spienza"
Rome, Italy

P.P. Stapleton
Royal College of Surgeons in Irland
Department of Surgery
Beaumont Hospital
Dublin 9, Ireland

J.A. Sturman
Department of Developmental Biochemistry
Institute for Basic Research in
Developmental Disabilities
1050 Forest Hill Road
Staten Island, NY 10314

K. Sugahara
Department of Chemistry
Kochi Medical School
Okocho, Nankoku
Kochi 783, Japan

Y. Sun
Department of Pharmacology
University of Arizona
College of Medicine
Tucson, AZ 85724

K. Takahahsi
Department of Clinical Evaluation of Medicines and Therapeutics
Faculty of Pharmaceutical Science
Osaka University
Osaka, Japan

Y. Tamari
Department of Chemistry
Konan University
Kobe 658, Japan

E. Tang
Department of Pharmacodynamics College of Pharmacy
University of Florida
JHMHC Box 100487
Gainesville, FL 23610

X.W. Tang
Department of Physiology and Cell Biology
University of Kansas
Lawrence, KS 66045

L. Tao
Department of Cardiovascular Pharmacology
Nanjing Medical University
Nanjing, P.R. China

M. Tappaz
INSERUM U 433
Faculté de Médecine Laennec
F-69372 Lyon, France

J. Tamarit-Rodriquez
Depto. Bioquìmica
Facultad de Medicina
Universidad Complutense
28040 Madrid, Spain

V. Theofanopoulos
St. John's University
College of Pharmacy and Allied Health Professions
8000 Utopia Parkway
Jamaica, NY 11439

K.F. Tipton
Department of Biochemistry
Trinity College
Dublin, Ireland

E. Trenkner
New York State Institute for Basic Research and Developmental Disabilities
1050 Forest Hill Road
Staten Island, NY 10314

D.R. Wallace
Department of pharmacology and Physiology
Oklahoma State University
Tulsa, OK 74107

X.B. Wang
Institute of Nutrition and Food Hygiene
Chinese Academy of Preventive Medicine
Beijing 10050, China

P. Wingenfeld
Department of Pediatrics
University of Cologne
Joseph-Stelzmann-Str. 9
D - 50924 Cologne, Germany

J.-Y. Wu
Department of Physiology and Cell Biology
University of Kansas
Lawrence, KS 66045

C.C. Yan
Department of Pharmacology
University of Arizona
College of Medicine
Tucson, AZ 85724

H. Yokogoshi
School of Food and Nutritional Sciences
The University of Shizuoka
52-1 Yada
Shizuoka 422, Japan

J.S. You
Department of Food and Nutrition
Inha University
Inchon, 402-751, Korea

G. Youmbi
Istituto di Scienze Farmacologiche
Università di Siena
Siena, Italy

X.H. Zhao
Institute of Nutrition and Food Hygiene
Chinese Academy of Preventive Medicine
Beijing 100050, China

# PREFACE

This volume is the selected, edited proceedings of the International Taurine Symposium held in Tucson, Arizona, in July 1997. The meeting was a satellite symposium of the 16th Biennial Meeting of the International Society for Neurochemistry, which was held in Boston immediately following the Tucson meeting. In view of the desert location of Tucson, the meeting was advertised tongue-in-cheek as being the hottest scientific meeting ever. As the weather lived up to its billing, the Symposium may well have earned the title.

The meeting was held in an atmospheric cluster of adobe buildings, old by the standards of the American southwest, at the Westward Look Resort in the Sonoran Desert foothills of Tucson, which is overlooked by the 9000' high Santa Catalina mountains. As is the norm for taurine symposia, participants formed a multinational group, with representatives from China, Korea, Japan, United States, Mexico, Venezuela, Ireland, England, Spain, Italy, Finland, France, Denmark, Germany, Norway and Armenia.

The meeting was organized around plenary lectures by Russell Chesney (University of Tennessee: *Taurine and infant nutrition*), Herminia Pasantes-Morales (National Autonomous University of Mexico: *Taurine: An osmolyte in mammalian tissues*) and Kinya Kuriyama (Kyoto Prefectural University of Medicine: *Interrelationship between taurine and GABA*).

A novel aspect of this meeting was a poster session highlighting student presentations from the University of Arizona. An International Awards Committee judged posters for scientific content and effectiveness of presentation. Three of the four awards were in the name of deceased scientists who made significant contributions to the taurine area and who had participated in former taurine symposia. The three scientists were Paul Mandel (University of Strasbourg), André Barbeau (Clinical Research Institute, Montreal) and Kanji Izumi (Kagoshima University). Reminiscences of these scientists were distributed at the awards ceremony. Award winners were Linda Tompkins, Jeremy Richman, Kim Mitchell and Yoshiaki Hosohata.

The various social activities included a southwestern banquet at the Arizona-Sonora Desert Museum and a classical music concert performed by Tannis Gibson (piano), Mark Rush (violin) and Judith Pearce (flute).

As always, we are indebted to the sponsoring organizations. These were: Taisho Pharmaceutical Co., Ltd., Red Bull GmbH, The International Society for Neurochemistry, Milupa, and Nestlé SA. The Departments of Pharmacology, Pharmacology and Toxicology, and Physiology at the University of Arizona helped contribute to the student awards.

The International Advisory Committee composed of W. Ewart Davies (England), Flavia Franconi (Italy), Byong-Kak Kim (Korea), and Dietrich V. Michalk (Germany) are acknowledged for their helpful contributions in the planning and organization of this international meeting.

I am profoundly grateful to J. Barry Lombardini and Steve Schaffer for carrying much of the burden of editing this volume. The advent of desktop computers has meant a shift of workload from publishers to editors. Consequentially, a volume such as this represents many hundreds of hours of editorial and secretarial work. The editors are grateful to Josie L. Aleman (Texas Tech University Health Sciences Center) for her retyping and formatting of manuscripts, and to Jane Ageton (University of Arizona) for her assistance in organizing the meeting. We also thank Ed French and Frank Porreca, who, together with myself, constituted the Local Organizing Committee.

Ryan J. Huxtable
Editor-in-Chief

CONTENTS

## Part 1: Chemistry, Biochemistry, and Metabolism of Taurine



## Part 2: Taurine and the Liver and Pancreas

## Part 5: Osmotic Actions of Taurine



## Part 6: Transport and General Studies on Taurine

**Part 7: Taurine and the Central Nervous System. I: The Brain**

# Part 1:
# Chemistry, Biochemistry, and Metabolism of Taurine

# HYPOTAURINE OXIDATION: AN HPLC-MASS APPROACH

Silvestro Dupré,[1] Alessandra Spirito,[1] Kazunori Sugahara,[2] and Hiroyuki Kodama[2]

[1] Dipartimento di Scienze Biochimiche and
Centro di Biologia Molecolare del C.N.R.
Università di Roma "La Sapienza"
Roma, Italy
[2] Department of Chemistry
Kochi Medical School
Okocho, Nankoku
Kochi 783, Japan

## INTRODUCTION

Hypotaurine reacts with various reactive oxygen species, although mainly with hydroxyl radicals. The compound *bis*-aminoethyl-α-disulfone ($H_2N-CH_2-CH_2-SO_2-SO_2-CH_2-CH_2-NH_2$) has been proposed as one of the intermediate products during the ultraviolet light-catalyzed oxidation of hypotaurine[13]. The disulfone has been proposed as a product of the hydroxyl radical quenching activity of hypotaurine[6] and has been demonstrated to be present *in vitro* after hypotaurine oxidation by Fenton's reagent or in the presence of enzymatically-generated hydroxyl radicals[7]. The sulfonyl radical $RSO_2\cdot$ has been hypothesized as a reaction intermediate[6,7]. In an attempt to understand in more detail the reaction of hypotaurine with Fenton's reagent, we studied this reaction by LC/APCI-MS which evaluates the time course of product formation and the production of intermediates. The presence of *bis*-aminoethyl-α-disulfone among the reaction products could not be demonstrated by this technique. Attempts to synthesize this compound by reported methods for the synthesis of other α-disulfones or by reaction involving chemically- or enzymatically-generated hydroxyl radicals were also unsuccessful.

## MATERIAL AND METHODS

### Chemicals

Hypotaurine, cystamine dihydrochloride, ferrous sulfate ($FeSO_4 \cdot 7H_2O$) and $H_2O_2$ were purchased from Sigma Chemical Co.

**Hypotaurine Oxidation**

Hypotaurine (10 μmol) was subjected to oxidation in the presence of $Fe^{++}$ (2.5 - 5 μmol $FeSO_4$) and 30 - 60 μmol $H_2O_2$ in water (1.0 ml final volume). The reaction course was followed by HPLC separation and monitored by mass spectrometry at the beginning of the reaction and after various reaction time intervals. Liquid chromatography-mass determination was performed with a Hitachi L-6200 high-performance liquid chromatography (HPLC) instrument, equipped with a 5 μm Inertsil ODS-2 column (150 cm x 4.6 mm i.d.) from Gasukuro Kogyo (Tokyo, Japan) and connected to a Hitachi M80B mass spectrometer/computer system through the APCI interface. The nebulizer and vaporizer temperatures were 255°C and 380°C respectively. Analyses were performed with a mobile phase of 100 mM $CH_3COONH_4$ : $CH_3CN$ (80:20 v/v) at a flow rate of 0.9 ml/min. Amino acid analysis was performed with a Hitachi L-8500 instrument using citrate buffer (0.068 M, pH 2.8).

## RESULTS AND DISCUSSION

Hypotaurine is reported to be oxidized by Fenton chemistry[6], an oxidation reaction involving hydroxyl radicals. The second-order rate constant for the reaction of hypotaurine and hydroxyl radicals[3] is $0.6 - 1.15 \cdot 10^{10}\ M^{-1}s^{-1}$. At millimolar concentrations of hypotaurine a rate of the order of $\mu Ms^{-1}$ is expected. Figure 1 depicts the time course of the reaction. Relative amounts of hypotaurine and taurine are reported which were calculated from mass spectra and from single ion monitoring by comparison with peak intensities of known amounts of pure compounds. The reaction is complete between 30 and 60 min and results

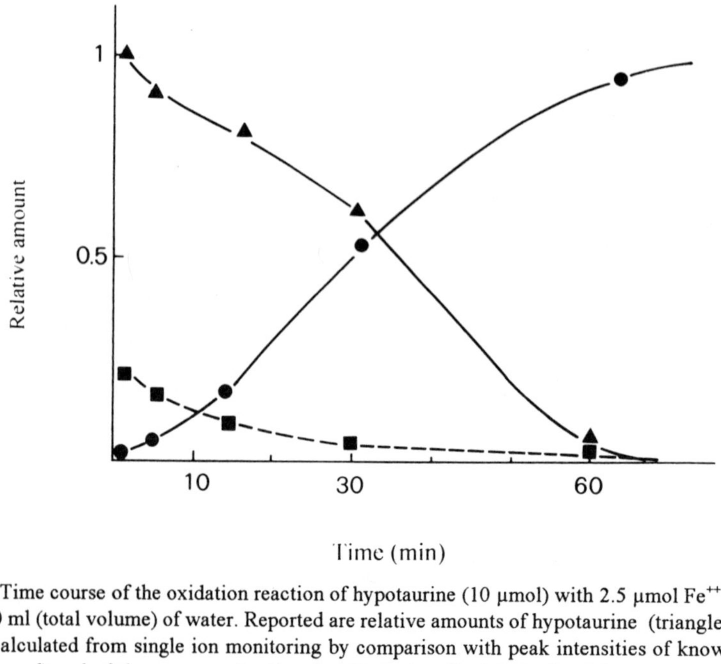


**Figure 1.** Time course of the oxidation reaction of hypotaurine (10 μmol) with 2.5 μmol $Fe^{++}$ and 60 μmol $H_2O_2$ in 1.0 ml (total volume) of water. Reported are relative amounts of hypotaurine (triangles) and taurine (circles) (calculated from single ion monitoring by comparison with peak intensities of known amounts of pure compound) and of the compound with m/z 153 (broken line) (calculated from mass spectra).

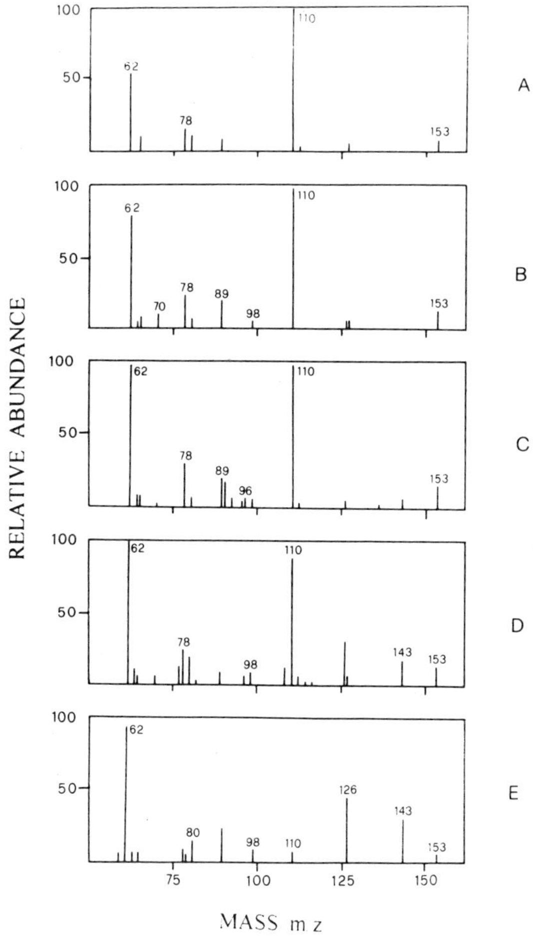


**Figure 2.** Mass spectra of the oxidation reaction. Oxidation conditions are as in Fig. 1. Total ion chromatography of the reaction mixture is shown in Fig. 3 (see TIC of Fig. 3). Mass spectra of samples [(32*43) - 18], scanned from 50 to 160, were recorded at time 0 (A), 10 min (B), 15 min (C), 30 min (D) and 60 min (E), respectively.

in the conversion of hypotaurine to taurine. The inspection of mass spectra taken at various times (Fig. 2) shows the presence of other compounds during the reaction and at the end of the reaction. A compound with m/z 62 is present in hypotaurine standard solutions[5] and is identified as protonated ethanolamine. A compound with m/z 153, present at a fairly low concentration until shortly before the end of the reaction (Fig. 1, broken line), is identified as a product ($H_2N$-$CH_2$-$CH_2$-$SO_2$-$CH_2$-$CH_2$-$NH_2$) of fragmentation ions[12]. Amino acid analysis (not shown) at the end of the reaction identifies taurine, low amounts of some other ninhydrin-positive compounds, and an appreciable amount of glycine. Glycine (m/z 76) was, however, not detected by mass analysis, and therefore an unidentified compound, running as glycine by ion-exchange chromatography, is present as a product of hypotaurine oxidation. Single-ion monitoring over the whole chromatographic run does not reveal formation of a any compound of molecular weight 216, corresponding to *bis*-aminoethyl-α-disulfone, nor the formation of other putative fragmentation derivatives of α-disulfone, such as the monoaminoethyl-α-disulfone fragment (m/z = 173) or partial

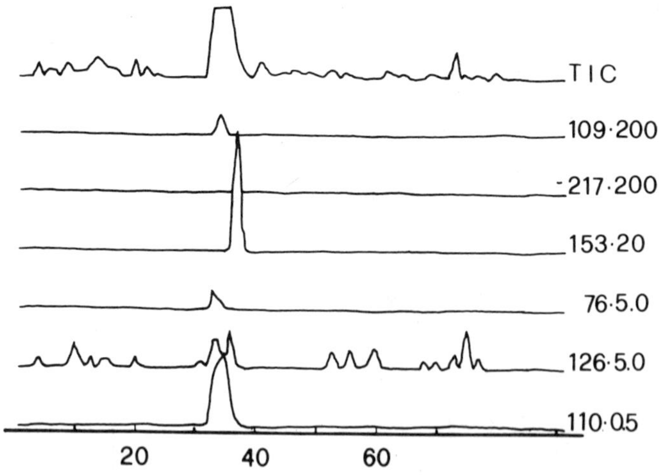


**Figure 3.** Total ion chromatography (TIC) and selected ion monitoring of the oxidation reaction. Oxidation conditions are as in Fig. 1. The recordings were taken 5 min after the initiation of the reaction's. Sensitivities of each recording are shown. Reported m/z values are: hypotaurine (110), taurine (126), glycine (76), $H_2N\text{-}CH_2\text{-}CH_2\text{-}SO_2\text{-}CH_2\text{-}CH_2\text{-}NH_2$ (153), *bis*-aminoethyl-α-disulfone (217), $H_2N\text{-}CH_2\text{-}CH_2\text{-}SO_2H^+$ (109).

or partial oxidation intermediates such as 2-aminoethyl 2-aminoethanethiolsulfonate (m/z = 185)[5]. By contrast utilizing the same experimental conditions Fellman et al.[7] observed the presence of *bis*-aminoethyl-α-disulfone shortly (probably a few minutes) after the beginning of the reaction of hypotaurine with Fenton's reagent. Figure 3 shows one of the single-ion monitoring records, taken 5 min after the initiation of the reaction. Ions with m/z 217 and 109 ($H_2N\text{-}CH_2\text{-}CH_2\text{-}SO_2H^+$) were monitored at the highest sensitivity allowed. If *bis*-aminoethyl-α-disulfone was present in our experiments, the quantity was below the detection limit of the techniques used in this study. Apparently the yield of *bis*-aminoethyl-α-disulfone is very low. Previously, this compound was accurately identified and characterized, but no yield of the chemical preparation was reported[7]. Alternatively, it is possible that the compound is highly unstable under the detection conditions of mass spectrometry and the molecular ion is undetectable by this analytical technique. The molecular ion of *bis*-aminoethyl-α-disulfone was also not detected by FAB-mass spectrometry[7]. However, the monoaminoethyl-α-disulfone fragment (m/z = 173) was present[7]. According to the literature[11], fragmentation of *bis*-aminoethyl-α-disulfone should have yielded $H_2N\text{-}CH_2\text{-}CH_2\text{-}SO_2^+$ (molecular weight 108) as the main peak, but this peak was not detected by mass spectrometry in our experiments.

Syntheses of various aliphatic α-disulfones have been reported[2,8,10,15]. A high yield of α-disulfone was also reported by sulfinate oxidation with acidic $KMnO_4$[1]. However, we were not able to obtain *bis*-aminoethyl-α-disulfone using this synthetic method. Also, repeated attempts to synthesize the disulfone using chemically- or enzymatically-generated hydroxyl radicals[7] were not successful in our laboratory. Therefore, we are not able to confirm the data presented in earlier reports.

Without pure samples of *bis*-aminoethyl-α-disulfone it is not possible to study the stability of this compound. Quantitative data on its stability are not available, although stability data for other α-disulfones are reported[2,16]; these compounds are stable in aqueous neutral solution. During the reported isolation procedure[7], *bis*-aminoethyl-α-disulfone is separated from hypotaurine and taurine by ion-exchange chromatography on a strong cationic resin, and is eluted with 2 N HCl in the last elution step. Thus, *bis*-aminoethyl-a-disulfone seems to be stable in strong acid.

We recently identified and characterized by proton and carbon NMR and by mass spectrometry the corresponding lower oxidation form $H_2N-CH_2-CH_2-SO_2-S-CH_2-CH_2-NH_2$[5]. This compound was observed as the molecular ion (m/z = 185) under the LC/APCI-MS conditions used in this work. Its mass spectrum also exhibits minor peaks at m/z 153 ($H_2N-CH_2-CH_2-SO_2-CH_2-CH_2-NH_2$) and 169. The signal at m/z 153 which was observed during Fenton oxidation of hypotaurine, could be due to a product arising from fragmentation of *bis*-aminoethyl-α-disulfone or of hypotaurine. However, as we have seen above, there are no signals corresponding to the main fragmentation ion reported in the literature for α-disulfones[11]. Therefore, we can reasonably conclude that the absence of the molecular ion of *bis*-aminoethyl-α-disulfone in LC/APCI-MS under the conditions used in this work is not due either to chemical unstability of the compound or to further fragmentation of the molecular ion.

The presence of intermediates and final products other than taurine (revealed by mass chromatography and by amino acid analysis) indicates that the reaction mechanism of hypotaurine oxidation by Fenton's reagent is not simple. It has been shown that sulfinates[9,14,15,17], and in particular hypotaurine[4], disproportionate under various conditions; hypotaurine yields cystamine and cystamine thiol sulfonate. The strong oxidative conditions used in the hypotaurine reaction with Fenton's reagent are not favourable for a disproportionation reaction, however, it cannot be excluded. Also, *bis*-aminoethyl-α-disulfone is reported to disproportionate[7] yielding one mole of hypotaurine and one mole of taurine. A radical reaction followed by disproportionation is probably only one of the possible oxidative pathways involving hypotaurine.

## ACKNOWLEDGMENTS

Financial support from MURST (40%) is acknowledged. We thank Mr. Cesare Paoloni for the artwork. S.D. gratefully acknowledges financial support from C.N.R. and from Kochi Medical School which enabled some of the experimental work to be performed in Japan. We also thank Professor F. Pinnen for many helpful discussions.

## REFERENCES:

1. Allen, P., Karger, L., Haygood, J.D., and Shrensel, J., 1951, Aliphatic sulfinic acids. III. Preparation of aliphatic alpha-disulfones, *J. Org. Chem.*, 16:767-770.
2. Allen, P. and Brook, J.W., 1962, Preparation of alkyl thiolsulfinates, thiolsulfonates and α-disulfones, *J. Org. Chem.*, 27:1019-1020.
3. Aruoma O.L., Halliwell B., Hoey B.M., and Butler J., 1988, The antioxidant action of taurine, hypotaurine and their metabolic precursors, *Biochem. J.*, 256: 251-255.
4. Duprè, S., Fontana, M., Pitari, G., and Cavallini, D., 1996, In vitro reactions of hypotaurine, in Adv. Exp. Med. Biol. "Taurine 2: Basic and Clinical Aspects", Huxtable, R.J., Azuma, J., Kuriyama, K., Nakagawa, M., and Baba, A., eds., Plenum Press, New York, Vol. 403, pp 3-8.
5. Duprè, S., Spirito, A., Pinnen, F. Sugahara, K., and Kodama, H., manuscript in preparation.
6. Fellman, J.H. and Roth, E.S., 1985, The biological oxidation of hypotaurine to taurine: hypotaurine as an antioxidant, in *Progr. Clin. Biol. Res.* "Taurine: Biological Actions and Clinical Perspectives", Oja, S.S., Ahtee, L., Kontro, P. and Paasonen, M.K., eds., Alan R. Liss, Inc., New York, Vol. 179, pp 71-82.

7. Fellman, J.H., Green, T.R., and Eicher, A.L., 1987, The oxidation of hypotaurine to taurine: *bis*-aminoethyl-α-disulfone, a metabolic intermediate in mammalian tissue, in Adv. Exp. Med. Biol. "The Biology of Taurine: Methods and Mechanism", Huxtable, R.J., Franconi, F., and Giotti, A., eds., Plenum Press, New York, Vol. 217, pp. 39-48.
8. Hoyle, J., 1990, The oxidation and reduction of sulphinic acids and their derivatives, in "The Chemistry of sulphinic acids, esters and their derivatives", Patai, S., ed., John Wiley & Sons, New York, pp. 463.
9. Kice, J.L. and Bowers, K.W., 1962, The mechanism of the disproportionation of sulfinic acids, *J. Amer. Chem. Soc.*, 84:605-610
10. Kice, J.L. and Kasparek G.J., 1969, Quantitative comparison of nucleophilic substitution at sulfonyl vs. sulfinyl sulfur. The hydrolysis of aryl α-disulfones in aqueous dioxane, *J. Amer. Chem. Soc.*, 91:5510-5516
11. Pihlaja, K., 1988, Mass spectra of sulfoxides and sulfones, in "The Chemistry of Sulphones and Sulphoxides", S. Patai, ed., Wiley and Sons, pp. 149.
12. Pihlaja, K., 1990, Mass spectra of sulfinic acids, in "The Chemistry of Sulphinic Acids, Esters and Their Derivatives", S. Patai, ed., Wiley and Sons, pp. 107-128
13. Ricci, G., Duprè, S., Federici, G., Spoto, G., Matarese, R.M., and Cavallini, D., 1978, Oxidation of hypotaurine to taurine by ultraviolet irradiation, *Physiol. Chem. & Phys.*, 10:435-441.
14. Schöberl, A. and Wagner, A. 1955a, Methoden zur Herstellung und Umwandlung von Thiosulfon- und Thiosulfinsäureestern, in "Methoden der Organischen Chemie (Houben-Weyl)", E. Müller, ed., G. Thieme Verlag, Stuttgart, 4th ed., Vol. 9, pp. 689-691.
15. Schöberl, A. and Wagner, A., 1955b, Methoden zur Herstellung und Umwandlung von Sulfonen und Sulfoximinen, in "Methoden der Organischen Chemie (Houben-Weyl)", E. Müller, ed., Vol. 9, 4th ed., G. Thieme Verlag, Stuttgart, pp. 254-255.
16. Toennies, G. and Lavine, T.F., 1934, The oxidation of cystine in non-aqueous media. III. Products of exhaustive oxygenation of cystine perchlorate in acetonitrile, *J. Biol. Chem.*, 105:107-115.
17. Wellish, E., Gibstein, E., and Sweeting, O.J., 1961, Thermal decomposition of sulfinic acids, *J. Org. Chem.*, 27:1810-1812.

# *IN VITRO* EVALUATION OF HYPOTAURINE ACTIVITY ON OXIDIZED LDL

Silvestro Dupré,[1] Alberto Macone,[1] Roberta Masella,[2]
Daniela Modesti,[2] Claudio Giovannini,[2] and Alfredo Cantafora[2]

[1] Dipartimento di Scienze Biochimiche and Centro di Biologia Molecolare del C.N.R.
Università di Roma "La Sapienza"
[2] Istituto Superiore di Sanità
Laboratorio di Metabolismo e Biochimica Patologica
Roma, Italy

## INTRODUCTION

Recent reports on the antioxidant activity of hypotaurine indicate that it behaves as a hydroxyl radical scavenger[7]. *In vitro* experiments have shown excellent reactivity with ·OH and HOCl based on rate constants calculated for these reactions[2]. Cystamine-α-disulfone has been proposed as a product of the radical reaction[12], as demonstrated by the isolation of this compound following an *in vitro* reaction of hypotaurine with Fenton's reagent[8]. Notably cystamine-α-disulfone has been also isolated from male sexual tissues[8]. Hypotaurine affects the initiation, propagation and termination phases of lipid peroxidation[15]. In a recent review[9] the antioxidant role of hypotaurine in the regenerating liver and male reproductive system, tissues in which hypotaurine is present in mM concentrations, has been discussed. In other tissues hypotaurine is present in such a minute amount (micromolar or less) as to cast doubts about its real capacity to act as an *in vivo* antioxidant. Target molecules for possible oxidation by highly reactive oxygen species (ROS, reactive oxygen species) are, among others, membrane lipids, protein structures and nucleic acids. A general protective effect of hypotaurine on membrane lipid structure is likely, in view of the protective action of hypotaurine on sperm motility[1] and capacitation[11].

We present data of *in vitro* experiments on the effects of hypotaurine on the process of lipid peroxidation. The protective activity of hypotaurine on $Cu^{++}$-catalyzed oxidation of low-density lipoproteins (LDL) isolated from pooled human plasma was followed monitoring diene formation at 234 nm. The length of the lag-phase preceeding the oxidation phase and the oxidation rate were studied as a function of hypotaurine concentration. Similar experiments were performed on whole human plasma, where the oxidation of plasma lipoproteins is monitored. The antioxidant property of hypotaurine was also tested by evaluating its protective action on the cytotoxic activity of oxidized

*Taurine 3*
Edited by Schaffer *et al.*, Plenum Press, New York, 1998

LDL incubated with cultured CaCo-2 cells. Hypotaurine appeared to be a mild protective agent, when compared to other known natural or synthetic antioxidants. Its *in vivo* activity is probably effective only in the millimolar concentration range.

## MATERIAL AND METHODS

### Materials

Hypotaurine, caffeic acid and tyrosol were purchased from Sigma Chemical Co. Other chemicals were analytical grade compounds. Protein concentration was determined by Bradford's method[5] using bovine serum albumin as standard.

### Low Density Lipoprotein (LDL) Preparation

Human EDTA-added plasma was pooled from samples collected from healthy subjects fasting 12 hours. The LDL fraction was isolated by sequential ultracentrifugation as described by Lindgren[10]. Plasma samples and the purified LDL fractions were dialyzed for 2 hours against a 200 fold volume of 150 mM NaCl and 10 μM EDTA, then for 60 hours against 150 mM NaCl which was changed at least two times. Experiments were performed within one week after the preparation of the LDL fractions.

### Conjugated Diene Determination

The formation of conjugated dienes was followed spectrophotometrically by recording the increase in absorbance of LDL at 234 nm[4] or plasma solutions at 245 nm[13]. Initially, the LDL fraction (50 μg) in 0.96 ml PBS, was incubated with $CuCl_2$ (1.6 μM final concentration), while 25 μM $CuCl_2$ was added to the plasma.

### Evaluation of Cytotoxic Activity of Oxidized LDL

CaCo-2 cells were grown in Dulbecco's modified Eagle medium (DMEM) (bioWhittaker, Maryland) containing 4.5 g glucose per liter, supplemented with 1% (vol/vol) non essential aminoacids, 0.2 mM L-glutamine, 5 U/ml penicillin and 5 μg/ml streptomycin and 10% (vol/vol) fetal bovine serum, and then cultured at 37°C in a humidified atmosphere of 5% $CO_2$ in air. Each experiment was conducted by seeding $6x10^4$ cells per well in 24 multiwell plates.

### Viability Studies

On the 5th day of culture, cells were washed twice with medium without serum before being placed in 500 μl of DMEM containing 2% Ultroser G (a lipoprotein-free serum substitute) (Gibco Brl, Scotland) and exposed to native LDL or oxidized LDL with and without polyphenols for 44 hrs. After this incubation period, 100 μl of the same medium containing 0.05 mCi of $^{14}C$ thymidine (sp. act. 56.5 mCi/mmol) per well was added. The experiments were performed as follows: after 4 hrs of incubation, the incorporation was stopped by 1 ml of 10% trichloroacetic acid at 0°C. The wells were washed twice with 1 ml of PBS, cells were fixed with 1 ml of methyl alcohol for 10 minutes, then the alcohol was removed and the cell monolayer dissolved in 0.5 ml of 0.1 N NaOH. Radioactivity was evaluated in 200 μl aliquots of NaOH extracts using a liquid scintillation counter.

## RESULTS AND DISCUSSION

Figure 1 shows a representative time course for the oxidation of the LDL fraction in the presence of 1.6 μM $Cu^{++}$ and variable amounts of hypotaurine, as determined by monitoring the rate of conjugated diene formation (absorbance at 234 nm). The presence of hypotaurine has a slight effect on the lag-phase of the reaction, and even a minor effect on the oxidation rate, as seen from the slope of the curve. A clear increase in the total amount of oxidized LDL, which is dependent upon hypotaurine concentration, is also observed.

Table 1 reports the average data of three experiments expressed as percent increase in the length of the lag-phase versus hypotaurine concentration. The data (Fig. 2) yielded a correlation coefficient of R = 0.823 for the two parameters. The protective activity of hypotaurine against $Cu^{++}$-induced LDL oxidation is rather low in comparison with other known antioxidants (Fig. 3). At a concentration of 3 mM, the hypotaurine effect (100% increment of the lag-phase as compared to the control) is the same as that obtained with other antioxidants at concentrations at least two orders of magnitude lower. The

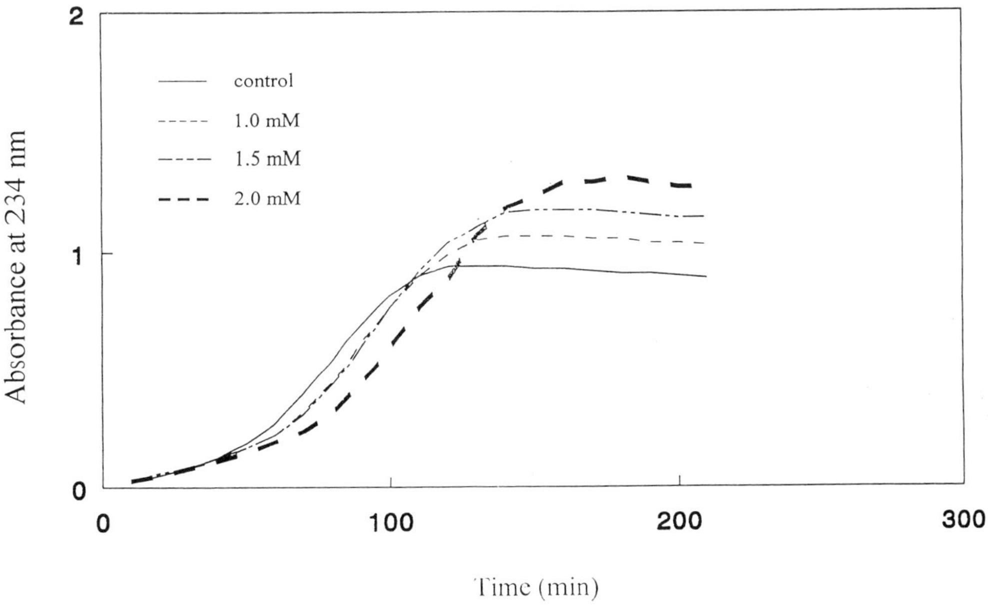


**Figure 1.** Time course of the oxidation of LDL (50 μg) solution in the presence of 1.6 μM $Cu^{++}$ at various concentrations of hypotaurine.

**Table 1.** Hypotaurine effect on lag-phase percentage increase during copper-induced LDL oxidation.

| Hypotaurine (concentrations) | Lag-phase (% increase) |
|---|---|
| 1.0 mM | 25 ± 8.5 |
| 1.5 mM | 40 ± 10 |
| 2.0 mM | 57 ± 24 |
| 2.5 mM | 76 ± 15 |
| 3.0 mM | 125 ± 18 |

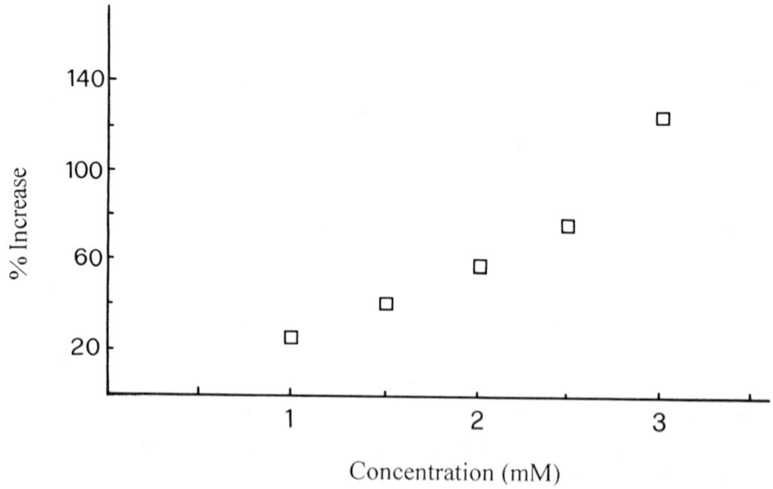


**Figure 2.** Correlation between hypotaurine concentration and percentage increase of lag-phase. The lag phase length was determined from the intercepts of tangents to the slow and fast increase of the absorption curve at 234 nm. Correlation coefficient R = 0.823, calculated from a line drawn by linear regression analysis.

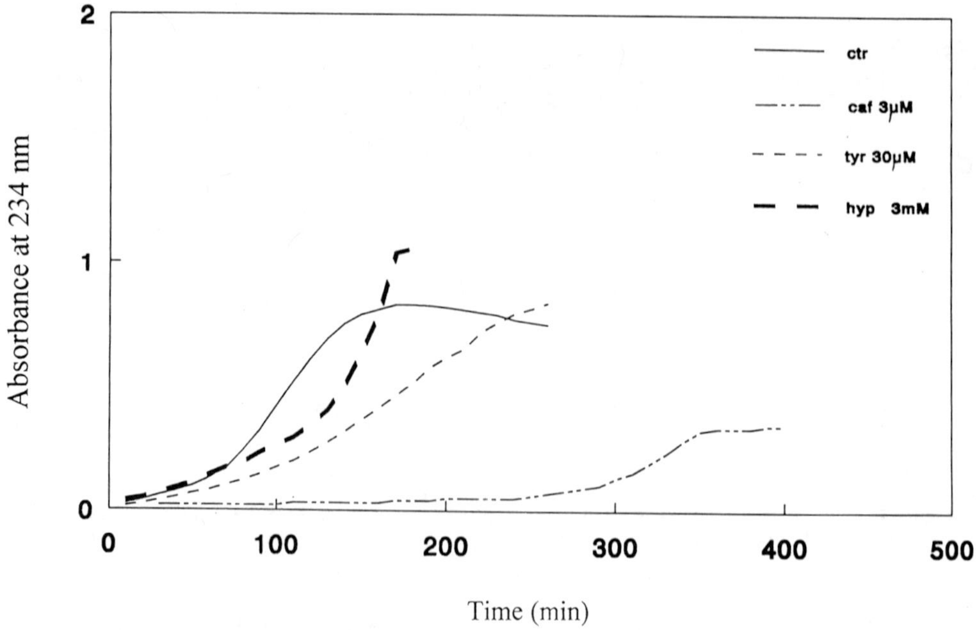


**Figure 3.** Time course of the oxidation of LDL (50 μg) performed with 1.6 μM $Cu^{++}$ (ctr) and in the presence of tyrosol (tyr), caffeic acid (caf) or hypotaurine (hyp).

mechanism of LDL oxidation is known in some detail (see ref. 6 for an extensive review). It involves different steps, only a few of them utilizing hydroxyl radicals. Hypotaurine is believed to be a good scavenger of hydroxyl radicals, but not of either superoxide ion or $H_2O_2$[2]. Its activity as a scavenger of other free radicals has not been studied in detail. The oxidation of LDLs involves the formation of lipid hydroperoxides, LOOH, and their corresponding free radical LOO·, as well as other free radicals. Several antioxidants are probably more efficient, in scavenging the other free radical products of lipid peroxidation than hypotaurine. Thus, a potent hydroxyl radical scavenger is not very effective in this specific oxidation reaction.

In the presence of higher concentrations of hypotaurine (2-5 mM) we observed also a slower rate of diene formation at the beginning of the lag-phase (data not shown), similar to the effect seen with other antioxidants[14,17], the so-called tocopherol-mediated peroxidation (TMP) kinetics. This phenomenon has been explained by a sudden rise in free radical concentration due to the presence of compounds, like tocopherols and probably hypotaurine, which easily form free radicals and initiate a new free radical chain reaction.

The antioxidant activity of hypotaurine was also evaluated by its ability to protect against the cytotoxic action of oxidized LDL towards cultured CaCo-2 cells. The effect of hypotaurine (2-12 mM) added to the culture medium is again rather low. About 10% protection is observed with 12 mM hypotaurine, whereas 100% protection is obtained with 0.5 mM tyrosol (Fig. 4). Prior incubation of cells with medium containing hypotaurine (12 mM) for a period of 24 hrs before the addition of oxidized LDL did not improve the degree of cell protection relative to cells exposed simultaneously to hypotaurine and oxidized LDL. No data are available, to our knowledge, on the intracellular transport or uptake of hypotaurine in this cell type. The above data however show that apparently very low uptake is experienced and the mild protective effect is likely attributed to an extracellular scavenging action of hypotaurine. It is conceivable that a greater uptake of hypotaurine would exert a larger protective effect on the cytotoxic activity of oxidized LDL, which presumably is internalized and exerts its lethal effect on the inside of the cell.

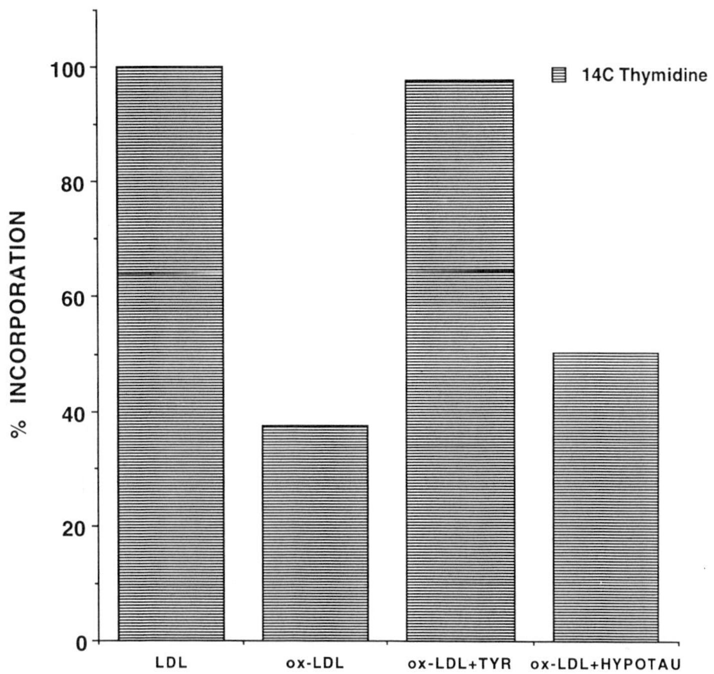


**Figure 4**. Effect of antioxidants on labeled thymidine incorporation. Viability of CaCo-2 cells was measured as reported in Materials and Methods; cells were incubated for 44 hours with (from left): native LDL (0.2 mg/ml), oxidized LDL (0.2 mg/ml), oxidized LDL + 0.5 mM tyrosol, oxidized LDL + 12 mM hypotaurine. Viability is expressed as percent incorporation.

Using a recently published method[13], we performed the same experiment using whole plasma in place of purified LDL fractions. When hypotaurine was tested on plasma samples of individuals, the results were scattered and this depends on the age of plasma and other individual factors which could not be controlled. This suggests that human plasma may contain substances capable of affecting both the mechanism and the kinetics of hypotaurine action. In fact, experiments with pooled plasma samples yielded similar results to those performed with LDL fractions purified from the very same plasma sample (not shown). Hypotaurine behaves as an antioxidant with LDL fractions and at high concentrations, as a pro-oxidant with plasma fractions. A slight pro-oxidant effect on the oxidation rate is also observed (Fig. 5). Whole plasma is a more complex system and the oxidation curve observed may be due to the oxidation of many lipoprotein fractions, each with different kinetics. We can not exclude the possibility that hypotaurine may exert its protective activity on the oxidation of other fractions, but this conclusion can not be ascertained from a comparison between whole plasma and the LDL oxidation data.

The conclusions we can draw from these data are:

i) hypotaurine exhibits *in vitro* a low antioxidant activity on LDL oxidation and on the cytotoxic activity of oxidized LDL. The activity may be observed at millimolar concentrations and the effect is about two order of magnitude lower than the same effect observed with known antioxidants, such as caffeic acid or polyphenols; ii) hypotaurine *in vivo* may not be an important antioxidant in plasma, due to the very low concentrations present, but could be very important in the seminal male tract, where concentrations are of the order of millimolar. In fact many recent papers point to this role of hypotaurine[3,16], correlating the protection of membrane lipids with the activity of sperm cells; iii) the rather low antioxidant activity of hypotaurine in these systems may be due to the specific reactivity of hypotaurine towards certain radicals formed during lipid peroxidation. According to the generally accepted mechanism of LDL oxidation oxygen free radicals are not the predominant oxidizing species; iv) experiments with plasma are not very informative because results show high variations due to many different factors, including the dietary habits of the patients.

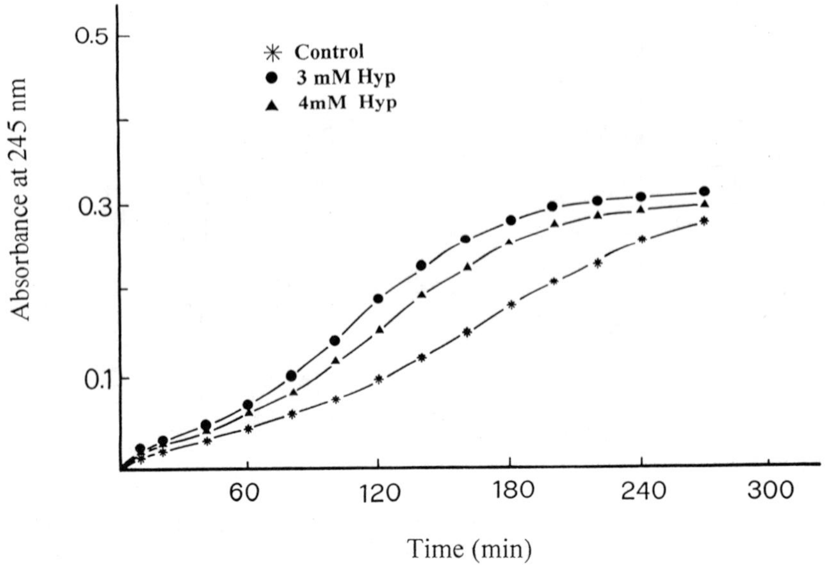


**Figure 5**. Time course of diene formation measuared at 245 nm. 50-fold diluted citrated plasma (20 μl) is added simultaneously with 25 μM $Cu^{++}$ (1 ml final volume) to initiate the reaction.

## ACKNOWLEDGMENTS

Financial support by MURST (40%) and C.N.R. is acknowledged. We thank Mr. Cesare Paoloni for the artwork.

## REFERENCES

1. Alvarez, J.G. and Storey, B.T., 1983, Taurine, hypotaurine, epinephrine and albumin inhibit lipid peroxidation in rabbit spermatozoa and protect against loss of motility, *Biol. Reprod.*, 29:548-555.
2. Aruoma O.L., Halliwell B., Hoey B.M., and Butler J., 1988, The antioxidant action of taurine, hypotaurine and their metabolic precursors, *Biochem. J.*, 256:251-255.
3. Baker, H.W., Brindle, J., Irvine, D.S., and Aitken, R.J., 1996, Protective effect of antioxidants on the impairment of sperm motility by activated polymorphonuclear leukocytes, *Fertil. Steril.*, 65:411-419.
4. Beuge, J.A and Aust, S.D., 1978, Microsomal lipid peroxidation, in Methods Enzymol. "Biomembranes", Fleischer, S. and Packer, L., eds., 52:308.
5. Bradford, M., 1976, A rapid and sensitive method for the quantitation of microgram quantities of protein utilizing the principle of dye binding, *Anal. Biochem.*, 72:248-254.
6. Esterbauer, H., Gebicki, J., Puhl, H., and Jürgens, G., 1992, The role of lipid peroxidation and antioxidants in oxidative modification of LDL, *Free Rad. Biol. and Med.*, 13:341-390.
7. Fellman, J.H. and Roth, E.S., 1985, The biological oxidation of hypotaurine to taurine: hypotaurine as an antioxidant, in Progr. Clin. Biol. Res. "Taurine: Biological actions and clinical Perspectives", Oja, S.S., Ahtee, L., Kontro, P., and Paasonen, M.K., eds., Alan R. Liss, Inc., New York, vol. 179, pp. 71-82.
8. Fellman, J.H., Green, T.R., and Eicher, A.L., 1987, The oxidation of hypotaurine to taurine: bis-aminoethyl-α-disulfone, a metabolic intermediate in mammalian tissue, in Adv. Exp. Med. Biol. "The Biology of Taurine: Methods and Mechanism", Huxtable, R.J., Franconi, F., and Giotti, A., eds., Plenum Press, New York, Vol. 217, pp. 39-48.
9. Huxtable, R.J., 1992, Physiological actions of taurine, *Physiol. Reviews*, 72:101-163.
10. Lindgren, F.T., 1975, Preparative ultracentrifugal laboratory procedures and suggestions for lipoprotein analysis, in: "Analysis of lipids and lipoproteins", Perkins E.G., ed., AOCS, Campaign Ill., pp. 204-224.
11. Meizel, S., Lui, C.W., Working, P.K., and Mrsny, R.J., 1980, Taurine and hypotaurine: their effects on motility, capacitation and the acrosome reaction of hamster sperm in vitro and their presence in sperm and reproductive tract fluids of several mammals, *Dev. Growth Differ.*, 22:483-494.
12. Ricci, G., Duprè, S., Federici, G., Spoto, G., Matarese, R.M., and Cavallini, D., 1978, Oxidation of hypotaurine to taurine by ultraviolet irradiation, *Physiol. Chem. & Phys.*, 10:435-441.
13. Schnitzer, E., Pinchuk, I., Fainaru, M., Schafer, Z., and Lichtenberg, 1995, Copper-induced lipid oxidation in unfractioned plasma:the lag phase preceding oxidation as a measure of oxidation resistance, *Biochem. Biophys. Res. Commun.*, 216:854-861.
14. Schnitzer, E., Pinchuk, I., Bor, A., Fainaru, M., and Lichtenberg, D., 1997, The effect of albumin on copper-induced LDL oxidation, *Biochim. Biophys. Acta*, 1344:300-311.
15. Tadolini, B., Pintus, G., Pinna, G.G., Bennardini, F., and Franconi, F., 1995, Effects of taurine and hypotaurine on lipid peroxidation, *Biochem. Biophys. Res. Commun.*, 213:820-826.
16. Wakayama, T., Suto, J.K., Imamura, K., Toyoda, Y., Kurohmaru, M., and Hayashi, Y., 1996, Effect of hypotaurine on in vitro fertilization and production of term offspring from in vitro-fertilized ova of the Japanese fiels vole, *Microtus montebelli*, *Biol. Reprod.*, 54:625-630.
17. Witting, P.K., Upston, J.M., and Stocker, R., 1997, Role of α-tocopheroxyl radical in the initiation of lipid peroxidation in human low-density lipoprotein exposed to horse radish peroxidase, *Biochemistry*, 36:1251-1258.

# HYPOTAURINE PROTECTION ON CELL DAMAGE BY $H_2O_2$ AND ON PROTEIN OXIDATION BY $Cu^{++}$ AND $H_2O_2$

Silvestro Dupré,[1] Mara Costa,[1] Alessandra Spirito,[1] Giuseppina Pitari,[2] Paola Rossi,[2] and Fernanda Amicarelli[2]

[1] Dipartimento di Scienze Biochimiche and
Centro di Biologia Molecolare del C.N.R.
Università di Roma "La Sapienza"
Roma, Italy
[2] Dipartimento di Biologia di Base ed Applicata
Università de L'Aquila
L'Aquila, Italy

## INTRODUCTION

The antioxidant activity of hypotaurine on cell structure has primarily been studied utilizing artificial liposomes and isolated lipoproteins[18]. Most studies with whole cells have utilized sperm cells to study the effect of oxidative stress on cellular function in the presence of varying concentrations of hypotaurine[1,4,11,14,19]. In this paper we will report experiments performed using two *in vitro* model systems.

Cultured cells suspended in PBS were subjected to oxidative stress by the addition of $H_2O_2$. The protective action of hypotaurine against oxidative damage was studied at two different concentrations of the antioxidant. Viability of cells, recorded 72 hours after washing out the oxidant, depended upon antioxidant activity. In other experiments cells were either preincubated with hypotaurine, to load the cells with the antioxidant, or hypotaurine was added also during the recovery step. The sulfinate has a clear protective activity which is observed only on the outside of the cell and only when hypotaurine is added before or simultaneously with $H_2O_2$.

In the second model system, protection by hypotaurine against the oxidation of bovine serum albumin in the presence of $Cu^{++}$ and $H_2O_2$ was studied. Under conditions where mannitol is without effect, indicative of site-specific production of hydroxyl radicals, hypotaurine exerts a slight but significant protective effect which can not simply be explained as a a scavenging action of $H_2O_2$.

## MATERIAL AND METHODS

### Chemicals

Hypotaurine, mannitol, cysteamine hydrochloride and $H_2O_2$ were purchased from Sigma Chemicals.

### Cell Cultures

Carl-1 human melanotic melanoma, Lipari glioblastoma and SK-N-BE(2) human neuroblastoma cells were cultured as monolayers in RPMI-1640 medium containing 2 mM glutamine, 20 mM Hepes, 100 IU/ml penicillin, 100 μg/ml streptomycin and supplemented with 10% fetal bovine serum (Flow Laboratories Irvine, Scotland). The cultures were incubated under 5% $CO_2$ moist air at a humidity of 80% and a temperature of 37°C. Cell detachment was carried out with 1 mM EDTA and 0.5% porcine trypsin in Earle's balanced medium; cells were counted by using a Burker chamber. After washing with PBS, cells in triplicate were incubated at 37°C in serum-free medium with two concentrations of $H_2O_2$ (0.05 and 0.2 mM) with or without two concentrations of hypotaurine (0.8 and 1.6 mM). After an incubation period of 1 hour, cells were washed, added again to complete culture medium, and viability measured after 24, 48 and 72 hours. In some experiments, a preincubation of the cells with 0.8 mM hypotaurine was performed before the incubation with $H_2O_2$.

### Protein Oxidation

Oxidation of bovine serum albumin (0.5 mg, 7.5 μM) in the presence of 0.2 mM $Cu^{++}$ and 2 mM $H_2O_2$ was followed fluorimetrically for one hour ($\lambda_{ex}$ = 290 nm, $\lambda_{em}$ = 340 nm)[13]. $H_2O_2$ concentration was measured by the peroxidase method with guaiacol as substrate[16]. The hypotaurine concentration was measured at 220 nm ($\varepsilon_{mM}$ = 1.72 at pH = 7.0)[17].

## RESULTS AND DISCUSSION

### Protection of Cultured Cells Against Oxidative Stress by $H_2O_2$

Three cell types were used in these experiments. Melanoma, glioblastoma and neuroblastoma cells were cultured and prepared as reported in the experimental section. After washing with PBS, cells were submitted to oxidative stress for 1 hour. Survival rates up to 72 hours after introduction of the stress have been used as a measure of oxidative damage and of the protective action of hypotaurine. Furthermore, the growth rate of cells indicates if the cells that survive have been partially damaged or if the oxidative damage has been completely lethal. In some experiments, preincubation of the cells with 0.8 mM hypotaurine was performed before the incubation with $H_2O_2$. Figure 1 shows the effects of oxidative stress and of the presence of hypotaurine on the growth rate of melanoma cells. Only one concentration each of $H_2O_2$ and hypotaurine are shown. Hypotaurine has a significant protective effect which is clearly evident at a 4:1 ratio with $H_2O_2$. The growth rate lines for cells exposed to $H_2O_2$ plus hypotaurine at various times are parallel to each other indicating that the surviving cells grow normally (not shown). Preincubation of the cells with 0.8 mM hypotaurine, followed by removal of the hypotaurine before the addition of $H_2O_2$, does not protect cells unless hypotaurine is again added simultaneously

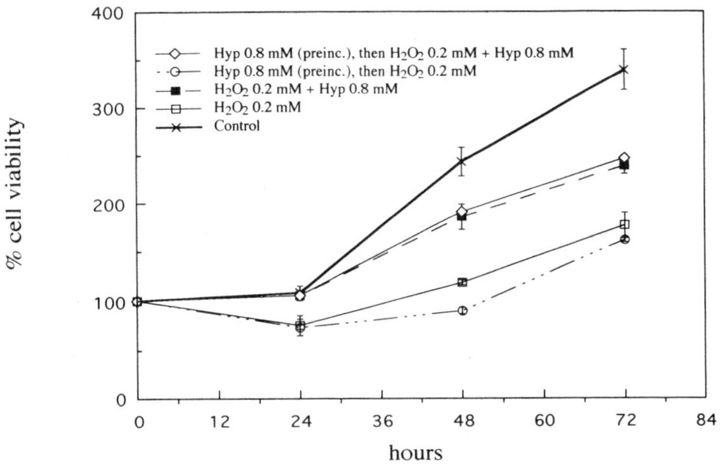


**Figure 1.** Growth rate of cultured melanoma cells under oxidative stress. Melanoma cells were cultured as described in the experimental section, submitted to oxidative stress as described in the figure, and then washed and cultured again for 72 hours to obtain growth rates. Counts were performed at 24, 48 and 72 hours to assess cell viability. Preincubation with hypotaurine was for 1 hour; cells were then washed before further additions. Incubation with $H_2O_2$ was for 1 hour. For experimental details see Methods.

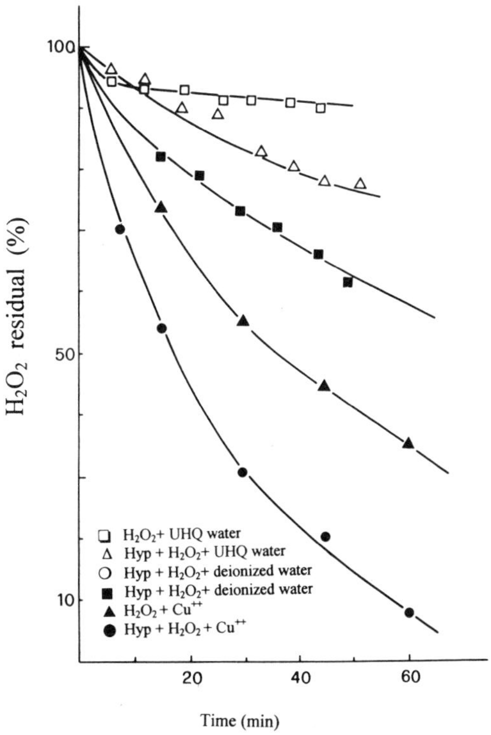


**Figure 2.** Disappearence kinetics of $H_2O_2$ under various experimental conditions. $H_2O_2$ (2.0 mM) and hypotaurine (10 mM) were incubated in a final volume of 1.0 ml of UHQ water, deionized water or in the presence of 0.2 mM $Cu^{++}$. Control curves represent the disappearence of $H_2O_2$ alone in UHQ water or in the presence of 0.2 mM $Cu^{++}$. (Conditions are described in the figure.) At various intervals samples of the reaction mixture were removed and $H_2O_2$ content was measured with peroxidase by the guaiacol method[16]. For experimental details see Methods.

with $H_2O_2$. Hypotaurine has been reported to react very slowly with $H_2O_2$[2,5]. Figure 2 shows the disappearance kinetics of $H_2O_2$ under various experimental conditions. A second-order rate constant of approximately $2 \cdot 10^{-2}$ $M^{-1}s^{-1}$ was calculated for this reaction. In the presence of trace amounts of metal the reaction is accelerated indicating that some other reaction involving activated oxygen free radicals must have occurred. The protective activity of hypotaurine cannot be explained by a simple scavenging of $H_2O_2$, to reduce its concentration since the reaction of hypotaurine and $H_2O_2$ at these concentrations is responsible for no more than 10% of the protection. We may, therefore, conclude that the observed antioxidant effect of hypotaurine is due to scavenging of reactive oxygen species other than $H_2O_2$. Melanoma cells are quite resistant to $H_2O_2$-mediated oxidative stress which is probably due to the levels of the antioxidative enzymes in this rather specialized cell type.

The same experiment was performed with neuroblastoma cultured cells. Results obtained reveal a minor protective activity for hypotaurine (Fig. 3). Since the survival rate of the neuroblastoma cells is much lower than that seen in melanoma cells, the effect of hypotaurine can be attributed mainly to the lowering of the $H_2O_2$ concentration through the reaction of $H_2O_2$ and hypotaurine. Addition of the same amount of hypotaurine after the incubation with $H_2O_2$, thereby assuring the presence of hypotaurine both during the oxidative stress step and after $H_2O_2$ has been removed, does not change the survival rate (data not shown). If we assume that $H_2O_2$-mediated damage to cells is a rapid process, and that the damage is due to lipid peroxidation initiated either outside or inside the cell, then the protective activity of hypotaurine due to scavenging either $H_2O_2$ or another reactive species should be noticeable at the beginning of the incubation period. According to this scenario, hypotaurine should have no effect during the growth period on either the degree of cell damage or on the survival rate. Pre-loading of cells with hypotaurine is without any effect unless hypotaurine is also present during initiation of the oxidative stress. The uptake of hypotaurine in mouse brain slices and neuroblastoma cells has been reported to be high[8,15], whereas it has, to our knowledge, not been studied in melanoma cells. In view of our results, that pre-loading with hypotaurine before $H_2O_2$ incubation has no influence on the survival rate, the uptake rate seems not to be important in the protective activity of hypotaurine.

Similar experiments performed with cultured glioblastoma cells show that hypotaurine is without any protective effect on this cell type. Glioblastoma cells are the most sensitive to oxidative stress and are almost completely destroyed by the presence of 0.05 mM $H_2O_2$.

**Protection of Protein Oxidation**

The antioxidant effect of hypotaurine has also been studied using a model system of protein oxidation. Bovine serum albumin is oxidized with $Cu^{++}$-$H_2O_2$[9,10] at pH 7.6, a process that can be followed by fluorescence ($\lambda_{ex}$ = 290 nm, $\lambda_{em}$ = 340 nm), monitoring the disappearance of aromatic amino acids[6]. The oxidation is dose-dependent with respect to $H_2O_2$ (Fig. 4) and becomes asymptotic after about one hour. The presence of 0.01 M hypotaurine has a low but significant protective effect on the oxidation, whereas 0.01 M mannitol is without effect (Fig. 5). Cysteamine at the same concentration protects more than 90%. These results are consistent with a site-specific oxidation of the protein. $Cu^{++}$ is chelated by the protein and a site-directed production of hydroxyl radicals is responsible for the oxidation[12]. Mannitol, a specific scavenger of free OH radicals[7], is completely without protective effect, whereas cysteamine, which reacts with $H_2O_2$ with a second-order rate constant of 1.7 $M^{-1}s^{-1}$ and is a strong $Cu^{++}$ chelator[2], behaves as a powerful antioxidant. Hydroxyl radicals produced with $Fe^{++}$ and $H_2O_2$, in the presence or absence of EDTA, are much less effective as protein oxidants, at least in the case of albumin which is not

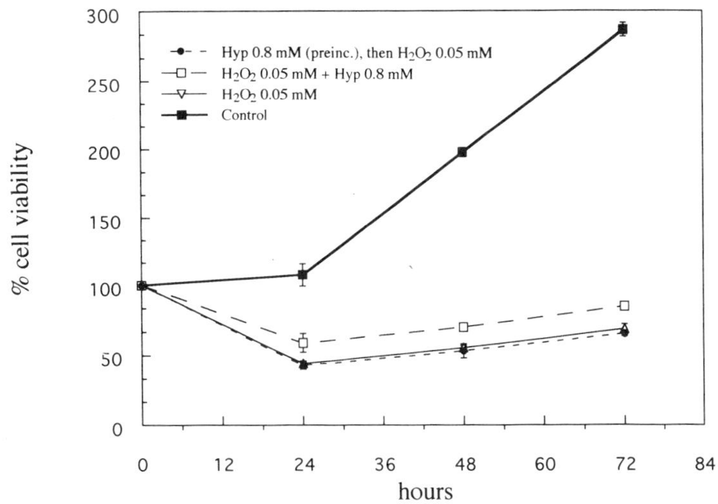


**Figure 3.** Growth rate of cultured neuroblastoma cells under oxidative stress. Neuroblastoma cells were cultured as described in the experimental section, submitted to oxidative stress as described in the figure, and then washed and cultured again for 72 hours to obtain growth rates. Counts were performed at 24, 48 and 72 hours to assess cell viability. Preincubation with hypotaurine was for 1 hour; cells were then washed before further additions. Incubation with $H_2O_2$ was for 1 hour. For experimental details see Methods.

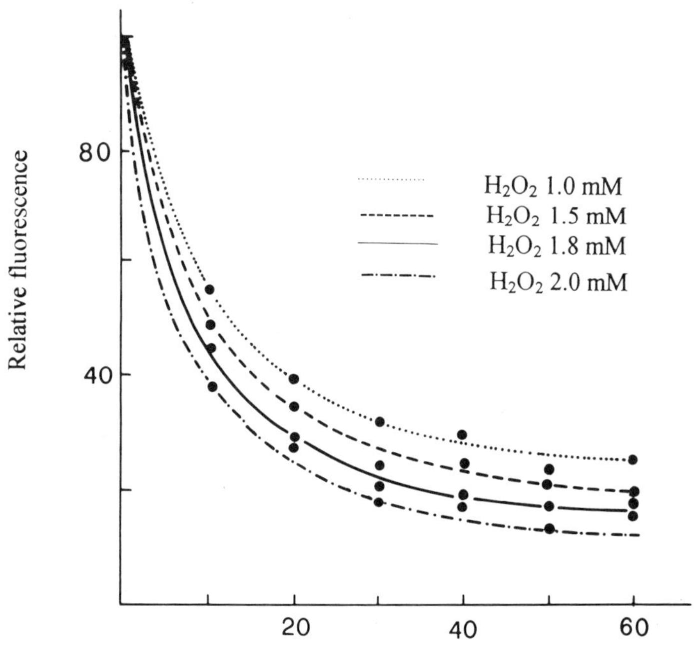


**Figure 4.** Oxidation of bovine serum albumin in the presence of $Cu^{++}$ and various $H_2O_2$ concentrations. Final reaction concentrations were: 7.5 μM bovine serum albumin, 0.2 mM $Cu^{++}$, and $H_2O_2$ 1.0 mM, 1.5 mM, 1.8 mM and 2.0 mM, respectively. (Conditions are described in figure.) The changes in the fluorescence properties of the protein were followed at appropriate time intervals and reported as arbitrary units. For experimental details see Methods.

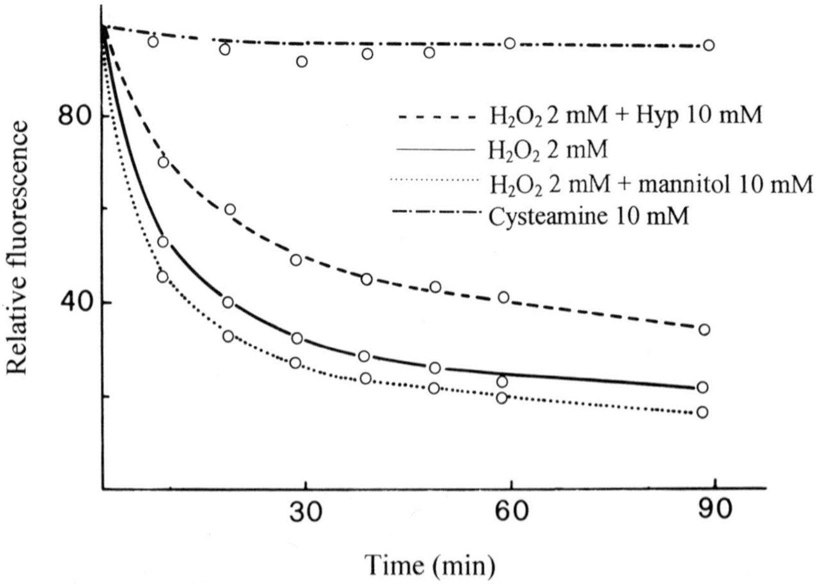


**Figure 5.** Effects of various compounds on bovine serum albumin oxidation. The decrease of fluorescence was measured in the absence or in the presence of 10 mM mannitol, 10 mM hypotaurine, and 10 mM cysteamine. (Conditions are described in the figure.) The final concentrations of bovine serum albumin, $Cu^{++}$ and $H_2O_2$ were 7.5 μM, 0.2 mM and 2.0 mM, respectively. For experimental details see Methods.

oxidized even after an incubation of 1 hour. $Fe^{++}$/EDTA binds weakly to albumin[12] and the free ·OH radicals that are generated are less efficient oxidizing agents. The effect in the presence of hypotaurine mimics the oxidation pattern observed in the presence of lower concentrations of $H_2O_2$. The reaction of hypotaurine with $H_2O_2$ is very slow but becomes faster in the presence of trace amounts of metal (see Fig. 2). Hypotaurine increases the disappearance rate of $H_2O_2$ in the presence of $Cu^{++}$ probably by scavenging ·OH. The antioxidant activity of hypotaurine is, therefore, not simply due to a reaction with $H_2O_2$ (too low to be consistent with the observed data) but mainly to a scavenging of hydroxyl radicals. This result implies that hypotaurine is also able to partially scavenge the so-called crypto hydroxyl radicals in the site-directed Haber-Weiss reaction. A similar effect has been reported with another crypto-radical generating system where the production of hydroxyl radicals was monitored using deoxyribose as substrate[3]. Site-directed protein oxidation has also been reported for fatty acid-free human albumin[6].

## ACKNOWLEDGMENTS

Financial support from MURST (40%) and C.N.R. is acknowledged. We thank Mr. Cesare Paoloni for the artwork.

## REFERENCES


1. Alvarez, J.G. and Storey, B.T., 1983, Taurine, hypotaurine, epinephrine and albumin inhibit lipid peroxidation in rabbit spermatozoa and protect against loss of motility, *Biol. Reprod.*, 29:548-555.
2. Aruoma, O.I., Halliwell, B., Hoey, B.M., and Butler, J., 1988, The antioxidant action of taurine, hypotaurine and their metabolic precursors, *Biochem. J.*, 256:251-255.

3. Aruoma, O.I., 1994, Deoxyribose assay for detecting hydroxyl radicals, in *Methods Enzymol.*, 233:57-66.
4. Baker, H.W., Brindle, J., Irvine, D.S., and Aitken, R.J., 1996, Protective effect of antioxidants on the impairment of sperm motility by activated polymorphonuclear leukocytes, *Fertil. Steril.*, 65:411-419.
5. Fiori, A. and Costa, M., 1969, Ossidazione dell'ipotaurina con $H_2O_2$, *Acta Vitaminol. Enzymol.*, 23:204-207.
6. Gutteridge, J.M.C. and Wilkins, S., 1983, Copper salt-dependent hydroxyl radical formation. Damage to proteins acting as antioxidants, *Biochim. Biophys. Acta*, 759:38-41.
7. Gutteridge, J.M.C., 1984, Reactivity of hydroxyl and hydroxyl-like radicals discriminated by release of thiobarbituric acid-reactive material from deoxy sugars, nucleosides and benzoate, *Biochem. J.*, 224:761-767.
8. Holopainen, I., 1984, Taurine and hypotaurine transport by a single system in cultured neuroblastoma cells, *Acta Physiol. Scand.*, 122:381-386.
9. Hunt, J.V., Simpson, J.A.. and Dean, R.T., 1988, Hydroperoxide-mediated fragmentation of proteins, *Biochem. J.*, 250:87-93.
10. Hunt, J.V., Dean, R.T., and Wolff, S.P., 1988, Hydroxyl radical production and autoxidative glycosilation. Glucose autoxidation as the cause of protein damage in the experimental glycation model of diabetes mellitus and ageing, *Biochem. J.*, 256:205-212.
11. Huxtable R.J., 1992, Physiological actions of taurine, *Physiol. Reviews*, 72:101-163.
12. Kocha, T., Yamaguchi, M., Ohtaki, H., Fukuda, T., and Aoyagi, T., 1997, Hydrogen peroxide-mediated degradation of protein: different oxidation modes of copper- and iron-dependent hydroxyl radicals on the degradation of albumin, *Biochim. Biophys. Acta*, 1337:319-326.
13. Marx G. and Chevion M., 1986, Site specific modification of albumin by free radicals. Reaction with copper(II) and ascorbate, *Biochem. J.*, 236:397-400.
14. Meizel, S., Lui, C.W., Working, P.K., and Mrsny, R.J., 1980, Taurine and hypotaurine: their effects on motility, capacitation and the acrosome reaction of hamster sperm in vitro and their presence in sperm and reproductive tract fluids of several mammals, *Dev. Growth Differ.*, 22:483-494.
15. Oja, S.S. and Kontro, P., 1982, Hypotaurine uptake in mouse brain slices, *Adv. Exp. Med. Biol.*, 139:115-126.
16. Pruitt, K.M., Kamau, D. N., Miller, K., Mansson-Rahemtulla, B., and Rahemtulla, F., 1990, Quantitative, Standardized Assays for Determining the Concentration of Bovine Lactoperoxidase, Human Salivary Peroxidase, and Human Myeloperoxidase, 191:278-286.
17. Ricci, G., Duprè, S., Federici, G., Spoto, G., Matarese, R.M., and Cavallini, D., 1978, Oxidation of hypotaurine to taurine by ultraviolet irradiation, *Physiol. Chem. & Phys.*, 10:435-441.
18. Tadolini, B., Pintus, G., Pinna, G.G., Bennardini, F., and Franconi, F., 1995, Effects of taurine and hypotaurine on lipid peroxidation, *Biochem. Biophys. Res. Commun.*, 213:820-826.
19. Wakayama, T., Suto, J.K., Imamura, K., Toyoda, Y., Kurohmaru, M., and Hayashi, Y., 1996, Effect of hypotaurine on in vitro fertilization and production of term offspring from in vitro-fertilized ova of the Japanese fiels vole, *Microtus montebelli*, *Biol. Reprod.*, 54:625-630.

# CYSTEINE SULFINATE DECARBOXYLASE (CSD): MOLECULAR CLONING, SEQUENCE AND GENOMIC EXPRESSION IN BRAIN

Marcel Tappaz,[1] Isabelle Reymond,[1] Marc Bitoun,[1] and Alain Sergeant[2]

[1] INSERM U 433
Faculté de Médecine Laennec
F-69372 Lyon
[2] INSERM U 412
Ecole Normale Supérieure
F-69634 Lyon

## INTRODUCTION

Cysteine sulfinate decarboxylase (CSD) is thought to be the rate limiting step of taurine biosynthesis[9]. It may thus represent a key enzyme in the function of taurine. While a CSD activity was detected in crude brain extracts more than four decades ago, its identification, characterization and cellular localization in brain proved difficult and led to controversial reports (For review see[14,15]). We previously established that there exists in brain an enzyme showing a strong affinity and narrow specificity for cysteine sulfinic acid (CSA) and cysteic acid (CA)[6]. Brain CSD appeared to be very similar to the CSD from liver: both proteins have similar physico-chemical and enzymatic properties[3,6] and share common antigenic epitopes[9]. Using a specific antiserum against liver CSD[1] we showed that CSD immunopositive cells in hippocampus and cerebellum were astrocytes through a quantitative double immunofluorescence analysis with appropriate astroglial markers[11]. In rat brain, CSD expression level reveals two distinct phenotypes in astrocytes that exhibit a different distribution pattern[10]. Given that taurine was reported to be localized predominantly in neurons[5,8], our findings suggest that the poorly understood functional role of taurine in the brain[4,7] should be investigated within the framework of astrocyte-neuron interactions. In addition, our results suggest that the CSD expression level may vary greatly among astrocytes according to their location and functional state. Disclosing the regulation of the genomic expression of CSD in various experimental situations could help provide insight into the physiological role of taurine in the brain. With this long-term goal in mind, we have carried out molecular cloning of CSD[12], determined its sequence and developed new appropriate tools, such as the quantification of CSD-mRNA through competitive reverse transcription coupled with polymerase chain reaction (RT-PCR). These molecular biology techniques will enable us to investigate the regulation of the genomic expression of CSD in brain, as well as in astrocytes in primary culture.

## RESULTS AND DISCUSSION

### Strategy for CSD Molecular Cloning

Our previous results indicated that CSD makes up less than 1/10000 of the total liver soluble protein content, and that it is about 20-fold enriched in the liver compared to the brain[1,9]. In order to enhance the chances of success, we used a rat liver cDNA library and designed a specific DNA probe for screening the library. This probe was a cloned PCR-amplified CSD-cDNA fragment generated from liver reverse transcribed poly(A+)-mRNA (Fig. 1).

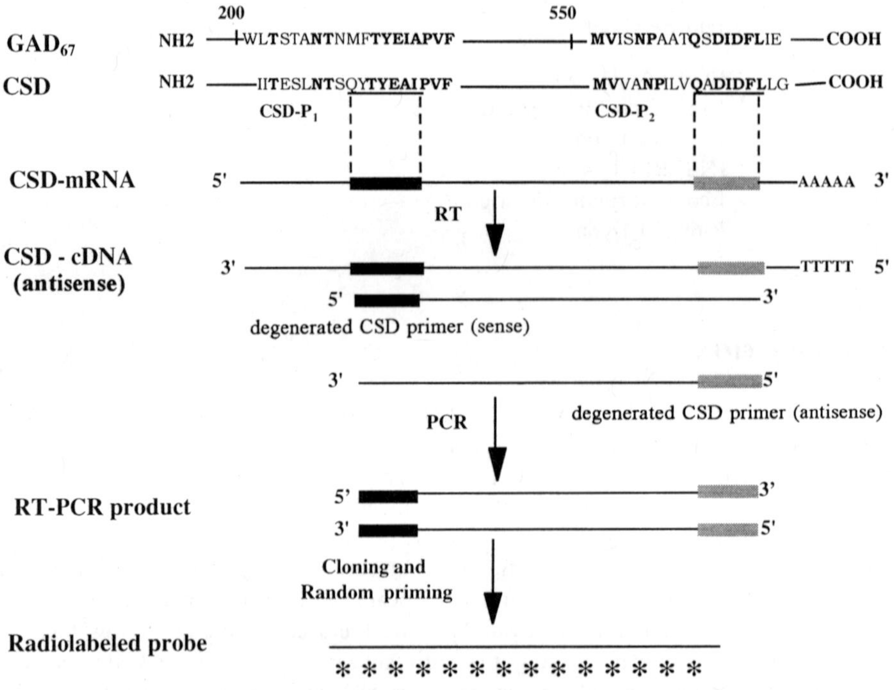


**Figure 1.** Strategy for the production of a specific DNA probe for molecular cloning of CSD. Through a classical purification procedure, a homogeneous CSD fraction was isolated from liver homogenate[9]. It was ascertained to be CSD by the production of an antiserum that immunoprecipitated CSD activity when used as an immunogen. From this protein, three peptides could be isolated and sequenced. Two of them, CSD-$P_1$ and CSD-$P_2$ showed a marked homology with the 67 kDa isoform of glutamic decarboxylase ($GAD_{67}$)(bold letters). According to this homology, CSD-$P_1$ and CSD-$P_2$ were postulated to be on the N-terminal and C-terminal side of the CSD protein. Sense and antisense-degenerated primers corresponding to the underlined amino acids were used to amplify a DNA fragment through PCR of CSD-cDNA obtained by reverse transcription of liver poly(A+)-mRNA. The fragment was cloned and sequenced in order to check if it encoded the expected amino-acids of CSD-$P_1$ and CSD-$P_2$. This cloned DNA fragment was radiolabeled using random primers and used to screen a rat liver cDNA library constructed in a lambda ZAP II vector. For experimental details see Reymond et al.[12].

## Liver CSD Molecular Cloning and Sequence Analysis

Three positive clones were isolated from a rat liver cDNA library containing about $10^5$ independent recombinants. Altogether a 1981 base pairs (bp) sequence was determined. An open reading frame (ORF) containing 1479 bp was found on this sequence that encoded a protein of 493 amino acids with a predicted molecular weight of 55.2 kDa, close to the experimental value for liver CSD[9,16]. The protein encoded by this ORF contained the amino-acid sequence NPHK which is considered as a putative pyridoxal phosphate (PLP) binding site found in many PLP-dependent decarboxylases, as well as the sequence of the three CSD-P1, -P2 and -P3 peptides obtained from a homogeneous liver CSD fraction. The fusion protein, obtained by cloning the ORF in an expression vector, decarboxylated CSA to form hypotaurine and showed the same specificity for various substrates as the native enzyme (Fig. 2). These data are the clearest evidence that the present sequence, reported in detail elsewhere[12] (Genbank accession number X 94152), corresponds to the cDNA of CSD.

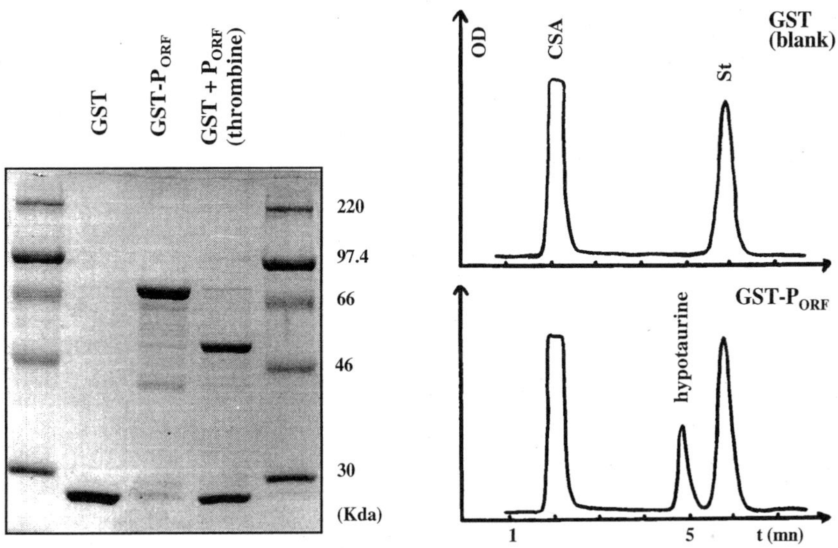


| Substrate | Reaction product | Native CSD | GST-$P_{ORF}$ |
|---|---|---|---|
| CSA | hypotaurine | +++ | +++ |
| CA | taurine | ++ | ++ |
| HCSA | homohypotaurine | - | - |
| HCA | homotaurine | - | - |
| Glutamate | GABA | - | - |

**Figure 2.** Analysis of the fusion protein GST-$P_{ORF}$. The ORF, amplified through PCR from a putative cloned CSD-cDNA using appropriate primers, was inserted after a glutathione-S-transferase (GST) gene of a fusion plasmid vector. SDS-PAGE shows, GST (MW: 27 kDa), the fusion protein GST-$P_{ORF}$ (MW: 75 kDa) purified by glutathione affinity chromatography from transformed bacteria, as well as the protein $P_{ORF}$ (MW: 52kDa) and GST (MW: 27 kDa) obtained from the fusion protein following site specific cleavage by thrombin protease. CSA was incubated with either GST (blank) or the fusion protein GST-$P_{ORF}$. HPLC analysis shows the formation of a reaction product following incubation with GST-$P_{ORF}$ that was identified as hypotaurine according to its retention time. St: HPLC internal standard. The fusion protein decarboxylated CA to form taurine, but did not transform the homoanalogues, HCSA and HCA, or glutamate. Thus the fusion protein showed the same substrate specificity as the native protein determined in a previous study[3].

## Molecular Identity of Brain and Liver CSD

The sequence of CSD from rat brain was determined through another approach. Using different pairs of primers defined from the sequence of liver CSD-cDNA, overlapping RT-PCR products were obtained from brain poly(A+)-mRNA, cloned and then sequenced. Altogether a sequence of 1687 bp was determined that showed nearly 100% identity with the sequence of liver CSD-cDNA. The sequence of the protein encoded by the corresponding ORF was strictly identical to liver CSD. Thus, these data show that there exists in brain a mRNA encoding a protein identical to liver CSD. They establish at the molecular level the identity of brain and liver CSD that we have suggested previously[9].

## CSD Sequence Versus Sequence of Other Decarboxylases

CSD belongs to a family of PLP-dependent enzymes which decarboxylate α-amino acids to form the corresponding amines. Some of these decarboxylases have already been cloned and sequenced. CSD shows a strong homology with each isoform of glutamic acid decarboxylase ($GAD_{65}$ and $GAD_{67}$) (Fig. 3A). It must be noted that their respective substrates, CSA and glutamate, share a common feature: both have an acidic group on the lateral chain, albeit of a different chemical nature, sulfinic (-SOOH) versus carboxylic (-COOH) group. CSD and GAD may thus derive from a common ancestor. Since GAD can decarboxylate CSA while CSD does not decarboxylate glutamate[6], CSD may have

A) Phylogenetic tree

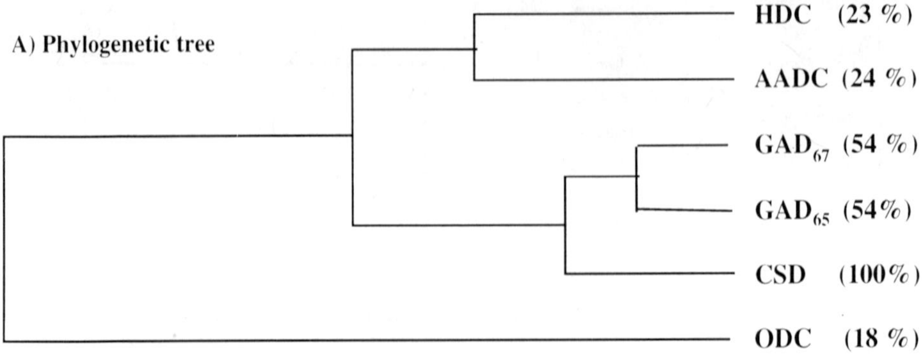


B) Region of PLP binding site

```
CSD     PLDAIADVCQRHGLWLHVDAAWGGSVLLSRTHRHLLDGIQRADSVAW  NPHK  LLAAGLQCSALLLRD
GAD67   PIQEIADICEKYNLWLHVDAAWGGGLLMSRKHRHKLSGIERANSVTW  NPHK  MMGVLLQCSAILVKE
GAD65   PLLAVADICKKYKIWMHVDAAWGGGLLMSRKHKWKLNGVERANSVTW  NPHK  MMGVPLQCSALLVRE
AADC    NLLEBGPICNQEGVWLHIDAAYAGSAFICPEFRYLLNGVEFADSFNF  NPHK  WLLVNFDCSAMWVKK
HDC     KLSELGPICAREGLWLHVDAAYAGTAFLRPELRGFLKGIEYADSFTF  NPSK  WMMVHFDCTGFWVKD
        ** * ****  ************* ** ** **   ************  ****  * ********  **
```

**Figure 3.** Sequence homology between the decarboxylases. The amino-acid sequence of the different decarboxylases, CSD, aromatic amino acid decarboxylase (AADC), glutamic acid decarboxylase (GAD), histidine decarboxylase (HDC) and ornithine decarboxylase (ODC) were compared through an appropriate computer program. a) Phylogenetic tree drawn after cross-comparison of the sequences. The percentage of homology with CSD is given in parenthesis. CSD shows the strongest homology with both isoforms of GAD ($GAD_{67}$ and $GAD_{65}$). b) Sequence alignment in the vicinity of the PLP binding site (framed) of CSD, AADC, GAD and HDC. Amino acids common to the five decarboxylases are in bold letters. Amino acids common to at least three decarboxylases are indicated by a star. All five decarboxylases show a strong sequence homology in the binding region of the PLP coenzyme.

evolved more from this common ancestor than GAD. The homology with the other decarboxylases is much weaker, except in the vicinity of PLP binding site (Fig. 3B). This region is likely to represent the site for molecular interactions between substrate, enzyme and PLP that lead to the reaction of decarboxylation through a similar mechanism while the more distant variable sequences may be involved in the specific recognition of their distinct substrates.

### Northern Blot of CSD-mRNA

Using a phospholabeled CSD-cDNA as a probe, a single band around 2.5 bp mRNA was detected in RNA extracts from liver, brain and kidney (Fig. 4). No band was detected in heart RNA extracts. CSD-mRNA in brain did not appear to be very abundant. Its detection on a Northern blot required the analysis of around 10 µg of poly(A+)-mRNA i.e. the amount extracted from roughly 1 g of crude brain tissue. Therefore, in order to be able to measure CSD-mRNA in smaller brain regions or astrocytes in primary culture, we have developed a quantitative assay for CSD-mRNA through competitive RT-PCR.

### Quantification of CSD-mRNA Through Competitive RT-PCR

This method relies on the *simultaneous* amplification of a fragment of endogenous CSD-mRNA and of a fragment of an exogenous standard CSD-RNA through RT-PCR. Each RT-PCR amplified fragment was quantified and the unkown amount of mRNA-CSD was determined in reference to known amounts of exogenous standard CSD-RNA (Fig. 5). The specificity of the amplification was established by checking that the sequence of the RT-PCR amplified product corresponded to that of CSD-mRNA. The PCR-amplification method improved the sensitivity of detection of CSD-mRNA about 1000-fold compared to Northern Blot. The standard, that was simultaneously coamplified with the primers, allowed for normalization of the data, thereby correcting for the inherent variability of PCR amplification. Thus, the intra-assay variability was less than 5%. In addition, since the standard CSD-RNA was added in known amounts, an absolute quantification of CSD-mRNA could be achieved by this method.

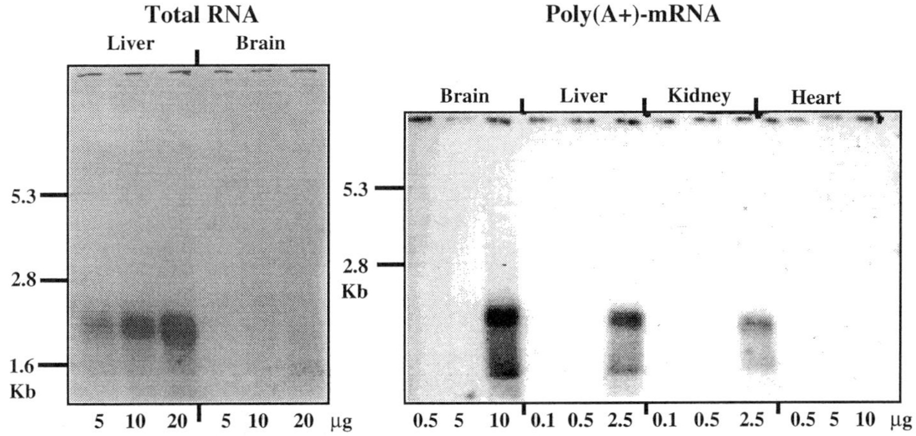


**Figure 4.** Northern blot of CSD-mRNA. Increasing amounts of total RNA or poly(A+)-mRNA as indicated were hybridized with random-primed phospholabeled CSD-cDNA probe. A main single band around 2.5 kb is detected. The light smear below the main band likely reflects some degradation of the poly(A+)-mRNA.

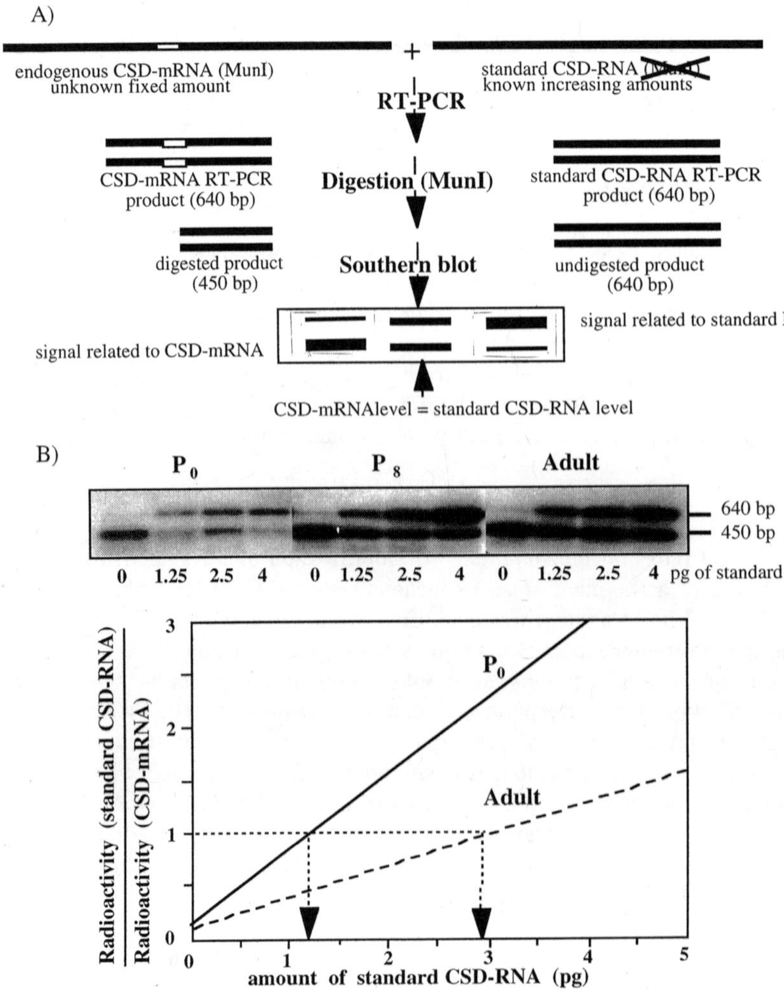


**Figure 5.** Quantitative assay of CSD-mRNA through competitive RT-PCR. A) Principle of the assay. Standard CSD-RNA, obtained by in vitro transcription of the antisense strand of cloned CSD-cDNA in which restriction site MunI was suppressed, is added in increasing amount to a fixed amount of total RNA extract. Endogenous CSD-mRNA and added standard CSD-RNA are RT reverse transcribed and PCR amplified simultaneously using the same primers. Two same-sized RT-PCR products are formed. The one derived from CSD-mRNA only is cleaved by restriction enzyme MunI. They are then separated and quantified through Southern blot. There is a competition for the formation of RT-PCR products. When both are formed in equal amount, the amount of endogenous CSD-mRNA in the initial extract is equal to the known amount of added standard CSD-RNA. B) Quantification of CSD-mRNA in rat brain RNA extract. One μg of brain total RNA from adult rats and neonates at birth ($P_0$) or 8 days post natal ($P_8$) was added to 0, 1.25, 2.5 and 4 pg of standard CSD-RNA. Following MunI digestion, RT-PCR products were separated by electrophoresis, electroblotted, hybridized with a phospholabeled oligonucleotide probe and autoradiographed. The radioactivity associated with each RT-PCR product was measured and the ratio was plotted against the amount of CSD-RNA standard. When this ratio was equal to 1 the amount of endogenous CSD-mRNA was equal to the amount of added standard (arrows).

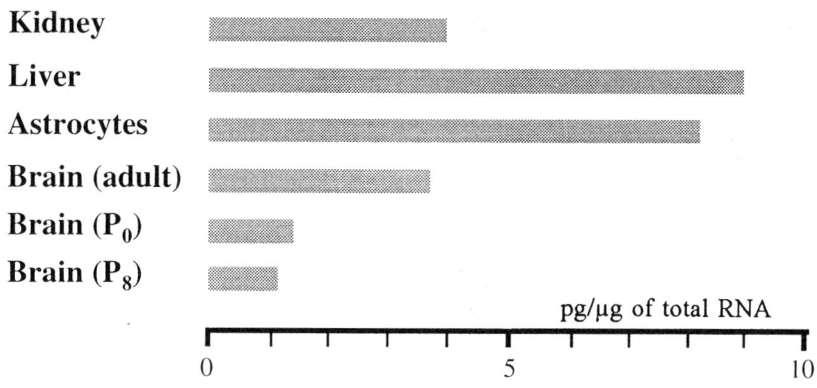


**Figure 6.** CSD-mRNA levels in various tissues. CSD-mRNA levels were determined through competitive RT-PCR using RNA extracts from liver, kidney, brain of adult rats, brain of neonates at birth ($P_0$) and 8 days postnatal ($P_8$) and from astrocyte three-weeks primary culture. Quantification was carried out in reference to a CSD-RNA standard. Intra-assay variability was less than 5%.

## Genomic Expression of CSD in Various Tissues

Absolute levels of CSD-mRNA were determined in RNA extracts from various tissues (Fig. 6). The highest level was found in liver RNA extracts, with a value close to 10 pg/µg of total RNA. Assuming that mRNA represents around 1% of total RNA, it can be estimated that CSD-mRNA represents between 0.1% and 0.02% of mRNA, at least according to the tissue extract. These figures confirm that CSD-mRNA is moderately abundant. CSD-RNA level is about 2.5 fold higher in liver RNA extracts compared to brain RNA extracts. The level of CSD protein was estimated to be about 20-fold higher in liver protein extracts compared to brain protein extracts[9]. Comparison of these figures thus suggests that regulation of post-transcriptional processes is different in liver and brain. CSD mRNA was enriched in astrocyte RNA extracts, in agreement with the strict location of CSD in these glial cells in brain[11]. During development CSD is already significantly expressed in brain at birth: CSD-mRNA levels in brain of neonates is nearly half the value found in adult rats. CSD thus appears to be an early marker of astrocyte differentiation. The pattern of CSD expression in brain during postnatal development was intermediate to that of other astrocyte markers, such as vimentin and glial acidic fibrillary protein (GFAP)[13].

## ACKNOWLEDGMENTS

This work was supported by INSERM and DRET (Grant 95/067). Marc Bitoun is the recipient of financial support by DRET.

## REFERENCES

1. Almarghini, K., Barbagli, B., and Tappaz, M., 1994, Production and characterization of a new specific antiserum against the taurine putative biosynthetic enzyme cysteine sulfinate decarboxylase, *J. Neurochem.*, 62:1604-1614.
2. De La Rosa, J. and Stipanuk, M. H., 1985, Evidence for a rate-limiting role of cysteine sulfinate decarboxylase activity in taurine biosynthesis in vivo, *Comp. Biochem. Physiol.*, 81B:565-571.
3. Do, K. Q. and Tappaz, M. L., 1996, Specificity of cysteine sulfinate decarboxylase (CSD) for sulfur-containing amino acids, *Neurochem. Int.*, 28:363-371.

4. Huxtable, R. J., 1989, Taurine in the central nervous system and the mammalian actions of taurine, *Prog. Neurobiol.*, 32:471-533.
5. Madsen, S., Ottersen, O. P., and Storm-Mathisen, J., 1985, Immunocytochemical visualization of taurine: neuronal localization in the rat cerebellum, *Neurosci. Lett.*, 60:255-260.
6. Oertel, W. H., Schmechel, D. E., Weise, V. K., Ranson, D. H., Tappaz, M., Krutzsch, H. C., and Kopin, I. J., 1981, Comparison of cysteine sulphinic acid decarboxylase isoenzymes and glutamic acid decarboxylase in rat liver and brain, *Neuroscience*, 6:2701-2714.
7. Oja, S. S. and Saransaari, P., 1996, Taurine as osmoregulator and neuromodulator in the brain, *Metab. Brain Dis.*, 11:153-164.
8. Ottersen, O. P., 1988, Quantitative assessment of taurine-like immunoreactivity in different cell types and processed in rat cerebellum: an electronmicroscopic study based on a postembedding immunogold labelling procedure, *Anat. Embryol.*, 178:407-421.
9. Remy, A., Henry, S., and Tappaz, M., 1990, Specific antiserum and monoclonal antibodies against taurine biosynthesis enzyme cysteine sulfinate decarboxylase (CSD): identity of brain and liver enzyme, *J. Neurochem.*, 54:870-879.
10. Reymond, I., Almarghini, K., and Tappaz, M., 1995, Immunocytochemical localization of cysteine sulfinate decarboxylase (CSD) in astrocytes: evidence for two distinct phenotypes, Abstr., '95 International Taurine Symposium (Osaka).
11. Reymond, I., Almarghini, K., and Tappaz, M., 1996, Immunocytochemical localization of cysteine sulfinate decarboxylase in astrocytes in the cerebellum and hippocampus: a quantitative double immunofluorescence study with glial fibrillary acidic protein and S-100 protein, *Neuroscience*, 75:619-633.
12. Reymond, I., Sergeant, A., and Tappaz, M., 1996, Molecular cloning and sequence analysis of the cDNA encoding rat liver cysteine sulfinate decarboxylase (CSD), *Biochim. Biophys. Acta*, 1307:152-156.
13. Sancho-Tello, M., Valles, S., Montoliu, C., Renau-Piqueras, J., and Guerri, C., 1995, Development pattern of GFAP and vimentin gene expression in rat brain and radial cultures, *Glia*, 15:157-166.
14. Tappaz, M., Almarghini, K., and Do, K. D., 1994, Cysteine sulfinate decarboxylase (CSD) in brain: identification, characterization and immunocytochemical location in astrocytes, *Adv. Exp. Med. Biol.*, 359:257-268.
15. Tappaz, M., Almarghini, K., Legay, F., and Remy, A., 1992, Taurine biosynthesis enzyme cysteine sulfinate decarboxylase (CSD) from brain - The long and tricky trail to identification, *Neurochem. Res.*, 17:849-859.
16. Weinstein, C. L. and Griffith, O. W., 1987, Multiple forms of rat liver cysteine sulfinate decarboxylase, *J. Biol. Chem.*, 262:7254-7263.

# EFFECT OF TAURINE AND GUANIDINOETHANE SULFONATE ON GLUTATHIONE METABOLISM IN THE RAT

Chong Chao Yan and Ryan J. Huxtable

Department of Pharmacology
University of Arizona
College of Medicine
Tucson, AZ 85724

## INTRODUCTION

Taurine is synthesized via the oxidation and decarboxylation of cysteine (Cys), a precursor for glutathione (GSH) biosynthesis. GSH is the major intracellular nonprotein sulfhydryl compound and is important in various biochemical processes, including detoxification of reactive oxygen species and electrophilic metabolites formed from xenobiotics. It is known that dietary sulfur amino acids, such as Cys or methionine, significantly influence the synthesis of taurine[9,27] and GSH[9,29]. However, the effect of dietary taurine on Cys or GSH metabolism has long been ignored.

We have examined the effect of taurine on GSH metabolism in liver, heart and brain in rats. Tissue taurine concentration was modified by the administration of either 1% taurine or 1% guanidinoethane sulfonate (GES) in drinking water for 4 weeks. GES depletes taurine content in various tissues and it has been used to study the functions of taurine in mammalian cells and tissues[2,3,5,8,10,15,16,20,23,25,26]. To explore the possible influence of taurine or GES on GSH metabolism, the activities of enzymes involved in GSH metabolism were also determined: γ-glutamylcysteine (γ-Glu-Cys) synthetase, the rate-limiting enzyme for GSH synthesis; GSH reductase and GSH peroxidase, involved in GSH and GSSG balance; γ-glutamyl transpeptidase, concerned with the degradation and translocation of GSH; and dipeptidase, an enzyme metabolizing cysteinylglycine (Cys-Gly).a product of γ-glutamyl transpeptidase.

## MATERIALS AND METHODS

### Chemicals

Taurine, glutathione (GSH), L-Cys, glycine (Gly), L-glutamic acid, DL-dithiothreitol

*Taurine 3*
Edited by Schaffer *et al.*, Plenum Press, New York, 1998

(DTT), γ-Glu-Cys, Cys-Gly, glycylglycine, N-ethylmorpholine, NADPH, GSH reductase, oxidized glutathione (GSSG), $H_2O_2$ and the kit for protein determination were purchased from Sigma (St. Louis, MO). Monobromobimane (MB; Thiolyte Reagent) was from Calbiochem (San Diago, CA). GES was synthesized according to the procedure of Huxtable *et al.*[16]

**Procedures**

Eighteen male Sprague-Dawley rats (body weight 60-75 g) were obtained from the University of Arizona Division of Animal Resources and housed 2-3 per cage in a room maintaining 12 hr cycles of light and dark. After 3 days for adaptation, the animals were randomly divided into 3 groups, with 6 animals in each group. Rats in the control group were given tap water, while rats in the taurine group and GES groups were maintained on 1% taurine and 1% GES in drinking water, respectively. Animals were allowed eating and drinking *ad libitum*. After 4 weeks, all animals were killed. To avoid differences due to diurnal variation in GSH concentration and enzyme activities, the animals were killed by decapitation between 9:00-10:00 am. Whole blood was collected for amino acid assay. Liver, heart, and brain were removed for determination of amino acid composition, GSH content and enzyme activities.

Tissue cytosol and microsome isolation, assays of tissue Cys and GSH concentration and the activities of γ-Glu-Cys synthetase, γ-glutamyl transpeptidase and dipeptidase were performed as previously described[30-33].

Cytosolic GSH reductase and GSH peroxidase activities were assayed by monitoring the disappearance of NADPH absorption at 340 nm[24,28].

Amino acid determination was conducted as previously reported[30].

**Statistics**

Statistics comparisions were obtained by MANOVA followed by a Student "t" test (Graphpad Software, version 1.14).

## RESULTS

As we have previously reported, administration of 1% taurine or GES in drinking water for 4 weeks modified taurine concentration in rat tissues[16]. Taurine treatment significantly increased taurine concentration in whole blood, liver and heart, but not in brain. On the other hand, supplement of GES significantly decreased taurine concentration in all examined tissues (Fig. 1).

Tissue Cys concentration was also influenced by taurine or GES administration. Compared to control animals, both taurine and GES treatment significantly lowered concentrations of Cys in liver and brain without altering Cys concentration in the heart (Fig. 2).

Compared to the control, administration of taurine significantly depressed hepatic GSH concentration. However, supplementation with GES significantly elevated threefold GSH concentration in heart. In addition, GES treatment resulted in a significantly higher GSH concentration in brain compared to the taurine treatment (Fig. 3). However, neither taurine nor GES produced a significant change from control in brain GSH concentration (Fig. 3).

To explore the mechanism of GSH changes following taurine or GES treatment, a simultaneous determination of the specific activities of the enzymes involved GSH meta-

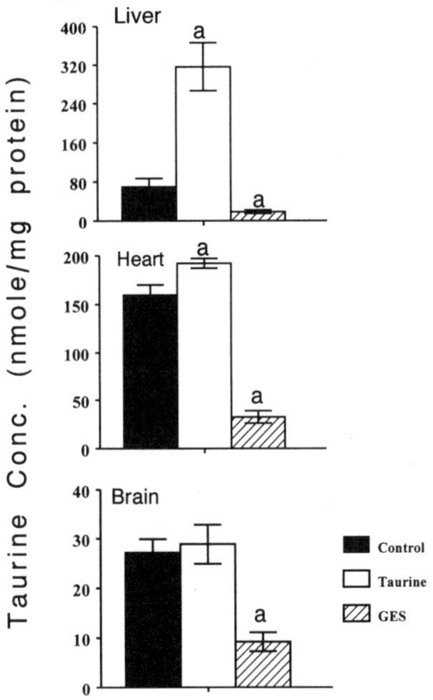


**Figure 1.** Taurine concentration in rat tissues after taurine or GES administration for 4 weeks. Data are means ± SD of 6 animals per group, [a] $p < 0.05$ vs the control.

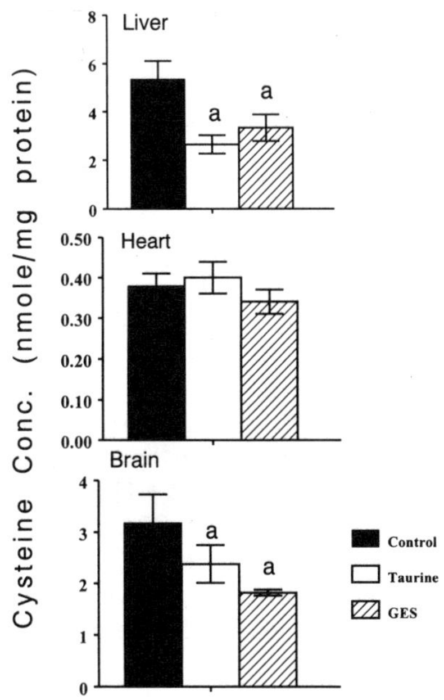


**Figure 2.** Cysteine concentration in rat tissues after taurine or GES administration for 4 weeks. Data are means ± SD of 6 animals per group, [a] $p < 0.05$ vs the control.

**Table 1.** Effect of taurine and GES administration on the activity of enzymes related to glutathione metabolism in rat liver

| Enzyme activity | Control | Taurine | GES |
|---|---|---|---|
| γ-Glutamylcysteine synthetase | 7.05 ± 0.94 | 7.26 ± 0.84 | 4.78 ± 0.68$^{ab}$ |
| GSH reductase | 10.76 ± 0.58 | 9.96 ± 0.70 | 11.82 ± 0.50 |
| GSH peroxidase | 6.73 ± 1.11 | 5.76 ± 0.41 | 6.13 ± 0.69 |
| γ-Glutamyl transpeptidase | 2.88 ± 0.58 | 2.47 ± 0.25 | 2.77 ± 0.54 |
| Dipeptidase | 4.48 ± 0.56 | 2.90 ± 0.71$^{a}$ | 4.93 ± 0.36 |

$^{a}p < 0.05$ versus control; $^{b}p < 0.05$ versus taurine-treated

**Table 2.** Effect of taurine and GES administration on the activity of enzymes related to glutathione metabolism in rat heart

| Enzyme activity | Control | Taurine | GES |
|---|---|---|---|
| γ-Glutamylcysteine synthetase | 3.65 ± 0.66 | 3.29 ± 1.11 | 5.85 ± 0.91$^{ab}$ |
| GSH reductase | 10.24 ± 0.98 | 9.21 ± 1.68 | 11.74 ± 0.75$^{b}$ |
| GSH peroxidase | 6.58 ± 0.45 | 7.21 ± 0.31 | 6.03 ± 0.56$^{b}$ |
| γ-Glutamyl transpeptidase | 0.68 ± 0.32 | 0.56 ± 0.16 | 0.58 ± 0.16 |
| Dipeptidase | 0.76 ± 0.09 | 1.19 ± 0.30$^{a}$ | 0.42 ± 0.19$^{b}$ |

$^{a}p < 0.05$ versus control; $^{b}p < 0.05$ versus taurine-treated

**Table 3.** Effect of taurine and GES administration on the activity of enzymes related to glutathione metabolism in rat brain

| Enzyme activity | Control | Taurine | GES |
|---|---|---|---|
| γ-Glutamylcysteine synthetase | 0.77 ± 0.07 | 0.83 ± 0.07 | 0.94 ± 0.18 |
| GSH reductase | 5.19 ± 0.22 | 4.91 ± 0.50 | 5.15 ± 0.40 |
| GSH peroxidase | 11.96 ± 1.51 | 11.00 ± 0.53 | 13.15 ± 0.58$^{b}$ |
| γ-Glutamyl transpeptidase | 2.88 ± 0.53 | 2.71 ± 0.65 | 2.50 ± 0.77 |
| Dipeptidase | 3.62 ± 0.47 | 2.50 ± 0.29$^{a}$ | 2.84 ± 0.63 |

$^{a}p < 0.05$ versus control; $^{b}p < 0.05$ versus taurine-treated

bolism was performed (Tables 1-3). Compared to the control, supplementation with taurine significantly depressed dipeptidase activity in liver and brain, but elevated dipeptidase activity in the heart. GES treatment decreased γ-Glu-Cys synthetase activity in the liver but increased this enzyme activity in the heart. Compared with the taurine group, the GES group showed significantly higher activities of γ-Glu-Cys synthetase and GSH reductase in the heart and higher GSH peroxidase activity in brain, but lower activity of hepatic γ-Glu-Cys synthetase and lower activities of GSH peroxidase and dipeptidase in the heart.

## DISCUSSION

Taurine has been shown to protect against toxic exposures in liver, heart and lung[14]. However, the mechanism is still unidentified. Biochemically, taurine is synthesized from

Cys by oxidation and decarboxylation. Cys is also the substrate for γ-Glu-Cys synthetase, the enzyme mediating the first step in GSH synthesis. Cys availability helps regulate GSH synthesis[1,4,6,7,9,22]. On the other hand, intracellular GSH serves as a reservoir for Cys. Thus, taurine, Cys and GSH have a close metabolic relationship, the study of which may yield insight into the mechanism by which taurine exerts some of its enantiostatic actions[13,14].

The present study establishes the following: (i) taurine concentrations in the tissues examined were significantly modified by taurine or GES administration, in confirmation of earler findings; (ii) Cys concentrations in liver and brain were significantly lower in treated animals than in controls; (iii) GSH concentrations were significantly lower in liver following taurine treatment but higher in heart and brain following GES administration; and (iv) enzyme activities related to GSH metabolism were influenced by both taurine and GES administration.

The mechanism by which taurine and GES depress tissue Cys concentration in liver (and brain) is unclear. The large increase in GSH concentration in heart following GES administration seems to be due to stimulation of GSH synthesis, as indicated by the increase in γ-Glu-Cys synthetase activity (Table 2). There may also be increased availability of Cys for GSH synthesis as a result of inhibition of taurine biosynthesis, as suggested by Ide and Murata[21] and Huxtable[11,17]. Although an increase in exogenous taurine decreases the proportion of whole body taurine obtained from biosynthesis, this is not due to inhibition of biosynthesis but to increased turnover of body taurine pools[12,18,19]: i.e. biosynthetic taurine is cleared faster. The decreased hepatic GSH in taurine-exposed animals may be partially due to lowered Cys levels, but the reason for these lowered levels is unclear.

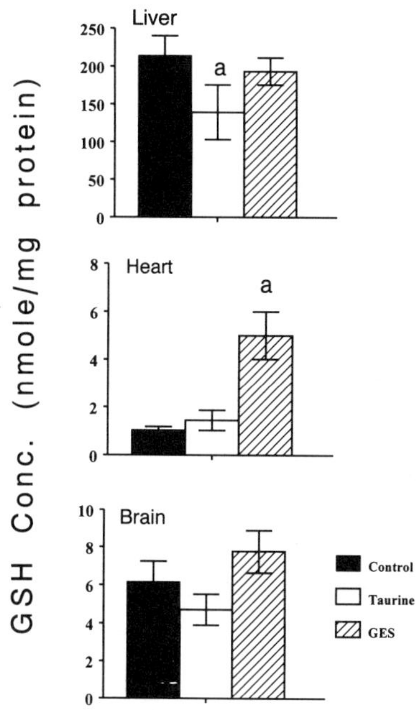


**Figure 3.** Glutathione concentration in rat tissues after taurine or GES administration for 4 weeks. Data are mean ± SD of 6 animals per group, [a] $p < 0.05$ vs the control.

## REFERENCES


1. Aebi, S. and Lauterburg, B.H. 1992, Divergent effects of intravenous GSH and cysteine on renal and hepatic GSH, *Am. J. Physiol. Regul. Integr. Comp. Physiol.*, 263:R348-R352.
2. Bonhaus, D.W., Lippincott, S.E., and Huxtable, R.J. 1983, Effect of pre- and postnatal exposure to guanidinoethane sulfonate upon taurine concentrations in the neonatal rat, *Federation Proceedings,* 42:1054.
3. Bonhaus, D.W., Pasantes-Morales, H., and Huxtable, R.J. 1985, Actions of guanidinoethane sulfonate on taurine concentration, retinal morphology and seizure threshold in the neonatal rat, *Neurochem. International,* 7:263-270.
4. Chung, T.K., Funk, M.A., and Baker, D.H. 1990, L-2-Oxothiazolidine-4-carboxylate as a cysteine precursor: Efficacy for growth and hepatic glutathione synthesis in chicks and rats, *J. Nutr.*, 120:158-165.
5. Cocker, S. and Lake, N. 1988, Electroretinographic alterations and their reversal in rats treated with guanidinoethyl sulfonate, a taurine depletor, *Exp. Eye Res.*, 45:977-987.
6. Dalhoff, K. and Poulsen, H.E. 1992, Effects of cysteine and acetaminophen on the syntheses of glutathione and adenosine 3'-phosphate 5'-phosphosulfate in isolated rat hepatocytes, *Biochem. Pharmacol.*, 44:447-454.
7. Dalhoff, K. and Poulsen, H.E. 1993, Synthesis rates of glutathione and activated sulphate (PAPS) and response to cysteine and acetaminophen administration in glutathione-depleted rat hepatocytes, *Biochem. Pharmacol.*, 46:1295-1297.
8. De La Rosa, J. and Stipanuk, M.H. 1984, Effect of guanidinoethane sulfonate adminis-tration on taurine levels in rat dams and their pups, *Nutrition Reports Inter.*, 30:1121.
9. Glazenburg, E.J., Jekel-Halsema, I.M.C., Scholtens, E., Baars, A.J., and Mulder, G.J. 1983, Effects of variation in the dietary supply of cysteine and methionine on liver concentration of glutathione and active sulfate (PAPS) and serum levels of sulfate, cystine, methionine and taurine: relation to the metabolism of acetaminophen, *J. Nutr.*, 113:1363-1374.
10. Hiramatsu, M., Niiya-Nishihara, H., and Mori, A. 1982, Effect of taurocyamine on taurine and other amino acids in the brain, liver and muscle of mice, *Neurosciences (Kobe),* 8:289-294.
11. Huxtable, R.J. 1982, Guanidinoethane sulfonate and the disposition of dietary taurine in the rat, *J. Nutr.*, 12:2293-2300.
12. Huxtable, R.J. 1986, Taurine and the oxidative metabolism of cysteine, in: "Biochemistry of Sulfur",. Huxtable, R.J., ed., Plenum Press, New York. pp. 121-198.
13. Huxtable, R.J. 1992, Physiological actions of taurine, *Physiol. Rev.*, 72:101-163.
14. Huxtable, R.J. 1996, Taurine: Past, present and future, in: "Taurine: Basic and Clinical Aspects",. Huxtable, R.J., Azuma, J., Nakagawa, M., Kuriyama, K., and Baba, A., eds., Plenum, New York. pp.
15. Huxtable, R.J., Bonhaus, D., Nakagawa, K., Laird, H.E., and Pasantes-Morales, H. 1985, Taurine and the actions of guanidinoethane sulfonate, in: "Guanidines: Historical, Biological, Biochemical and Clinical Aspects of the Naturally-Occurring Guanidino Compounds",. Mori, A., Cohen, B.D., and Lowenthal, A., eds., Plenum Press, New York. pp. 213-226.
16. Huxtable, R.J., Laird, H.E., and Lippincott, S.E. 1979, The transport of taurine in the heart and the rapid depletion of tissue taurine content by guanidinoethyl sulfonate, *J. Pharmacol. Exptl. Therap.*, 211:465-471.
17. Huxtable, R.J. and Lippincott, S.E. 1981, Comparative metabolism and taurine depleting effects of guanidinoethane sulfonate in cats, mice, and guinea pigs, *Arch. Biochem. Biophys.*, 210:698-709.
18. Huxtable, R.J. and Lippincott, S.E. 1982a, Relative contribution of diet and biosyn-thesis to the taurine content of the adult rat, *Drug-Nutrient Inter.*, 2:153-168.
19. Huxtable, R.J. and Lippincott, S.E. 1982b, Sources and turnover rates of taurine in newborn, weanling, and mature rats, in: "Taurine in Nutrition and Neurology",. Huxtable, R.J. and Pasantes-Morales, H., eds., Plenum Press, New York. pp. 23-45.
20. Huxtable, R.J., Murphy, J., and Lleu, P.-L. 1997, Guanidinoethane sulfonate in the investigation of taurine-phospholipid interactions in the developing brain, in: "Guanidino Compounds in Biology & Medicine:2",. De Deyn, P.P., Marescau, B., Qureshi, I.A., and Mori, A., eds., John Libbey, London. pp. 359-371.
21. Ide, T. and Murata, M. 1994, Reduction by guanidinoethane sulfonate of the activities of enzymes involved in taurine synthesis in rat liver, *Biosci. Biotechnol. Biochem.*, 58:1584-1588.
22. Jain, A., Madsen, D.C., Auld, P.A.M., Frayer, W.W., Schwartz, M.K., Meister, A., and Mårtensson, J., 1995, L-2-Oxothiazolidine-4-carboxylate, a cysteine precursor, stimulates growth and normalizes tissue glutathione concentrations in rats fed a sulfur amino acid-deficient diet, *J. Nutr.*, 125:851-856.

23. Lake, N. 1981, Depletion of retinal taurine by treatment with guanidinoethyl sulfonate, *Life Sci.*, 29:445-448.
24. Lawrence, R.A., Sunde, R.A., Schwartz, G.L., and Hoekstra, W.G. 1974, Glutathione peroxidase activity in rat lens and other tissues in relation to dietary selenium intake, *Exp. Eye Res.*, 18:563-569.
25. Lehmann, A., Huxtable, R.J., and Hamberger, A. 1987, Taurine deficiency in the rat and cat: effects on neurotoxic and biochemical actions of kainate, in: "The Biology of Taurine: Methods and Mechanisms",. Huxtable, R.J., Franconi, F., and Giotti, A., eds., Plenum Press, New York. pp. 331-339.
26. Lombardini, J.B. 1981, Combined effects of guanidinoethanesulfonate, a depletor of tissue taurine levels, and isoproterenol or methoxamine on rat tissues, *Biochem. Pharmacol.*, 30:1698-1701.
27. Lombardini, J.B. and Medina, E.V. 1978, Effects of dietary inorganic sulfate, taurine, and methionine on tissue levels of taurine in the growing rat, *J. Nutr.*, 108:428-433.
28. Massey, V. and Williams, C.H. 1965, On the reaction mechanism of yeast glutathione reductase, *J. Biol. Chem.*, 240:4470-4480.
29. Takeuchi, F., Izuta, S., Tsubouchi, R., and Shibata, Y. 1991, Glutathione levels and related enzyme activities in vitamin B-6-deficient rats fed a high methionine and low cystine diet, *J. Nutr.*, 121:1366-1373.
30. Yan, C.C. and Huxtable, R.J. 1995a, Fluorimetric determination of monobromobimane and o-phthalaldehyde adducts of γ-glutamylcysteine and glutathione: Application to assay of γ-glutamylcysteinyl synthetase activity and glutathione concentration in liver, *J. Chromatogr. B Biomed. Appl.*, 672:217-224.
31. Yan, C.C. and Huxtable, R.J. 1995b, Relationship between glutathione concentration and metabolism of the pyrrolizidine alkaloid, monocrotaline, in the isolated, perfused liver, *Toxicol. Appl. Pharmacol.*, 130:132-139.
32. Yan, C.C. and Huxtable, R.J. 1996a, Effect of the pyrrolizidine alkaloid monocrotaline on taurine and sulfur amino acid metabolism in the rat liver, *Adv. Exp. Med. Biol.*, 403:135-145.
33. Yan, C.C. and Huxtable, R.J. 1996b, Effects of monocrotaline, a pyrrolizidine alkaloid, on glutathione metabolism in the rat, *Biochem. Pharmacol.*, 51:375-379.

# DETERMINATION OF CYSTEINYL-CONTAINING PEPTIDES AND ASSOCIATED ENZYME ACTIVITIES IN RAT TISSUES BY REVERSE PHASE HPLC

Chong Chao Yan and Ryan J. Huxtable

Department of Pharmacology
College of Medicine
University of Arizona
Tucson, Arizona 85724

## INTRODUCTION

We have developed a sensitive chromatographic method using fluorescence detection for the rapid analysis of monobromobimane-derivatized cysteine and cysteine-containing peptides. The method allows determination of these substances and assay of the enzymes involved in their formation and breakdown: γ-glutamylcysteine synthetase (γ-Glu-Cys synthetase), GSH synthetase, γ-glutamyl transpeptidase (γ-Glu transpeptidase) and dipeptidase. In the assay of GSH synthetase, 90% of the endogenous interfering GSH can be removed by preincubation with thiopropyl Sepharose resin. The distribution of the four enzymes listed above has been measured in liver, heart, lung and kidney of the rat. Some kinetic constants are: hepatic γ-Gly-Cys synthetase: $K_m$ (glutamate) 3.36 ± 0.33 mM, $V_{max}$ 85.0 ± 2.3 nmol/min/mg protein ($K_i$ for buthionine sulfoximine 1.84 ± 0.11 μM); hepatic GSH synthetase: high affinity $K_m$ (glycine) 0.67 ± 0.30 mM, $V_{max}$ 7.86 ± 1.17 nmol/min/mg protein; renal γ-Glu transpeptidase: $K_m$ (glycylglycine) 8.63 ± 0.07 mM, $V_{max}$ 23.6 ± 0.7 μmol/min/mg protein ($K_i$ for AT-125 2.64 ± 0.17 mM); renal dipeptidase: $K_m$ (cysteinylglycine) 0.55 ± 0.06 mM, $V_{max}$ 100.6 ± 4.5 nmol/min/mg protein.

Glutathione (γ-L-glutamyl-L-cysteinylglycine; GSH) is present in all cells and functions directly or indirectly in many important biological phenomena, including transport, regulation of enzyme activity, metabolism and protection of cells, and redox processes involved in the synthesis of proteins and DNA[17]. GSH is synthesized intracellularly by the consecutive actions of γ-Glu-Cys synthetase (EC 6.3.2.2) and GSH synthetase (EC 6.3.2.3).

These enzymes catalyze, respectively, the following reactions:

$$\text{L-glutamate} + \text{L-cysteine} + \text{ATP} \rightarrow \text{L-}\gamma\text{-Glu-Cys} + \text{ADP} + P_i$$

$$\text{L-}\gamma\text{-Glu-Cys} + \text{glycine} + \text{ATP} \rightarrow \text{GSH} + \text{ADP} + P_i$$

The breakdown of GSH is catalyzed by γ-Glu transpeptidase (EC 2.3.2.2), which catalyzes the transfer of the γ-glutamyl moiety to acceptor molecules such as cystine, glutamine, and methionine, certain peptides, GSH itself and water[18]:

$$\text{GSH + L-amino acid} \rightarrow \text{L-}\gamma\text{-Glu-L-amino acid + L-Cys-Gly}$$

The product of γ-Glu transpeptidase, cysteinylglycine (Cys-Gly), is hydrolyzed by dipeptidase (EC 3.4.13.11):

$$\text{Cys-Gly} \rightarrow \text{Cys + Gly}$$

Although there are many methods for the quantitative determination of GSH and biological thiols, relatively few assays are available for determining enzyme activities related to GSH metabolism. Assays for GSH-synthesizing enzymes are based principally on the determination of the inorganic phosphate released from ATP during the reactions. Such assays have limited sensitivity in tissues having low enzyme activities, although the sensitivity can be enhanced by the utilization of [$\gamma$-$^{32}$P]ATP[25]. γ-Glu-Cys synthetase and GSH synthetase activities can also be assayed by measurement of the formation of ADP in the presence of pyruvate kinase, lactate dehydrogenase, phosphoenolpyruvate, and NADH[24]. Although these assays are appropriate for isolated enzyme preparations, their utility in tissue homogenates and subcellular fractions are limited by lack of selectivity. Radiometric assays, which measure the incorporation of [$^{14}$C]Glu or [$^{14}$C]Gly into γ-Glu-Cys or GSH, are sensitive, but separation from excess substrates and the quantitation of the products are cumbersome[1,2,19,27].

HPLC separation of the thiol derivatives formed with dithionitrobenzene or monobromobimane (MB) has been used in the assay of GSH synthesis[3,9,20,23]. Of these methods, only the method of Hamel *et al.*[9] has been applied to the assay of γ-Glu-Cys synthetase and GSH synthetase activities in different tissues of the rat. Their assay yields an unidentified peak between cysteine (Cys) and GSH, necessitating a somewhat longer HPLC run.

Many methods are available for determining γ-Glu transpeptidase activity, including both spectrometric and HPLC assays[10,11,26,28,29]. Few of these methods, however, utilize the natural substrate, GSH, as an assay substrate. Furthermore, there are few studies on the distribution of Glu transpeptidase and dipeptidase activities in rat other than in the kidney. In the present study, therefore, we have developed a gradient reverse phase-HPLC system for separating the MB derivatives of Cys and Cys-containing peptides, and have applied this method to the determination of enzymes involved in the synthesis and degradation of GSH in various tissues.

## MATERIALS AND METHODS

### Chemicals

GSH, Cys, glycine, L-glutamic acid, DL-dithiothreitol (DTT), γ-Glu-Cys, cysteinylglycine (Cys-Gly), glycylglycine (Gly-Gly), N-ethylmorpholine, buthionine sulfoximine, AT-125 (avicin) and the kit for protein determination were purchased from Sigma (St. Louis, MO). Monobromobimane (MB) under the name Thiolyte Reagent was purchased from Calbiochem (San Diego, CA). Thiopropyl Sepharose 6B resin was supplied from Pharmacia LKB Biotechnology AB (Uppsala, Sweden). Resin (1 g) was prepared for use by being suspended in 0.85% NaCl solution and rinsed thoroughly. β-Mercaptoethanol (0.1 ml) was added to the resin slurry and left for 30 min at room temperature. The resin was washed three times with 0.85% NaCl solution, and then suspended in 10 ml 0.85% NaCl[30].

**High Performance Liquid Chromatography**

HPLC was performed using a 110B solvent delivery module liquid chromatograph equipped with a programmable 421 A Controller, Beckman 157 fluorescence detector (Beckman, San Ramon, CA) and a Varian 4290 integrator. The excitation range of the detector was 303-395 nm and the detection range was 420-470 nm.

Separation was achieved using a Beckman Ultrasphere™ ODS (250 x 4.6 mm I.D., 5 μm particle size) reversed-phase stainless-steel column (Beckman, San Ramon, CA) maintained at room temperature. The analytical column was protected by a supelguard (5 x 4.6 mm, I.D.), pellicular reversed-phase cartridge precolumn (Supelco, Bellefonte, PA). Elution was carried out using a gradient program with 0.25% acetic acid as solvent A (pH 3.9) and methanol: 0.25% acetic acid (30:70, v/v) as solvent B (pH 3.9). The flow rate was 1.5 ml/min and the starting conditions were 35% B for 1 min followed by a linear gradient to 62% B within 6 min. This was maintained for another 2 min. The column was then washed with 100% B for 2 min and maintained under starting conditions for 5 min before reuse. Retention times are shown on Fig. 1.

**Preparation of Derivatized Standards and Samples**

MB itself is only weakly fluorescent but selectively reacts with thiols to yield stable, highly fluorescent thioethers which can be easily detected at the pmol level. Freshly prepared standard solutions of Cys, Cys-Gly, γ-Glu-Cys and GSH were added to a microfuge tube containing 20 μl of 50 mM N-ethylmorpholine (pH 8.7) and 10 μl of 50 mM MB in acetonitrile. The mixtures were incubated in the dark at room temperature for at least 15 min and diluted with water. 20 μl of the dilution was injected into the HPLC. Tissue incubation samples (100 μl) were mixed in a microfuge tube with 5% sulfosalicylic acid (50 μl) and centrifuged for 3 min to precipitate protein. Supernatants (100 μl) were derivatized as described for the standards. Derivatized samples can be stored at 4°C for 24 hr in the dark, or at -20°C for several weeks, without loss of fluorescence[31]. Analyses reported in this paper were performed within 12 hr of derivatization.

**Tissue Cytosol and Microsomal Preparation**

Male Sprague-Dawley rats (body weight 200-250 g) were obtained from the University of Arizona Division of Animal Resources and housed 2 to a cage in a room maintaining 12 hr cycles of light and dark. They were allowed food and tap water *ad libitum*. To avoid differences due to diurnal variation in enzyme activities related to tissue GSH metabolism, all rats were killed by decapitation between 9:00-10:00 am. Tissues were removed and washed 2 times in saline at 4°C. Each tissue was minced and homogenized in 4 vols of 0.25 M sucrose and 20 mM Tris-HCl containing 1 mM EDTA (pH 7.4). The homogenate was centrifuged at 3000 g for 10 min at 4°C and then the supernatant was centrifuged at 10,000 g for 20 min at 4°C. Finally, the supernatant was centrifuged again at 105,000 g for 60 min at 4°C. The supernatant from this final spin (cytosol) was used to determine the activities of γ-Glu-Cys synthetase and glutathione synthetase and the 105,000 g microsomal pellet were used to determine the activities of dipeptidase and γ-Glu transpeptidase.

### Enzyme Assays

The enzyme activity determinations were based on the formation of product in the presence of their substrates, protein, coenzymes and metal ions at 37°C. The reactions were stopped by adding 100 μl incubation to 50 μl 5% sulfosalicylic acid and centrifuging for 3 min to precipitate protein. The supernatant was derivatized with MB as described above and injected into the HPLC. Peaks were identified by comparison with the retention times of authentic standards (Fig. 1).

### Cytosolic γ-Glu-Cys Synthetase and GSH Synthetase

Cytosolic γ-Glu-Cys synthetase determination was based on the formation of γ-Glu-Cys. Unless otherwise stated, incubations were performed in 1.5 ml microfuge tubes containing 10 mM Glu, 5 mM cysteine, 10 mM ATP, 0.1 M Tris-HCl (pH 8.2), 0.15 M KCl, 2 mM EDTA and 20 mM $MgCl_2$. After preincubation for 3 min at 37°C, cytosolic protein (about 1 mg) was added to initiate the enzyme reaction. The total volume was 1.0 ml and the incubation time was 15 min.

The conditions for determination of GSH synthetase activity were the same as for γ-Glu-Cys synthetase except that Glu and Cys were replaced by 20 mM glycine and 5 mM γ-Glu-Cys. Enzyme activity was followed by measuring GSH formation. Endogenous GSH must be removed as much as possible. Prior incubation of cytosol (50 μl) with thiopropyl Sepharose 6B resin (500 μl of 1 g resin suspended in 5 ml 0.9% NaCl) at 4°C for 3 hr removed 90% of GSH. The mixture was placed in a microfuge tube and gently shaken in an incubator bath. After 3 hr, resin was removed by centrifugation. The content of GSH remaining in the cytosol was subtracted from the final GSH content to calculate the newly synthesized GSH. In general, sample values were around 10x blank values.

### Microsomal γ-Glutamyl Transpeptidase and Dipeptidase

γ-Glu transpeptidase was assayed based on the transfer of glutamyl from GSH to Gly-Gly with the concomitant release of Cys-Gly from GSH. The standard assay solution (1.0 ml) contained 0.1 M Tris-HCl buffer (pH 8.0), 75 mM NaCl, 20 mM Gly-Gly, 5 mM GSH, 50 μM DTT and microsomal enzyme (0.5-1.0 mg). The solution was incubated at 37°C for 20 min. To correct for hydrolytic release of Cys-Gly from GSH by γ-Glu transpeptidase, a blank tube lacking Gly-Gly was incubated at the same time[18]. For the kidney, the hydrolytic activity of transpeptidase was also assayed with 0.1 mM GSH and 0.1 M Tris-HCl (pH 8.0). After equilibration at 37°C, the reaction was initiated by addition of enzyme. Transpeptidase activity was 5-6x higher than hydrolytic release of Cys-Gly catalyzed by the same enzyme.

Microsomal dipeptidase was assayed based on Cys release from Cys-Gly. The incubation system contained 2 mM Cys-Gly and 0.1 M Tris-HCl (pH 8.0) in a final volume of 1 ml. The solution was brought to 37°C and incubated for 20 min. To correct for nonenzymatic hydrolysis of Cys-Gly, a blank tube lacking microsomal protein was prepared at the same time. Cys was determined by HPLC and taken to represent dipeptidase activity. For kidney, this procedure determines combined dipeptidase and aminopeptidase M activities. Cys-Gly is contaminated with Cys, leading to a relatively high sample/blank ratio averaging 3.5.

Protein determination was performed according to Lowry[14] using bovine serum albumin as a calibration standard.

## Kinetic Analysis

Enzyme rate data ($K_m$ and $V_{max}$) were fitted by nonlinear regression to the Michaelis-Menton equation using SigmaPlot7 software. Inhibition constants ($K_i$) were derived using the Pharm/PCS (version 4) software (Springer-Verlag, NY). Data are reported as means ± SD.

# RESULTS

## Chromatographic Separation

The MB method has not hitherto been applied to the determination of dipeptidase or γ-Glu transpeptidase activities. In order to assay enzymes involved in the biosynthesis of GSH, therefore, it was necessary to develop HPLC techniques for the separation and determination of their products. Under the conditions described in the Materials and Methods section, good separations were obtained within 16 min of the monobromobimane derivatives of Cys, Cys-Gly, γ-Glu-Cys and GSH (Fig. 1). For all four substances, chromatographic peak areas were linearly dependent on substrate up to at least 400 pmol. Amounts as low as 5 pmol could be easily detected and quantified. This separation procedure was used to determine the activities of four enzymes related to GSH metabolism: γ-Glu-Cys synthetase, GSH synthetase, γ-Glu transferase, and dipeptidase. Selected kinetic constants for these enzymes are reported on Table 1.

## γ-Glu-Cys Synthetase Activity

In cytosol of rat liver, Glu-Cys formation is proportional to incubation time up to at least 30 min, and is proportional to protein content up to 2 mg protein.

Glu (Fig. 2) and Cys are both substrates for the enzyme. Cys concentrations above 5 mM interfere with the assay due to competition for MB. This results in a significantly decreased peak area of γ-Glu-Cys on the chromatogram (Data not shown). This interference can be overcome by increasing the amount of MB used. For this reason, under our standard incubation conditions, Cys concentration was fixed at 5 mM. The use of 10 μl of 50 mM MB with 100 μl incubation supernatant is sufficient to derivatize all the γ-Glu-Cys under these conditions. Due to this problem, kinetic constants were not determined for cysteine.

Buthionine sulfoximine is a well-known inhibitor of γ-Glu-Cys synthetase[7]. We find buthionine sulfoximine to have an inhibition constant ($IC_{50}$) for the enzyme of 1.84 ± 0.11 μM (Fig. 2).

## GSH Synthetase Activity

Determination of GSH synthetase activity is based on measurement of the newly synthesized GSH in the incubation system. As endogenous GSH in the liver is a confounding variable, as much of this should be removed as much as possible. We have found a convenient and reproducible method of doing this; namely, the incubation of cytosol with thiopropyl Sepharose 6B resin at 4°C for 3 hr. This sulfhydryl-containing resin binds GSH by disulfide bond formation. As shown in Fig. 3, cytosolic GSH decreases exponentially with length of incubation, exhibiting a $t_{½}$ of 43.1 ± 2.5 min ($r^2$ 0.983). In the absence of resin, GSH levels fall only slightly (probably due to aerial oxidation to GSSG).

**Table 1.** Kinetic constants of enzymes involved in GSH synthesis and breakdown.

| Enzyme | Tissue | Substrate | $K_m$ (mM) | $V_{max}$ (nmol/min/mg protein) | Correlation coefficient ($r^2$) |
|---|---|---|---|---|---|
| γ-Glutamylcysteine Synthetase[a] | Liver | Glutamate | 3.36 ± 0.33 | 85.0 ± 2.3 | 0.997 |
| GSH Synthetase[b] | Liver | Glycine | 0.67 ± 0.30; 80.8 ± 57.3 | 7.86 ± 1.17; 32.3 ± 2.8 | 0.999 |
| γ-Glutamyl Transpeptidase[c] | Kidney | Glycylglycine | 8.63 ± 0.07 | 23580 ± 700 | 0.997 |
| Dipeptidase[d] | Kidney | Cysteinylglycine | 0.55 ± 0.06 | 100.6 ± 4.5 | 0.988 |

[a]see Figure 1; [b]see Figure 4; [c]see Figure 5; [d]see Figure 6

**Table 2.** Tissue distribution of enzymes involved in GSH synthesis and breakdown

| Tissues | γ-Glutamylcysteine Synthetase | GSH Synthetase | γ-Glutamyl Transpeptidase | Dipeptidase |
|---|---|---|---|---|
| | | nmol/min/mg | | |
| Liver | 7.02 ± 0.76 | 4.62 ± 0.38 | 0.90 ± 0.08 | 0.53 ± 0.05 |
| Heart | 1.42 ± 0.18 | 1.02 ± 0.12 | 26.7 ± 2.0 | 0.62 ± 0.08 |
| Lung | 0.87 ± 0.02 | 1.26 ± 0.09 | 12.2 ± 1.6 | 1.45 ± 0.13 |
| Kidney | 132 ± 8 | 39.2 ± 3.6 | 8360 ± 680 | 69.9 ± 3.9 |

Data are means ± SD for 5 samples per assay

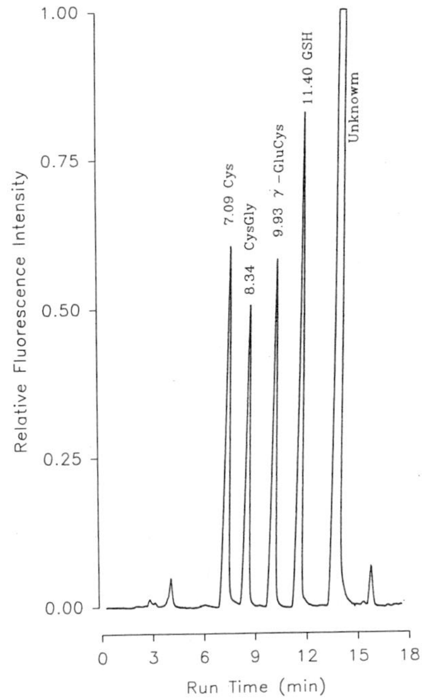


**Figure 1.** A chromatogram of standards of MB-derivatized cysteine and cysteine-containing peptides. The chromatographic conditions are described under Materials and Methods.

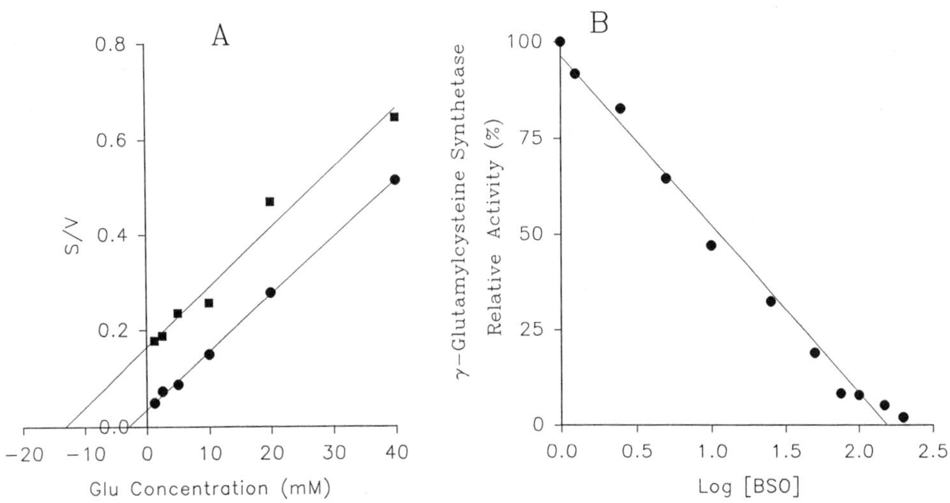


**Figure 2.** A: Hanes-Woolf plot of the effect of glutamate (Glu) concentration on production of γ-Glu-Cys by γ-Glu-Cys synthetase. Liver cytosol of male rat was used as enzyme source. Cys concentration was 5 mM. (●): control; (■): preincubation with buthionine sulfoximine (10 μM) for 5 min. B: Inhibition of γ-Glu-Cys synthetase activity by preincubation with buthionine sulfoximine. The $IC_{50}$ for buthionine sulfoximine is 1.84 ± 0.11 μM. See Table 1 for kinetic parameters.

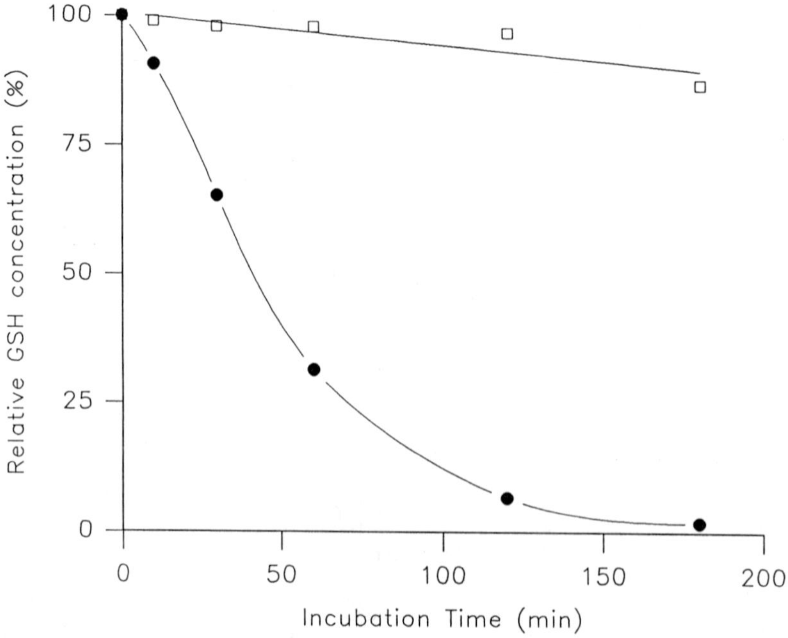


**Figure 3.** The effect of incubation with thiopropyl Sepharose 6B resin on cytosol GSH concentration. ●: 50 μl liver cytosol of male rat was incubated with 85 mg treated resin. The content of GSH remaining at the indicated time was determined by HPLC. The half-life for removal of GSH is 43.1 ± 2.5 min ($r^2$ 0.983). □: 50 μl liver cytosol incubated in the absence of resin.

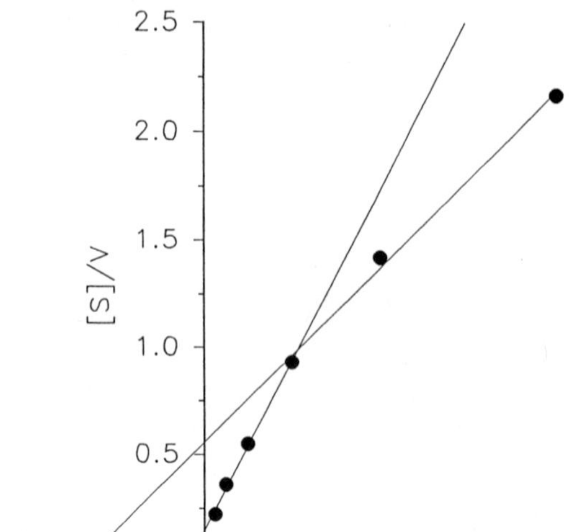


**Figure 4.** Hanes-Woolf plot of the effect of glycine concentration on rate of GSH formation by GSH synthetase. Liver cytosol of male rat was used as enzyme source. γ-Glu-Glu-Cys concentration was 5 mM. See Table 1 for kinetic parameters.

Extrahepatic tissues contain lower concentrations of GSH. For assay of these tissues, we found removal of endogenous GSH was in general not necessary. However, the initial GSH content in the incubation mixture still needs to be determined.

GSH synthetase activity in liver preparations is proportional to incubation time up to 30 min and proportional to protein content up to 3 mg (data not shown). γ-Glu-Cys concentrations above 10 mM interfered with the determination of GSH synthetase activity by competition with GSH for derivatization. Therefore, in this experiment, γ-Glu-Cys was fixed at 5 mM. At this concentration, all the GSH formed could be derivatized by the standard derivation process. We found two affinity sites for Gly in hepatic cytosol (Table 1).

### γ-Glutamyl Transpeptidase Activity

GSH degradation is initiated by γ-Glu transpeptidase located on the cell membrane. Many amino acids or peptides can serve as γ-glutamyl acceptors. However, Gly-Gly is among the most active dipeptide acceptors. The enzyme assay is based on the release of Cys-Gly from GSH on incubation of the enzyme with Gly-Gly. Cys-Gly can also be hydrolyzed by dipeptidase. However, the rate of this reaction is low as the concentration of Cys-Gly achieved by hydrolysis of GSH is well below the $K_m$ of the dipeptidase (see below). Under the conditions used here, Cys-Gly release was proportion to protein content up to 2 mg and proportional to time of incubation up to 30 min (Fig. 5).

No significant dipeptidase activity was seen when hepatic microsomal protein was used as the enzyme source. When kidney γ-Glu transpeptidase was determined (Fig. 5), a lower protein content (10-20 μg) and a shorter incubation time (5-10 min) was necessary. Under these conditions, we did not find significant interference by aminopeptidase or dipeptidase activities, which result in a Cys peak appearing on the chromatogram (see Fig. 1). AT-125 (5.6 μM) inhibited Cys-Gly release, showing a $K_i$ of 2.64 ± 0.17 mM (Fig 5).

Under the conditions of the assay, hydrolysis of GSH to Cys-Gly by γ-Glu transpeptidase (i.e. transfer of glutamyl to water rather than to Gly-Gly) was 0.625 μmol/min/mg protein in the kidney (determined in the absence of Gly-Gly as an acceptor).

### Dipeptidase Activity

Cys-Gly, the degradative product of GSH formed by γ-Glu transpeptidase, can be further metabolized by a membrane-bound dipeptidase. We have determined directly the release of Cys from L-Cys-Gly by incubation with microsomal enzymes. However, in kidney (Fig. 6), both dipeptidase and aminopeptidase M catalyze Cys-Gly hydrolysis. Therefore, the method used here determines the combined activities of dipeptidase and aminopeptidase in kidney microsomes. Dipeptidase is present largely on the surface of renal brush border membranes. However, the microsomal preparation we use includes at least 70% of brush border membrane activities[16].

Under the conditions used, Cys release in the presence of hepatic microsomal protein was proportional to protein content up to 1.5 mg, and proportional to incubation time up to 20 min (data not shown).

### Distribution of Enzyme Activities in Rat Tissues

The procedures described above were used to determine enzyme activities in selected tissues of the rat (Table 2). γ-Glu-Cys synthetase and GSH synthetase activities were assayed in a cytosolic fraction (105,000 g supernatant) and γ-Glu transpeptidase and dipeptidase activities were assayed in a microsomal fraction (105,000 g pellet). The enzyme distributions found are highly asymmetric. γ-Glu transpeptidase and dipeptidase activities were found in liver, heart and lung although these activities were much lower than in the kidney (Table 2). This reflects the importance of these tissues in GSH catabolism.

## DISCUSSION

MB was introduced by Kosower *et al.*[12] for the analysis of low molecular weight biological thiols. Others have utilized bimane derivatives for the separation of the four cysteine-related substances reported herein[5]. However, the modifications employed here - primarily the use of a gradient - results in an appreciably faster analysis: 16 min as compared to 25 min for Fahey and Newton[5]. To the best of our knowledge, this is the first application of such a separation procedure to the determination of four enzyme activities (Table 2). The simultaneous determination of these four enzymes in four tissues of the rat minimizes differences due to technique or laboratory, and allows a comparison under identical conditions of the distribution among various organs of enzymes involved in GSH biochemistry.

Fluorometric detection is both more selective and more sensitive than uv detection. The calibration curves for Cys, Cys-Gly, γ-Glu-Cys and GSH are linear with 0-400 pmol with variations of only 4.5%-7.0%. Both the separation procedure we report (Fig. 1) and that of Hamel *et al.*[9] produce a sizeable unknown peak. It is a relative advantage of our procedure, however, that this peak emerges at the end of the separation, thereby reducing possible interference and loss of sensitivity[9]. The total time required for one separation, including re-equilibration, is reduced in our procedure (16 min) compared with other published procedures, such as those of Dennda and Kula[3] (23 min) and Hamel *et al.*[9] (18 min). The gradient programs used by us or others are necessary for good chroma-tography. Otherwise, resolution is poor[20] or the single run time is much longer (>30 min) (Yan and Huxtable, unpublished observations).

The high levels of endogenous GSH in liver interfere with determination of GSH synthetase. It is, thus, necessary when employing such assays to lower the endogenous GSH levels. Hamel *et al.*[9] achieved this by a continuous-flow dialysis technique 4°C for 2.5 hr, with a flow rate of 1 liter/hr. We have developed a more convenient method based on the binding of GSH to thiopropyl Sepharose resin by disulfide formation. As shown in Fig. 3, cytosolic GSH can be removed in a reproducible manner, depending on the time of preincubation with the resin. This method has two advantages compared with a dialysis technique: (i) no dialysis equipment is needed and no large volume of buffer is used; (ii) less than 50 μl sample can be simply processed.

The resin-treatment procedure is especially useful for assaying GSH synthetase activity in the liver. As GSH concentrations in other tissues are much lower than in the liver, depletion of endogenous GSH is not so necessary in order to achieve accurate assay of GSH synthetase. It is still necessary, however, to determine the GSH content present initially in the incubation mixture so that the newly synthesized GSH can be calculated accurately.

There appears to be no earlier report on simultaneous determination of the four enzymes in four tissues reported on Table 2. The kinetic constants obtained by us (Table 1) for the various enzymes are in line with values reported by others. Our $IC_{50}$ value for inhibition of γ-Glu-Cys synthetase by buthionine sulfoximine (Fig. 2) is comparable to the reported $IC_{50}$ against rat kidney γ-Glu-Cys synthetase of 1 μM[8]. GSH synthetase shows two affinities in hepatic cytosol (Fig. 4). The $K_m$ for the higher affinity activity is comparable to the $K_m$ of 0.76 mM reported for the rat kidney enzyme[21]. The lower affinity

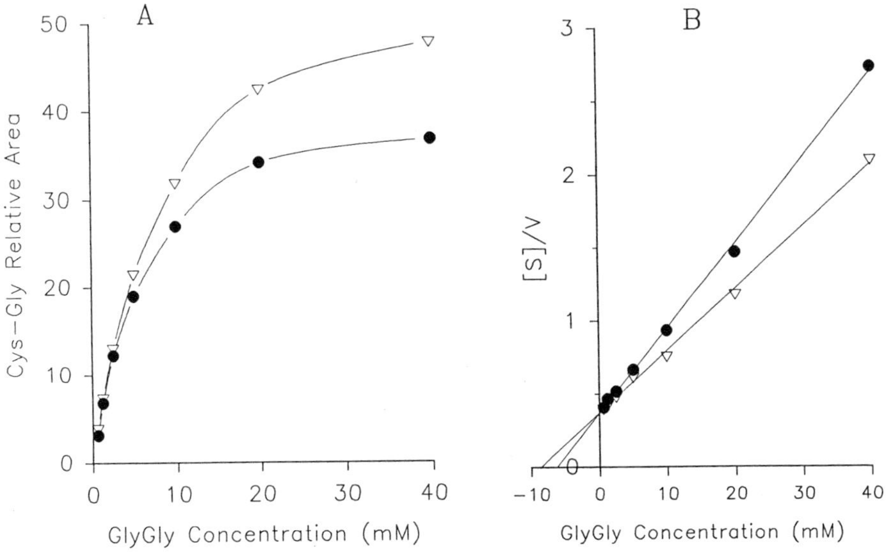


**Figure 5.** **A:** Dependency of γ-Glu transpeptidase activity on Gly-Gly concentration. (▽): Control; (●): Enzyme preincubated with 5.6 μM AT-125. Liver microsomes of male rat were used as enzyme source. GSH concentration was 10 mM. **B:** Hanes-Woolf plot transformation of **A**. See Table 1 for kinetic parameters. AT-125 had a $K_i$ of 2.64 ± 0.17 mM.

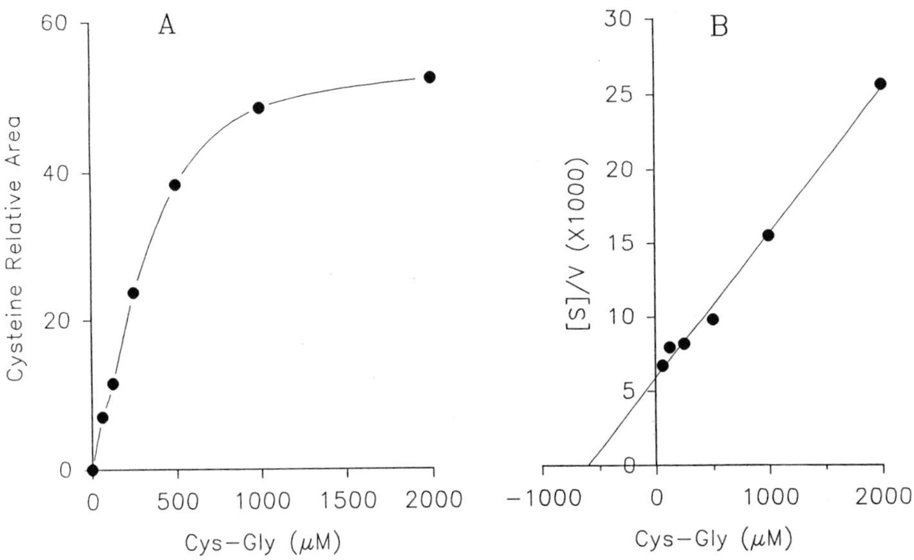


**Figure 6.** A: Dependency of dipeptidase activity on Cys-Gly concentration. Liver microsomes of male rat were used as enzyme source. B: Hanes-Woolf plot transformation of A. See Table 1 for kinetic parameters.

system has not been reported before, probably because the range of glycine concentrations examined did not extend high enough. This low affinity system has no biochemical significance as the binding site remains unoccupied under the range of glycine concentrations normally found in liver (around 2 μmol/g wet weight[6]).

γ-Glu transpeptidase activity has been assayed by many methods. Typically, however, synthetic substrates have been used, rather than the endogenous substrate, GSH[28]. The use of GSH is complicated by its oxidation to glutathione disulfide. However, oxidation can be retarded by including EDTA (1 mM) or DTT in the reaction mixture without changing enzyme activity[4,28].

Cys-Gly, released from GSH by the action of γ-Glu transpeptidase, can be metabolized by membrane-bound dipeptidase. Dipeptidase activity can be determined by the ninhydrin-cyanide procedure, which nonspecifically measures amino acid release from most peptides[15,22]. Other assay procedures for this enzyme include the use of [$^{3}$H]glycine-labeled L-Cys-Gly and mercurial-agarose to separate the substrate from the labeled product[16], and the use of L-amino acid oxidase to determine the released amino acid. We have determined directly the release of cysteine from L-Cys-Gly in incubations with microsomal enzymes. Compared with the above methods, this method is sensitive, easy and specific in the liver. However, in kidney both dipeptidase and aminopeptidase M catalyze cleavage of Cys-Gly. If used without modification, therefore, our method determines the combined activities of dipeptidase and aminopeptidase in kidney microsomes. Selective inhibition of dipeptidase and aminopeptidase M (by thiols and bestatin, respectively) allows one to estimate the contribution of each to the hydrolysis of Cys-Gly[13].

We did not observe inhibition of dipeptidase and aminopeptidase activities by high substrate concentrations[13]. The apparent $K_m$ and $V_{max}$ we find for activity in the kidney (Fig. 6) are close to the values of 0.46 mM and 46 nmol/min/mg, respectively, reported by McIntyre and Curthoys[16] for isolated rat renal brush border membranes.

## REFERENCES


1. Beutler, E. and Gelbart, T., 1986, Improved assay of the enzymes of glutathione synthesis: γ-glutamyl-cysteine synthetase and glutathione synthetase, *Clin. Chim. Acta,* 158:115-123.
2. Davis, J.S., Balinsky, J.B., Harington, J.S., and Shepherd, J.B., 1973, Assay, purification, properties and mechanism of action γ-glutamylcysteine synthetase from the liver of the rat and *Xenopus laevis, Biochem J.,* 133:667-678.
3. Dennda, G. and Kula, M.R., 1986, Assay of the glutathione-synthesizing enzymes by high-performance liquid chromatography, *Biotechnology and Applied Biochemistry,* 8:459-464.
4. Elce, J.S. and Broxmeyer, B., 1976, γ-Glutamyltransferase of rat kidney, *Biochem J.,* 153:223-232.
5. Fahey, R.C. and Newton, G.L., 1987, Determination of low-molecular-weight thiols using monobromobimane fluorescent labeling and high-performance liquid chromatography, *Meth. Enzymol.,* 143:85-96.
6. Galindo, J.D., Cremades, A., Monserrat, F., and Peñafiel, R., 1992, The effect of glycine administration on taurine concentration in the rat liver, *Comp. Biochem. Physiol. [A],* 102A:147-149.
7. Griffith, O.W., 1981, Depletion of glutathione by inhibition of biosynthesis, *Meth. Enzymol.,* 77:59-63.
8. Griffith, O.W., 1982, Mechanism of action, metabolism and toxicity of buthionine sulphoximine and its higher homologues, potent inhibitors of glutathione synthesis, *J. Biol. Chem.,* 257:13704-13712.
9. Hamel, D.M., White, C., and Eaton, D.L., 1992, Determination of γ-glutamylcysteine synthetase and glutathione synthetase activity by HPLC, *Toxicology Methods,* 1:273-288.
10. Hoskins, J.A. and Davies, F.F. 1986, The analysis of low levels of γ-glutamyltransferase activity by high-performance liquid chromatography, *Anal. Biochem.,* 152:314-318.
11. Kiuchi, K., Nagatsu, T., Togari, A., and Kumagai, H., 1986, Highly sensitive assay for γ-glutamyl-transpeptidase activity by high-performance liquid chromatography with electrochemical detection, *J. Chromatogr.,* 357:191-198.
12. Kosower, E.M., Pazhenchevsky, B., and Hershkowitz, E., 1978, 1,5-diazabicylo[3.3.0]octadiene-diones (9,10-dioxabimanes). Strongly fluorescent syn isomers, *J. Amer. Chem. Soc.,* 100: 6516-6518.

13. Kozak, E.M. and Tate, S.S., 1982, Glutathione-degrading enzymes of microvillus membranes, *J. Biol. Chem.*, 257:6322-6327.
14. Lowry, O.J., Rosebrough, N.J., Farr, A.L., and Randall, R.J., 1951, Protein measurement with folin phenol reagent, *J. Biol. Chem.*, 193:265-275.
15. Matheson, A.T. and Tattrie, B.L., 1964, A modified Yemm and Cocking ninhydrin reagent for peptidase assay, *Canadian Journal of Biochemistry*, 42:95-103.
16. McIntyre, T. and Curthoys, N.P., 1982, Renal catabolism of glutathione: Characterization of a particulate rat renal dipeptidase that catalyzes the hydrolysis of cysteinylglycine, *J. Biol. Chem.*, 257:11915-11921.
17. Meister, A. and Anderson, M.E., 1983, Glutathione, *Ann. Rev. Biochem.*, 52:711-760.
18. Meister, A., Tate, S.S., and Griffith, O.W., 1981, γ-Glutamyl transpeptidase, *Meth. Enzymol.*, 77:237-251.
19. Minnich, V., Smith, M.B., Brauner, M.J., and Majerus, P.W., 1971, Glutathione biosynthesis in human erythrocytes, *J. Clin. Invest.*, 50:507-513.
20. Nardi, G., Cipollaro, M., and Loguercio, C., 1990, Assay of γ-glutamylcysteine synthetase and glutathione synthetase in erythrocytes by high-performance liquid chromatograhy with fluorimetric detection, *J. Chromatog.*, 530:122-128.
21. Oppenheimer, L., Wellner, V.P., Griffith, O.W., and Meister, A., 1979, Glutathione synthetase: Purification from rat kidney and mapping of the substrate binding sites, *J. Biol. Chem.*, 254:5184-5190.
22. Patterson, E., 1976, A dipeptidase from *Escherichia coli* B[1], *Meth. Enzymol.*, 45:377-386.
23. Reeve, R., Kuhlenkamp, J., and Kaplowitz, N., 1980, Estimation of glutathione in rat liver by reversed-phase high- performance liquid chromatography: separation from cysteine and γ-glutamylcysteine, *J. Chromatogr.*, 194:424-428.
24. Seelig, G. and Meister, A., 1985, Glutathione biosynthesis; γ-glutamylcysteine synthetase from rat kidney, *Meth. Enzymol.*, 113:379-399.
25. Sekura, R. and Meister, A., 1977, γ-Glutamylcysteine synthetase: Further purification, "half of the sites" reactivity, subunits and specificity, *J. Biol. Chem.*, 252:2599-2605.
26. Smith, G.D., Ding, J.L., and Peters, T.J., 1979, A sensitive fluorimetric assay for γ-glutamyl transferase, *Anal. Biochem.*, 100:136-139.
27. Snoke, J.E. and Bloch, K., 1955, Studies on the mechanism of action of glutathione synthetase, *J. Biol. Chem.*, 213:825-835.
28. Tate, S.S. and Meister, A., 1974, Interaction of gamma-glutamyl transpeptidase with amino acids, dipeptides, and derivatives and analogs of glutathione, *J. Biol. Chem.*, 249:7593-7602.
29. Wapnir, R.A., Mancusi, V.J., and Goldstein, L.A., 1982, Comparative ontogenesis of γ-glutamyl transpeptidase in rat tissues, *Experientia*, 38:647-648.
30. Yan, C.C. and Huxtable, R.J., 1994, Quantitation of the hepatic release of metabolites of the pyrrolizidine alkaloid, monocrotaline, *Toxicol. Appl. Pharmacol.*, 127:58-63.
31. Yan, C.C. and Huxtable, R.J., 1995, Fluorometric determination of monobromobimane and orthophthaldehyde adducts of γ-glutamylcysteine and glutathione: Application of assay to γ-glutamylcysteine synthetase activity and glutathione concentration in the liver, *Journal of Chromatography, B: Biomedical Applications*, 672:217-224.

# THE EFFECTS OF AGING ON TAURINE CONTENT AND BIOSYNTHESIS IN DIFFERENT STRAINS OF RATS

Baerbel Eppler and Ralph Dawson, Jr.

Department of Pharmacodynamics
College of Pharmacy
University of Florida
JHMHC Box 100487
Gainesville, FL 23610

## INTRODUCTION

Taurine has been postulated to have enantiostatic properties to stabilize cellular functions when challenged with external or internal stressors such as perturbations in calcium, free radicals or osmolality[11]. Taurine content has been found to decline with advancing age in a number of tissues in rodents[8]. A decrease in intracellular taurine during senescence would likely further exacerbate the well documented age-related declines in antioxidant defense systems, calcium regulation, and membrane integrity[3,10]. We have previously found that taurine content was decreased in a number of tissues in Fischer 344 rats (F344) and that this decline was not attributable to a failure in the renal conservation of taurine[9]. This has lead us to examine the biosynthesis of taurine to determine if it declines during aging and whether altered biosynthesis could be a possible mechanism contributing to the age-related decline in tissue taurine content.

Sulfur metabolism in mammals is complex and involves pathways for the synthesis of both glutathione and taurine[4]. The rate-limiting step in taurine synthesis in the liver appears to be cysteine sulfinate decarboxylase (CSD, EC 4.1.1.29)[1,4,15]. CSD is a pyridoxal phosphate requiring enzyme that catalyzes the conversion of cysteine sulfinic acid (CSA) to hypotaurine which is subsequently oxidized to taurine[11]. CSD activity is under inhibitory regulation by thyroid hormone, estrogen, and high protein diets[15]. Cysteine dioxygenase (CDO, EC 1.13.11.20) is also a key hepatic enzyme for regulating taurine production[2,4,21]. CDO catalyzes the oxygenation of cysteine to CSA. CDO is also involved in the regulation of the cellular levels of cysteine, methionine and glutathione[1,4]. CDO activity is increased by glucocorticoids and sulfur amino acids and inhibited by cAMP[21]. Recent studies suggest that measurement of both CDO and CSD activity are necessary to estimate taurine production via this enzymatic pathway[1,4]. While many studies have examined glutathione levels in the aging animal, the activity of CSD and CDO in the liver during aging has not been examined extensively despite the interrelated nature of taurine and glutathione synthesis. The exact mechanisms responsible for the loss of intracellular taurine concentrations in aging are poorly understood. In the present study, we examined the hepatic synthesis of taurine in 3 strains of rats that have widely been used as experimental models of aging. Taurine content in specific organs is known to vary substantially due to disease, endocrine status and other variables[14]. It is important to distinguish between alterations in taurine associated with aging and other factors that are not age-related. We therefore compared taurine levels in a number of tissues from these 3 different rat strains. We also speculate about potential mechanisms that may account for tissue specific reductions in taurine content due to aging.

## MATERIALS AND METHODS

### Animals

Male rats obtained from Harlan Sprague-Dawley, Inc. (Indianapolis, IN) were used in all the studies. F344 rats were 10 (n=6) and 26-28 (n=10-14) months of age when sacrificed. These rats were fed Purina diets (5012) or a taurine free diet composed of soy protein (Purina 5729C) which was otherwise identical to the standard diet (5012). One group was also fed 5012 diet and given 1.5% taurine in the drinking water as previously described[9]. The Sprague-Dawley (SD) rats were 5, 20, and 25 months of age (n=4 per age group) and were fed Purina chow (5001). FBNF1 rats (F344 and Brown Norway crosses) used in these studies were 13-15 (n=4-5) and 31-36 (n=3-7) months of age.

### Enzyme Activity Measurements

The liver and cerebellum were homogenized in 50 mM potassium phosphate buffer (pH=6.8) using a polytron. The homogenates were used for the assay of CSD and CDO. CSD activity was assayed by the method of Stipanuk and coworkers[1,4] using (final concentrations): 25 mM cysteine sulfinic acid (CSA), 0.8 mM pyridoxal phosphate, 0.55 mM dithiothreitol and 15 mM glutamate in potassium phosphate buffer (pH=7.0). Glutamate was used to inhibit glutamic acid decarboxylase activity (GAD)[20]. The taurine formed was assayed by HPLC as previously described[17]. CDO activity was assayed by the method of Bagley et al.[2] using a final cysteine concentration of 5 mM. The CSA and cysteic acid generated were determined using a modification of a previously described HPLC assay[8]. Protein content of the homogenates was assayed by the method of Bradford[5]. The taurine content in the liver and cerebellar homogenates was also assayed by HPLC[17].

### Tissue Content of Taurine

Taurine content was also determined in perchloric acid extracts from the heart, testes, kidney and skeletal muscle. The serum taurine from the SD rats was extracted in 100% methanol. The samples were prepared and assayed as previously described[8].

## RESULTS AND DISCUSSION

CSD activity was significantly decreased in the livers of aged F344 rats, but was not altered in either FBNF1 or SD rats (Fig. 1). SD rats had 3-4 times higher hepatic CSD activity than either FBNF1 or F344 rats (Fig. 1). Interestingly, the F344 rats have more

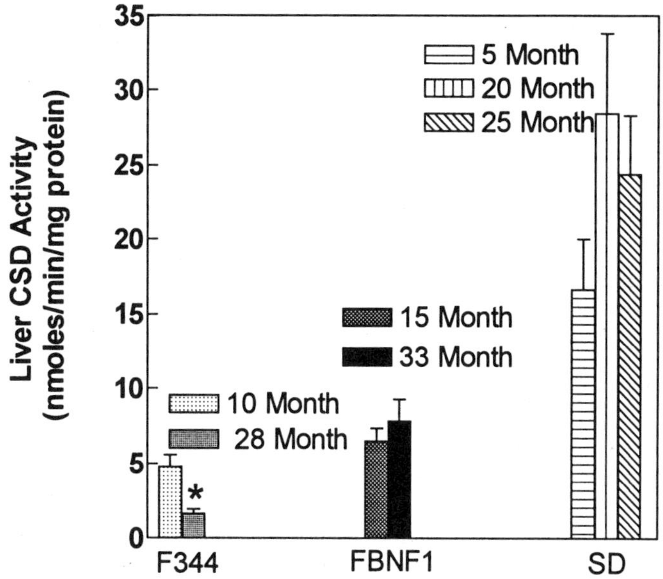


**Figure 1**. Liver CSD activity in 3 rat strains of adult and aged rats. *p<0.05 versus 10 month old rats.

than 10 fold higher levels of CDO activity than the SD rats (Fig. 2). Bagley et al.[2] found Wistar rats to have nearly 10 fold lower CDO activity than SD rats. Therefore, strain differences are prominent in CDO activity. CDO activity was significantly decreased in aged F344 rats (Fig. 2). The taurine content of the liver was significantly decreased in the aged F344 rats and there was also a tendency for a decline in the FBNF1 rats (Table 1). SD rats had very low liver taurine content which was not further reduced by aging. CDO activity was reduced approximately 44% in both the 20 and 25 month old SD rats when compared to the 5 month old rats.

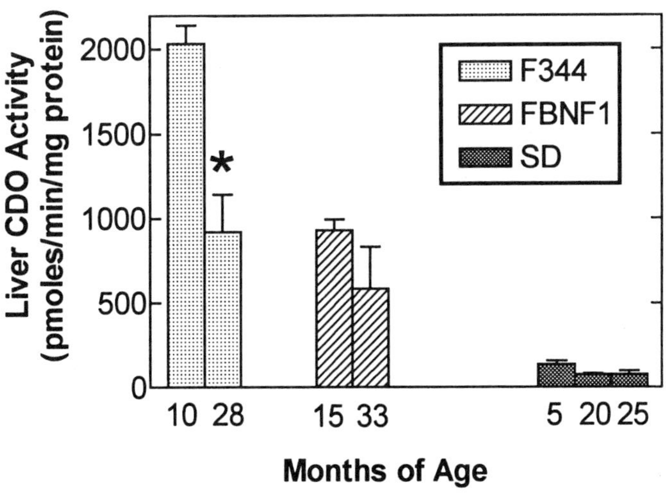


**Figure 2**. CDO activity in the liver of adult and aged rats. *p<0.05 versus 10 month old rats.

Table 1. Taurine content in tissues from 3 strains of adult and aged rats.

| | | Liver | Heart | Kidney | T. Muscle | Testes | Serum |
|---|---|---|---|---|---|---|---|
| F344 | 10 Month | 5.9 ± 1.8 | 28.9 ± 1.2 | 7.24 ± 0.46 | 8.5 ± 0.5 | 2.49 ± 0.07 | 324 ± 33 |
| | 28 Month | 2.4 ± 0.5* | 26.3 ± 1.4 | 4.78 ± 0.61* | 8.4 ± 0.6 | 11.75 ± 1.57* | 172 ± 21* |
| FBNF1 | 15 Month | 12.0 ± 1.5 | 24.1 ± 3.1 | ND | ND | ND | ND |
| | 33 Month | 7.0 ± 2.0 | 29.2 ± 4.5 | | | | |
| SD | 5 Month | 1.8 ± 0.3 | 26.5 ± 2.7 | 3.13 ± 0.07 | 34.0 ± 2.5 | 3.08 ± 0.14 | 324 ± 20 |
| | 20 Month | 2.5 ± 0.4 | 22.3 ± 4.8 | 3.31 ± 0.20 | 23.5 ± 3.0* | 3.75 ± 0.28 | 310 ± 16 |
| | 25 Month | 1.6 ± 0.1 | 25.3 ± 3.0 | 3.34 ± 0.22 | 32.5 ± 0.8 | 3.42 ± 0.07 | 236 ± 27* |

Tissue content data are presented as μmoles/g tissue weight. Serum values are nmoles/ml. *$p<0.05$ versus adult value. T=temporalis

The effects of dietary taurine supplementation or restriction was evaluated in adult (10 month) and aged (28 month) F344 as previously described[9]. CSD activity was decreased in aged F344 rats compared to the CSD activity in adult rats (Table 2). A diet that did not contain any taurine did not increase CSD activity in aged rats, however, taurine supplementation appeared to suppress CSD activity although this effect was not statistically significant. Compared to the other aged groups, CDO activity was also decreased in aged F344 rats and increased dietary taurine intake appeared to modestly suppress CDO activity (Table 2). Taurine supplementation significantly increased liver content of taurine and the absence of taurine in the diet or the soy protein diet seemed to moderate the effects of aging (Table 2). It is unclear why the taurine free diet did not result in an increase in either CDO or CSD activity to compensate for the lack of a dietary source of taurine. Taurine supplementation did result in what appeared to be a modest feedback (end product) mediated down regulation of CSD activity. The taurine free diet was a soy protein based diet which has been shown to markedly reduce age-related nephropathy and improve the general health of F344 rats[22]. All of the other groups received standard Purina diets that contained fish meal as the protein source. The positive effects of a soy protein diet in retarding renal pathology in F344 rats confounds the interpretation of the effects of the taurine free diets, although dietary cysteine and methionine levels were comparable in the diets.

Cerebellar CSD activity and taurine content were evaluated in F344 rats and SD rats (Table 3). Unlike the liver, cerebellar CSD activity was similar in F344 and SD rats. There was also no effect of aging on cerebellar CSD activity (Table 3). Cerebellar taurine content was reduced 15% in aged F344 rats and was significantly lower in aged (20 & 25 month old) SD rats. Dietary restriction or supplementation of aged F344 rats had no effect

**Table 2**. Effects of manipulating dietary taurine on the liver of F344 rats.

| | Adult Control | Aged Control | Aged No Taurine | Aged 1.5% Taurine |
|---|---|---|---|---|
| CSD | 4.79 ± 0.78 | 1.66 ± 0.30* | 1.14 ± .18* | 0.41 ± 0.10* |
| CDO | 2033 ± 106 | 922 ± 221* | 672 ± 193* | 544 ± 164* |
| Taurine | 5.88 ± 1.80 | 2.38 ± 0.46* | 5.17 ± 1.25 | 12.26 ± 1.52$^{+}$ |

CSD activity expressed as nmoles/min/mg protein. CDO activity expressed as pmoles/min/mg protein. Taurine content expressed as μmoles/g tissue weight. *$p<0.05$ versus adult control group; $^{+}p<0.05$ versus all other groups.

**Table 3.** Cerebellar CSD activity and taurine content in F344 and SD rats.

| | F344 Adult | F344 Aged | SD 5 Months | SD 20 Months | SD 25 Months |
|---|---|---|---|---|---|
| CSD | 4.28 ± 0.50 | 5.06 ± 0.27 | 5.97 ± 0.80 | 5.92 ± 0.35 | 5.20 ± 0.24 |
| Taurine | 4.39 ± 0.13 | 3.74 ± 0.25 | 5.70 ± 0.13 | 5.38 ± 0.19* | 4.57 ± 0.14* |

CSD activity expressed as nmoles/min/mg protein. Taurine content is expressed as µmoles/g tissue weight. * $p<0.05$ versus SD 5 month controls.

on cerebellar CSD activity (data not shown). Dietary supplementation significantly increased cerebellar taurine content in the aged F344 rats (6.82 ± 1.00 µmoles/g) when compared to aged rats on either control (Table 3) or taurine free diets (4.13 ± 0.18 µmoles/g). These findings suggest that the decline in cerebellar taurine may be an age-related effect that is not due to a decline in CSD activity. A defect in taurine transport or a decrease in CDO activity[12] could account for a loss of cerebellar taurine content. The decline in cerebellar taurine content in aged rats could also reflect a loss of taurine containing neurons or glia that do not express CSD activity.

There was a clear decline in liver's biosynthetic capacity for taurine in aged F344 rats. This may in part explain the decline in tissue taurine concentration in this particular rat strain. Therefore, it would not appear that decreased hepatic taurine synthesis is a universal aspect of aging since no such changes were seen in either SD or FBNF1 rats. Furthermore, the age-related decline in CSD activity was not evident in the cerebellum of F344 rats. The loss of CSD and CDO activity in aged F344 rats may be related to the commonly noted hepatic lesions seen in aged F344 rats[7]. Endocrine factors (thyroid status, testicular atrophy, adrenal function etc.) may also contribute to the large strain differences in CDO and CSD activity[14].

Taurine content in tissues from adult and aged rats are presented Table 1. Serum taurine was decreased in aging F344 and SD rats. The taurine content of the testes in aged F344 rats was significantly elevated while no such change was noted in SD rats (Table 1). This finding is most likely due to the high incidence (>70%) of testicular interstitial cell tumors in the aged F344 rats[7] which was not noted at necropsy in SD rats. At present, it is unknown how testicular interstitial cell tumors and tissue taurine content are related in F344 rats. Liver and renal tissue content of taurine were much lower in SD rats than F344 rats and did not show an age-related decline (Table 1). The taurine content in heart muscle was not altered by aging in the 3 rat strains examined (Table 1). In contrast, temporalis muscle content of aged SD rats was over 3 fold higher than in F344 rats (Table 1).

Skeletal muscle taurine content was examined in detail in aging FBNF1 rats (Fig. 3). There were no age-related declines in taurine content in FBNF1 muscles and taurine content was significantly increased in the tibialis anterior (TA) muscle of aged rats (Fig. 3). Skeletal muscle taurine content is known to increase after denervation[13] and aging is associated with a loss of cholinergic innervation and a decline in function involving the neuromuscular junction[6]. Skeletal muscle contains the major body pool of taurine[19], however, little is known about the dynamics of taurine uptake or synthesis in the skeletal muscle pool of aged rats. These preliminary findings would suggest that taurine may be sequestered to a greater extent in skeletal muscle of aged rats. The increase in TA taurine content in aged FBNF1 rats could reflect a response to age-related denervation or other adaptations to aging. Further studies are needed to examine the significance of this observation, especially given that skeletal muscle chloride channel conductance impairments in aged rats can be attenuated by dietary taurine treatment[18]. Aged skeletal muscle may require a higher taurine content to maintain contractile function. An age associated increase

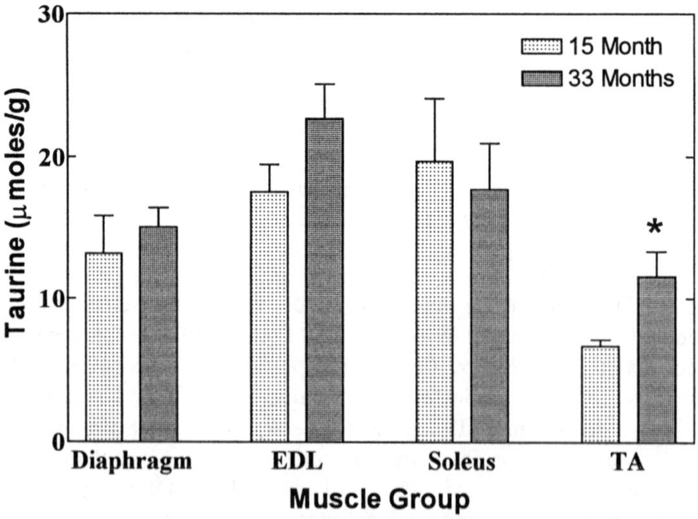


**Figure 3.** Taurine content in muscle groups from adult and aged FBNF1 rats. EDL=extensor digitorum longus, TA=tibialis anterior *p<0.05 versus 33 month group.

in skeletal muscle taurine uptake could contribute to the decline in serum and tissue taurine content seen with advanced age. If aged skeletal muscle was more efficient than other tissues at sequestering taurine from the plasma pool, then other tissues could over time exhibit a decrease in intracellular taurine concentrations.

In summary, our studies have demonstrated that significant differences exist in the biosynthetic capacity for taurine by the CDO and CSD pathways in different strains of rats. Furthermore, these differences are generally maintained in aging, although F344 rats show marked declines in hepatic CSD and CDO activity with advanced age. These changes in CSD activity in F344 rats may be more related to pathological changes in the liver than to aging[7], since these changes were not seen in the cerebellum or in the other rat strains. There was a trend for CDO activity to decline with age in all the strains examined. This could have implications for both hepatic taurine and glutathione production in aging due to the pivotal role that CDO plays in cysteine metabolism[1,2,4]. As previously noted, there did not appear to be a strong correlation between the activity of CDO or CSD and hepatic taurine content[16]. Our studies also suggest that it is important to consider factors such as the strain of rat used in aging studies before important generalizations about basic mechanisms related to aging can be clearly differentiated from other factors (pathogenic, hormonal, dietary etc.).

## ACKNOWLEDGMENTS

The authors gratefully acknowledge Dr. Martha H. Stipanuk and Deborah L. Bella for their assistance with various aspects of the CSD and CDO assays. The authors also wish to thank Dr. Thomas Nosek for his gift of tissues from FBNF1 rats and Dr. Stephen Borst for his gift of aged SD rats. The authors thank Steve Liu, Joel Cherdack and Seena Salyani for their technical assistance. This work was supported in part by a grant from Taisho Pharmaceutical Co., LTD.

## REFERENCES


1. Bagley, P.J. and Stipanuk, M.H., 1994, The activities of hepatic cysteine dioxygenase and cysteinesulfinate decarboxylase are regulated in a reciprocal manner in response to dietary casein level in rats, *J. Nutr.*, 124:2410-2421.
2. Bagley, P., Hirschberger, L.L., and Stipanuk, M.H., 1995, Evaluation and modification of an assay procedure for cysteine dioxygenase activity: high-performance liquid chromatographic method for measurement of cysteine sulfinate and demonstration of physiological revelveance of cysteine dioxygenase activity in cysteine catabolism, *Analyt. Biochem.*, 227:40-48.
3. Beckman, K.B. and Ames, B.N., 1997, Oxidants, antioxidants, and aging, in "Oxidative Stress and the Molecular Biology of Antioxidant Defenses", Cold Spring Harbor Laboratory Press, pp. 201-246.
4. Bella, D. and Stipanuk, M.H., 1996, High levels of dietary proteins or methionine have different effects on cysteine metabolism in rat hepatocytes, in Adv. Exp. Med. Biol. "Taurine 2 Basic and Clinical Aspects", Huxtable, R.J., Azuma, J., Kuriyama, K., Nakagawa, M., and Baba, A., eds., Plenum Press, New York, Vol. 403, pp. 73-84.
5. Bradford, M.M., 1976, A rapid and sensitive method for the quantitation of microgram quantities of protein utilizing the principle of protein-dye binding, *Anal. Biochem.*, 72:248-254.
6. Carmeli, E. and Reznick, A.Z., 1994, The physiology and biochemistry of skeletal muscle atrophy as a function of age, *Proc. Soc. Exp. Biol.*, 206:103-113.
7. Coleman, G.L., Barthold, S.W., Osbaldiston, G.W., Foster, S.J., and Jonas, A.M., 1977, Pathological changes during aging in barrier-reared Fischer 344 male rats, *J. Gerontol.*, 32:258-278.
8. Dawson, R. and Wallace, D.R., 1992, Taurine content in tissues from aged Fischer 344 rats, *Age*, 15:73-81.
9. Dawson, R., Eppler, B., Patterson, T.A., Shih, D., and Liu, S., 1996, The effects of taurine in a rodent model of aging, in Adv. Exp. Med. Biol. "Taurine 2 Basic and Clinical Aspects", Huxtable, R.J., Azuma, J., Kuriyama, K., Nakagawa, M., and Baba, A., eds., Plenum Press, New York, Vol. 403, pp. 37-50.
10. Halliwell, B., 1992, Reactive oxygen species in the central nervous system, *J. Neurochem.*, 59:1609-1623.
11. Huxtable, R.J., 1992, Physiological actions of taurine, *Physiol. Rev.*, 72: 101-163.
12. Ida, S., Ohkuma, S., Kimori, M., Kuriyama, K., Morimoto, N., and Ibata, Y., 1985, Regulatory role of cysteine dioxygenase in cerebral biosynthesis of taurine. Analysis using cerebellum from 3-acetylpyridine-treated rats, *Brain Res.*, 344:62-69.
13. Iwata, H., Obara, T., Kim, B.K., and Baba, A., 1986, Regulation of taurine transport in rat skeletal muscle, *J. Neurochem.*, 47:158-163.
14. Jacobsen, J. and Smith, L.H., 1968, Biochemistry and physiology of taurine and taurine derivatives, *Physiol. Rev.*, 48:424-511.
15. Kaisaki, P.J., Jerkins, A.A., Goodspeed, D.C., and Steele, R.D., 1995, Cloning and characterization of rat cysteine sulfinic acid decarboxylase, *Biochim. Biophys. Acta*, 1262:79-82.
16. Kuo, S-M and Stipanuk, M.H., 1984, Changes in cysteine dioxygenase and cysteinesulfinate decarboxylase activities and taurine levels in tissues of pregnant or lactating rat dams and their fetuses or pups, *Biol. Neonate*, 46:237-248.
17. Patterson, T.A., Dawson, R., Iyer, S.N., and Katovich, M.J., 1995, Plasma tryptophan is altered in the Dahl model of hypertension, *Biogenic Amines*, 11:109-121.
18. Pierno, S., De Luca, A., Huxtable, R.J., and Conte Camerino, D., 1996, Effects of chronic taurine treatment on the electrical and contractile properties of skeletal muscle fibers of aged rats, in Adv. Exp. Med. Biol. "Taurine 2 Basic and Clinical Aspects", Huxtable, R.J., Azuma, J., Kuriyama, K., Nakagawa, M., and Baba, A., eds., Plenum Press, New York, Vol. 403, pp. 249-255.
19. Quesada, O., Lu, P., and Sturman, J.A., 1993, Taurine distribution in different cat muscles as visualized by immunohistochemistry: changes with stimulus state, *Cytobios*, 73:143-154.
20. Tappaz, M., Almarghini, K., Legay, F., and Remy, A., 1992, Taurine biosynthesis enzyme cysteine sulfinate decarboxylase (CSD) from brain: the long and tricky trail to identification, *Neurochem. Res.*, 17:849-859.
21. Tsuboyama, N., Hosokawa, Y., Totani, M., Oka, J., Matsumoto, A., Koide, T., and Kodama, H., 1996, Structural organization and tissue-specific expression of the gene encoding rat cys-teine dioxygenase, *Gene*, 181:161-165.
22. Weindruch, R. and Masoro, E.J., 1991, Concerns about rodent models for aging research, *J. Gerontol.*, 46:B87-B88.

# Part 2:
# Taurine and the
# Liver and Pancreas

# TAURINE LEVELS AND LOCALIZATION IN PANCREATIC ISLETS

Julián Bustamante,[2] Francisco J. Alonso,[1] María V.T. Lobo[1], Elena Giné,[3] Jorge Tamarit-Rodriguez,[3] José M. Solís,[1] and Rafael Martín del Río[1]

[1] Servicio de Neurobiología
Depto. de Investigación
Hospital Ramón y Cajal
Crta Colmenar Km 9
[2] Depto. Fisiología
[3] Depto. Bioquímica
Facultad de Medicina
Universidad Complutense
28040 Madrid, Spain

## INTRODUCTION

Taurine is present in most mammalian tissues including the pancreas. In this organ, taurine appears to be specifically concentrated in the islets as was reported 25 years ago using semi-quantitative techniques[2]. Experiments confirming these data or studies on the implications of taurine on pancreatic physiology have been scarce. It has been shown, for instance, that taurine administration strongly suppresses glucose-stimulated secretion of insulin from isolated mice islets[15], and also that intraperitoneal injection of taurine inhibits the increase in serum insulin induced by glucose administration[8]. On the contrary, it has been reported very recently that taurine enhances glucose-stimulated insulin release by cultured rat fetal islets[4]. Clinical studies linking taurine and pancreatic pathological situations are almost absent. It has been shown, however, that taurine plasma levels are low in diabetic patients and that supplementation with this amino acid reduces the increased tendency towards platelet aggregation in these patients[5]. These data indicate possible functional roles of taurine in the pancreas, and thus, a re-evaluation of its presence and cellular distribution in the islet needs to be carried out. The aim of this study was to determine taurine levels in the pancreatic islets by using a quantitative biochemical method, and also to localize the taurine distribution among the different cell types of the rat pancreas by immunohistochemical techniques.

*Taurine 3*
Edited by Schaffer *et al.*, Plenum Press, New York, 1998

## MATERIALS AND METHODS

Quantitative analyses of the taurine concentration in isolated intact islets and in the exocrine pancreas were carried out by high performance liquid chromatography (HPLC)[9]. The integrity of the isolated islets, obtained by collagenase digestion of the pancreatic tissue, was confirmed by histological analysis.

To perform histological studies, albino Wistar rats (200-250 g) were anesthetized with a mixture of ketamine, xilacine and atropine and perfused through the ascending aorta with 4% paraformaldehyde and 0.3% to 4% glutaraldehyde in phosphate buffer. Tissue samples were embedded in paraffin or araldite by conventional methods. After xylene deparaffinization or after removing epoxy resin from the araldite sections with sodium ethoxide, serial sections were incubated with the rabbit anti-taurine antibody (Chemicon Int. Inc., Temecula, CA) and the immunoreactivity was revealed by the streptavidin-peroxidase method. The reaction product of diaminobenzidine was intensified with nickel nitrate to obtain a dark black colour of immunostained antigens (10 ml of nickel nitrate (8%) in 1ml DAB-$H_2O_2$ solution). Some paraffin sections were also immunolabelled using conventional immunofluorescence procedures. Controls were performed as described in detail previously[11]. Since it has been demonstrated that poor fixation may lead to a considerable redistribution of taurine[10], cerebellar sections from the same rats were immunostained simultaneously with pancreatic tissue samples to evaluate this possibility. The primary antibody that was used in these experiments has been extensively characterized to confirm its specificity by the ELISA method, demonstrating that it does not cross-react with other related amino acids[3]. The distribution of taurine in pancreatic samples from adult dogs was also studied because of the important histological differences of their insular apparatus compared with those from rats. Pancreatic samples from adult dogs, killed under sodium penthobarbital anesthesia, were generously provided by the Veterinary Military Center (Madrid, Spain). Pancreata were embedded in paraffin and the immunostaining was performed as described above.

## RESULTS AND DISCUSSION

Quantitative analysis of taurine in the rat pancreas demonstrated that this amino acid is highly concentrated in the islets (62.0 ± 8.6 nmol/mg prot.) whereas it is present at a much lower concentration in the exocrine pancreas (4.2 ± 0.9 nmol/mg prot.). These values for taurine content in isolated islets are higher than those reported in previous studies[2]. This discrepancy could be due to the different sensitivity of the chromatography procedures and/or to intrinsic differences between the species that were studied.

We found using histological techniques that within the exocrine pancreas taurine was mainly localized in vascular cells (Fig. 1A). Smooth muscle cells and endothelial cells from muscular vessels were intensely immunostained. In the continuous capillaries, taurine appeared to be homogeneously distributed within the nucleus and cytoplasm of endothelial cells (Fig. 1B). Some long, polymorphic cells, frequently showing a spindle-shaped morphology and located adjacent to the wall of nonmuscular microvessels, were also stained. These cells were identified by their size, morphology and distribution as pericytes and fibroblasts. All acinar cells were unstained (Fig. 1A, B). Some ducts of the exocrine pancreas were immunolabelled, whereas other ducts were weakly stained.

In the rat islet of Langerhans, the different cell types have specific locations. The β-cells are centrally situated forming the inner insular core and are surrounded by non-β-cells (α-,δ- and PP-cells), which form a thin layer in three-dimensional reconstructions[1]. In the endocrine pancreas, taurine was mainly distributed within non-β-cells (Fig. 1A, B).

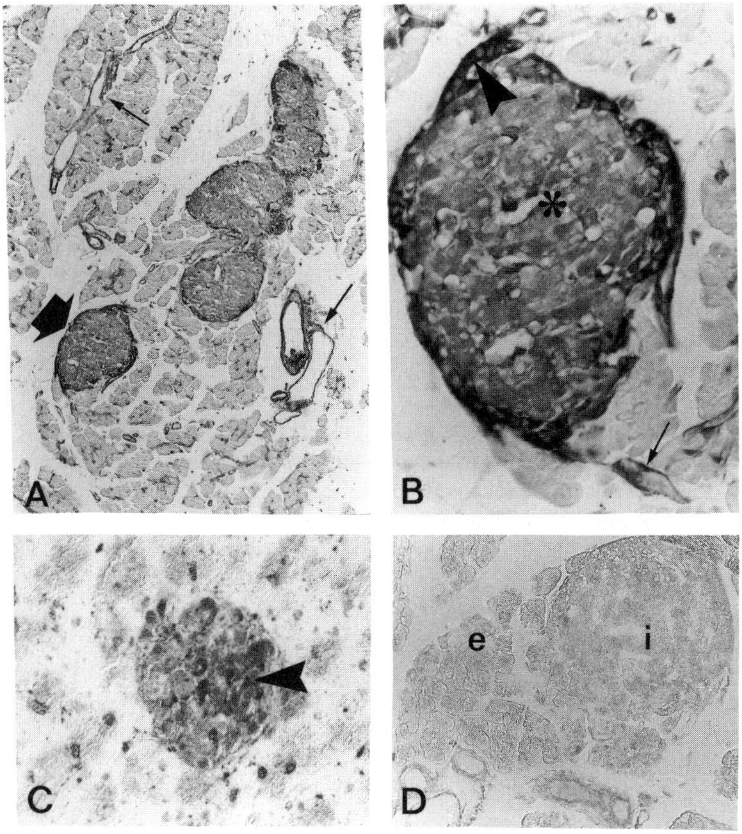


**Figure 1.** Immunohistochemical localization of taurine. (A) Overall view of the staining in the rat pancreas. Within the islet, immunoreactivity is seen in non-β-cells surrounding the inner core of weakly stained β-cells. Thin arrows, vessels from the exocrine pancreas. (B) Enlarged view of an islet from A (thick arrow). Arrowhead, non-β-cell. Asterisk, fenestrated capillar. Thin arrow, continuous capillar. (C) Dog islet. Arrowhead, immunostain non-β-cells. (D) Control of immunostaining from a rat pancreatic section. (i) islet; (e) exocrine pancreas.

The islets frequently showed a thin ring of immunolabelled cells surrounding the weakly stained or unstained core of β-cells. Occasionally, small islets were found, showing mostly insular cells that were immunostained. However, serial sections demonstrated that tangential sections from the pancreatic islets were mostly composed of non-β-cells. Endothelial cells from the fenestrated capillaries of the endocrine pancreas were also intensely immunostained (Fig. 1B).

Dog pancreatic samples were studied because their islets are similar to those of humans but differ from those found in the rat, being smaller and with a different distribution of insular cells. Within the dog islets, non-β-cells are also located in the central region of the islets among the numerous β-cells. Despite these and other differences, taurine appears to be localized, in the dog pancreas, with a similar pattern of distribution to that found in the rat. In the endocrine pancreas, β-cells were unstained whereas the fenestrated capillaries and non-β-cells showed a positive immunolabelling (Fig. 1C). In the exocrine pancreas, the continuous capillaries were stained whereas acinar cells were unstained.

No staining was found in negative controls (Fig. 1D). Positive controls showed a taurine distribution within cells of rat cerebellar sections similar to that previously reported in several studies[3,10] (not shown), suggesting that the normal pattern of taurine distribution in the rat pancreas was not altered.

These data suggest interesting relations between the distribution of taurine and GABA in the islet of Langerhans. Both are highly concentrated in the endocrine pancreas showing levels comparable with those found in the central nervous system. GABA as well as the enzymes implicated in its synthesis and degradation are confined to β-cells[12,14], whereas taurine is mainly located in non-β-cells. The role of GABA in the endocrine pancreas is unclear, mainly because physiological studies have failed to demonstrate GABA release from β-cells. However, it has been frequently proposed that GABA could have a paracrine role in the islet since GABA receptors are localized on α- and δ-cells but not on β-cells[13]. Recent studies seem to support this hypothesis as they demonstrated that GABA is secreted by a β-cell line in response to glucose and that GABA attenuates glucagon secretion by an α-cell line[7].

From the results presented here and the previously reported modulatory effect of exogenous taurine on glucose-stimulated insulin secretion by β-cells, it can be suggested that the tissue pool of taurine located in non-β-cells may have a paracrine and/or autocrine function in the physiology of the insular apparatus since taurine interacts with $GABA_A$ receptors on α- or δ-cells as has been reported in other tissues[6].

## ACKNOWLEDGMENTS

This work was supported by the "Ministerio de Educación y Cultura" of Spain (Grant PM95-170). We thank A. Latorre for technical assistance. The presentation of this work was sponsored by "Red Bull España".

## REFERENCES

1. Brelje, T.C., Scharp, D.W., and Sorenon R.L., 1989, Three-dimensional imaging of intact isolated islets of Langerhans with confocal microscopy, *Diabetes*, 38:808-814.
2. Briel, G., Gylfe, E., Hellman, B., and Neuhoff, V., 1972, Microdetermination of free amino acids in pancreatic islets isolated from obese-hyperglycemic mice, *Acta Physiol. Scand.*, 84:247-253.
3. Campiston, G., Geffard, M., and Buijs, R.M., 1986, Immunological approach to the detection of taurine and immunocytochemical results, *J. Neurochem.*, 46:862-868.
4. Cherif, H., Reusens, B., Dahri, S., Remacle, C., and Hoet, J-J. 1996, Stimulatory effects of taurine on insulin secretion by fetal rat islets cultured *in vitro*, *J. Endocrinol.*, 151:501-506.
5. Franconi F., Bennardini F., Mattana A., Miceli M., Ciuti M. Mian, M., Gironi, A., Anichini R., and Seghieri G., 1995, Plasma and platelet taurine are reduced in subjects with insulin-dependent diabetes mellitus: effects of taurine supplementation. *Am. J. Clin.. Nutr.*, 61:1115-9.
6. Galarreta, M., Bustamante, J., Martín del Río, R., and Solís, J.M., 1996, Taurine induces a long-lasting increase of synaptic efficacy and axon excitability in the hippocampus, *J. Neurosci.* 16:92-102.
7. Gaskins, H.R., Baldeón M.E., Selassie, L., and Beverly, J.L., 1995, Glucose modulates γ-aminobutyric acid release from the pancreatic bTC6 cell line, *J. Biol. Chem.*, 270:30286-30289.
8. Kulakowski, E.C., and Maturo, J., 1984, Hypoglycemic properties of taurine: not mediated by enhanced insulin release, *Biochem. Pharmacol.*, 33:2835-2838.
9. Lerma, J., Herranz, A.S., Herreras, O., Abraira, V. and Martín del Río, R., 1986, "*In vivo*" determination of extracellular concentration of amino acids in the rat hippocampus. A method based on brain dialysis and computerized analysis, *Brain Res.*, 384:145-155.

10. Madsen, S., Ottersen, O.P., Storm-Mathisen, J., and Sturman, J.A., 1990, Immunocyto-chemical localization of taurine: methodological aspects, in: "Taurine: Functional Neurochemistry, Physiology, and Cardiology", Pasante-Morales, H., Martin, D.L., Shain, W., Martín del Río, R., eds., Wiley-Liss, pp. 37-44.
11. Merighi, A., 1992, Post-embedding electron microscopic immunocytochemistry, in: "Electron Microscopic Immunocytochemistry", Polak, J.M. and Priestley, J.V., eds., Oxford University Press, pp. 51-85.
12. Michalik, M., and Erecinska, M., 1992, GABA in pancreatic islets: metabolism and function, *Biochem. Pharmacol.,* 44:1-9.
13. Rorsman, P., Berggren, P-O., Bokwist, K., Ericson, H., Möhler, H., Östenson, C-G., and Smith, P.A., 1989, Glucose-inhibition of glucagon secretion involves activation of $GABA_A$-receptor chloride channels, *Nature,* 341:233-236.
14. Sorenson, R.L., Garry, D.G., and Brelje, T.C., 1991, Structural and functional considerations of GABA in islets of Langerhans, Beta-cells and nerves, *Diabetes,* 40:1365-1374.
15. Tokunaga, H., Yoneda, Y., and Kuriyama, K., 1983, Streptozotocin-induced elevation of pancreatic taurine content and suppressive effect of taurine on insulin secretion, *Eur. J. Pharmacol.,* 87:237-243.

# PYRROLIZIDINE ALKALOID-INDUCED DEPLETION OF TAURINE FROM RAT LIVER *IN VIVO* AND *IN VITRO*

Y. Sun, S.J. Hasal, C.C. Yan, K. Brendel, and R.J. Huxtable

Department of Pharmacology
University of Arizona
College of Medicine
Tucson, AZ 85724

## INTRODUCTION

Pyrrolizidine alkaloids (PAs) are a group of plant hepatotoxins[13,19]. Certain PAs can also cause toxicity in extra-hepatic organs, including heart, lung, kidney and the CNS. PAs produce hepatotoxicity as a result of bioactivation to the corresponding dehydroalkaloids. Dehydroalkaloids are highly reactive molecules, with half-lives of a few seconds in water[7]. It has been suggested that alkylation of cell macromolecules by electrophilic dehydro-alkaloids is the biochemical basis of toxicity[7,25]. Dehydroalkaloids can also conjugate with glutathione to form 7-glutathionyl-6,7-dihydroxymethyl-5H-pyrrolizine (GSDHP), a polar detoxification product released in high concentration into the bile[20]. Taurine has been shown to protect against toxic exposures in liver[8,23], lung[5,10,16,18,31-33] and heart[2-4,9]. We have reported that pretreatment with taurine reduced the cardiac toxicity of the PA, monocrotaline, in rats and reduced the $LD_{50}$ from 80 mg/kg to 120 mg/kg[40]. In this study, we report the effect of PA exposures on hepatic taurine concentration in acute, subacute and pair-feeding studies *in vivo*, and perfused liver and liver slice incubations *in vitro*.

## METHODS

### Animals

Male Sprague-Dawley rats were obtained from the University of Arizona Division of Animal Resources. They were maintained 2-3 to a cage and allowed food and water *ad libitum* except in the pair-feeding study. To avoid differences due to diurnal variation of liver taurine concentration, animals were killed between 9:00-11:00 am.

*Taurine 3*
Edited by Schaffer *et al.*, Plenum Press, New York, 1998

### Acute Effects of Retrorsine *In Vivo*

Rats were administered the pyrrolizidine alkaloid, retrorsine (20.4 mg/kg, 60% of $LD_{50}$ dose, i.p.) and killed by decapitation at 0 (as control), 2, 4, 8, 12, 16, 24, and 48 hrs later. Hepatic cytosolic fraction was isolated by ultracentifugation for taurine measurement by HPLC.

### Acute Effects of Monocrotaline *In Vivo*

Rats were injected monocrotaline (65 mg/kg, 60% of $LD_{50}$ dose, i.p.), while control animals were given i.p. saline. After 24 hrs, rats were killed and the hepatic cytosolic fraction was isolated for taurine determination.

### Subacute Effects of Monocrotaline *In Vivo*

Rats were injected monocrotaline (65 mg/kg, 60% of $LD_{50}$ dose, i.p.), while control animals were given i.p. saline. After 2 weeks, rats were killed and the hepatic cytosolic fraction was isolated for taurine determination.

### Pair-feeding Study with Monocrotaline

One group was given monocrotaline (65 mg/kg, 60% of $LD_{50}$ dose, i.p.) and another group was injected with saline as a control. The food intake in the monocrotaline group was recorded daily for each cage. The control group was given the same weight of food consumed by the monocrotaline group on the previous day. Both groups were allowed tap water *ad libitum*. After 2 week, the study was terminated. Hepatic taurine concentrations were measured.

### *In Vitro* Perfused Rat Liver

Rat livers were surgically removed under anesthesia and the bile duct was cannulated as previously described[35]. Livers were perfused as previously described[34] with a Krebs-Henseleit bicarbonate buffer containing 0.5 mM monocrotaline. Samples of perfusate were taken at 10 min intervals for taurine analysis. After 60 min perfusion, hepatic taurine concentrations were measured.

### *In Vitro* Incubation of Precision-Cut Rat Liver Slices

Rats were decapitated, the liver removed, and cylindrical tissue cores were bored from the liver. The cores were used for the preparation of liver slices in a Brendel/Vitron slicer under ice-cold Sack's preservation solution[24]. Slices were incubated in Krebs medium containing retrorsine concentrations of (mM): 0 (as control), 0.125, 0.5 and 1, at 37°C under 95% $O_2$/5% $CO_2$ for 0, 2, 4, 8, and 12 hr. At the end of the incubation, taurine content in the slices was determined.

## RESULTS AND DISCUSSION

We have investigated in six *in vivo* or *in vitro* systems the influence of pyrrolizidine alkaloid exposure on the concentration of taurine in rat liver. In acute studies *in vivo*, hepatic taurine concentration is significantly decreased within 2 hr of an i.p. injection of

60% of the $LD_{50}$ dose of the pyrrolizidine alkaloid, retrorsine. Taurine concentrations remained markedly depressed throughout the 48 hr period studied (Table 1). Similarly, hepatic taurine concentrations fell to 25% of control within 24 hr in rats treated with an equitoxic dose of the pyrrolizidine alkaloid, monocrotaline (Fig 1). However, after 2 weeks, hepatic taurine concentration is comparable with the control rats (Fig. 2). In a two-week pair-feeding study, taurine concentration in the liver of rat treated with monocrotaline (65 mg/kg, i.p.) also showed no significant alteration compared to the control (Fig. 2).

In the isolated rat liver after a one hour perfusion, perfusate taurine level in the monocrotaline-perfused liver is only about 50% of the control (Fig. 3). However, the concentration of taurine in bile and liver (Fig. 3) remains unchanged. In the precision-cut liver slices incubated with retrorsine (0.125 mM to 1 mM) for 12 hr, taurine concentration is dose-dependently decreased from 8 hr onwards (Fig. 4). In control slices, taurine concentration increases over the first 3 hr of incubation before stabilizing. This is in keeping with the observed general rise in free amino acid concentrations in the slice and medium over this period, which presumably reflects an increase rate of protein breakdown. The rise in taurine must be due to substrate-controlled stimulation in taurine synthesis, this being the only source of extra taurine in the system.

Monocrotaline and retrorsine are representatives of the *Crotalaria*-type and *Senecio*-type pyrrolizidines, respectively[1,26]. These types differ in macrocyclic ring size, the former being a 11-membered and the latter a 12-membered macrocycle. The two alkaloids also differ in toxicity, monocrotaline producing right ventricular hypertrophy of the heart[15] and pulmonary arterial hypertension[17] in addition to hepatotoxicity. Except in sucrose-loaded animals[21], retrorsine is hepatotoxic only[26]. Both alkaloids have been responsible for human poisonings. For example, monocrotaline appears to have been the primary agent in Caribbean veno-occlusive disease[13,28], while retrorsine was the major alkaloid responsible for the gordolobo poisonings in Arizona[11,12,27].

Taurine is both hepato- and cardioprotective. However, our studies on the two pyrrolizidines under a variety of conditions concur in establishing that taurine levels fall sharply and rapidly in the pyrrolizidine-exposed liver, and remain depressed for at least the first few days. After two weeks, hepatic taurine content has returned to normal. The reason for this depression in taurine content is unclear. It may relate to the stimulation in GSH synthesis produced by both monocrotaline and retrorsine[37,39] and an associated diversion of cysteine from taurine synthesis.

Table 1. Acute depletion of hepatic taurine after retrorsine (20.4 mg/kg, i.p.) administration in rats

| Time (hr) | Hepatic Taurine Concentration (nmol/g liver) |
|---|---|
| 0 | 970 ± 10* |
| 2 | 570 ± 50* |
| 4 | 580 ± 20* |
| 8 | 590 ± 30* |
| 12 | 500 ± 20* |
| 16 | 550 ± 20* |
| 48 | 540 ± 20* |

Data are mean ± SD for 5 samples per assay.
[a] $p < 0.05$ vs the tiem 0 hr. Taken from [36].

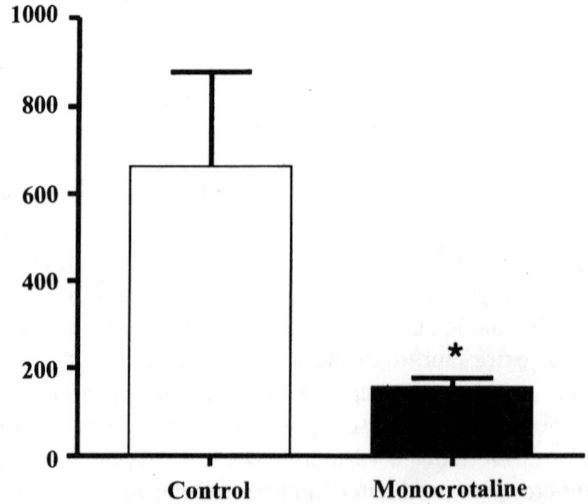


**Figure 1.** Depletion of hepatic taurine concentration 24 hr after monocrotaline administration (65 mg/kg, i.p.).

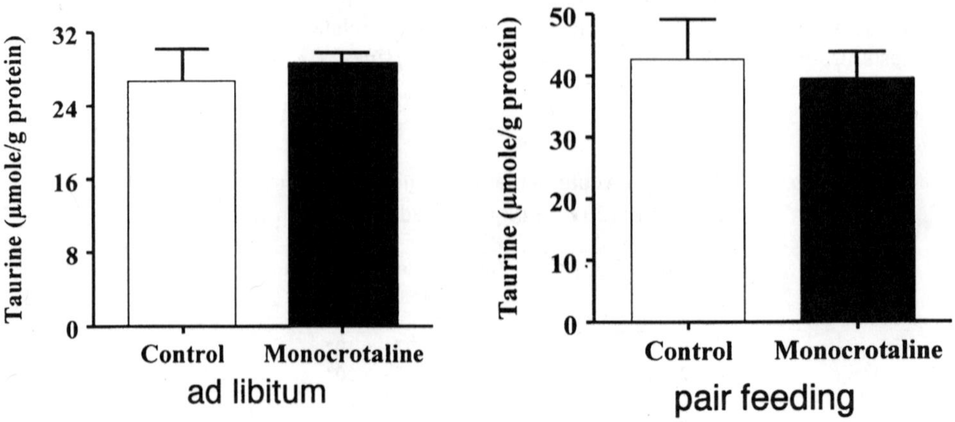


**Figure 2.** Hepatic taurine concentration after monocrotaline (65 mg/kg, i.p.) administration for two weeks. The pair-feeding data were reported at the Osaka Taurine Symposium[38].

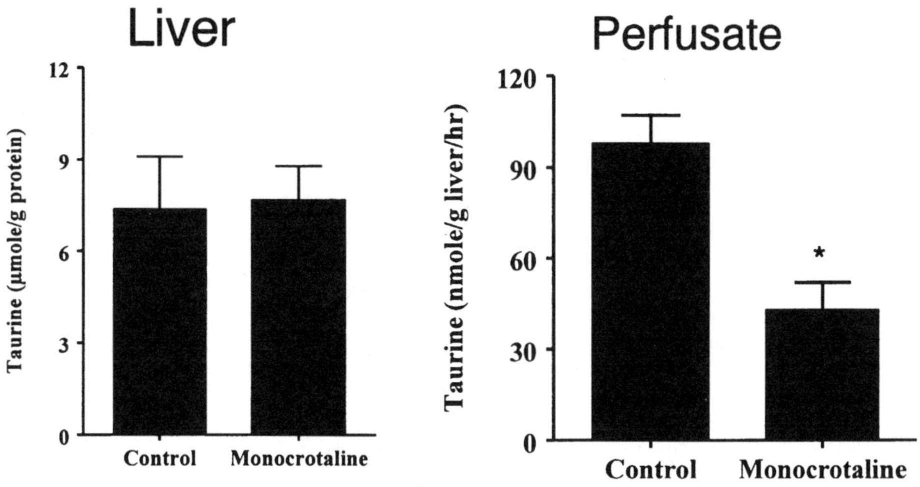


**Figure 3.** Taurine concentration in liver and perfusate after 0.5 mM monocrotaline perfusion for 1 hr.

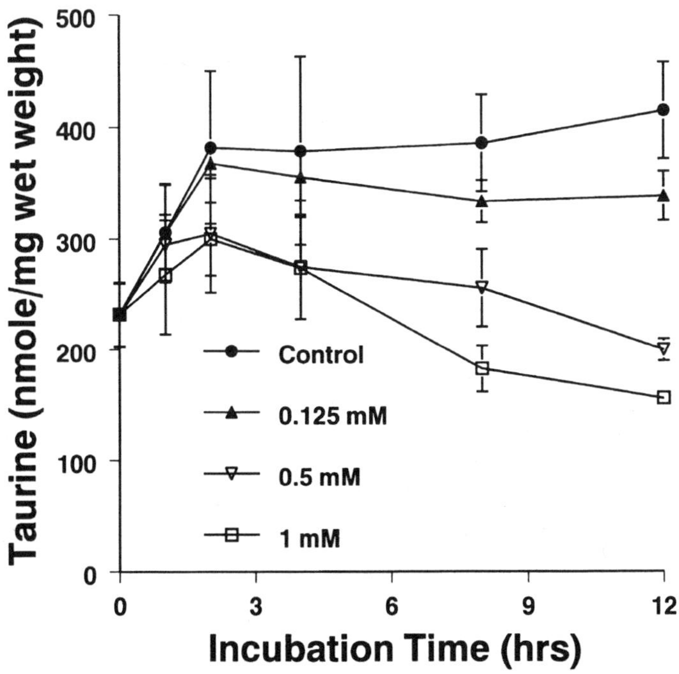


**Figure 4.** Taurine concentration in liver slices incubated with retrorsine at the indicated concentrations (0-1 mM).

It would be of interest to examine if hepatic taurine levels were lowered in pyrrolizidine-exposed human populations[6,14,22,29,30], as this would suggest that oral taurine supplementation may be of value in such populations.

## ACKNOWLEDGMENTS

This work was supported by a grant from Taisho Pharmaceutical Co., Ltd.

## REFERENCES


1. Adams, R. and Rogers, E.F., 1939, The structure of monocrotaline, the alkaloid in Crotalaria spectabilis and Crotalaria retusa, *J. Amer. Chem. Soc.*, 61:2815-2819.
2. Azari, J., Brumbaugh, P., and Huxtable, R.J., 1980, Prophylaxis by taurine in the hearts of cardiomyopathic hamsters, *J. Mol. Cell. Cardiol.*, 12:1353-1366.
3. Azuma, J., Sawamura, A., Awata, N., Ohta, H., Hamaguchi, T., Harada, H., Takihara, K., Hasegawa, H., Yamagami, T., Ishiyama, T., Iwata, H., and Kishimoto, S., 1985, Therapeutic effect of taurine in congestive heart failure: a double-blind crossover trial, *Clin.Cardiol.*, 8:276-282.
4. Azuma, J., Takahisa, K., Awata, N., Ohta, H., Sawamura, A., Harada, H., and Kishimoto, S., 1984, Beneficial effect of taurine on congestive heart failure induced by chronic aortic regurgitation in rabbits, *Res. Comm. Chem. Pathol. Pharmacol.*, 45:261-270.
5. Banks, M.A., Porter, D.W., Martin, W.G., and Castranova, V., 1991, Ozone-induced lipid peroxidation and membrane leakage in isolated rat alveolar macrophages: Protective effects of taurine, *J. Nutr. Biochem.*, 2:308-313.
6. Chauvin, P., Dillon, J.-C., Moren, A., Talbak, S., and Barakaev, S., 1993, Heliotrope poisoning in Tadjikistan, *Lancet*, 341:1663-1663.
7. Cooper, R.A. and Huxtable, R.J., 1996, A simple procedure for determining the aqueous half-lives of pyrrolic metabolites of pyrrolizidine alkaloids, *Toxicon*, 34:604-607.
8. Dokshina, G.A., Yartsev, E.I., Kolesnikov, Y.A., and Kostesha, N.Y., 1974, Influence of taurine on the recovery of tissue respiration and oxidative phosphorylation in liver mitochondria in irradiated rats, *Radiology*, 14:44-48.
9. Elizarova, E.P., Orlova, T.R., and Medvedeva, N.V., 1993, Effects on heart membranes after taurine treatment in rabbits with congestive heart failure, *Arzneimittelforschung*, 43:308-312.
10. Gordon, R.E., Shaker, A.A., and Solano, D.F., 1986, Taurine protects hamster bronchioles from acute $NO_2$-induced alterations, *Am.J.Pathol.*, 125:585-600.
11. Huxtable, R., Stillman, A., and Ciaramitaro, D., 1977, Characterization of alkaloids involved in human senecio (pyrrolizidine) poisoning, *Proc. Western Pharmacol. Soc.*, 20:455-459.
12. Huxtable, R.J., 1980, Herbal teas and toxins: novel aspects of pyrrolizidine poisoning in the United States, *Perspect.Biol.Med.*, 24:1-14.
13. Huxtable, R.J., 1989, Human health implications of pyrrolizidine alkaloids and herbs containing them, in: "Toxicants of Plant Origin, Vol I: Alkaloids",. Cheeke, P.R., ed., CRC Press, Boca Raton, Florida. pp. 41-86.
14. Huxtable, R.J., 1997, Pyrrolizidine alkaloids, in: "Comprehensive Toxicology, Vol. 9. Hepatic and Gastrointestinal Toxicology", McCuskey, R.S. and Earnest, D.L., eds., Pergamon, pp. 423-431.
15. Huxtable, R.J., Paplanus, S., and Laugharn, J., 1977, The prevention of monocrotaline-induced right ventricular hypertrophy, *Chest*, 71S:308-310.
16. Izumi, K., Nagata, R., Motoya, T., Yamashita, J., Hirokane, T., Nagata, T., Satoh, Y., Sawada, Y., Ishibashi, M., Yoshida, H., and Fukuda, T., 1989, Preventive effect of taurine against acute paraquat intoxication in beagles, *Jpn. J. Pharmacol.*, 50:229-233.
17. Kean, P.M. and Kay, J.M., 1965, Lung angiotensin converting enzyme activity in monocrotaline pulmonary hypertension, *Thorax*, 40:159
18. Lehnert, B.E., Archuleta, D.C., Ellis, T., Session, W.S., Lehnert, N.M., Gurley, L.R., and Stavert, D.M., 1994, Lung injury following exposure of rats to relatively high mass concentrations of nitrogen dioxide, *Toxicology*, 89:239-277.
19. Mattocks, A.R., 1986. Chemistry and Toxicology of Pyrrolizidine Alkaloids. Academic Press, London.

20. Mattocks, A.R., Croswell, S., Jukes, R., and Huxtable, R.J., 1991, Identity of a biliary metabolite formed from monocrotaline in isolated, perfused rat liver, *Toxicon,* 29:409-415.

21. Mattocks, A.R. and White, I.N.H., 1973, Toxic effects and pyrrolic metabolites in the liver of young rats given the pyrrolizidine alkaloid retrorsine, *Chem.Biol.Inter.,* 6:297-306.

22. Mayer, F. and Lüthy, J., 1993, Heliotrope poisoning in Tadjikistan, *Lancet,* 342:246-247.

23. Nakashima, T., Takino, T., and Kuriyama, K., 1982, Therapeutic effect of taurine administration on carbon tetrachloride-induced hepatic injury, *Jpn. J. Pharmacol.,* 32:583-590.

24. Parrish, A.R., Gandolfi, A.J., and Brendel, K., 1995, Precision-cut tissue slices: Applications in pharmacology and toxicology, *Life Sci.,* 21:1887-1991.

25. Robertson, K.A., 1982, Alkylation of N2 in deoxyguanosine by dehydroretronecine, a carcinogenic metabolite of the pyrrolizidine alkaloid monocrotaline, *Cancer Res.,* 42:8-14.

26. Schoental, R., 1959, Liver lesions in young rats suckled by mothers treated with the pyrrolizidine (*Senecio*) alkaloids, lasiocarpine, and retrorsine, *J.Path.Bact.,* 77:485-495.

27. Stillman, A.E., Huxtable, R., Consroe, P., Kohnen, P., and Smith, S., 1977, Hepatic veno-occlusive disease due to pyrrolizidine poisoning in Arizona, *Gastroenterology,* 73:349-352.

28. Stuart, K.L. and Bras, G., 1956, Veno-occlusive disease of the liver in Barbados, *West Indian Med.J.,* 5:33-36.

29. Tandon, B.N., Tandon, R.K., Tandon, H.D., Narndranathan, M., and Joshi, Y.K., 1976, An epidemic of veno-occlusive disease of liver in central India, *Lancet,* ii:271-272.

30. Tandon, H.D., Tandon, B.N., and Mattocks, A.R., 1978, An epidemic of veno-occlusive disease of the liver in Afghanistan, *Am.J.Gastroenterology,* 70:607-613.

31. Venkatesan, N. and Chandrakasan, G., 1994, In vivo administration of taurine and niacin modulate cyclophosphamide- induced lung injury, *Eur. J. Pharmacol. Environ. Toxicol. Pharmacol.,* 292:75-80.

32. Wang, Q., Giri, S.N., Hyde, D.M., and Li, C., 1991, Amelioration of bleomycin-induced pulmonary fibrosis in hamsters by combined treatment with taurine and niacin, *Biochem. Pharmacol.,* 42:1115-1122.

33. Wang, Q., Giri, S.N., Hyde, D.M., and Nakashima, J.M., 1989, Effects of taurine on bleomycin-induced lung fibrosis in hamsters, *Proc. Soc. Exp. Biol. Med.,* 190:330-338.

34. Yan, C.C. and Huxtable, R.J., 1994, Quantitation of the hepatic release of metabolites of the pyrrolizidine alkaloid, monocrotaline, *Toxicol. Appl. Pharmacol.,* 127:58-63.

35. Yan, C.C. and Huxtable, R.J., 1995a, Effect of the pyrrolizidine alkaloid, monocrotaline, on bile composition of the isolated, perfused rat liver, *Life Sci.,* 57:617-626.

36. Yan, C.C. and Huxtable, R.J., 1995b, Effects of the pyrrolizidine alkaloid, retrorsine, on sulfur metabolism in the liver, *Proc. Western Pharmacol. Soc.,* 38:37-40.

37. Yan, C.C. and Huxtable, R.J., 1995c, The effect of the pyrrolizidine alkaloids, monocrotaline and trichodesmine, on tissue pyrrole binding and glutathione metabolism in the rat, *Toxicon,* 33:627-634.

38. Yan, C.C. and Huxtable, R.J., 1996a, Effect of the pyrrolizidine alkaloid monocrotaline on taurine and sulfur amino acid metabolism in the rat liver, *Adv. Exp. Med. Biol.,* 403:135-145.

39. Yan, C.C. and Huxtable, R.J., 1996b, Effects of monocrotaline, a pyrrolizidine alkaloid, on glutathione metabolism in the rat, *Biochem. Pharmacol.,* 51:375-379.

40. Yan, C.C. and Huxtable, R.J., 1996c, Effects of taurine and guanidinoethane sulfonate on right ventricular hypertrophy induced by the pyrrolizidine alkaloid, monocrotaline, *Biochem. Pharmacol.,* 51:321-329.

# EFFECTS OF TAURINE IN PRECISION-CUT LIVER SLICES EXPOSED TO THE PYRROLIZIDINE ALKALOID, RETRORSINE

Steve J. Hasal, Yun Sun, Chong Chao Yan, Klaus Brendel, and Ryan J. Huxtable

Department of Pharmacology
University of Arizona
College of Medicine
Tucson, AZ 85724

## INTRODUCTION

Pyrrolizidine alkaloids (PA) comprise a major group of toxins derived from plants. They are botanically and geographically widespread and cause many public health and economic problems[5]. PAs cause hepatomegaly, veno-occlusive disease of the liver, and cirrhosis[7]. The toxicity of PAs can also be expressed in other organ systems, including the heart, lung, and central nervous system.

Retrorsine is a PA found in certain *Senecio*, a genus in the Asteraceae. Retrorsine is hepatotoxic only after bioactivation to the dehydroalkaloid, dehydroretrorsine. Dehydroretrorsine is a highly reactive molecule with a half-life of 1.06 sec in aqueous solution[1]. Dehydroretrorsine can alkylate cell macromolecules, conjugate with glutathione (GSH), or hydrolyze to 6,7-dihydro-7-hydroxy-1-hydroxymethyl-5*H*-pyrrolizine (DHP).

Taurine can protect the liver[8], heart[6], and lungs[4] from toxic insult. Taurine raises the $LD_{50}$ value of the PA, monocrotaline, in rats[9]. In the hopes of gaining insight into the mechanism of the protective effect on taurine against PA toxicity, we have examined the effect of taurine on the toxicity of the PA retrorsine in precision-cut rat liver slices. Three indices of cell damage were used: (i) intracellular potassium content, a secondary measure of cell energy status and a measure of membrane permeability; (ii) ATP content, a primary measure of cell energy status; and (iii) GSH content, an indicator of oxidative stress.

## METHODS

Male Sprague-Dawley rats (200-250 g) were fed Wayne Lab Chow and water *ad libitum*. They were housed 3 to a cage on wood shavings on a 12 hour light/dark cycle.

*Taurine 3*
Edited by Schaffer *et al.*, Plenum Press, New York, 1998

### Slice Preparation

Animals were killed by $CO_2$ exposure. Each liver was excised through a midventral incision and immediately placed in ice-cold, modified V7 cold preservation solution. V7 solution was made without glutathione to minimize exogenous sulfhydryl content. The livers were cored with an 8 mm sharpened stainless steel coring tool, forming tissue cylinders. Tissue slices were made from these tissue cylinders using a Brendel-Vitron tissue slicer under ice-cold V7 solution according to the method of Fisher *et al.*[3]

### Slice Organ Culture

Culture media included: (i) Krebs/bicarbonate buffer; and (ii) Waymouth's medium without phenol red supplemented with 10% fetal bovine serum and 10 ml Fungi-bact solution. All media were filter-sterilized and supplemented with 50 μg/ml gentamycin. Liver slices were floated onto Teflon/Viton/Titanium rollers. These rollers were then blotted and placed in glass scintillation vials containing 1.7 ml of medium. The vials were closed with polypropylene caps having a central hole of 2 mm diameter and placed in a dynamic roller culture incubator at 37°C. The incubators were gassed with 95% $O_2$ / 5% $CO_2$ for the duration of the experiment. Slices were preincubated for 1 hr in either control medium or medium supplemented with 20 mM taurine. Then rollers from the control medium were placed into either control medium or medium containing 0.5 mM retrorsine, and the rollers from the medium supplemented with 20 mM taurine were placed into medium containing either 20 mM taurine or both 20 mM taurine and 0.5 mM retrorsine. The slices were incubated for 2, 4, or 6 hr.

### Measures of Slice Viability

At the appropriate time point, the slice was removed from the roller, weighed on an analytical balance, homogenized in 10% trichloroacetic acid and snap-frozen in liquid nitrogen. At the time of the assays, the homogenate was centrifuged and the supernatant was assayed.

For potassium retention, 500 μl of the supernatant was assayed for $K^+$ using a Bacharach flame photometer (Model Ca-51). Data are presented as nmol $K^+$/mg wet weight.

For ATP content, 10 μl of the supernatant from each sample was aliquoted into 12 X 75mm plastic test tubes. The aliquots were then diluted with 2 ml HEPES buffer. 200 μl of this diluted sample was aliquoted into round plastic cuvettes, and ATP was measured by measuring the luminescence of the sample when 100 μl of a luciferin/luciferase solution is added to the cuvette in a luminometer. Data are presented as pmol ATP/mg wet weight.

GSH content was assessed by the method of Ellman[2]. Data are presented as nmol GSH/mg wet weight.

### Statistical Analysis

The program Graphpad Prism was used to analyze the data. One-way analysis of variance (ANOVA) was performed with the Newman-Keuls post-test.

## RESULTS AND DISCUSSION

In Krebs-bicarbonate buffer, retrorsine lowered slice potassium content slightly but significantly only after 6 hr of incubation (Fig. 1A). A sharp decrease in slice ATP content occurred at the same time (Fig. 1B). GSH content was significantly depressed from 2 hr of incubation onwards (Fig. 1C). Taurine was without effect on any of these actions of retrorsine (Fig. 1). Slices incubated with taurine showed a significantly lower content of potassium at 6 hr relative to control slices (Fig. 1A).

Slices handled exposure to retrorsine better in the more complex Waymouth's solution with fetal bovine serum. Over 6 hr incubation, no effect was seen on potassium content (Fig. 2A). ATP content was mildly but significantly decreased at 6 hr (Fig. 2B). GSH content was depressed relative to control slices from 4 hr onwards (Fig. 2C). Taurine was without effect on these retrorsine-induced changes.

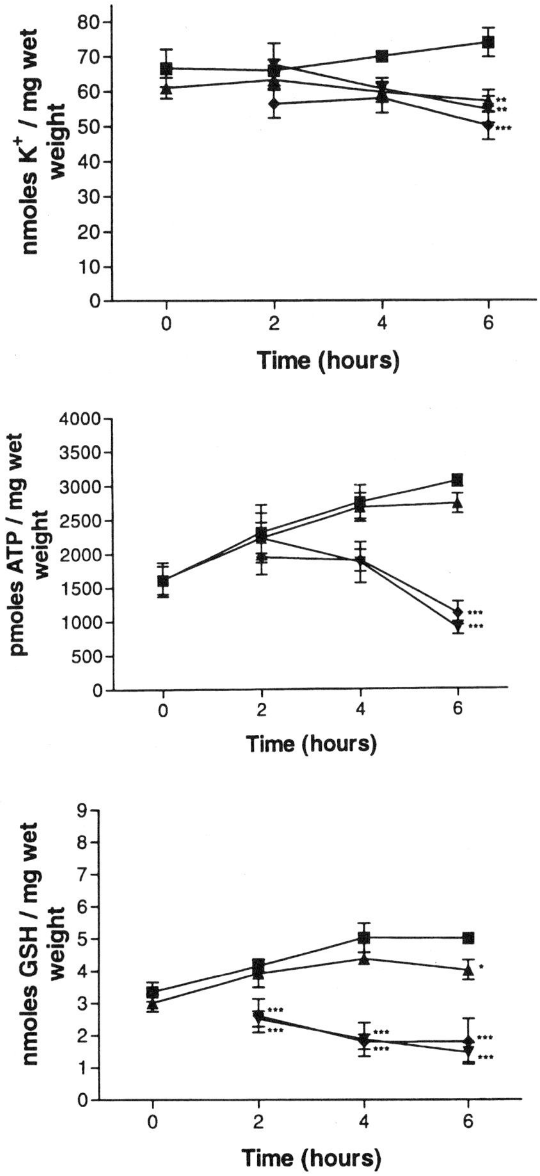


**Figure 1.** (A) top - Intracellular potassium content in rat liver slices incubated in Krebs-bicarbonate buffer following a 1 hr preincubation. Control (■), 20 mM taurine (▲), 0.5 mM retrorsine (▼), and 20 mM taurine + 0.5 mM retrorsine (◆). Each value is the mean ± SEM of 3 experiments in which three slices were used for each condition. Data are expressed as nmol $K^+$/mg wet weight. **p < 0.01 compared to control, ***p < 0.001 as compared to control. (B) middle - ATP content in rat liver slices incubated in Krebs-bicarbonate buffer following a 1 hr preincubation. Control (■), 20 mM taurine (▲), 0.5 mM retrorsine (▼), and 20 mM taurine + 0.5 mM retrorsine (◆). Each value is the mean ± SEM of 3 experiments in which three slices were used for each condition. Data are expressed as pmol ATP/mg wet weight. ***p < 0.001 as compared to control. (C) bottom - GSH content in rat liver slices incubated in Krebs-bicarbonate buffer following a 1 hr preincubation. Control (■), 20 mM taurine (▲), 0.5 mM retrorsine (▼), and 20 mM taurine + 0.5 mM retrorsine (◆). Each value is the mean ± SEM of 3 experiments in which three slices were used for each condition. Data are expressed as nmol GSH/mg wet weight. *p < 0.05 compared to control, ***p < 0.001 as compared to control.

High intracellular potassium concentration is maintained by the $Na^+/K^+$-ATPase. A marked decrease in ATP content usually precedes a fall in cell potassium. Changes in ATP content were only seen after 6 hr of incubation in both culture media. The drop in ATP content was greater in slices cultured in Krebs-bicarbonate buffer. The difference in ATP content in the two media may account for the difference in potassium content in the slices incubated in Krebs bicarbonate buffer.

Retrorsine was less toxic to slices incubated in a Waymouth's/10% fetal bovine serum than to slices incubated in Krebs bicarbonate medium.

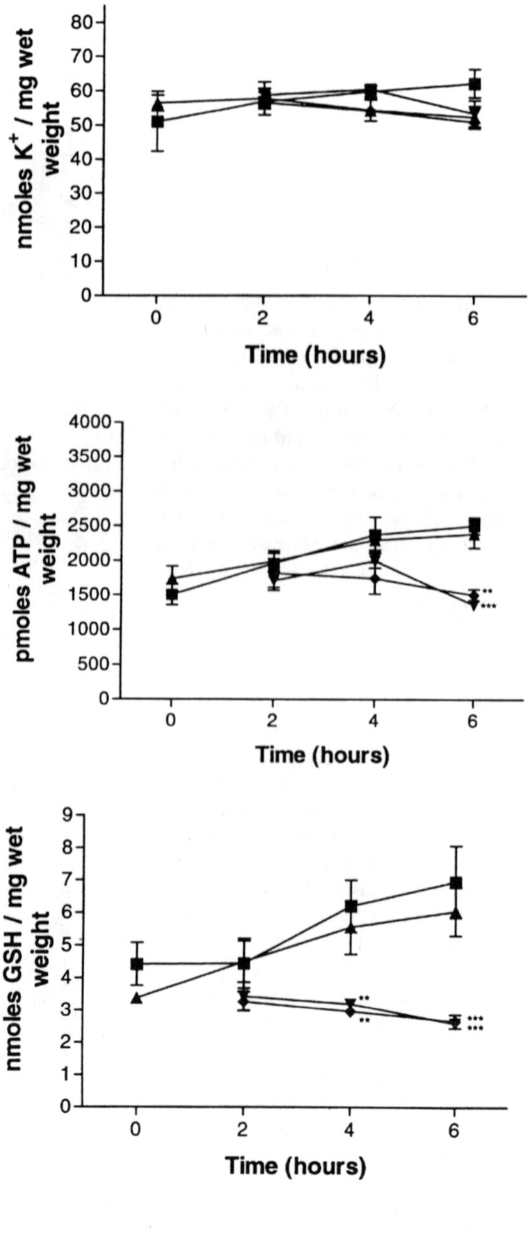


**Figure 2.** (A) top - Intracellular potassium content in rat liver slices incubated in Waymouth's + 10% fetal bovine serum following a 1 hr preincubation. Control (■), 20 mM taurine (▲), 0.5 mM retrorsine (▼), and 20 mM taurine + 0.5 mM retrorsine (◆). Each value is the mean ± SEM of 3 experiments in which three slices were used for each condition. Data are expressed as nmol $K^+$/mg wet weight. (B) middle - ATP content in rat liver slices incubated in Waymouth's + 10% fetal bovine serum following a 1 hr preincubation. Control (■), 20 mM taurine (▲), 0.5 mM retrorsine (▼), and 20 mM taurine + 0.5 mM retrorsine (◆). Each value is the mean ± SEM of 3 experiments in which three slices were used for each condition. Data are expressed as pmol ATP/mg wet weight. **$p < 0.01$ compared to control, ***$p < 0.001$ as compared to control. (C) bottom - GSH content in rat liver slices incubated in Waymouth's + 10% fetal bovine serum following a 1 hr preincubation. Control (■), 20 mM taurine (▲), 0.5 mM retrorsine (▼), and 20 mM taurine + 0.5 mM retrorsine (◆). Each value is the mean ± SEM of 3 experiments in which three slices were used for each condition. Data are expressed as nmol GSH/mg wet weight. **$p < 0.01$ compared to control, ***$p < 0.001$ as compared to control.

Our preliminary results indicate the value of the precision-cut liver slice technique for investigating the actions of hepatotoxins. Advantages of the technique include its parsimony for both toxin and animal tissue. In other studies, we have demonstrated the different metabolic profiles of PAs in precision-cut rat liver slices (Sun, Yan and Huxtable, unpublished observations).

The studies we report herein with taurine give no insight as to its protective mechanisms against PA toxicity. As reported elsewhere in this volume, however, taurine increases the biliary excretion of detoxified conjugates of dehydromonocrotaline, the toxic metabolite of the PA, monocrotaline. These findings suggest that alteration in the metabolic profile of PAs in the liver may contribute more to the protective action of taurine than an ameliorating effect on the mechanisms whereby the dehydroalkaloid metabolites of PAs produce toxicity.

## ACKNOWLEDGMENTS

Supported in part by a grant from Taisho Pharmaceutical Co., Ltd. Tokyo.

## REFERENCES

1. Cooper, R.A. and Huxtable, R.J., 1996, A simple procedure for determining the aqueous half-lives of pyrrolic metabolites of pyrrolizidine alkaloids, *Toxicon,* 34:604-607.
2. Ellman, G.L., 1958, A colorimetric method for determining low concentrations of mercaptans, *Arch. Biochem. Biophys.,* 74:443-450.
3. Fisher, R.L., Hasal, S.J., Sanuik, J.T., Gandolfi, A.J., and Brendel, K., 1995, Determination of optimal incubation media and suitable slice diameters in precision-cut liver slices: Optimization of tissue slice culture, II, *Toxicology Methods,* 5:115-130.
4. Gordon, R.E., Shaker, A.A., and Solano, D.F., 1986, Taurine protects hamster bronchioles from acute $NO_2$-induced alterations, *Am. J. Pathol.,* 125:585-600.
5. Huxtable, R.J., 1989, Human health implications of pyrrolizidine alkaloids and herbs containing them, in: "Toxicants of Plant Origin, Vol I: Alkaloids",. Cheeke, P.R., ed., CRC Press, Boca Raton, Florida. pp. 41-86.
6. Lake, N., 1993, Loss of cardiac myofibrils: Mechanism of contractile deficits induced by taurine deficiency, *Am. J. Physiol. Heart Circ. Physiol.,* 264:H1323-H1326
7. McLean, E.K., 1970, The toxic actions of pyrrolizidine (*Senecio*) alkaloids, *Pharm.Rev.,* 22:429-483.
8. Watanabe, A., Hobara, N., Kobayashi, M., and Nagashima, H., 1985, Effect of taurine on blood acetaldehyde elevation following alcohol ingestion, *Res. Comm. Sub. Abuse,* 6:247-250.
9. Yan, C.C. and Huxtable, R.J., 1996, Effects of taurine and guanidinoethane sulfonate on right ventricular hypertrophy induced by the pyrrolizidine alkaloid, monocrotaline, *Biochem. Pharmacol.,* 51:321-329.

# EFFECT OF TAURINE ON BILIARY METABOLITES OF GLUTATHIONE IN LIVER PERFUSED WITH THE PYRROLIZIDINE ALKALOID, MONOCROTALINE

Chong Chao Yan and Ryan J. Huxtable

Department of Pharmacology
University of Arizona
College of Medicine
Tucson, AZ 85724

## INTRODUCTION

Pyrrolizidines are a group of hepatotoxic plant alkaloids[6,9]. Certain pyrrolizidines are also toxic to lung, heart, the central nerve system, and other organs. Hepatic bioactivation is required for the expression of toxicity[8].

The initial step in the metabolism of pyrrolizidine alkaloids is a cytochrome P450-mediated dehydrogenation to the corresponding dehydroalkaloid[5]. Dehydroalkaloids are highly reactive electrophiles. They alkylate cell macromolecules, such as protein and nucleic acids, and are readily hydrolyzed, exhibiting half-lives in aqueous solutions of only a few seconds[4]. The alkylating ability of the primary metabolite is presumably the molecular basis of the cellular toxicity of pyrrolizidines. Dehydroalkaloids can also be released from the liver into the circulation to cause extrahepatic toxicity. Dehydroalkaloids are detoxified by conjugation with hepatic glutathione to form 7-glutathionyl-1-hydroxymethyl-6,7-dihydro-5*H*-pyrroline (GSDHP) (Fig. 1)[13,16]. GSDHP is released in high concentration into the bile[12]. Our studies have showed that variations in hepatic GSH levels regulate the production of GSDHP.

Taurine is known to protect against a number of insults in the heart, lung and liver[7,10,11]. Pretreatment of rats with taurine for two weeks reduces the toxicity of the pyrrolizidine alkaloid, monocrotaline (Fig. 1), and reduces the lethality from an $LD_{50}$ of 80 mg/kg to 120 mg/kg[17]. However, the mechanism of the protective action of taurine is not clear.

We now report the effects of taurine and its analog, guanidinoethane sulfonate (GES), on biliary glutathione metabolites in the monocrotaline-perfused isolated rat liver.

## METHODS AND MATERIALS

Monocrotaline was isolated from the seed of *Crotalaria spectabilis*, purified by

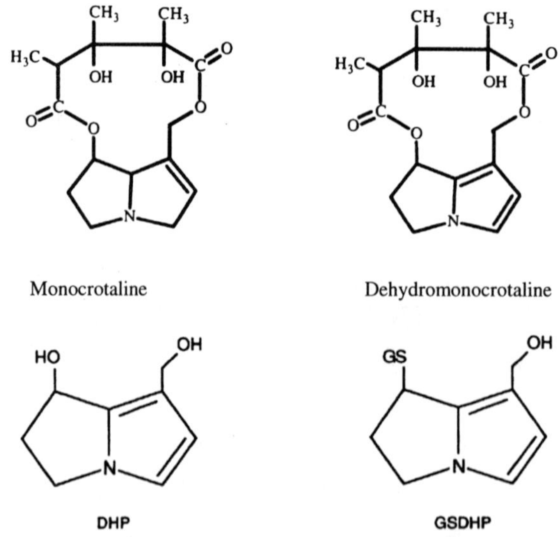


**Figure 1.** Structure of the pyrrolizidine alkaloid, monocrotaline, and its metabolites: the initially formed and highly toxic dehydromonocrotaline; the less toxic glutathionyl conjugate formed from dehydromonocrotaline, 7-glutathionyl-1-hydroxymethyl-6,7-dihydro-5*H*-pyrroline (GSDHP); and the hydrolytic product of dehydromonocrotaline, 6,7-dihydro-7-hydroxy-1-hydroxymethyl-5*H*-pyrrolizine (DHP).

counter-current distribution and identified by mass spectrometry. Bovine serum albumin (fraction V) was purchased from Sigma Chemical Co. (St. Louis, MO).

Twenty-four male Sprague-Dawley rats (body weight 150-175) were obtained from the University of Arizona Division of Animal Resources and were housed to a cage in the room kept on 12 hr cycles of light and dark. They were randomly divided into four groups receiving, respectively tap water (control and monocrotaline groups), 1% taurine (taurine group) or 1% GES (GES group) for 4 weeks. All rats were allowed food and water *ad libitum*.

After 4 weeks, rats were administered 2000 units heparin per kg body weight by a single i.p. injection. Fifteen min later, rats were anesthetized with sodium pentobarbital (50 mg/kg, i.p.). The common bile duct and the portal vein were cannulated, and the livers were perfused for 1 hr in a recirculating system. Liver perfusions were conducted with Krebs-Henseleit buffer containing 120 mM NaCl, 4.7 mM KCl, 2.5 mM $CaCl_2$, 1.2 mM $MgSO_4$, 1.2 mM $KH_2PO_4$, 25 mM $NaHCO_3$, 5.5 mM glucose and 2% bovine serum albumin. The buffer was equilibrated with 95% $O_2$-5% $CO_2$ and the pH adjusted to 7.40. At the beginning of the experiment, monocrotaline (final concentration 0.5 mM) was dissolved in the reservoir (100 ml total volume) for the monocrotaline, taurine and GES groups, but was not added to the reservoir for the control group. The duration of the perfusion was 60 min. At 15 min intervals, the bile was collected in a cold preweighed microfuge containing 0.25 ml 5% sulfosalicylic acid. The microfuge tube was kept on ice. Bile volume was determined gravimetrically, assuming a density of 1.0 gm/ml. The livers from control animals were perfused with monocrotaline-free solution.

GSDHP in bile and perfusate were assayed as previously reported[13]. GSH and its biliary metabolites, cysteine and cysteinylglycine, were analyzed by HPLC after being derivatized with monobromobimane as reported previously[15].

## RESULTS AND DISCUSSION

As previous reported[15], pretreatment of rats with 1% taurine or 1% GES in the drinking water significantly altered the taurine concentration in the liver. In the present study, after a 4 week treatment, taurine concentrations in the liver in the control, taurine and GES groups were 70 ± 17, 316 ± 50 and 19 ± 4 μmol/g protein, respectively. Bile flows in isolated perfused livers from the control, monocrotaline, taurine and GES group were 59.1 ± 2.3, 56.7 ± 6.5, 75.7 ± 4.7 and 91.4 ± 10.9 μl/g liver. Bile flows in the taurine and GES groups were significantly higher than either the control or monocrotaline-alone groups.

Neither taurine nor GES pretreatment changed the production of 6,7-dihydro-7-hydroxy-1-hydroxymethyl-5*H*-pyrrolizine (DHP) in the perfusate (Fig. 2, left). DHP (Fig. 1) is formed from nonenzymatic hydrolysis of dehydromonocrotaline (Fig. 1). However, both taurine and GES pretreatment (Fig. 2, right) significantly increased biliary GSDHP output. This finding suggests that one basis for the protective effect of these agents *in vivo* is an increased excretion of the detoxified GSDHP conjugate.

The release of GSH itself into the bile is markedly increased in monocrotaline-perfused livers (Fig. 3), confirming our earlier report[14]. GSH release into the bile increased even further in monocrotaline-perfused livers from rats pretreated with taurine or GES (Fig. 3).

The biliary releases of the GSH metabolites, cysteine and cysteinylglycine, were also increased in monocrotaline-perfused livers from rats pretreated with taurine or GES (Fig. 4).

It has been established that about 20% of hepatic GSH release is into bile and about 80% is into the blood. Although the detailed mechanism of GSH release into bile is unclear, the indications are that it probably involves an active carrier-mediated transport system that is independent of bile flow rate[1,3]. Exogenous agents having no effect on hepatic GSH concentration can alter GSH release into bile[2].

GSH, transported across the canalicular membrane in canalicular space and bile ducts, serves as substrate for the membrane-bound enzyme, (γ-glutamyltransferase. This enzyme hydrolyzes GSH to cysteinylglycine. The latter is in turn degraded by dipeptidase into

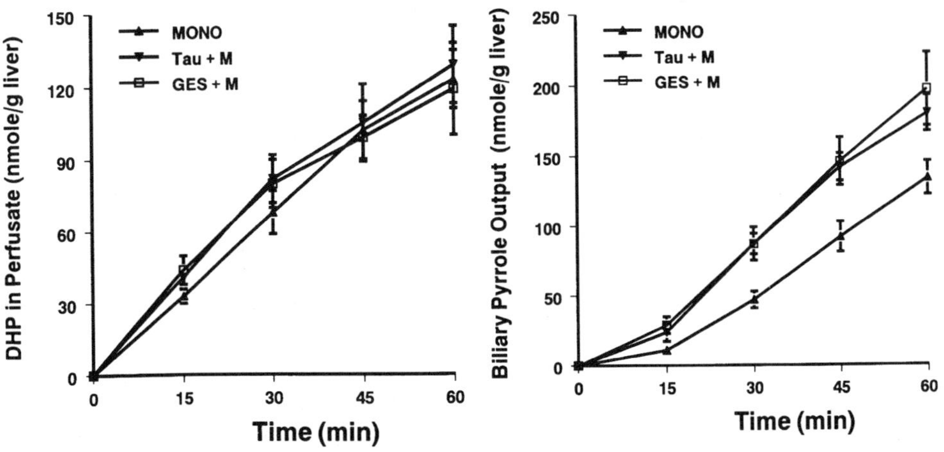


**Figure 2.** Effect of *in vivo* pretreatment with taurine (Tau + M) or guanidinoethane sulfonate (GES + M) on: (Left) the *in vitro* release into perfusate of DHP from the monocrotaline-perfused liver; (Right) the *in vitro* release into bile of GSDHP from the monocrotaline-perfused liver.

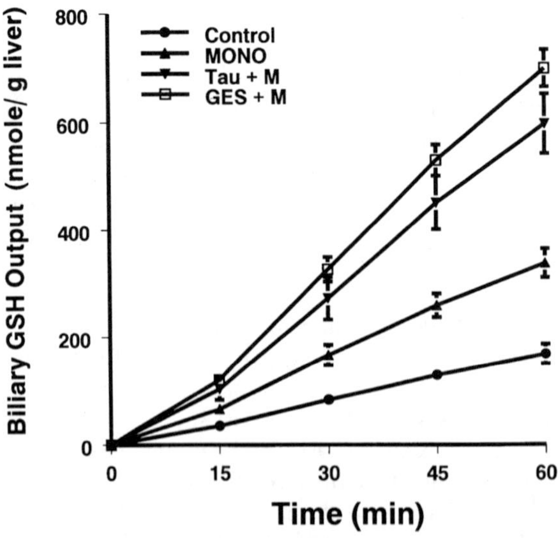


**Figure 3.** Biliary release of GSH in isolated perfused rat liver. Mono: Livers perfused with 0.5 mM monocrotaline; Tau + M: Livers from rats pretreated with taurine and perfused with 0.5 mM monocrotaline; and GES + M: Livers from rats pretreated with guanidinoethane sulfonate and perfused with 0.5 mM monocrotaline.

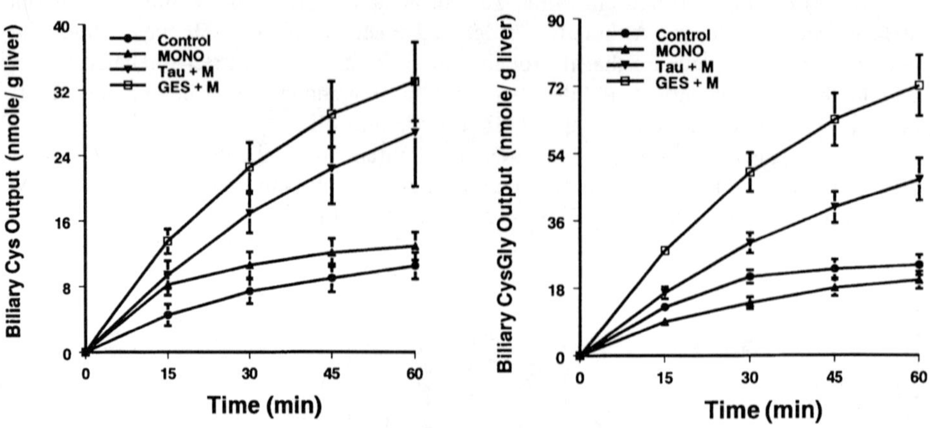


**Figure 4.** (Left) Biliary output of cysteine; and (Right) Biliary output of cysteinylglycine in the isolated perfused liver. Legend as for Figure 2.

cysteine and glycine[2]. Determination, therefore, of biliary GSH, cysteine and cysteinylglycine serves as a measure of the total biliary secretion of GSH from the liver. Total secretion is significantly higher in monocrotaline-perfused liver, and higher again in monocrotaline-perfused livers from animals pretreated *in vivo* with taurine or GES.

The physiological significance of biliary GSH is unknown. However, it may help prevent, or repair, damage to canalicular and bile duct membranes caused by the oxidant properties of bile. The increased GSH release in taurine- and GES-pretreated animals may, therefore, be one aspect of the hepatoprotective effects of these agents in monocrotaline-exposed rats[17].

In conclusion, as reported elsewhere in this volume, both taurine and GES modify sulfur metabolism in the monocrotaline-exposed liver, stimulating the biliary excretion of both GSH and GSDHP, a detoxified product of the highly toxic dehydropyrrolizidine metabolite of monocrotaline. Both or these effects may relate to the protective effects of GES and taurine in monocrotaline-exposed rats[17].

## REFERENCES


1. Abbott, W.A. and Meister, A., 1986, Intrahepatic transport and utilization of biliary glutathione and its metabolites, *Proc. Natl. Acad. Sci.*, 83:1246-1250.
2. Ballatori, N. and Clarkson, R.W., 1985, Biliary secretion of glutathione and of glutathione-metal complexes, *Fund. Appl. Toxicol.*, 5:816-831.
3. Ballatori, N., Jacob, R., and Boyer, J.L., 1986, Intrabiliary glutathione hydrolysis - a source of glutamate in bile, *J. Biol. Chem.*, 261:7860-7865.
4. Cooper, R.A. and Huxtable, R.J., 1996, A simple procedure for determining the aqueous half-lives of pyrrolic metabolites of pyrrolizidine alkaloids, *Toxicon,* 34:604-607.
5. Glowaz, S.L., Michnika, M., and Huxtable, R.J., 1992, Detection of a reactive pyrrole in the hepatic metabolism of the pyrrolizidine alkaloid, monocrotaline, *Toxicol. Appl. Pharmacol.,* 115:168-173.
6. Huxtable, R.J., 1989, Human health implications of pyrrolizidine alkaloids and herbs containing them, in: "Toxicants of Plant Origin, Vol I: Alkaloids",. Cheeke, P.R., ed., CRC Press, Boca Raton, Florida. pp. 41-86.
7. Huxtable, R.J., 1996, Taurine: Past, present and future, in: "Taurine: Basic and Clinical Aspects",. Huxtable, R.J., Azuma, J., Nakagawa, M., Kuriyama, K., and Baba, A., eds., Plenum, New York. pp. 641-650.
8. Mattocks, A.R., 1968, Toxicity of pyrrolizidine alkaloids, *Nature (London),* 217:723-728.
9. Mattocks, A.R., 1986, Chemistry and Toxicology of Pyrrolizidine Alkaloids. Academic Press, London.
10. Nakashima, T., Takino, T., and Kuriyama, K., 1982, Therapeutic effect of taurine administration on carbon tetrachloride-induced hepatic injury, *Jpn. J. Pharmacol.,* 32:583-590.
11. Waterfield, C.J., Turton, J.A., Scales, M.D.C., and Timbrell, J.A., 1991, Taurine, a possible urinary marker of liver damage: A study of taurine excretion in carbon tetrachloride-treated rats, *Arch. Toxicol.,* 65:548-555.
12. Yan, C.C., Cooper, R.A., and Huxtable, R.J., 1995, The comparative metabolism of the four pyrrolizidine alkaloids, seneciphylline, retrorsine, monocrotaline, and trichodesmine in the isolated, perfused rat liver, *Toxicol. Appl. Pharmacol.,* 133:277-284.
13. Yan, C.C. and Huxtable, R.J., 1994, Quantitation of the hepatic release of metabolites of the pyrrolizidine alkaloid, monocrotaline, *Toxicol. Appl. Pharmacol.,* 127:58-63.
14. Yan, C.C. and Huxtable, R.J., 1995a, Effect of the pyrrolizidine alkaloid, monocrotaline, on bile composition of the isolated, perfused rat liver, *Life Sci.,* 57:617-626.
15. Yan, C.C. and Huxtable, R.J., 1995b, Fluorimetric determination of monobromobimane and ophthalaldehyde adducts of gamma-glutamylcysteine and glutathione: Application to assay of gamma-glutamylcysteinyl synthetase activity and glutathione concentration in liver, *J. Chromatogr. B Biomed. Appl.,* 672:217-224.
16. Yan, C.C. and Huxtable, R.J., 1995c, Relationship between glutathione concentration and metabolism of the pyrrolizidine alkaloid, monocrotaline, in the isolated, perfused liver, *Toxicol. Appl. Pharmacol.,* 130:132-139.
17. Yan, C.C. and Huxtable, R.J., 1996, Effects of taurine and guanidinoethane sulfonate on right ventricular hypertrophy induced by the pyrrolizidine alkaloid, monocrotaline, *Biochem. Pharmacol.,* 51:321-329.

# ANTAGONISM BY TAURINE ON THE RUTHENIUM RED-INDUCED AND 6-HYDROXYDOPAMINE PLUS 1-METHYL-4-PHENYLPYRIDINIUM-INDUCED $Ca^{2+}$ RELEASE FROM RAT LIVER MITOCHONDRIA

M. Palmi,[1] G. Youmbi,[1] F. Fusi,[1] M. Frosini,[1] G.P. Sgaragli,[1]
L. Della Corte,[2] L. Bianchi,[2] and K.F. Tipton[3]

[1] Istituto di Scienze Farmacologiche
Università di Siena
Italy
[2] Dipartimento di Farmacologia
Università di Firenze
Italy
[3] Department of Biochemistry
Trinity College
Dublin, Ireland

## INTRODUCTION

Mitochondria have been implicated in the maintenance of the $Ca^{2+}$ "set-point" in cells[2,13,15], where control of $Ca^{2+}$ levels plays a significant role in enzymatic regulation and energy production[14,9,12,16,10]. Pathological conditions that result in increased tissue $Ca^{2+}$ concentration, e.g., ischemia, oxidative stress, excito- and neuro-toxicity, involve alterations in the continued ability of $Ca^{2+}$-regulating organelles such as sarcolemma, mitochondria and sarcoplasmic reticulum, to provide long-term control of cellular $Ca^{2+}$ levels (for review see 19, 20). The disruption of $Ca^{2+}$ homeostasis with loss of accumulated $Ca^{2+}$ possibly by opening a specific mitochondrial pore may account to some extent for the toxicity of 1-methyl-4-phenylpyridinium (MPP) and 6-hydroxydopamine (6-HD) that cause lesions similar to idiopathic Parkinson's disease in primates[6].

Recently, we reported that taurine (2-aminoethanesulphonic acid) a naturally occurring amino acid present in high (mM) concentration in many mammalian tissues[11] stimulates the rate of $Ca^{2+}$ accumulation by isolated liver mitochondria through a specific and saturable process that was hypothesized to involve the electrogenic uniport system responsible for mitochondrial $Ca^{2+}$ accumulation[18]. To substantiate this hypothesis and check whether taurine might exert a protective effect against toxicity induced by MPP and

*Taurine 3*
Edited by Schaffer *et al.*, Plenum Press, New York, 1998

6-HD, in this report we have used isolated, functionally intact mitochondria from rat liver pre-loaded with $Ca^{2+}$ as a model to study the effects of taurine on the release of $Ca^{2+}$ by ruthenium red (RR), a specific inhibitor of the uniport system[8]. This has been compared with the release of $Ca^{2+}$ in the presence of a combination of RR and the neurotoxins 6-HD and MPP.

## MATERIALS AND METHODS

### Mitochondria Preparation

Male Wistar rats (200-250 g) were lightly anaesthetized with a mixture of Ketavet® and Rompun®. Liver was rapidly removed and homogenized in a medium containing 250 mM sucrose, 5 mM Tris, 1 mM EGTA (STE), at pH 7.4 and mitochondria were separated by differential centrifugation[5]. After centrifugation, the mitochondrial pellet was resuspended in 1-2 ml of a medium containing 150 mM KCl, 25 mM sucrose, 5 mM $Mg^{2+}$, 2.7 mM Hepes, for determinations of $Ca^{2+}$ uptake. The mitochondrial protein content was determined by the biuret method[7] using bovine serum albumin as a standard.

### $Ca^{2+}$ Uptake

$Ca^{2+}$ uptake was determined spectrophotometrically as previously described[17] by following the differential absorbance ($A_{710\text{-}790}$) decrease of the $Ca^{2+}$-antipyrylazo III complex after addition of 50 μM $CaCl_2$ to a suspension containing 1mg/ml of mitochondrial protein.

### $Ca^{2+}$ Release

After mitochondria were loaded with $Ca^{2+}$ by treatment with 50 μM $CaCl_2$ for a period of 20 min, they were treated with 50 or 200 nM RR, a specific inhibitor of $Ca^{2+}$ uptake. Addition of RR terminated $Ca^{2+}$ uptake and stimulated the release of $Ca^{2+}$. This was followed by a time resolvable increase in $A_{710\text{-}790}$ as a result of $Ca^{2+}$ efflux from mitochondria and its complexation by antipyrylazo. Compounds to be tested were added shortly after RR. The rates of $Ca^{2+}$ release were determined at 2, 4, 6, 20 and 60 min following addition of RR.

### Mitochondria Swelling

To check possible permeability changes in the mitochondrial inner membrane during experiments with 6-HD and MPP, mitochondrial swelling was determined following the method of Broekemeier[3] by monitoring the decrease in the light-scattering of a mitochondrial suspension. Following preparation in STE without EGTA as described above, the mitochondria were suspended in the same medium used to determine $Ca^{2+}$ transport and placed in a 3 ml stirred cuvette at 30°C. The final suspension containing 1mg/ml of mitochondria was monitored spectrophotometrically at 540 nm. After the attenuance stabilized, 50 μM $Ca^{2+}$ was added followed 20 min later by 1.5 mM 6-HD and 1.5 mM MPP together. The results showed that either in the presence or in the absence of taurine, addition of 6-HD and $MPP^+$ together did not result in significant mitochondrial swelling (Fig. 5).

### Statistics

The means ± SEM from 3 to 5 separate experiments were determined and the % differences between these and controls were then assessed. The statistical significance of the differences was determined by using Student's t test for unpaired data followed by Welch's t test.

## RESULTS

As shown in Figure 1, the addition of 50 nM RR to $Ca^{2+}$ loaded mitochondria induced a rapid release of $Ca^{2+}$ whose rate declined with time. The rate of $Ca^{2+}$ release determined 2 min after RR addition was 11.03 ± 0.91 nmol/min/mg of mitochondrial protein and progressively decreased to a value of 1.37 ± 0.071 nmol/min/mg of mitochondrial protein one hour later. Figure 2 shows that 1, 5, 10 and 20 mM taurine significantly antagonized the RR-stimulated $Ca^{2+}$ release with an effect that was both time and concentration-dependent. At 2 min, 1mM taurine reduced $Ca^{2+}$ efflux by 10.6 ± 0.3 % ($p>0.05$) of controls, whereas 5, 10 and 20 mM taurine reduced it by 29.3 ± 0.7 % ($p<0.05$), 36.9 ± 1.7 % ($p<0.01$), and 44.2 ± 0.55 % ($p<0.01$), respectively. Similar values of percentage inhibition were obtained at 4 and 6 min, (Fig. 2a) whereas at 20, 40 and 60 min the % inhibitions were constant and not related to the concentration of taurine (range 15.7 ± 0.95 - 20.9 ± 3.7 % of controls) (Fig. 2b).

In the presence of a higher (200 nM) RR concentration, $Ca^{2+}$ efflux was similar (10.95 ± 1.32 nmol/min/mg of mitochondrial protein) to that induced by the lower RR concentration, however, this was not changed in a statistically significant way by treatment with taurine at any concentration in the 1-20 mM range. In this case the % reduction of $Ca^{2+}$ release by taurine was between 8.85 ± 2.55 and 17.1 ± 3.81 % of controls with no significant differences related to time (Fig. 3a and b).

Pro-oxidant neurotoxins such as MPP and 6-HD are reported to induce $Ca^{2+}$ loss as a possible consequence of a change in the permeability of the mitochondrial inner

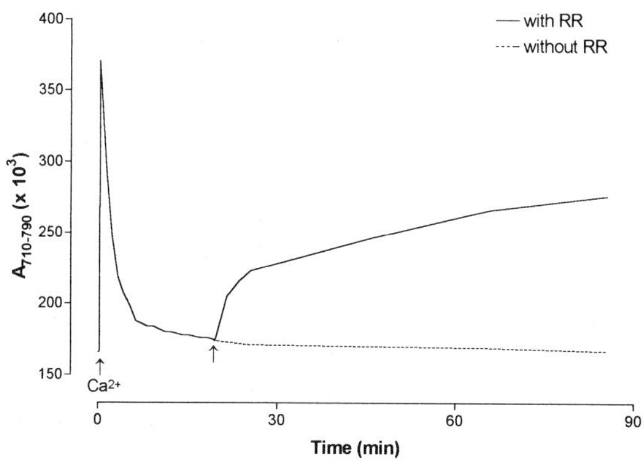


**Figure 1.** Representative spectrophotometrical tracings of kinetics of $Ca^{2+}$ uptake (downward deflection) and release (upward deflection) in rat liver mitochondria. After mitochondria were loaded with 50 μM $CaCl_2$ for 20 min, the release was initiated by addition of 50 nM ruthenium red (RR) to a cuvette containing 1 mg/ml mitochondria. The addition of RR terminated $Ca^{2+}$ uptake and stimulated its release. Rates of $Ca^{2+}$ transport in nmol/min/mg of mitochondrial protein were determined spectrophotometrically from the differential absorbance changes ($A_{710-790}$) using the $Ca^{2+}$-indicator antipyrylazo III.

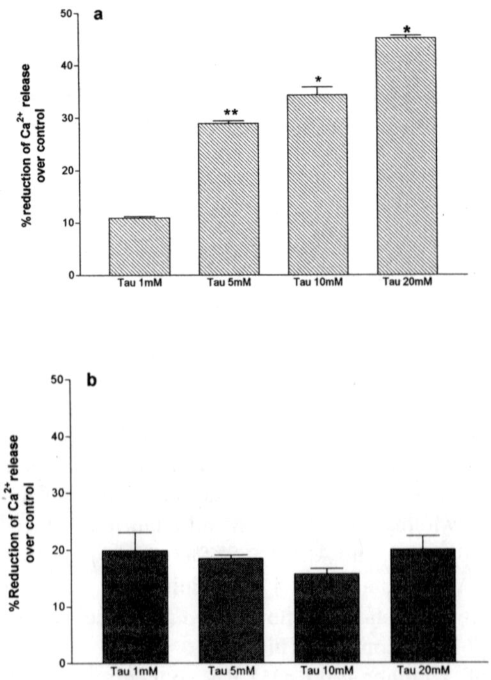


**Figure 2.** Effect of taurine on 50 nM ruthenium red (RR)-stimulated $Ca^{2+}$ release from preloaded mitochondria. After the $Ca^{2+}$ loading procedure was terminated (cf Fig. 1), 50 nM RR alone (control) or in combination with various (1, 5, 10, 20 mM) taurine concentrations was added to a suspension containing 1 mg/ml mitochondria. Columns in the figure (mean of duplicate determinations from 3 to 5 separate experiments with SEM.) represent % reduction of $Ca^{2+}$ release (nmol/min/mg of protein) over controls as a function of taurine concentrations. Panels in the figure represent rate values obtained at different intervals from taurine addition: group data from 2, 4 and 6 min intervals are represented in panel a and group data from 20, 40 and 60 min in panel b. Within panels the asterisk in each column indicate statistical significance of that value with respect to the preceding value. Statistical significances at *$p<0.05$ and **$p<0.01$ levels were determined by the Student's t test followed by Welch's t test.

membrane[20]. To verify this hypothesis, as well as to study the effect of taurine on the neurotoxin-induced $Ca^{2+}$ release, we studied the effects of the combination of MPP and 6-HD on RR-induced $Ca^{2+}$ release and on mitochondrial swelling. These effects were also investigated in the presence of taurine. The results showed that $Ca^{2+}$ efflux stimulated by RR was higher in the presence of 1.5 mM 6-HD and 1.5 mM MPP together than that stimulated by RR alone (13.3 ± 1.45 and 11.03 ± 0.91 nmol/min/mg of mitochondrial protein, respectively) and that taurine antagonized the rate of neurotoxin-induced $Ca^{2+}$ release but to a limited extent reducing it significantly ($p<0.05$) only at concentrations as high as 5 and 10 mM and when the rate was determined within the first 2 min interval. In that case, the % inhibition values (peak 28.1 ± 2.7 %, obtained with 10 mM taurine) did not change with increasing taurine concentrations (Fig. 4a). Similar effects (peak value 21.5 ± 1.74 at 10 mM taurine) were observed when % inhibition rates were determined at longer (20-60 min) intervals (Fig. 4b). In addition, in this case there was no statistical difference from controls at any of the times and taurine concentrations used.

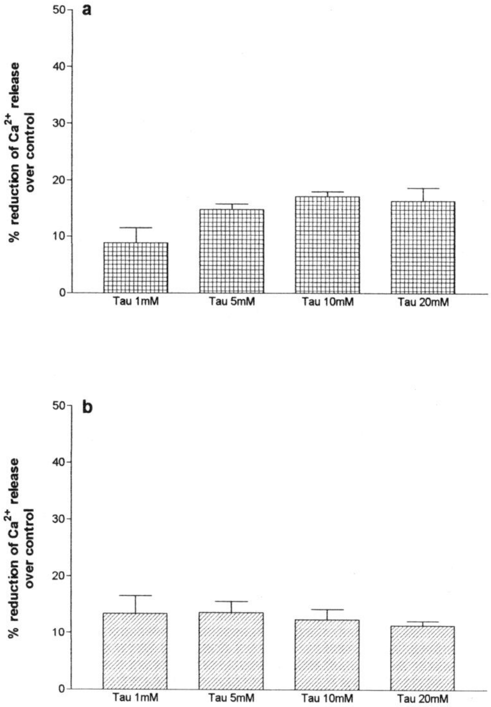


**Figure 3.** Effect of taurine on 200 nM ruthenium red (RR)-stimulated $Ca^{2+}$ release from $Ca^{2+}$ preloaded mitochondria. After the $Ca^{2+}$ loading procedure was terminated, 200 nM RR alone (control) or in combination with various concentrations of (1, 5, 10, 20 mM) taurine was added to a suspension containing 1 mg/ml mitochondria. (cf Fig. 2).

When the ability of 6-HD and MPP to induce mitochondrial swelling was tested, we observed an increase in attenuance at 540 nm shortly after addition of the neurotoxins together (Fig. 5). This effect, which was probably due to the reported spontaneous autoxidation of 6-HD[8] stabilized within 2-3 min and remained constant for the following 20 min, indicating that under these conditions, the mitochondria were not swelling. The same trend was observed in the presence of 20 mM taurine (Fig. 5).

## DISCUSSION

Mitochondria from liver possess at least three possible mechanisms for maintenance of cellular $Ca^{2+}$ homeostasis. These include a membrane potential-dependent uniport system for $Ca^{2+}$ uptake into the mitochondrial matrix, a $H^+/Ca^{2+}$ exchange pathway, the primary $Ca^{2+}$-efflux mechanism of mitochondria from liver, kidney, lung and smooth muscle and a $Na^+/Ca^{2+}$ exchange, the primary $Ca^{2+}$-efflux mechanism of mitochondria from excitable tissue (for review see ref 8).

Recently, we reported that taurine stimulates the initial rate of $Ca^{2+}$ accumulation in isolated liver mitochondria through a specific and saturable process[18]. The hypothesis that this might be due to increased $Ca^{2+}$ uptake via stimulation of the uniporter is supported by the present data showing that taurine concentration-dependently antagonizes $Ca^{2+}$ loss

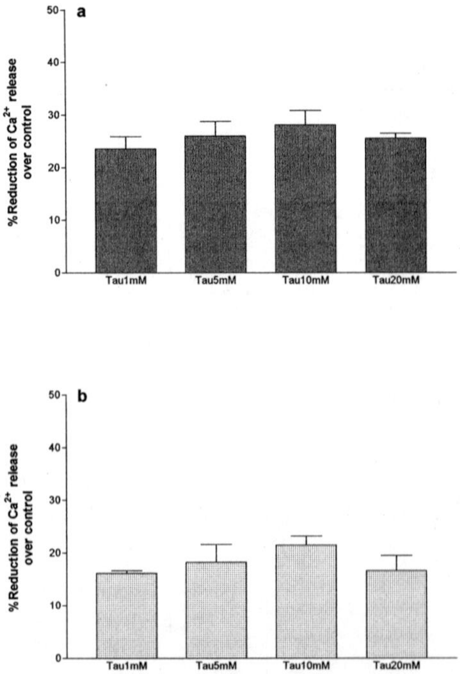


**Figure 4.** Effect of taurine on $Ca^{2+}$ released by a combination of 6-hydroxydopamine (6-HD), 1-methyl-4-phenylpyridinium (MPP) and ruthenium red (RR). After the $Ca^{2+}$ loading procedure was terminated, 50 nM RR, 1.5 mM 6-HD and 1.5 mM MPP were added to a suspension containing 1 mg/ml mitochondria (see cf. Fig. 2).

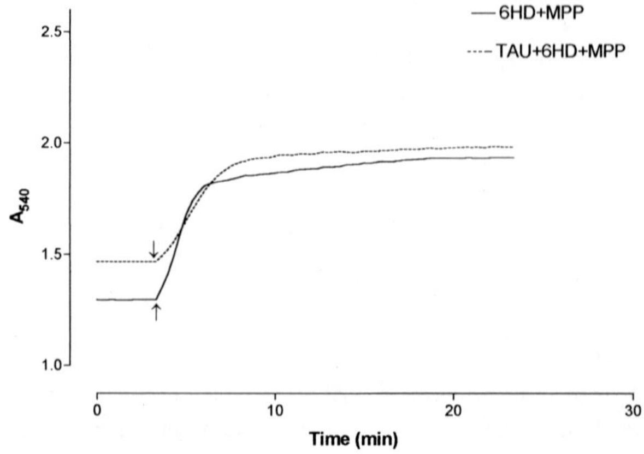


**Figure 5.** Effect of 6-hydroxydopamine (6-HD) and 1-methyl-4-phenylpyridinium (MPP) alone or in the presence of taurine on mitochondrial swelling. Representative spectrophotometric tracings of changes in attenuance at 540 nm to measure variations in volume of mitochondria (1 mg/ml) are represented. After the mitochondrial loading procedure was terminated, a combination of 1.5 mM 6-HD and 1.5 mM MPP alone or with 20 mM taurine were added (indicated by the arrow) and changes in mitochondria volume followed for a 20 min period.

induced by 50 nM RR, a specific inhibitor of the uniporter in mitochondria[8] and that this effect is seen when the efflux rates were determined within the 2-6 min interval but not at longer times.

This observation is necessary but not sufficient to demonstrate our hypothesis since an increased $Ca^{2+}$ accumulation might be the consequence not only of stimulated uptake but also of inhibition of the electroneutral ion ($Na^+/Ca^{2+}$ and/or $H^+/Ca^{2+}$) exchangers responsible for $Ca^{2+}$ extrusion. To distinguish between these two possibilities, we tested the effect of taurine on $Ca^{2+}$ efflux induced by a higher concentration of ruthenium red (RR). The observation that with 200 nM RR, taurine, even at the highest concentration used, failed to inhibit $Ca^{2+}$ release indicates that its action involves, at least mainly, the uniport system. If this were not the case, the % inhibition values would have been similar with both of the RR concentrations.

In agreement with earlier reports[19,4] that combinations of 6-HD and MPP stimulate $Ca^{2+}$ efflux from mitochondria we observed potentiation of $Ca^{2+}$ efflux when these neurotoxins were added to RR in $Ca^{2+}$-loaded mitochondria. However, this effect was not statistically significant. Since in our study the rate of $Ca^{2+}$ release was about 2-3 times higher than that observed in the previous work[19], it is conceivable that our conditions approximated the maximal rate of release which could not be significantly enhanced further by the neurotoxins.

Taurine antagonized the $Ca^{2+}$ loss induced by the neurotoxins to a limited extent as it reduced only the initial rate (determined at 2 min interval) of $Ca^{2+}$ release, and its effect was not concentration-dependent. If it is assumed that the total $Ca^{2+}$ released by RR plus the neurotoxins is the net result of the combination of two different mechanisms, the inhibition of the uniporter plus an, as yet undefined process induced by the neurotoxins. It is possible that taurine, by acting primarily on the former process, only partially counteracts the $Ca^{2+}$ release induced by the association of RR and the neurotoxins. Earlier findings showing that taurine reacts poorly with pro-oxidants, including superoxide, peroxide and the hydroxyl radical[1], seem to be consistent with such an hypothesis.

It may be that the mechanism of loss of accumulated $Ca^{2+}$ by 6-HD and MPP together involves on one hand, the mitochondrial membrane-pore opening induced by 6-HD autoxidation, consequent to the production of superoxide and hydroxyl radicals, and on the other, the MPP potentiation of these effects[19,4]. Although further experiments such as measurements of membrane potential and sucrose permeation are needed to assess the presence of a permeability transition, our data, by showing lack of swelling in mitochondria treated with the neurotoxins, would seem to exclude the involvement of this mechanism in the neurotoxin-induced $Ca^{2+}$ release and toxicity.

In conclusion we have shown that taurine potentiates the initial rate of $Ca^{2+}$ uptake in mitochondria and that this is due to activation of the uniporter present in the mitochondrial inner membrane. This effect might explain the partial inhibition by taurine on the $Ca^{2+}$ loss induced by combination of the pro-oxidant neurotoxins 6-HD plus MPP in the presence of RR.

Our data showing the effects of taurine on $Ca^{2+}$ uptake and pro-oxidant mediated $Ca^{2+}$ release suggest that this amino acid might be of importance in those pathological conditions, such as ischaemia and oxidative stress, characterized by cell loss of $Ca^{2+}$ homeostasis.

## ACKNOWLEDGMENTS

This work was supported by grants from the European Community (BIOMED 1, BHMT1 CT-94-1402) and from Italian MPI 60%.

## REFERENCES


1. Aruoma, O.I., Halliwell, B., Hoey, B.M., and Butler, J., 1988, The antioxidant action of taurine, hypotaurine and their metabolic precursors, *Biochem. J.*, 256:251-255.
2. Becker, G.L., 1981, Regulation of free $Ca^{2+}$ by cardiac mitochondria and skinned myocytes, in: "Calcium and Phosphate Transport Across Biomembranes", Bronner, F., and Peterlik, M., eds., Academic, New York, pp. 79-82.
3. Broekemeier, K.M., Dempsey, M.E., and Pfeiffer, D.R., 1989, Cyclosporin A is a potent inhibitor of the inner membrane permeability transition in liver mitochondria, *J. Biol. Chem.*, 264:7826-7830.
4. Cohen, G. and Heikkila, R.E., 1974, The generation of hydrogen peroxide, superoxide radical, and hydroxyl radical by 6-hydroxydopamine, dialuric acid, and related cytotoxic agents, *J. Biol. Chem.*, 249:2447-2452.
5. Chappell, J.B. and Hansford, R.G., 1972, Preparation of mitochondria from animal tissues and yeast, in: "Subcellular Components: Preparation and Fractionation", 2nd Ed., Birmie, G.D, ed., Butterworth, London, pp. 77-91.
6. Gerlach, M., Riederer, P., Przunteck, H., and Youdim, M.B.H., 1992, MPTP mechanisms of neurotoxicity and their implications for Parkinson's disease, *Eur. J. Pharmacol.*, 208, 273-286.
7. Gornall, A.G., Bardawill, C.J., and David, M.M., 1949, Determination of serum protein by means of the biuret reaction, *J. Biol. Chem.*, 177:751-766.
8. Gunter, T.E. and Pfeiffer, D.R., 1990, Mechanism by which mitochondria transport calcium., *Am. J. Physiol.*, 258:C755-C786.
9. Hansford, R.G., 1985, Relation between mitochondrial calcium transport and control of energy metabolism, *Biochem. Pharmacol.*, 102:1-72.
10. Hansford, R.G., 1994, Physiological role of mitochondrial $Ca^{2+}$ transport, *J. Bioenerg. Biomembr.*, 26:495-508.
11. Huxtable, R.J., 1992, Physiological actions of taurine, *Physiol. Rev.*, 72:101-159.
12. McCormack, J.G., Halestrap, A.P., and Denton, R.M., 1990, Role of calcium ions in the regulation of mammalian intramitochondrial metabolism, *Physiol. Rev.*, 70:391-425.
13. Murphy, E., Coll, K., Rich, T.L., and Williamson, J.R., 1989, Hormonal effects on calcium homeostasis in isolated hepatocytes, *J. Biol. Chem.*, 255:6600-6608.
14. Murphy, M.P., 1989, Slip and leak in mitochondrial oxidative phosphorylation, *Biochem. Biophys. Acta*, 977:123-141.
15. Nicholls, D.G., 1978, The regulation of extramitochondrial free calcium ion concentration by rat liver mitochondria, *Biochem. J.*, 176:463-474.
16. Nicholls, D.G. and Ferguson, S.J., 1992, Secondary transport, in: "Bioenergetics 2", Nicholls, D.G. and Ferguson, S.J., eds., Academic Press, London, pp. 207-233.
17. Palade, P., 1987, Drug-induced $Ca^{2+}$ release from isolated sarcoplasmic reticulum by use of pyrophosphate to study caffeine-induced $Ca^{2+}$ release, *J. Biol. Chem.*, 262:6135-6141.
18. Palmi, M., Fusi, F., Youmbi, G., Frosini, M., Bianchi, L., Della Corte, L., Sgaragli, G.P., and Tipton, K.F., 1996, Effects of taurine and structurally related analogues on $Ca^{2+}$ uptake and respiration rate in rat liver mitochondria, *Adv. Exp. Med. Biol.*, 403:117-124.
19. Reichman, N., Porteous, C.M., and Murphy, M.P., 1994, Cyclosporin A blocks 6-hydroxydopamine-induced efflux of $Ca^{2+}$ from mitochondria without inactivating the mitochondrial inner membrane pore, *Biochem. J.*, 297:151-155.
20. Richter, C. and Frei, B., 1988, $Ca^{2+}$ release from mitochondria induced by pro-oxidants, *Free Radic. Biol. Med.*, 4:365-375.

# EFFECT OF TAURINE SUPPLEMENTATION ON THE LIPID PEROXIDE FORMATION AND THE ACTIVITIES OF GLUTATHIONE-RELATED ENZYMES IN THE LIVER AND ISLET OF TYPE I AND II DIABETIC MODEL MICE

Eunyoung Lim, Sungyoun Park, and Harriet Kim

Department of Food and Nutrition
Seoul National University
Seoul, 151-742, Korea

## ABSTRACT

Effects of taurine supplementation on lipid peroxide formation and the activities of glutathione (GSH) dependent enzymes in diabetic model mice were investigated. Type I diabetes mellitus was induced by injecting alloxan to ICR mice while type II diabetes mellitus was produced by high calorie diet feeding to genetically hyperglycemic KK mice. Taurine was given in drinking water at the level of 5% (w/v) for seven days. The malondialdehyde (MDA) levels of liver and the islets of type I diabetes were significantly increased compared to the control group but the levels were significantly decreased by taurine supplementation. In the type II diabetic model, the concentrations of MDA were not changed by taurine treatment. The activity of hepatic and islet GSH-peroxidase (GPX) was increased in the type I diabetic group, but in type II animals it was decreased. Hepatic GPX activity of both type I and II diabetics was not altered by taurine supplementation but was increased in the islets of the type II animals. No effect on the activity of GSH S-transferase (GST) was observed in both types of diabetes (I and II) following taurine supplementation. These results suggest that taurine supplementation protects type I diabetic mice from lipid peroxide formation.

## INTRODUCTION

Diabetes mellitus (DM) is a chronic metabolic disease that has been extensively studied throughout the entire life span of the disease[18]. Type I DM can be induced by destruction of the pancreatic beta-cells with drugs or immunologic reactions. Although the majority of diabetics are type II (NIDDM), the complexity of the genetic and environmental factors, along with ethical problems, has prevented human experimentation. Nonetheless, progress in studying NIDDM has been hampered by the paucity of animal models reproducing the human NIDDM condition[12]. The KK mouse was originally developed by Kondo et al.[13] for the study of NIDDM. The animals in this model develop obesity, polyuria, and hyperglycemia. However, the incidence of DM in this mouse is only about 50 to 70%, although a high calorie-diet can increase the incidence rate and thereby shorten the induction period of DM. The KK mouse has been exploited as an animal model of NIDDM because it develops obesity and insulin-resistance[4], beginning 4 to 12 months after birth.

Malondialdehyde (MDA), the widely used indicator of lipid peroxidation, is increased in the plasma of diabetic patients[10], while levels of glutathione (GSH), another measure of oxidant stress, is 20-30% lower than normal in sera and red cells of NIDDM patients[3]. Moreover, the infusion of GSH increases the ratio of GSH/GSSG, and lowers blood glucose levels. These data suggest that the ratio of GSH/GSSG plays an important role in maintaining glucose homeostasis[13]. In light of the important role of GSH as an antioxidant which regulates the production of free radicals, many investigators have been interested in the changes provoked by supplementation of antioxidative agents to animals with DM.

Taurine (2-aminoethanesulfonic acid), a sulfur containing amino acid, is of interest because it is not involved in the synthesis of cellular proteins or for energy production. The best known function for taurine is in its conjugation with bile acids[8]. Brain, heart, eye, muscle, and liver contain high levels of taurine. In these tissues, taurine is recognized for its involvement in the coordination of nerve function[2,15], stabilization of the cell membrane, detoxification[5], anti-oxidant reactions[14], and modulation of osmotic pressure[10]. It has been reported that taurine supplementation reduces blood glucose levels and reverses the extent of pancreatic β-cells injury in IDDM[9]. Taurine, which decreases lipid peroxidation, helps to maintain the normal rod cell shape in the retina and also preserve rabbit sperm cell mobility[1]. The concentration of taurine in the brain has been demonstrated to decrease as the animal ages. The object of this paper is to measure the effects of taurine as an antioxidant in diabetic animals. The content of lipid peroxidation products in the liver and the islets after taurine supplementation were measured in the two types of diabetic animals. The activities of the antioxidant enzymes, glutathione peroxidase and glutathione S-transferase, were measured.

## MATERIALS AND METHODS

Weaned ICR mice and KK mice from the Experimental Animal Breeding Center of Seoul National University were maintained on a regular pellet diet *ad libitum* for two weeks. The onset of spontaneous hyperglycemic diabetes in the KK mice was induced by a high fat diet (20 g/100 g diet). Water was supplied *ad libitum*; temperature was 20-25°; humidity was 55%; and the light-dark cycle was set at 12 hour intervals. After fasting for 12 hours to induce type I diabetes, alloxan (100 mg/kg b.w.) was injected intraperitoneally into the ICR mice. After access to a high fat diet for 8 weeks to induce type II DM, fasting blood glucose levels higher than 120 mg/dl were used as the criteria to identify type II diabetes. Taurine was added to the drinking water at 5% (v/w). Thus, the experimental groups were (n=7) used the normal ICR mouse (CON. I), Diabetic ICR mouse (DM I), normal KK mouse (CON. II), Diabetic KK mouse (DM II) and four groups treated with taurine. Animals were fasted for 18 hours prior to the determination of blood glucose levels. Following decapitation, the livers and islets were removed, washed in cold physiological saline, and then frozen at -60°C. Blood glucose was measured with a test strip (Blood glucose sensor: ExactTech, MediSense. Inc., Cambridge, MA) using the glucose oxidase (*Aspergillus niger*) method.

To measure lipid peroxides, liver and islet homogenates were prepared in 1:4 (w/v) ice-cold 1.15% KCl solution. Lipid peroxide content was measured by the assay of thiobarbituric acid reactive substances (TBARS), according to the method of Wills[20]. The standard was 1,1,3,3-tetraethoxypropane. The activity of glutathione peroxidase was assayed in the cytosol of liver and islets by the method of Tappel[19]; cumene-hydroperoxide was used as the substrate. The activity of glutathione S-transferase was assayed spectrophotometrically at 340 nm by measuring the formation of the conjugate of glutathione and CDNB (1-chloro-2,4-dinitrobenzene) by the method of Habig[9]. SAS was used for statistical analyses. Means and standard deviations were calculated by ANOVA ($P<0.05$) and the Duncan's multiple-range test was used to establish significance between each group.

## RESULTS AND DISCUSSION

### Blood Glucose Levels

Taurine supplementation for 7 days at a concentration of 5% (w/v) had little effect on fasting blood glucose levels of either the control or diabetic groups (types I and II) (Table 1). However, Hisashi et al.[11] has reported that taurine lowers blood glucose levels in type I DM animals.

### Effects on Lipid Peroxides

Taurine-induced changes in liver and islets malondialdehyde (MDA) levels of normal and diabetic type I and II animals are shown in Figure 1. Alloxan treatment of the ICR mice resulted in almost a 3-fold increase in hepatic MDA levels. However this 3-fold increase was reduced to almost control level when taurine was added to the drinking water. It is interesting to note that this effect of taurine on MDA levels was not observed in type II diabetic KK mice. Alloxan treatment caused pancreatic damage to the ICR mouse although Hisashi et al.[11] reported that structural and functional pancreatic damage in STZ-treated mice was reversed by taurine treatment. Presumably, alloxan treatment results in rapid accumulation of hydrogen peroxide and the generation of hypochlorite, the latter which can be scavenged by taurine.

**Table 1.** Blood glucose levels of experimental groups with or without taurine supplementation (n=7)

| | Fasting blood glucose (mg/dl) | |
|---|---|---|
| | (-) Taurine | (+) Taurine |
| Con. I | $93.4 \pm 8.9^{c}$ | $111.7 \pm 2.6^{c}$ |
| DM I | $258.2 \pm 5.0^{ab}$ | $281.0 \pm 10.8^{ab}$ |
| Con. II | $104.0 \pm 10.2^{c}$ | $98.0 \pm 9.2^{c}$ |
| DM II | $161.0 \pm 22.7^{bc}$ | $163.0 \pm 33.9^{bc}$ |

Con. I: Control group for type I; diabetic DM I: type 1 diabetic model; Con. II: Control group for type II; DM II: type II diabetic model. Values are mean ± S.D. Values with different letters are significantly different at P <0.05 by ANOVA and the Duncan's multiple-range test.

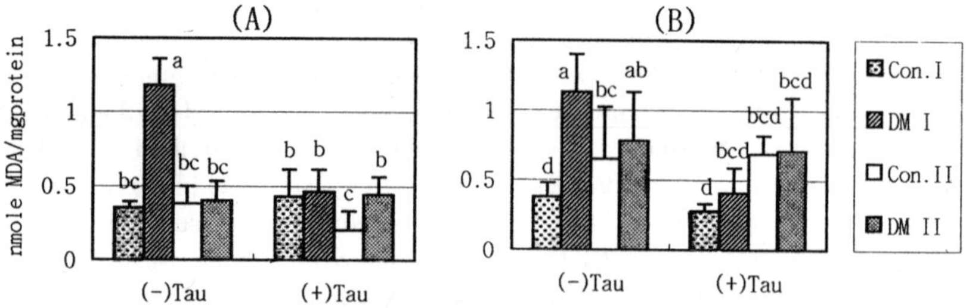


**Figure 1.** Effects of taurine supplementation on liver (A) and islet (B) concentrations of malondialdehyde in diabetic mice.

When excess taurine is available to the hepatocyte or the β-cell, it is likely that hypochlorite is converted to chloramine, thereby, reducing the degree of oxidative damage to the tissues. However, this protective effect of taurine in reducing lipid peroxide formation in liver, as well as in the islet (Fig. 1), was observed only in type I diabetic mice.

## GSH - Related Enzymes

The activity of GSH-related enzymes, glutathione peroxidase (GPX), and glutathione - S-transferase (GST), in liver and islets of type I and II diabetic animals with or without taurine treatment was measured. The results are shown in Figures 2 and 3. Taurine supplementation had very little effect on the activities of GPX and GST in the liver and islets of diabetic mice. However, the GPX activity of liver and islets of type I diabetic mice was significantly higher than the control animals (Fig. 2). Rushmore et al.[17] observed

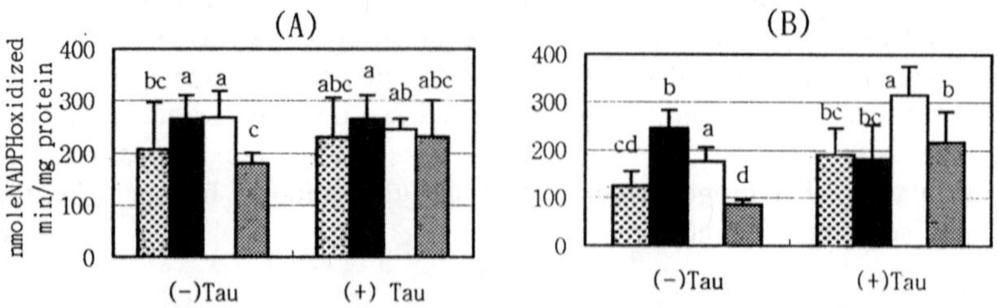


**Figure 2.** Effects of taurine supplementation on liver (A) and islet (B) activities of glutathione peroxidase in diabetic mice.

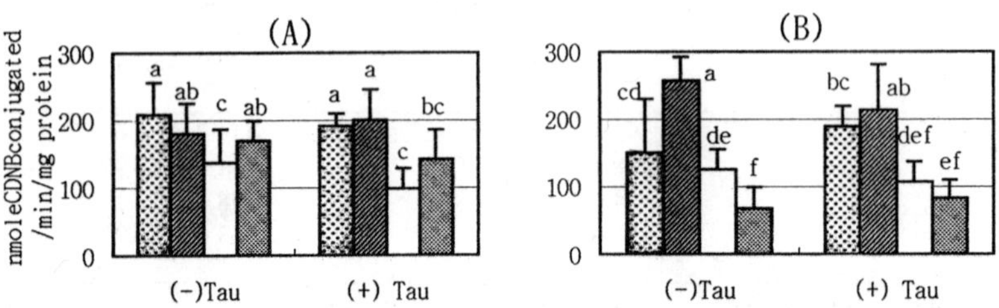


**Figure 3.** Effects of taurine supplementation on liver (A) and islet (B) activities of glutathione S-transferase in diabetic mice.

an increase in GST activity in hepatoma cells that were cultured in the presence of hydrogen peroxide. Recently, Giugliano et al.[8] reported on the normalizing effect of the antioxidants, gluatathione and vitamin E, in type II diabetic patients. These investigators measured the rheology of blood after the administration of glycerylnitrate, followed by the infusion of glutathione and vitamin E. The data suggested that the increase in oxidative stress in diabetic patients may explain the link between hyperglycemia and vascular complications. However, taurine supplementation appears to have a negligible effect on the activities of glutathione-related enzymes in type I and II diabetic mice.

## REFERENCES:


1. Alvarez, J.G. and Storey, B.T., 1983, Taurine, hypotaurine, epinephrine and albumin inhibit lipid peroxidation in rabbit spermatozoa and protect against loss of mobility, *Biol. Reproduction.*, 29:54829.
2. Aruoma, O.I., Halliwell, B., Hoey, B.M., and Butler, J., 1988, The antioxidant action of taurine, hypotaurine and their metabolic precursor, *Biochem. J.*, 256:251.
3. Arzate, M.E. and Moran, J.H., 1989, Inhibitory effect of taurine on 4-aminopuridine-stimulated release of labelled dopamine from straital synaptosomes. *Neuropharmacology*, 25: 689.
4. Awadallah, R., Sessoukey, E.A., Doss, H., Khalifa, K., and Hawry, Z., Blood-reduced glutathione, serum ceruloplasmin and mineral changes in juvenile diabetes. *Endocrinology*, 17: 79.
5. Dillard, C.J., Kunert, K.J., and Yappel, A.L., 1983, Effects of vitamin E, ascorbic acid and mannitol on alloxan induced lipid peroxidation in rats, *Biochem. J.*, 216:204.
6. Dulin, W. E. and Wyse, B. M., 1970, Diabetes in the KK mouse, *Diabetologia*, 6:317.
7. Emudianughe, T.S., Caldwell, J., and Smith, R.L, 1983, The utilization of exogenous taurine for the conjugation of xenobiotic acids in the ferrot, *Xenobiotica*, 13:133.
8. Giugliano, D., Paolisso, G., Varricchio, M., D'amore, A., Galzerano, D., D'onofrio, F., and Balbi, V., 1993, Daily vitamin E supplements improve metabollic control but not insulin secretion in elderly type II diabetic patients, *Diabetes Care*, 16:1433.
9. Habig, W.H., Pabst, M.J., and Jakoby, W.B., 1974, Glutathione S-transferase, *J. Biol. Chem.*, 249:7130.
10. Hayes, K. C. and D.V.N., 1976, A review on biological function of taurine, *Nutr. Rev.*, 34:161.
11. Hisashi, T., Yukio, Y., and Kinya, K., 1979, Protective action of taurine against streptozotocin induced hyperglycemia, *Biochem. Pharm.*, 28:2807.
12. Hoffman, E.K. and Lamvert, I.H., 1983, Amino acid transport and cell volume regulation in Ehilich ascites tumor cells. *J. Physiol. Lond.*, 338:613.
13. Kondo, K.K., Oxawa, T., Tomida, T., and Ezaki, K. 1957, Inbred strains resulting from Japanese mice, *Bull. Esp. Animals*, 6:107.
14. Mordes, J.P. and Rosin, A.A., 1980, Animal models of diabetes, *Am. J. Med.*, 70,353.
15. Palolisso, G., Maro, G.D., Pizza, G., and D'onnofrio, F., 1992, Plasma GSH/GSSG affects glucose homeostasis in healthy subjects and non-insulin-dependent diabetics, *Am. J. Phsiol.*, 263:E435.
16. Pasantes, M.H., Martin, D.L., and Sain, W., 1990, Taurine:Function Neurochemistry, Physiology and cardiology, Wiley-Liss. New York, 191.
17. Rushmore, T.H., Morton, M.R., and Pickette, C.B., 1991, The anti-oxidant responsive element, *J. Biol. Chem.*, 266:11632.
18. Sato, Y., Hotta, N., Sakamoto, N, Matsuoka, S., Oshishi, N., and Yagi, K, 1979, Lipid peroxidation and glutathione peroxidase activities in patients with diabetes mellitus, *Horm. Metab. Res.*, 19:89.
19. Tappel, A.L., 1978, Glutathione peroxidase and hydroperoxides, in: "Methods in Enzymology", Fleisher, S. and Packer, L., eds., pp. 52:506.
20. Wills, E.D., 1966, Mechanisms of lipid peroxide formation in animal tissue, *Biochem. J.* 99:667.
21. Zimmet, P., 1983, Epidemiology of diabetes mellitus, in: "Diabetes mellitus, theory and practice", Ellenberg, M., and Rifkin, H., eds., Medical Examination Publ., New Hyde Park, 451.

# TAURINE PROTECTS THE LIVER AGAINST LIPID PEROXIDATION AND MEMBRANE DISINTEGRATION DURING RAT HEPATOCARCINOGENESIS

Jeong Soon You and Kyung Ja Chang
Department of Food and Nutrition
Inha University
Inchon, Korea 402-751

## ABSTRACT

The purpose of this study was to determine the effects of taurine supplementation on both hepatic morphological changes and the extent of hepatic lipid peroxidation and membrane disintegration during rat hepatocarcinogenesis. Sprague Dawley rats were fed high fat diets containing 15% corn oil and were maintained on drinking water with or without 1% taurine. Two weeks after the appropriate feeding regimen, hepatocarcinogenesis was induced by a modification of the Solt and Farber method. This involved a 8 week protocol, including diethylnitrosamine initiation, 3 weeks of 2-acetylaminofluorene feeding and finally a 70% partial hepatectomy. Morphological changes of the hepatocyte were observed by transmission electron microscopy (TEM). Hepatocytes of the carcinogen-treated rat not exposed to taurine contained normal nuclei, but the endoplasmic reticulum (ER) and the mitochondria (Mi) were almost destroyed. By contrast, although the hepatocytes from the taurine supplemented group contained some irregular contour nuclei, the ER and Mi were normal. In the carcinogen-treated groups, lipid peroxidation was decreased because of the activation of several detoxifying enzymes. Glutathione S-transferase (GST) activity increased in the carcinogen-treated groups but less so in the group supplemented with taurine before treatment with the carcinogen. In the group supplemented with taurine prior to treatment with the carcinogen, glutathione peroxidase (GPx) activity was higher than in the carcinogen-treated group lacking taurine exposure. Consistent with the severe destruction to the membrane in the carcinogen-treated rats, hepatic glucose-6-phosphatase (G6Pase) activity, an index of membrane stability, was also decreased. However, both the fall in G6Pase activity and the degree of membrane damage was reduced in the carcinogen-treated animals receiving oral taurine. These results suggest that taurine appears to inhibit lipid peroxidation, to alter the activity of the defense enzymes and to protect the liver against membrane disintegration during rat hepato-carcinogenesis.

*Taurine 3*
Edited by Schaffer *et al.*, Plenum Press, New York, 1998 

## INTRODUCTION

Taurine (2-aminoethansulfonic acid) is a free ß-amino acid that is normally present in high concentrations in many tissues of man and other animal species. There are many established and proposed roles for taurine, including conjugation with bile acids, modulation of calcium levels, maintenance of osmolarity, antioxidation and stabilization of membranes[11]. Also, taurine has a number of protective functions which have been observed both *in vitro* and *in vivo* in several organs, particularly in the lungs, heart and liver. When administered prophylactically to the hamster, taurine completely protects the lungs against inflammation and morphological changes caused by exposure to nitrogen dioxide. Moreover, taurine administered (1% solution) prior to dosing reduces pulmonary pathology in hamsters caused by the administration of paraquat or bleomycin[8]. Taurine administered orally to chicks has been reported to protect against isoprenaline-induced cardiac damage[4]. The amino acid also exerts a protective effect in hepatocytes by decreasing lipid peroxidation due to hyperoxidation, thereby preventing cell death[15]. As expected, hypotaurine is more potent than taurine in blocking lipid peroxidation[2], a process that can be initiated by oxygen radicals and has been implicated as a major cause of cancer. Various carcinogens have been reported to increase the extent of lipid peroxidation[3,14], which in turn decreases microsomal membrane integrity and adversely influences G6Pase activity. It has also been reported that taurine prevents the loss in motility and inhibits lipid peroxidation in rabbit spermatozoa[1]. On the other hand, in rod outer segments, taurine protects against the loss of morphology[19] but fails to block chemical- or ischemia-induced lipid peroxidation or ischemia-induced glutathione depletion[23]. Paradoxically, Nakashima et al.[16] have shown that taurine reduces the extent of hepatic lipid peroxidation induced by $CCl_4$ if administered after onset of $CCl_4$ toxicity in the liver, but pretreatment with taurine actually facilitates $CCl_4$-mediated lipid peroxidation[17]. Thus, the effect of taurine is dependent upon the experimental conditions employed.

The purpose of the present study was to examine the effect of taurine administration on liver morphology, the extent of lipid peroxidation and the degree of membrane disintegration during rat hepatocarcinogenesis.

## MATERIALS AND METHODS

Male Sprague-Dawley rats (50-70 g) were acclimatized to their environment for 3 days prior to their use in an experiment. All rats were kept at 23 - 27°C with a 12 hr light-dark cycle, were fed a high fat diet of 15% corn oil and were maintained ad libitum on drinking water with or without 1% taurine.

Rats were randomly divided into the following groups: CO (control, without taurine), CT (control, with taurine during the experimental period), EO (carcinogen-treated, without taurine), EB (carcinogen-treated, with taurine before treatment with the carcinogen), EA (carcinogen-treated, with taurine after treatment with the carcinogen) and ET (carcinogen-treated, with taurine throughout the experimental protocol). Two weeks after feeding, hepatocarcinogenesis was induced using a modification of the method of Solt and Farber[20]. Rats received a single intraperitoneal injection of diethylnitrosamine (DEN, 200 mg/kg body weight), which was dissolved in saline solution (Fig. 1). After 2 weeks on the appropriate experimental diet, the animals were fed diets containing 0.02% 2-acetylaminofluorene (2-AAF) for a period of 10 days and then were subjected to a 70% partial hepatectomy (PH). Thereafter, the rats were fed a diet containing 2-AAF for a period of 10 days and then switched to the basal diet for 3 weeks before initiation of the

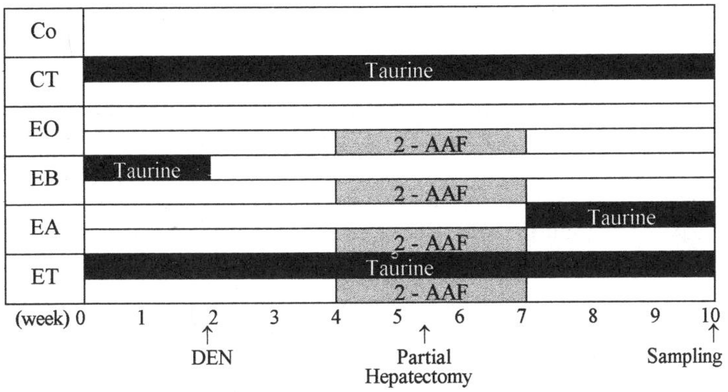


CO : Control group
CT : Control group with taurine
EO : Carcinogen-treated group
EB : Carcinogen-treated group + pre 1% taurine
EA : Carcinogen-treated group + post 1% taurine
ET : Carcinogen-treated group + all 1% taurine

**Figure 1.** Experimental design

experiment. The control group was injected with a saline solution instead of DEN and was subjected to a sham operation instead of the partial hepatectomy. Animals were killed by decapitation after a 12 hr fast and the livers were immediately removed. A portion of the liver was reserved for transmittance electron microscopy (TEM). These liver samples were prefixed in 2.5% paraform-glutaraldehyde for 18 hr. and then postfixed in 1% osmium tetroxide for 2 hr. Following dehydration, the samples were treated with Epon 812. Thin sections were stained with uranyl acetate and lead citrate and examined with a Jeol-100CX electron microscope. The portion of the liver not used in the electron microscope studies were used for the preparation of microsomal and cytosolic fractions. These samples were homogenized and subjected to differential centrifugation[7]. Microsomal and cytosolic fractions were stored at -70°C until use.

The content of microsomal lipid peroxides in the liver was determined by measuring the formation of thiobarbituric acid reactive substances (TBARS)[6]. Cytosolic glutathione S-transferase (GST) activity was assayed by the method of Habig et al.[9] using 1-chloro-2,4-dinitrobenzene (CDNB) and glutathione as substrates. Cytosolic glutathione peroxidase (GPx) activity was estimated by Tappel's method using cumene hydroperoxide as the substrate[2]. Microsomal glucose 6-phosphatase (G6Pase) activity was determined by measurement of the amount of inorganic phosphate liberated from glucose-6-phosphate[5]. The protein concentration of hepatic cytosol and microsomes was estimated by the Lowry method using bovine serum albumin as the standard[13]. All values are given as the mean $\pm$ S.E. All statistical analyses were carried out by Duncan's multiple range test using the SAS program. A p values of $< 0.05$ was selected as a limit of statistical significance.

## RESULTS AND DISCUSSION

Morphological changes of rat hepatocytes were observed by transmission electron microscopy (Fig. 2). Control hepatocytes contained round nuclei, abundant endoplasmic reticulum (ER) and rod shaped mitochondria (Mi). However, hepatocytes from the carcinogen-treated group not exposed to taurine contained normal nuclei, but the ER and

Mi were virtually 100% destroyed. By comparison, hepatocytes from the EB group, in which taurine was administered prior to carcinogen exposure, contained irregular contour nuclei, but normal Mi. Also, in the taurine supplemented EB group, normal ER were concentrated on one side while lipid granules were scattered. Interestingly, glycogen granules were prevalent in the EB group. In this regard, it is significant that taurine influences blood glucose and insulin levels, increases glycogen synthesis and inhibits glycogenolysis[12].

TBARS content of the control CT group, which had been exposed to taurine were higher than those of the taurine free CO control group (Table 1). By contrast, TBARS content of the carcinogen-treated, taurine free group (EO) were lower than those of the control. Surprisingly, in the carcinogen-treated groups, there was no difference between taurine exposure before treatment with the carcinogen and the EO group with no taurine exposure. However, TBARS content of the group exposed to taurine after treatment with the carcinogen (EA) was higher than all other groups. Previously, it has been reported that TBARS content increases in the carcinogen-treated group[3,14]. However, our data is consistent with another study showing a decrease in TBARS content[18]. Enzymes, such as

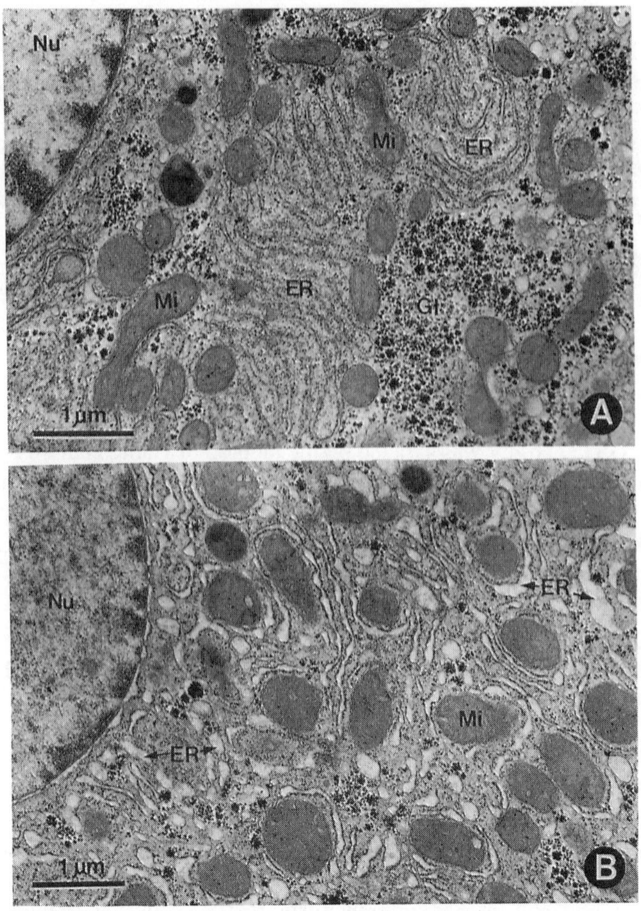


**Figure 2AB.** Effect of taurine on hepatocyte morphology of the carcinogen-treated rat. A. Control group (CO) X 23,000; B. Control group with taurine (CT) X 21,000.

**Table 1.** Effect of taurine supplementation on hepatic microsomal thiobarbituric acid reactive substances content.

| Group | Thiobarbituric acid reactive substance (nmole TBARS/mg protein) |
|---|---|
| CO | 0.176 ± 0.005[bc] |
| CT | 0.208 ± 0.016[a] |
| EO | 0.140 ± 0.009[d] |
| EB | 0.144 ± 0.006[d] |
| EA | 0.190 ± 0.009[ab] |
| ET | 0.163 ± 0.003[cd] |

Values are mean ± S.E. Means with different letters are significantly different at $p < 0.05$ by Duncan's multiple range test.

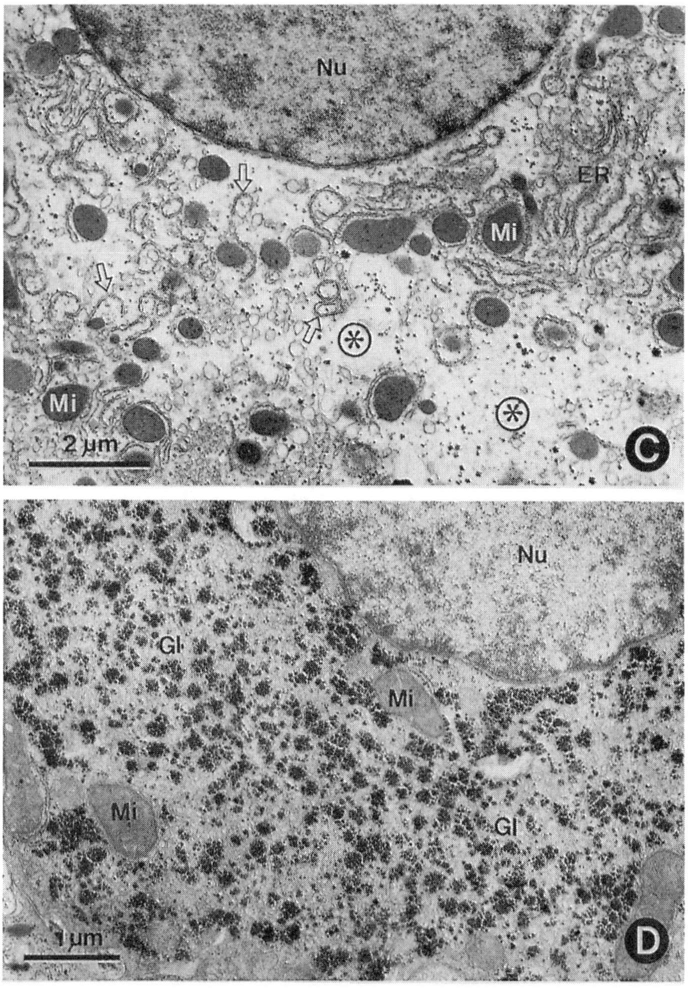


**Figure 2CD.** Effect of taurine on hepatocyte morphology of the carcinogen-treated rat. C. Carcinogen treated group (EO) X 13,000; D. Carcinogen treated group + pre 1% taurine (EB) X 12,000.

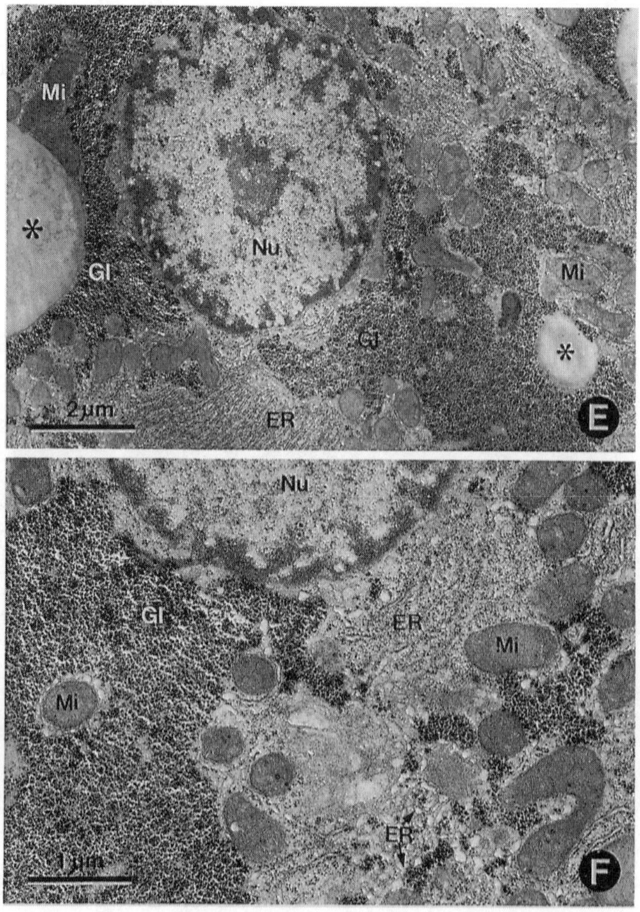


**Figure 2EF.** Effect of taurine on hepatocyte morphology of the carcinogen-treated rat. E. Carcinogen treated group + post 1% taurine (EA) X 20,000; F. Carcinogen treated group + all 1% taurine (ET) X 24,000.

glutathione S-transferase, play a central role in drug detoxification[10]. It has been assumed that TBARS content decreases because of the activation of detoxifying enzymes.

Glutathione-S-transferase (GST) activity of the carcinogen-treated groups was significantly increased compared to those of the control groups (Table 2). This result agrees with many previous reports on GST activity during hepatocarcinogenesis[10,18]. By contrast, the GST activity of groups exposed to taurine before treatment with the carcinogen or throughout the experimental procedure were only slightly depressed. The reason for this result might be that taurine supplementation before DEN injection slowed or weakened the carcinogenic response resulting from DEN administration.

In agreement with previous studies[10,18,21], we found that glutathione peroxidase (GPx) activity of the carcinogen-treated groups was decreased compared to those of the control groups (Table 2). Interestingly, GPx activity of the carcinogen-treated groups supplemented with taurine were higher than those of the taurine free groups.

**Table 2.** Effect of taurine supplementation on hepatic glutathione-S-transferase, glutathione peroxidase and glucose-6-phosphatase activity

| Group | Glutathione S-transferease activities (nmole CDNB conjugated /mg protein) | Glutathione peroxidase activities (nmole NADPH oxidized/ /mg protein) | Glucose 6-phosphatase activities (nmole Pi liberated /min/mg protein) |
|---|---|---|---|
| CO | 277.887 ± 8.278[c] | 59.923 ± 2.353[a] | 1016.557 ± 54.674[a] |
| CT | 249.418 ± 3.973[d] | 55.880 ± 3.931[ab] | 1108.628 ± 20.360[a] |
| EO | 391.672 ± 5.381[a] | 41.429 ± 0.940[d] | 459.238 ± 37.275[d] |
| EB | 352.539 ± 7.095[b] | 52.790 ± 1.824[bc] | 498.598 ± 25.339[cd] |
| EA | 394.292 ± 1.560[a] | 49.213 ± 1.077[c] | 581.414 ± 39.219[c] |
| ET | 343.696 ± 2.729[b] | 49.375 ± 1.445[c] | 681.967 ± 27.385[b] |

Values are mean ± S.E. Means with different letters are significantly different at $p < 0.05$ by Duncan's multiple range test.

Glucose-6-phosphatase (G6Pase) activity was measured as an index of membrane stability of the microsomal fraction (Table 2). G6Pase activity of the carcinogen-treated groups was significantly decreased compared to those of the control groups. However, G6Pase activity of the EA group supplemented with taurine after treatment with the carcinogen or the ET group exposed to taurine throughout the experimental protocol were higher than those of the carcinogen-treated group without taurine. This result was confirmed by morphological results analyzed by electron microscopy. We conclude that taurine stabilizes biomembranes during hepatocarcinogenesis.

Therefore, these results suggest that taurine inhibits lipid peroxidation, alters the activity of several defense enzymes and protects the membrane against disintegration during rat hepatocarcinogenesis. Nonetheless, further studies are required in order to determine the dose-response effect.

## ACKNOWLEDGMENTS

This work was supported by grants from Inha University. We thank Dong-A Pharmaceutical Co. for the donation of taurine.

## REFERENCES

1. Alvarez, J.G. and Storey, B.T., 1983, Taurine, hypotaurine, epinephrine and albumin inhibit lipid peroxidation in rabbit spermatozoa and protect against lossof motility, *Biol. Reproduction*, 29:548-555.
2. Aruoma, O.I., Halliwell, B., Honey, B.W., and Butler, J., 1988, The antioxidant action of taurine, hypotaurine and their metabolic precursors, *Biochem. J.*, 256:251-255.
3. Astrom, A., Maner, S., and DePierre, J.W., 1986, Induction of cytochrome p450 and related drug metabolizing activities in the livers of different rodent species by 2-acetylaminoflurorene and 3-methylcholanthrene, *Biochem. Pharmacol.*, 35:2703-2713.
4. Azuma, J., Hamaguchi, T., Ohta, H., Takihara, K., Awata, N., Sawamura, A., Harada, H., Tanaka, Y., and Kishimoto, S., 1987, Calcium overload-induced myocardial damage caused by isoproterenol and by adriamycin: possible role of taurine in its prevention, *Adv. Exp. Med. Biol.*, 217:167-179.
5. Baginski, E.S., Foa, P.P., and Zak, B., 1983, Glucose 6-phosphatase, *Method. Enz. Anal.*, 2:876-880.
6. Buege, J.A. and Aust, S.D., 1978, Microsomal lipid peroxidation, *Meth. Enzymol.*, 52:302-310.
7. Fleisher, S. and Kervina, M., 1974, Subcellular fraction of rat liver, *Meth. Enzymol.*, 31:6-13.
8. Gordon, R.E., Richard, F.H., and Rachel, F.H., 1992, Taurine protection of lungs in hamster weeks of oxidant injury: a morphologic time study of paraquat and bleomycin treatment, *Adv. Exp. Med. Biol.*, 315:319-328.

9. Habig, W.H., Pabst, M.J., and Jakoby, W.B., 1974, Glutathione S-transferase, *J. Biol. Chem.*, 249:7130-7139.
10. Howie, A.F., Forrester, L.M., Glancey, M.J., Schlager, J.J., Powis, G., Beckett, G.J., Hayes, J.D., and Wolf, C.R., 1990, Glutathione S-transferase and glutathione peroxidase expression in normal and tumor human tissues, *Carcinogenesis*, 11:451-458.
11. Huxtable, R.J., 1992, Physiological actions of taurine, *Physiol. Rev.*, 72:101-163.
12. Kulakowski, E.C. and Matura, J., 1984, Hypoglycemic properties of taurine: not mediated by enhanced insulin release, *Biochem. Pharmac.*, 33:2835-2838.
13. Lowry, O.H., Rosebrough, N.J., Farr, A.L., and Randall, R.J., 1951, Protein measurement with Folin phenol reagent, *J. Biol. Chem.*, 193:265-275.
14. Mukai, F.H., Goldstein, B.D., 1981, Mutagenicity of malondialdehyde, a decomposition product of peroxidized polyunsaturated fatty acids, *Science*, 197:868-869.
15. Nakashima, T., Seto, Y., Toshikazu, N., Shima, T., Iwai, Y., Zeiao, K., Okanque, T., and Kashima, K., 1990, Calcium-associated cytoprotective effect of taurine on the calcium and oxygen paradoxes in isolated rat hepatocytes, *Liver*, 10:167-172.
16. Nakashima, T., Takino, T., and Kuriyama, K., 1982, Therapeutic effect of taurine administration on carbon tetrachloride-induced hepatic injury, *Japan J. Pharm.*, 32:583-589.
17. Nakashima, T., Takino, T., and Kuriyama, K., 1983, pretreatment with taurine facilitates hepatic lipid peroxide formation associated with carbon tetrachloride intoxidation, *Japan J. Pharm.*, 33:515-523.
18. Park, K.A., 1994, Effect of different fats and proteins on the hepatic lipid peroxidation and membrane stability in phenobarbital-induced hepatocarcinogenesis-Casein vs Soy protein-, Master thesis of Seoul National University.
19. Pasantes-Morales, H. and Cruz, C., 1985, Taurine and hypotaurine inhibit light-induced lipid peroxidation and protect rod outer segment structure, *Brain Res.*, 330:154-157.
20. Solt, D. and Farber, E., 1976, new principle for the analysis of chemical carcinogenesis, *Nature*, 263:701-703.
21. Stout, D.L. and Becker, F.F., 1986, Xenobiotic metabolizing enzymes in genetically and chemically initiated mouse liver tumors, *Cancer Res.*, 46:2693-2696.
22. Tappel, A.L., 1978, Glutathione peroxidase and hydroperoxides, *Meth. Enzymol.*, 52:506-513.
23. Zhou, M., Ma, T., and Tseng, M.T., 1991, Effects of taurine and ketamine on bovine retinal membrane lipid peroxidation, *Neuroscience*, 45(2):461-465.

# Part 3:
# Taurine in Muscle and the Cardiovascular System

# EFFECT OF TAURINE ON EXCITATION-CONTRACTION COUPLING OF EXTENSOR DIGITORUM LONGUS MUSCLE OF DYSTROPHIC MDX MOUSE

Annamaria De Luca,[1] Sabata Pierno,[1] Claudia Camerino,[2]
Ryan J. Huxtable,[2] and Diana Conte Camerino[1]

[1] Unit of Pharmacology
Department of Pharmacobiology
Faculty of Pharmacy
University of Bari
Via Orabona 4, 70125
Bari, Italy
[2] Department of Pharmacology
College of Medicine
University of Arizona
Tucson, AZ 85724

## INTRODUCTION

Dystrophin is a cytoskeletal protein associated with the sarcolemma. Genetic depletion of the protein leads to Duchenne muscular dystrophy (DMD) and dystrophic conditions in animals such as the mdx mouse[9]. Mechanical weakness of the sarcolemma during the stress of contraction can account for an abnormal calcium entry in the cell that in turn triggers proteolysis-dependent dystrophic muscle necrosis[5,7,9]. This mechanism can also account, in a positive feedback loop, for the increased activity of calcium-selective leak ion channels recorded in dystrophinless muscle fibers[15]. The increased cytosolic calcium is also responsible for longer calcium transients in the dystrophinless fibers[7,14], which may affect muscle contractile properties. In skeletal muscle, contraction is triggered mainly by the calcium ions released from the sarcoplasmic reticulum due to depolarization of the sarcolemma and transverse tubules[11]. Nevertheless, higher cytosolic calcium levels can affect the rate of calcium release and reuptake[1,14], thus modifying the excitation-contraction coupling mechanism.

Intracellular calcium movements can also be altered in dystrophic fibers by other mechanisms. As a consequence of membrane weakness the dystrophic fibers can loose cytoplasmic components. The urinary excretion of taurine is increased in DMD patients[4], leading to the hypothesis that the intracellular muscle concentration of this amino acid, high in normal subject's[6], may decrease due to fiber degeneration. Furthermore, the striated fibers of mdx mouse have a decreased resistance to osmotic stress, and it is well recognized that taurine controls osmoregulation in many tissues[6,10]. It has been found that taurine exerts a pivotal role for the excitation-contraction coupling mechanism in heart[8]. Recently, we have demonstrated a similar physiological function for taurine in skeletal muscle, probably through a control of calcium availability. Thus, taurine depletion in rats shifts the voltage threshold for contraction of striated fibers towards more negative potentials as a result of increased cytosolic calcium[1].

In light of the above findings, the present study evaluates the excitation-contraction coupling mechanism of extensor digitorum longus (EDL) muscle fibers of dystrophic mdx mouse by measuring the voltage threshold for mechanical activation by the "point" voltage clamp method. The possible relationship between this process and taurine was investigated by testing the effect of *in vitro* application of taurine.

## MATERIALS AND METHODS

The extensor digitorum longus (EDL) muscles of 6 month-old mdx and wild type (C57BL/10) mice were removed under urethane anesthesia and placed in a muscle bath at their resting length[1]. The preparation was perfused with a physiological solution of the following composition (mM): NaCl 148; KCl 4.5; $CaCl_2$ 2.0; $MgCl_2$ 1.0; $NaH_2PO_4$ 0.44; $NaHCO_3$ 12 and glucose 5.55, continuously bubbled with 95% $O_2$ and 5% $CO_2$ and maintained at 30 ± 1°C (pH = 7.2). Taurine (Teofarma, Italy), 60 mM, was daily prepared in the above solution and incubated for the appropriate time before recordings. The mechanical threshold was determined using a two microelectrode point voltage clamp method as previously described[1]. In brief, a voltage-sensing electrode (3 M KCl) and a current-passing electrode (2 M potassium citrate) were inserted within 50 μm of each other into the central region of a randomly-selected superficial fiber which was continuously viewed using a stereomicroscope (100 x magnification). The holding potential was set at -90 mV and depolarizing command pulses of variable duration were given at a rate of about 0.3 Hz. Tetrodotoxin (3 μM) was continuously present during recordings to prevent action potential generation[1,3]. As a standard protocol the command-pulse duration was usually set sequentially to each of the following values: 500, 50, 5, 100 and 10 ms. At each duration, the command voltage was increased using an analogue control until contraction was just visible, and then backed down until the contraction just disappeared. A digital sample-and-hold millivoltmeter stored the value of the threshold membrane potential at this point. The threshold membrane potential V (mV) for each fiber was averaged at each pulse duration t (ms). A fit estimate of the rheobase voltage (R) and of the time constant to reach the rheobase was obtained by a nonlinear least square algorithm using the following equation:

$$V= [H-R \exp (t/\tau)]/[1-\exp (t/\tau)]$$

where H is the holding potential (mV), R is the rheobase (mV) and $\tau$ is the time constant[1]. The mechanical threshold values are expressed as the fitted rheobase (R) parameter ± standard error (S.E.) which was determined from the variance-covariance matrix in the nonlinear least squares fitting algorithm. The statistical significance of the differences between the mean values of threshold potential or the fitted values of R and $\tau$ was estimated by a Student's t distribution. For the fitted R and $\tau$ values, standard errors used were estimated from the fit and the number of degrees of freedom equal to the total number of values determining the curves minus the number of means minus two for the free parameters[1].

## RESULTS

The threshold potential for contraction of EDL muscle fibers from both wild type and mdx mice showed the typical dependence on command pulse duration; i. e. it was the more negative the longer the duration of the pulse. Under the experimental conditions used in the present study, a constant rheobase value was almost fully reached at the longest pulses used, a behavior which has been commonly seen with mammalian muscle fibers[1]. However, the mechanical threshold of EDL muscle fibers of mdx mice was significantly different from that of wild type ones. Indeed at each pulse duration the mdx muscle fibers needed significantly less depolarization to contract (Table 1). In fact the voltage at rheobase estimated from the fit was -62.2 ± 0.46 mV for wild type and -69.8 ± 0.62 mV for mdx mice and the difference between these values was significantly different with a $p < 0.001$. The time constant ($\tau$) to reach the rheobase was slightly but not significantly longer in the EDL muscle fibers of mdx mice, being 7.4 ± 0.4 ms vs. 6.9 ± 0.25 ms of wild types.

The *in vitro* application of 60 mM taurine partially but significantly restored the mechanical threshold of mdx muscle fibers; i. e. at each pulse duration the contraction occurred at more positive membrane potentials, whereas no effect of taurine was observed on the mechanical threshold of wild type fibers (Table 1). In fact, the rheobase value after taurine application was 65.2 ± 0.4 mV in mdx ($p<0.01$ vs. related control) and 62.4 ± 0.5 mV in wild type (no significance). Interestingly, the $\tau$ value of mdx fibers was significantly shortened by taurine application to 6.2 ± 0.3 ms.

**Table 1.** Effect of *in vitro* application of taurine on the mechanical threshold of extensor digitorum longus muscle fibers of dystrophic mdx mouse.

| | Duration (ms) | | | | |
|---|---|---|---|---|---|
| | 5 | 10 | 50 | 100 | 500 |
| wild type | -38.7 ± 1.0 (17) | -46.1 ± 1.1 (15) | -58.7 ± 1.3 (18) | -61.9 ± 0.9 (16) | -64.0 ± 0.6 (23) |
| wild type + 60 mM taurine | -39.8 ± 1.2 (12) | -49.0 ± 2.2 (11) | -60.9 ± 1.4 (13) | -61.8 ± 1.0 (15) | -63.8 ± 0.8 (15) |
| mdx | -50.1 ± 0.9[a] (13) | -55.7 ± 1.5[a] (15) | -67.5 ± 1.1[a] (9) | -70.1 ± 1.2[a] (13) | -72.8 ± 1.0[a] (13) |
| mdx + 60 mM taurine | -47.4 ± 1.1[a] (14) | -51.2 ± 1.4[a,b] (11) | -61.9 ± 1.0[b] (18) | -65.6 ± 1.1[a,b] (15) | -67.9 ± 0.8[a,b] (19) |

The mechanical threshold has been measured in extensor digitorum longus muscle fibers of healthy wild type (C57BL/10) and dystrophic mdx mice in the absence and presence of 60 mM taurine. For each experimental condition are shown the threshold membrane potential values (in mV) obtained with depolarizing command pulse of duration ranging from 5 ms up to 500 ms. The values are expressed as mean ± S.E. from the number of fibers shown in brackets below each value.

[a] Significantly different with respect to wildtype ($p < 0.05$)

[b] Significantly different with respect to related control without taurine ($p < 0.05$)

## DISCUSSION

The present results show that the rheobase voltage for mechanical threshold of muscle fibers of dystrophic mdx mice is shifted towards more negative potentials with respect to control animals. It is known that the rheobase is the voltage at which the release of calcium from the sarcoplasmic reticulum equals the speed at which the calcium is buffered to stop contraction[1,3]. Any pathological or pharmacological intervention that increases the release of calcium or reduces the uptake would shift the rheobase voltage in a negative direction[1,3]. No direct alterations of either the calcium release or calcium uptake mechanisms have been found in the muscle of mdx mice[13-15]. Furthermore, no changes in membrane electrical properties occur in mdx EDL muscle fibers at the age at which we measured the mechanical threshold[2]. However, it has been observed that the increased cytosolic calcium levels of dystrophinless fibers can produce a prolongation of calcium transients due to a slowdown of the calcium reuptake mechanism from sarcoplasmic reticulum after contraction[5,7,14]. In addition, the mdx muscle fibers show a more prominent contracture by caffeine[13], suggesting an abnormal leakage of calcium from sarcoplasmic reticulum occurring under particular conditions. The present results also suggest that a decrease of taurine content, as a consequence of fiber degeneration, can play a role in the alteration of excitation-contraction coupling observed. In fact we have previously found that a pharmacological depletion of taurine content by chronic administration of guanidinoethane sulfonate (GES) to rats produces alteration of mechanical threshold similar to that observed here in mdx muscle fibers[1]. In support of this hypothesis we found that *in vitro* application of the amino acid is able to counteract significantly the alteration observed, whereas no effects were found when taurine is applied on the muscle from wild type animals. This is in line with previous findings that taurine does not affect the excitation-contraction coupling of striated fibers when its intracellular concentration is in the physiological range[1]. Furthermore, an increased urinary excretion of taurine has been observed in DMD patients[4].

In conclusion, the spontaneous lowering of the voltage threshold for contraction in dystrophic muscle can be an index of the abnormalities of calcium movements and may in turn worsen the dystrophic condition that is primarily related to the mechanical stress on the sarcolemm[9]. Taurine significantly improved *in vitro* the altered mechanical threshold of mdx mice, supporting a potential pharmacological benefit of this safe amino acid in the dystrophinopathies. Moreover taurine has been found to provide protection against myocardial injury in hamster genetically affected by a cardiomyopathy caused by the absence of a dystrophin-related protein[4,9,12]. However, the real involvement of taurine in Duchenne muscular dystrophy and its animal model needs to be better established.

## ACKNOWLEDGMENTS

The support of Telethon-Italy to the project # 322 is gratefully acknowledged.

## REFERENCES

1. De Luca, A., Pierno, S., and Conte Camerino, D., 1996, Effect of taurine depletion on excitation contraction coupling and $Cl^-$ conductance of rat skeletal muscle, *Eur. J. Pharmacol.,* 296: 215-222.
2. De Luca, A., Pierno, S., and Conte Camerino, D., 1997, Electrical properties of diaphragm and EDL muscles during the life of dystrophic mice, *Am. J. Physiol.* 272 (Cell Physiol. 41): C333-C340.
3. Dulhunty, A., 1988, Internal citrate ions reduce the membrane potential for contraction threshold in mammalian skeletal muscle fibers, *Biophys. J.,* 53: 609-615.

4. Engel, A.G., Yamamoto, M., and Fischbeck, K.H., 1994, Dystrophinopathies, in: Myology, Engel, A.G. and Franzini-Armstrong, C. eds., Mc Graw-Hill, Inc., New York, Vol. 2, pp. 1133-1187.
5. Franco-Obregon, A., and Lansman, J.B., 1994, Mechanosensitive ion channels in skeletal muscle from normal and dystrophic mice, *J. Physiol.*, 481.2: 299-309.
6. Huxtable, R.J., 1992, The physiological actions of taurine, *Physiol. Rev.* 72, 101-163.
7. Imbert, N., Cognard, C., Duport, G., Guillon, C., and Raymond, G., Abnormal calcium homeostasis in Duchenne muscular dystrophy myotubes contracting in vitro, *Cell Calcium*, 18: 177-186.
8. Lake, N., 1992, Effects of taurine deficiency on arrhythmogenesis and excitation-contraction coupling in cardiac tissues, in Adv. Exp. Med. Biol. Taurine, Nutritional Value and Mechanism of Action, Lombardini J.B., Schaffer S.W. and Azuma J. eds., Plenum Press, New York, Vol. 315, pp. 173-176.
9. Matsumura, K., and Campbell, K.P., 1994, Dystrophin-glycoprotein complex: its role in the molecular pathogenesis of muscular dystrophies, *Muscle Nerve* , 17: 2-15
10. Menke, A. and Jockusch, H., 1991, Decreased osmotic stability of dystrophin-less muscle cells from the mdx mouse, *Nature,* 349: 69-71.
11. Rios, E., and Pizarro, G., 1991, Voltage sensor of excitation-contraction coupling in skeletal muscle, *Physiol. Rev.,* 71: 849-860.
12. Schaffer, S.W and Azuma, J., 1992, Review: myocardial physiological effects of taurine and their significance, in Adv. Exp. Med. Biol. Taurine, Nutritional Value and Mechanism of Action, Lombardini J.B., Schaffer S.W. and Azuma J. eds., Plenum Press, New York, Vol. 315, pp. 105-120.
13. Takagi, A., Kojima, S., Ida M., and Araki, M., 1992, Increased leakage of calcium ion from the sarcoplasmic reticulum of the mdx mouse, *J. Neurol. Sci.*, 110: 160-164.
14. Turner, P.R., Fong, P., Denetclaw, W.F., and Steinhardt, R.A., 1991, Increased calcium influx in dystrophic muscle, *J. Cell Biol.,* 115: 1701-1712.
15. Turner, P.R., Schultz, R., Ganguly, B., and Steinhardt, R.A., 1993, Proteolysis results in altered leak channel kinetics and elevated free calcium in mdx muscle, *J. Membr. Biol.* 133:243-251.

# CARDIAC ACTIONS OF TAURINE AS A MODULATOR OF THE ION CHANNELS

Hiroyasu Satoh

Department of Pharmacology
Division of Molecular and Cellular Biology
Nara Medical University
Kashihara, Nara 634
Japan

## ABSTRACT

During ischemia, hypoxia and cardiac failure, the heart undergoes several adverse changes, including a reduction in taurine (2-aminoethanesulfonic acid)[9,10]. Oral administration of taurine under these disease conditions would be expected to act like a mild cardiac glycoside. Taurine would exert improvement in the accumulation of $[Na]_i$ and the loss of $\alpha$-amino acids. Nonetheless, when intracellular taurine content is raised, there would be the benefit of increased $Ca^{2+}$ release from the sarcoplasmic reticulum and increased $Ca^{2+}$ sensitivity of the contractile proteins, as well as possible changes in the action potential associated with the actions of taurine on ion channels. In fact, intracellular application of taurine produces the opposite actions to extracellularly administration of the amino acid.

From our previous experiments, the electrophysiological actions of taurine on cardiac muscle cells include the following[3,28]: (a) Prolongation of action potential duration (APD) at high $[Ca]_i$ and shortening of APD at low $[Ca]_i$. In multicellular preparations, however, taurine did not always prevent $[Ca]_o$-induced effects. (b) Stimulation of spontaneous activity at low intracellular and extracellular $Ca^{2+}$ concentrations ($[Ca]_i$ and $[Ca]_o$), and vice versa. (c) Inhibition of the L-type $Ca^{2+}$ current ($I_{Ca(L)}$) at high $[Ca]_i$, and vice versa. (d) Enhancement of the T-type $Ca^{2+}$ current ($I_{Ca(T)}$). (e) Inhibition of fast $Na^+$ current ($I_{Na}$). (f) Enhancement of TTX-insensitive slow $Na^+$ current. (g) Inhibition of delayed rectifier $K^+$ current ($I_{Krec}$) at high $[Ca]_i$, and vice versa. (h) Enhancement of the transient outward current ($I_{to}$). (i) Inhibition of the ATP-sensitive $K^+$ current ($I_{K(ATP)}$).

Since taurine acts on so many ion channels and transporters, it is clearly non-specific. Although it is very difficult to understand the diversity of taurine's actions, it is possible that taurine can exert its potent cardioprotective actions under the conditions of low $[Ca]_i$, as well as $Ca^{2+}$ overload. Thus, although taurine-induced modulation of ion channels located on the cardiac cell membrane is complex, the multiple effects may combine to yield useful therapeutic results.

*Taurine 3*
Edited by Schaffer *et al.*, Plenum Press, New York, 1998

## INTRODUCTION

Taurine is classified as a β-amino acid, and because of its sulfonic acid moiety it is not incorporated into proteins or membranes. Heart cells are abundant in taurine (around 10-20 mM), but the plasma has relatively low levels. In ischemic heart failure, taurine levels in the cytosol are reduced. Thus, external taurine might be expected to play an important role as a regulator of cardiac function. Taurine has been reported to possess several beneficial cardiac actions. It mediates a positive inotropic effect at low $Ca^{2+}$, but reduces contraction at high $Ca^{2+}$ levels. Therefore, taurine might be useful for the treatment of congestive heart failure, and to exert protective actions against cellular $Ca^{2+}$ overload.

Recently, several pharmacological actions of taurine have been described[28]. Taurine produces a positive inotropic effect without affecting the resting potential, and modulates ionic currents across the cell membrane[7,15,33]. We have also reported that taurine exerts electrophysiological actions which are dependent on $[Ca]_i$ and $[Ca]_o$. Taurine inhibits $I_{Ca(L)}$ and $I_{Krec}$ at pCa 7, whereas taurine enhances both currents at pCa 10. Thus, the effects of taurine on the cardiac cell are complex. Taurine increases APD at pCa 6, but decreases APD at pCa 8. Simultaneously, taurine enhances $I_{krec}$ at pCa 8, and inhibited $I_{krec}$ at pCa 6. Taurine stimulates a TTX-insensitive $Na^+$ current, but depresses the fast (TTX-sensitive) $Na^+$ current[26].

The inwardly rectifying $K^+$ currents ($I_{K1}$) contributes the resting potential and helps stabilize the membrane. The effects of taurine on $I_{K1}$ are discussed here.

## METHODS AND MATERIALS

Cells were prepared from tissue taken from ventricular muscle of adult guinea pig hearts, using methods similar to those described previously[17,22]. Under anesthesia with sodium pentobarbital (30 mg/kg), the chest was opened and the aorta was cannulated *in situ*. The heart was dissected out and perfused with normal Tyrode solution on a Langendorff apparatus. After a washout of blood, the heart was perfused with $Ca^{2+}$-free Tyrode solution, causing spontaneous beating to cease. Then, the perfusate was switched to low-$Ca^{2+}$ (30 to 60 μM) Tyrode solution containing 0.4 mg/ml collagenase (Type I, Sigma Chemical) for about 20 min. The heart was washed with a high-$K^+$ and low-$Cl^-$ solution (KB solution), and was dissected with scissors. The temperature of all solutions was maintained at 36°C.

Whole-cell voltage-clamp recordings were performed using a Axopatch patch-clamp amplifier (Axon Instruments) and standard techniques. Patch pipettes were fabricated using a two-stage puller, and a resistance of 3-5 MΩ for whole-cell experiments, and 1-3 MΩ for cell-attached (or inside-out) patch-clamp experiments was used. The tip of the electrode was coated with Sylgard (KE106, Shin-etsu Chemical). Experiments were carried out at a temperature of 36°C. The data were stored and analyzed on an IBM-AT micro-computer, using the PCLAMP analysis program (Axon Instruments). Current traces were filtered using a cut-off frequency of 1 KHz. All values are given as means ± S.E.M. Differences in mean values were analyzed by ANOVA and the Student *t*-test for paired data. A p value less than 0.05 was considered significant.

The composition of the Tyrode solution was (in mM): NaCl 137, KCl 5.4, $CaCl_2$ 1.8, $MgCl_2$ 1.0, $NaH_2PO_4$ 0.3, glucose 5.0, and HEPES 5.0. The pH was adjusted to 7.4 with NaOH. Taurine (Sigma Chemical) was dissolved at the desired concentration directly into the bath solution before superfusion. The pipette (intracellular) solution was determined by the desired ion current. The concentration of calcium in the pipette solution was determined according to calculations.

## RESULTS AND DISCUSSION

Taurine, an amino acid largely obtained from the diet, exerts many important physiological functions, including many cardiovascular actions that have already been described[7,28,30,32]. The electrical and mechanical actions induced by taurine are independent of known intracellular regulators, such as cAMP and cGMP levels, (Na,K)-ATPase activity and calmodulin-dependent protein kinase action, although taurine may inhibit protein kinase C activity[11,30].

The actions of taurine on ion channels are presented here. Also, a comparison between the effects of taurine on embryonic and adult cardiomyocytes are included (Table 1).

### $Na^+$ Current

In young embryonic chick cells, the TTX-sensitive fast $Na^+$ current ($I_{Na}$) was enhanced by low concentrations (1-5 mM) of taurine, but was inhibited by high concentrations of taurine (10-20 mM)[26]. Intracellular application of taurine at 10 and 20 mM enhanced $I_{Na}$. In adult cells (guinea pig), taurine also inhibited $I_{Na}$ slightly more than in embryonic cells[25]. Simultaneously, taurine decreased the time constant of inactivation in adult cells. Amino acids, such as taurine, alanine and glutamine, and $Na^+$ can cotransport across the membrane[34]. Actually, taurine shifts the reversal potential of the fast $Na^+$ current ($I_{Na}$) in the hyperpolarizing direction and reduces cellular $Na^+$ content in

**Table 1.** Comparison with the actions of taurine on the ionic currents in embryonic and adult heart cells

| | Embryo | | | Adult | | |
|---|---|---|---|---|---|---|
| | n | 10 mM | 20 mM | n | 10 mM | 20 mM |
| $I_{Na}$ | | | | | | |
| Amplitude | 10 | -38.2 ± 4.3** | -49.5 ± 4.6*** | 9 | -39.0 ± 4.1** | -56.1 ± 4.7*** |
| $\tau$ | 8 | +0.1 ± 1.8 | +1.8 ± 1.0 | 8 | -26.1 ± 3.1* | -21.8 ± 2.3* |
| $I_{Ca}$ | | At pCa 10 | | | At pCa 8 | |
| Amplitude | 5 | +7.2 ± 3.1 | +47.3 ± 2.4** | 7 | +10.1 ± 3.2* | +41.7 ± 2.1*** |
| $\tau_f$ | 4 | -3.1 ± 2.4 | +3.2 ± 2.7 | 8 | +1.7 ± 1.4 | +56.9 ± 5.1*** |
| $\tau_s$ | 4 | +14.5 ± 2.9* | +66.2 ± 2.8*** | 8 | +9.4 ± 3.2 | +24.0 ± 3.4** |
| | | At pCa 7 | | | At pCa 6 | |
| Amplitude | 5 | -56.3 ± 2.8** | -73.2 ± 2.6*** | 8 | -14.4 ± 2.0** | -31.5 ± 2.2*** |
| $\tau_f$ | 4 | -0.9 ± 2.5 | -6.5 ± 2.6 | 7 | +140.2 ± 5.5*** | +204.8 ± 6.7*** |
| $\tau_s$ | 4 | -4.8 ± 2.9 | +16.7 ± 2.8** | 7 | +41.1 ± 5.3** | +72.3 ± 5.1*** |
| $I_{Krec}$ | | At pCa 10 | | | At pCa 8 | |
| Amplitude | 5 | +35.7 ± 3.6** | +66.7 ± 5.2*** | 6 | +13.4 ± 3.2* | +22.4 ± 3.1** |
| | | At pCa 7 | | | At pCa 6 | |
| Amplitude | 5 | -11.4 ± 2.8* | -39.8 ± 3.5*** | 6 | -16.5 ± 2.7* | -27.1 ± 2.8** |

Values (%) represent mean ± S.E.M. Amplitude was measured at voltage which the maximum current was activated. *: P<0.05, **: P<0.01, ***: P<0.001, with respect to control value.

embryonic chick cardiomyocytes, but not in adult (guinea pig) heart cells[3,18,24,26]. This cotransport is considered to be a simple symport of 1 $Na^+$- 1 taurine[4]. A change in cellular $Na^+$ content may alter $[Ca]_i$ levels via the Na-Ca exchange.

These differences might be explained mostly by the developmental changes in the $I_{Na}$ channel of heart cells. The channel density is much lower in embryonic hearts than in adult hearts[6]. The poor development of $I_{Na}$ channel may indirectly cause a shift in $E_{Na}$. In young cardiomyocytes (developing cells), the TTX- and $Mn^{2+}$-insensitive $Na^+$ channel (but not usually a fast $Na^+$ channel) plays an important role in the physiological function of the cell[2,33]. This current might be closely involved with taurine action since taurine might alter the charge density on the inner surface of the membrane.

## $Ca^{2+}$ Current

Taurine inhibits $I_{Ca(L)}$ at high $[Ca]_i$ and $[Ca]_o$, and enhances $I_{Ca(L)}$ at both low $[Ca]_i$ and $[Ca]_o$ in embryonic chick cardiomyocytes[27]. Intracellular application of taurine inhibits $I_{Ca(L)}$ even at low pCa. In contrast, T-type $Ca^{2+}$ current ($I_{Ca(T)}$) is stimulated by taurine. On the other hand, in adult cells, at pCa 8 taurine application increases $I_{Ca(L)}$, whereas at pCa 6 taurine decreases $I_{Ca(L)}$. After washing out taurine, the current amplitude is not reversed.

Time decay of the inactivation phase for $I_{Ca(L)}$ is one of the factors regulating $[Ca]_i$. The inactivation phase is composed of fast and slow exponential components ($\tau_f$ and $\tau_s$). $\tau_f$ is $[Ca]_i$-dependent, while $\tau_s$ is voltage-dependent. Both $\tau_f$ and $\tau_s$ increase during exposure to taurine in both embryonic chick cells[27] and adult cells[25]. In these studies, $[Ca]_i$ was buffered with either 10mM EGTA or BAPTA, eliminating an effect of $[Ca]_i$ on ion channels. Therefore, the slowing of inactivation induced by taurine appears on involve a direct action of the amino acid.

Taurine decreased the open probability ($P_o$) of the unitary $Ca^{2+}$ channel at 5.4 mM $[Ca]_o$, and increased it at 0.9 mM $[Ca]_o$ without affecting channel conductance[25]. Both dwell (open and closed) times fit the two exponential component mode. At both low and high $[Ca]_o$, the time constant of the open time increased slightly after taurine exposure whereas those of the closed time decreased. Thus, the enhancement of $I_{Ca(L)}$ by taurine at low pCa is caused by the increase in the open time and the decrease in the closed time of the $Ca^{2+}$ channel. On the other hand, at high pCa, taurine inhibits $I_{Ca(L)}$, which cannot be explained by the decrease in the closed time and the increase in the open time of the $Ca^{2+}$ channel. Therefore, we conclude that the modulation $Ca^{2+}$ channels by taurine involves primarily changes in the open probability.

## Delayed Rectifier $K^+$ Currents

The repolarization of action potentials is mostly regulated by the outward $K^+$ current and partly by the inward $Na^+$ and $Ca^{2+}$ currents. The effects of taurine on the delayed rectifier $K^+$ current ($I_{Krec}$) are $[Ca]_i$-dependent. In embryonic cells, taurine inhibited $I_{Krec}$, at pCa 7 but enhanced $I_{Krec}$ at pCa 10[19]. By contrast, taurine increased $I_{Krec}$ at pCa 8, but decreased $I_{Krec}$ at pCa 6 in adult cells (unpublished data). Taurine modulated the rapid component of $I_{Krec}$ ($I_{Kr}$), but not the slow one ($I_{Ks}$). However, the effects of taurine on action potential duration (APD) are controversial. In adult (guinea pig) cells, taurine shortened APD at pCa 8, whereas taurine prolonged APD at pCa 6. In embryonic chick myocytes, taurine shortened APD at pCa 10, but had no effect at pCa 7. By comparison, taurine prolonged APD at low $[Ca]_o$ and shortened APD at high $[Ca]_o$ in rabbit SA node cells[16]. Thus, the actions of taurine are dependent on both $[Ca]_i$ and $[Ca]_o$ levels, and vary with different tissue isoforms.

## Inwardly Rectifying K$^+$ Current

The inwardly rectifying K$^+$ current ($I_{K1}$) is considered to be mainly related with the resting potential of cardiac myocytes. Effects of taurine on $I_{K1}$ in isolated guinea pig ventricular cardiomyocytes were examined using a whole-cell and a cell-attached patch voltage-clamp system (Fig. 1). Holding potential was maintained at -30 mV. Taurine application at a concentration of 10 mM and 20 mM inhibited $I_{K1}$ at −120 mV by 28.3 ± 1.1 % (n = 6, P<0.05) and 36.0 ± 2.1 % (n = 6, P<0.01), respectively.

In cell-attached patch-clamp experiments, taurine decreased the mean patch $I_{K1}$ current (open probability) in a concentration-dependent manner. The amplitude was 2.2 ± 0.3pA (n = 9) at +50 mV, when using a 140 mM K$^+$ solution in both the bath and the pipette. The current was activated with inwardly rectification, and was completely blocked by $Ba^{2+}$. The channel conductance was 31 ± 3pS (n = 9). Taurine did not affect the channel conductance (approximately 34 pS), but markedly decreased the open probability (as a mean patch current for 30 sec) of the channel by 21.5 ± 2.4 % (n = 8, P<0.01) and 56.7 ± 3.8 % (n = 8, P<0.001) at 10 mM and at 20 mM respectively. The responses were reversible. These results suggest that taurine directly modulates the open probability of $I_{K1}$ current, thereby altering cell function.

Although, taurine markedly decreased $I_{K1}$ in adult guinea pig cardiomyocytes, it had no significant effect on resting potential, an observation consistent with the results in multicellular muscle preparations (guinea pig)[15]. The lack of change in the resting potential would be the result of the resting potential being close to the K$^+$ equilibrium potential of cardiac cells. If taurine were to decrease $I_{K1}$ in diseased hearts it could lower the resting potential which would diminish the activation of $I_{Na}$ and/or $I_{Ca(T)}$, which represents a cardioprotective action. This effect might also be antiarrhythmic or antihyper-

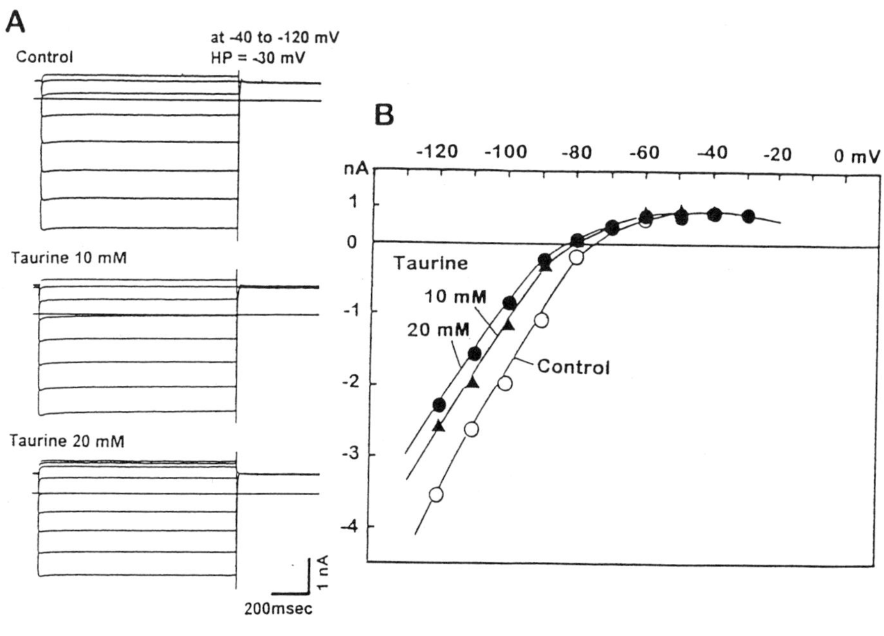


**Figure 1**. Modulation by taurine of the inwardly rectifying K$^+$ current in guinea pig ventricular myocytes. A: Current traces in the absence and presence of 10 and 20 mM taurine. Test pulses for 1 s were applied between -40 to -120 mV from a holding potential of -30 mV. The horizontal line on the current records represents the zero current level. B: Current-voltage (I-V) curves. Symbols used are control (open circles), 10 mM (triangles) and 20 mM (filled circles) of taurine.

tensive. Therefore, this action of taurine can be considered one of its important cardioprotective effects.

### ATP-Sensitive $K^+$ Current

Since ATP-sensitive $K^+$ channel openers shorten the APD and mediate negative chronotropic and inotropic effects, this current plays an important role in cell function[14]. The ATP-sensitive $K^+$ current ($I_{K(ATP)}$) is activated by a decrease in intracellular ATP concentration (to less than 1 mM)[13] and is thus activated during ischemia and cardiac failure. By hyperpolarizing the membrane[14], openers inhibit $I_{Ca(L)}$, resulting in a reduction in $Ca^{2+}$ influx into the cell. These actions would help prevent $Ca^{2+}$ overload-related arrhythmias.

The ATP-sensitive $K^+$ current ($I_{K(ATP)}$) in adult guinea pig myocytes is inhibited by taurine[5,22], with the opening of the unitary $I_{K(ATP)}$ channel significantly reduced at 10 mM, and almost completely blocked at 20 mM taurine, thereby resulting in a prolongation of APD[22]. Under ischemic conditions, $I_{K(ATP)}$ is presumably fully activated. The question is: will taurine exert a cardioprotective action in this setting. It is known that glibenclamide (a blocker of $I_{K(ATP)}$ channels) prevents arrhythmias generated during ischemia, presumably by preventing $K^+$ efflux from whole heart during ischemia. Therefore, taurine may act as an antiarrhythmic during ischemia by blocking $I_{K(ATP)}$.

### Transient Outward Current

Transient outward current ($I_{to}$) also plays an important role in repolarizing the membrane. Taurine enhances the 4-AP sensitive current in single embryonic chick heart cells[19], where normal $I_{to}$ amplitude is approximately 22 pA/pF at +70 mV (HP = -60 mV). Addition of 4-acetamido-4'-isothiocyanotostilbene-2', 2'-disulfonic acid (SITS), a specific $Cl^-$ current inhibitor, to the bath in the presence of 4-AP had little or no effect on the amplitude of $I_{to}$. These results indicate that in the embryonic cell the $Cl^-$ current is only a minor contributor to $I_{to}$ current, and the effect of taurine on the $Cl^-$ current-dependent $I_{to}$ is presently unclear.

### Pump and Exchange Currents

Taurine had no effect on Na/K-pump current[9], but Na/Ca exchange current is reportedly either decreased or not affected in guinea pig cardiomyocytes[5]. Finally, taurine enhances a background current in rabbit atrial myocytes.

### Cardioprotective Actions of Taurine

Our recent investigation shows that taurine possesses potent cardioprotective activity [15,26-28,32,33]. It enhances $I_{Ca(L)}$ at low pCa, and inhibits $I_{Ca(L)}$ at high pCa, indicating that one of taurine's actions is to normalize $Ca^{2+}$ movement through the sarcolemma, dampening the effects of variations in perfusate $Ca^{2+}$ concentration and functioning to keep $[Ca]_i$ levels relatively constant, thereby affecting contractile function and cell viability.

Taurine also inhibits $I_{Na}$ (TTX-sensitive) current[26], which may be a way of protecting heart muscle against arrhythmias. A decline in cellular $Na^+$ content should decrease $[Ca]_i$ levels which should minimize cellular calcium overload during ischemia. Since calcium overload can trigger arrhythmias, it is not surprising that taurine can abolish arrhythmias due to triggered activity[15,16]. Moreover, taurine promotes the sudden emergence of a dormant T-type $Ca^{2+}$ current ($I_{Ca(T)}$)[20], which contributes to the generation of pacemaker activity during diastole[1]. Automaticity is depressed in the calcium overload state[29]. In rabbit SA nodal cells, taurine enhances the automaticity at low $[Ca]_o$, whereas taurine inhibits it at high $[Ca]_o$. Thus, the stimulation of $I_{Ca(T)}$ may also represent a protective action of taurine. In a related effect, taurine decreases the prolonged APD seen at low pCa. Since extreme prolongation of APD may elicit arrhythmias, this may be another means by which taurine can serve as an antiarrhythmic agent[15,16].

In addition, taurine attenuates the development of hypertension in SHR[12]. These results strongly suggest that taurine possesses potent cardioprotective activity, presumably related to its ability to regulate $Ca^{2+}$ movement in the cytoplasm. Although many of taurine's actions on cardiomyocytes still remain unclear, further studies should reveal possible therapeutic uses of taurine as a sulfur amino acid.

## REFERENCES


1. Armstrong, C.M. and Matteson, D.R., 1985, Two distinct populations of calcium channels in a clonal line of pituitary cells, *Science*, 227:65-67.
2. Bkaily, Gg., Chahine, M., Sperelakis, N., and Yamamoto, T., 1988, Taurine increases one type of slow $Na^+$ and $Ca^{2+}$ currents in embryonic heart, *J. Physiol.* (Lond.), 406:91.
3. Chapman, R.A., Suleiman, M.S., and Earm, Y.E., 1993, Taurine and the heart, *Cardiovasc. Res.*, 27:358-363.
4. Christensen, H.N. and Kilberg, M.S., 1987, Amino acid transport across the plasma membrane: role of regulation in interorgan flows, in "Amino Acid Transport in Animal Cells", Yudilevich, D.L. and Boyd, C.A.R., eds., Manchester Univ Press, Manchester, pp. 10-45.
5. Earm, Y.E., Ho, W.K., So, I., Leem, C.H., and Han, J., 1993, Effect of taurine on the activitity of background current in cardiac myocytes of the rabbit, in "Ionic Channels and Effect of Taurine on the Heart", Noble, E. and Earm, Y.E., eds., Kluwer Academic Publishers, Boston, pp. 119-138.
6. Fujii, S., Ayer, R.K., DeHaan, R.L., 1988, Development of the fast sodium current in early embryonic chick heart cells, *J. Membr. Biol.*, 101:209-223.
7. Huxtable, R.J. and Sebring, L.A., 1983, Cardiovascular actions of taurine, in "Sulfur Amino Acids: Biochemical and Clinical Aspects", Kuriyama, K., Huxtable, R.J., Iwata, H., and Liss, A.R., eds., Biochem Clin Aspec, New York, pp. 5-37.
8. Khatter, J.C., Soni, P.L., Hoeschen, R.J., Alto, L.E., and Dhalla, N.S., 1981, Subcellular effects of taurine on guinea pig heart, in "The Effect of Taurine on Excitable Tissues", Schaffer, S.W., Baskin, S.I., and Kocsis, J.J., eds., Spectrum Publications, New York, pp. 281-294.
9. Kramer, J.H., Chovan, J.P., and Schaffer, S.W., 1981, Effect of taurine on calcium paradox and ischemic heart failure, *Am. J. Physiol.*, 240:H238-H246.
10. Lombardini, J.B., 1980, Effect of ischemia on taurine levels, in "Natural Sulfur compounds", Cavallini, D., Gaull, G.E., and Zappia, V., eds., Plenum Press, New York, pp. 255-261.
11. Lombardini, J.B., 1992, Effects of taurine on protein phosphorylation in mammalian tissues, in Adv. Exp. Med. Biol. "Taurine: Nutritional Value and Mechanism of Actions", Lombardini, B., Schaffer, S. and Azuma, J., eds., Plenum Press, New York, Vol. 315, pp. 309-318.
12. Meldrum, M.J., Tu, R., Patterson, T., Dawson, R., and Petty, T., 1994, The effect of taurine on blood pressure, and uninary sodium, potassium and calcium excretion, in Adv. Exp. Med. Biol. "Taurine in Health and Disease", Huxtable, R.J. and Michalk, D., eds., Plenum Press, New York, Plenum Press, New York, Vol. 359, pp. 207-215.
13. Noma, A., 1983, ATP-regulated $K^+$ channels in cardiac muscle, *Nature*, 305:147-148.
14. Satoh, H., 1993, Effects of ATP-sensitive $K^+$ channel openers on the pacemaker activity in isolated rabbit single sino-atrial node cells, *J. Cardiovasc. Pharmacol.*, 22:863-868.
15. Satoh, H., 1994, Cardioprotective actions of taurine against intracellular and extra-cellular calcium-induced effects, in Adv. Exp. Med. Biol. "Taurine in Health and Disease", Huxtable, R.J. and Michalk, D., eds., Plenum Press, New York, Vol. 359, pp. 181-196.
16. Satoh, H., 1995, Electrophysiological actions of taurine on spontaneously beating rabbit sino-atrial nodal cells, *Jpn. J. Pharmacol.*, 67:29-34.
17. Satoh, H., 1995, Regulation by taurine of the spontaneous activity in young embryonic chick cardiomyocytes, *J. Cardiovasc. Pharmacol.*, 25:3-8.

18. Satoh, H., 1995, Tauirne-induced hyperpolarization shift of the reversal potential of the fast $Na^+$ current in embryonic chick cardiomyocytes, *Gen. Pharmacol.*, 26:517-521.
19. Satoh, H., 1995, A dual action of taurine on the delayed rectifier $K^+$ current in embryonic chick cardiomyocytes, *Amino Acids*, 9:235-246.
20. Satoh, H., 1995, Role of T-type $Ca^{2+}$ channel inhibitors in the pacemaker depolarization in rabbit sino-atrial nodal cells, *Gen. Pharmacol.*, 26:581-587.
21. Satoh, H., 1995, Identification of and developmental changes in transient outward current in embryonic chick cardiomyocytes, *Reprod. Fertil. Dev.*, 7:1369-1374.
22. Satoh, H., 1996, Direct inhibition by taurine of the ATP-sensitive $K^+$ channel in guinea pig ventricular cardiomyocytes, *Gen. Pharmacol.*, 27:625-627.
23. Satoh, H., 1996, Electrophysiological and electropharmacological actions of taurine on cardiac cells, in Adv. Exp. Med. Biol. "Taurine 2; Basic and Clinical Aspects". Huxtable, R.J., Azuma J., Kuriyama, K. et al. eds., Plenum Press, New York, Plenum Press, New York, Vol. 403, pp. 285-296.
24. Satoh, H., 1997, Inhibition of the fast $Na^+$ current by taurine in guinea pig ventricular myocytes, *Gen. Pharmacol.*, (in press).
25. Satoh, H. and Horie, M., 1997, Actions of taurine on the L-type $Ca^{2+}$ channel current in guinea pig ventricular cardiomyocytes, *J. Cariovasc. Pharmacol.*, 31:(in press).
26. Satoh, H. and Sperelakis, N., 1992, Taurine inhibition of $Na^+$ current in embryonic chick ventricular myocytes, *Eur. J. Pharmacol.*, 218:83-89.
27. Satoh, H. and Sperelakis, N., 1993, Effects of taurine on $Ca^{2+}$ currents in young embryonic chick cardiomyocytes, *Eur. J. Pharmacol.*, 231:443-449.
28. Satoh, H. and Sperelakis, N., 1997, Review of some actions of taurine on ion channels of cardiac muscle cells and others, *Gen. Pharmacol.*, 30:(in press).
29. Satoh, H., Tsuchida, K., and Hashimoto, K., 1989, Electrophysiological actions of A23187 and X-537A in spontaneously beating and in voltage-clamped rabbit sino-atrial node preparations, *Naunyn-Schmiedeberg's Arch. Pharmacol.*, 339:320-326.
30. Schaffer, S.W., Kramer, J., and Chovan, J.P., 1980, Regulation of calcium homeostasis in the heart by taurine, *Fed. Proc.*, 39:2691-2694.
31. Segawa, T., Nomura, Y., and Shimazaki, I., 1985, Possible involvement of calmoduline in modulatory role of taurine in rat cerebral (β-adrenergic neurones, in "Taurine: Biological Actions and Clinical Perspectives", Oja, S.S., Athee, L., Kontro, P., and Paasonen, M.K., eds., Alan R. Liss, New York, pp. 321-330.
32. Sperelakis, N. and Satoh, H., 1993, Taurine effects on ion channels of cardiac muscle, in "Ionic Channels and Effect of Taurine on the Heart", Noble, D. and Earm, Y.E., eds., Kluwer Academic Publishers, Boston, pp. 93-118.
33. Sperelakis, N., Satoh, H., and Bkaily, G., 1992, Taurine's effects on ionic currents in myocardial cells, in Adv. Exp. Med. Biol. "Taurine: Nutritional Value and Mechanical of Actions", Lombardini, B., Schaffer, S., and Azuma, J., eds., Plenum Press, New York, Vol. 315, pp. 129-143.
34. Suleiman, M.S., Rodrigo, G.C., and Chapman, R.A., 1992, Interdependence of intracellular taurine and sodium in guinea pig heart, *Cardiovasc. Res.*, 26:897-907.

# TAURINE IMPROVES ANGIOTENSIN II-INDUCED HYPERTROPHY OF CULTURED NEONATAL RAT HEART CELLS

Kyoko Takahahsi,[1] Mitsuhiro Azuma,[1] Akemichi Baba,[2]
Stephen Schaffer,[3] and Junichi Azuma[1]
[1] Department of Clinical Evaluation of Medicines and Therapeutics
[2] Department of Pharmacology
Faculty of Pharmaceutical Science
Osaka University
Osaka, Japan
[3] Department of Pharmacology
University of South Alabama
School of Medicine
Mobile, Alabama
USA

## ABSTRACT

The effect of taurine on angiotensin II-induced hypertrophy of cultured neonatal rat heart cells (myocytes and nonmyocytes) was examined. Angiotensin II (1-100 nM) alone caused an increase in the rate of protein synthesis and the surface area of myocytes without altering the rate of DNA synthesis or cell number. It also mediated an increase in DNA synthesis and in cell number of nonmyocytes. Exposure of the cells to taurine (20 mM) in the absence of angiotensin II had no effect on either hyperplastic growth or hypertrophy of the two types of cultured cardiac cells. However, myocytes pretreated with 20 mM taurine exhibited reduced responsiveness to angiotensin II. Following a 24 hr pretreatment with 20 mM taurine, the stimulation in protein synthesis by angiotensin II (1 nM) was significantly suppressed. Similarly, taurine treatment of nonmyocytes reduced the degree of angiotensin II-induced promotion of hyperplastic growth (DNA synthesis and cell number). Finally, taurine partially prevented the rise in $[Ca^{2+}]_i$ mediated by angiotensin II in cardiac cells.

The present results indicate that taurine is an effective inhibitor of angiotensin II action. The possibility that the beneficial effects of taurine in the treatment of heart failure might be related to its suppression of angiotensin II-mediated cellular responses is discussed.

*Taurine 3*
Edited by Schaffer *et al.*, Plenum Press, New York, 1998 

## INTRODUCTION

The development of heart failure is thought to proceed through several phases, one of the most important being cardiac hypertrophy. Recently, it has been established that one of the important factors contributing to the development of cardiac hypertrophy and heart failure is angiotensin II[4,6,9]. We have previously reported that taurine suppresses angiotensin II-induced IEG expression in both myocytes and nonmyocytes of cultured neonatal rat heart and inhibits angiotensin II-mediated hyperplastic growth of nonmyocytes[14]. These findings provide an explanation for the clinical observation that taurine therapy benefits patients with congestive heart failure[2].

The present study was designed to test the idea that taurine may act to modulate angiotensin II activity and to improve cardiac hypertrophy. To examine this idea, the experiments focused on the interaction between taurine and angiotensin II in cultured myocytes and cardiac fibroblasts (nonmyocytes) prepared from neonatal rats. The parameters examined were DNA synthesis, cell number, cell size, protein synthesis and changes in intracellular free calcium concentration ($[Ca^{2+}]_i$) of fura-2 loaded cardiac cells.

## MATERIALS AND METHODS

### Preparation of Cardiac Myocyte and Nonmyocyte Culture

The preparation of primary cardiac myocyte and nonmyocyte cultures from 1 day old Wistar rats has been described by Sadoshima et al[12]. For selective enrichment of cardiac myocytes, dissociated cells were preplated for 1 hr, during which time the nonmyocytes attached readily to the bottom of the culture dish[6]. This procedure yielded cultures with 90-95% myocytes, as assessed by microscopic observation of cell beating. Cells were kept in serum-containing culture medium for 48 hr, followed by serum-free medium. All experiments were performed 24 hr after transferring the cells to serum-free medium.

### Evaluation of DNA Synthesis and Cell Density

Cells were grown in serum-free medium for 24 hr and then stimulated with angiotensin II (1-100 nM). After 24 hr, methyl-[$^3$H]-thymidine (1 $\mu C_i$/ml) was added. Following a 6 hr incubation in methyl-[$^3$H]-thymidine containing buffer, the cells were washed 3 times with phosphate buffer and harvested with 0.1 N NaOH. Aliquots were counted by a scintillation counter. The results were expressed as counts/min/μg protein. The protein concentration was determined by the method of Lowery[7], using bovine serum albumin as a standard. The cells were detached with 1 ml of trypsin-EDTA and dissociated by titration. Cell density was counted using a hemocytometer.

### Evaluation of Protein Synthesis and Cell Size

As an index of protein synthesis, cellular L-2, 6-[$^3$H]-phenylalanine accumulation was measured as described by Sadoshima et al[12]. After incubation in serum-free medium for 24 hr, the cells were stimulated with angiotensin II (1 nM) for 48 hr in the presence of [$^3$H]-phenylalanine (5 $\mu C_i$/ml). The cells were then washed 3 times with phosphate buffer and harvested with 0.1 N NaOH. Aliquots were prepared and radioactivity measured as described above.

Cell size was quantified by measuring cell surface area. The shape and location of each myocyte was recorded before initiating the experiment by taking a photograph. The surface area in the same cell was estimated by planimetry enlarged photographs using NIH image software.

## Measurement of $[Ca^{2+}]_i$

The method of measuring $[Ca^{2+}]_i$ has been previously described by Takahashi et al[15]. An INTERDAC #M-1000 fluorescent spectromicroscope system (Osaka, Japan) was used to evaluate $[Ca^{2+}]_i$. The results were generally presented as fura-2 ratios. During the course of one cycle of contraction and relaxation, the 340/380 fluorescence ratio increased to a maximum value during systole [Max], and fell to a minimum value during diastole [Min]. The calcium transient (Ca-T) represents the difference between the Max and Min values. All data are expressed as Max, Min and Ca-T and represent the mean of 10 beats.

## Statistics

Statistics significance was determined by the Student's t-test and the $X^2$-test. Each value was expressed as the mean ± S.E.M. Differences were considered statistically significant when the calculated p value was less than 0.05.

# RESULTS AND DISCUSSION

## The Effect of Taurine on Angiotensin II-Induced Hypertrophy of Cardiac Cells

Since cardiac myocytes are terminally differentiated and lose their ability to duplicate soon after birth, they respond to a rise in workload by increasing cell size (hypertrophy) rather than increasing cell number (hyperplasia)[8]. In agreement with this notion, we found that angiotensin II neither increased DNA synthesis (as measured by $[^3H]$-thymidine uptake) nor cell number when added to cardiac myocytes in culture (Fig. 1). However, angiotensin II (1 nM) caused a significant increase (50%) in cell size and mediated an increase (22%) in protein synthesis of the myocyte, as measured by $[^3H]$-phenylalanine incorporation over a 48 hr period (Figs. 2 and 3). By contrast, in cardiac nonmyocyte cultures dominated by fibroblasts, angiotensin II (1 and 100 nM) treatment caused a significant increase (1.2 - 1.3 fold) in $[^3H]$-thymidine uptake over a 24 hr period (Fig. 1). The mitogenic effect of angiotensin II on nonmyocytes was also confirmed by counting the number of cells, although the magnitude of the increase in $[^3H]$-thymidine uptake did not

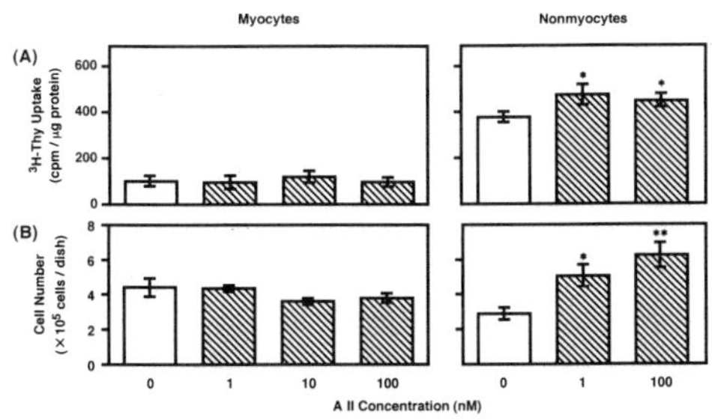


**Figure 1.** Angiotensin II-induced growth responses in cultured neonatal rat heart cells. (A) Cells were stimulated with angiotensin II (1-100 nM). After 24 hrs, $^3H$-thymidine (1 μCi/ml) was added for 6 hrs. Data are means ± SEM obtained from 9-21 samples. *$p<0.05$ vs control. (B) Cells were stimulated with angiotensin II (1-100 nM) for 24 hrs. Data are means ± SEM obtained from 6 samples. *$p<0.05$ and **$p<0.01$ vs control.

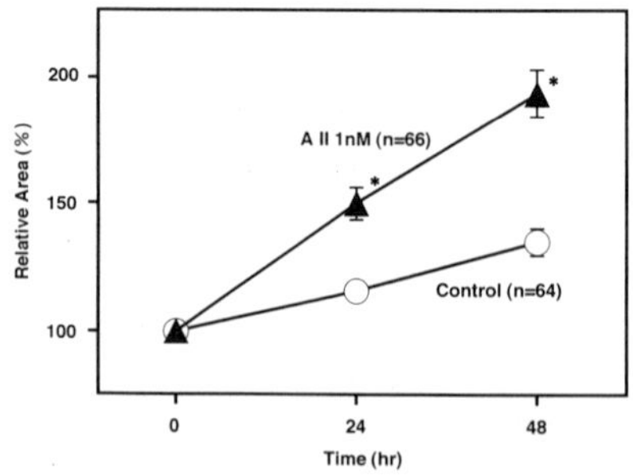


**Figure 2**. Effect of angiotensin II on cell size of myocytes from neonatal rat heart. Cells were stimulated with angiotensin II (1 nM) for either 24 or 48 hrs. Data are means ± SEM obtained from 66-64 cells. *$p<0.05$ vs control.

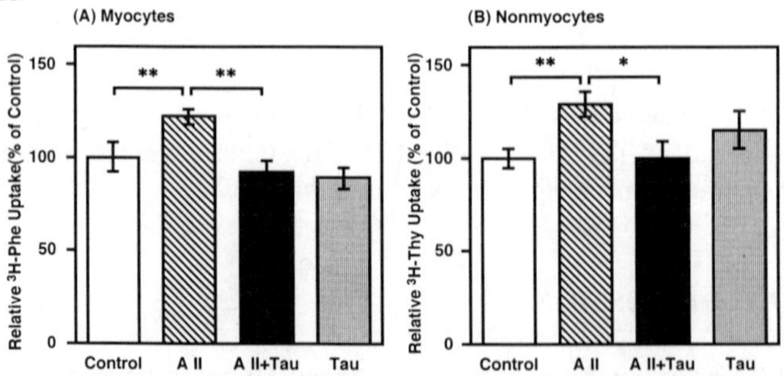


**Figure 3**. Effect of taurine on angiotensin II-induced hypertrophic responses of myocytes and nonmyocytes from neonatal rat heart. Cells were stimulated with angiotensin II (1 nM) for either 24 or 48 hrs. Data are means ± SEM obtained from 9-34 cells. *$p<0.05$, **$p<0.01$ vs control or angiotensin II.

parallel the elevation in cell number (Fig. 1). The effect of taurine treatment on angiotensin II-induced hypertrophy of myocytes and nonmyocytes was examined using cells prepared in serum-free medium containing 20 mM taurine before stimulation with 1 nM angiotensin II for a period of either 24 or 48 hr in the presence of 20 mM taurine (Fig. 3). While angiotensin II (1 nM) significantly increased both [$^3$H]-thymidine uptake in nonmyocytes and [$^3$H]-phenylalanine incorporation into myocytes, taurine reversed these effects, decreasing [$^3$H]-thymidine uptake by nonmyocytes in the presence of angiotensin II from 129% to 100% and [$^3$H]-phenylalanine incorporation into angiotensin II-exposed myocytes from 122% to 92%, respectively. By contrast, in the absence of angiotensin II, taurine had no effect on either [$^3$H]-thymidine uptake or [$^3$H]-phenylalanine incorporation into the two types of cardiac cells. Previously, we have reported that 20 mM taurine prevented the 60% increase in nonmyocyte cell number seen following angiotensin II (1 nM) exposure[14].

## The Effect of Taurine on Angiotensin II - Mediated Cellular Responses

It has been established that most hypertrophic stimuli, including angiotensin II, exhibit similar signal transduction pathways, in which $[Ca^{2+}]_i$ rises and protein kinase C is activated[1,11,12]. Therefore, we examined the effect of taurine on angiotensin II-induced changes in $[Ca^{2+}]_i$. The control, unstimulated myocytes formed monolayer sheets of synchronously beating cells which exhibited typical calcium transients (Ca-T), defined as the difference between peak $[Ca^{2+}]_i$ (Max) and baseline $[Ca^{2+}]_i$ (Min). In the absence of taurine, angiotensin II (1 nM) immediately increased the magnitude of Ca-T (Fig. 4). Within 5 min of angiotensin II exposure, Ca-T increased from 1.50 ± 0.61 to 1.74 ± 0.68 and remained at the elevated level for at least 5 minutes. The elevation in Ca-T was observed in 93-100% of the measured cells. By contrast, myocytes treated for 24 hr with 20 mM taurine exhibited little change in Ca-T upon addition of angiotensin II (1 nM); Ca-T rose only 5 - 9% compared to 17% in the absence of taurine. Interestingly, about 30% of the cells treated with taurine exhibited no significant change in Ca-T following exposure to angiotensin II (Fig. 4).

Figure 5 shows that taurine treatment also altered angiotensin II-induced elevations in $[Ca^{2+}]_i$ of the nonmyocyte cells. After 4 min of exposure, angiotensin II promoted a transient increase in $[Ca^{2+}]_i$ of about 30%. Taurine (20 mM) completely blocked this rise in $[Ca^{2+}]_i$. As expected, taurine alone had no effect on $[Ca^{2+}]_i$ in the two types of cells.

The most important finding of this study is the observed interaction between taurine and angiotensin II. While exposure of the cells to taurine (20 mM) in the absence of angiotensin II had no effect on $[Ca^{2+}]_i$ or either hypertrophic or hyperplastic growth of cultured cardiac cells, the cultured myocytes pretreated with 20 mM taurine exhibited reduced angiotensin II responsiveness. Following a 24 hr pretreatment period with 20 mM taurine, the stimulation in protein synthesis by angiotensin II was significantly suppressed. Moreover, taurine pretreatment also inhibited angiotensin II-mediated hyperplastic growth in culture nonmyocytes. As predicted, taurine attenuated the elevation in $[Ca^{2+}]_i$ induced by angiotensin II in both types of cardiac cells.

The mechanism underlying the attenuation of angiotensin II-induced actions by taurine remains an area of considerable interest. The signal transduction pathway responsible for the hypertrophic effects of angiotensin II begins with the interaction of

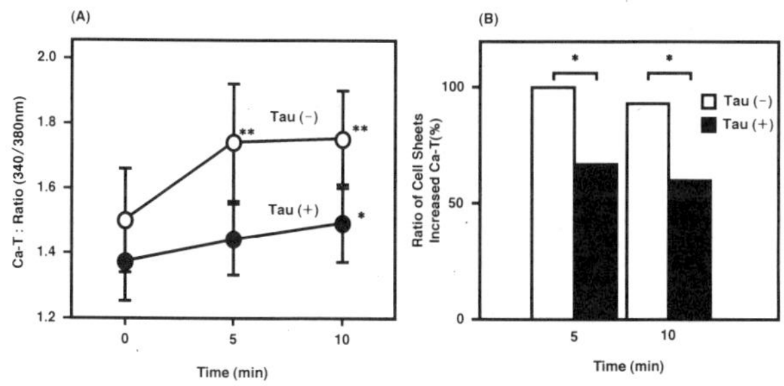


**Figure 4**. Effect of taurine on angiotensin II-induced changes in $[Ca^{2+}]_i$ transients of myocytes from neonatal rat heart. (A) Each value represents the mean ± SEM obtained from 15 samples. *p<0.05, **p<0.001 vs o min. (B) The ratio of cell sheets with increased Ca-T represents the number of cell sheets with increased Ca-T/total number of cell sheets (n=15).

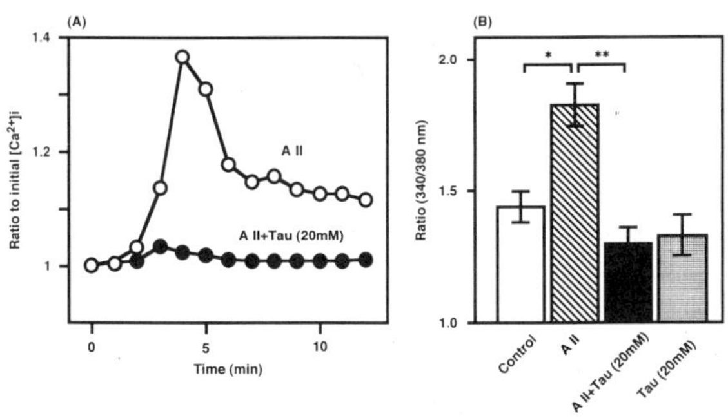


**Figure 5**. Effect of taurine on angiotensin II-induced $[Ca^{2+}]_i$ alterations of nonmyocytes from neonatal rat heart. (A) Typical time course pattern. (B) Data are means ± SEM obtained from 5-8 samples. *p<0.05, **p<0.01 vs control or angiotensin II.

angiotensin II with its receptor. The angiotensin II receptor is coupled through a G protein to a membrane bound phospholipase C isozyme[2], which is subsequently activated. The resulting products of phospholipase C action, $IP_3$ and diacylglycerol, are important second messengers in hypertrophic growth[1,2,11]. $IP_3$ promotes the release of $Ca^{2+}$ from the sarcoplasmic reticulum, thereby increasing $[Ca^{2+}]_i$. Moreover, angiotensin II promotes calcium influx through the calcium channel and calcium handling by the $Na^+$-$Ca^{2+}$ exchanger[5]. The subsequent rise in $[Ca^{2+}]_i$ plays a potentiating role in the growth process by activating several signaling steps[11,12].

One of the major functions of taurine in the heart is modulation of calcium transport[3,13]. Depending on the status of the myocardium, taurine is capable of either increasing or decreasing $[Ca^{2+}]_i$. This complicated biphasic pattern has been difficult to resolve mechanistically. The only rational explanation for this phenomenon assumes that taurine exerts multiple effects on calcium transport, some which promote myocyte calcium accumulation while others enhance cellular calcium loss. Thus, it is not surprising that taurine attenuates the rise in $[Ca^{2+}]_i$ mediated by angiotensin II. Since $[Ca^{2+}]_i$ is thought to be an important hypertrophic messenger, the modulation of the angiotensin II-mediated changes in $[Ca^{2+}]_i$ provides a rational explanation for the reversal of some angiotensin II effects by taurine.

The present results indicate that taurine is an effective inhibitor of angiotensin II action. It is plausible that the beneficial effect of taurine in the treatment of heart failure could be related to its suppression of angiotensin II-mediated cellular responses.

## REFERENCES


1. Allos, S.N., Carl, L.L., and Morgan, H.E., 1992, Acceleration of growth of cultured cardiomyocytes and translocation of protein kinase C, *Am. J. Physiol.*, 263:C319-C325.
2. Azuma, J., Sawamura, A., Awata, N., Hasegawa, H., Ogura, K., Harada, H., Ohta, H., Yamauchi, K., Kishimoto, S., Yamagami, T., Ueda, E., and Ishiyama, T., 1983, Double-blind randomized crossover trial of taurine in congestive heart failure, *Curr. Therap. Res.*, 34:543-557.
3. Azuma, J. and Schaffer, S.W., 1995, Protective effect of taurine on the failing heart and its possible mechanism of action, in: The Failing Heart, Dnalla, N.S., Beamish, R.E., Takeda, N., and Nagano, M., eds., Lippncutt-Raven Publishers, Philadelphia, pp. 421-431.
4. Baker, K.M., Booz, G.W., and Dostal, D.E., 1992 Cardiac actions of angiotensin II: Role of an intra-cardiac renin-angiotensin system, *Ann. Rev. Physiol.*, 54:227-241.
5. Ballard, C. and Schaffer, S.W., 1996, Stimulation of the $Na^+/Ca^{2+}$ exchanger by phenylephrine, angio-tensin II and endothelin 1, *J. Mol. Cell. Cardiol.*, 28:11-17.
6. Lindpainter, K. and Ganten, D., 1991, The cardiac renin-angiotensin system, *Circ. Res.*, 68:905-921.
7. Lowry, O.H., Rosebrough, N.J., Farr, A.L., and rAndall, R.J., 1951, Protein measurement with the Folin phenol regent, *J. Biol. Chem.*, 193:265-275.
8. Komuro, I. and Yazaki, Y., 1993, Control of cardiac gene expression by mechanical stress, *Annu. Rev. Physiol.*, 55:55-75.
9. Sadoshima, J. and Izumo, S., 1993, Molecular characterization of angiotensin II-induced hypertrophy of cardiac myocytes and hyperplasia of cardiac fibroblasts, *Circ. Res.*, 73:413-423.
10. Sadoshima, J. and Izumo, S., 1993, Signal transduction pathways of angiotensin II-induced c-fos gene expression in cardiac myocytes *in vitro*, *Circ. Res.*, 73:424-438.
11. Sadoshima, J. and Izumo, S., 1993, Mechanical stretch rapidly activates multiple signal transuction pathways in cardiac myocytes: potential involvement of an autocrine/paracrine mechanism, *The EMBO J.*, 12:1681-1692.
12. Sadoshima, J., Jahn, L., Takahashi, T., Kulik, T.J., and Izumo, S., 1992, Molecular characterization of the stretch-induced adaptation of cultured cardiac cells, *J. Biol. Chem.*, 267:10551-10560.
13. Takahashi, K., Azuma, J., Awata, N., Sawamura, A., Kishimoto, S., Yamagami, T., Kishi, T., Harada, H., and Schaffer, S.W., 1988, Protective effect of taurine on the irregular beating pattern of cultured myocardial cells induced by high and low extracellular calcium ion, *J. Mol. Cell. Cardiol.*, 20:397-403.

14. Takahashi, K., Hashimoto, H., Baba, A., Schaffer, S.W., and Azuma, J., 1996, Effect of taurine on angiotensin II-induced expression of immediate early response genes in primary cultured neonatal rat heart cells, in: Taurine 2; Basic and Clinical aspects, Huxtable, R.J., Azuma, J., Kuriyama, K., Nakagawa, M., and Baba, A., eds., Plenum Press, New York, pp. 297-304.
15. Takahashi, K., Schaffer, S.W., and Azuma, J., 1997, Taurine prevents intracellular calcium overload during calcium paradox of cultured cardiomyocytes, *Amino Acids*, 13:1-11.

# EFFECTS OF TAURINE ON SIGNAL TRANSDUCTION STEPS INDUCED DURING HYPERTROPHY OF RAT HEART MYOCYTES

M.R. Rao and L. Tao
Department of Cardiovascular Pharmacology
Nanjing Medical University
Nanjing, P.R. China

## ABSTRACT

Angiotensin II plays an important role in the development of cardiac hypertrophy. One factor thought to contribute to the trophic activity of angiotensin II in fibroblasts is the elevation in $[Ca^{2+}]_i$. Although this theory has received considerable support in cardiac fibroblasts, it is much more controversial in cardiac myocytes. Therefore, the aim of this report was to examine the effect of several $Ca^{2+}$ modulators on protein synthesis in neonatal cardiac myocytes. We found that angiotensin II increased both $[Ca^{2+}]_i$ and the rate of protein synthesis in isolated myocytes. Both effects were blocked by nifedipine, but only the angiotensin II-mediated increase in $[Ca^{2+}]_i$ was inhibited by taurine in a dose-dependent manner. These data support the notion that $Ca^{2+}$ plays only a permissive role in angiotensin II-mediated stimulation of protein synthesis. By contrast, the ability of taurine to attenuate the positive chronotropic effect, the prolongation of the action potential and the proarrhythmic activity of angiotensin II appear to be linked directly to changes in $[Ca^{2+}]_i$. We conclude that taurine reverses these actions of angiotensin II by altering $Ca^{2+}$ flux across the cell membrane.

## INTRODUCTION

Hypertrophy of the left ventricle (LVH) occurs as a pathophysiologic adaptation to the rise in afterload that accompanies hypertension. Consequently, over a period of many years, the left ventricle of patients with essential hypertension increases in mass and wall thickness in proportion to the elevation in total peripheral resistance and arterial pressure. This process initially allows the heart to adapt to the pressure overload without succumbing to congestive heart failure, which inevitably results when the left ventricle cannot successfully sustain its afterload. Nonetheless, clinical epidemiological studies have shown that subjects with ECG evidence of LVH are at increased risk of compounding factors, such as elevated blood pressure, myocardial ischemia, heart failure, arrhythmias and sudden cardiac death. To date, LVH has been internationally listed as an independent risk factor for cardiovascular death.

Several stressful conditions, such as hypertension, chronic pressure overload and elevations in circulating levels of trophic factors, can cause the heart to undergo major changes, including an increase in the size of preexisting myocytes (hypertrophy) and an accumulation of fibrillar collagen (remodeling). Experimental and clinical evidence suggest that angiotensin II acts as a growth factor and is capable of causing both LVH and myocardial fibrosis. In order to evaluate the direct effect of angiotensin II on several hypertrophy-linked reactions, a cardiomyocyte model of hypertrophy produced by chronically treating cells with angiotensin II was used. Since abnormalities in $[Ca^{2+}]_i$ are thought to contribute to the development of LVH[4,5], the effects of several $Ca^{2+}$ modulators were examined in angiotensin II treated myocytes.

## RESULTS

As observed previously, we found that one of the major effects of angiotensin II (100 nM) on cultured myocytes was to increase the rate of protein synthesis (Fig. 1A). This effect was inhibited by inclusion of the $Ca^{2+}$ antagonist, nifedipine (10 μM), in the culture medium. However, the angiotensin II effect on protein synthesis was not blocked by either taurine (20 mM) or the ACE inhibitor, enalapril (1 mM). Enalapril and taurine also failed to inhibit the increase in cellular protein content induced by angiotensin II (Fig. 1B). However, taurine was found to be an effective inhibitor of the angiotensin II-mediated increase in myocyte beating rate (Table 1). It also reversed the elevation in action potential amplitude, prolongation of the action potential and the rise in the number of arrhythmias.

It has generally been assumed that most myocardial actions of taurine are linked to changes in $[Ca^{2+}]_i$. Thus, the observation that taurine reverses the electrophysiological effects of angiotensin II but not its trophic effects was unexpected. To provide information necessary in clarifying this apparent paradox, $[Ca^{2+}]_i$ was monitored before and after treatment with taurine. Figure 2 shows that angiotensin II (100 nM) increased $[Ca^{2+}]_i$ of neonatal rat myocytes from 84 ± 11 nM to 294 ± 17 nM. As expected, both nifedipine (10 μM) and taurine (20 mM) significantly reduced the angiotensin II-mediated rise in $[Ca^{2+}]_i$. However, enalapril (100 μM) had no effect of the angiotensin II response, supporting the notion that enalapril's only function is to block the formation of angiotensin II.

With the aim of providing information on the mechanism underlying taurine's action, the interaction between taurine and two established modulators of $[Ca^{2+}]_i$ was examined. In agreement with previous studies, both ryanodine (10 μM) and ouabain (200 nM) elevated $[Ca^{2+}]_i$ in the normal cardiomyocyte (Fig. 3). Taurine (20 mM) attenuated the rise in $[Ca^{2+}]_i$ induced by ouabain, but had no effect on the ryanodine-mediated change in $[Ca^{2+}]_i$ (Fig. 3). Interestingly, while taurine had no effect on ryanodine's ability to release $Ca^{2+}$ from intracellular stores, the amino acid significantly reduced $[Ca^{2+}]_i$ in the presence of both angiotensin II and ryanodine (data not shown). These data suggest that the taurine-sensitive, angiotensin II-mediated elevation in $[Ca^{2+}]_i$ involves a modulation of $Ca^{2+}$ movement across the cell membrane.

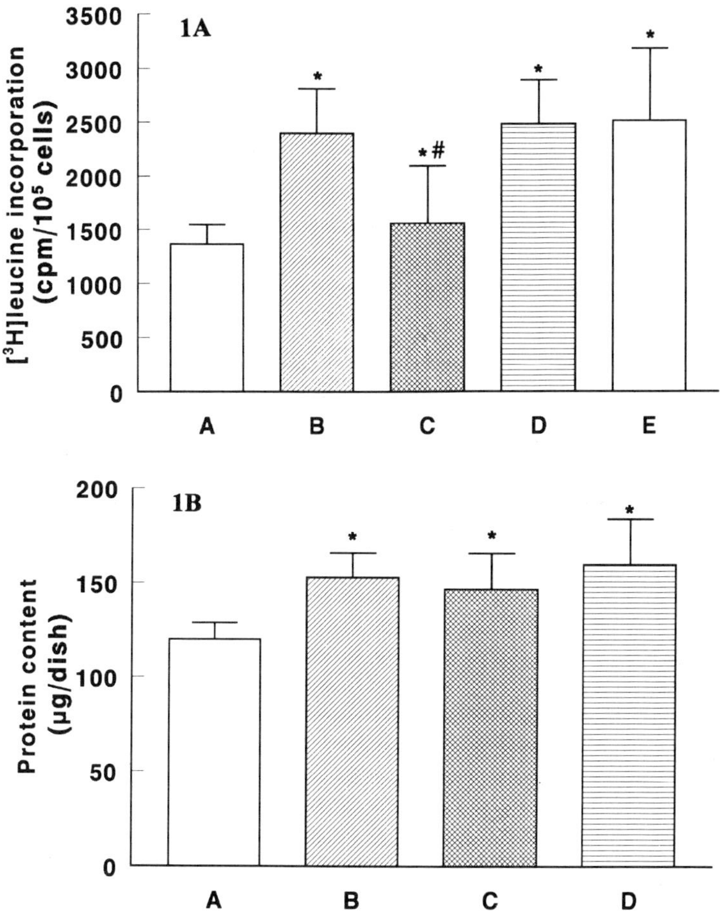


**Figure 1.** Effect of angiotensin II, taurine and enalapril on the rate of protein synthesis and protein content of cardiac myocytes. **A.** Rat neonatal cardiac myocytes were incubated with medium supplemented with the following: (A) no additions, (B) 100 nM angiotensin II, (C) Nifedipine (10 µM) + Angiotensin II (100 nM), (D) Enalapril (1 mM) + Angiotensin II (100 nM) or (E) Taurine (20 mM) + Angiotensin II. After a 72 hour incubation, the rate of $^{3}$H-leucine incorporation into protein was determined. Values shown represent the means ± S.E.M. of 10 cells. Asterisks denote significant differences ($p < 0.01$) from the control (A) while # indicates a significant difference from the angiotensin II-treated group (B). **B.** Rat neonatal cardiac myocytes were incubated for 7 days with medium supplemented with the following: (A) no additions, (B) Angiotensin II (100 nM), (C) Enalapril (1 mM) and Angiotensin II (100 nM) or (D) Taurine (20 mM) and Angiotensin II (100 nM). Values shown represent the means ± S.E.M. of 10 cells. Asterisks denote a significant difference ($p < 0.01$) from the control (A).

## DISCUSSION

An early step in the cascade of events triggered by either the induction of volume overload or an ischemic insult is a dramatic elevation in cardiac renin mRNA expression[3,8]. Since transgenic mice overexpressing the renin gene develop cardiac hypertrophy[7], a clear link between angiotensin II and cardiac hypertrophy has been established. Also supporting the cause-effect relationship between angiotensin II and hypertrophy is evidence that angiotensin II increases both protein synthesis and the expression of specific fetal genes, which are thought to contribute to the remodeling of the myocardium.

**Table 1.** Changes in beating rate, action potential characteristics and arrhythmia frequency of neonatal rat myocytes after exposure to angiotensin II (100 nM) or angiotensin II (100 nM) + taurine.

| Group | Beating rate (bpm) | APA(mV) | $APD_{50}$(ms) | $APD_{90}$(ms) | MDP(mV) | SCL(ms) | % rate of arrhythmias |
|---|---|---|---|---|---|---|---|
| | | Action Potential Characteristics | | | | | |
| Normal | 89 ± 9 | 66 ± 8 | 77 ± 9 | 155 ± 9 | 54 ± 4 | 661 ± 40 | 10 (1/10) |
| Ang II | 138 ± 18* | 77 ± 5* | 91 ± 12* | 168 ± 10* | 55 ± 4 | 441 ± 60 | 78 (7/9)* |
| Ang II + Taurine | | | | | | | |
| 5 mM | 118 ± 14# | 72 ± 5# | 85 ± 5# | 166 ± 5 | 55 ± 5 | 509 ± 34# | n.d. |
| 10mM | 117 ± 11# | 72 ± 3# | 85 ± 6# | 165 ± 5 | 53 ± 5 | 514 ± 51# | n.d. |
| 20mM | 100 ± 9# | 63 ± 4# | 79 ± 5# | 157 ± 8# | 56 ± 8 | 595 ± 49# | 20 (2/10)# |

Neonatal rat cardiomyocytes were maintained in culture for 7 days in medium containing no additions (Normal), 100 nM angiotensin II (Ang II) or the combination of 100 nM angiotensin II and the indicated amount of taurine (Ang II + Taurine). Values shown represent the means ± S.E.M. of 6-10 cells. Asterisks denote significant differences between the normal and the angiotensin II-treated group while # indicates a significant difference between the angiotensin II + taurine and the angiotensin II groups ($p < 0.05$). APA: action potential amplitude; $APD_{50}$: action potential duration to 50% repolarization; $APD_{90}$: action potential duration to 90% repolarization; MDP: maximal diastolic potential; SCL sinus cycle length.

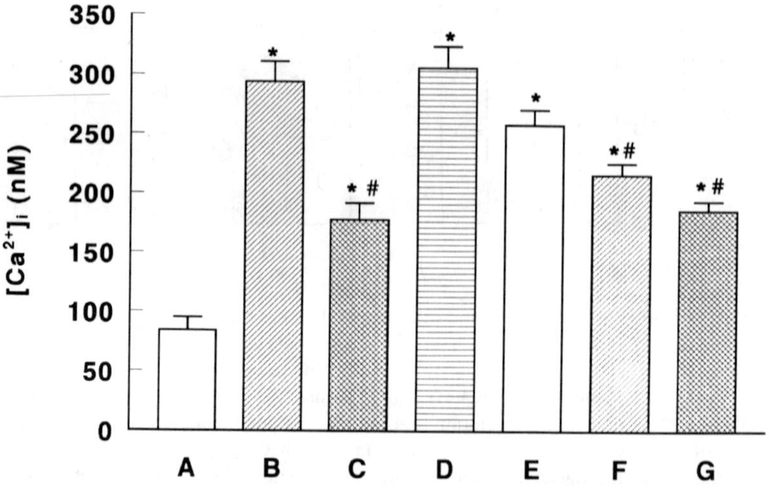


**Figure 2.** Effect of angiotensin II, nifedipine, enalapril and taurine on $[Ca^{2+}]_i$ of neonatal rat cardiomyocytes. Changes in $[Ca^{2+}]_i$ were monitored in cultured neonatal rat myocytes following exposure to the following agents: (A) No additions, (B) Angiotensin II (100 nM), (C) Nifedipine (10 nM) + Angiotensin II (100 nM), (D) Enalapril (100 µM) + Angiotensin II (100 nM), (E) Taurine (5 mM) + Angiotensin II (100 nM), (F) Taurine (10 mM) + Angiotensin II (100 nM) and (G) Taurine (20 mM) + Angiotensin II (100 nM). Values shown represent the means ± S.E.M. of 6 cells. Asterisks denote significant differences from the control (A) while # indicates a significant difference between the angiotensin II group and groups containing angiotensin II and another agent ($p < 0.01$).

Nonetheless, the signal transduction pathway for angiotensin II-induced cellular hypertrophy remains to be definitively defined. Several reports implicate changes in $[Ca^{2+}]_i$ in this process[4,5]. Booz et al.[4] found that chelation of intracellular $Ca^{2+}$ using the chelator BAPTA prevented the increase in cell proliferation seen 48 hours after treatment with 1 µM [$Sar^1$]-angiotensin II. Based on their data, they concluded that the hypertrophic

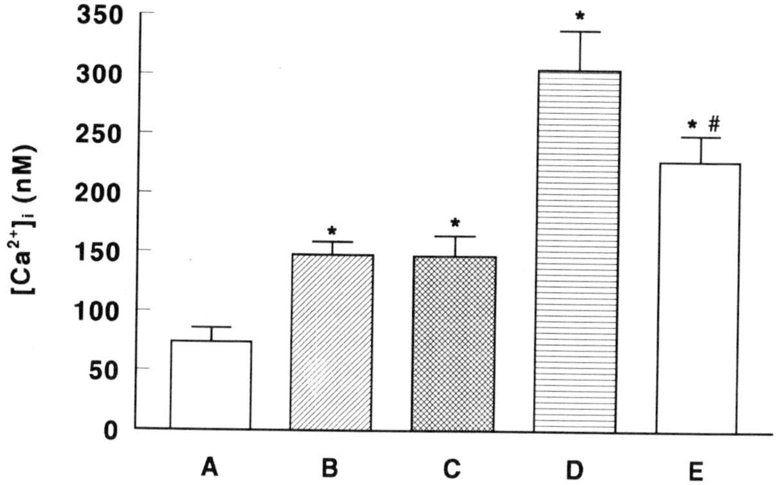


**Figure 3.** Effect of ryanodine, ouabain and taurine on $[Ca^{2+}]_i$ of cultured neonatal myocytes. Rat neonatal myocytes were incubated for 72 hours in medium containing the following: (A) no additions, (B) Ryanodine (10 μM), (C) Ryanodine (10 μM) + Taurine (20 mM), (D) Ouabain (200 nM) or (E) Ouabain (200 nM) + Taurine (20 mM). Values shown represent the means ± S.E.M. of 8 cells. Asterisks denote significant differences from the control while # indicates a significant difference between the taurine + ouabain and the ouabain groups ($p < 0.01$).

response was initiated by a series of protein kinase steps, including the activation of MAP kinase and ribosomal S6 protein kinase, both which phosphorylate transcriptional factors, such as c-jun, c-myc and c-fos. This process is apparently $Ca^{2+}$ dependent since $Ca^{2+}$ ionophores partially mimic the actions of angiotensin II in the cardiac fibroblast by elevating $[Ca^{2+}]_i$ and stimulating MAP kinase activity[10]. In apparent contrast to the studies using isolated fibroblasts, Kinugawa et al.[5] reported a decline in $[Ca^{2+}]_i$ following exposure of isolated neonatal rat myocytes to 10 nM angiotensin II. However, since reductions in $[Ca^{2+}]_i$ to less than 100 nM prevented the induction of c-fos by angiotensin II, Kinugawa et al.[5] concluded that a minimal amount of intracellular $Ca^{2+}$ is required for the hypertrophic response of angiotensin II. This conclusion was in agreement with an earlier study by Baker and Aceto[1], who found that the angiotensin II antagonist, [Sar[1], Ile[8]]-angiotensin II, elevated $[Ca^{2+}]_i$ but blocked the angiotensin II-mediated increase in protein synthesis.

The present study supports the notion that $[Ca^{2+}]_i$ is not the sole factor responsible for initiating protein synthesis. We found that the $Ca^{2+}$ channel antagonist, nifedipine, attenuated the angiotensin II-mediated elevation in both $[Ca^{2+}]_i$ and the rate of protein synthesis (Figs. 1 and 2). On the other hand, 20 mM taurine had no effect on angiotensin II-mediated stimulation in protein synthesis (Fig. 1), but dramatically reduced $[Ca^{2+}]_i$ in the angiotensin II exposed myocyte.

In agreement with two previous studies[2,12], we found that an interaction must exist between the signaling pathways of taurine and those of angiotensin II. However, the site of this interaction cannot be ascertained from the data because of differences in the experimental approaches used. In the present study, taurine was found to only reverse the actions of angiotensin II related to changes in cation transport, such as the elevation in $[Ca^{2+}]_i$, increased beating rate, the alterations in the action potential and augmented dysarrhythmia generation (Fig. 2 and Table 1). Taurine was ineffective in preventing $Ca^{2+}$ independent events, such as angiotensin II-mediated increases in protein synthesis and in total myocyte protein content (Fig. 1). By comparison, the study by Ballard-Croft et al.[2] focused on the interaction between taurine and angiotensin II relative to contractile function and the activity of the $Na^+$-$Ca^{2+}$ exchanger. Since these actions of taurine are related to changes in $[Ca^{2+}]_i$, the observation that taurine served as an effective inhibitor of these actions of angiotensin II are consistent with our data. However, Takahashi et al.[12] reported that both angiotensin II-mediated upregulation of the immediate early response genes, c-fos and c-jun, and its positive inotropic effect were suppressed by taurine. Although the immediate early response gene expression data are seemingly inconsistent with the present study, some major differences exist between our study and the work of Takahashi et al.[12]. First, the present study examined the rate of $^3$H-leucine incorporation into protein 72 hours after the initial exposure of the myocytes to taurine and angiotensin II. This contrasts with the 24 hour incubation period chosen in the study by Takahashi et al.[12]. Second, we examined the effect of angiotensin II on protein synthesis whereas Takahashi et al.[12] focused on the upregulation of immediate early response gene expression. Although several years ago it was assumed that the upregulation of the immediate early response genes, c-fos and c-jun, was a necessary step in the angiotensin II signaling pathway leading to the development of cardiac hypertrophy, recent data by Takano et al.[13] suggest that this assumption may be incorrect. Takano et al.[13] reported that the immunosuppressant, rapamycin, completely inhibited angiotensin II-induced phenylalanine incorporation into proteins without altering angiotensin II-induced c-fos expression, suggesting that immediate early response gene expression may belong to a different angiotensin II signaling pathway than the elevation in protein synthesis. Therefore, taurine may act specifically to block the c-fos pathway without affecting the rapamycin sensitive protein synthesis pathway. It would be attractive to suggest that since rapamycin suppresses the angiotensin II-mediated activation of the 70 kD form of S6 kinase, taurine may modulate this signaling step.

Another interesting finding of this study is that taurine blocks ouabain-induced elevations in $[Ca^{2+}]_i$ without affecting the actions of ryanodine. To date, there is no evidence that taurine specifically alters either the binding of ouabain to its receptor or the activity of the $Na^+$,$K^+$ ATPase. Instead, taurine appears to modulate the transport of $Na^+$ or $Ca^{2+}$ by voltage sensitive transporters[11]. One transporter that plays a central role in the elevation in $[Ca^{2+}]_i$ by ouabain is the $Na^+$-$Ca^{2+}$ exchanger. Recently, Ballard-Croft et al.[2] found that taurine inhibits the activation of the exchanger by angiotensin II. If taurine were also effective in inhibiting the ouabain-mediated stimulation of $Ca^{2+}$ flux by the $Na^+$-$Ca^{2+}$ exchanger, it could account for the observed reversal of the ouabain effect. It could also explain the failure of taurine to block the rise in $[Ca^{2+}]_i$ in cells treated with ryanodine, an agent capable of only releasing $Ca^{2+}$ from intracellular stores.

## REFERENCES


1. Baker, K.M. and Aceto, J.F., 1990, Angiotensin II stimulation of protein synthesis and cell growth in chick heart cells, *Am. J. Physiol.*, 259:H610-H618.
2. Ballard-Croft, C., Mozaffari, M.S., Azuma, J., and Schaffer, S., 1997, Interaction between taurine and angiotensin II: modulation of calcium transport and myocardial contractile function, *Amino Acids*, 13:105-114.
3. Boer, P.H., Ruzicka, M., Lear, W., Harmsen, E., Rosenthal, J., and Leenen, F.H.H., 1994, Stretch-mediated activation of cardiac renin gene, *Am. J. Physiol.* 267:H1630-H1636.
4. Booz, G.W., Dostal, D.E., Singer, H.A., and Baker, K.M., 1994, Involvement of protein kinase C and $Ca^{2+}$ in angiotensin II-induced mitogenesis of cardiac fibroblasts, *Am. J. Physiol.*, 267:C1308-1318.
5. Kinugawa, K-I, Takahashi, T., Kohmoto, O., Yao, A., Ikenouchi, H., and Serizawa, T., 1995, $Ca^{2+}$-growth coupling in angiotensin II-induced hypertrophy in cultured rat cardiac cells, *Cardiovasc. Res.*, 30:419-431.

6. Neyses, L., Nouskas, J., Luyken, J., Fronhoffs, S., Oberdorf, S., Pfeifer, U., Williams, R.S., Sukhatme, V.P., and Vetter, H., 1993, Induction of immediate-early genes by angiotensin II and endothelin-1 in adult rat cardiomyocytes, *J. Hypertension* 11:927-934.
7. Nickenig, G., Laufs, U., Schnabel, P., Knorr, A., Paul, M. and Bohm, M., 1997, Down-regulation of aortic and cardiac $AT_1$ receptor gene expression in transgenic (mRen-2) 27 rats, *Br. J. Pharmacol.*, 121:134-140.
8. Passier, R.C.J.J., Smits, J.F.M., Verluyten, M.J.A., and Daemen, M.J.A.P., 1996, Expression and localization of renin and angiotensinogen in rat heart after myocardial infarction, *Am. J. Physiol.*, 271:H1040-1048.
9. Sadoshima, J. and Izumo, S., 1995, Rapamycin selectively inhibits angiotensin II-induced increase in protein synthesis in cardiac myocytes in vitro, *Circ. Res.*, 77:1040-1052.
10. Sadoshima, J., Qui, Z., Morgan, J.P., and Izumo, S., 1995, Angiotensin II and other hypertrophic stimuli mediated by G protein-coupled receptors activate tyrosine kinase, mitogen-activated protein kinase and 90-kD S6 kinase in cardiac myocytes, *Circ. Res.*, 76:1-15.
11. Sperelakis, N., Katsube, Y., and Kusaka, M., 1996, Some actions of taurine on ionic current of myocardial cells and myometrial cells, in "Taurine 2: Basic and Clinical Aspects", Huxtable, R.J., Azuma, J., Kuriyama, K., Nakagawa, M., and Baba, A., eds., Plenum Press, New York and London, pp. 275-284.
12. Takahashi, K., Hashimoto, H., Baba, A., Schaffer, S.W., and Azuma, J., 1996, in "Taurine 2: Basic and Clinical Aspects", Huxtable, R.J., Azuma, J., Kuriyama, K., Nakagawa, M. and Baba, A., eds., Plenum Press, New York and London, pp. 297-304.
13. Takano, H., Komuro, I., Zou, Y., Kudoh, S., Yamazaki, T., and Yazaki, Y., 1996, Activation of p70 S6 protein kinase is necessary for angiotensin II-induced hypertrophy in neonatal rat cardiac myocytes, *FEBS Lett.*, 379:255-259.

# EFFECT OF TAURINE DEPLETION ON ANGIOTENSIN II-MEDIATED MODULATION OF MYOCARDIAL FUNCTION

Stephen W. Schaffer,[1] Cherry Ballard-Croft,[1] Kyoko Takahashi,[2] and Junichi Azuma[2]

[1] Department of Pharmacology
University of South Alabama School of Medicine
Mobile, AL 36688
[2] Department of Clinical Evaluation of Medicines and Therapeutics
Faculty of Pharmaceutical Science
Osaka University
Osaka, Japan

## ABSTRACT

Taurine depletion was induced by either incubation of isolated myocytes with 5 mM ß-alanine or feeding rats with water containing 3% ß-alanine. Hearts of taurine depleted rats exhibited an impairment in myocardial relaxation, associated with a decrease in $Na^+$-$Ca^{2+}$ exchanger activity. Exposure of the heart to angiotensin II, an activator of the $Na^+$-$Ca^{2+}$ exchanger, eliminated the relaxation defect. In agreement with the contractile results, taurine depletion prolonged the calcium transient, an effect which was partially eliminated by exposure to angiotensin II. Although peak systolic $[Ca^{2+}]_i$ was modestly depressed in the taurine depleted myocyte, peak ventricular pressure was normal. This may be related to an elevation in $pH_i$ induced by taurine depletion. Angiotensin II had little effect on contractility of the taurine depleted heart, presumably because of two opposing effects, a reduction in $pH_i$ and an increase in $[Ca^{2+}]_i$. Thus, taurine depletion alters contractile function and ion transport and both of these effects are modulated by exposure of the heart to angiotensin II.

## INTRODUCTION

Remodeling is considered one of the most important steps in the development of congestive heart failure. Traditionally, the term "remodeling" has referred to the anatomical changes occurring in the ventricular wall of the heart during development of hypertrophy and heart failure[7]. However, more recently the science of cardiac remodeling

*Taurine 3*
Edited by Schaffer *et al.*, Plenum Press, New York, 1998

has focused on the biochemical and physiological changes of heart failure. One of these biochemical changes is the accumulation of the osmolyte, taurine[8,14,15]. Interestingly, this accumulation of taurine strictly depends upon the severity of congestive heart failure[14].

Since taurine exerts multiple actions on the heart[2,17], it has been proposed that taurine accumulation may influence the progression of heart failure. Although a definitive role for taurine in the development of structural and physiological remodeling has not been established, there seems little doubt that taurine therapy improves the status of congestive heart failure. The first report describing the benefit of taurine therapy in heart failure utilized the cardiomyopathic hamster model of heart failure[12]. The major effect of taurine treatment in this genetic model was a decline in intracellular calcium levels, resulting in a reduction in the number of myocardial lesions. In another study testing the effects of taurine therapy, Takihara et al.[20] induced congestive heart failure in rabbits by experimentally producing aortic regurgitation. Although this procedure of inducing heart failure is very different from the genetic hamster model, taurine therapy was equally effective, as evidenced by the significant increase in the survival rate of the rabbits. Based on the favorable response to taurine therapy in these animal studies, Azuma et al.[1] conducted a clinical trial examining the efficacy of taurine as a therapeutic agent in patients suffering from congestive heart failure. Significantly, they found that taurine was superior to placebo in managing congestive heart failure, even among patients receiving alternative therapy, such as digitalis glycosides and diuretics.

The mechanism underlying the beneficial effect of taurine therapy in congestive heart failure is at the present time poorly understood. This paucity of information on taurine action led us to test a new hypothesis; namely, that taurine acts by inhibiting the actions of angiotensin II, a neurohumoral agent that plays a central role in the development of congestive heart failure. The rationale behind this hypothesis was two-fold. First, since taurine inhibits the actions of several $G_q$-coupled agents[3,6], it might be expected to inhibit the $G_q$-linked reactions of angiotensin II. Second, taurine mimics some of the actions of the ACE inhibitors, whose main function is to block the production of angiotensin II[2,17]. Thus, in the present study we began testing the validity of the hypothesis by examining the effects of taurine depletion on angiotensin II-induced changes in contractile function and ion transport.

## MATERIALS AND METHODS

### Procedure for Producing Taurine Depleted Hearts

Male Wistar rats (240-280 g) were maintained for 4 weeks on water containing either 0 or 3% ß-alanine. Taurine content of hearts obtained from ß-alanine treated and untreated rats was 51 $\pm$ 3 wt and 106 $\pm$ 4 µmol/g dry wt, respectively. Hearts from control and taurine depleted rats were used in the preparation of isolated membrane or in isolated heart experiments.

### Measurement of $Na^+$-$Ca^{2+}$ Exchanger Activity

Isolated sarcolemma was prepared according to the method of Pitts[12]. To assay for $Na^+$-$Ca^{2+}$ exchanger activity, the vesicles from taurine depleted and normal hearts were loaded with buffer containing 160 mM NaCl, 20 mM MOPS, 100 µM Gpp(NH)p and 1 mM $MgCl_2$. Following a preincubation period at 37°C, either 0 or 2 nM angiotensin II was added to the medium. The $Na^+$-$Ca^{2+}$ exchange reaction was initiated after 10 minutes of angiotensin II exposure. To assay $Na^+$-$Ca^{2+}$ exchanger activity, the membranes were added to a phosphate buffer containing 160 mM KCl, 20 mM MOPS, 40 μM $^{45}CaCl_2$ and 5 μM valinomycin. The reaction was terminated after 2-5 sec by the addition of 3 ml of ice cold MOPS buffer containing 160 mM KCl and 1 mM $LaCl_3$, followed by rapid filtration. The filters were washed 5 times with the $LaCl_3$ containing buffer before being dried and counted.

## Measurement of Hemodynamic Parameters using the Isolated Perfused Heart

Hearts from 240-280 g male Wistar rats were perfused with Krebs-Henseleit buffer (37°C) supplemented with 10 mM glucose and 2.0 mM calcium. Following a 15 min stabilization period, some hearts were titrated with increasing concentrations of angiotensin II. Prior to exposure to a new angiotensin II concentration, the hearts were allowed to stabilize for at least 5 minutes. Left ventricular pressure, the rates of left ventricular development and decay, and heart rate were monitored before and after the addition of each concentration of angiotensin II. The hemodynamic parameters were measured with a Statham p23Gb pressure transducer by inserting a 22 gauge needle through the wall of the ventricle.

## Preparation and Culturing of the Neonatal Rat Cardiomyocyte

Neonatal cardiomyocytes were isolated by enzymatic digestion using the procedure described previously[13]. The cells were suspended in minimal essential medium containing 10% newborn calf serum and 0.1 mM 5-bromo 2'-deoxyuridine. The cells were allowed to plate on dishes precoated with 0.1% gelatin. They were then placed in serum free medium containing 56 U/l insulin, 10 μg/ml transferrin, 30 nM selenium, 0.25 mM ascorbate and supplemented with either 0 or 5 mM ß-alanine.

In some experiments, the cells were incubated for 45 minutes with buffer containing various concentrations of $^{125}$I-angiotensin II. The binding reaction was terminated by aspiration of the medium and rinsing with ice cold incubation buffer. The cells were removed from the dishes by the addition of 0.5 ml of 0.25 N NaOH and 0.5 % sodium dodecyl sulfate. After a 10 minute incubation, an aliquot was removed and counted. Angiotensin II binding was expressed as fmol/mg protein.

## Determination of $pH_i$ and $[Ca^{2+}]_i$

To measure $[Ca^{2+}]_i$, the cells were first loaded with indo-1 using an incubation medium containing 2.5 μM indo-1-AM. After a 20 minute incubation at 37°C, the cells were washed 3 times with indo 1 free medium and finally placed in normal serum free medium. Fluorescence at wavelengths 410 nm and 490 nm were monitored using a confocal microscope. A calcium transient was generated during each spontaneous depolarization and contraction. The data were expressed as the fluorescence ratio, $F_{410}/F_{490}$, obtained from the fluorescence at 410 nm and 490 nm.

Intracellular pH ($pH_i$) was measured in cells loaded with the pH probe, carboxy-seminaphthorhodafluor (C-SNARF-1). Cells were incubated for 15 minutes at 37°C with buffer containing 2 μM C-SNARF-1/AM and then washed 3 times with C-SNARF-1 free medium. After being placed in normal serum free medium, 7 mM 2,3 butanedione monoxime was added to inhibit contractile function. Fluorescence at wavelengths 580 nm and 640 nm were obtained using a confocal microscope. Intracellular pH was calculated from the fluorescence ratio, $F_{580}/F_{640}$, and a standard curve generated using the procedure of Blank et al.[4].

## RESULTS

Angiotensin II has previously been shown to mediate a modest positive inotropic effect in isolated adult rat hearts perfused with buffer containing 0.6 mM calcium and paced at 300 beats/minute[3]. However, in the unpaced rat heart perfused with buffer containing 2.5 mM calcium, the major effect of angiotensin II was to elevate heart rate without significantly affecting pressure development (Fig. 1A). Interestingly, the effect of angiotensin II on heart rate was less pronounced in hearts made taurine deficient using ß-alanine. Another major angiotensin-induced difference between the unpaced, taurine deficient and the unpaced, normal heart was noted in the relaxation phase of the cardiac cycle. Although relaxation was significantly prolonged in the taurine deficient heart, this defect was eliminated by the addition of angiotensin II to the perfusion medium II (Fig. 1B). By contrast, angiotensin II did not alter relaxation in the normal heart (data not shown).

Since many of the known myocardial actions of taurine and angiotensin II are caused by alterations in calcium transport[11,17], we examined the effect of angiotensin II on $[Ca^{2+}]_i$ of control and taurine depleted neonatal rat myocytes. We found that the major effect of taurine depletion on $[Ca^{2+}]_i$ was a prolongation in the decay phase of the calcium transient (Fig. 2). Taurine deficiency also mediated a modest decrease in the amplitude of the calcium transient, but it had no effect on the initial upstroke of the calcium transient. Exposure of the taurine deficient myocyte to 100 nM angiotensin II shortened the calcium transient, largely eliminating differences between the relaxation properties of the taurine deficient and normal cell (Table 1). Although angiotensin II had no effect on calcium transient duration of the normal myocyte, it significantly increased the amplitude of the calcium transient. Yet, the change in amplitude was larger in the taurine deficient myocyte than the normal cell (Table 1).

The observation that only some of the actions of angiotensin II were greater in the taurine deficient myocardium than the control heart suggests that the interaction between taurine deficiency and angiotensin II probably does not reside with the $AT_1$ receptor. Nonetheless, to confirm this notion, the binding of $^{125}$I-angiotensin II to taurine deficient and control rat neonatal myocytes was examined. Two sets of angiotensin II binding sites were observed in the two cell types, with the higher affinity site, the $AT_1$ receptor, exhibit-

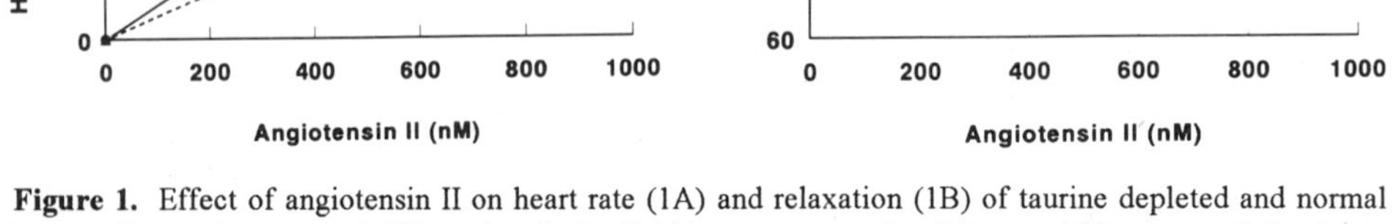


**Figure 1.** Effect of angiotensin II on heart rate (1A) and relaxation (1B) of taurine depleted and normal hearts. Hearts from normal (●) taurine depleted (▲) rats were perfused for a stabilization period prior to being exposed to varying concentrations of angiotensin II. Values shown represent the means ± S.E.M. of 4-5 hearts. Asterisks denote significant differences between the taurine depleted and normal hearts ($p < 0.05$).

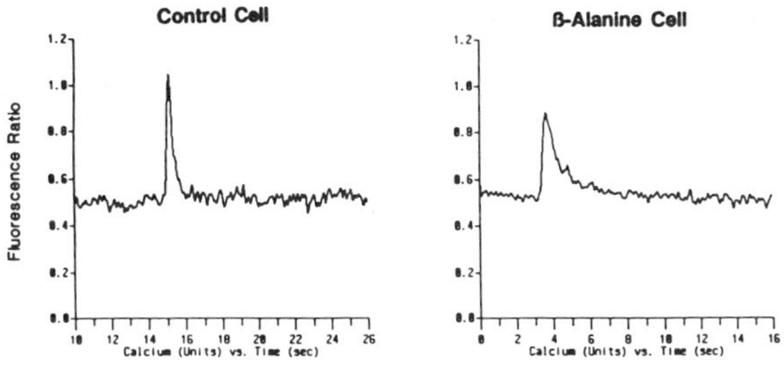


**Figure 2.** Effect of taurine depletion on $Ca^{2+}$ transients of isolated myocytes. Isolated neonatal rat myocytes were incubated for 3 days in medium containing either no addition (Control cell) or 5 mM ß-alanine (ß-alanine cell). The cells were then loaded with indo-1 and calcium transients measured as a ratio of fluorescence at two wavelengths ($F_{410}/F_{490}$).

**Table 1.** Effect of Taurine Depletion and Angiotensin II on Calcium Transients of Isolated Myocytes.

| Condition | Systolic $Ca^{2+}$ | Diastolic $Ca^{2+}$ | Peak Amplitude | $PR_{90}$ |
|---|---|---|---|---|
| Control | $1.04 \pm 0.02$ | $0.51 \pm 0.03$ | $0.54 \pm 0.05$ | $0.7 \pm .07$ |
| Control + Ang II | $1.15 \pm 0.05^{\#}$ | $0.53 \pm 0.02$ | $0.61 \pm 0.06$ | $1.1 \pm .31$ |
| Taurine depleted | $0.90 \pm 0.06^{*}$ | $0.51 \pm 0.03$ | $0.40 \pm 0.05^{*}$ | $1.8 \pm .29^{*}$ |
| Taurine depleted + Ang II | $1.11 \pm 0.05^{\#}$ | $0.56 \pm 0.08$ | $0.56 \pm 0.02$ | $1.2 \pm .13^{\#}$ |

Systolic $Ca^{2+}$ correspond to peak $[Ca^{2+}]_i$, diastolic $Ca^{2+}$ represents basal $[Ca^{2+}]_i$ in the resting myocyte, and peak amplitude refers to the difference between systolic and diastolic $Ca^{2+}$. All $Ca^{2+}$ values are expressed as the fluorescence ratio ($F_{410}/F_{490}$). $PR_{90}$ refers to the time in seconds necessary for the calcium transient to decay from the peak to 10% of peak amplitude. Values shown represent means ± S.E.M. of 4-5 separate myocyte preparations. Asterisks denote significant differences between the taurine depleted and control cells ($p < 0.05$). The pound symbol denotes a significant difference between the angiotensin II treated cells and the untreated cells ($p < 0.05$).

ing a $K_d$ of approximately 0.3 nM in both the taurine deficient and normal cell. The number of the high affinity binding sites on each myocyte was approximately 23 fmol/mg protein for the two cell types. Thus, taurine deficiency had no effect on either the $K_d$ for angiotensin II binding or the number of high affinity angiotensin II binding sites located on the cell surface.

While numerous studies have explored the basis underlying angiotensin II-mediated alterations in the amplitude of the calcium transient[9-11], the shortening of the decay phase of the calcium transient by angiotensin II has remained unexplored. Therefore, we examined the effect of taurine deficiency on the activity of the sarcolemmal $Na^+$-$Ca^{2+}$ exchanger, a major determinant of the decay phase of the calcium transient. Figure 3 shows that while $Na^+$-$Ca^{2+}$ exchanger activity was significantly depressed in the taurine depleted heart, angiotensin II dramatically increased the activity of the $Na^+$-$Ca^{2+}$ exchanger.

Another mechanism by which angiotensin II might influence contractile function is through the modulation of myofilament calcium sensitivity. Several inotropic agents have been shown to alter the contractile state of the heart by affecting $pH_i$[9,10]. Moreover, metabolic modulation may influence contractile function in part by changing $pH_i$[19]. Thus, because of the close association between $pH_i$ and contractile function, we examined the effect of taurine depletion on angiotensin II-induced changes in $pH_i$. Prior to angiotensin II exposure, the $pH_i$ of the nonbeating, control and taurine deficient cell was 7.09 $\pm$ 0.01 and 7.22 $\pm$ 0.02, respectively (Fig. 4). Addition of 100 nM angiotensin II to the medium caused a 0.05 $pH_i$ unit decrease in the control myocyte but an insignificant 0.01 $pH_i$ unit decline in the taurine deficient cell (Fig. 4). Thus, exposure of the cell to 100 nM angiotensin II widens the difference in $pH_i$ between the control and taurine deficient myocyte.

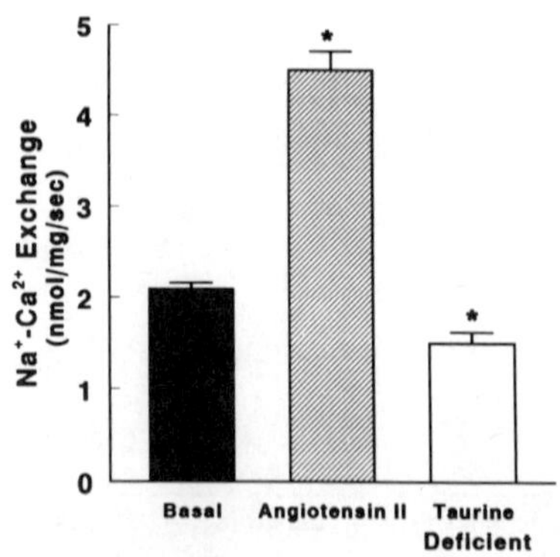


**Figure 3.** Effect of taurine depletion and angiotensin II on sarcolemmal $Na^+$-$Ca^{2+}$ exchanger activity. Isolated sarcolemma prepared from taurine depleted and normal rats were assayed for $Na^+$-$Ca^{2+}$ exchanger activity following exposure to 2 nM angiotensin II. Values shown represent the means $\pm$ S.E.M. of 4-5 preparations. Asterisks denote significant differences from the basal group ($p < 0.05$).

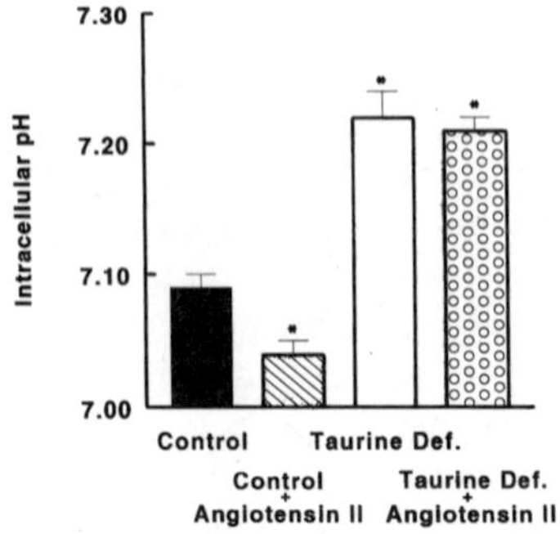


**Figure 4.** Effect of taurine depletion and angiotensin II on $pH_i$ of isolated myocytes. Neonatal rat myocytes were incubated for 3 days in medium containing (Taurine def.) or lacking (Control) 5 mM ß-alanine. The cells were loaded with C-SNARF-1 and $pH_i$ was determined from the fluorescence ratio, $F_{580}/F_{640}$, before and after addition of 100 nM angiotensin II. Values shown represent the means $\pm$ S.E.M. of 19 cells. Asterisks denote significant differences from the control cells ($p < 0.05$).

## DISCUSSION

The present study supports the hypothesis that taurine deficiency modulates the actions of angiotensin II. However, taurine deficiency does not affect all angiotensin II actions equally, suggesting that taurine does not act at an initial step in the signal transduction pathway of angiotensin II. This conclusion is supported by the observation that taurine deficiency does not alter the $K_d$ and $B_{max}$ of the high affinity ($AT_1$) receptor.

The present study further suggests that the effects of taurine deficiency on angiotensin II action are probably related to changes in cation transport. Figure 2 shows that myocytes made taurine deficient by treatment with 5 mM ß-alanine experience two major changes in its spontaneously-induced calcium transients: (1) a reduction in the amplitude and (2) a prolongation in the decay process. Significantly, the lengthening of the calcium transient in the taurine deficient myocyte correlates with the prolongation of the contraction cycle (Figure 1), implying that impaired calcium movement accounts for the relaxation defect of the taurine deficient heart. The present study further suggests that the relaxation defect is caused in part by depressed $Na^+$-$Ca^{2+}$ exchanger activity (Fig. 3).

In contrast to the relaxation phase of the cardiac cycle, taurine deficiency had no effect on either peak left ventricular pressure or maximal cardiac work. Nonetheless, myocytes treated with 5 mM ß-alanine exhibited a 15% reduction in systolic $[Ca^{2+}]_i$. Since this change in $[Ca^{2+}]_i$ is large enough to produce a significant decline in contractile function, some other change must have occurred within the cell to prevent the negative inotropic effect. The most probable compensatory mechanism involves a rise in $pH_i$, which increases the calcium sensitivity of the myofibrils, thereby elevating contractile function[9].

Changes in $pH_i$ may also play a role in some of the actions of angiotensin II . In agreement with a previous study by Kohmoto et al.[10], we observed a reduction in $pH_i$ following exposure of the neonatal myocyte to angiotensin II. Although the basis underlying this decline in $pH_i$ has not been definitively established, it is interesting that the decline is eliminated by removal of glucose from the incubation medium, suggesting that anaerobic metabolism may be responsible for the angiotensin II-mediated change in $pH_i$[10]. Consistent with this conclusion is our observation that termination of cellular contraction, which dramatically reduces anaerobic metabolism, also reduces the extent of the angiotensin II-induced $pH_i$ decline (data not shown). In this regard, it is also significant that the $pH_i$ change induced by angiotensin II is attenuated in the taurine deficient cell (Fig. 4). This may be related to the fact that both taurine deficiency and angiotensin II promote glucose uptake and metabolism[5,13]. Since the rate of glucose metabolism is already very high in the taurine deficient myocyte, the ability of angiotensin II to further increase glucose metabolism in the taurine deficient myocyte is restricted.

Another important effect of angiotensin II is the activation of the $Na^+$-$Ca^{2+}$ exchanger. This activity is important because the $Na^+$-$Ca^{2+}$ exchanger functions to extrude calcium from the cell during diastole. Because the taurine deficient cell exhibits reduced $Na^+$-$Ca^{2+}$ exchanger activity, angiotensin II, by activating the exchanger, shortens the calcium transient and accelerates the relaxation process in the taurine deficient myocyte. Thus, although angiotensin II widens the difference in $pH_i$ between the taurine deficient and normal cell, it lessens the differences between the calcium transients of the two cell types.

## ACKNOWLEDGMENTS

This work was supported in part by a grant from the Taisho Pharmaceutical Co., Ltd.

## REFERENCES


1. Azuma, J., Sawamura, A., Awata, N., Hasegawa, H., Ogura, K., Harada, H., Ohta, H., Yamauchi, K., Kishimoto, S., Yamagami, T., Ueda, E., and Ishiyama, T., 1983, Double-blind randomized crossover trial of taurine in congestive heart failure, *Curr. Therapeut. Res.*, 34:543-557.
2 Azuma, J. and Schaffer, S.W., 1995, Protective effect of taurine on the failing heart and its possible mechanism of action, in: "The Failing Heart", Dhalla, N.S., Beamish, R.E., Takeda, N. and Nagano, M., eds., Lippincott-Raven Publishers, Philadelphia, pp. 421-431.
3. Ballard-Croft, C., Mozaffari, M.S., Azuma, J., and Schaffer, S., Interaction between taurine and angiotensin II: Modulation of calcium transport and myocardial contractile function, *Amino Acids*, in press.
4. Blank, P.S., Silverman, H.S., Chung, O.Y., Hogue, B.A., Stern, M.D., Hansford, R.G., Lakatta, E.G., and Capogrossi, M.C., 1992, Cytosolic pH measurements in single cardiac myocytes using carboxy-seminaphthorhodafluor-1, *Am. J. Physiol.*, 263:H276-H284.
5. DeTata, V., Del Roso, A., Gori, Z., and Bergamini, E., 1990, Effects of the administration of angiotensin II on cardiac glycogen metabolism in the rat, *Eur. J. Physiol.*, 416:673-678.
6. Failli, P., Fazzini, A., Franconi, F., Stendardi, I., and Giotti, A., 1992, Taurine antagonizes the increase in intracellular calcium concentration induced by α-adrenergic stimulation in freshly isolated guinea-pig cardiomyocytes, *J. Mol. Cell. Cardiol.*, 24:1253-1265.
7. Gerdes, A.M. and Capasso, J.M., 1995, Structural remodeling and mechanical dysfunction of cardiac myocytes in heart failure, *J. Mol. Cell. Cardiol.*, 27:849-856.
8. Huxtable, R. and Bressler, R., 1974, Elevation of taurine in human heart failure, *Life Sci.*, 14:1353-1359.
9. Ikenouchi, H., Barry, W.H., Bridge, J.H.B., Weinberg, E.O., Apstein, C.S., and Lorell, B.H., 1994, Effects of angiotensin II on intracellular $Ca^{2+}$ and pH in isolated beating rabbit hearts and myocytes loaded with the indicator indo-1, *J. Physiol.*, 480:203-215.
10. Kohmoto, O., Ikenouchi, H., Hirata, Y., Shin-Ichi, M., Serizawa, T., and Barry, W.H., 1993, Variable effects of endothelin-1 on $[Ca^{2+}]_i$ transients, $pH_i$, and contraction in ventricular myocytes, *Am. J. Physiol.*, 265:H793-H800.
11. Lindpaintner, K. and Ganten, D., 1991, The cardiac renin-angiotensin system, *Circ. Res.*, 68:905-921.
12. McBroom, M.J. and Welty, J.D., 1977, Effects of taurine on heart calcium in the cardiomyopathic hamster, *J. Mol. Cell. Cardiol.*, 9:853-858.
13. Mozaffari, M.S., Tan, B.H., Lucia, M.A., and Schaffer, S.W., 1986, Effect of drug-induced taurine depletion on cardiac contractility and metabolism, *Biochem. Pharmacol.*, 35:985-989.
14. Newman, W.H., Frangakis, C.J., Grosso, D.S., and Bressler, R., 1977, A relation between myocardial taurine content and pulmonary wedge pressure in dogs with heart failure, *Physiol. Chem. Physics*, 9:259-263.
15. Peterson, M.B., Mead, R.J., and Welty, J.D., 1973, Free amino acids in congestive heart failure, *J. Mol. Cell. Cardiol.*, 5:139-147.
16. Pitts, B.J.R., 1979, Stoichiometry of sodium-calcium exchange in cardiac sarcolemmal vesicles. *J. Biol. Chem.*, 254:6232-6235.
17. Schaffer, S.W. and Azuma, J., 1992, Physiological effects of taurine in the heart, in: "Taurine: nutritional value and mechanisms of action", Lombardini, J.B., Schaffer, S.W. and Azuma, J., eds., Plenum Press, New York, pp. 105-120.
18. Schaffer, S.W., Ballard-Croft, C., Boerth, S., and Allo, S.N., 1997, Mechanisms underlying depressed $Na^+/Ca^{2+}$ exchanger activity in the diabetic heart, *Cardiovasc. Res.*, 34:129-136.
19. Schaffer, S.W., Safer, B., Ford, C., Illingworth, J., and Williamson, J.R., 1978, Respiratory acidosis and its reversibility in perfused rat heart: regulation of citric acid cycle activity, *Am. J. Physiol.*, 234:H40-H51.
20. Takihara, K., Azuma, J., Awata, N., Ohta, H., Hamaguchi, T., Sawamura, A., Tanaka, Y., Kishomoto, S., and Sperelakis, N., 1986, Beneficial effect of taurine in rabbits with chronic congestive heart failure, *Am. Heart J.*, 112:1278-1284.

# Part 4:
# Antioxidant and Anti-Ischemic Actions of Taurine

# TAURINE INHIBITION OF IRON-STIMULATED CATECHOLAMINE OXIDATION

Ralph Dawson, Jr., Elisa Tang, Debbie Shih, Hunter Hern, Ming Hu, Deron Baker, and Baerbel Eppler

Department of Pharmacodynamics
College of Pharmacy
JHMHC Box 100487
University of Florida
Gainesville, FL 32610

## INTRODUCTION

Taurine (2-aminoethane sulfonic acid) is a sulfur containing amino acid that is present in mammalian tissues in millimolar concentrations[10]. Taurine is involved in a diverse array of biological functions which include: osmoregulation, membrane stabilization, neuromodulation, bile salt conjugation, and calcium modulation[10]. A role for taurine in cellular antioxidant defense mechanisms has been observed under a number of conditions[3,17,21,25]. Although taurine appears to have little intrinsic reactivity with oxidizing species[1], its high concentration may still allow it to directly scavenge free radicals. Furthermore, taurine may bind reactive quinones involved in redox cycling and indirectly inhibit free radical formation[11,22]. Taurine could also moderate the effects of free radical damage by blunting the ability of prooxidants to increase intracellular free calcium[10]. The biosynthesis of taurine generates intermediates (hypotaurine and cysteamine) that also have free radical scavenging properties[1,10,25]. The synthesis of taurine also utilizes cysteine which is an excitatory amino acid that undergoes autoxidation reactions[13]. Taurine is present in tissues that contain high concentrations of catecholamines which are known to be cytotoxic and increase oxidative stress[8,9,16,20]. Taurine content declines in brain and peripheral tissues in aged rodents[4]. The age-related decline in tissue taurine content could increase the susceptibility of cells to free radical-mediated damage. Catecholamines are known to undergo autoxidation reactions to generate free radicals ($H_2O_2$, hydroxyl, superoxide) and cytotoxic quinones[8,11,22]. The oxidation of catecholamines is catalyzed by metals such as iron[13]. Iron-stimulated autoxidation of catecholamines is thought to play a role in neurodegenerative diseases[26]. One study has reported that chronic treatment with l-dopa caused a 13% decline in brain taurine content[23]. At present, it is unclear whether taurine under *in vivo* conditions serves as a scavenger for reactive quinones and free radicals derived from catecholamine autoxidation.

The goals of the present series of experiments was to determine: 1) could taurine directly inhibit catecholamine autoxidation, 2) could taurine prevent lipid peroxidation or protein oxidation by catecholamine autoxidation products, and 3) would *in vivo* assays confirm *in vitro* findings of taurine's antioxidant properties.

## MATERIALS AND METHODS

### Measurement of Catecholamine Autoxidation

The rate of catecholamine autoxidation was monitored by measuring the increase in aminochrome formation[8]. Aminochromes (quinoidal oxidation products) were detected at 490 nm using a Beckman DU 7000 diode array spectrophotometer. Experiments were conducted in phosphate buffered saline (PBS, pH=7.4) at ambient temperatures (21-24° C) except for some experiments conducted at 37° C. The autoxidation reactions were conducted in an assay volume of 3 ml and the rate of quinone formation was determined by measuring absorbance changes every 20 sec for a total of 5 min. The rate of quinone formation was linear with time and reaction rates were calculated by linear regression. Catecholamine oxidation rates were measured in the presence and absence of iron (ferric and ferrous forms) and the effects of taurine and other amino acids were assessed for their ability to inhibit catecholamine autoxidation. Specific experimental conditions and reagent concentrations are given in figure captions and tables.

### Protein Carbonyl Measurements

Oxidative damage to proteins was measured by determining carbonyl content in oxidatively modified proteins by the method of Levine et al.[12]. Carbonyl content was assessed by reaction of protein samples with 2,4-dinitrophenylhydrazine and detection of the derivatives spectrophotometrically at 370 nm. Bovine γ-globulin (BGG, 1 mg) was oxidatively damaged using either hydrogen peroxide (0.15%, 44 mM) or ferric chloride (250 μM) and l-dopa (250 μM). The assays were conducted in total volume of 1 ml and oxidation reactions were carried for either 30 min or 60 min in a shaking water bath at 37°C. The oxidation reaction was stopped by precipitating the protein with 10% trichloroacetic acid and the iron was chelated by adding desferroxamine (50 μM). Protein carbonyl content was determined in the precipitated proteins. The effects of 10-20 mM taurine to prevent oxidative damage was assessed and glutathione (10 mM) was used as a positive control in one experiment.

Total tissue content of protein carbonyls was determined in whole kidney homogenates from male Fischer 344 rats fed normal diets, diets free of taurine or diets supplemented with 1.5% taurine in the drinking water[6]. Kidney tissue was homogenized in 50 volumes (w/v) 100 mM phosphate buffer containing 0.1 mM $Na_2EDTA$, aprotinin (0.2 mg/L), and phenylmethanesulfphonyl fluoride (230 μM). Precipitation of nucleic acids in the kidney homogenates with 1% streptomycin sulfate was performed to remove nucleic acids containing reactive carbonyl groups[11]. Tissue carbonyl content was assayed by the method of Levine et al.[12].

### Lipid Peroxidation

Lipid peroxidation was estimated using the thiobarbituric acid test (TBARS). TBARS were assayed by a modification of the method described by Wills[24]. Malondialdehyde (TBAR) formation was quantitated using 1,1,3,3-tetraethoxypropane as a standard. The absorbance of TBARs was read at 530 nm in a Beckman spectrophotometer. The ability of taurine (1-20 mM) to directly inhibit ferric chloride-stimulated lipid peroxidation was examined in adult Sprague-Dawley cerebral cortex. Rat cortex was homogenized in ice cold saline (pH=7.0) and incubated (200 mg protein/tube) in the presence and absence of either ferric chloride (200 μM) and/or taurine (1-20 mM) for 15 min at 37°C in a shaking water bath. The reaction was stopped by addition of 1 ml of a solution containing 0.8 M HCl, 12.5% TCA and O.1 mM desferrioxamine. The samples were then assayed for TBAR formation in duplicate. Kidney and adrenal tissue from the Fischer 344 rats described above were also assayed for basal levels of lipid peroxidation.

## Renal Cell Culture and the Effects of Iron-stimulated Catecholamine Oxidation on Cell Viability

LLC-$PK_1$ cells are a porcine-derived renal epithelial cell line with characteristics of proximal tubular cells. LLC-$PK_1$ cells have a high capacity for dopamine synthesis and contain dopamine transporters and receptors[5]. LLC-$PK_1$ cells were grown and maintained as previously described[5]. The cells were treated with saline containing 75 μM ferric chloride and 250 μM l-dopa for 60 min and then the cells were washed and fresh culture media (DMEM) was added. The cells were allowed 24 hours to recover and cytotoxicity (viability) was determined using an assay based on tetrazolium salt (MTT) reduction[15]. In addition to the MTT assay, total protein and DNA was measured to confirm the MTT results. The ability of 10 and 20 mM taurine to attenuate cytotoxicity was evaluated.

## RESULTS AND DISCUSSION

Ferric chloride stimulated l-dopa autoxidation in a concentration dependent manner and 20 mM taurine significantly ($p < 0.01$) inhibited quinone formation at all ferric chloride concentrations tested (Fig. 1). Ferrous sulfate stimulated l-dopa autoxidation, but only about 33% of the rate seen with ferric chloride (data not shown). Addition of ascorbic acid (0.1 mM) to ferrous sulfate increased l-dopa autoxidation slightly, but not to the level seen with ferric chloride (data not shown). Taurine did not attenuate l-dopa autoxidation stimulated by the ferrous form of iron in the presence or absence of ascorbic acid. The ferric form of iron is preferentially bound by the ortho hydroquinone moiety of catechol-

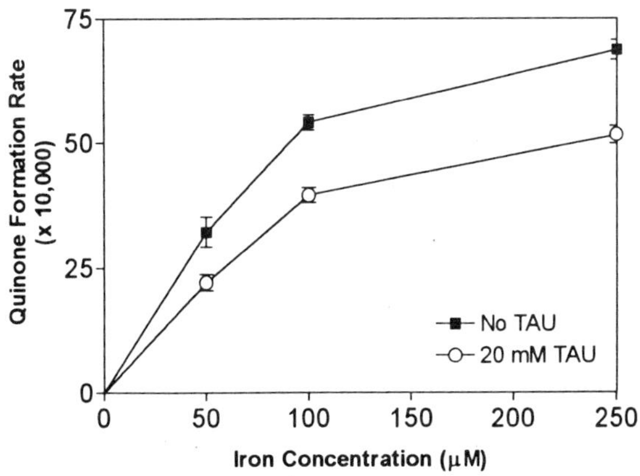


**Figure 1.** Ferric chloride stimulated 1-dopa oxidation. n=6 determinations. p<0.01 no TAU versus 20 mM TAU for all iron concentrations tested.

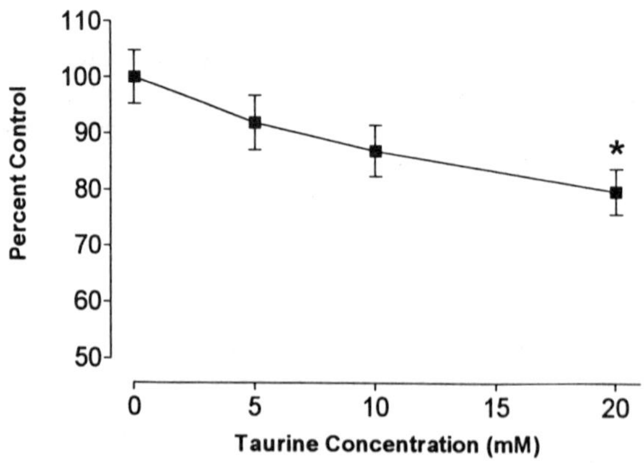


**Figure 2.** Taurine inhibition of ferric chloride (500 μM) stimulated l-dopa (500 μM) oxidation. Data are based on 5-15 determinations. *$p<0.01$ versus control.

amines, which leads to the reduction of $Fe^{3+}$ to $Fe^{2+}$ with concomitant oxidation of the catecholamine[13]. The lower oxidation rates of l-dopa seen with the ferrous form of iron may preclude the detection of the antioxidant properties of taurine using the quinone formation assay. Taurine caused a concentration related decrease in quinone formation (Fig. 2). Taurine does not appear to inhibit the oxidation of catecholamines due to metal chelation since the stability constant for metal complexation by taurine is neglible[25]. Taurine was also effective across a range of ferric chloride concentrations (Fig. 1). In unpublished studies, we have found that taurine inhibits l-dopa oxidation in borate buffer (pH=10), but was less effective in 50 mM Tris buffer (pH=7.0) perhaps due to the metal chelating and free radical scavenging properties of Tris buffers[13].

Hypotaurine, cystine, glutathione, and ascorbic acid inhibited iron-stimulated l-dopa oxidation (Table 1). β-Alanine had no effect on l-dopa oxidation, but the taurine precursors, methionine and cysteine stimulated l-dopa oxidation (Table 1). Lysine, which has been reported to scavenge quinones[11], had only a modest effect on iron-stimulated quinone formation. A final experiment was performed to examine taurine-induced inhibition of l-dopa oxidation over a longer time course and to determine the λmax for the l-dopa oxidation products formed. The iron-stimulated (250 μM ferric chloride) oxidation of l-dopa (500 μM) was followed for 15 minutes at 37°C in the presence and absence of tau-

**Table 1.** The Effect of Amino Acids on Iron-Stimulated L-dopa Oxidation

| Test Compound | Number of Replications | % Control |
|---|---|---|
| Taurine (20 mM) | 5 | 75 ± 3* |
| Taurine (20 mM)† | 12 | 83 ± 3* |
| Hypotaurine (20 mM) | 14 | 87 ± 3* |
| β-Alanine (20 mM) | 15 | 96 ± 6 ns |
| Cystine (20 mM) | 5 | 2 ± 2* |
| Glutathione (20 mM)† | 10 | 4 ± 3* |
| Cysteine (20 mM) | 9 | 222 ± 18* |
| Methionine (20 mM) | 9 | 121 ± 5* |
| Lysine (20 mM)† | 12 | 89 ± 4 ns |
| Ascorbic Acid (0.5 mM) | 5 | 77 ± 3* |

Ferric chloride (500 μM)-stimulated l-dopa (500 μM) oxidation. † Ferric chloride was 250 μM for these experiments.*$p< 0.05$ versus control, ns=not significant.

rine and a wavelength scan (200-800 nm) was conducted at 0, 5 and 15 minutes after iron addition. The presence of taurine during the 15 minute oxidation period caused a 38% decrease in the absorbance values for l-dopa oxidation products at 490 nm and the λmax absorbance at 550 nm. These experiments suggest that taurine can inhibit quinone formation from oxidized catecholamines despite its slow rate constant for reactions with free radicals[1] and reactive quinones[22].

Taurine protected BGG from the oxidative damage caused by either 30 or 60 minute exposures to ferric chloride (250 μM) and l-dopa (250 μM). Taurine (20 mM) significantly ($p<0.05$) inhibited protein carbonyl formation 47 ± 3% (n=6) after 30 min of exposure, but 10 mM taurine only reduced the damage by 6% (data not shown). Taurine (10 or 20 mM) or glutathione (10 mM) significantly decreased the carbonyl content of BBG treated with l-dopa and iron for 60 min (Fig. 3). Taurine did not inhibit protein carbonyl formation induced by $H_2O_2$ which was more than 2-fold greater than the damage produced by iron and l-dopa (data not shown).

Dietary taurine supplementation significantly reduced protein carbonyl content in aged rats (n=8) when compared to aged rats on normal diets (n=8) (Fig. 4). Aged rats (28 months old) had significantly ($p<0.05$) lower renal taurine content (8.5 ± 0.5 μmoles/g) than 10 month old adult rats (11.1 ± 0.4 μmoles/g). Renal taurine content was increased nearly two-fold in aged rats supplemented with 1.5% taurine in the drinking water. The aged rats were on the experimental diets for 10 months and taurine supplementation

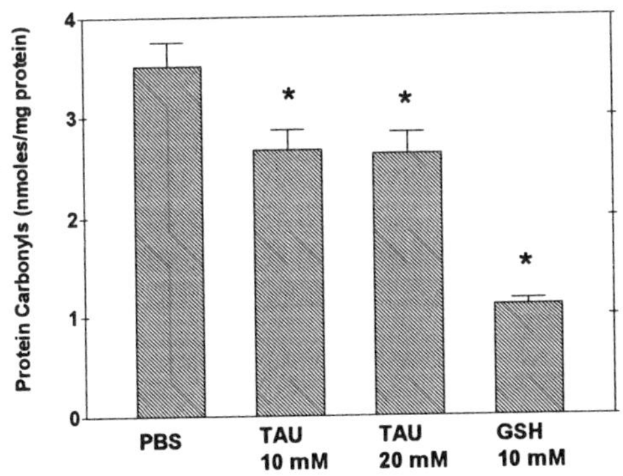


**Figure 3.** Taurine inhibition of protein carbonyl formation induced by l-dopa and ferric chloride. n=5-6 determinations. *p<0.05 versus PBS.

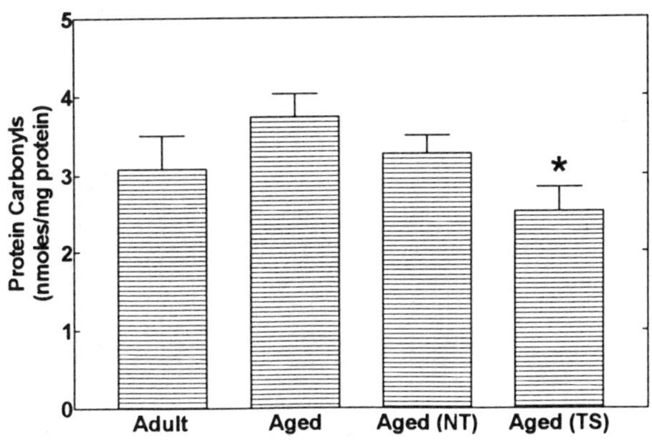


**Figure 4.** Protein carbonyl content in kidneys from adult and aged F322 rats on normal diets, no taurine diets (NT) or taurine supplemented diets (TS). *p<0.05 vs aged rats on normal diets.

reduced total protein carbonyl content. Previous studies have shown that free radical scavenging agents can reduce the age-related increases in tissue carbonyl content[2].

TBARs in the kidneys of taurine supplemented (TS) rats were 15% lower than aged rats on normal diets, but this was not statistically significant (data not shown). Adrenal TBAR levels in aged F344 rats were also not significantly reduced by taurine supplementation (data not presented). Taurine (1-20 mM) did not inhibit ferric chloride (400 μM) stimulated TBAR formation in rat cerebral cortex homogenates (data not shown). These data are consistent with a previous study that found taurine to be ineffective in inhibiting iron-stimulated lipid peroxidation[17].

Treatment of LLC-$PK_1$ cells with l-dopa (250 μM) and ferric chloride (75 μM) reduced cell viability about 50% (*p<0.01) and taurine (10 or 20 mM) significantly (+p<0.05) decreased cytotoxicity as indexed by the MTT assay (Figure 5). Similar results were found when the protein or DNA content of the cells were determined (data not shown). Taurine alone had no effect on cell viability, protein, or DNA content (data not shown). Ferric chloride stimulated l-dopa oxidation was determined under identical conditions as the cell culture experiments were conducted and taurine significantly (p<0.001) decreased quinone formation by 23% (10 mM) and 34% (20 mM) (data not shown).

Taurine was clearly shown to directly inhibit l-dopa oxidation that was catalyzed by ferric chloride. Furthermore, this inhibition of l-dopa oxidation resulted in decreased oxidative damage to protein (BGG) and a reduction in cytotoxicity in cultured renal epithelial cells. The high intracellular concentration and water solubility of taurine may function to protect vital cytosolic proteins from oxidative damage arising from metal-stimulated catecholamine oxidation. Metal-catalyzed oxidation of proteins have been implicated in aging[2,19] and l-dopa protein adducts can also serve as redox cycling systems to generate further free radicals[7]. A number of studies have shown that catecholamines (dopamine and l-dopa) can produce oxidative stress and cytotoxicity that are related to metal catalyzed autoxidation reactions which can be prevented by antioxidants[9,16,20]. Extensive oxidative damage to DNA can also be produced by l-dopa and its metabolites in the presence of trace levels of metals[18]. Preliminary studies in our laboratory have found that taurine can also inhibit oxidative damage to DNA by iron stimulated l-dopa oxidation.

Thus, the ability of taurine to inhibit metal-stimulated catecholamine oxidation could have therapeutic implications in diseases that are associated with catecholamine-induced oxidative stress and free radicals.

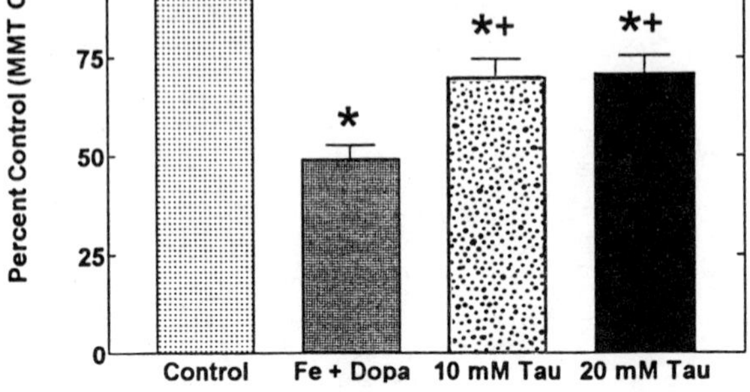


**Figure 5.** Taurine mediated attenuation of the cytotoxic effects of ferric chloride (75 μM) and 1-dopa (250 μM) on LLC-$PK_1$ cell viability. n=10-13 experiments. *p<0.01 versus control, †p<0.05 versus Fe + Dopa.

As previously suggested[10], the synthesis of taurine may have a general cytoprotective effect by converting reactive cysteine and methionine to the antioxidants hypotaurine and taurine. Cysteine can produce oxidative damage depending on cellular conditions[13,14]. Thiols interacting with metals can also catalyze oxidative protein modifications[19]. Furthermore, detoxification of free radicals and reactive intermediates by cysteine or glutathione may result in cytotoxic quinone thioether derivatives[14]. Detoxification of reactive species by taurine may be a biologically "safer" route that generates less reactive detoxification products. Further studies are needed to fully understand the mechanism of taurine-induced inhibition of ferric iron stimulated l-dopa oxidation. The conclusion from our studies is that taurine can directly inhibit the iron-mediated oxidation of l-dopa and related catecholamines and this can result in an attenuation in oxidatively damaged proteins and the cytotoxicity associated with l-dopa oxidation products.


## ACKNOWLEDGMENTS

This project was funded by a grant from Taisho Pharmaceutical Co. LTD.


## REFERENCES


1. Aruoma, O.I., Halliwell, B., Hoey, B.M., and Butler, J., 1988, The antioxidant action of taurine, hypotaurine and their metabolic precursors, *Biochem. J.*, 256:251-255.
2. Carney, J.M., Starke-Reed, P.E., Oliver, C.N., Landum, R.W., Cheng, M.S., Wu, J.F., and Floyd, R.A., 1991, Reversal of age-related increases in brain protein oxidation, decrease in enzyme activity, and loss in temporal and spatial memory by chronic administration of the spin-trapping compound N-tert-butyl-α-phenylnitrone, *Proc. Natl. Acad. Sci.*, 88:3633-3636.
3. Cozzi, R., Ricordy, R., Bartolini, Ramadori, L., Perticone, P., and De Salvia, R., 1995, Taurine and ellagic acid: two differently-acting natural antioxidants, *Environ. Mol. Mutagen.*, 26:248-254.
4. Dawson, R. and Wallace, D.R., 1992, Taurine content in tissues from aged Fischer 344 rats, *Age*, 15:73-81.
5. Dawson, R., Felheim, R. and Phillips, M.I., 1994, Characterization of the synthesis and release of dopamine in LLC-$PK_1$ cells, *Renal Physiol. Biochem.*, 17:85-100.
6. Dawson, R., Eppler, B., Patterson, T.A., Shih, D., and Liu, S., 1996, The effects of taurine in a rodent model of aging, in Adv. Exp. Med. Biol. "Taurine 2 Basic and Clinical Aspects", Huxtable, R.J., Azuma, J., Kuriyama, K., Nakagawa, M., and Baba, A., eds., Plenum Press, Vol. 403, pp. 37-50.
7. Dean, R.T., Gieseg, S., and Davies, R.T., 1993, Reactive species and their accumulation on radical damaged proteins, *Trends Biochem. Sci.*, 18:437-441.
8. Graham, D.G., Tiffany, S.M., Bell, W.R., and Gutknecht, W.F., 1978, Autoxidation versus covalent binding of quinones as the mechanism of toxicity of dopamine, 6-hydroxydopamine, and related compounds toward C1300 neuroblastoma cells *in vitro*, *Mol. Pharmacol.*, 14:644-653.
9. Han, S-K., Mytilineou, C., and Cohen, G., 1996, L-DOPA up-regulates glutathione and protects mesencephalic cultures against oxidative stress, *J. Neurochem.*, 66:501-510.
10. Huxtable, R.J., 1992, Physiological actions of taurine, *Physiol. Rev.*, 72:101-163.
11. Kalyanaraman, B., Premovic, P.I., and Sealy, R.C., 1987, Semiquinone anion radicals from addition of amino acids, peptides, and proteins to quinones derived from oxidation of catechols and catecholamines, *J. Biol. Chem.*, 262:11080-11087.
12. Levine, R.L, Garland, D., Oliver, C.N., Amici, A.A., Climent, I., Lenz, A., Ahn, B., Shaltiel, S., and Stadtman, E.R., 1990, Determination of carbonyl content in oxidatively modified proteins, *Meth. Enzymol.*, 186:464-478.
13. Miller, D.M., Buettner, G.R., and Aust, S.D., 1990, Transition metals as catalysts of "autoxidation" reactions, *Free Rad. Biol. Med.*, 8:95-108.
14. Monks, T.J. and Lau, S.S., 1992, Toxicology of quinone-thioethers, *Crit. Rev. Toxicol.*, 22:243-270.
15. Mosmann, T., 1983, Rapid colorimetric assay for cellular growth and survival: application to proliferation and cytotoxicity assays, *J. Immunol. Meth.*, 65:55-63.

16. Pardo, B., Mena, M.A., Casarejos, M.J., Paino, C.L., and De Yebenes, J.G., 1995, Toxic effects of l-dopa on mesencephalic cell cultures: protection with antioxidants, *Brain Res.*, 682:133-143.
17. Pasantes-Morales, H., Wright, C.E., and Gaull, G.E., 1985, Taurine protection of lymphoblastoid cells from iron-ascorbate induced damage, *Biochem. Pharmacol.*, 34:2205-2207.
18. Spencer, J.P.E., Jenner, A., Aruoma, O.I., Evans, P.J., Kaur, H., Dexter, D.T., Jenner, P., Lees, A.J., Marsden, D.C., and Halliwell, B., 1994, Intense oxidative DNA damage promoted by l-dopa and its metabolites implications for neurodegenerative disease, *FEBS Lett.*, 353:246-250.
19. Stadtman, E.R. and Oliver, C.N., 1991, Metal-catalyzed oxidation of proteins, *J. Biol. Chem.*, 266:2005-2008.
20. Tanaka, M., Sotomatsu, A., Kanai, H., and Hirai, S., 1991, Dopa and dopamine cause cultured neuronal death in the presence of iron, *J. Neurol. Sci.*, 101:198-203.
21. Trachtman,H., Futterwiet, S., and Bienkowski, R.S., 1993, Taurine prevents glucose-induced lipid peroxidation and increased collagen production in cultured rat mesangial cells, *Biochem. Biophys. Res. Comm.*, 191:759-765.
22. Tse, D.C.S., McCreery, R.L., and Adams, R., 1976, Potential oxidative pathways of brain catecholamines, *J. Med. Chem.*, 19:37-40.
23. Tyce, G.M. and Owen, C.A., 1973, The effect of L-3,4-dihydroxyphenylalanine administration on glucose metabolism in brain, *J. Neurochem.*, 20:1563-1573.
24. Wills, E.D., 1987, Evaluation of lipid peroxidation in lipids and biological membranes, in "Biochemical Toxicology a Practical Approach" Snell, K. and Mullock, B., eds., IRL Press, Oxford, pp. 127-152.
25. Wright, C.E., Tallan, H.H., Lin, Y.Y., and Gaull, G.E., 1986, Taurine: biological update, *Ann. Rev. Biochem.*, 55:427-453.
26. Youdim, M.B.H., Ben-Shachar, D., Eshel, G, Finberg, J.P.M., and Riederer, P., 1993, The neurotoxicity of iron and nitric oxide relevance to the etiology of Parkinson's disease, *Adv. Neurol.*, 60:259-265.

# EFFECTS OF TAURINE SUPPLEMENTATION ON LIPID PEROXIDATION, BLOOD GLUCOSE AND BLOOD LIPID METABOLISM IN STREP-TOZOTOCIN-INDUCED DIABETIC RATS

Jeong Soon You and Kyung Ja Chang
Department of Food Nutrition
Inha University
Inchon, Korea 402-751

## ABSTRACT

The purpose of this study was to determine the effect of taurine on several complications of diabetes, including oxidative stress, glucose intolerance and blood lipid profile. Sprague Dawley male rats were fed an experimental diet for 7 weeks, at which time they were maintained on drinking water with or without 1% taurine. The experimental period was 7 weeks and the rats were administered streptozotocin (STZ) to induce diabetes. Thiobarbituric acid reactive substances (TBARS) content was increased following the STZ injection, but was lowered by prior treatment with taurine. The primary diabetic symptoms, such as polydipsia and polyuria, were ameliorated in rats supplemented with taurine before the STZ injection. Plasma triglyceride (TG) levels of the diabetic group were decreased by taurine supplementation, although plasma total cholesterol (T-chol) and HDL cholesterol (HDL-chol) were not different among the groups. LDL cholesterol (LDL-chol) levels of the control group were significantly decreased by taurine supplementation, however, the time of taurine administration affected the response of the diabetic group; only diabetic rats treated with taurine after the administration of STZ showed a decrease in LDL cholesterol. Therefore, taurine inhibits lipid peroxidation and decreases blood TG and LDL-chol levels, however, the time and dose of taurine supplementation are variables that need to be considered in the treatment of diabetes.

## INTRODUCTION

Taurine (2-aminoethanesulfonic acid) is a free ß-amino acid that is present in high concentrations in mammalian tissue. It is available from dietary sources and can also be biosynthesized from cysteine. Many biological functions have been attributed to taurine, including bile acid conjugation, neuromodulation, detoxification, osmoregulation,

*Taurine 3*
Edited by Schaffer *et al.*, Plenum Press, New York, 1998 

membrane stabilization and antioxidant properties[1]. Hyperglycemia is the major etiologic factor in the complication of diabetes, which is significant because elevated glucose concentration directly injures cells and induces lipid peroxidation. It was found that taurine suppressed insulin secretion in nondiabetic pancreatic islets, suggesting that taurine might serve as a physiological regulator of insulin secretion and plasma glucose levels[12]. It has been reported that taurine can protect against streptozotocin via membrane stabilization by binding to the membrane at or near the insulin receptor[6]. Franconi et al.[4] reported recently that plasma and platelet taurine concentration is reduced in subjects with insulin-dependent diabetes mellitus, although this defect can be reversed. Moreover, taurine supplementation (5% w/v) increased fasting glucose levels in alloxan-induced diabetic mice[7]. Therefore, the purpose of this study was to determine the effect of taurine on lipid peroxidation, blood glucose levels and blood lipid metabolism in the diabetic rat.

## MATERIALS AND METHODS

Male Sprague-Dawley rats weighing 90-100 g were acclimatized for 1 week prior to their use in experiments. The rats were kept at 25 - 30°C with a 12 hr light-dark cycle. For 7 weeks the rats were fed an experimental diet and were maintained ad libitum on drinking water with or without 1% taurine. Three weeks after feeding the experimental diet, diabetes was induced by a single thigh subcutaneous injection of streptozotocin (50 mg/kg body weight). As shown in Figure 1, rats were randomly divided into the following groups: CO (control, without taurine); CT (control, with taurine during all experimental procedures); EO (diabetic, without taurine); EB (diabetic, with taurine before the STZ injection); EA (diabetic, with taurine after the STZ injection) and ET (diabetic, with taurine during all experimental procedures). Three weeks after inducing diabetes, metabolic cages were used to collect urine for urinary volume measurements. Four weeks after inducing

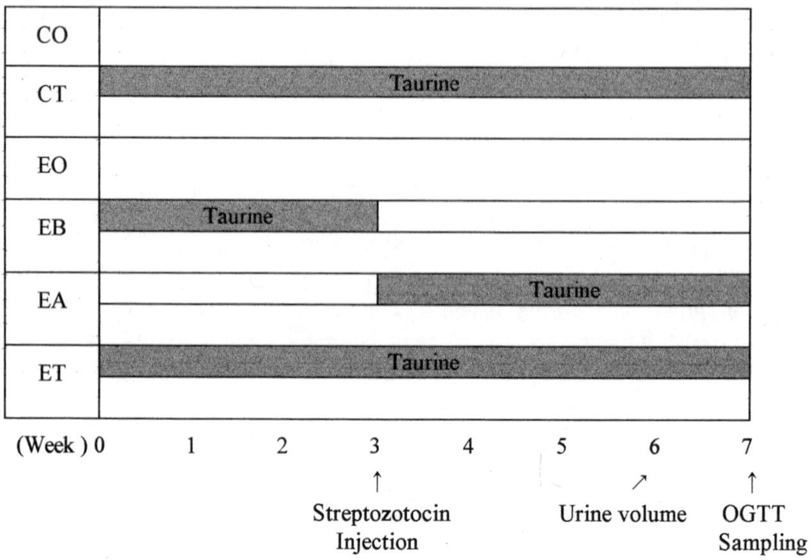


CO : Control group
CT : Control group with taurine
EO : Diabetic group
EB : Diabetic group + pre 1% taurine
EA : Diabetic group + post 1% taurine
ET : Diabetic group + all 1% taurine

**Figure 1.** Experimental design.

diabetes, oral glucose tolerance tests were conducted. Initially, a fasting blood glucose level was obtained. After an oral administration of glucose (0.1 g/100 g body weight), blood samples were collected every 30 min for 2 hours, followed by hour intervals. Blood glucose concentration was determined with a reagent strip and a blood glucose sensor (Exactech, Medisense, Inc., U.S.A.). The next day, liver and plasma samples of the rats were obtained. Upon removal, the liver was homogenized and a microsomal fraction prepared by differential centrifugation. The microsomal fraction and the plasma were stored at -70°C until use.

The content of lipid peroxides in liver microsomes was determined by measuring the formation of thiobarbituric acid reactive substances (TBARS)[1]. Hepatic microsomal protein concentration was estimated by the Lowry method using bovine serum albumin as a standard[8]. Plasma TG, T-chol and HDL-chol levels were measured with a biochemical automated analyzer (Hitachi 747). Plasma LDL-chol content was calculated from measured levels. All values represent the mean ± S.E. All statistical analyses were carried out by Duncan's multiple range test using the SAS program. A p value of $< 0.05$ was selected as a limit of statistical significance.

## RESULTS AND DISCUSSION

The observation that urinary volume of the diabetic group was higher than that of the control group is consistent with other reports[10,11]. In the diabetic group, urinary volume of the rats receiving taurine supplemention before the STZ injection was lower than untreated diabetic rats, although taurine supplementation increased urinary output if administered following STZ administration (Table 1). Thus, the diabetic symptoms might be ameliorated by taurine supplementation before inducing diabetes but might deteriorate if supplementation occurs after STZ injection.

Glucose levels of the diabetic group supplemented with taurine after the STZ injection was higher than that of the other groups (Fig. 2). Glucose levels of the diabetic group supplemented with taurine before the STZ injection was significantly lower than that of the diabetic group without taurine. The results of the urinary volume and OGTT measurements suggest that taurine supplementation after STZ administration might provoke a deterioration in the diabetic symptoms.

**Table 1.** Effect of taurine supplementation on urinary volume and TBARS content

| Group | Urine volume (ml/100g B.W/day) | Thiobarbituric acid reactive substance (nmole TBARS/mg protein) |
|---|---|---|
| CO | 2.61 ± 0.39$^{d}$ | 0.145 ± 0.006$^{c}$ |
| CT | 3.82 ± 1.07$^{d}$ | 0.127 ± 0.007$^{c}$ |
| EO | 56.10 ± 12.21$^{bc}$ | 0.218 ± 0.016$^{a}$ |
| EB | 30.64 ± 9.13$^{cd}$ | 0.163 ± 0.007$^{bc}$ |
| EA | 96.08 ± 11.70$^{a}$ | 0.199 ± 0.010$^{ab}$ |
| ET | 70.69 ± 10.04$^{ab}$ | 0.197 ± 0.029$^{ab}$ |

Values are mean ± S.E. Means with different letters are significantly different at $p < 0.05$ by Duncan's multiple range test.

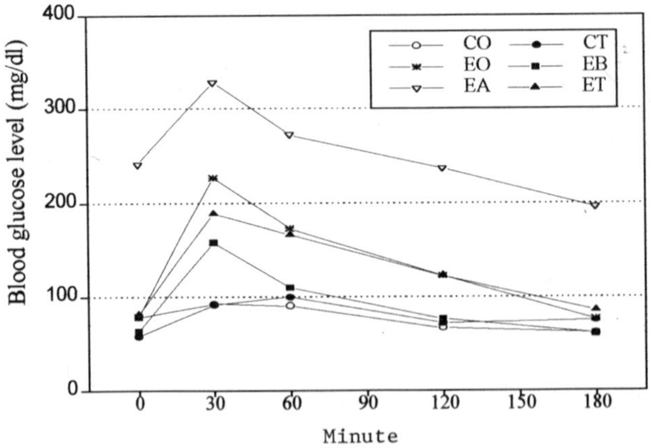


**Figure 2.** Effect of taurine supplementation on the oral glucose tolerance test.

TBARS content was increased following the injection of STZ (Table 1). In the diabetic group supplemented with taurine before the STZ injection, TBARS content was significantly lower than the diabetic group without taurine treatment. By comparison, the TBARS content of the diabetic group supplemented with taurine either after the STZ injection or throughout all of the experimental procedures tended to be lower, although the effect was not significant. It has previously been reported that the TBARS content of liver and pancreatic islets from type I diabetic rats was significantly increased compared to the control group, an effect which was significantly attenuated by taurine supplementation[7]. From these results, we suggest that taurine supplementation protects the liver against lipid peroxidation in the type I diabetic rat. However, further studies related to the time and dose of taurine supplementation are warranted.

Consistent with several other reports[2,7,13], but not that of Park et al.[10], we found that plasma TG levels of the diabetic group without taurine supplementation were significantly higher than the control group. Interestingly, plasma TG levels of the diabetic group were decreased by taurine supplementation (Table 2). Nonetheless, plasma T-chol levels and HDL-chol levels were not different among the various groups. As seen in Table 3, LDL chol levels of the control group and the diabetic group treated with taurine after the STZ injection were significantly decreased by taurine supplementation. However, taurine was ineffective in lowering LDL-chol in the other diabetic groups.

It has previously been reported that serum TG, T-chol, HDL-chol and LDL/VLDL-chol levels in the diabetic rat are significantly increased compared with corresponding levels in the normal rat[9]. Moreover, taurine supplementation significantly reduces the extent of the diabetes-induced increase[9], suggesting that taurine inhibits lipid peroxidation and decreases blood TG and LDL-chol levels in the diabetic. However, since the time and dose of taurine supplementation affects the results, further work is required in order to determine whether the effect of taurine on prevention and treatment of diabetes is a competitive effect or a dose effect.

**Table 2.** Effect of taurine supplementation on triglyceride and total cholesterol levels.

| Group | Plasma triglyceride level (mg/dl) | Plasma total cholesterol level[N.S.] (mg/dl) |
|---|---|---|
| CO | 60.67 ± 12.68[bc] | 90.00 ± 12.74 |
| CT | 72.33 ± 6.64[ab] | 78.67 ± 5.61 |
| EO | 86.20 ± 9.11[a] | 68.88 ± 5.51 |
| EB | 42.00 ± 4.72[c] | 68.14 ± 6.23 |
| EA | 49.00 ± 7.08[bc] | 64.50 ± 6.12 |
| ET | 53.50 ± 12.77[bc] | 78.14 ± 6.69 |

Values are mean ± S.E. Means with different letters are significantly different at $p < 0.05$ by Duncan's multiple range test N.S.: Not significant ($p<0.05$).

**Table 3.** Effect of taurine supplementation on HDL cholesterol and LDL cholesterol levels

| Group | Plasma HDL-cholesterol level[N.S.] (mg/dl) | Plasma LDL-cholesterol level[N.S.] (mg/dl) |
|---|---|---|
| CO | 47.33 ± 6.23 | 30.53 ± 4.68[a] |
| CT | 53.67 ± 2.85 | 10.53 ± 2.47[cd] |
| EO | 42.50 ± 3.37 | 13.13 ± 1.67[cd] |
| EB | 41.29 ± 3.99 | 18.97 ± 2.97[bc] |
| EA | 43.86 ± 4.33 | 7.90 ± 2.02[d] |
| ET | 43.50 ± 3.75 | 24.23 ± 4.06[ab] |

Values are mean ± S.E. Means with different letters are significantly different at $p < 0.05$ by Duncan's multiple range test N.S.: Not significant ($p < 0.05$).

## ACKNOWLEDGMENTS

This work was supported by grants from Inha University and Miwon Research Institute of the Korean food and dietary culture. We thank Dong-A Pharmaceutical Co. for the donation of taurine.

## REFERENCES

1. Buege, J.A. and Aust, S.D., 1978, Microsomal lipid peroxidation, *Meth. Enzymol.*, 52:302-310.
2. Ceriello, A., Giugliano, D., Quatraro, A., Donzella, C., Dipalo, G., and Lefebvre, P.J., 1991, Vitamin E reduction of protein glycosylation in diabetes, *Diabetes Care*, 14:68-72.
3. Flesiher, S. and Kervina, M., 1974, Subcellular fraction of rat liver, *Meth. Enzymol.*, 31:6-13.
4. Franconi, F., Bennardini, F., Mattana, A., Miceli, M., Ciuti, M., Mian, M., Gironi, A., Anichini, R., and Seghieri, G., 1995, Plasma and platelet taurine are reduced in subjects with insulin-dependent diabetes mellitus: effects of taurine supplementation, *Am. J. Clin. Nutr.*, 61:1115-1119.
5. Huxtable, R.J., 1992, Physiological actions of taurine, *Physiol Rev.*, 72:101-163.
6. Kulakowski, E.C. and Matura, J., 1984, Hypoglycemic properties of taurine: not mediated by enhanced insulin release, *Biochem. Pharmac.*, 33:2835-2838.
7. Lim, E.Y. and Kim, H., 1995, Effect of taurine supplement on the lipid peroxide formation and the activity of glutathione-dependent enzyme in the liver and islet of diabetic model mice, *J. Korean Soc. Food Nutr.*, 24(2):195-201.

8. Lowry, O.H., Roserbrough, N.J., Farr, A.L., and Randall, R.J., 1951, Protein measurement with Folin phenol reagent, *J. Biol. Chem.*, 193:265-275.
9. Nanami, K., Oda, H., and Yokogoshi, H., 1996, Antihypercholesterolemic action of taurine on streptozotocin-diabetic rats fed a high cholesterol diet, *Adv. Exp. Med. Biol.*, 403:561-568.
10. Park, S.H., Lee, Y.K., Lee, H.S., 1994, The effect of dietary fiber feeding on gastrointestinal functions and lipid and glucose metabolism in streptozotocin-induced diabetic rats, *Korean J. Nutrition*, 27(4):311-322.
11. Park, S.Y., 1995, Effect of vitamin E and taurine supplement on the lipid peroxidation and the activities of glutathione-related antioxidative enzymes according to blood glucose content in KK mice, Mater thesis of Seoul National University.
12. Tokunaga, H., Yoneda, Y., and Kuriyama, K., 1983, Streptozotocin-induced elevation of pancreatic taurine content and suppressive effect of taurine on insulin secretion, *Eur. J. Pharmacol.*, 87:237-243.
13. Winkler, R. and Moser, M., 1992, Alterations of antioxidant tissue defense enzumes and related metabolic parameters in streptozocin-diabetic rats-effects of iodine treatment, *Wien-Klin-Wochenschr.*, 104(4):409-413.

# UPTAKE OF TAURINE AND TAURINE CHLORAMINE IN MURINE MACROPHAGES AND THEIR DISTRIBUTION IN MICE WITH EXPERIMENTAL INFLAMMATION

C. Kim,[1] J.-K. Chung,[2] J.M. Jeong,[2] Y.S. Chang,[2] Y.J. Lee,[2]
Y.J. Kim,[2] M.C. Lee,[2] C.-S. Koh,[2] and B.-K. Kim[3]

[1] Indiana University School of Medicine
Herman B Wells Center for Pediatric Research
1044 W Walnut
Indianapolis, IN 46202
[2] Department of Nuclear Medicine
College of Medicine
[3] Department of Microbial Chemistry
College of Pharmacy
Seoul National University
Seoul, 110-744
Korea

## INTRODUCTION

Taurine is an abundant amino acid in inflammatory cells, where it acts as a trap for toxic hypochlorous acid ($HOCl/OCl^-$) and protects tissue from damage resulting from overt inflammatory reactions as demonstrated *in vivo* and *in vitro* models of inflammation[7,10,11]. High concentrations of taurine are present in the cytosol of leukocytes (20-50 mM)[4,8]. Under physiological conditions the formation of taurine chloramine (Tau-Cl) is catalyzed by a reaction of the polymorphonuclear leukocytes (PMN) derived from myeloperoxidase (MPO) and hypochlorous acid with taurine. Formation of Tau-Cl may also be catalyzed directly by a halide-dependent myeloperoxidase reaction not involving formation of $HOCl/OCl^-$. Although Tau-Cl inherently possesses oxidative potential, it is more stable and less toxic than HOCl. The addition of exogenous taurine strongly enhances chloramine formation. Tau-Cl plays a role in inflammation by inhibiting the production of nitric oxide, TNF-α, IL-1, IL-6, IL-8 and prostaglandin E2 in macrophages and production of superoxide anion ($O_2^-$) in PMN[7, 10, 11, 12].

A close correlation between endogenous taurine concentration and autoradiograms of injected taurine, except brain and lung, was reported[14]. As a normal constituent of mammalian tissue, taurine is found in particularly high concentration in the brain, liver, lung, kidney, muscle, intestine, heart and adrenal gland[5,17]. Injected [[35]S]taurine mainly

*Taurine 3*
Edited by Schaffer *et al.*, Plenum Press, New York, 1998 

accumulates in the liver, kidney, intestine, eye, hypophysis, thymus, stomach and adrenal glands[14]. Transport of taurine is mediated by a taurine transporter which has been cloned from rat brain, mouse brain and human placenta[9,13,15]. The taurine transporter is $Na^+$- and $Cl^-$-dependent with a molecular mass of approximately 70 kDa[1, 19]. The affinity of the transporter for taurine in different cell lines is significantly different. $K_m$ values vary from 0.9 ± 0.1 μM in Caco-2 cells to 10.6 ± 0.3 μM in HT-29 cells[3].

This study was designed based on the concept that the accumulation of taurine and Tau-Cl in inflammatory lesions is essential for the anti-inflammatory role of taurine and Tau-Cl. We determined the uptake of Tau-Cl and its precursor taurine in resident and lipopolysaccharide (LPS)-stimulated macrophages. Transport of taurine and Tau-Cl into murine macrophages was determined by using various concentrations of radiolabeled [$^3$H]taurine and [$^3$H]Tau-Cl. In addition, the biodistribution was examined in ICR mice having chronic muscle inflammation induced by turpentine.

## MATERIALS AND METHODS

### Preparation of Macrophages

Male ICR mice (8-14 weeks old) obtained from Seoul National University Hospital were injected *i.p.* with 1 ml of 3% thioglycollate (Sigma, St. Louis, MO). Cells were removed five days after thioglycollate injection by peritoneal lavage with HBSS (w/o $Ca^{+2}$, $Mg^{+2}$ and phenol red; Gibco, Gaithersburg, MD), suspended in Krebs-Ringer-phosphate-glucose solution (KRPG; containing 16.2 mM sodium phosphate buffer, pH 7.4, 118.5 mM NaCl, 3.0 mM KCl, 0.88 mM $CaCl_2$, 1.18 mM $MgSO_4$, and 10.0 mM D-glucose). For the activation of macrophages, cells were treated for 1 hr with LPS (10 μg/ml, from *E. coli* 0111:B4, Sigma) before onset of the transport study.

### Taurine and Tau-Cl Transport

Taurine and Tau-Cl transport were measured by the modified method of Tallan *et al.*[19]. Various concentrations (12.5 - 500 μM) of [$^3$H]taurine (Du Pont NEN, Boston, MA) and [$^3$H]Tau-Cl (synthesized in our laboratory) were incubated with 1 x $10^6$ cells at 37.5°C for 20 min in a total volume of 1.1 ml of KRPG. Immediately after incubation each tube received 10 ml of cold KRPG and was gently swirled at 4°C for 5 min. The cells were washed three times with 5 ml of cold KRPG. Cells were lysed overnight in Soluene 350 (Packard, Downers Grove, IL) and the radioactivity was counted with 3 ml of liquid scintillation solution in a liquid scintillation analyzer (RPI-CAPB, Packard). Uptake results are expressed as pmol amino acid taken up per minute per 1 x $10^6$ cells.

### Biodistribution in Mice With Inflammation

Biodistribution studies were performed in male ICR mice (30-35 g) having turpentine induced inflammation. Turpentine (0.1 ml) was injected *i.m.* into the right thigh of mice and the distribution study was performed after 7 days. Groups of five mice were injected intravenously with 1 μCi of [$^3$H]taurine or [$^3$H]Tau-Cl in 0.1 ml and were sacrificed 1 hr after injection. Tissues of interest were removed, weighed and dissolved overnight in 1 ml of Soluene 350. Radioactivity was determined with 10 ml of liquid scintillation solution in a liquid scintillation analyzer. The injected dose (%ID) per gram or %ID per organ were calculated by comparing the activity in each sample with an injection standard of an appropriate count.

## Stability Study

The stability of taurine (Sigma) and Tau-Cl was determined by measuring their UV spectra and separating them on thin layer chromatography (TLC). Two mM taurine and Tau-Cl were incubated at 37°C for 1hr. After incubation the UV spectrum at 250 nm and 293 nm was determined using a spectrophotometer (8452A, Hewlett Packard, German).

For the TLC analysis, 1 μl of 200 mM taurine and Tau-Cl were spotted on Whatman paper (No. 1) and developed using n-propanol:water (1:1) as the solvent. The chromatogram was sprayed with 0.2% ninhydrin in water-saturated butanol. In addition, radiolabeled taurine and Tau-Cl were incubated with human serum at 37°C for 1 hr in a $CO_2$ incubator and TLC analysis was performed. The chromatogram was exposed to BAS film for 7-10 days and analyzed by a Bio-imaging Analyzer System (Fuji film BAS-2500, Japan).

# RESULTS AND DISCUSSION

## Transport of Taurine and Tau-Cl

Transport of taurine and Tau-Cl into murine macrophages was determined by using various concentrations of radiolabeled [$^3$H]taurine and [$^3$H]Tau-Cl. Murine macrophages possess an uptake system for taurine and Tau-Cl (Fig. 1 and Fig. 2). The apparent $K_m$ and $V_{max}$ of resident macrophages were 28.1 μM and 90.9 pmol/min/$10^6$ cells for taurine and 23.3 μM and 51.3 pmol/min/$10^6$ cell for Tau-Cl, respectively. The apparent $K_m$ and $V_{max}$ of LPS-treated macrophages were 17.3 μM and 116.3 pmol/min/$10^6$ cells for taurine and 45.9 μM and 82.6 pmol/min/$10^6$ cell for Tau-Cl, respectively (Table 1).

It has been previously reported that the taurine and Tau-Cl transport systems are separate and both uptake systems are temperature, sodium, and energy dependent in RAW 264.7 cells[11]. In this study, taurine and Tau-Cl revealed different uptake patterns. These findings indicate that taurine and Tau-Cl transport systems are separate in murine macrophages. $V_{max}$ of LPS-treated macrophages was higher than that of resident macrophages in the taurine and Tau-Cl transport system. The affinity of the system for taurine was increased in LPS-treated macrophages, whereas that for Tau-Cl was decreased. Activated macrophages transport more taurine and Tau-Cl than resident macrophages and it is thus concluded that the transport of taurine and Tau-Cl play a role in inflammation.

**Table 1.** [$^3$H]taurine and [$^3$H]Tau-Cl transport into murine peritoneal macrophages

| Cell | Taurine | | Tau-Cl | |
|---|---|---|---|---|
| | $K_m$ (μM) | $V_{max}$ (pmol/min/$10^6$ cell) | $K_m$ (μM) | $V_{max}$ (pmol/min/$10^6$ cell) |
| Macrophages | 28.1 | 91.7 | 23.3 | 56.5 |
| Activated-macrophages | 17.3 | 116.3 | 45.9 | 82.6 |

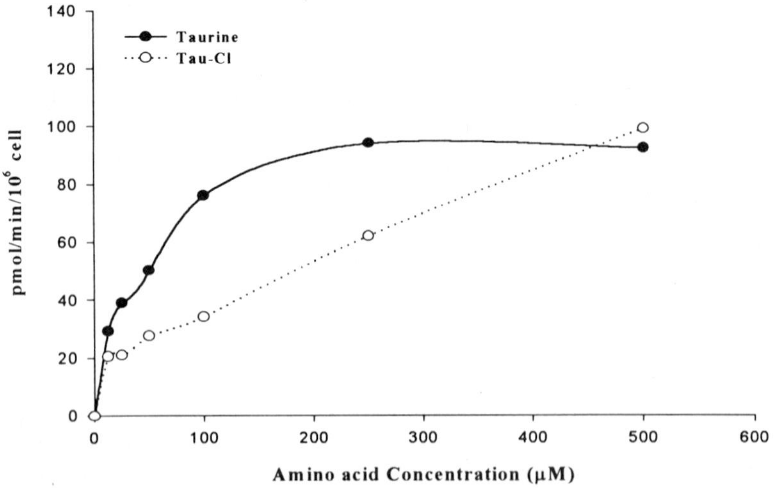


**Figure 1.** Uptake of [$^3$H]taurine and [$^3$H]Tau-Cl into murine peritoneal macrophages. The uptake was measured at 37 °C for 20 min. The lysate was mixed with 3 ml liquid scintillation solution and was measured in a β-counter for one min.

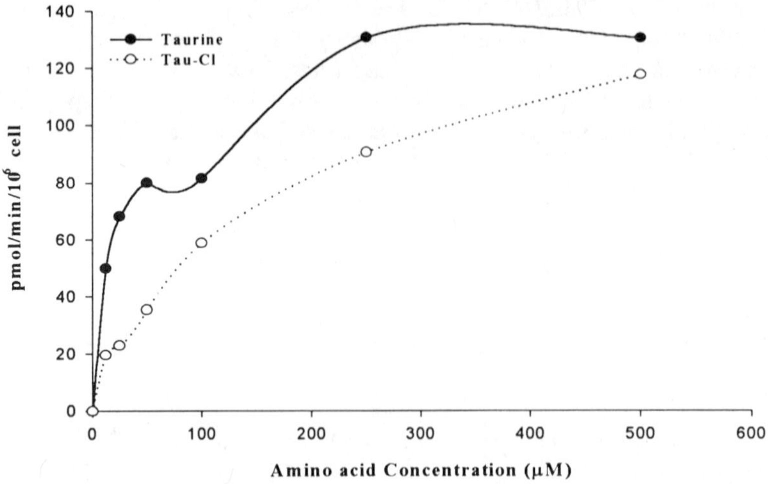


**Figure 2.** Uptake of [$^3$H]taurine and [$^3$H]Tau-Cl into LPS-treated murine peritoneal macrophages. LPS (10 μg/ml) was added 1 hr before uptake and maintained throughout the uptake. The uptake was measured at 37 °C for 20 min. The lysate was mixed with 3 ml liquid scintillation solution and was measured in a β-counter for one min.

## Biodistribution Studies in Mice

Biodistribution of taurine and Tau-Cl was determined one hour after intravenous injection of either 1 μCi of [$^3$H]taurine or [$^3$H]Tau-Cl. The biodistribution patterns were similar. Rapid clearance of taurine and Tau-Cl from the blood was observed with high retention in the lung, liver, spleen, stomach, intestine and kidney (Fig. 3 and Fig. 4). Significant accumulation of [$^3$H]taurine and [$^3$H] Tau-Cl were observed based on the inflammatory lesion (ratio of accumulation: inflammation/blood = 6.43 and 4.84, respectively) (Table 2). The inflammatory lesions in the muscle had ratios of 1.42 and 1.67 (Table 2). In addition, [$^3$H]taurine and [$^3$H]Tau-Cl were taken up by the inflammatory lesion without influencing the overall biodistribution. However, brain accumulation of taurine and Tau-Cl was negligible: this phenomenon was most likely due to the blood-brain barrier[6]. It was also reported that injected taurine demonstrate a rapid uptake by the kidney, liver, spleen and bone marrow whereas there is a considerably slower uptake by the heart and muscle[5].

Boquet and Fromageot reported the rate of turnover of taurine was tissue-dependent and the rate of cellular uptake and the retention varied with the cell type or organ[2].

### *In Vitro* Stability

The stability of taurine and Tau-Cl was monitored by measuring their UV spectra. Tau-Cl and NaOCl had maximum optical densities at 250 nm and 293 nm, respectively. The optical densities of taurine and Tau-Cl were not changed after a 1 hr incubation at 37°C. Taurine was visualized by ninhydrin after TLC, whereas Tau-Cl was not. The $R_f$ values of taurine and Tau-Cl were 0.75 and 0.83, respectively; these values did not change after 1 hr incubation at 37°C. These results suggest that taurine and Tau-Cl are stable throughout the time of transport into the macrophages.

Analyses of radiolabeled taurine and Tau-Cl after TLC were performed. The $R_f$ values of taurine and Tau-Cl changed from 0.75 to 0.65 and from 0.83 to 0.59, respectively, after a 1 hr incubation with human serum at 37°C in a $CO_2$ incubator. The changes in the $R_f$ values after a 1 hr incubation with human serum do not necessarily signify an instability in the compounds, however, further studies are required. Taurine is present as a free amino acid in all tissues and as a conjugate with bile acids in the liver[3,17]. No metabolism of injected taurine in the organs has been described[2]. In the future the reaction of taurine and Tau-Cl with serum components needs to be investigated.

**Table 2.** The inflammation to organ radioactivity ratio at 1 hr after *i.v.* injection of [$^3$H]taurine and [$^3$H]Tau-Cl (n=5)

| Ratio | % injected dose/g | |
|---|---|---|
| | Taurine | Tau-Cl |
| Inflammation/Blood | 6.43 ± 1.75* | 4.48 ± 1.86 |
| Inflammation/Muscle | 1.42 ± 0.27 | 1.66 ± 0.15 |
| Inflammation/Brain | 4.64 ± 0.52 | 5.46 ± 0.97 |
| Inflammation/Liver | 0.23 ± 0.03 | 0.21 ± 0.04 |
| Inflammation/Spleen | 0.36 ± 0.07 | 0.45 ± 0.08 |

* : mean ± S.D.

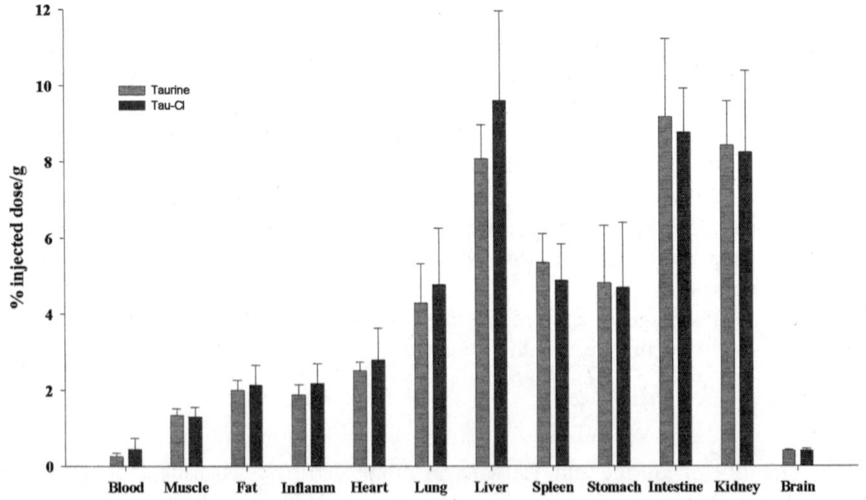


**Figure 3**. Distributions of [$^3$H]taurine and [$^3$H]Tau-Cl in ICR mice having turpentine induced inflammation at 1 hr post injection. The data are expressed in percent injected dose per gram of tissue (n=5 for each value). No significant difference was observed between the distribution of [$^3$H]taurine and [$^3$H]Tau-Cl. (Inflamm = Inflammation)

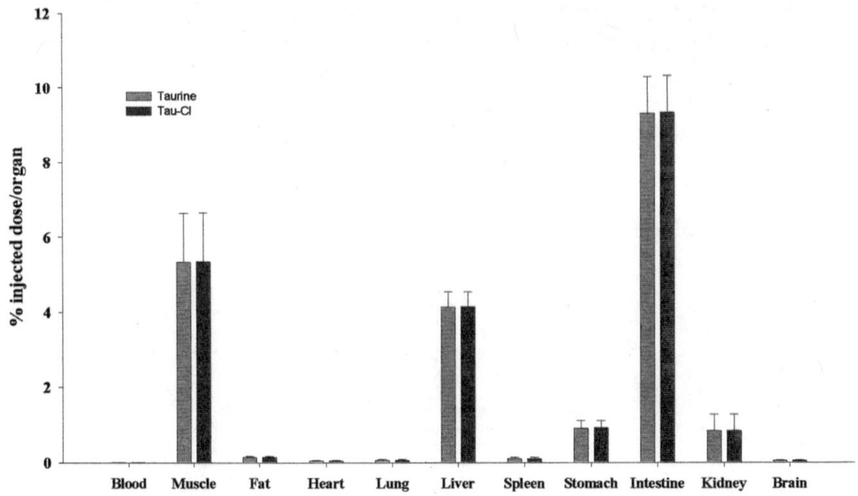


**Figure 4.** Organ distributions of [$^3$H]taurine and [$^3$H]Tau-Cl in ICR mice having turpentine induced inflammation at 1 hr post injection. The data are expressed in percent injected dose per organ (n=5 for each value). No significant difference was observed between the distribution of [$^3$H]taurine and [$^3$H]Tau-Cl .

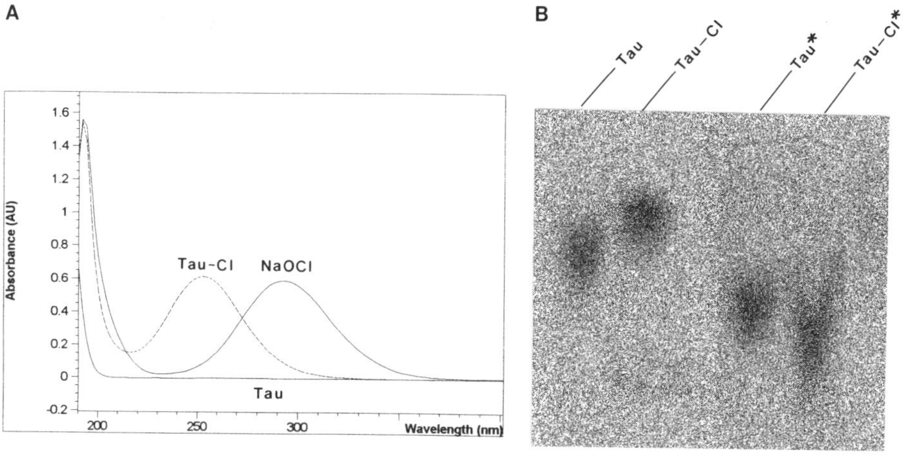


**Figure 5.** UV spectrum (A) and instant image of taurine (Tau) and Tau-Cl after TLC (B).
* Represents the incubation with human serum at 37°C for 1 hr in a $CO_2$ incubator.

We compared the uptake of taurine and Tau-Cl into macrophages and the biodistribution of taurine and Tau-Cl in mice with inflammation. Taurine and Tau-Cl have a separate transport system and are actively transported into inflammatory cells and inflammatory lesions. The LPS-stimulated macrophages revealed higher taurine and Tau-Cl transport than resident macrophages. Radiolabeled taurine and Tau-Cl were in inflammatory lesions after intravenous injection. These phenomena may be attributed to the anti-inflammatory role of Tau-Cl and its precursor, taurine. In inflammatory lesions, elevated taurine acts as a scavenger of the toxic substance, hypochlorous acid. Tau-Cl also inhibits the production of inflammatory mediators[7,12]. Taurine and Tau-Cl were stable after a 1 hr incubation at 37°C either alone or with human serum. Even though taurine and Tau-Cl were transported into macrophages with different distribution patterns, the biodistribution of taurine and Tau-Cl in mice was very similar. From these results, we may conclude that taurine and Tau-Cl are distributed in the same manner in the body. We may also discount the possibility of the dissociation of Tau-Cl into taurine and $Cl^-$. However, the stability and reactivity of taurine and Tau-Cl in vivo must be further studied.

## ACKNOWLEDGMENTS

This research was supported in part by a grant from the Korean Taurine Society and we acknowledge their support. Chaekyun Kim would like to thank Dr. Eunkyue Park and Dr. Georgia Schuller-Levis at the New York State Institute for Basic Research in Developmental Disabilities.

## REEFERENCES

1. Blakely, R.D., 1992, Advances in molecular biology of neurotransmitter transporters, *Curr. Opin. Psychiatry*, 5:69-73.

2. Boquet, P.L. and Fromageot, P., 1965, Renouvellement de la taurine tissulaire chez le rat. *Biochim. Biophys. Acta*, 97:222-232.
3. Ganapathy, V. and Leibach, F.H., 1994, Expression and regulation of the taurine transporter in cultured cell lines of human origin, *Avd. Exp. Med. Biol.*, 359:51-57.
4. Green, T., Feldman, J.H., Eicher, A.L. and Pratt, K.J., 1991, Antioxidant role and subcellular location of hypotaurine and taurine in human neutrophils, *Biochem. Biophys. Acta*, 1073:91-97.
5. Huxtable, R.J., 1981, Sources and turnover rates of taurine in nursing and weaned rat pups, *J. Nutr.*, 111:1275-1286.
6. Jhiang, S.M., Fithian, L., Smanik, P., McGill, J., Tong, Q. and Mazzaferri, E.L., 1993, Cloning of the human taurine transporter and characterization of taurine uptake in thyroid cells, *FEBS Letters*, 318:139-144.
7. Kim, C., Park, E., Quinn, M.R. and Schuller-Levis, G., 1996, The production of superoxide anion and nitric oxide by cultured murine leukocytes and the accumulation of TNF-$\alpha$ in the conditioned media is inhibited by taurine chloramine, *Immunopharmacol.*, 34:89-95.
8. Learn, D.B., Fried, V.A. and Thomas, E.L., 1990, Taurine and hypotaurine content of human leukocytes, *J. Leuko. Biol.*, 48:174-182.
9. Liu, Q.R., Lopez-Corcurea, B., Nelson, H., Mandiyan, S. and Nelson, N., 1992, Cloning and expression of a cDNA encoding the transporter of taurine and $\beta$-alanine in mouse brain, *Proc. Natl. Acad. Sci.*, 89: 12145-12149.
10. Marcinkiewicz, J., Grabowska, A., Bereta, J. and Stelmaszynska, T., 1995, Taurine chloramine, a product of activated neutrophils, inhibits in vitro the generation of nitric oxide and other macrophage inflammatory mediators, *J. Leuk. Biol.*, 58:667-674.
11. McLoughlin, D.M., Stapleton, P.P. and Bloomfield, F.J., 1991, Influence of taurine and a substituted respiratory burst pathway in the inflammatory response, *Biochem. Soc. Trans.*, 19:73-78.
12. Park, E., Quinn, M.R., Wright, C. and Schuller-Levis, G., 1993, Taurine chloramine inhibits the synthesis of nitric oxide production and the release of TNF-$\alpha$ in activated RAW 264.7 cells, *J. Leuko. Biol.*, 54:119-124.
13. Ramamoorthy, S., Ainol, L.S., Leibach, F.H., Blakely, R.D. and Ganapathy, V., 1993, Molecular cloning, expression, chromosomal localization, and regulation of the human placental taurine transporter, *Placenta*, 14:A.60.
14. Shimada, M., Shimono, R., Watanabe, M., Imahayashi, T., Ozaki, H.S., Kihara, T., Yamaguchi, K. and Niizeki, S., 1984, Distribution of $^{35}$S-taurine in rat neonates and adults, *Histochem.*, 80:225-230.
15. Smith, K.E., Borden, L.A., Wang, H.D., Hartig, P.R., Branchek, T.A. and Weinshank, R.L., 1992, Cloning and expression of a high affinity taurine transporter from rat brain, *Mol. Pharmacol.*, 42:563-569.
16. Stapleton, P.P., Mahon, T.M., Nowlan, P. and Bloomfield., 1994, Effects of in-vivo administration of taurine and HEPES on the inflammatory response in rats, *J. Pharm. Pharmacol.*, 46:745-750.
17. Sturman, J.A., 1993, Taurine in development, *Physiol. Rev.*, 73:119-147.
18. Sturman, J.A., Rassin, D.K., Gaull ,G.E., 1977, Taurine in developing rat brain transfer of [$^{35}$S]taurine to pups via the milk, *Pediatr. Res.*, 11:28-33.
19. Tallan, H.H., Jacobson, E., Wright, C.E., Scheidman, K., Gaull, G.E., 1983, Taurine uptake by cultured human lymphoblastoid cells, *Life Sci.*, 33:1853-1860.

# TAURINE CHLORAMINE INHIBITS THE PRODUCTION OF SUPEROXIDE ANION, IL-6 AND IL-8 IN ACTIVATED HUMAN POLYMORPHONUCLEAR LEUKOCYTES

Eunkyue Park,[1] James Alberti,[1] Michael R. Quinn,[2] and Georgia Schuller-Levis[1]

[1] Departments of Immunology and
[2] Developmental Biochemistry
NY State Institute for Basic Research in Developmental Disabilities
Staten Island, NY 10314

## INTRODUCTION

Polymorphonuclear leukocytes (PMN) are the initial cells recruited to the site of inflammation where foreign invaders such as microorganisms, toxic gases and chemicals elicit an inflammation reaction[4,6,12]. Activated PMN produce various oxygen and hydroxyl radicals which kill bacteria and fungi but may also induce indiscriminate cellular damage[14]. For example, $HOCl/OCl^-$ produced by the halide-dependent myeloperoxidase of PMN is highly toxic. Taurine attenuates the damage caused by $HOCl/OCl^-$ by forming taurine chloramine (Tau-Cl), a relatively non-toxic and long-lived oxidant[7,9,13,15]. Previously we demonstrated that Tau-Cl suppressed the production of nitric oxide (NO), tumor necrosis factor-α (TNF-α) and prostaglandin $E_2$ ($PGE_2$) in activated murine peritoneal macrophages and/or RAW 264.7, a murine macrophage cell line[8,10,11]. Tau-Cl also inhibits the production of superoxide anion in activated murine peritoneal PMN[5]. Since human PMN produce Tau-Cl by halide-dependent myeloperoxidase, we investigated the effect of Tau-Cl on the production of various cytokines and superoxide anion by activated human PMN. In this report we demonstrate that Tau-Cl down-regulates the production of superoxide anion, IL-6 and IL-8 by activated human PMN.

## MATERIALS AND METHODS

### Preparation of PMN

Human peripheral blood PMN were separated from whole blood of healthy volunteers using neutrophil isolation media (Cardinal Associates, Inc., Santa Fe, NM). Cell suspensions were > 95% PMNs by morphology as determined by Wright's Giemsa stain.

For production of chloramines, PMN in HBSS were activated with phorbol myristate acetate (PMA, 1 μg/ml) for 1 hr in the presence or absence of various amino acids. For superoxide anion studies, PMN were suspended in phenol red-free HBSS, added to 96 well flat-bottom microculture plates ($2x10^5$ cells/well) (Costar, Cambridge, MA), and were activated with phorbol myristate acetate (PMA, 500 ng/ml). For cytokine production, cells ($2 \times 10^5$/well) were suspended in RPMI 1640 (GIBCO BRL) containing 10% fetal calf serum and antibiotics in 96 well plates and activated with 1 μg/ml of lipopolysaccharide. Female Balb/c (4-8 wks old) mice (Jackson Laboratories, Bar Harbor, ME) and female Sprague-Dawley rats (Taconic Farms, Germantown, NY) were used for preparing murine and rat peritoneal PMN, respectively. Peritoneal PMN were removed by lavaging the peritoneal cavity with HBSS one day after intraperitoneal injection of 3% thioglycollate (Difco Laboratories, Detroit, MI). Each population of PMN was more than 90% pure as determined by differential staining. For production of Tau-Cl, $3 \times 10^6$ cell were activated with PMA (1 μg/ml). The viability of cells was unaffected, as determined by trypan blue dye exclusion.

### Oxidation Potential of Chloramine

The oxidation potential of chloramines was measured using a potassium iodide (KI) method as described by Witko et al.[16]. Briefly, 500 μl samples were reacted with 25 μl of 1.16 M KI for 2 min at room temperature before stopping the reaction with 50 μl of glacial acetic acid and measuring the absorbance at 340 nm. The oxidation potential of each sample was calculated from a standard curve using chloramine-T.

### Preparation of Tau-Cl and Other Assays

Tau-Cl was prepared by adding equimolar amounts of NaOCl dropwise into a taurine solution at pH 8.3 as previously described[9]. Superoxide anion production was determined by the superoxide anion dismutase (SOD) inhibitable cytochrome c reduction assay[5]. IL-6 and IL-8 production was measured by enzyme linked immunoabsorbent assay (ELISA)[8,10]. Chemical reagents were purchased from Sigma Chemical Co. (St. Louis, MI) unless noted otherwise. Antibodies for ELISA were obtained from Genzyme (Cambridge, MA).

## RESULTS AND DISCUSSION

We previously demonstrate that Tau-Cl inhibits production of NO, TNF-α and $PGE_2$ in activated murine peritoneal macrophage and/or RAW 264.7 cells[8,10] and inhibits superoxide anion production in activated murine peritoneal PMN[5]. In the present study the effects of Tau-Cl on activated human leukocytes was examined and the production of Tau-Cl by PMNs obtained from mice, rats and human subjects was compared. PMN from all three species produced Tau–Cl in amounts that were dependent on extracellular taurine concentration (Fig. 1). Human PMN produced the most Tau-Cl with maximum amounts generated in the presence of 25 mM taurine (Fig. 1) under the conditions examined. We also evaluated the ability of taurine, serine, α-alanine and β-alanine to form stable chloramines as indicated by maintenance of oxidation potential. Tau-Cl maintained the highest oxidation potential of the chloramines evaluated (Fig. 2). The production of Tau-Cl increased in proportion to the number of the human PMN present in the assay (Fig. 3). We also investigated the effects of Tau-Cl on the generation of selected proinflammatory mediators produced by activated human PMN. Superoxide was measured by SOD inhibitable cytochrome c reduction method, while IL-6 and IL-8 were determined by

ELISA. Production of superoxide anion by PMN activated with PMA (500 μg/ml) was decreased by Tau-Cl in a concentration-dependent manner ($IC_{50}$~0.1 mM) (Fig. 4). Secretion of IL-6 and IL-8 by PMN activated with LPS (1 μg/ml) was suppressed by Tau-Cl (Figs. 5 and 6). Inhibition of IL-6 ($IC_{50}$~0.2 mM) was more dramatic than that of IL-8 ($IC_{50}$~0.4 mM). The effective range of Tau-Cl (0.1 mM~0.4 mM) in human leukocytes was much lower than that of murine leukocytes (0.3 mM~1.0 mM)[5,8,11]. Taurine has antioxidant as well as antiinflammatory activities in several animal models of inflammation[1-3,13]. The present results suggest that the ability of Tau-Cl to modulate production of inflammatory mediators is not species specific and extends to human leukocytes.

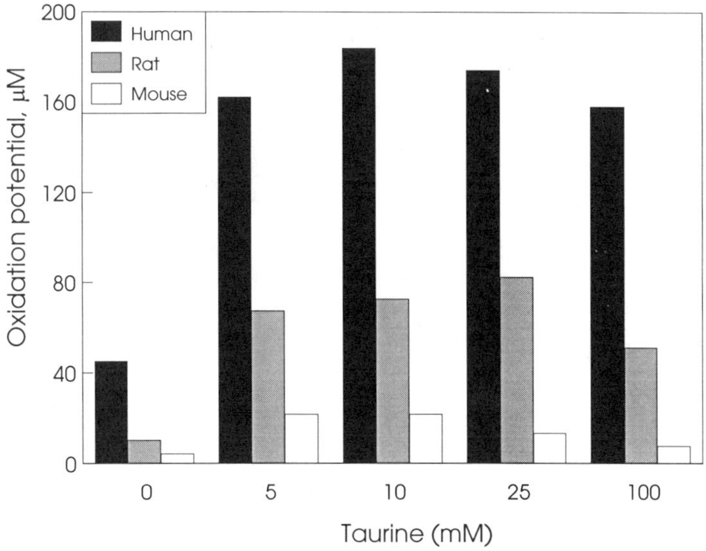


**Figure 1.** Production of Tau-Cl by PMN (3 x $10^6$) from mouse, rat and human. Data represent the mean of duplicate samples. Similar results were obtained in one additional experiment.

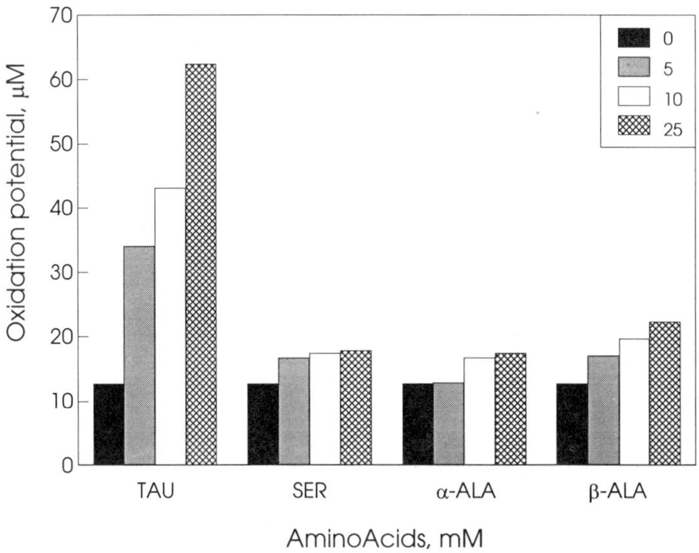


**Figure 2.** Production of chloramines by human PMN (3 x $10^6$) using various amino acids. Data represent the mean of duplicate samples. Similar results were obtained in one additional experiment.

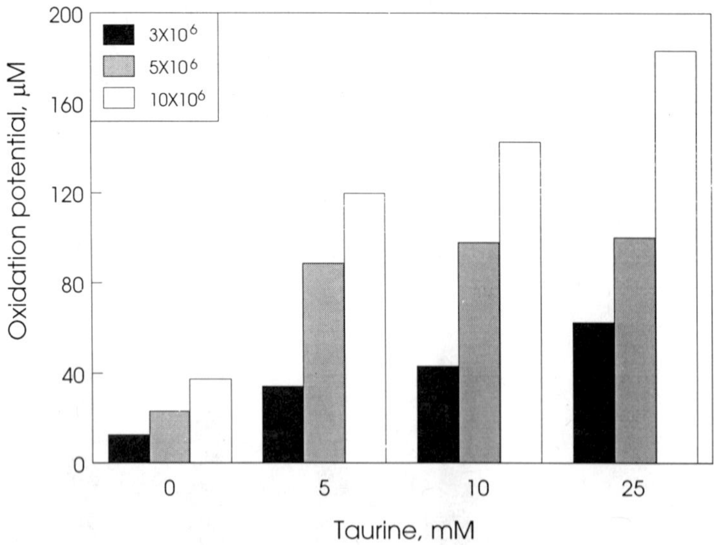


**Figure 3.** Production of Tau-Cl by various numbers of human PMN. Data represent the mean of duplicate samples. Similar results were obtained in one additional experiment.

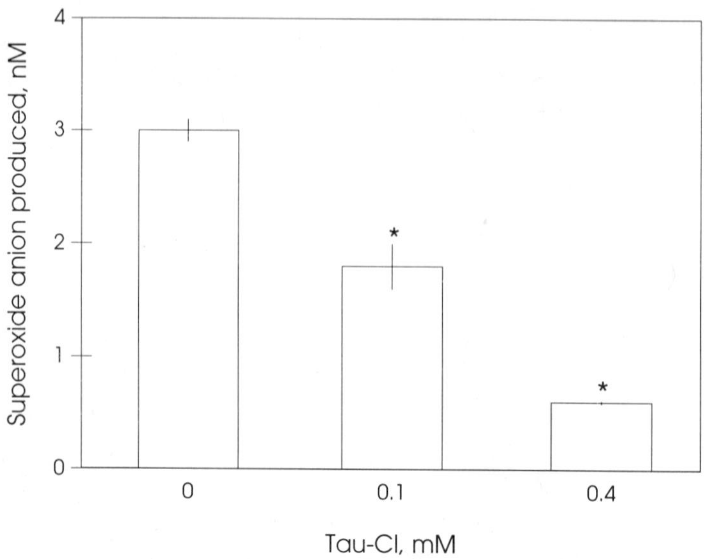


**Figure 4.** Inhibition of superoxide anion production by Tau-Cl in PMA-activated human PMN. Data represent the mean ± SD. Statistical significance was considered as $p < 0.05$. Similar results were obtained in one additional experiment.

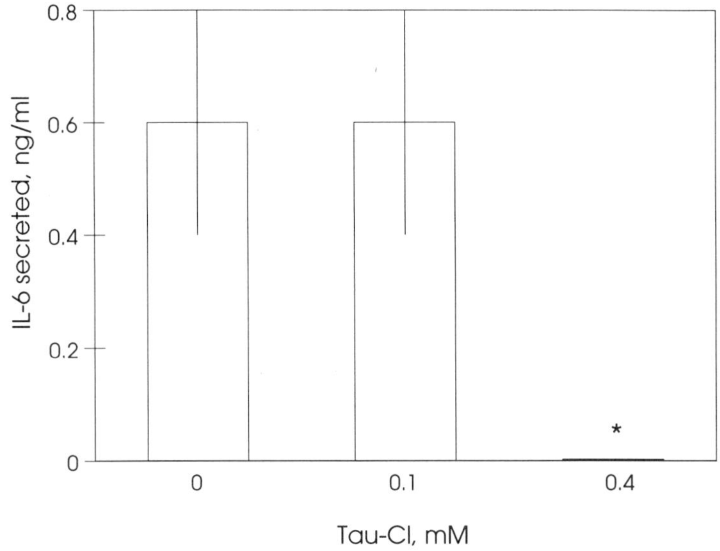


**Figure 5.** Inhibition of IL-6 in LPS-activated human PMN by Tau-Cl. Data represent the mean ± SD. Statistical significance was considered as $p < 0.05$. Similar results obtained in two additional experiments.

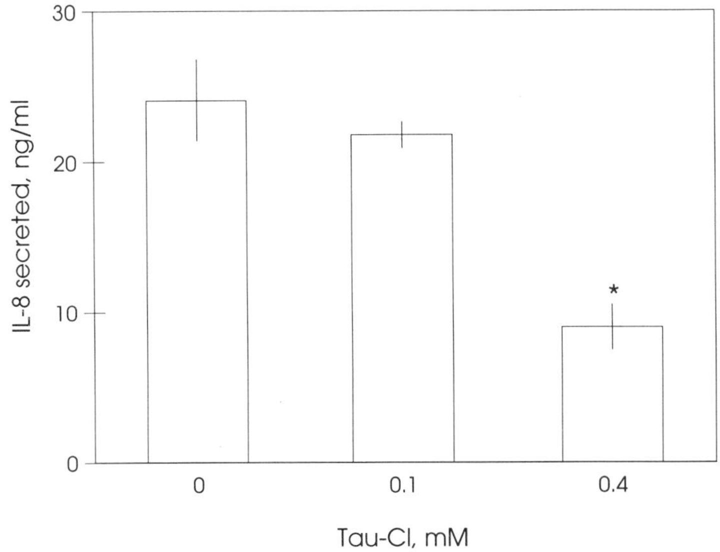


**Figure 6.** Inhibition of IL-8 in LPS-activated human PMN by Tau-Cl. Data represent the mean ± SD. Statistical significance was considered as $p < 0.05$. Similar results obtained in two additional experiments.

## ACKNOWLEDGMENTS

This study was supported by the Office of Mental Retardation and Developmental Disabilities of the State of New York and USPHS Grant HL-49942. We thank Joanne Lopez for secretarial assistance.

## REFERENCES


1. Gordon, R.E., Shaked, A.A., and Solano, D.F., 1986, Taurine protects hamster bronchioles from acute $NO_2$-induced alterations, a histologic, ultrastructural and freeze-fracture study, *Am. J. Pathol.*, 125:585-600.
2. Green T., Fellman, J.H., Eicher, A.I., and Pratt, K.J., 1991, Antioxidant role and subcellular location of hypotaurine and taurine in human neutrophils, *Biochem. Biophys. Acta.*, 1073:93-97.
3. Gurujeyalakshmi, G., Iyer, S.N., Hollinger, M.A., and Geri, S.N., 1996, Procollagen gene expression is down-regulated by taurine and niacin at the transcriptional level in the bleomycin model of lung fibrosis, *J. Pharmacol. Exp. Ther.*, 277:1152-1157.
4. Harada, A., Sekido, N., Akahoshi, T., Wada, T., Mukaida, N., and Matsushima, K., 1994, Essential involvement of interleukin-8 (IL-8) in acute inflammation, *J. Leukoc. Biol.*, 56:559-564.
5. Kim, C., Park, E., Quinn, M.R., and Schuller-Levis, G., 1996, The production of superoxide anion and nitric oxide by cultured murine leukocytes and the accumulation of TNF-$\alpha$ in the conditioned media is inhibited by taurine chloramine, *Immunopharmacology* 34:89-95.
6. Lloyd, A. and Oppenheim, J., 1992, Poly's lament: the neglected role of the polymorphonuclear neutrophil in the afferent limb of the immune response, *Immunology Today*, 13:169-172.
7 Marques, L.A. and Dunford, H.B., Chlorination of taurine by myeloperoxidase: kinetic evidence for an enzyme-bound intermediate, *J. Biol. Chem.*, 269:7950-7956.
8. Park, E., Quinn, M.R., Wright, C.E., and Schuller-Levis, G., 1993, Taurine chloramine inhibits the synthesis of nitric oxide and release of tumor necrosis factor in activated RAW 264.7 cells, *J. Leukoc. Biol.*, 54:119-124.
9. Park, E., Schuller-Levis, G., Jia, J., and Quinn, M.R., 1997, Preactivation exposure of RAW 264.7 cells to taurine chloramine attenuates subsequent production of nitric oxide and expression of iNOS mRNA, *J. Leukoc. Biol.*, 61:161-166.
10. Park, E., Schuller-Levis, G., and Quinn, M.R., 1995, Taurine chloramine inhibits production of nitric oxide and TNF-$\alpha$ in activated RAW 264.7 cells by mechanisms that involve transcriptional and translational events, *J. Immunol.*, 154:4778-4784.
11. Quinn, M.R., Park, E., and Schuller-Levis, G., 1996, Taurine chloramine inhibits prostaglandin $E_2$ production in activated RAW 264.7 cells by post-transcriptional effects on inducible cyclooxygenase expression, *Immunol. Lett.*, 50:185-188.
12. Schleimer, R., Freeland, H., Peter, S., Brown, K., and Derse, C., 1989, An assessment of the effects of glucocorticoid on degranulation, chemotaxis, binding to vascular endothelium and formation of leukotriene $B_4$ by purified human neutrophils, *J. Pharmacol. Exp. Ther.*, 250:598-602.
13. Schuller-Levis, G., Gordon, R.E., Park, E., Pendino, K., and Laskin, D., 1995, Taurine protects rat bronchioles from acute ozone-induced lung inflammation and hyperplasia, *Exp. Lung Res.*, 21:877-888.
14. Weiss, S.J. 1989, Tissue destruction by neutrophils, *New Engl. J. Med.*, 320:365-375.
15. Weiss, S.J., Kline, R., Slivka, A., and Wei, M., 1982, Chlorination of taurine by human neutrophils, evidence for hypochlorous acid generation, *J. Clin. Invest.*, 70:598-607.
16. Witko, V., Nguyen, A.T., and Descamps-Latscha, B., 1992, Microtiter plate assay for phagocyte-derived taurine chloramine, *J. Clin. Lab. Anal.*, 6:47-53.

# MYELOPEROXIDASE (MPO) MAY MEDIATE NEUTROPHIL ADHERENCE TO THE ENDOTHELIUM THROUGH UPREGULATION OF CD11B EXPRESSION - AN EFFECT DOWNREGULATED BY TAURINE

Philip P. Stapleton, H. Paul Redmond, and David J. Bouchier-Hayes

Royal College of Surgeons in Ireland
Department of Surgery
Beaumont Hospital
Dublin 9
Ireland

## ABSTRACT

Extracellular myeloperoxidase (MPO) is a macrophage modulator which stimulates release of the pro-inflammatory cytokine TNFα in addition to reactive oxygen species (ROS) generated by these cells. MPO-induced macrophage secretion of pro-inflammatory mediators indirectly upregulates neutrophil pro-inflammatory capacity through contributing to neutrophil priming for respiratory burst activity. However, to date the question concerning a direct influence on the neutrophil by MPO or the MPO-derived product hypochlorous acid (HOCl) remains to be elucidated. Taurine, the most abundant free amino acid in human neutrophils acts as an antioxidant through the formation of taurine-chloramine by sequestering HOCl. Zinc also has antioxidant properties and taurine-zinc complexes have been shown to have greater efficacy than either agent alone in protection against ROS-mediated tissue damage. The aims of this study were: (a) to determine if extracellular MPO modulates the inflammatory response through autocrine feedback on the neutrophil and to investigate if taurine either directly or indirectly through taurine-chloramine formation may further influence this pathway and (b) to evaluate the efficacy of a taurine-zinc combination in modulating MPO-induced CD11b receptor expression.

## INTRODUCTION

Neutrophil infiltration in response to chemotactic stimuli is an early and characteristic feature of the inflammatory response. These cells are central to host defense against pathogenic insult although paradoxically they have been implicated in the pathogenesis of host tissue destruction following activation[23]. Neutrophils exert protective effects through mobilisation of endogenous anti-bacterial defence mechanisms by release of an array of toxic ROS through respiratory burst activity in addition to proteolytic enzyme cascades. However, extracellular release of ROS plays a major role both in the initiation and perpetuation of host tissue damage following neutrophil activation. The principal ROS produced by the respiratory burst pathway are the superoxide anion ($O_2^-$), hydrogen peroxide ($H_2O_2$) and hypochlorous acid (HOCl)[22]. While much recent research interest has focused on methods to attenuate the production of both $O_2^-$, and $H_2O_2$ in order to reduce host tissue damage, these studies have met with limited success.

With the exception of some noteworthy studies[2,3] little attention has centered on HOCl, perhaps the most toxic of the ROS, and a product of the $H_2O_2$-myeloperoxidase (MPO)-chloride component of the respiratory burst pathway. HOCl is produced by MPO through oxidation of chloride ions ($Cl^-$) at the expense of $H_2O_2$ following activation of the respiratory burst[22]. MPO is a heme-enzyme found in neutrophil azurophilic granules and has been associated with both pro-[6] and anti-inflammatory properties[20]. Taurine, the most concentrated free amino acid in human granulocytes[8] functions as an anti-oxidant in response to NAD(P)H oxidase-dependent activation of the respiratory burst[14] pathway through sequestration of HOCl to form its stable chlorinated derivative taurinechloramine[19]. We have previously reported that in addition to sequestration of HOCl, taurine upregulates MPO activity in human neutrophils which is associated with a concomitant decrease in the levels of the chemotactic agent $LTB_4$[14]. We proposed that taurine had anti-inflammatory properties through two mechanisms, the direct removal of HOCl and the indirect depression of $LTB_4$ levels through upregulation of MPO.

Although a pivotal component of the respiratory burst pathway, the ability of MPO to interact with the inflammatory response is not restricted to an intracellular environment. Extracellular MPO upregulates TNF$\alpha$ and IFN$\alpha/\beta$ secretion by mouse macrophages in addition to ROS generation by these cells[11]. Furthermore pre-incubation with human recombinant MPO enhances mouse macrophage-mediated cytotoxicity towards the TNF$\alpha$ sensitive cell line WEHI 164, and titers of systemic TNF$\alpha$ and IFN$\alpha/\beta$ are elevated in response to intravenous MPO administration. MPO is therefore an immunomodulator through direct interaction with mouse macrophages. It is not known whether extracellular MPO can influence neutrophil action in a similar manner. If it does, the effect of taurine on MPO may further modify neutrophil function which could be important as plasma taurine levels are reduced in many pathological states.

Zinc like taurine is a known antioxidant[1]. Supraphysiological concentrations of zinc *in vitro* have antioxidant-like effects in organelle-based systems and isolated cell systems. Administration of pharmacological doses of zinc *in vivo* protects against general and liver specific pro-oxidants. Increased susceptibility to oxidative damage has been shown in the membranes of some tissues of subjects following a zinc deficient diet. A protective effect of taurine and zinc on peroxidation-induced damage in photoreceptor outer segments was shown by Pasantes-Morales and Cruz[17] who postulated that a naturally occurring taurine-zinc complex may act in a physiological protective capacity.

The aim of the present study was to assess the ability of MPO to act as an autocrine stimulator of the neutrophil. We were further interested to elucidate a role if any for taurine or a taurine-zinc complex in regulation of MPO-mediated neutrophil functional capacity.

## MATERIALS AND METHODS

### Reagents

Human leukocyte myeloperoxidase was purchased from Sigma Chemical Co., Ltd., Poole, Dorset. U.K. Zinc was purchased from Merck, Darmstedt Germany. NaOCl was from BDH Ltd., Poole, Dorset, U.K.

### Materials for Cell Isolation

Dulbecco's Phosphate Buffered Saline (PBS) was supplied by Gibco Limited, Paisley, Scotland. Ficoll-Paque® was obtained from Pharmacia, Uppsala, Sweden. Lithium heparin vacutainers were purchased from Sarstedt, Ireland. Dextran 150 Injection in 0.9% NaCl was supplied by Sigma Chemical Co., Ltd., Poole, Dorset, U.K.

### Isolation of Neutrophils

The procedure for the isolation and counting of human neutrophils has been previously described[14].

### Measurement of Human Neutrophil Respiratory Burst Activity and Phagocytosis

Neutrophil respiratory burst activity and phagocytosis were assessed using the BURSTTEST and Phagotest kits respectively both supplied by Orpegen, Heidelberg Germany. The experimental procedure was as described in the manufacturer's protocol. For both procedures whole blood (50 μl) was incubated with MPO (50-200 mU) for 1hr at 37°C in a humidified 5% $CO_2$ incubator prior to assay. Analysis was on a FACScan flow cytometer (Becton Dickenson). A minimum number of 10,000 cells was collected and analysed using the software Lysis II.

### Measurement of CD11b Receptor Expression

Whole blood (50 μl) was incubated with MPO (50-200 mU) for the indicated time period after which it was stained with 10 μl of FITC-conjugated anti-Leu15 (anti-CR/CD11b) mouse anti-human mAb (Becton Dickenson) or 10 μl FITC-conjugated isotype IgG2a control mAb (Becton Dickenson) and incubated for a further 15 min at room temperature. CD11b receptor expression on neutrophils was analysed on a FACScan flow cytometer (Becton Dickenson). The mean channel fluorescence (MCF) intensity of stained cells was detected with FL1 on the basis of a minimum number of 10,000 cells collected and analysed using the software Lysis II.

### Taurine/Zinc Modulation of MPO-Induced CD11b Receptor Expression

Briefly these experiments were set up as follows: whole blood (50 μl) was incubated with (a) MPO (50 mU) plus taurine (10 μl of either a 0.1M or 0.2M solution) and incubated for 2 hr at 37°C in a humidified 5% $CO_2$ incubator (b) taurine, for 1hr prior to addition of MPO which after gently vortexing was incubated for a further 1hr period or (c) MPO, for 1hr prior to addition of taurine and further incubated for 1hr. After the required incubation period receptor expression was determined as described above. Where the effect of zinc or a taurine-zine combination was investigated the experimental procedure was as above with the exception that either zinc alone (10 μl of a 100 ppm solution) or the taurine-zinc combination replaced taurine.

### Chloramine Formation and Identification

Taurinechloramine (0.2M) was synthesised by reacting taurine with NaOCl (sodium hypochlorite) at room temperature[24]. Identification was by U.V. spectrophotometry. Taurinechloramine absorbs maximally at 250-254 nm. Taurine alone does not absorb in the U.V. region while hypochlorite absorbs maximally at 295-302 nm[14].

### Hypochlorite and Taurinechloramine Modulation of MPO-Induced CD11b Receptor Expression

Whole blood (50 μl) was incubated with NaOCl (10 μl of a 0.2 M solution) or taurine (10 μl of a 0.2M solution) or taurinechloramine (10 μl of a 0.2 M solution) and incubated for 30 min at 37°C in a humidified 5% $CO_2$ incubator. After the required incubation period receptor expression was determined as described above.

### Statistical Analysis

Statistical differences were determined using Student's t-test for paired data (not significant, $P>0.05$)

## RESULTS AND DISCUSSION

MPO increased both respiratory burst activity and phagocytosis although the observed increases did not reach statistical significance (data not shown). CD11b levels were significantly increased, however, compared to control cells preincubated with the saline vehicle (Fig. 1). Taurine had no significant effect on receptor expression (Fig. 2).

Taurine modulated MPO-induced CD11b receptor expression in a concentration dependent manner. Interestingly, we observed that at the higher concentration, taurine maintained receptor density at levels comparable to controls even when introduced into the system 1 hr after MPO addition (Fig. 3). From earlier experiments (Fig. 1) we have demonstrated that there is a significant MPO-induced increase in CD11b receptor density after this time suggesting that taurine modulates receptor expression by reverting levels expressed back to similar levels expressed by untreated cells. This may be a further mechanism through which taurine functions as an anti-inflammatory agent *in vivo*.

We used NaOCl to mimic the effects of endogenously produced HOCl to study the effects of the $H_2O_2$-MPO-$Cl^-$ system on CD11b expression. Receptor expression was significantly downregulated by NaOCl (Fig. 4). Taurinechloramine demonstrated a similar profile to that of taurine, augmenting the physiological importance of taurine as a cellular protective agent.

Zinc alone did not influence receptor expression. When whole blood was preincubated with taurine or zinc plus MPO, CD11b expression was not affected suggesting that the protective effect of taurine and zinc was more influential than the ability of MPO to induce expression of this receptor (Fig. 5). Similarly when blood was incubated with taurine and zinc prior to addition of MPO the upregulatory effect of MPO was negated. Surprisingly taurine and zinc in combination were unable to attenuate the ability of MPO to enhance CD11b expression when added after MPO (Fig. 5). This result was unexpected as taurine alone at this concentration abrogated MPO-mediated receptor upregulation and although zinc was unable to negate the MPO affect it was anticipated that their combination would attenuate MPO-induced upregulation of CD11b.

Modern therapeutic strategies for the management of a chronic inflammatory response are limited by varying inter-person efficacy and toxic side effects. While much research has focused on modulation of the deleterious effects of an overactive respiratory burst pathway few significant advances have been made in recent times. Several independent lines of experimental evidence have clarified a role for neutrophils in the pathophysiology

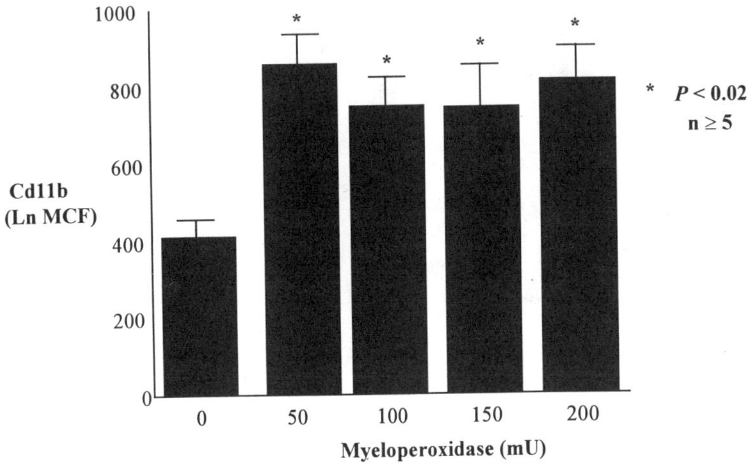


**Figure 1**. The effect of myeloperoxidase on neutrophil CD11b receptor expression.

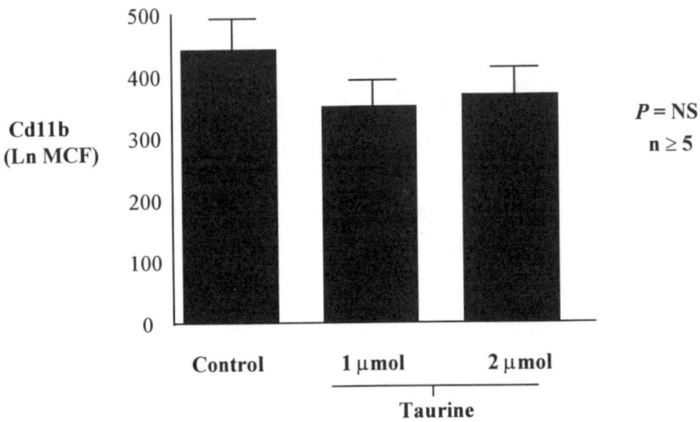


**Figure 2**. The effect of taurine on human neutrophil CD11b receptor expression.

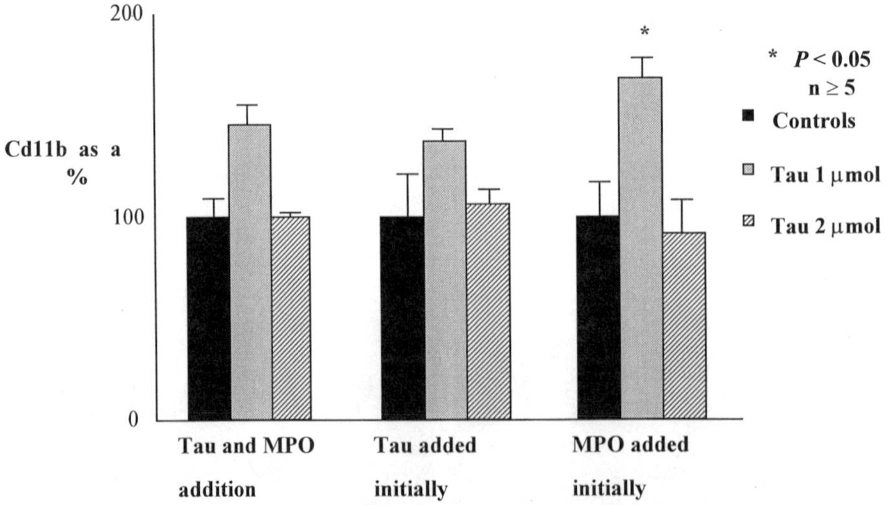


**Figure 3**. The effect of taurine on MPO-induced neutrophil CD11b receptor expression.

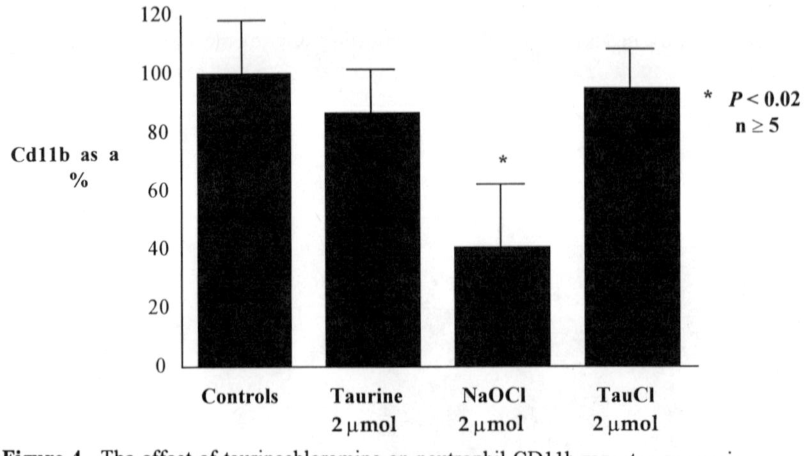


**Figure 4**. The effect of taurinechloramine on neutrophil CD11b receptor expression.

of inflammation[9,23]. Adhesion to endothelium primes neutrophils for physiological activation in response to cytokines and other stimuli. The $\beta_2$-integrins are a family of cell surface adhesive receptors that play a pivotal role in neutrophil adherence to endothelial surfaces. Bound neutrophils have been shown to respond differently to pro-inflammatory stimuli than circulating neutrophils. The initial response is slower but once initiated is greater in magnitude increasing potential for host tissue damage. Attenuation of the $\beta_2$-integrin, CD11b/CD18 (Mac-1) by the pharmacological agent NPC 15669 attenuates neutrophil-mediated endothelial damage through inhibition of cell surface CD11b/CD18 expression[18]. Furthermore, anti-CD11b antibodies have proved to be beneficial in reducing the clinical signs of inflammation in animal models[4].

In our studies we have investigated if MPO acts on the neutrophil to influence its functional capacity. An inevitable consequence of inflammatory cell activation is extracellular release of intracellular components, a process referred to as degranulation. Extracellular MPO is a known macrophage modulator stimulating release of the pro-inflammatory cytokine TNFα in addition to ROS generation by these cells. The ability of MPO to influence macrophage secretion of pro-inflammatory mediators is by itself indirectly upregulating neutrophil pro-inflammatory capacity through contributing to neutrophil priming by these agents. However, to date the question concerning a direct influence on the neutrophil by MPO or the MPO derived product HOCl remains to be elucidated. We have shown that MPO significantly upregulates human neutrophil CD11b receptor expression whilst unaffecting either *E.coli*-induced respiratory burst activity or phagocytosis. The ability of MPO to enhance surface CD11b receptor density in our system may indicate an important role for extracellular MPO *in vivo*. Increased ability of neutrophils to attach to and transcend endothelial barrier will as a consequence increase focal concentration of activated cells at the point of inflammation thereby enhancing expression of ROS produced locally and potentiating host tissue damage.

When we examined the ability of taurine to influence CD11b receptor expression no difference was found between treated and non-treated cells. This was as expected since taurine is a naturally occurring compound which over the past decade has been used in numerous clinical situations with minimal or no adverse effects. However, the ability of taurine to modulate MPO-mediate CD11b expression is a new and exciting development. Taurine maintained surface receptor expression at levels comparable to those expressed by untreated cells when administered simultaneously with MPO, prior to MPO, or even when administered after incubation with the enzyme for 1 hr. These data suggest a direct effect of taurine on CD11b. We know that the concentration of MPO used in these experiments induces significantly elevated surface expression over the chosen time period yet this is reversed by subsequent addition of taurine. In this situation, taurine may be functioning as an enantiostatic agent by maintaining CD11b levels within a normal physiological range sufficient to ensure neutrophil margination whilst mediating against enhanced neutrophil infiltration at local inflammatory sites.

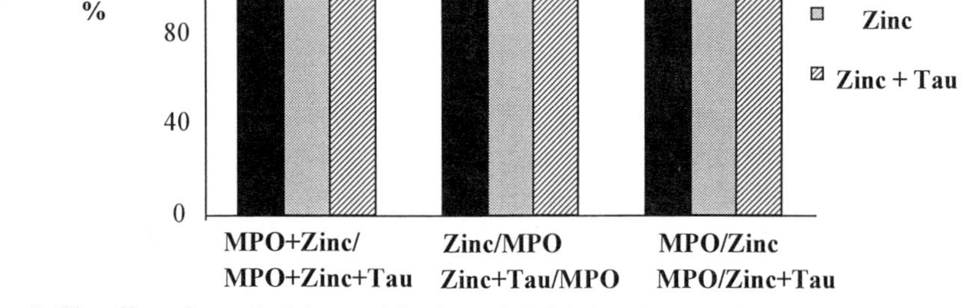


**Figure 5**. The effect of a taurine/zinc combination on MPO-induced neutrophil CD11b receptor expression.

We have previously reported taurine to upregulate MPO activity in neutrophils[14] an effect we proposed to be intrinsically anti-inflammatory as increased MPO activity correlated with decreased levels of the chemotactic agent $LTB_4$. In these current studies where we have demonstrated modulation of MPO-induced CD11b expression, we once again propose this to indicate an anti-inflammatory role for taurine. Our current and previous observations detailing effects mediated through interaction of taurine with MPO would appear to be contradictory. However, in this study we also assessed the ability of taurine to regulate MPO activity in a cell-free system and have found taurine to have no effect on MPO (data not shown). This indicates that taurine alone does not modulate MPO activity and that previous findings suggest that within a cellular system taurine upregulates MPO activity not directly but through interaction with at least one other component.

Zinc is a known anti-oxidant and Bray and Bettger[1] outlined two mechanisms of zinc protection; a) the protection of sulphydryl groups (SH) against oxidation and the inhibition of production of reactive oxidant species. Chvapil[5] reviewed the importance of zinc as a membrane stabiliser. This ability was said to arise from the interaction of zinc with sulphydryl groups on membrane proteins to form stable compounds. This effect was observed irrespective of temperature and therefore a non-enzymic mechanism was proposed. Increased susceptibility to oxidative damage has been shown in the membranes of some tissue of subjects following a zinc deficient diet. It has been reported that taurine deficient cats exhibit disruption and disorganisation of the membrane surrounding the tapetal rods. There is a corresponding depletion of zinc on the periphery of these rods in these animals. It was suggested that taurine and zinc act in cooperation to maintain stability of this membrane[17]. A synergistic relationship between taurine and zinc has been proposed by other groups. Several lines of experimental data suggest that in combination zinc and taurine function more effectively than either agent alone. A membrane stabiliser action was proposed as the mechanism underlying the protective effect of taurine and zinc against retinol-induced damage in human lymphoblastoid cells[16]. The plausable benefits of a physiological association between taurine and zinc has generated much interest since Pasantes-Morales et al[17] postulated that a naturally occurring taurine-zinc complex acts in a physiological protective mechanism against ROS-induced host tissue damage.

Zinc alone had no effect on neutrophil CD11b expression. Furthermore, when added simultaneously with MPO (Fig. 5a) or when incubated with whole blood prior to MPO (Fig. 5b) no elevation in CD11b expression was recorded. These effects may possibly be explained by membrane stabilizing effects of zinc. We were interested to investigate the possible effects of a zinc-taurine combination. Results were similar for either agent alone when added with MPO or addition prior to the enzyme, however cells cultured with MPO initially induced CD11b expression which was not attenuated by the addition of zinc and taurine in combination (Fig. 5c). The result would suggest that the ability of taurine to modulate post-expressed levels of CD11b is attenuated by zinc.

The $H_2O_2$-MPO-$Cl^-$ system is a critical component of oxygen-dependent neutrophil-mediated host defense against infection, but it is also involved in the initiation and promotion of host tissue damage. Neutrophils are very well equipped for their defence role. They contain an array of proteolytic enzymes and also produce reactive oxidants and the potent anti-bacterial agent HOCl following phagocytic stimulation. These products, although effective in eliminating pathogenic insults are the primary source of tissue damage in chronic inflammation. HOCl is a powerful oxidising agent which will indiscriminately attack tissue from both host and pathogen origin. Activated human leukocytes utilise the $H_2O_2$-MPO-$Cl^-$ system to generate taurinechloramine[10]. A pivotal role for taurine in maintaining cellular viability following stimulation was supported by Thomas et al[21] having shown taurine to preserve leukocyte integrity by competing for HOCl. This view is shared by Lin et al[12] who intimated a buffering role for taurine in preventing oxidative damage to cells from attack by HOCl and other oxidants.

Neutrophil CD11b receptor expression was significantly depressed following incubation with NaOCl which correlated with decreased viability. The ability of the chlorinated derivative to preserve cellular viability and maintain functional capacity indicates the physiological importance of maintaining adequate endogenous extracellular taurine. In pathological conditions where plasma taurine levels decrease, the potential to protect against HOCl-mediated host tissue damage may be attenuated. Recent evidence has suggested that taurinechloramine is by itself an agent of potential importance. Taurinechloramine protects against inhibition of collagenase activity[7] and it has been shown to inhibit the production of nitric oxide and TNF$\alpha$ by RAW 264.7 cells[15] in addition to attenuation of $PGE_2$ and IL-6 generation by mouse macrophages[13]. Our results while supporting a protective capacity for taurine through sequestration of HOCl also show that taurinechloramine maintains cellular capacity in a manner similar to that shown by taurine.

In conclusion, these studies have shown that extracellular MPO can act in an autocrine fashion to upregulate CD11b receptor density on neutrophils. Increased receptor expression will augment neutrophil population attachment to endothelial surfaces *in vivo*. Overexpression of neutrophil migration will lead to higher focal concentrations of activated inflammatory cells which enhances the possibility of local tissue damage. Furthermore, crosslinked $\beta_2$-integrins prime neutrophils for activation of the respiratory burst by the inflammatory mediators TNF$\alpha$, fMLP and GM-CSF. The ability of taurine to maintain receptor density at levels comparable to those expressed by control cells suggests a possible role for taurine *in vivo* in maintaining neutrophil-endothelium interaction within a window which affords normal neutrophil functioning whilst controlling against possible elevated neutrophil influx across the endothelium and enhanced neutrophil respiratory burst activity. The concentration of MPO released extracellularly is difficult to assess, circulating MPO is not currently used as a diagnostic marker. Therefore, how the levels used in this study compare to levels circulating extracellularly is difficult to ascertain. However, circulating levels will be dependent upon factors such as extent and duration of the inflammatory response. We propose that we have identified a novel and important role for taurine as an anti-inflammatory agent. Questions arise from these studies not least the mechanism(s) through which MPO induces Cd11b expression, and those through which taurine modulated this effect. The results provided by these studies warrant further investigation into possible other MPO-mediated effects through autocrine stimulation of neutrophils and the role taurine may play in modulating these effects.

## ACKNOWLEDGMENT

This work was supported by the Charitable Infirmary Charitable Trust, Dublin, Ireland.

## REFERENCES

1. Bray, T.M. and Bettger, W.J., 1990, The Physiological Role of Zinc as an Antioxidant, *Free Radical Biology Medicine*, 8:281-291.
2. Carey, P.D., Byrne, K., Jenkins, J.K., Sielaff, T.D., Walsh, C.J., Fowler, A.A., and Sugerman, H.J., 1990, Ibuprofen attenuates hypochlorous acid production from neutrophils in porcine acute lung injury, *J Surg. Tres.*, 49:262-270.
3. Carey, P.D., Wakefield, C.H., Thayeb, A., Monson, J.R.T., Darzi, A., and Guillou, P.J., 1994, Effects of minimally invasive surgery on hypochlorous acid production by neutrophils, *Br. J. Surg.*, 81:557-560.

4. Chopp, M., Zhang, R.L., Chen, H., Li, Y., Jiang, N., and Rusche, J.S., 1994, Postischemic administration of an anti-Mac-1 antibody reduces ischemic cell damage after transient middle cerebral artery occlusion in rats, *Stroke*, 25:869-875.
5. Chvapil, M., 1973, New aspects in the biological role of zinc: A stabilizer of macromolecules and biological membranes, *Life Sci.*, 13:1041-1049.
6. Couser, W.G., 1993, Mediation of immune glomeruli injury, *Clin. Investig.*, 71:808-811.
7. Davies, J. M. S., Horwitz, D. A., and Davies, K. J. A., 1994, Inhibition of collagenase activity by N-chlorotaurine, a product of activated neutrophils, *Arthritis and Rheumatism*, 37:424-427.
8. Fukuda, K., Hirai, Y., Yoshida, H., Nakajima, T., and Usai, T., 1982, Free Amino Acid Content of Lymphocytes and Granulocytes Compared, *Clin Chem*, 28:1758-1761.
9. Langdale, L.A., Flaherty, L.C., Liggitt, H.D., Harlan, J.M., Rice, C.L., and Winn, R.K., 1993, Neutrophils contribute to hepatic ischemia-reperfusion injury by a CD18-independent mechanism, *J. Leuko. Biol.*, 53:511-517.
10. Lampert, M.B. and Weiss, S.J., 1983, The chlorinating potential of the human monocyte, *Blood*, 62:645-651.
11. Lefkowitz, D.L., Mills, K.C., Moguilevsky, N., Bollen, A., Vaz, A., and Lefkowitz, S., 1993, Regulation of macrophage function by human recombinant myeloperoxidase, *Immunology Letters* 36:43-50.
12. Lin, Y. Y., Wright, C. E., Zagorski, M., and Nakanishi, K., 1988, 13C-NMR study of taurine and chlorotaurine in human cells, *Biochim. Biophys. Acta*, 969:242-248.
13. Marcinkiewicz, J., Grabowska, A., Bereta, J., and Stelmaszynska, T., 1995, Taurine chloramine, a product of activated neutrophils, inhibits in vitro the generaton of nitric oxide and other macrophage inflammatory mediators, *J. Leuko. Biol.*, 58:667-674.
14. McLoughlin, D.M., Stapleton, P.P., and Bloomfield, F.J., 1991, Influence of taurine and a substituted taurine on the respiratory burst pathway in the inflammatory response, *Biochem. Soc. Trans.*, 19:73-78.
15. Park, E., Quinn, M. R., Wright, C. E., and Schuller-Levis, G., 1993, Taurine chloramine inhibits the synthesis of nitric oxide and the release of tumor necrosis factor in activated RAW 264.7 cells, *J. Leuko. Biol.*, 54:119-124.
16. Pasantes-Morales, H., Wright, C.E., and Gaull, G.E., 1984, Protective effect of taurine, zinc and tocopheral on retinol-induced damage in human lymphoblastoid cells, *J. Nutr.*, 114:2256-2261.
17. Pasantes-Morales, H., and Cruz C., 1984, Protective effect of taurine and zinc on peroxidation-induced damage in photoreceptor outer segments, *J. Neurosci. Res.*, 11:303.
18. Pou, S., Gunther, M.R., Pou, W.S., Cao, G.L., Bater, J.M., Cohen, M.S., Burch, R. M., and Rosen, G.M., 1993, Effect of NPC 15669, an inhibitor of neutrophil recruitment and neutrophil-mediated inflammation, on neutrophil function *in vitro*, *Biochem. Pharmacol.*, 45:2123-2127.
19. Stapleton, PP. and Bloomfield, FJ., 1993, Effect of zwitterions on the respiratory burst, *J. Biomed. Sci.*, 3:79-84.
20. Stendahl, O., Coble, B-I., Dahlgren, C., Hed, J., and Molin, L., 1984, Myeloperoxidase modulates the phagocytic activity of polymorphonuclear neutrophil leukocytes. Studies with cells from a myeloperoxidase-deficient patient, *J. Clin. Invest.* 73:366-373.
21. Thomas, E. L., 1979, Myeloperoxidase-hydrogen peroxide-chloride antimicrobial system: Effect of exogenous amines on anti-bactericidal action against Escherichia coli, *Infect. Immun.*, 25:110-116.
22. Weiss, S.J., Klein, R., Slivka, A., and Wei, W., 1982, Chlorination of taurine by human neutrophils evidence for hypochlorous acid generation, *J. Clin. Invest.*, 70:598-607.
23. Weiss, S.J., 1989, Tissue destruction by neutrophils, *New Engl. J. Med.*, 320:365-376.
24. Wright, C.E., Tallan, H.H., Lin, Y.Y., and Gaull, G.E., 1986, Taurine: Biological update, 55:427-453.

# NO BENEFICIAL EFFECTS OF TAURINE APPLICATION ON OXYGEN FREE RADICAL PRODUCTION AFTER HEMORRHAGIC SHOCK IN RATS

F. Niessen, W. Isselhard, and T. Minor

Institute for Experimental Medicine
University of Cologne
Robert-Koch-Str. 10
50931 Cologne
Germany

## ABSTRACT

Oxygen free radical generation contributes to the reinfusion damage after hemorrhagic shock. Taurine has been proposed to have radical scavenging properties under certain experimental conditions. Therefore the present study was undertaken to investigate if taurine would be able to attenuate adverse effects of shock/resuscitation in male rats (fasted over night).

Under pentobarbital anesthesia, hemorrhagic shock (HS) was induced for 1h by bleeding of the animal [mean arterial blood pressure (MAP) = 40 mm Hg] followed by shed blood reinfusion and another 1h period of resuscitation. Rats were divided into two groups: Treated rats (n=6) were injected with taurine (40 mg/kg body mass) prior to withdrawal of shed blood; untreated rats (n=9) received respective volumes of a normal saline solution. In untreated animals, free radical induced lipid peroxidation was documented by an increase of malondialdehyde (MDA) in the systemic circulation (nmol/ml; HPLC measurement) from 1.06 ± 0.08 during normotension (NT) to 1.35 ± 0.18** 1h after resuscitation (RS).

Accordingly, plasma levels of alanine aminotransferase (ALT) (11 ± 2; 35 ± 12; 94 ± 44 U/l, NT; HS; RS) and ammonia (120 ± 39; 532 ± 161; 224 ± 101 µg/dl) changed significantly during the experimental protocol. Hepatic ATPase-content as an indicator of energetic status of the liver fell from 4.8 ± 0.83 to 0.56 ± 0.27 after HS and recovered to only 2.7 ± 1.6 µmol/g after RS. Leukocyte infiltration of the liver was followed by tissue levels of myeloperoxidase (MPO) which did not change during HS, but rose during RS (37.9 ± 18.5; 38.6 ± 16.4; 77.5 ± 24; arbitrary units), documenting an inflammatory reaction after HS. Taurine treated rats showed levels of MDA not different from untreated rats after RS; also no differences were observed concerning enzyme concentrations and ammonia levels. The liver tissue levels of ATP and MPO revealed no differences between

*Taurine 3*
Edited by Schaffer *et al.*, Plenum Press, New York, 1998

the two groups during the various periods of the experiment. Liver tissue perfusion, as measured by Laser Doppler flowmetry, also did not show significant differences between both groups. MAP was significantly higher in the taurine-treated rats during the first 40 min of resuscitation.

It is concluded that even a relatively high dose of taurine failed to attenuate the impact of oxygen free radicals and did not improve the recovery of the rats during the early resuscitation period.

## INTRODUCTION

During hemorrhagic shock, reduced tissue perfusion occurs in various organs including the liver. Liver perfusion at reduced flow rates has been shown to produce zones of virtual anoxia in pericentral areas[1]. Reoxygenation of the hypoxic liver may lead to oxygen free radical generation and parenchymal tissue alteration as has been demonstrated *in vitro*[2] and *in vivo*[3]. The oxygen free radical scavenger superoxide dismutase was protective in reducing functional and morphological impairment of the liver upon resumption of normal flow conditions *in vitro*[2], as well as in preventing multiple organ failure after polytrauma in a clinical pilot study[4]. Taurine protects against oxidative damage under many conditions, decreasing rates of malondialdehyde formation from unsaturated membrane lipids[5]. Preoperative infusion of taurine during myocardial revascularization reduced reperfusion injury[6].

This study was designed to investigate if systemic treatment with taurine would be beneficial in reducing liver injury during hemorrhagic shock and in subsequent reinfusion *in vivo*.

## MATERIALS AND METHODS

### Animals

Male Wistar rats were housed under controlled conditions, including a 12 h light/dark cycle. The animals had free access to tap water until the day of surgery. Standard pellet food was removed the night before the experiment. The study was performed in accordance with the German legislation on protection of animals after approval by the local ethics committee. The rats were anesthetized with pentobarbital (50 mg/kg body mass i.p.) and then placed on a thermostated operating table. The rectal temperature of the animals was controlled using a thermistor probe and kept constant between 35° and 37°C. A tracheotomy was performed and the animals were mechanically ventilated with room air (Hoek Loos aIV Mk3 Amsterdam Infant Ventilator; Hoek Loos medical equipment, Schidam, NL).

The right carotid artery was cannulated, preserving the vagal nerve, and connected to a Stadtham P 23 Db pressure transducer (Gould, Oxnard, Cal) for continous monitoring of systemic blood pressure. The left jugular vein was also cannulated for volume replacement and application of the drug.

### Experimental Model

The rats were randomly assigned to either treatment with taurine (Sigma Chemical Co., St. Louis; 40 mg/kg body mass in 1 ml of saline solution administered i.v. prior to induction of hemorrhagic shock) or to the control group (1 ml of saline solution i.v.).

After the surgical manipulations, a stabilization period followed that lasted for 30 min. The rats were then bled from the carotid artery to a mean artery blood pressure of 40 mm Hg. The shed blood was kept in a heparinized syringe. Blood pressure was kept constant at 40 mm Hg for 60 min by additional blood withdrawal or reinfusion of a saline solution. After 60 min of hemorrhagic shock, the shed blood was reinfused for 5 min and the animals were observed for another 60 min. Animals were sacrificed for biochemical analysis of the liver tissue and blood plasma: 1) after the initial stabilization period (baseline), 2) at the end of 60 min of hemorrhagic shock, and 3) at the end of 60 min of reinfusion.

### Determination of Tissue-Associated Activity of Myeloperoxidase (MPO) in the Liver

Leucocyte infiltration into the liver was assessed using the measurement of MPO based on the $H_2O_2$-dependent oxidation of tetramethylbenzidine[7] from frozen liver tissue after extraction as described earlier[8].

### Assay of Lipid Peroxides in Plasma

Free radical-mediated tissue injury was approximated from the release of lipid peroxidation products into the systemic circulation, determined as plasma concentrations of malondialdehyde (MDA) by high-performance liquid chromatography (HPLC) as described previously[9,10].

### Assay of Enzyme Activities in Plasma

Parenchymal liver injury was monitored by the plasma activities of alanine aminotransferase (ALT) and lactic dehydrogenase (LDH), determined photometrically using commercialized standard kits (Fa. Boehringer, Mannheim, FRG).

### Statistics

Statistical significance between sequential measurements within the groups was determined after one-way analysis of variance and comparison of the means with the Bonferroni t-test. Significant differences between treated and untreated animals were calculated using the Student's t-test. Results were considered to be significantly different if $P<0.05$.

## RESULTS

### Hemodynamics

Initial mean arterial pressure (MAP) showed no differences between treated and untreated rats during baseline conditions (Fig. 1). Following 60 min of hemorrhagic shock with MAP kept at 40 mm Hg in either group and reinfusion of the shed blood, a rise of MAP to preshock values above 100 mm Hg was observed. The MAP in taurine-treated rats was significantly higher during the first 40 min of the reinfusion period. Liver tissue perfusion, as measured by Laser Doppler flowmetry, showed a decrease in erythrocyte flux in the livers during the shock period to approximately 60% of the baseline values, which was not affected by the application of taurine (Fig. 2). During the resuscitation period the liver perfusion values were restored to normal, which were not significantly different from

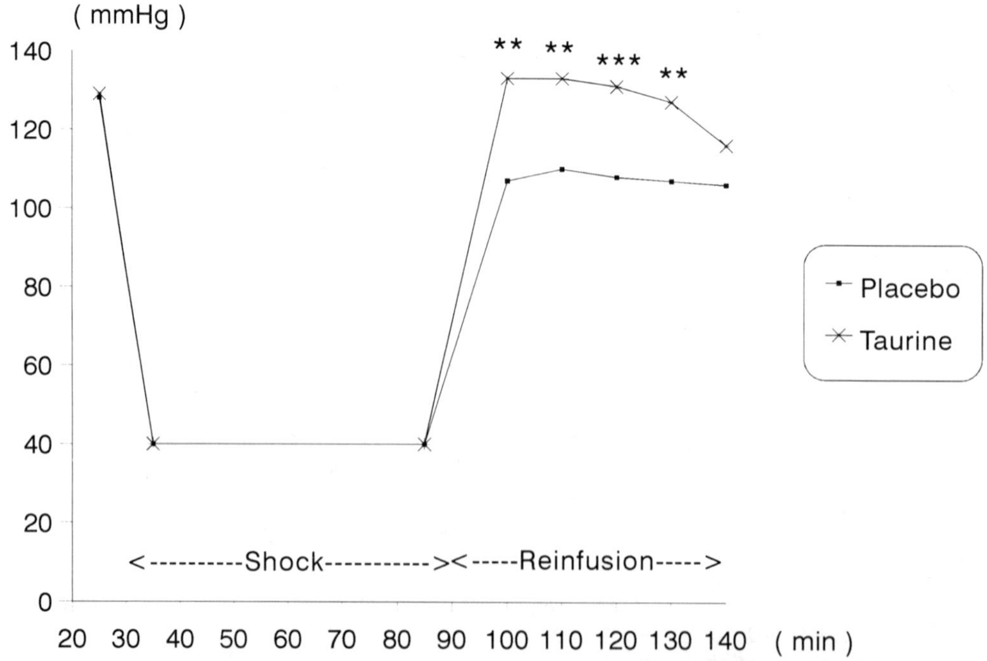


**Figure 1**. Mean arterial blood pressure (MAP) at baseline conditions, during hemorrhagic shock and during resuscitation in untreated and taurine-treated animals (Values are given as mean ± SEM of n>6 observations; indicators of SEM have been omitted for the sake of clarity; *:p< 0.05 vs baseline).

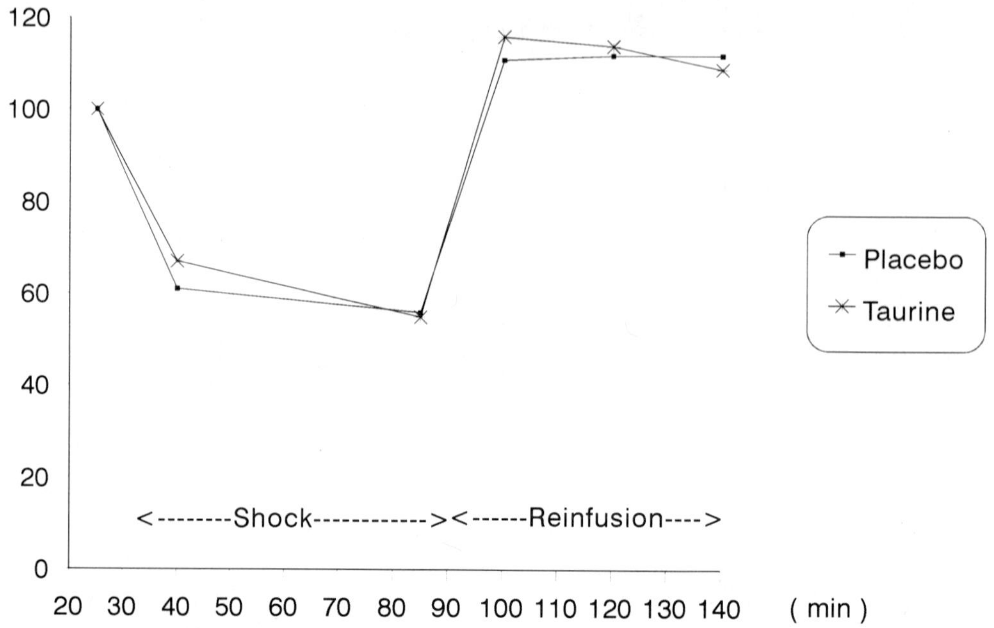


**Figure 2**. Hepatic tissue perfusion as evaluated by Laser Doppler flowmetry during the respective periods of the experiment in untreated and taurine-treated animals. (Values are given as mean ± SEM of n>6 observations; indicators of SEM have been omitttted for the sake of clarity).

the respective baseline measurements. There was no difference between treated and untreated rats.

**Tissue Analysis**

Hepatic tissue levels of myeloperoxidase were determined in order to estimate inflammatory leukocyte infiltration into the liver (Table 1). The mean activity of myeloperoxidase in the liver tissue did not change significantly during 60 min of hemorrhagic shock in either group with respect to baseline values. Upon resuscitation, however, tissue-associated activity of MPO increased concordantly in both groups, but no differences were observed between taurine-treated and untreated rats.

**Plasma Analysis**

Plasma concentrations of the lipid peroxidation product MDA are depicted in Figure 3. During hemorrhage, the systemic levels of MDA remained unchanged in either group as compared to the baseline values. After resuscitation, an increase in plasma MDA concentration was seen, which was most pronounced in the untreated group (P<0.05 vs. baseline). This increase was less pronounced and did not reach statistical significance in the taurine-treated group, however, there was no significant difference between treated and untreated groups at this time.

Table 1 summarizes the variations of liver enzymes in plasma during and after hemorrhage. A slight but not significant increase of plasma activities of ALT as well as LDH were observed during hemorrhagic shock as compared to the baseline values. Resuscitation of the animals led to a two- to three-fold rise in systemic activities of both enzymes in the two groups, with no particular benefit of taurine treatment. Blood

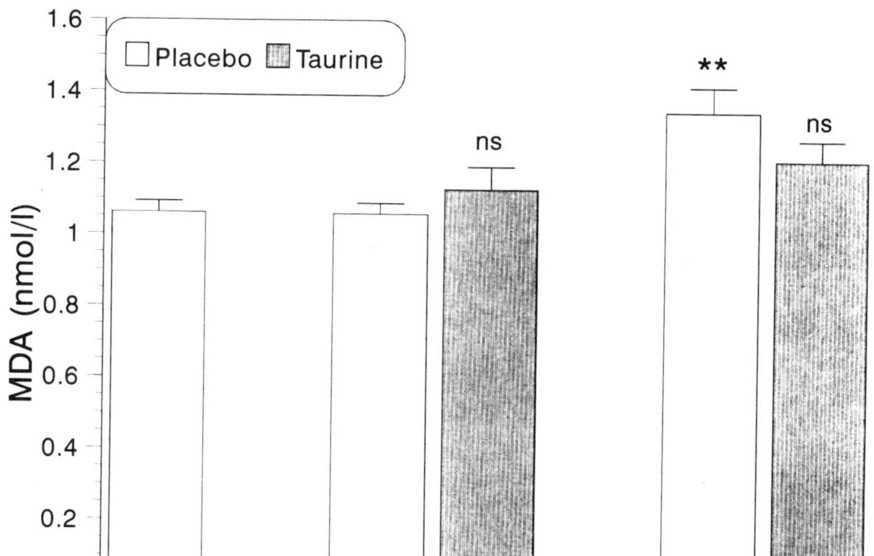


**Figure 3**. Plasmatic levels of malondialdehyde (MDA) at baseline conditions, at the end of hemorrhagic shock (HS) and after resuscitation in untreated and taurine-treated animals. (Values are given as mean ± SEM of n>6 observations; *: p< 0.05 vs baseline).

ammonia, a powerful toxin supposed to be cleared from circulation by the intact liver[11], showed significantly elevated levels during hemorrhagic shock with no differences between treated and untreated animals. Upon resumption of normal blood pressure during resuscitation the elevated ammonia concentrations were markedly reduced in all animals.

## DISCUSSION

The model of hemorrhagic shock used in this study resulted in a significant reduction of hepatic tissue perfusion to about 60% of the reference values, corresponding to the severe hypotension induced by the withdrawal of circulating blood. The observed perturbation of hepatic tissue perfusion is in keeping with the data previously reported by Vollmar et al.[12], showing a reduction of the sinusoidal perfusion rate to an average of 73 to 84% after 1 h of hemorrhagic shock as evaluated by intravital microscopy and to about 52% of the erythrocyte flux measured by Laser Doppler flowmetry. Although the circulatory disturbances in the liver were fully reversible in our experiments, as was the macrohemodynamic status of the animal, our data give indirect evidence for significant tissue alterations after resumption of normovolemia.

A considerable amount of evidence implicates oxygen derived free radicals in tissue alteration after shock or ischemia[2,3,13,14]. In our experiments we demonstrated an increase in the plasma levels of MDA, not during hemorrhagic shock, but after normalization of the systemic circulation of the animal. It can thus be presumed that peroxidative injury occurred during resumption of adequate tissue oxygenation after reinfusion of the blood. This is in agreement with previous data obtained using the isolated rat liver *in vitro*, when generation of superoxide anion radicals were observed and hepatic perfusion returned to physiological values after a period of low flow hypoxia[2].

During circulatory shock, the functional capacity of the liver to clear toxic ammonia from the systemic circulation was reduced and ammonia blood levels rose significantly. Significant parenchymal morphological damage, as judged from the release of ALT or LDH, appeared and was pronounced after reinfusion of the shed blood. These results may be interpreted as indicative of reperfusion-associated damage to cellular membranes leading to increased membrane permeability.

It is of interest that the occurrence of free radical-mediated lipid peroxidation and parenchymal enzyme release was paralleled by a two-fold rise in myeloperoxidase activity of the liver. Activated leukocytes, invading postischemic tissue, have been shown to play a major role in organ dysfunction after ischemia or shock[12,15], and they are likely sources of peroxidative alterations due to the release of oxygen free radicals upon activation. However, site specific oxidant injury to organs may also be promoted by xanthine oxidase, present in the circulation after hemorrhagic shock, after binding to the vascular endothelium[14].

Taurine protects against oxidative damage under many conditions in part by decreasing rates of malondialdehyde formation from unsaturated membrane lipids[5]. Preoperative infusion of taurine during myocardial revascularization reduces reperfusion injury[6]. Taurine has demonstrated protective actions against ischemia-reperfusion liver injury in rats[16]. However, no protective effect of taurine was demonstrated after ischemia-reperfusion on bovine retinal membrane lipid peroxidation[17]. Very interesting effects of taurine on malondialdehyde levels were observed on perfused heart muscle after reperfusion.

In this study, elevated malondialdehyde levels were observed if taurine was administered before anoxia and reperfusion, but decreased malondialdehyde levels were noted if taurine was administered after anoxia[18]. In accordance with our study, anoxia did

**Table 1**. Plasma levels of ALT, LDH and ammonia and liver content of myeloperoxidase (MPO) under baseline conditions, after 60 min of hemorrrhagic shock (HS) and after resuscitation (RS) in untreated and taurine treated animals.

| | ALT (U/l) | LDH (U/l) | $NH_3$ (μg/dl) | MPO (arb U) |
|---|---|---|---|---|
| Baseline | 11 ± 2 | 148 ± 88 | 120 ± 39 | 37.9 ± 18.5 |
| HS | | | | |
| Untreated | 35 ± 12* | 456 ± 170* | 532 ± 161* | 38.6 ± 16.4 |
| Taurine | 24 ± 8 | 363 ± 205 | 728 ± 244 | 53.7 ± 19.6 |
| RS | | | | |
| Untreated | 94 ± 44* | 1191 ± 420* | 224 ± 101* | 77.5 ± 24.0* |
| Taurine | 164 ± 103 | 1365 ± 861 | 194 ± 133 | 108.6 ± 41.7 |

Values are mean ± SD of n>6 experiments per group. (*:P<0.05 vs baseline-ANOVA and Boferroni t-test.)

not affect MDA-levels significantly. Another study demonstated a reduction in liver taurine levels during hemorrhagic shock[19].

In conclusion, the reported data, taken together with the literature, suggest that the possible antioxidative effect of taurine seems to be largely dependent on the respective experimental conditions. In our model of circulatory shock *in vivo*, a monotherapy with taurine does not seem to be of value.

## REFERENCES


1. Biasi, F., Chiarpotto, E., and Lafranco, G., et al., 1994, Oxidative stress in the development of human ischemic hepatitis during circulatory shock, *Free Radical Biol. Med.*, 17:225-233.
2. Chen, Y.X., 1993, [Protective action of taurine on ischemia-reperfusion liver injury in rats and its mechanism], *Chung Hua I Hsueh Tsa Chih.*, 73(5):276-9, 318-319.
3. Huxtable, R.J., 1992, Physiological actions of taurine, *Physiol. Rev.*, 72 (1):135-136.
4. Kaplan, B., Aricioglu, A., Erbas, D., Erbas, S., and Turkozkan, N., 1993, The effects of taurine on perfused heart muscle malondialdehyde levels, *Gen. Pharmacol.*, 24(6):1411-1413.
5. Kobelt, F., Schreck, U., and Henrich, A., 1994, Involvement of liver in the decompensation of hemorrhagic shock, *Shock*, 2:281-288.
6. Marotto, M.E., Thurman, R.G., and Lemasters, J.J., 1988, Early midzonal cell death during low-flow hypoxia in the isolated perfused rat liver: Protection by allopurinol, *Hepatology*, 8:585-890.
7. Marzi, I., Rose, S., and Bühren, V., 1991, Prävention des Multiorganversagens nach Trauma durch Einsatz des Sauerstoffradikalfängers rh-Superoxid Dismutase, *Intensivmed*, 28:443 (abstract).
8. Marzi, I., Bauer, M., Secchi, A., Bahrami, S., Redl, H., and Schlag, G., 1995, Effect of anti-tumor necrosis factor alpha on leukocyte adhesion in the liver after hemorrhagic shock: an intravital microscopic study in the rat, *Shock*, 3:27-33.
9. Matsumoto, S. and Yokoyama, H., 1994, Changes in liver tissue amino acids during hemorrhagic shock, *Masui*, 43(3):334-338.
10. Milei, J., Ferreira, R., Llesuy, S., Forcada, P., Covarrubias, J., and Boveris, A., 1992, Reduction of reperfusion injury with preoperative rapid intravenous infusion of taurine during myocardial revascularization, *Am. Heart J.*, 123(2):339-345.
11. Minor, T., Isselhard, W., Kunz, G., and Saad, S., 1991, Involvement of oxygen in harvesting injury of liver. An experimental study including substrate free organ persufflation to evaluate a specific therapeutic approach, *Res. Exp. Med.*, 191:167-175.
12. Minor, T., Isselhard, W., and Berghaus, K., 1993, Parenchymal and vascular endothelial cell injury in the hypoxic and reperfused rat liver: Evidence for superoxide anion generation by perfusion with ferricytochrome, *Biomed. Pharmacother*, 47:213-218.
13. Minor, T., Yamaguchi, T., and Isselhard, W., 1995, Treatment of preservation reperfusion injury by platelet-activating factor antagonism in the rat liver graft, *Transplant Proc.*, 27:522-523.

14. Minor, T., Sturz, J., Klauke, H., and Isselhard, W., 1995, Reduction of oxidative tissue injury and endothelial dysfunction by graduated reperfusion after cardioplegic arrest in isolated rat hearts, *Free Radical Biol. Med.*, 18:621-626.
15. Suzuki, K., Ota, H., Sasagawa, S., Sakatani, T., and Fujikura, T., 1983, Assay method for myeloperoxidase in human polymorphonuclear leukocytes, *Anal. Biochem.*, 132:345-352.
16. Tan, S., Yokoyama, Y., Dickens, E., Cash, T.G., Freeman, B.A., and Parks, D.A., 1993, Xanthine oxidase activity in the circulation of rats following hemorrhagic shock, *FRBM*, 15:407-414.
17. Vollmar, B., Menger, M., Lang, G., Post, S., and Messmer, K., 1993, Hepatic micro-circulation in hemorrhagic shock, *Prog. Appl. Microcirc.*, 19:85-105.
18. Wong, S.H.Y., Knight, J.A., Hopfer, S.M., Zahria, O., Leach, C.N.J., and Sunderman, F.W., 1987, Lipoperoxides in plasma as measured by liquid chromatographic separation of malondialdehyde-thiobarbituric acid adduct, *Clin. Chem.*, 33:214-220.
19. Zhou, M., Ma, T., and Tseng, M.T., 1991, Effects of taurine and ketamine on bovine retinal membrane lipid peroxidation, *Neuroscience*, 45(2):461-465.

# INFLUENCE OF TAURINE SUPPLEMENTATION ON ISCHEMIC PRESERVATION OF THE ISOLATED RAT KIDNEY

Christoph Licht,[1] Eckhard Kriegesmann,[2] Thomas Minor,[2]
Peter Wingenfeld,[1] Wolf Isselhard,[2] and Dietrich Volker Michalk[1]

[1] Department of Pediatrics
University of Cologne
Joseph-Stelzmann-Str. 9
D - 50924 Cologne, Germany
[2] Institute for Experimental Medicine
University of Cologne
Robert-Koch-Str. 10
D - 50924 Cologne, Germany

## INTRODUCTION

Taurine (2-aminoethanesulfonic acid) is an endogenous amino acid with unique functions as a modulator of transmembraneous calcium transport, an osmoregulator and putatively a free radical scavenger[4,7,12].

In previous experiments with cell cultures of renal tubular epithelium and colon epithelium, we showed that taurine supplementation of preservation medium (1 and 10 mmol/l, respectively) reduced tissue alterations during hypoxia and reoxygenation[7,13].

In a subsequent project, these results were confirmed using a rat liver model. We found that addition of taurine to the preservation medium improved the recovery of cold stored livers during postischemic reperfusion[9,10].

The aim of the present study was to examine if taurine had a similar protective effect on the preservation of rat kidney.

## MATERIALS AND METHODS

Kidneys of male Wistar rats (300-500 g) were preserved either in pure University of Wisconsin (UW) medium or in UW medium supplemented with 1 mmol/l of taurine (Sigma Chemical Co., St. Louis, MO, USA). Under sodium pentobarbital anesthesia (50 mg/kg i.p.) and artificial respiration via a tracheotomy the left kidney was isolated, separated from the blood circulation, immediately perfused by gravity (100 mm Hg) with the respective preservation medium, removed, placed in the appropriate medium and then subjected to ischemia at 4°C for 15 hours. Afterwards, the kidney was mounted on a Langendorff apparatus and reperfused for 1 hour with oxygen-saturated Krebs-Henseleit medium at 37°C.

As parameters for functional recovery during this period, renal vascular resistance [kPa/ml x g x min] and oxygen consumption [mmol/g x min] were measured every 10 minutes over a period of 1 hour, with the first measurement occurring after 1 minute. Renal vascular resistance was calculated according to the formula TVR=(AP-VP)/Q with VP assumed to be zero (TVR: total vascular resistance, AP: arterial pressure, VP: venous pressure, Q: flow). Oxygen consumption was calculated by the difference of arterial oxygen saturation and that of the venous effluent (acid-base laboratory ABL-2, Radiometer, Copenhagen, Denmark).

Furthermore, renal tissue ATP content [mmol/g] was determined by high-pressure liquid chromatography (HPLC)[8] and the ECP (energy charge potential) was calculated according to Atkinson using the equation ECP=(ATP+1/2 ADP)/(ATP+ADP+AMP)[1]. We also measured lactate tissue concentration [mmol/g] by standard enzymatic methods according to Gercken[3]. The measurements were carried out at the endpoint of the 1 hour reperfusion period.

Results are given as means ± SD. Statistical analysis of the data was performed by unpaired and two-sided Student's t-test. Differences were considered to be significant if $p<0.05$.

## RESULTS

### Renal Vascular Resistance

Renal vascular resistance showed elevated values during initial reperfusion, but decreased afterwards in both the taurine and the control group to reach a minimal level after about 30 minutes (Fig. 1).

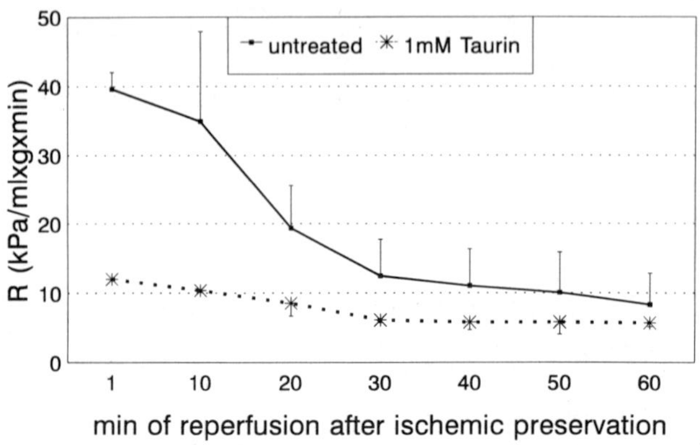


**Figure 1.** Renal vascular resistance during isolated reperfusion. Data are given as means ± SD.

## Oxygen Consumption

Renal vascular resistance of the taurine group was lower than that of the control group during the whole reperfusion period of 1 hour, with the greatest difference found during the initial period. Oxygen consumption increased in both the taurine and the control group during the initial 30 minutes of reperfusion (and then decreased in both groups) (Fig. 2). Oxygen consumption of the taurine group was larger than that of the control group, with the greatest difference occurring during the initial period.

## Renal Tissue Content of ATP and ECP

Renal tissue content of ATP (Fig. 3), as well as ECP (Fig. 4), showed a significantly ($p<0.05$) higher level in the taurine group in comparison to the control group.

## Lactate Tissue Concentration

We found no difference between tissue lactate concentration of the taurine and the control group.

# CONCLUSION

In kidney transplantation in man, an increased vascular resistance is assumed to be a sign of severe postischemic damage or acute rejection[5,11]. It is assumed that hypoxia causes inter alia an imbalance between endothelial derived relaxing factor (EDRF) and endothelin, caused by decreased release of EDRF and augmented release of endothelin[2,6]. Corresponding to this, renal vascular resistance in our experiments was elevated in both the taurine supplemented and the untreated control group after 15 hours of ischemic preservation. The resistance of the taurine group, however, was lower than that of the control group, with the main difference seen at the beginning of the reperfusion period. During the reperfusion period, there was a decrease in resistance of both groups. A mini-

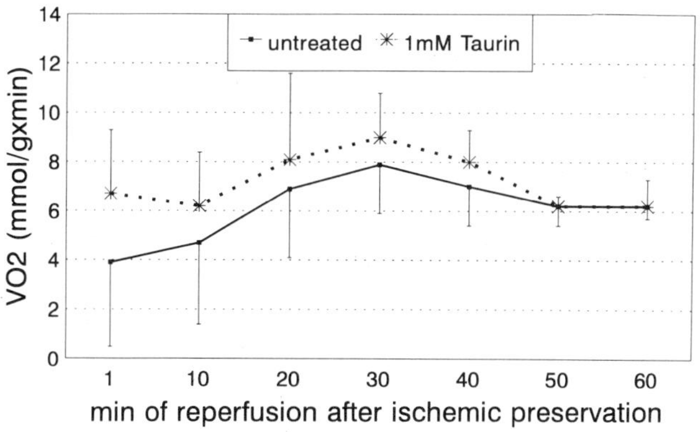


**Figure 2.** Oxygen consumption during isolated reperfusion. Data are given as means ± SD.

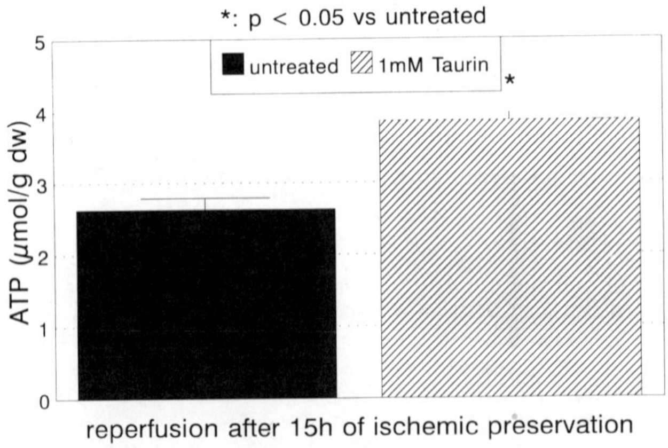


**Figure 3.** Renal tissue content of ATP after isolated reperfusion for 1 hour. Data are given as means ± SD. *:p<0.05 vs. untreated.

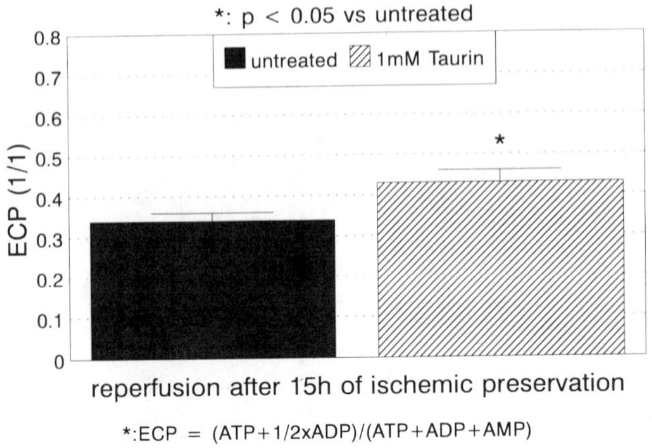


**Figure 4.** Energy charge potential after isolated reperfusion for 1 hour. Data are given as means ± SD. *:p<0.05 vs. untreated.

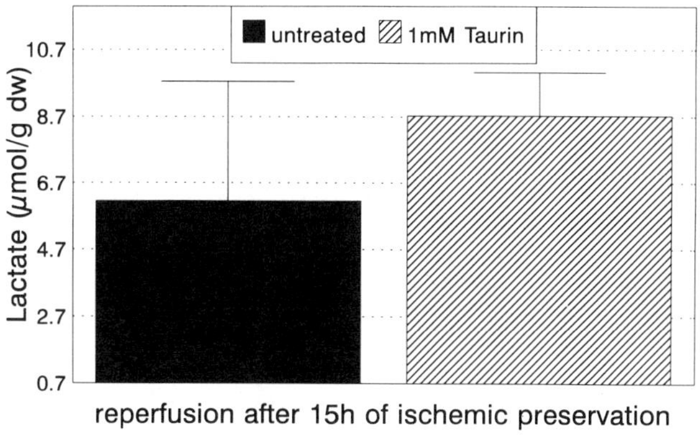


**Figure 5.** Lactate tissue concentration after isolated reperfusion for 1 hour. Data are given as means ± SD.

mum resistance level was reached after about 30 minutes. This result suggests a possible cytoprotective effect of taurine during ischemic preservation.

Oxygen consumption of the taurine group during postischemic reperfusion was higher than that of the control group, with the main effect apparent after about 30 minutes, suggesting that taurine supplementation during ischemic preservation causes a higher level of aerobic metabolism during postischemic reperfusion, although the observed difference did not reach statistical significance in these experiments.

Renal tissue content of ATP and the valve of ECP showed a significantly higher level of postischemic energy metabolism in the taurine group ($p<0.05$). Lactate tissue concentration exhibited no significant difference between the taurine and the control group.

Further experiments using a modified protocol will be carried out in order to test the results of this pilot study in the isolated rat kidney. We will extend the ischemic period from 15 to 24 hours, a common time limit for organ transplantation in man. Furthermore, we will utilize a taurine concentration of 10 mmol/l, a concentration with promising results in previous cell culture experiments.

## REFERENCES


1. Atkinson, D.E., 1968, The energy charge of the adenylate pool as a regulatory parameter. Interaction with feedback modifiers, *Biochemistry,* 7:4030-4034.
2. Bacha, E.A., Hervé, P., Murakami, S., Chapelier, A., Mazmanian, G.-M., de Montpreville, V., Détruit, H., Libert, J.-M., and Dartevelle, P., 1996, Lasting beneficial effect of short-term inhaled nitric oxide on graft function after lung transplantation, *J. Thorac. Cardiovasc. Surg.*, 112:590-598.
3. Gercken, G., 1961, Glukose-, Milchsäure-, ATP-, ADP-Konzentrationen im arteriellen und venösen Coronarblut des Hundes bei Sauerstoff-Mangel, *Pflügers Arch.*, 272:323.
4. Huxtable, R.J., 1992, Physiological actions of taurine, *Physiol. Rev.*, 72:101-163.
5. Hoyer, P.F., Offner, G., and Brodehl, J., 1983, Acute rejection episodes after renal transplantation in children, *Clin. Nephrol.*, 19 (2):61-66.
6. Lerman, A. and Burnett, J.C., 1992, Intact and altered endothelium in regulation of vasomotion, *Circulation* 86 (suppl. III):III-12 - III-19.

7. Michalk, D.V., Wingenfeld, P., Licht, C., Ugur, T., and Farrokh Siar, L., 1996, The mechanisms of taurine mediated protection against cell damage induced by hypoxia and reoxygenation 223-232, in: "Taurine 2. Advances in experimental medicine and biology", Huxtable et al., eds., Vol. 403, Plenum Press, New York and London.
8. Minor, T., Wingenfeld, P., Gehrmann, U., Strübind, S., Isselhard, W., and Michalk, D.V., 1994, Energetic status and viability of porcine kidney cells after hypoxic, hypothermic preservation in UW solution and subsequent reoxygenation: the influence of taurine, *Pathophysiology,* 1:247-250.
9. Minor, T., Yamaguchi T., and Isselhard, W., 1995, Effects of taurine on liver preservation in UW solution with consecutive ischemic rewarming in the isolated perfused rat liver, *Transpl. Int.*, 8:174-179.
10. Minor, T., Yamaguchi, T., Klauke, H., Wingenfeld, P., Michalk, D.V., and Isselhard, W., 1996, Taurine reduces experimental liver injury after cold ischemic preservation and a period of rewarming prior to reperfusion, 157-161, in: Taurine 2. Advances in Experimental Medicine and Biology, Huxtable et al., eds., Vol. 403, Plenum Press, New York and London.
11. Perrella, R.R., Duerinckx, A.J., Tessler, F.N., Danovitch, G.M., Wilkinson, A., Gonzalez, S., Cohen, A.H. and Grant E.G., 1990, Evaluation of renal transplant dysfunction by duplex doppler sonography: a prospective study and review of the literature, *Am. J. Kidney Dis.*, 15 (6):544-550.
12. Schaffer, S.W., Ballard, C., and Azuma, L., 1994, Mechanisms underlying physiological and pharmacological actions of taurine on myocardial calcium transport, 171-196, in: Taurine in Health and Disease. Advances in Experimental Medicine and Biology, Huxtable, R.J. and Michalk, D.V., eds., Vol. 359, Plenum Press, New York and London.
13. Wingenfeld, P., Minor, T., Gehrmann, U., Strübind, S., Isselhard, W., and Michalk, D.V., 1995, Hypoxic cellular deterioration and its prevention by the amino acid taurine in a transplantation model with renal tubular cells (LLC PK1), *In Vitro Cell. Dev. Biol.*, 31:483-486.

# Part 5:
# Osmotic Actions of Taurine

# TAURINE: AN OSMOLYTE IN MAMMALIAN TISSUES

H. Pasantes-Morales, O. Quesada, and J. Morán

Institute of Cell Physiology
National University of Mexico
Mexico City, Mexico

One of the distinctive features of taurine is its presence at high levels in most animal tissues[5,3]. Although differences exist among cells and species, taurine is consistently found in mM concentrations. It is noteworthy that excitable tissues including brain, striatal muscle and heart contain large, inert, and intracellular taurine pools[3,9]. The most prominent example in this respect is the retina. In all species so far examined, retinal taurine levels exceed 20 mM and in some species its concentration is as high as 60 mM[39]. In addition to these high levels, taurine in excitable tissues has a very slow turnover rate. In rat organs, values for taurine turnover rate fit into three main groups: those with a relatively fast rate calculated in less than 1 day and include liver, kidney and pancreas; a second group, with a medium rate of about 2-3 days comprises lung, spleen, intestine, testes, bone marrow and the third group, with the slowest rate of more than 3 days and up to 7 days, is represented by brain, heart and muscle[30].

Taurine in cells is found largely in a soluble form in the cytosol, and except in the liver, its involvement in metabolic cell function is null. All these properties of taurine are difficult to ascribe to a functionally active compound, but they fit very well into a role as an osmolyte. A molecule acting as an osmolyte should be present in sufficiently high amounts to contribute to water movement, be readily accessible to ensure a rapid and efficient mobilization and be as dispensable as possible for cell function in a way that its concentration may change with minimal disturbance of cell metabolism. In addition, an osmolyte candidate should be able to respond to changes in cell volume caused by fluctuations in external or internal osmolarity by modifying its concentration in the necessary direction to restablish the osmotic equilibrium of the cell. Osmolytes in cells consists of the inorganic ions, K, Cl and Na, and a variety of organic molecules, including amino acids and derivatives, sugars, polyalcohols, and amines.

## TAURINE RELEASE ASSOCIATED WITH CHANGES IN CELL VOLUME

Cells in culture from both primary cultures or cell lines are convenient models to initially examine the mechanisms subserving the fluxes of taurine occurring in connection

*Taurine 3*
Edited by Schaffer *et al.*, Plenum Press, New York, 1998

to changes in cell volume. More integrated preparations should be subsequently examined and all results observed *in vitro* must be validated *in vivo*. The observations summarized in this chapter correspond only to nervous tissue preparations: astrocytes and neurons in culture, retina, brain slices and brain *in vivo*.

When cultured brain cells, astrocytes as well as neurons, are exposed to 50% hyposmotic media, they first behave as perfect osmometers, following the predictions of the Boyle-van't Hoff relationship, and swell to increase volume up to 70-80%. Immediately thereafter, a volume regulatory response ensues which leads to cell volume recovery despite the persistence of the hyposmotic condition. The regulatory volume decrease exhibits a rapid phase lasting 2-6 min after swelling and a more slow phase during the following 10-15 min[18,19]. This cell volume regulation is accomplished by the extrusion of osmotically active intracellular solutes. In these cells, and in fact in most cells so far examined, taurine is one of these osmolytes. In astrocytes and neurons, cell swelling elicited by hyposmotic medium results in the immediate activation of taurine release, with a peak at about 1-2 min after the stimulus followed by a slower phase of inactivation. The time course of taurine efflux closely parallels the decrease in cell volume[20]. Other amino acids (as well as the inorganic ions Cl and K) are also released from cultured astrocytes and neurons in response to cell swelling[17]. Chloride efflux (measured with $^{125}$I as a tracer) activates and inactivates more rapidly than taurine, whereas K (measured as $^{86}$Rb) is somewhat delayed[24,25] (Fig. 1). During the volume regulatory decrease that occurs in cells exposed to 50% hyposmotic medium, more than 80% of the cellular taurine pool is released.

The efflux of taurine associated with swelling induced by exposure to hyposmotic solutions is also observed in more integrated preparations. Figure 2 show the release of taurine from the excised chick retina and from rat hippocampal slices. The time course, onset and offset of the efflux are similar to those observed in cultured cells. Essentially all of the intracellular pool is released from the tissue upon hyposmotic stimulus. Results in Figure 2 are similar to those reported by Law[7] and by Oja et al.[14] in brain cortical slices. Hyposmotic-induced release of taurine has been observed also *in vivo*. Using microdialysis

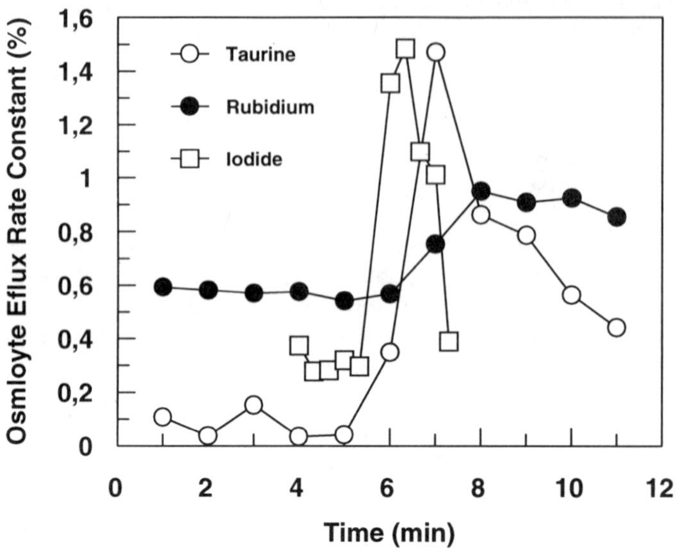


**Figure 1.** Time course of the swelling-activated efflux of osmolytes from cultured cerebellar astrocytes. Cells were preloaded with $^{3}$H-taurine, $^{125}$I and $^{86}$Rb (as tracers for Cl and K, respectively) and superfused with isosmotic medium. At min 5 superfusion was continued with 50% hyposmotic medium. Fractional efflux of osmolytes is expressed as radioactivity released at each fraction as percent of total radioactivity accumulated at loading.

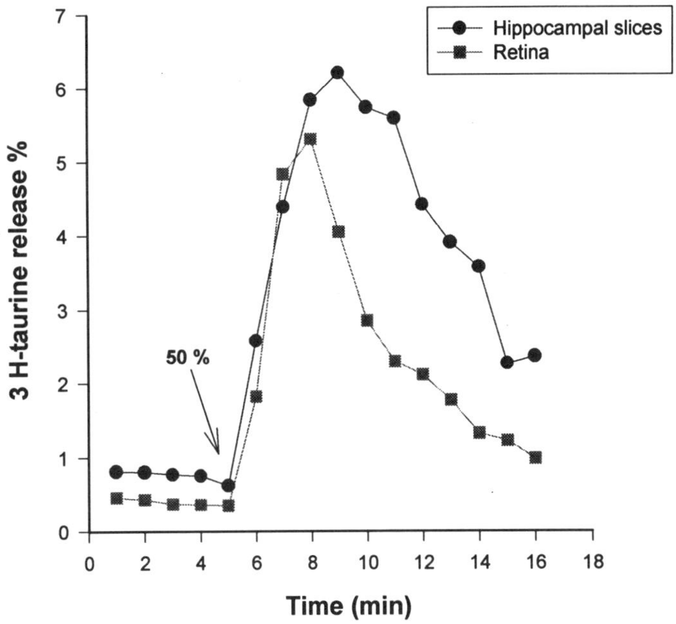


**Figure 2.** Swelling-activated efflux of taurine from chick retina and rat hippocampal slices. Tissues preloaded with $^3$H taurine were superfused with isosmotic medium during 10 min and subsequently with 50% hyposmotic medium (at the arrow). Results are expressed as in Figure 1.

techniques, Solis et al.[29] and Lehman et al.[8] described a prominent efflux of taurine when the hippocampus was perfused with media of decreased osmolarity. It is noteworthy that in the hippocampus, taurine is the amino acid exhibiting the lowest efflux threshold. Whereas the excitatory amino acid glutamate is released only after drastic reductions in osmolarity, the efflux of taurine is more osmotically sensitive[29]. The increase in taurine observed in the perfusate is a reflection of the decrease in tissue taurine. Taurine is part of a group called "idiogenic" osmolytes, a term coined to describe the compounds playing a crucial role in the adaptive response of the brain to changes in cell volume in the face of severe hypo- and hypernatremia. In fact, more than 90% of the brain taurine content decreases during experimental hyponatremia in rats[38]. This decrease in brain taurine during hyponatremia was reported in 1980 by Thurston et al.[33] who were the first to propose an osmoregulatory role for taurine in brain and heart. Interestingly, Law reported on the decease on brain taurine levels during the physiological mild hyponatremia accompanying pregnancy[6].

Taurine is also functioning as an osmolyte during cell volume regulation subsequent to cell shrinkage in hypernatremia. In cultured astrocytes and neurons grown in hyperosmotic media, taurine levels markedly rise and together with other organic osmolytes, compensate for the new osmotic gradient, thereby avoiding cell shrinkage[26]. Accumulation of organic osmolytes, including taurine, occurs slowly, reaching a maximal concentration only 12 h after the hyperosmotic stimulus. The same adaptive response is observed in the brain *in vivo* under conditions of experimental hypernatremia or during pathological states leading to increased plasma osmolarity[2]. The intracellular concentration of taurine increases and contributes in this way, the creation of a new pool of organic osmolytes capable of equilibrating the intracellular osmotic pressure with the imposed external gradient[2]. This new condition has to be taken into consideration when handling patients with chronic hypernatremia or under conditions when the plasma or the extracellular space becomes hyperosmotic. A too rapid correction of a hyperosmotic condition may result in a slow rate of intracellular organic osmolyte accumulation resulting in a transient phase of cell edema, which in the brain at least, may be fatal[34,38].

The fluxes of taurine activated by hypo- or hyperosmolarity have markedly different time courses, suggestive of the involvement of different pathways for taurine movement under the two conditions. Taurine release activated by cell swelling occurs immediately after the stimulus and inactivates within a few minutes. In contrast, taurine uptake stimulated by cell shrinkage is a slow process requiring several hours. Studies *in vitro* in cultured astrocytes have contributed to clarify the mechanisms involved in the response of taurine to changes in cell volume. The increase in taurine concentration observed after shrinkage seems to result from an increased activity of the Na-dependent taurine carrier, which normally transports the amino acid from the external medium into the cell. The $K_m$ of the transporter is unchanged but the number of transporter molecules is enhanced, most likely due to a *de novo* synthesis of the carrier. In astrocytes grown in hyperosmotic media, an increase in the mRNA expression of the carrier has been found in response to cell shrinkage[36].

The Na-dependent carrier does not seem to be involved in the efflux of taurine associated with cell swelling. Rather, current ideas favour a diffusional pathway as the mechanism of taurine release[16]. Taurine release activated by swelling is Na-independent and nonsaturable, with the change in volume not affecting the saturable component of taurine uptake. In contrast, the diffusional coefficient value of the uptake curve is markedly increased upon swelling. In addition, it has been shown that upon swelling, taurine movement occurs following the direction of the concentration gradient. Under normal conditions, i.e. when the intracellular taurine concentration largely exceeds that in the medium, activation of the volume-sensitive pathway drives taurine outside, promoted by the concentration gradient. The opposite occurs when the gradient is experimentally reversed and external levels are higher than intracellular content. Under these conditions, taurine accumulates in the cell and blocks RVD[27]. This suggests the involvement of a channel-like molecule in taurine efflux activated by swelling. This taurine leak pathway is ubiquitous since it has been described in essentially all cell types examined so far. It has been proposed that this taurine pathway could be an anion channel since known blockers of these channels are potent inhibitors of volume-activated taurine release[16]. Since only a small proportion of taurine in the cell is present as an anion, the proposed channel should allow the passage of zwitterions, such as taurine and other amino acids. In fact, it has been shown that, besides taurine, a number of amino acids with similar pharmacological sensitivity are also released from brain cells upon swelling. Electrophysiological studies have confirmed that swelling activates anion channels which are permeable to amino acids, including taurine, with a $P_{Cl}/P_{aa}$ ranging from 0.1 to 0.2[22]. The opposite is observed in the anion channel, phospholemman, which exhibits a much higher permeability for taurine than Cl[12]. The properties of this channel are described in detail in this book. Phospholemman is present in a variety of cell types that exhibit taurine efflux activated by swelling, including astrocytes and neurons in culture. Other candidates for taurine efflux are the recently cloned Cl channels, ClC2, ClC5, ClC6 and ClC7, a channel of the family of the porins and P-glycoprotein, which is an ubiquitous large pore. It has been proposed that the channel responsible for taurine and Cl fluxes activated during swelling may be a novel protein called VSOAC[31]. These putative taurine transporting channels are described in more detail in other chapters. Besides the molecular identity of the taurine leak pathway, other questions remain obscure, such as the signal transducing the change in cell volume to the activation of taurine efflux, the involvement of $Ca^{++}$ in the process and the sensing of volume recovery which leads to the inactivation of taurine efflux.

## TAURINE AS AN OSMOLYTE IN EXCITABLE TISSUES: WHY?

The presence of molecules functioning as osmolytes is required for cells in species naturally exposed to fluctuations in external osmolarity. In terrestrial animals, cells are surrounded by media of strictly controlled osmolarity and the requirement of these molecules is not so obvious. Nevertheless, in essentially all animal cells, the mechanisms adjusting cell volume in response to changes in osmolarity have been carefully preserved, and cell volume regulation in brain, muscle or kidney in terrestrial vertebrates are essentially identical to those of clams or worms living in the river deltas. Cell volume control may be necessary in a number of pathological situations, but also perhaps during normal functioning of the cell.

The response of brain cells to changes in external osmolarity in a number of pathological conditions has been investigated. Clinical states of hyponatremia or hypernatremia, diabetes and hepatic encephalopathy result in brain cell edema or shrinkage, which is rapidly corrected by activation of cell volume control mechanisms. More local changes in cell volume in brain occur in ischemia and in epilepsies. In these later, the importance of volume spaces has been recently emphasized by experiments showing that hypersynchronization of neuronal activity, which is a prominent feature of epileptiform activity, critically depends on the size of the extracellular space, reduced by cell swelling occurring as consequence of neuronal hyperexcitability[23]. These results support the observation by Tratchman et al.[35] on the higher susceptibility to seizures of taurine deficient cats.

Besides these changes in cell volume associated with various pathological conditions, numerous aspects of the normal functioning of cells generate microscopic osmotic gradients, leading to local volume changes of intracellular compartments. In highly specialized and organized tissues, such as brain, muscle or photoreceptors, uncontrolled changes in intracellular water may result in disruption of the cytoarchitecture, incompatible with normal cell function. These changes, which are result of normal cell function, such as ion gradients in brain cells, $Ca^{++}$ movement in the sarcoplasmic reticulum, or ion fluxes in response to light in photoreceptors, have to be carefully controlled. Thus, it might not be casual that it is in these tissues where taurine is present in the highest concentration. The photoreceptor outer segment contains no other osmolyte but taurine and is not unlikely that the dramatic consequences of taurine deficiency in these cells i.e. membrane disorganization and cell dysfunction[1], are somehow related to inadequate control of discal and intradiscal spaces, and water content.

Another interesting example of cell volume control in which taurine appears to play a crucial role is described in the work by Ottersen et al.[13], who showed that during hyponatremia, taurine from the Purkinje cells is released and rapidly accumulated by astrocytes, which ordinarily contain no taurine. As a consequence, neurons are spared from swelling. This redistribution of taurine is easily explained by the hyposmolarity-sensitive difusional pathway (described above), which is present in both astrocytes and neurons. This phenomenon may be part of the known nursing role that astrocytes play on neurons.

Taurine levels are quite different among tissues and species and the reason for this variability has been correlated with the water content of cells. Two such successful correlations have been found in brain. One reported by Lehman et al.[21] showing that taurine content in three species containing high (mouse), medium (rat) and low (gerbil) concentrations of taurine is highly correlated with the amount of water in the cells.

Another interesting observation in this respect was made by van Gelder[37], who pointed out a correlation between taurine levels and the activity of glucose metabolism in different species. During the glucose oxidation reaction new water, known as oxidation water or metabolic water, is formed at the proportion of 0.6 g of water per 1 g of glucose. Oxidation of other metabolites, such as fat yields even more oxidation water than carbohydrates. It would be interesting to examine whether a correlation with total oxidation water could explain the cell and species variations in taurine levels.

## PROTECTIVE EFFECTS OF TAURINE AND ITS ROLE AS AN OSMOLYTE.

One of the most active areas of taurine research is related to its protective effects in a variety of cells and tissues deteriorated by a number of different insults. Beneficial effects of taurine have been described in ischemic and diabetic heart and during cold storage and hypoxia in liver, renal and human colon cells[11,32,40,41]. Major damage in these tissues occurs sometimes as a consequence of reoxygenation, and the beneficial effects of taurine seems to continue at this stage as well. Taurine also exhibits dramatic protective effects in retinal photoreceptors exposed to light and to peroxidative damage[15] and it is also helpful in lung cells damaged by exposure to ozone[28]. In most cases, taurine seems to prevent ion overload through damaged membranes. Membrane lipid peroxidation is often associated with ischemia-hypoxia and other injury models. It is consistently reported that taurine prevents excessive ion influx through these membranes, although it is controversial whether taurine acts as an antioxidant or modulates membrane permeability by some other mechanism, which may involve a direct effect on Cl fluxes. In the ischemic heart, the cardioprotective effect of taurine involves attenuation of the $Ca^{++}$ paradox-induced injury

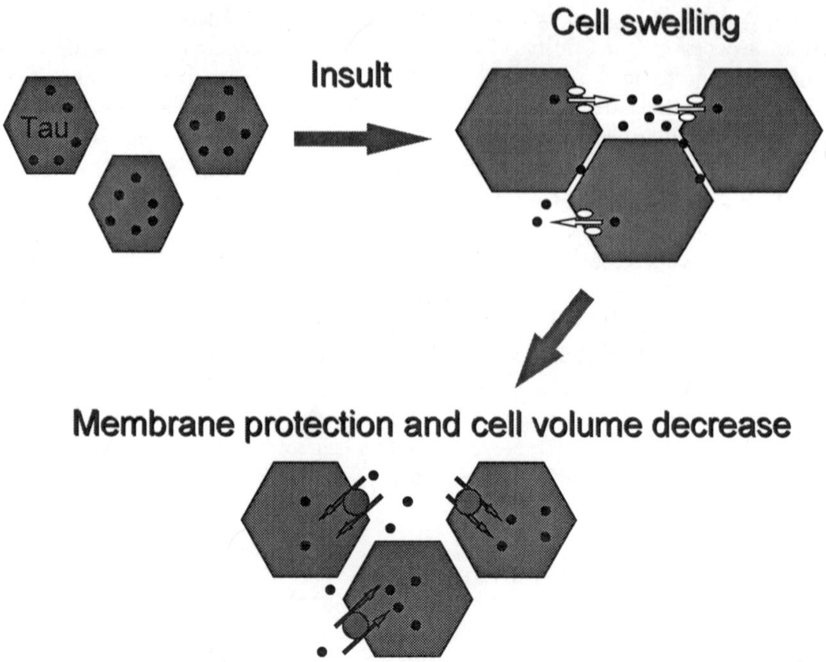


**Figure 3.** Cell swelling activates taurine leak raising its concentration at the extracellular space to levels required for its protective action. Once the normal conditions are restored, the energy-dependent carrier replenishes the intracellular pool.

occurring after reperfusion. This effect may be related to the well known effect of taurine on $Ca^{++}$ homeostasis in cardiac cells[4].

The main criticism consistently made of these studies was that the concentrations of taurine required to exhibit its protective actions were in the mM range, far above those present in plasma and in the extracellular space. However, quite often cell swelling is a prominent feature of the injury condition which is attenuated by taurine. Therefore, it is possible that taurine efflux subsequent to swelling raises its extracellular concentration to a level required for the beneficial effects observed on the cell membranes (Fig. 3). It is in this way that two important avenues of taurine research converge to suggest a unique function of this amino acid in the biology of mammalian tissues.


## ACKNOWLEDGMENTS

The experimental results of this work were supported in part by grants number 400 346-5-2262PN from the CONACYT and IN202093 from DGAPA, UNAM, Mexico.


## REFERENCES:


1. Hayes, K.C., Carey, R.E., and Schmidt, S.Y., 1975, Retinal degeneration associated with taurine deficiency in the cat, *Science*, 188:949-951.
2. Heilig, C.W., Stromski, M.E., Blumenfeld, J.D., Lee, J.P., and Gullans, S.R., 1989, Characterization of the major brain osmolytes that accumulate in salt-loaded rats, *Am. J. Physiol.*, 257:F1108-F1116.
3. Huxtable, R.J., 1992, Physiological actions of taurine, *Physiol. Rev.*, 72:101-163.
4. Huxtable, R.J. and Sebring, L.A., 1983, Cardiovascular actions of taurine, in Prog. Clin. Biol. Res. "Sulfur Amino Acids: Biochemical and Clinical Aspects", Kuriyama, K., Huxtable, R. J. and Iwata, H., eds., Alan R. Liss, New York, Vol. 125, pp 5-37.
5. Jacobsen, J.G and Smith, L.H.J., 1968, Biochemistry and physiology of taurine and taurine derivatives, *Physiol. Rev.*, 48:424-511.
6. Law, R.O., 1989, Effects of pregnancy on the contents of water, taurine and total amino nitrogen in rat cerebral cortex, *J. Neurochem.*, 53:300-302.
7. Law, R.O., 1994, Taurine efflux and the regulation of cell volume in incubated slices of rat cerebral cortex, *Biochim. Biophys. Acta*, 1221:21-28.
8. Lehmann, A., 1990, Derangements in cerebral homeostasis: a common denominator for stimulation of taurine and phosphoethanolamine release, in Prog. Clin. Biol. Res. "Taurine: Functional Neurochemistry, Physiology and Cardiology", Pasantes-Morales, H., Martin, D.L., Shain, W. and Del Río, R.M., eds., Wiley-Liss, Inc., New York, Vol. 351, pp 337-347.
9. Mandel, P. and Pasantes-Morales, H., 1978, Taurine in nervous tissue, in: "Reviews of Neurosciences", Ehrenpreis, S. and Kopin, I.J., eds., Raven Press, New York, Vol. 3, pp 157-194.
10. McManus, M.L. and Churchwell, K.B., 1994, Clinical significance of cellular osmoregulation, in: "Cellular and Molecular Physiology of Cell Volume Regulation", Strange, K., ed., CRS Press, Boca Ratón, pp 63-67.
11. Minor, T., Yamaguchi, T., Klauke, H., Wingenfeld, P., Michalk, D., and Isselhard, W., 1996, Taurine reduces experimental liver injury after cold ischemic preservation and a period of rewashing prior to reperfusion, in Adv. Exp. Med. Biol., "Taurine 2", Huxtable, R.J., Azuma, J., Kuriyama, K., Nakagawa, M. and Baba, A., eds., Plenum Press, New York, Vol. 403, pp 157-161.
12. Moorman, J.R., Ackerman, S.J., Kowdley, G.C., Griffin, M.P., Mounsey, J.P., Chen, Z., et al., 1995, Unitary anion currents through phospholemman channel molecules, *Nature*, 377:737-740.
13. Nagelhus, E.A., Lehmann, A., and Ottersen, O.P., 1993, Neuronal-glial exchange of taurine during hypo-osmotic stress: a combined immunocytochemical and biochemical analysis in rat cerebellar cortex, *Neuroscience*, 54:615-631.
14. Oja, S.S. and Saransaari, P., 1992, Cell volume changes and taurine release in cerebral cortical slices, in Adv. Exp. Med. Biol., "Taurine: Nutritional Value and Mechanisms of Action", Lombardini, J.B., Schaffer, S.W. and Azuma, J., eds., Plenum Press, New York, Vol. 315, pp 369-374.

15. Pasantes-Morales, H., 1986, Current concepts on the role of taurine in the retina, in: "Progress in Retinal Research", Osborne, N. and Chader, G., eds., Pergamon Press, Oxford, Vol. 5, pp 207-229.
16. Pasantes-Morales, H., 1996, Volume regulation in brain cells: cellular and molecular mechanisms, *Met. Brain Dis.*, 11:187-204.
17. Pasantes-Morales, H., Alavéz, S., Sánchez-Olea, R., and Morán, J., 1993, Contribution of organic osmolytes to volume regulation in rat brain cells in culture, *Neurochem. Res.*, 18:445-452.
18. Pasantes-Morales, H., Chacón, E., Murray, R.A., and Morán, J., 1994, Properties of osmolytes fluxes activated during regulatory volume decrease in cultured cerebellar granule neurons. *J. Neurosci. Res.*, 37:720-727.
19. Pasantes-Morales, H., Murray, R.A., Lilja, L.. and Morán, J., 1994, Regulatory volume decrease in cultured astrocytes, *Am. J. Physiol.*, 266:C165-C171.
20. Pasantes-Morales, H. and Schousboe, A., 1988, Volume regulation in astrocytes: a role for taurine as osmoeffector, *J. Neurosci. Res.*, 20:505-509.
21. Puka, M., Sundell, K., Lazarewicz, J.W., and Lehmann, A., 1991, Species differences in cerebral taurine concentrations correlate with brain water content, *Brain Res.*, 548:267-272.
22. Roy, G.,1995, Amino acid current through anion channels in cultured human glial cells, *J. Membr. Biol.* 147:35-44.
23. Saly, V. and Andrew, R.D., 1993, CA3 neuron excitation and epileptiform discharge are sensitive to osmolarity, *J. Neurophysiol.*, 69:1-9.
24. Sánchez-Olea, R., Morales-Mulia, M., García, O., and Pasantes-Morales, H., 1996, Chloride channel blockers and polyunsaturated fatty acids similarly inhibit the volume activated pathways for Cl and taurine in cultured cerebellar granule neurons, *Am. J. Physiol.*, 270:C1703-C1708.
25. Sánchez-Olea, R., Morán, J., Martínez, A.. and Pasantes-Morales, H., 1993, Volume-activated Rb transport in astrocytes in culture, *Am. J. Physiol.*, 33:C836-C842.
26. Sánchez-Olea, R., Morán, J., and Pasantes-Morales, H., 1992, Changes in taurine transport evoked by hyperosmolarity in cultured astrocytes, *J. Neurosci. Res.*, 32:86-92.
27. Sánchez-Olea, R., Morán, J., Schousboe, A., and Pasantes- Morales, H., 1991, Hyposmolarity-activated fluxes of taurine in astrocytes are mediated by diffusion, *Neurosci. Lett.*, 130:233-236.
28. Shuller-Levis, G., Quinn, M.R., Wright, C., and Park, E., 1994, Taurine protects against oxidant-induced lung injury: possible mechanism(s) of action, in Prog. Clin. Biol. Res. "Taurine: Functional Neurochemistry, Physiology and Cardiology", Pasantes-Morales, H., Martín, D.L., Shain, W. and Del Río, R.M., eds., Willey-Liss, Inc., New York, Vol. 351, pp 31-39.
29. Solís, J.M., Herranz, A.S., Herreras, O., Lerma, J., and Martin del Río, R., 1988, Does taurine act as an osmoregulatory substance in the rat brain?, *Neurosci. Lett.*, 91:53-58.
30. Spaeth, D.G. and Schneider, D.L., 1974, Turnover of taurine in rat tissues, *J. Nutr.*, 104:179-186.
31. Strange, K., Emma, F., and Jackson, P.S., 1996, Cellular and molecular physiology of volume-sensitive anion channels, *Am. J. Physiol.*, 270:C711-C730.
32. Tatsumi, T., Matoba, S., Kawahara, A., Kobara, M., et al., 1996, Cardioprotective effect of taurine on calcium paradox in streptozotocin-induced diabetic rat hearts, in Adv. Exp. Med. Biol., "Taurine 2", Huxtable, R.J., Azuma, J., Kuriyama, K., Nakagawa, M. and Baba, A., eds., Plenum Press, New York, Vol. 403, pp 539-549.
33. Thurston, J.H., Hauhart, R.E., and Dirco, J.A., 1980, Taurine: a role in osmotic regulation in mammalian brain and possible clinical significance, *Life Sci.*, 26:1561-1568.
34. Trachtman, H., 1992, Cell volume regulation: a review of cerebral adaptive mechanisms and implications for clinical treatment of osmolal disturbances: II, *Pediatr. Nephrol.*, 6:104-112.
35. Trachtman, H., Barbour, R., Sturman, J.A., and Finberg, L., 1988, Taurine and osmoregulation: taurine is a cerebral osmoprotective molecule in chronic hypernatremic dehydration, *Pediatr. Res.*, 23:35-39.
36. Uchida, S., MooKwon, H., Yamauchi, A., Preston, A.S., Marumo, F., and Haudler, J.S., 1992, Molecular cloning of the cDNA for an MDCK cell $Na^+$ and $Cl^-$-dependent taurine transporter that is regulated by hypertonicity, *Proc. Natl. Acad. Sci.*, 89:8230-8234.
37. van Gelder, N.M., Neuronal discharge hypersyncrony and the intracranial water balance in relation to glutamic acid and taurine redistribution: migraine and epilepsy, in Prog. Clin. Biol. Res. "Taurine: Functional Neurochemistry, Physiology and Cardiology", Pasantes-Morales, H., Martin, D.L., Shain, W. and Del Río, R.M., eds., Willey-Liss, Inc., New York, Vol. 351, pp 1-20.
38. Verbalis, J.G. and Gullans, S.R., 1991, Hyponatremia causes large sustained reduction in brain content of multiple organic osmolytes in rats, *Brain Res.*, 567:274-282.
39. Voaden, M.J., Draedn, A.C.I., Marshall, J., and Lake, N., 1981, Taurine in the retina, in: "The Effects of Taurine on Excitable Tissues", Schaffer, S.W., Baskin, S.I., and Kocsis, J.J., eds., Spectrum Press, New York, pp 145-160.

40. Wingenfeld, P., Michalk, D., Sonntag, A., Paas, S., Minor, T., and Isselhard, W., 1996, Protective effect of taurine on hypoxia and reoxygenation-induced damage of human colon cells (HT29), in Adv. Exp. Med. Biol., "Taurine 2", Huxtable, R.J., Azuma, J., Kuriyama, K., Nakagawa, M. and Baba, A., eds., Plenum Press, New York, Vol. 403, pp 214-222.
41. Wingenfeld, P., Strübind, S., Gehrmann, U., Minor, T., Isselhard, W., and Michalk, D., 1994, Protective effect of taurine against hypoxic cell damage in renal tubular cells cultured in different transplant preservation solutions, in Prog. Clin. Biol. Res. "Taurine: Functional Neurochemistry, Physiology and Cardiology", Pasantes-Morales, H., Martín, D.L., Shain, W. and Del Río, R.M., eds., Willey-Liss, Inc., New York, Vol. 351, pp 159-169.

# PHOSPHOLEMMAN: A CARDIAC TAURINE CHANNEL INVOLVED IN REGULATION OF CELL VOLUME

J. Randall Moorman[1] and Larry R. Jones[2]

[1] University of Virginia
Charlottesville, Virginia
[2] Indiana University School of Medicine
Indianapolis, IN

## A NEW SUPERFAMILY OF MEMBRANE PROTEINS WITH SINGLE TRANSMEMBRANE DOMAINS

The story begins with the study of purified plasma membrane vesicles from canine heart[10,11]. Phosphorylation assays showed that these vesicles contained endogenous PK-A [16] and PK-C[24] activities. PK-A phosphorylated several proteins in cardiac SL vesicles, the major one of which migrated with an apparent $M_r$=15,000 on sodium dodecyl sulfate polyacrylamide gel electrophoresis (SDS-PAGE). PK-C, in contrast, phosphorylated only one substrate in cardiac SL vesicles, also of $M_r$=15,000[16,24]. Based on these results, this major SL protein substrate of PK-A and PK-C was provisionally named the "15-kDa protein." Phosphorylation by PK-A and PK-C was additive, suggesting that at least two different sites of the protein were phosphorylated. In rodent hearts perfused with β-adrenergic[23] and α1-adrenergic[15] agonists, the "15-kDa protein" was again the major SL substrate phosphorylated, and $^{32}$P-labeling of the protein in actively contracting myocardium correlated with the positive inotropic response induced by either agonist. Thus, results with intact myocardial tissue strongly supported a functional role for "15-kDa protein" phosphorylation by PK-A and PK-C. Phosphorylation of the "15-kDa protein" in cardiac membrane preparations and in intact cells was confirmed in other laboratories[12,17,18,26]. The protein was purified, cloned, and sequenced[22], and was named ***PHOSPHOLEMMAN*** (PLM). This name denotes the protein's location within the plasma membrane and its characteristic multisite phosphorylation. The mature protein (that present in SL vesicles) contained only 72 amino acids (Fig. 1), whose sequence is as follows:

E$^{1}$APQEHDPFTYEYQSL*R*I$^{18}$GGLIIAGILFILGILIVLS$^{37}$*RR*C*R*C*K*FNQQQ*R*TGEPDEEE
GTF*R*S$^{62}$SI*RR*LS$^{68}$T*RRR*$^{72}$

This is preceded by a 22 amino acid signal sequence, which is cleaved from the nascent protein prior to delivery to the SL. The calculated $M_r$ of the mature protein is

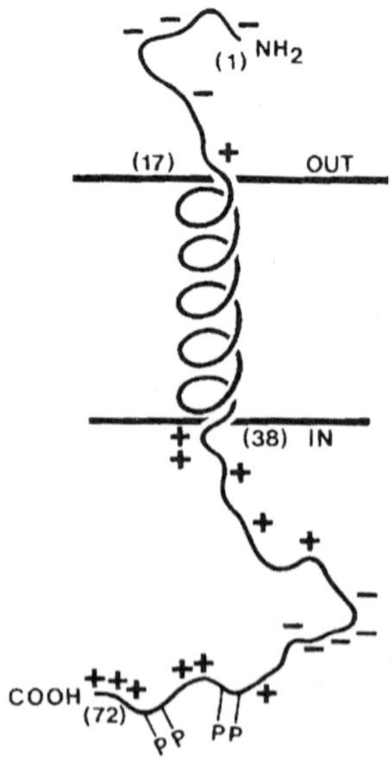


**Figure 1.** Phospholemman (PLM), a major membrane substrate for protein kinases, is a 72 amino acid phosphoprotein with a single transmembrane domain and four phosphorylationsites.

8,409, a value obviously much smaller than that of 15,000 estimated from the protein mobility on SDS-PAGE. Plasma membrane phosphoproteins with similar mobilities on SDS-PAGE have been identified in other tissues[2-4,29,31], and our Northern and Western blot analysis reveals that the protein has a widespread tissue distribution. For example, the protein is also concentrated in skeletal muscle, smooth muscle, liver, and brain.

The amino acid sequence of PLM shows several interesting features. Amino acids 18-37 (underlined) are uncharged and hydrophobic. These residues are postulated to span the SL membrane once as an α-helix. Residues 38-72 of PLM project into the cytoplasm and are mostly polar. At the C-terminal end of the molecule, there is a very high density of positive charge (italicized residues). This region contains the phosphorylated residues. $Ser^{68}$ is an obvious candidate for the residue phosphorylated by PK-A, and $Ser^{62}$, $Ser^{63}$, and $Thr^{69}$ are all likely candidates for residues phosphorylated by PK-C. Phosphoamino acid analysis reveals that PK-A phosphorylates only serine residues of PLM, whereas PK-C phosphorylates both serine and threonine residues[22,28]. The amino terminal end (residues 1-17) of PLM has an overall negative charge, and is postulated to project into the extracellular space.

While the amino acid sequence of PLM revealed limited homology to only the gamma-subunit of the Na-K-ATPase, subsequent cloning efforts have uncovered three homologues.

*Mat-8* (Mammary-associated tumor) is a gene expressed in mouse breast cancers initiated by the transgenic oncogenes c-*neu* or v-Ha-*ras* but not by c-*myc*[21]. It is thought to be a marker of a cell type preferentially transformed by Neu or Ras oncoproteins. All of 16 primary human breast cancers and 8 of 11 human breast cancer cell lines express Mat-8. This work from the Leder laboratory demonstrated a correlation between Mat-8 expression and initiating events of mouse breast carcinogenesis.

*RIC* (Related Ion Channel) is one of 12 genes induced by the oncoprotein E2a-PbxI

in NIH3T3 fibroblasts[6]. E2a-PbxI arises from the chromosome t(1;19) translocation found in some cases of childhood acute leukemia, and activates gene transcription. It contains the transactivation domain of E2a and the DNA-binding homeodomain of PbxI. RIC, which is not normally expressed in these cells, is 161 amino acids in length -- 128 extracellular (by far the longest extracellular domain on the family), 19 transmembrane, and 14 intracellular. It is most homologous to Mat-8, with most of the homology occurring in and around the transmembrane domain. It is strongly upregulated by Ras and tyrosine kinases but not by Neu. When expressed in *Xenopus* oocytes, it induces hyperpolarization-activated currents similar to those induced by PLM and other hydrophobic peptides (X. Fu, M. Kamps, and J.R. Moorman, not shown).
*CHIF* (Channel Inducing Factor) was isolated from colons of rats after mineralocorticoid treatment[1] and is constituitively expressed in kidney, where it is up-regulated by aldosterone[30]. Interestingly, its expression in *Xenopus* oocytes leads to depolarization-activated K currents[1].

## ELECTROPHYSIOLOGICAL STUDIES OF PLM

### In Oocytes Expressing

To test the hypothesis that PLM was an ion channel, we studied membrane currents in RNA-injected *Xenopus* oocytes under two-microelectrode voltage clamp[14,20]. We found that oocytes injected with PLM RNA had large, very slowly activating, non-inactivating inward currents on hyperpolarizing steps. Ion substitution experiments identified $Cl^-$ as the major permeant ion. We hereafter refer to the current as $I_{Cl(PLM)}$, to signify that it is a $Cl^-$ current induced after expression of PLM. The appearance of $I_{Cl(PLM)}$ was correlated with oocyte expression of PLM as detected by immunoblotting with polyclonal antibodies to either N- or C-termini. The simplest interpretation was that PLM was an ion channel. An alternative explanation was that PLM regulated an endogenous oocyte $Cl^-$ channel.

### In Lipid Bilayers Containing Purified Recombinant PLM

To distinguish between these possibilities, we measured currents across a lipid bilayer after adding highly purified protein. Mass spectroscopy of the purified recombinant protein by Dr. David Cafiso, Department of Chemistry, University of Virginia, shows that the preparations we use in the bilayer experiments is completely pure. We have reproducibly recorded very long-lived, very high-conductance anion currents, as shown in Figure 2. The upper traces are uncorrected data, and show currents during a voltage ramp from -75 to 75 mV over 10 sec. The solutions are 200 mM KCl on one side and 50 mM KCl on the other. In this experimental configuration, cation currents reverse at negative voltages, and anion currents reverse at positive voltages. The upper left-hand trace was obtained before addition of protein; it shows a transient capacitance current and a leak conductance of 10 pS. The upper right-hand trace was obtained a few minutes after adding purified recombinant PLM. There is the abrupt initiation of a conductance at about -50 mV. The lower traces were obtained after subtracting the mean of several traces obtained before addition of protein. The lower left-hand trace represents a corrected trace with no channel activity. The lower right-hand represents the corrected current trace, and allows calculation of conductance (700 pS) and reversal potential (17 mV). The positive value establishes anion-selectivity; at 17 mV, $P_K/P_{Cl} \cong 0.3$.

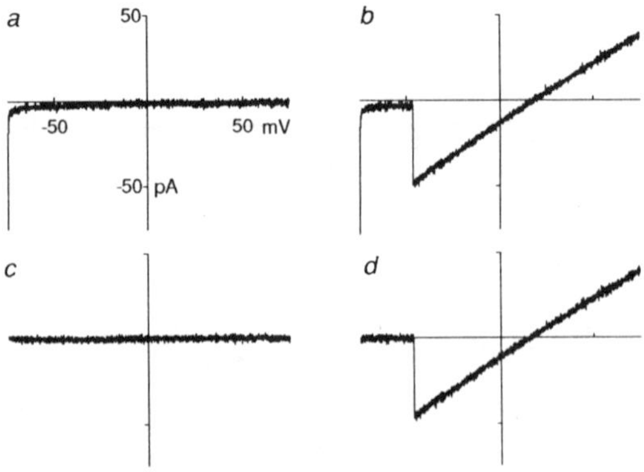


**Figure 2.** Addition of purified recombinant PLM to a phospholipid bilayer results in anion-selective currents. Raw data, top. Corrected for passive elements of leak and capacitance, bottom. The positive reversal potential establishes anion-selectivity. Tracings a and c lack PLM while b and d were obtained from PLM containing bilayers.

Currents through PLM channels in bilayers have two features not previously shown in any other kind of ion channel. Both are related to its selectivity properties. The first is a *very high selectivity for taurine* over any of the other 18 or 20 ions we tested. Moreover, this selectivity was very dependent on the nature of the group on the number 2 carbon. Substitution of the amino group by hydroxyl (isethionate) or methyl (propane-sulfonic acid) groups, or by Br (2-bromoethanesulfonic acid) drastically reduced permeation[19].

The second unique feature is *switching among conformations with drastically different selectivities for anions and cations*. In the left-hand panel of Figure 3, a voltage ramp from -30 to 30 mV elicits a 700 pS current with a reversal potential at 12.5 mV (right arrows) which makes instantaneous transitions to and from a 500 pS conductance with a reversal potential at -14 mV (left arrows). These represent transitions between anion-selective and cation-selective unitary currents. Our interpretation is that the PLM $Cl^-$ channel has the ability to switch among cation- and anion-selective states [13].

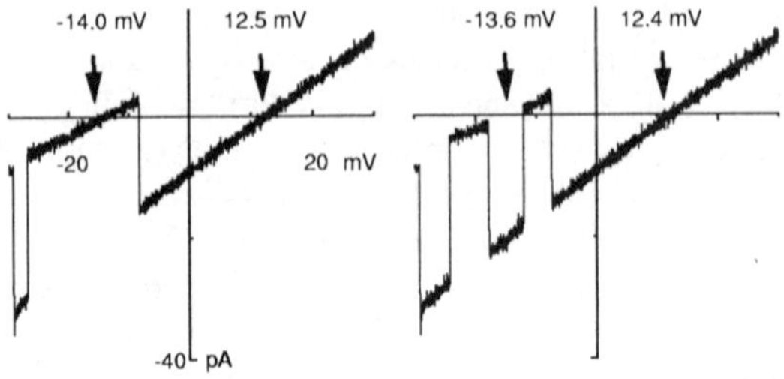


**Figure 3.** Voltage-dependent changes in PLM channel conformation. As before, a positive reversal potential signifies anion-selectivity. Here, a single channel makes several instantaneous changes between cation- and anion-selective conformations. This is consistent with the finding that PLM channels select for taurine over other ions.

### What Is the Physiological Role of PLM?

The sulfonic amino acid taurine is present in high concentration in many tissues[7]. It does not diffuse passively through lipid membranes and acts as an osmolyte in the regulation of cell volume. Taurine flux through swelling-activated anion-selective channels has been observed[8,25,27], suggesting that taurine efflux through membrane ion channels may be a mechanism for the regulatory volume decrease seen in many types of cells responding to swelling under hypotonic conditions. The most notable biophysical feature of the taurine molecule is its zwitterionic character, which makes it extremely polar. The nearby and oppositely charged amino and sulfonate groups, however, complicate the problem of selective transport of taurine through ion channels, which employ charged binding sites to select among and between cations and anions. A possible solution is that a taurine-selective pore might contain binding sites for both the negatively and positively charged moieties. Thus these two highly unusual features of an ion channel -- selectivity for the zwitterionic taurine, and the presence of binding sites for both anions and cations -- are in fact consistent with each other, and mutually support the hypothesis that PLM is a membrane mechanism for taurine flux. If that is true, then what is the physiological role of PLM? We have hypothesized that PLM is involved in regulation of cell volume, and, specifically is the taurine efflux mechanism necessary to effect RVD.

## A ROLE FOR PLM IN REGULATION OF CELL VOLUME

### PLM Expression Enhances Regulatory Volume Decrease

We have begun to test this idea by measuring regulation of volume of human embryonic kidney cells expressing transfected PLM. The phenomenon of interest is RVD, whereby cells swollen by hypotonic solution actively reduce their size by releasing intracellular osmolytes such as taurine. We have adapted a method described by Alvarez-Leefmans and co-workers[5]. The scheme involves loading cells with the fluorescent marker *calcein*, imaging them with a fluorescence microscope and recording the images with a videocamera, digitizing the images, and measuring the change in fluorescence intensity as cells are exposed to solutions of varying composition. The parameter which relates to cell volume is the fluorescence intensity of a small and arbitrarily chosen section of the cell. As cell volume increases, the calcein is diluted and local fluorescence intensity falls.

Figure 4 shows four fluorescence images of one cell from the PLM-expressing C8 line. For the first 10 min, the cells are perfused with a solution of osmolality 330 mOsm/kg $H_2O$ at 37° C. For the second 10 min, the solution is switched to one with identical ionic strength but osmolality 200 mOsm. For the final 10 min, the solution is returned to the original 330 mOsm solution. The first image was made at minute 10, the end of the first perfusion with the 330 mOsm solution. The second image was made at minute 12, at the peak of the cell swelling. The fluorescence intensity is diminished and the cell appears slightly larger. The third image was obtained at minute 20, at the end of the perfusion with the 200 mOsm solution. The cell appears smaller, and both the image and the finding that relative fluorescence of the PLM-expressing cells was greater than 1.0 at 20 minutes suggests that cell size diminished below the baseline size despite the persistent reduction in extracellular osmotic pressure. The fourth image was made at minute 22, after the return to the 330 mOsm solution. The fluorescence intensity has increased further, and the cell is still smaller.

Figure 5 demonstrates the important finding that *cells expressing PLM have enhanced RVD*. The plot is of relative fluorescence (the value at 10 minutes is set to 1.0), and a fall reflects an increase in cell volume. The cells swell after the hypotonic solution is started at minute 10. The slow return to normal is called *regulatory volume decrease (RVD)*, and reflects efflux of osmolytes through swelling-activated mechanisms. In this experiment, sham-transfected cells (open diamonds, n=8) are compared with cells from the C3 (filled circles, n=5) and C8 (filled squares, n=4) transfected cell lines. Bars are standard error of the mean. We tested the significance of the difference in RVD among the cell lines by comparing the value of relative fluorescence after 10 min in the 200 mOsm solution. For sham-transfected cells, the median value was 0.95. For the transfected HEK cells expressing PLM, the median values were 1.17 and 1.12. The difference between sham and either line of transfected HEK cells expressing PLM was significant ($p<0.05$, one way ANOVA on ranks).

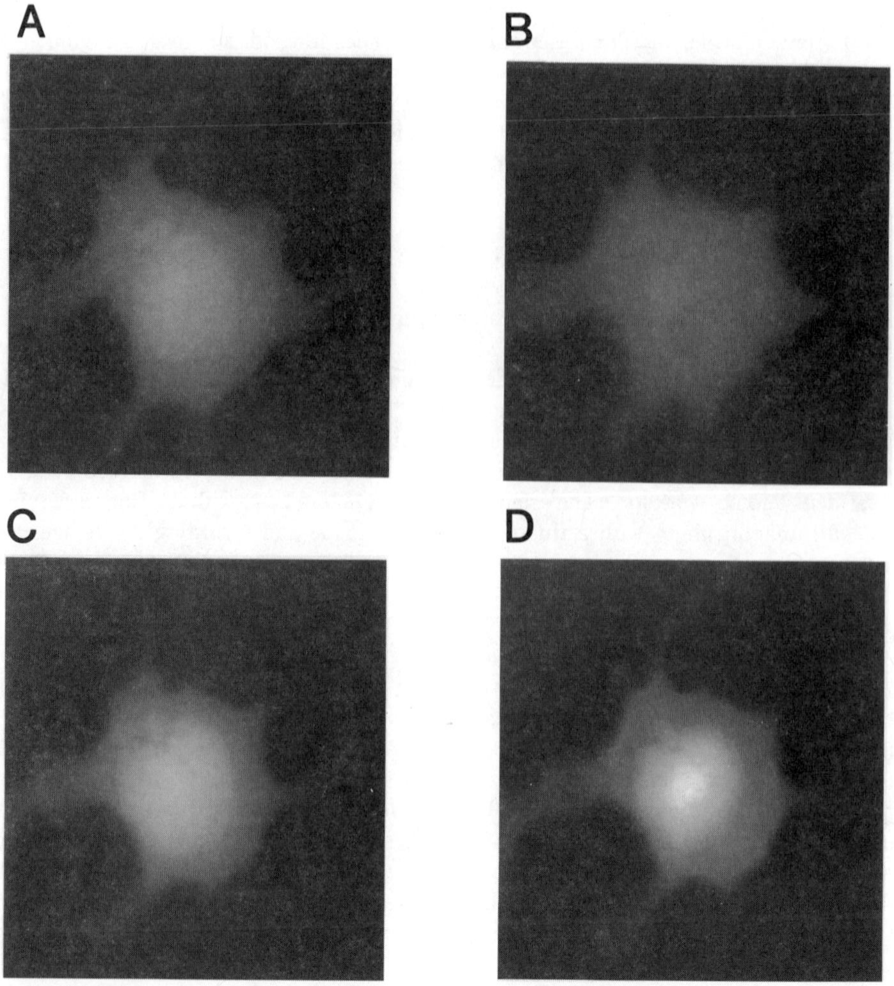


**Figure 4.** Regulatory volume decrease (RVD) in a transfected human embryonic kidney (HEK293) cell over-expressing PLM. At minute 10, the bathing solution was changed to one with lower osmotic pressure. The initial response (minute 12) is an increase in cell volume, followed by a reduction in volume as osmolytes are lost (minute 20). The cell shrinks further when the osmotic pressure is restored (minute 22).

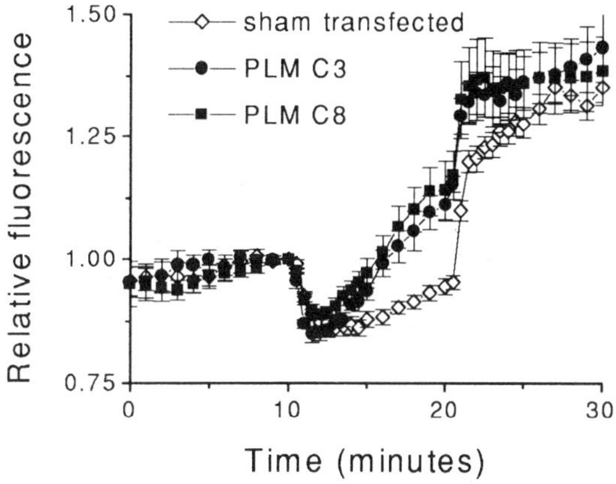


**Figure 5.** Enhanced RVD in cells over-expressing PLM. The plot shows relative fluorescence of calcein-loaded cells: a higher number reflects volume loss.

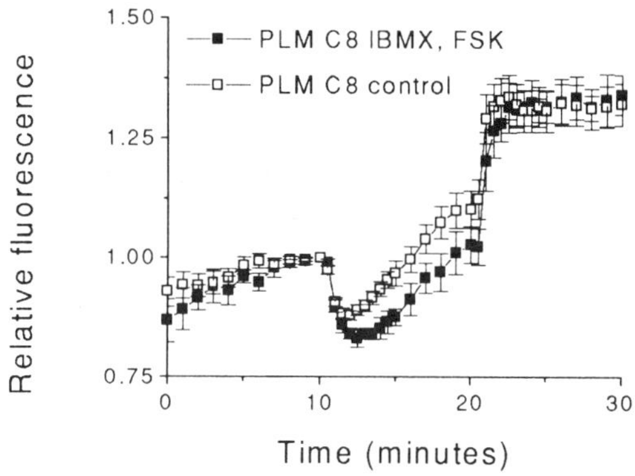


**Figure 6.** Modulation of the PLM effect on RVD by phosphorylation. Incubation in agents which increase phosphorylation by cAMP-dependent protein kinase reduces RVD in cells over-expressing PLM.

The response is modulated by phosphorylation. Figure 6 compares fluorescence measurements for PLM-expressing HEK cells incubated overnight in forskolin and IBMX, and cells incubated in the vehicle (0.05% ethanol) alone. Forskolin (FSK; 5 μM) activates PKA directly, and 3-isobutyl-1-methylxanthine (IBMX; 200 μM) activates PKA indirectly by blocking its degradation by phosphodiesterase. Data from the 3 treated cells are shown as filled squares, and data from the 10 control cells are shown as open squares. Incubation in ethanol had no effect on RVD (not shown). Incubation of sham-transfected cells in IBMX and FSK had no effect (not shown). As shown in Figure 6, the effect of these maneuvers to activate PKA was to diminish the RVD.

Taken together, these experiments are consistent with the hypothesis that osmolyte efflux through PLM channel molecules plays a role in RVD. The inhibitory effect of FSK and IBMX on RVD suggests that phosphorylation of PLM may reduce osmolyte efflux during sympathetic stimulation.

## PLM Expression Enhances Swelling-Activated Anion Currents

We have recently tested the effect of phospholemman (PLM) over-expression on swelling-activated currents in HEK 293 cells. We tested two of the cell lines described in the application -- non-transfected cells, and one of the cell lines expressing PLM (C8 cells). The experimental design was to elicit currents in the whole-cell configuration of the patch clamp technique using a hypertonic solution in the pipette, which would result in cell swelling. After compensation for capacitance and series resistance (80%), the voltage protocol was to step from 50 to 120 mV in 10 mV increments for 4 seconds, resting 10 seconds in-between at a holding potential of -50 mV. The bath solution (300 mOsm) contained (mM) 120 NaCl, 5.4 KCl, 0.8 MgC12, 1.0 CaC12, 5.6 dextrose, and 10.0 HEPES. The pipette solution (402 mOsm) contained 30 TEA-Cl, 10.0 HEPES, 1.0 EGTA, 5.0 K-ATP, 2.0 MgC12, 0.5 CaC12, 100.0 NMDG-aspartate.

Panels A and B (Fig. 7) show exemplary current recordings. Non-transfected HEK293 cells respond to cell-swelling by developing an outward current which decays with a single exponential time course at large depolarizations. The waveform is very similar to that reported by Jackson and Strange in C6 glioma cells[9], where it underlies swelling-induced flux of ions and taurine[8]. Cells over-expressing PLM had the same current, but with larger amplitudes. Panel C shows the current density-voltage relationship for 11 non-transfected and 12 C8 cells. The increase in current of the C8 cells is statistically significant ($p<0.001$, ANOVA; bars are S.E.M.). Cell capacitance was not significantly different, though there was a trend for increased cell size in the PLM-expressing cells ($14.6 \pm 2.9$ pF compared with $13.7 \pm 2.9$ pF). Panel D shows the proportion of current that inactivates during a 4 sec pulse (peak current minus steady state/peak) as a function of test potential. There was no significant difference, suggesting that the currents arise from channels with similar properties in both kinds of cells. Further ion substitution experiments similar to those described in the application (but using a different voltage protocol) also show anion-selectivity. In 129 mM $[Cl]_o$, the reversal potential was $-5.1 \pm 0.7$ mV (S.D.; n=3). In 9 mM $[Cl]_o$, the reversal potential was 21.6 $\pm$ 5.8 mV (n=4).

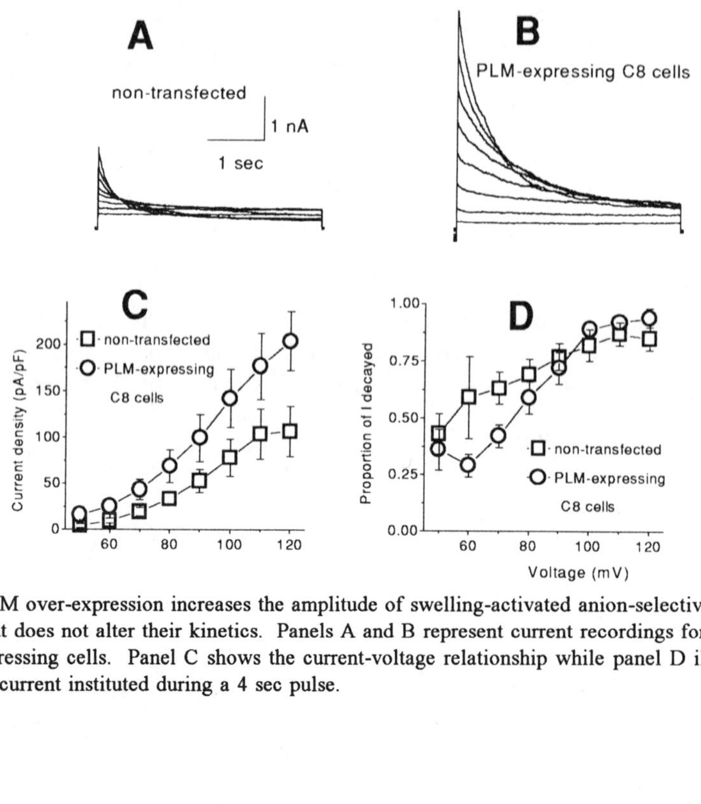


**Figure 7.** PLM over-expression increases the amplitude of swelling-activated anion-selective currents in HEK cells, but does not alter their kinetics. Panels A and B represent current recordings for control and PLM overexpressing cells. Panel C shows the current-voltage relationship while panel D illustrates the proportion of current instituted during a 4 sec pulse.

Thus, PLM over-expression augments the swelling-activated volume-regulatory anion current in HEK293 cells. Our interpretation is that PLM is part of the membrane mechanism for regulatory volume decrease.

## REFERENCES


1. Attali, B., Latter, H., Rachamim, N., and Garty, H, 1995, A corticosteroid-induced gene expressing an "IsK-like" $K^+$ channel activity in *Xenopus* oocytes, *Proc. Natl. Acad. Sci.*, 92:6092-6096.
2. Boulanger-Saunier, C., Kattenberg, D.M., Stoclet, J., 1985, Cyclic AMP-dependent phosphorylation of a 16-kDa protein in a plasma membrane-enriched fraction of rat aortic myocytes, *Febs Lett*, 193:283-288.
3. Boulanger-Saunier, C. and Stoclet, J., 1987, A 16-kDa protein substrate for protein kinase C and its phosphorylation upon stimulation of vasopressin receptors in rat aortic myocytes, *Biochem. Biophys. Res. Comm.*,143:517-524.
4. Cooper, R.H., Kobayashi, K., and Williamson, J.R., 1984, Phosphorylation of a 16-kDa protein by diacylglycerol-activated protein kinase C *in vitro* and by vasopressin in intact hepatocytes, *Febs Lett*, 166:125-130.
5. Crowe, W. E., J. Altamirano, L. Huerto, and F. J. Alvarez-Leefmans. 1995. Volume changes in single N1E-115 neuroblastoma cells measured with a fluorescent probe. *Neuroscience*, 69:283-296.
6. Fu, X. and Kamps, M.P., 1997, E2a-Pbx1 induces aberrant expression of tissue-specific and developmentally regulated genes when expressed in NIH3T3 fibroblasts, *Mol. Cell. Biol*,. 17:
7. Huxtable, R. J., 1992, Physiological actions of taurine, *Physiol. Rev.*, 72:101-163.
8. Jackson, P.S. and Strange, K., 1993, Volume-sensitive anion channels mediate swelling-activated inositol and taurine efflux, *Am. J. Physiol.*, 265:C1489-C1500.
9. Jackson, P.S. and Strange, K., 1995, Characterization of the voltage-dependent properties of a volume-sensitive anion conductance, *J. Gen. Physiol.*, 105:661-677.
10. Jones, L.R., 1988, Rapid preparation of canine cardiac sarcolemmal vesicles by sucrose flotation, *Methods Enzymol.*, 157:85-91.
11. Jones, L.R., Besch, H.R., Jr., Fleming, J.W., McConnaughey, M.M., and Watanabe, A.M., 1979, Separation of vesicles of cardiac sarcolemma from vesicles of cardiac sarcoplasmic reticulum: Comparative biochemical analysis of component activities, *J. Biol. Chem.*, 254:530-539.
12. Karczewski, P., Bartel, S., and Krause, E.-G., 1990, Differential sensitivity to isoprenaline of troponin I and phospholamban phosphorylation in isolated rat hearts. *Biochem. J.*, 266:115-122.
13. Kowdley, G. C., Ackerman, S.J., Chen, Z., Szabo, G., Jones, L.R., and Moorman, J.R., 1997, Anion, cation, and zwitterion-selectivity of phospholemman channel molecules, *Biophys. J.*, 72:141-145.
14. Kowdley, G. C., Ackerman, S.J., John, J.E., Jones, L.R., and Moorman, J.R., 1994, Hyperpolarization-activated chloride currents in *Xenopus* oocytes, *J. Gen. Physiol.*, 103:217-230.
15. Lindemann, J.P., 1985, α-Adrenergic stimulation of sarcolemmal protein phosphorylation and slow responses in intact myocardium, *J. Biol. Chem.*, 261:4860-4867.
16. Manalan, A.S. and Jones, L.R., 1982, Characterization of the intrinsic cAMP-dependent protein kinase activity and endogenous substrates in highly purified cardiac sarcolemmal vesicles, *J. Biol. Chem.*, 257:10052-10062.
17. Meij, J.T., Bezstarosti, A.K., Panagia, V., and Lamers, J.M.J., 1991, Phorbol ester and the actions of phosphatidylinositol 4, 5-bisphosphate specific phospholipase C and protein kinase C in microsomes prepared from cultured cardiomyocytes, *Mol. Cell. Biochem.*, 105:37-47.
18. Miyakoda, G., Yoshida, A., Takisawa, H., and Nakamura, T., 1987, β-adrenergic regulation of contractility and protein phosphorylation in spontaneously beating isolated rat myocardial cells, *J. Biochem.*, 102:211-224.
19. Moorman, J.R., Ackerman, S.J., Kowdley, G.C., Griffin, M.P., Mounsey, J.P., Chen, Z., Cala, S.E., O'Brian, J.J., Szabo, G., and Jones, L.R., 1995, Unitary anion currents through phospholemman channel molecules, *Nature*, 377:737-740.
20. Moorman, J.R., Palmer, C.J., John, J.E., Durieux, M.E., and Jones, L.R., 1992, Phospholemman expression induces a hyperpolarization-activated chloride current in *Xenopus* oocytes, *J. Biol. Chem.*, 267:14551-14554.
21. Morrison, B.W. and Leder, P., 1994, *neu* and *ras* initiate murine mammary tumors that share genetic markers generally absent in c-*myc* and *int*-2-initiated tumors. *Oncogene*, 9:3417-3426.

22. Palmer, C. J., Scott, B.T., and Jones, L.R., 1991, Purification and complete sequence determination of the major plasma membrane substrate for cAMP-dependent protein kinase and protein kinase C in myocardium, *J. Biol. Chem.*, 266:11126-11130.
23. Presti, C.F., Jones, L.R., and Lindemann, J.P., 1985, Isoproterenol-induced phosphorylation of a 15-kilodalton sarcolemmal protein in intact myocardium, *J. Biol. Chem.*, 260:3860-3867.
24. Presti, C.F., Scott, B.T., and Jones, L.R., 1985, Identification of an endogenous protein kinase C activity and its intrinsic 15-kilodalton substrate in purified canine cardiac sarcolemmal vesicles, *J. Biol. Chem.*, 260:13879-13889.
25. Szucs, G., Heinke, S., De Greef, C., Raeymaekers, L., Eggermont, J., Droogmans, G., and Nilius, B., 1996, The volume-activated chloride current in endothelial cells from bovine pulmonary artery is not modulated by phosphorylation, *Pflugers Archiv. - European. Journal of Physiology*, 431:540-548.
26. Talosi, L. and Kranias, E.G., 1992, Effect of α-adrenergic stimulation on activation of protein kinase C and phosphorylation of proteins in intact rabbit hearts, *Circ. Res.*, 70:670-678.
27. Voets, T., Szucs, G., Droogmans, G., and Nilius, B., 1995, Blockers of volume-activated Cl- currents inhibit endothelial cell proliferation, *Pflugers Archiv. - European. Journal of Physiology*, 431:132-134.
28. Walaas, S.I., Czernik, A.J., Olstad, O.K., Sletten, K., and Walaas, O., 1994, Protein kinase C and cyclic AMP-dependent protein kinase phosphorylate phospholemman, an insulin and adrenaline-regulated membrane phosphoprotein, at specific sites in the carboxy terminal domain, *Biochem. J.*, 304:635-640.
29. Walaas, S.I., Horn, R.S., Albert, K.A., Adler, A., and Walaas, A., 1988, Phosphorylation of multiple sites in a 15,000 dalton proteolipid from rat skeletal muscle sarcolemma, catalyzed by adenosine 3', 5'-monophosphate-dependent and calcium/phospholipid-dependent protein kinases, *Biochimica et Biophysica Acta*, 968:127-137.
30. Wald, H., Goldstein, O., Asher, C., Yagil, Y., and Garty, H., 1996, Aldosterone induction and epithelial distribution of CHIF, *Am. J. Physiol.*, 271:F322-9.
31. Widmaier, E.P., Osawa, S., and Hall, P.F., 1986, Phosphorylation of three proteins in the plasma membrane of Y-1 adrenal cells by a membrane-bound adenosine 3', 5'-monophosphate-dependent protein kinase, *Endocrinology*, 118:701-708.

# TAURINE EFFLUX AND INTRACELLULAR pH DURING ASTROCYTE VOLUME REGULATION

James E. Olson,[1,2] Robert W. Putnam,[2] Julie A. Evers,[1] and Nicholas Munoz[2]

[1] Department of Emergency Medicine
[2] Department of Physiology and Biophysics
Wright State University
Dayton, Ohio 45401

## ABSTRACT

Cytotoxic cerebral edema is characterized by enlarged astroglial cells. In tissue culture, osmotically swollen astrocytes return toward normal volume over a period of 15-30 min in a process termed regulatory volume decrease (RVD). RVD is due, in part, to net efflux of taurine and other amino acids. Our objective in these studies was to examine changes in astrocyte intracellular pH ($pH_i$) which may be related to taurine loss during RVD. We hypothesized net efflux of anionic taurine abandons a proton inside the cell, thus lowering $pH_i$. Primary cultures of cerebral astrocytes were prepared from neonatal rat pups and grown on glass coverslips. Confluent cells were loaded at 37°C with the fluorescent pH indicator BCECF. Fluorescence intensity ratios for excitation wavelengths of 440 nm and 494 nm (530 nm emission) were computed every 2 sec. Intensity ratios were calibrated to $pH_i$ at the end of each experiment using 140 mM KCl plus 8.6 μM nigericin at pH 7.4. $pH_i$ was measured in isoosmotic Hepes-buffered saline (290 mOsm) and then in hypoosmotic Hepes-buffered saline (200 mOsm) in the presence of 0.5 mM amiloride. Some solutions also contained 150 μM niflumic acid (NA). Cellular taurine content was determined in parallel studies using HPLC. Changes in $pH_i$ were compared between groups using Student's t-test with Bonferroni correction. Significance was assumed if $p<0.05$. In isoosmotic saline, mean ± SEM $pH_i$ was 7.58 ± 0.04 and decreased to 7.35 ± 0.09 after adding amiloride. Hypoosmotic exposure caused a further drop in $pH_i$ of 0.29 ± 0.03 within 15 min. Recovery of $pH_i$ in isoosmotic saline was amiloride-sensitive. Subsequent hypoosmotic exposure after recovery in isoosmotic saline produced a change in $pH_i$ which was 81 ± 9% of the change measured during the initial hypoosmotic exposure. Taurine content decreased from 147 ± 6 nmol/(mg protein) to 116 ± 7 nmol/(mg protein) during the 15 min hypoosmotic exposure in 0.5 mM amiloride. NA significantly reduced the hypoosmotically induced change in $pH_i$ to 0.17 ± 0.02 while completely blocking taurine loss. Assuming an intracellular buffering power of 13 mM, the NA-sensitive intracellular acidification of cells during hypoosmotic exposure in the presence of 0.5 mM amiloride corresponds to 1.6 mequiv/l additional intracellular $H^+$. This increase in intracellular $H^+$ content is equivalent to approximately 32% of the NA-sensitive loss of taurine. The association of changes in $pH_i$ with taurine efflux is supported by these data; however, efflux of other weak acids may contribute to intracellular acidification during astrocyte RVD and a significant portion of taurine must leave the cell with a proton.

## INTRODUCTION

Astrocytes exposed suddenly to hypoosmotic conditions (175 mOsm - 250 mOsm) swell rapidly and then demonstrate a regulatory volume decrease (RVD) which is mediated by net loss of intracellular osmolytes,[6,16]. While rates of unidirectional efflux of potassium and chloride are accelerated during RVD of astrocytes[9], no change occurs in the cellular content of these ions. Thus, net loss of potassium and chloride does not contribute to astrocyte RVD in response to moderate osmotic swelling[12,14]. In contrast, net loss of taurine and other amino acids occurs during RVD and, in the case of taurine, is quantitatively correlated with the magnitude of the volume response[22]. Taurine leaves osmotically swollen astrocytes by a passive process[19]. The diffusion rate of taurine from the cell interior may be controlled by osmotically sensitive anion channels capable of moving amino acids and possibly other organic osmolytes through the cell membrane[4,5,18]. Identification of an osmotically sensitive anion pathway capable of conducting taurine has recently been described in rat astrocytes from primary culture[13]. However, to measure significant taurine current in these electrophysiological studies, intracellular pH ($pH_i$) must be alkalinized to unphysiological levels. Based on these experiments, we cannot be certain that anionic taurine leaves the cell by this pathway in conditions of normal intracellular and extracellular pH. However, if we hypothesize taurine leaves the cell as an anion under conditions of normal $pH_i$, the taurine molecule must abandon a proton in the cytoplasm of the cell, leading to intracellular acidification. We tested this hypothesis by measuring $pH_i$ of rat astrocytes in primary culture during hypoosmotic volume regulation and correlating the resulting changes in $pH_i$ with hypoosmotically induced alterations in taurine content.

## MATERIALS AND METHODS

Astrocytes from the cerebral cortices of two to four day-old rat pups were grown in primary culture on 35 mm plastic petri dishes as previously described[2,15]. Some dishes contained two 9 × 18 mm glass coverslips. Cells attached and grew equally on the glass or plastic surfaces. After two weeks *in vitro*, the cells had nearly reached confluence.

A coverslip containing cells was removed from a petri dish after 2-4 weeks *in vitro* and the cells incubated for 15 min at 37°C with 2 μM of the acetoxymethylester derivative of the pH-sensitive fluorescent dye, 2',7'-bis-(2-carboxyethyl)-5-(6)-carboxyfluorescein (BCECF), in isoosmotic buffered saline (ISO) containing 100 mM NaCl, 2.7 mM KCl, 1 mM $CaCl_2$, 0.5 mM $MgCl_2$, 3.2 mM $Na_2HPO_4$, 0.5 mM $KH_2PO_4$, 5.5 mM glucose and 10 mM HEPES (pH=7.3) plus sufficient sucrose (≈90 mM) to raise the osmolality to 290 mOsm. The coverslips then were submerged in a 3 ml cuvette with constant perfusion of 37°C ISO at 1 ml/min[17]. The cuvette was placed in the holder of an SLM DMX 1000 spectrofluorometer and maintained at 37°C. For exposure to hypoosmotic buffered saline (HYPO), cells were perfused with a solution identical to that described above, but without sucrose (≈200 mOsm). Cells were illuminated alternately with 440 nm and 494 nm light and the emission intensity was measured at 530 nm for each excitation wavelength. Every two seconds, the ratio of the emission intensity for 494 nm excitation light ($F_{494}$) was divided by that measured with 440 nm excitation light ($F_{440}$) to give a fluorescence intensity ratio ($R_{fl}=F_{494}/F_{440}$). At the end of each experiment, $R_{fl}$ was determined while cells were perfused with 140 mM KCl, 10 mM HEPES (pH=7.4) plus 8.6 μM nigericin[21]. From these data, $R_{fl}$ values measured during experimental manipulations were converted to $pH_i$ using the method described by Shrode and Putnam[20].

Astrocyte amino acid and potassium content were measured in parallel studies as previously described[11]. Cells grown on plastic dishes were exposed to the same series of solutions as the cells used for $pH_i$ studies. At the appropriate times, dishes with cells were rinsed with 10 mM Tris buffer (pH=7.3) plus sufficient sucrose to match the osmolality of the experimental solution. Cells then were scraped into 0.6 M $HClO_4$ plus 4 mM CsCl and 5 μM homoserine (an internal standard for HPLC). Each suspension was centrifuged and the content of amino acids and potassium measured in the supernatant by HPLC and atomic absorption spectroscopy, respectively. Protein was determined in the pellet by the method of Lowry[8] after solubilization in 1 M NaOH.

## RESULTS

The basal $pH_i$ of cells in ISO was 7.58 ± 0.04 (mean ± SEM, N=29). Within 3 min of changing the perfusion solution to HYPO, $pH_i$ began falling gradually (Fig. 1A). After 15 min in this solution, the mean ± SEM $pH_i$ had fallen by 0.24 ± 0.05 units. $pH_i$ recovered if the solution perfusing the cells was returned to ISO. To examine whether the observed acidification was affected by alkalinizing $pH_i$ recovery mediated by Na/H exchange, changes in $pH_i$ during hypoosmotic exposure were measured in the presence of 0.5 mM amiloride, an inhibitor of Na/H exchange (Fig. 1B). Adding amiloride to cells in ISO caused acidification within 2 min to a value of 7.35 ± 0.09. Perfusion of these cells with HYPO in the continued presence of 0.5 mM amiloride resulted in a further drop in $pH_i$ of 0.29 ± 0.03 units (Fig. 1B), a value similar to that observed during hypoosmotic exposure without amiloride (Fig. 1A). Removing amiloride from the hypoosmotic perfusion solution or returning the cells to isoosmotic buffered saline in the continued presence of amiloride did not affect $pH_i$ (data not shown). However, a return towards the original $pH_i$ was observed if cells were perfused with ISO containing no amiloride (Figs. 1A-C). After 20 min recovery in ISO, subsequent exposure to HYPO plus 0.5 mM amiloride caused a mean ± SEM decrease in $pH_i$ which was only 81 ± 9% of that observed with the initial exposure to this same hypoosmotic solution ($p<0.05$) [data not shown].

Addition of 150 μM niflumic acid to cells in ISO containing 0.5 mM amiloride caused $pH_i$ to fall approximately 0.3 units to a value of 7.09 ± 0.07 (Fig. 1C). In these cells, exposure to HYPO containing the same amiloride and niflumic acid concentrations caused a $pH_i$ drop of 0.17 ± 0.02, a value significantly smaller than that obtained during hypoosmotic exposure in the absence of the drug ($p<0.05$).

Astrocyte content of alanine (Ala), asparagine (Asn), aspartate (Asp), glutamine (Gln), glutamate (Glu), and taurine (Tau) were unaffected by exposure of the cells in ISO to 0.5 mM amiloride or 0.5 mM amiloride plus 150 μM niflumic acid (Fig. 2). In contrast, hypoosmotic exposure caused a 31% loss of taurine over 15 min (Fig. 2A). Smaller percent changes in content of Ala, Asp, Gln, and Glu were observed in control dishes. While the content of Ala, Asp, Gln, and Glu were not affected by HYPO, the combined content of all four amino acids decreased significantly by 46 nmol/(mg protein) during HYPO exposure. By comparison, the content of taurine alone fell by 45 nmol/(mg protein) following 15 min of HYPO. With hypoosmotic exposure in the presence of 0.5 mM amiloride, only taurine demonstrated a significant decrease, equivalent to 21% of the value

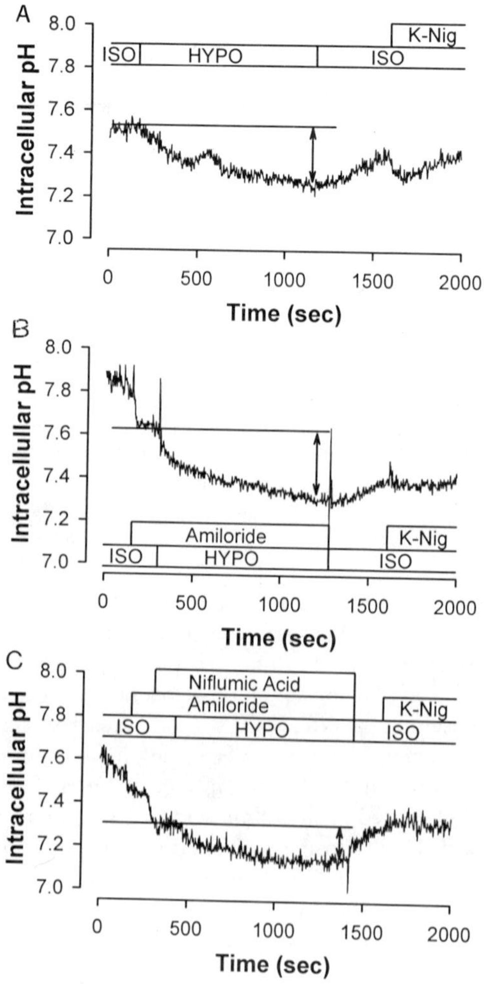


**Figure 1.** Astrocyte intracellular pH during exposure to isoosmotic and hypoosmotic buffered saline. A. Control cells. B. Cells exposed to 0.5 mM amiloride. C. Cells exposed to 0.5 mM amiloride plus 150 μM niflumic acid. Changes in the perfusion solutions were made at the times indicated. ISO, isoosmotic buffer, HYPO, hypoosmotic buffer; K-Nig, 140 mM KCl plus 8.6 μM nigericin (pH=7.4). A horizontal line is drawn through $pH_i$ values measured in ISO prior to exposure to HYPO. The vertical line indicates the magnitude of the change in $pH_i$ caused by 15 min exposure to HYPO.

measured in isoosmotic conditions (Fig. 2B). This loss of taurine during hypoosmotic exposure in the presence of 0.5 mM amiloride was completely inhibited by 150 μM niflumic acid (Fig. 2C). Potassium content (mean ± SEM) of astrocytes in ISO was 940 ± 35 nmol/(mg protein) and was unchanged by exposure to 0.5 mM amiloride, 150 μM niflumic acid and/or HYPO.

## DISCUSSION

These data demonstrate intracellular acidification of astrocytes during hypoosmotic exposure which is correlated with the loss of cellular taurine. Subsequent hypoosmotic exposure after recovery in isoosmotic buffer results in a smaller acidification, consistent with the decreased cellular taurine content following the initial hypoosmotic exposure. The acidification in hypoosmotic conditions does not appear to activate Na/H exchange suggesting astrocyte swelling inhibits this transporter. While inhibition of Na/H exchange by amiloride causes a fall in $pH_i$, the acidification observed during hypoosmotic swelling is clearly not due to inhibition of basal $H^+$ efflux mediated by this transporter.

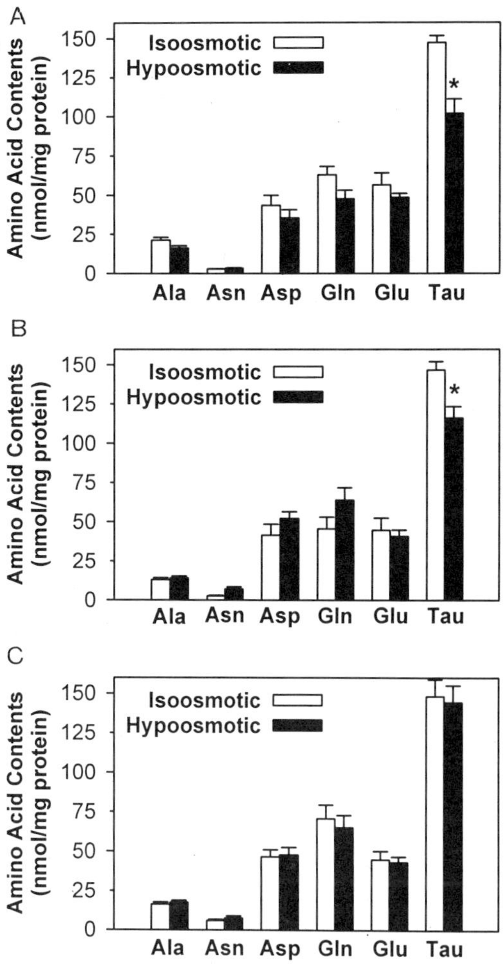


**Figure 2.** Astrocyte amino acid content in isoosmotic buffered saline and after 15 min in hypoosmotic buffered saline. A. Control cells. B. Cells exposed to 0.5 mM amiloride. C. Cells exposed to 0.5 mM amiloride plus 150 μM niflumic acid. Values are the mean ± SEM of 10-12 independent determinations. * Indicates mean values significantly different from those measured in isoosmotic conditions in the presence of the same additives.

These results suggest a significant portion of the taurine leaves the cells in an anionic form during hypoosmotic exposure, thus abandoning a proton in the cell interior. The pathway for anionic flux of taurine may be the volume-activated organic osmolyte channel or other similar anion channels demonstrated in these and other cell types[5,13,18]. When activated by cell swelling, these channels are capable of conducting taurine and other amino acids in their anionic form. In astrocytes, this pathway is inhibitable by 200 μM niflumic acid[13]. Similarly, we observed inhibition of swelling-induced taurine efflux and intracellular acidification with a comparable concentration of this drug.

Taurine may leave the astrocyte either as an anion or as a zwitterion. While loss of the zwitterion would produce little change in $pH_i$, efflux of each taurine anion abandons a proton, and thus, causes intracellular acidification. We can estimate the proportion of taurine which leaves the cell as an anion by comparing our measurements of the change in $pH_i$ with the change in total taurine content during hypoosmotic exposure. Assuming an intracellular buffering power of 13 mM[20] the observed intracellular acidification which is inhibitable by niflumic acid (0.12 units, Table 1) is equivalent to 1.6 mequiv/liter. Using a volume of isoosmotic cells of 5.4 μl/(mg protein)[1], the niflumic-acid inhibitable loss of taurine from astrocytes exposed to hypoosmotic conditions (28 nmol/mg protein, Table 1) is equivalent to 5.2 mM. Thus, if all of the niflumic acid-inhibitable change in $pH_i$ is due to efflux of anionic taurine, approximately 32% of the total taurine lost from

**Table 1.** Summary of astrocyte taurine content and $pH_i$ during isoosmotic and hypoosmotic exposure.

| Solution Additives | Taurine Content in Isoosmotic Buffer[a] | Taurine Content in Hypoosmotic Buffer | Hypoosmotic-induced Change in pH[b] |
|---|---|---|---|
| None | 147 ± 4 | 102 ± 9* | -0.24 ± 0.05 |
| 0.5 mM Amiloride | 147 ± 6 | 116 ± 7* | -0.29 ± 0.03 |
| 0.5 mM Amiloride plus 150 μM Niflumic Acid | 148 ± 11 | 144 ± 11 | -0.17 ± 0.02 |

Taurine contents are in units of nmol/(mg protein). Values are the means ± SEM of 11-12 (taurine content) or 8-9 change in pH), independent samples.

[a] Cells measured prior to hypoosmotic exposure.

[b] Cells measured after 15 min in hypoosmotic buffered saline.

* Indicates mean value which is significantly different from that measured in isoosmotic conditions with the same additives ($p<0.05$).

osmotically swollen astrocytes must leave as an anion. By comparison we can calculate the proportion of taurine in anionic and zwitterionic forms in the cell cytoplasm. The apparent pKa for taurine is approximately 8.6[3]. Thus, at a $pH_i$ of 7.4 (in ISO containing 0.5 mM amiloride), only 6.3% of the taurine is expected to be negatively charged. The comparison of the proportion of anionic taurine lost from the osmotically swollen astrocyte (32%) with the proportion of anionic taurine present under basal conditions (6.3%) suggests the anionic form of taurine is more permeable through the cell membrane of the swollen astrocyte than the zwitterion. Simultaneous measurements of taurine current and $pH_i$ are currently underway to more firmly characterize the relation between taurine efflux mechanisms and changes in astrocyte $pH_i$.

In addition to the loss of anionic taurine, other processes during RVD may lead to intracellular acidification. At normal $pH_i$, glutamate and aspartate are permeable through volume-activated anion channels described in other cell types[18]. However, since at normal $pH_i$ these amino acids exist predominately as univalent anions capable of losing or gaining a proton, their loss is not expected to significantly alter $pH_i$. Cell swelling also causes depolarization of the astrocyte cell membrane[7,13] which may lead to alkalization of the cell interior via bicarbonate influx mediated by a $Na^+$:$2HCO_3^-$ co-transporter[10]. This is not expected to play a major role in $pH_i$ changes in these experiments since the perfusion solution used was essentially bicarbonate-free. Intracellular calcium can increase with cell swelling[9]. Thus, proton mobilization from calcium binding sites or activation of cell metabolism[15] also may play a role in the cell acidification observed with hypoosmotic swelling. Clarification of the relative importance of these mechanisms for intracellular acidification must await further studies.

The majority of anionic taurine leaving the swollen astrocytes would acquire a proton from the extracellular space. Thus, taurine efflux stimulated by cell swelling would transfer a base equivalent in addition to an osmole from the cell interior to the exterior. This effect may be important in pathological conditions such as ischemia which lead to glial swelling and which are characterized by generalized brain tissue acidification. In this context, acidification of the astrocyte interior may proceed as extracellular space is protected from pH changes by the additional buffering power provided by anionic taurine lost from astrocytes.

## ACKNOWLEDGMENTS

This research was supported by funds from the Wright State University School of Medicine Biomedical Research Seed Grants and from the STREAMS program (NIH HL 07805).

## REFERENCES:

1. Bender, A.S. and Norenberg, M.D., 1994, Calcium dependence of hypoosmotically induced potassium release in cultured astrocytes, *J. Neurosci.*, 14:4237-4243.
2. Beetsch, J.W. and Olson, J.E., 1996, Hyperosmotic exposure alters total taurine quantity and cellular transport in rat astrocyte cultures, *Biochim. Biophys. Acta*, 1290:141-148.
3. Budavari, S., 1989, The Merck Index, 11th ed., Merck and Company, Rahway, NJ, pp. 1606.
4. Goldstein, L., Davis-Amaral, E.M., and Musch, M.W., 1996, Organic osmolyte channels: Transport characteristics and regulation, *Kidney Int.*, 49:1690-1694.
5. Jackson, P.S. and Strange, K., 1993, Volume-sensitive anion channels mediate swelling-activated inositol and taurine efflux, *Am. J. Physiol.*, 265:C1489-C1500.
6. Kimelberg, H.K. and Frangakis, M.V., 1985, Furosemide- and bumetanide-sensitive ion transport and volume control in primary astrocyte cultures from rat brain, *Brain Res.*, 361:125-134.
7. Kimelberg, H.K. and O'Connor, E., 1988, Swelling of astrocytes causes membrane potential depolarization, *Glia*, 1:219-224.
8. Lowry, O.H., Rosebrough, N.J., Farr, A.l., and Randall, R.J., 1951, Protein measurement with the Folin phenol reagent, *J. Biol. Chem.*, 193:265-275.
9. O'Connor, E.R. and Kimelberg, H.K., 1993, Role of calcium in astrocyte volume regulation and in the release of ions and amino acids, *J. Neurosci.*, 13:2638-2650.
10. O'Connor, E.R., Sontheimer, H., and Ransom, B.R., 1994, Rat hippocampal astrocytes exhibit electrogenic sodium-bicarbonate co-transport, *J. Neurophysiol.*, 72(6):2580-2589.
11. Olson, J.E. and Goldfinger, M.D., 1990, Amino acid content of rat cerebral astrocytes adapted to hyperosmotic medium in vitro, *J. Neurosci. Res.*, 27:241-246.
12. Olson, J.E. and Kimelberg, H.K., 1995, Hypoosmotic volume regulation and osmolyte transport in astrocytes is blocked by an anion transport inhibitor, L-644,711, *Brain Res.*, 682:197-202.
13. Olson, J.E. and Li, G.-Z., 1997, Increased potassium, chloride, and taurine conductances in astrocytes during hypoosmotic swelling, *Glia*, 20:254-261.
14. Olson, J.E., Alexander, C., Feller, D.A., Clayman, M.L., and Ramnath, E.M., 1995, Hypoosmotic volume regulation of astrocytes in elevated extracellular potassium, *J. Neurosci. Res.*, 40:333-342.
15. Olson, J.E., Holtzman, D., Sankar, R., Lawson, C., and Rosenberg, R., 1989, Octanoic acid inhibits astrocyte volume control: implications for cerebral edema in Reye's Syndrome, *J. Neurochem.*, 52:1197-1202.
16. Olson, J.E., Sankar, R., Holtzman, D., James, A., and Fleischhaker, D., 1986, Energy-dependent volume regulation in primary cultured cerebral astrocytes, *J. Cell. Physiol.*, 128:209-215.
17. Putnam, R.W. and Grubbs, R.D., 1990, Steady-state $pH_i$, buffering power, and effect of $CO_2$ in a smooth muscle-like cell line., *Am. J. Physiol.*, 258:C461-C469.
18. Roy, G. and Banderali, U., 1994, Channels for ions and amino acids in kidney cultured cells (MDCK) during volume regulation, *J. Exp. Zool.*, 268:121-126.
19. Sanchez-Olea, R., Morán, J., Schousboe, A., and Pasantes-Morales, A., 1991, Hyposmolarity-activated fluxes of taurine in astrocytes are mediated by diffusion, *Neurosci. Lett.*, 130:233-236.
20. Shrode, L.D. and Putnam, R.W., 1994, Intracellular pH in primary rat astrocytes and C6 glioma cells, *Glia*, 12:196-210.
21. Thomas, J.A., Buchsbaum, R.N., Zimniak, A., and Racker, E., 1979, Intracellular pH measurements in Ehrlich ascites tumor cells utilizing spectroscopic probes generated in situ, *Biochem. J.*, 18:2210-2218.
22. Vitarella, D., DiRisio, D.J., Kimelberg, H.K., and Aschner, M., 1994, Potassium and taurine release are highly correlated with regulatory volume decrease in neonatal primary rat astrocyte cultures, *J. Neurochem.*, 63:1143-1149.

# ADAPTATIONS OF CEREBRAL VOLUME-REGULATORY TAURINE TRANSPORT TO CHRONIC DISTURBANCES OF PLASMA OSMOLALITY

R.O. Law

Department of Cell Physiology and Pharmacology
University of Leicester
United Kingdom.

## INTRODUCTION

Numerous studies on cultured glia and neurons have demonstrated the pattern of volume-regulatory responses of brain cells acutely exposed to anisosmotic media[1,5,6,20-22]. Following a phase of rapid swelling (typically 1-2 minutes), cells subjected to hyposmotic media effect partial or complete volume recovery by shedding inorganic and organic solutes, notably taurine. This process is known as regulatory volume decrease (RVD). Cells transiently shrunken in hyperosmotic media recover their volume uptake of solute from the external medium (regulatory volume increase (RVI). RVI, which has been somewhat less extensively investigated than RVD, is usually a more prolonged process.

It is clear that these responses cannot provide an accurate reflection of the behaviour of cells in the intact animal, in which systemic anisosmotic stress typically arises only gradually and is usually less severe than the stress applied in experiments *in vitro*. For example, cerebral swelling in response to hyponatraemia must be severely constained by the presence of the bony capsule surrounding the brain; any swelling arising from osmotic or traumatic causes will be reflected in a sharp rise in intracranial pressure, compromising the cerebral microcirculation and posing the risk of herniation. The constancy of brain cell water in long-term hypo- or hypernatraemic animals suggest near perfect regulation of cell volume under these conditions (for review see[4,9]), and analysis of brain tissue suggests that even during acute experimental anisosmotic shock in intact animals cell volume is relatively well maintained[3]. Several years ago it was shown that renal proximal cells exposed to step-wise alterations in extracellular osmolality do not significantly change in volume[14,15], and comparable isovolumetric responses to anisosmolality have recently been demonstrated in C6 glioma cells[17] and a renal distal tubular cell line[28].

While the underlying principles of brain cell volume regulation, viz. the loss or uptake of osmotically active solute, are presumably the same during acute and chronic osmotic disturbance, there can be less certainty about the co-identity of the membrane transport process involved. To take a single example, a considerable body of evidence points to the existence of a population of swelling-activated anionic channels that also permit the diffusive efflux of low molecular weight organic solutes[24,25]. But it is difficult to envisage the playing a role in the response to brain cells to pathological hyponatraemia in which little, if any, swelling actually takes place. A further distinguishing factor when considering the responses of cultured cells, as opposed to cells in the intact brain or in tissue slices, is the absence of an intact interstitium in the former model. As has recently been emphasized[18], interstitial re-uptake of solute shed under hyposmotic conditions may significantly influence net solute loss, and hence cellular volume-regulatory behaviour.

The present chapter briefly reports data relating to the differences that can be observed between the volume-regulatory efflux of taurine from cells in rat cerebral superficial cortical minislices incubated under mildly anisosmotic conditions (i.e. within the range encountered clinically) and prepared from (i) normal rats, and (ii) rats rendered chronically (4 day) hyper- or hyponatraemic.

## MATERIALS AND METHODS

Details of the methodology for slice preparation (3-6 mg initial wet weight) and incubation, measurements of the rate constants for the slow phase of efflux of pre-loaded taurine, estimates of steady-state cell volumes (slice equilibrium non-inulin space), as well as the statistical treatment of results, have previously been published in full[10]. The details of the induction of hypernatraemia by provision of 2% saline in lieu of drinking water has also been previously published[13].

Hyponatraemia was induced over 4 days by (i) provision of Na-free diet, (ii) replacement of drinking water with 5% dextrose, and (iii) daily administration of lysine vasopressin (Sigma Chemical Co., 40 μg/kg body weight, i.p.). In a pilot study it was found that this produced plasma $Na^+$ at 122 ± 1.2 mM (12) and this level was routinely checked at sacrifice.

In previously published work from this laboratory the effects of the anion transport inhibitor niflumic acid (50 μM) have been reported. In more recent studies 4,4'-diisothiocyanatostilbene-2,2'-disulphonic acid (DIDS, 25 μM) has been used. Where necessary, results obtained using niflumic acid have been re-examined using DIDS; no significant differences have been observed between the effects of these inhibitors at the concentrations stated.

Variation in incubation media $Na^+$ concentrations were achieved by altering the NaCl content. Further experimental details are given in the relevant passages of the subsequent text.

## RESULTS

### Hypernatraemia

Findings relating to cellular taurine efflux and volume regulation in chronically hyper-natraemic rats, with brief data on the slow re-establishment of the "normal" efflux pattern following the restoration or normonatraemia, have already been published in detail[13], and those relevant to the present report are given here in summary only. Plasma $Na^+$ in these animals was within the range 161-172 mM, mean = 169 mM.

When slices from normonatraemic rats (plasma $Na^+$ ~142 mM) were acutely exposed

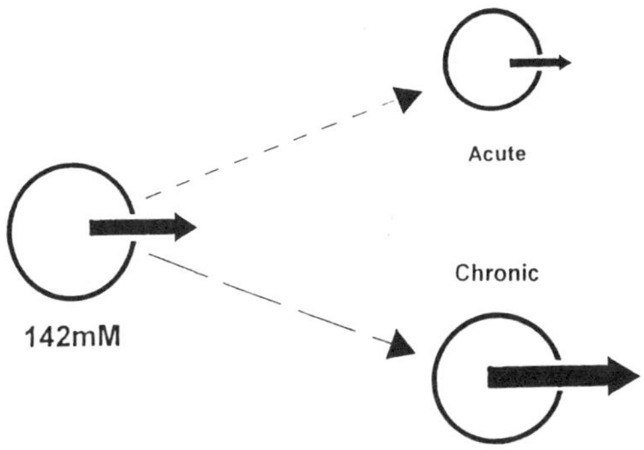


**Figure 1.** The effects of acutely and chronically imposed increase in external $Na^+$ on cell volume and taurine efflux in cerebral cortical minislices. This diagram, as well as those comprising Figs. 2 and 3, are semi-quantitative only. Increases and decreases in the rates of taurine efflux are represented by broadening and narrowing of the solid arrows. Increases and decreases in the diameter of circles indicates cell swelling and shrinkage. Precise mean values for these variables are provided in the accompanying text. Note that comparison of variables between Figs. 1-3 are not necessarily valid.

to media containing 169 mM the mean rate constant for cellular taurine efflux fell from $0.71 \times 10^{-2} min^{-1}$ to $0.47 \times 10^{-2} min^{-1}$ accompanied by a reduction of mean cell volume from 2.38 to 1.96 µl/mg dry wt. Conversely, similar treatment of cells from hypernatraemic rats resulted in acceleration of mean taurine efflux ($0.96 \times 10^{-2} min^{-1}$) and near perfect preservation of cell volume (2.28 µl/mg). These findings are represented in Figure 1.

In hyperosmotically incubated cells from normal animals niflumic acid (50 µM) further depresses taurine efflux and attenuates cell shrinkage; in hypernatraemic animals it is without effect on efflux[13]. Conversely, the calmodulin (CaM) activation inhibitor trifluoperazine (TFP, 25 µM), which has no effect on either variable in hyperosmotically incubated cells from normal rats, leads to marked depression of taurine efflux ($0.63 \times 10^{-2}$ $min^{-1}$) and cell swelling (2.66 µl/mg) in cells from hypernatraemic animals[13].

## Hyponatraemia

Like brain cells during chronic hypernatraemia, those in hyponatraemic rats display near-perfect volume regulation (mean = 2.33 µl/mg). Taurine efflux is slightly retarded, although this did not attain statistical significance in the present study ($0.1>P>0.05$). This contrasts markedly with the responses of cells from normal rats when suddenly exposed to 122 mM $Na^+$, in which there is acceleration of efflux ($1.14 \times 10^{-2} min^{-1}$) and cell swelling (2.69 µl/mg). These responses are represented in Figure 2.

DIDS, like niflumic acid, has a relatively powerful inhibitory effect on taurine efflux from hyposmotically incubated cells from normal rats (~ -43%), with significant cell swelling. These affects are attenuated in cells from hyponatraemic rats, with efflux slowed by only ~ -16% and negligible change in cell volume. TFP, which potently inhibits taurine efflux in hyposmotically incubated cells from normal rats (see Discussion), with pronounced cell swelling, has no effect on efflux in hyponatraemic rats. (Slight cell swelling occurs, but since this is also observed in cells that have not been pre-loaded with taurine presumably represents disruption of some other CaM-dependent process).

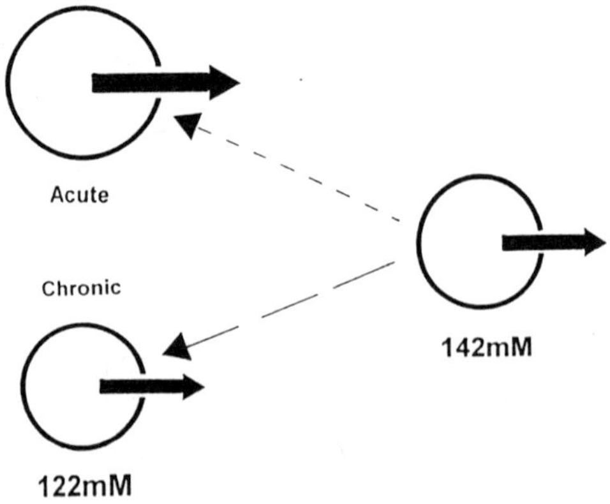


**Figure 2.** The effects of acutely and chronically imposed reductions in external $Na^+$ on cell volume and taurine efflux in cerebral cortical minislices. For explanation of the format of this diagram see the legend to Figure 1.

In a further series of experiments the ability of cells from hyponatraemic rats to regulate their volume was examined following acute reincubation in media containing 142 mM $Na^+$. This amounts, in effect, to sudden imposition of a relative hyperosmotic stress. Paradoxically, cells displayed a consistent, marked and significant increase in mean cells volume (2.81 µl/mg vs 2.33µl/mg in hyponatraemic controls, see above). A comparable response is not observed in cells pre-loaded with the carboxylic amino acids GABA or glutamate. This swelling response was abolished in the presence of the Na/anion cotransport inhibitor bumetanide (50 µM). Neither with or without bumetanide was there any significant variation in the rate of cellular taurine efflux. The swelling response was not observed in cells that had not been pre-loaded with taurine, and bumetanide had no effect in these cells. (These cells in fact swelled slightly, but this response is typical of cells incubated in 142 mM $Na^+$ but lacking the osmoprotective effect of pre-loaded taurine). These results are summarized in Figure 3.

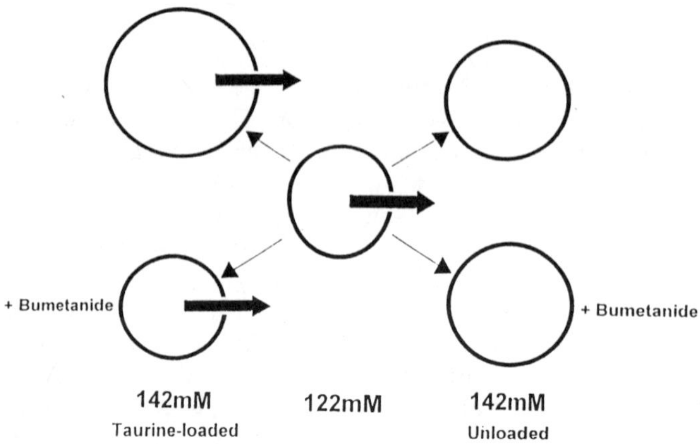


**Figure 3.** The volume responses of cells from hyponatraemic rats acutely exposed to media containing 142 mM $Na^+$ in the presence and absence of bumetanide (50µM) and pre-loaded taurine. For explanation of the format of this diagram, see the legend to Figure 1.

## DISCUSSION

Three principal questions arise of these studies. Firstly, why is it that during the near-perfect volume regulation that is observed under chronically anisosmotic conditions, changes in the rates of cellular taurine efflux are qualitatively the reverse of those occurring during acute osmotic perturbation (Figs. 1 and 2)? Secondly, how can the apparent differences in sensitivity to anion transport and CaM activation inhibition during chronic and acute osmotic stress be explained? Lastly, why do cells chronically adapted to hyponatraemia swell when acutely exposed to anisosmotic media?

The most straightforward explanation of the reversal of taurine efflux rates would be that whereas under acute hyperosmotic conditions (raised external $Na^+$ and $Cl^-$) (Fig. 1) transport is down-regulated with the aim of conserving intracellular osmotic potential, thereby minimizing cell shrinkage, acceleration of efflux during sustained hypernatraemia is a response to increased $Na^+$ and/or $Cl^-$-dependent taurine uptake; a new uptake/efflux balance is achieved that stabilizes cell volume at "normonatraemic" levels. Analogous arguments apply to hyponatraemia (Fig. 2). It has been shown in cultured astrocytes that prolonged (1-3 day) exposure to hyperosmotic media increases Vmax (though not $K_m$) for $Na^+$-dependent taurine uptake[22]. Similarly, synaptosomes prepared from rats that had been subjected to moderate (2 day) hypernatraemic dehydration (though not to dehydration per se) showed elevated rates of $Na^+$-dependent taurine uptake[27]. In cerebral cortical slices from adult mice, hyperosmotic media increased the maximal velocity or saturable low-affinity cellular taurine uptake, whereas hyposmotic media (or omission of external $Cl^-$) attenuated low-affinity and abolished high-affinity uptake[18,19]. Although taurine efflux in the present studies occurred into nominally taurine-free media, this consideration applies only to the bulk phase. Cells loaded for 2 hours in medium containing 1mM taurine accumulate taurine at 4-5mM[10] and it may reasonably be assumed that during net efflux the interstitial fluid contained concentrations of taurine sufficient to activate significant $Na^+$ and/or $Cl^-$-dependent re-uptake, measured rates of efflux thus reflecting the net balance between oppositely directed transport processes.

The apparent alterations in sensitivity to the inhibitory effects of DIDS (or niflumic acid) and TFP are difficult to interpret. It has been proposed that taurine efflux occurs by 2 mechanisms, an anion exchanger which operates under all osmotic conditions, and a CaM-dependent process, possibly a diffusion channel, that is activated only in hyposmotic media[10,11]. (The existence of an anionic exchanger, rather than an anion channel, is controversial, but is supported by the mechanism's dependence on the nature of external anions and its sensitivity to the chloride/bicarbonate exchange inhibitor pyrodoxal 5-phosphate (1mM)[12]. Moreover, recent evidence suggests that taurine transport occurs by anion/anion exchange in rat red cells[2,16]. Whatever the nature of this pathway, the strongly additive effects of DIDS and TFP on taurine efflux[10] are compatible with dual transport processes). The findings reported here were based on the use of unique concentrations of these agents, and a more rigorous pharmacological investigation would be required to establish the underlying causes of these changes. The CaM-dependent mechanism, quiescent in acutely hyperosmotically incubated cells, becomes the dominant pathway during chronic hypernatraemia. During chronic hyponatraemia, the marked inhibitory effects of DIDS and TFP on acutely stimulated taurine efflux become, respectively, attenuated and abolished. It is possible that during chronic adaptations to anisosmolality the populations of functional transport pathways differ from those subserving rapid volume-regulatory taurine efflux under acute conditions, and that these distinct populations possess different inhibitor sensitivities. But such a possibility would be very difficult to test using the present methodology.

The results obtained in hyposmotically adapted cells suddenly restored to isosmotic media further support the view that re-uptake of interstitial taurine plays a role in determining the volume-regulatory effectiveness of taurine efflux. If the rate of increase of $Na^+$ and/or $Cl^-$-dependent taurine uptake on transference from 122 to 142 mM $Na^+$ exceeds the rate of restoration of efflux (i.e. if efflux remains at its hyponatraemically down-regulated level), this would explain the observed cell swelling (Fig. 3). It is significant that no comparable swelling occurs in cells that have not been pre-loaded with taurine. The swelling response was inhibited by the NaCl cotransport inhibitor bumetanide, but only in taurine-loaded cells. It has recently been shown[23] that the NaCl cotransport inhibitor furosemide impairs fractional taurine reabsorption in the loops of Henle, but this effect may be ascribed to structural similarities between taurine and furosemide; no inhibition was found with bumetanide, which is not structurally related. However, the present findings are compatible with inwardly-directed Na/Cl/taurine cotransport in rat brain cells, analogous to the mechanism that has been demonstrated in ascites tumour cells[7,8], and which would serve to unify the ionic dependences that have been described in various brain cell preparations[18,20,22,26].

It might be argued that if bumetanide is inhibiting taurine uptake from the interstitium, this should result in an increase in the rate of net taurine efflux. This was not observed (Fig. 3). But it must be born in mind that in these experiments only the slow phase of cellular efflux was measured; there is a rapid cellular phase ($t_{1/2}$=2-3 min)[10] which might have subsumed any increase.

Finally, the observation that cerebral oedema resulting from hyponatraemia in cats is alleviated by chronic taurine deficiency[26] is also consistent with the view that interstitial re-uptake is a significant factor in cell volume regulation.

## ACKNOWLEDGMENT

Supported in part by the Wellcome Trust, Grant No. 043327/Z/94/Z.

## REFERENCES

1. Beetsch, J.W. and Olson, J.E., 1996, Hyperosmotic exposure alters total taurine quantity and cellular transport in rat astrocyte cultures, *Biochim. Biophys. Acta*, 1290:141-148.
2. Conejero, C., 1997, Anion exchanger AE1 as a candidate pathway for taurine exchange in rat erythrocytes, *Am. J. Physiol.*, 272:C1457-1464.
3. Cserr, H.F. and Patlak, C.S., 1991, Regulation of brain cell volume under isosmotic and anisosmotic conditions, in "Volume and Osmolality Control in Animal Cells", Gilles, R., Hoffmann, E.K. and Bolis, L., eds, Springer-Verlag, Berlin, pp 61-80.
4. Gullans, S.R. and Verbalis, J.G., 1993, Control of brain volume during hyperosmolar and hypoosmolar conditions, *Annu. Rev. Med.*, 44:289-301.
5. Hoffmann, E.K. and Simonsen, L.O., 1989, Membrane mechanisms in volume and pH regulation in vertebrate cells, *Physiol. Rev.*, 69:315-382.
6. Kimelberg, H.K., Goderie, S.K., Higman, S., Pang, S., and Waniewski, R.A., 1990, Swelling-induced release of glutamate, aspartate, and taurine from astrocyte cultures, *J. Neurosci.*, 10:1583-1591.
7. Lambert, I.H., 1985, Taurine transport in Ehrlich ascites tumour cells; specificity and chloride dependence, *Mol. Physiol.*, 7:323-332.
8. Lambert, I.H. and Hoffmann, E.K., 1993, Regulation of taurine transport in Ehrlich ascites tumor cells, *J. Memb. Biol.*, 131:67-79.
9. Law, R.O., 1994, Regulation of mammalian brain cell volume, *J. Exp. Zool.*, 268;90-96.
10. Law, R.O., 1994, Taurine efflux and the regulation of cell volume in incubated slices of rat cerebral cortex, *Biochim. Biophys. Acta*, 1221:21-28.

11. Law, R.O., 1994, Effects of extracellular bicarbonate ions and pH on volume-regulatory taurine efflux from rat cerebral cortical slices *in vitro*; evidence for separate neutral and anionic transport mechanisms, *Biochim. Biophys. Acta*, 1224:377-383.
12. Law, R.O., 1994, Taurine efflux and cell volume regulation in rat cerebral cortex following chronic (7 day) hypernatraemia, *J. Physiol. (Lond)*, 479:57P-58P (Abstract).
13. Law, R.O., 1995, Taurine efflux and cell volume regulation in cerebral cortical slices during chronic hypernatraemia, *Neurosci. Lett.*, 185:56-59.
14. Lohr, J.W., 1989, Cell volume maintenance in $S_2$ segments during gradual decrease in osmolality, *Clin. Res.*, 37:496A (Abstract).
15. Lohr, J.W., and Grantham, J.J., 1986, Isovolumetric regulation of isolated $S_2$ proximal tubules in anisotonic media, *J. Clin. Invest.*, 78:1165-1172.
16. Martin del Rio, R., Galarreta, M., Menendez, N., Conejero, C., and Solis, J.M., 1996, Taurine is a substrate of the anion exchanger transport system, in "Taurine 2", Huxtable, R.J., Azuma, J., Kuriyama, K., Nakagawa, M., and Baba, A., eds., Plenum Press, New York, pp. 401-407.
17. Mountian, I. and van Driessche, W., 1997, Isovolumetric regulation of C6 rat glioma cells in hyperosmotic media, *Am. J. Physiol.*, 272:C318-C323.
18. Oja, S.S. and Saransaari, P., 1996, Kinetic analysis of taurine influx into cerebral cortical slices from adult and developing mice in different incubatory conditions, *Neurochem. Res.*, 21:161-166.
19. Oja, S.S. and Saransaari, P., 1996, Taurine as osmoregulator and neuromodulator in the brain, *Metab, Brain Dis.*, 11:153-164.
20. Pasantes-Morales, H., 1996, Volume regulation in brain cells: cellular and molecular mechanisms, *Metab. Brain Dis.*, 11:187-204.
21. Pasantes-Morales, H. and Schousboe, A., 1988, Volume regulation in astrocytes: a role for taurine as an osmoeffector, *J. Neurosci. Res.*, 20:505-509.
22. Sanchez-Olea, R., Moran, J., and Pasantes-Morales, H., 1992, Changes in taurine transport evoked by hyperosmolarity in cultured astrocytes. *J. Neurosci. Res.*, 32:86-92.
23. Silbernagl, S., Volker, K., Lang, H-J., and Dantzler, W.H., 1997, Taurine reabsorption by a carrier inter-acting with furosemide in short and long Henle's loops of rat nephrons, *Am. J. Physiol.*, 272:F205-F213.
24. Strange, K., Emma, F., and Jackson, P.S., 1996, Cellular and molecular physiology of volume-sensitive anion channels, *Am. J. Physiol.*, 272:C1766-C1775.
25. Strange, K. and Jackson, P.S., 1995, Swelling-activated organic osmolyte efflux: a new role for anion channels, *Kidney Int.*, 48:994-1003.
26. Trachtman, H., del Pizzo, R., and Sturman, J.A., 1990, Taurine and osmoregulation. III. Taurine deficiency protects against cerebral edema during acute hyponatraemia, *Pediatr. Res.*, 27:85-88.
27. Trachtman, H., Futterweit, S., and del Pizzo, R., 1992, Taurine and osmoregulation. IV. Cerebral taurine is increased in rats with hypernatraemic dehydration. *Pediatr. Res.*, 32:118-124.
28. Van Driessche, W., de Smet, P., Li, J., Allen, S., Zizi, M., and Mountian, I., 1997, Isovolumetric regulation in a distal nephron cell line (A6), *Am. J. Physiol.*, 272:C1890-C1898.

# Part 6:
# Transport and General Studies on Taurine

# EFFECTS OF STEROID HORMONES AND CYCLOSPORINE A ON TAURINE-TRANSPORTER ACTIVITY IN THE RAW264.7 CELL LINE

Ha Won Kim,[1] Eun Jin Lee,[2] Mi Ja Shim,[1] and Byong Kak Kim[2]

[1] Department of Life Science
The Seoul City University of Seoul
Seoul 130-743
[2] Department of Microbial Chemistry
College of Pharmacy
Seoul National University
Seoul 151-742, Korea

## ABSTRACT

The activity of the taurine transporter is affected by various extracellular stimuli, such as ions, hormones and stress. To assess the effects of steroid hormones and cyclosporine A (CsA) on taurine-transporter activity, the murine monocytic cell line, RAW264.7, was stimulated with dexamethasone (DM), triamcinolone (TA), cortisone (CS), hydrocortisone (HCS), prednisone (PSN), prednisolone (PSL) and methylprednisolone (MPSL) in the presence of 12-O-tetradecanoylphorbol 13-acetate (TPA). Treatment of the cell with TPA led to a significant reduction in taurine-transporter activity. However, in the case of the stimulation of the cells with steroid hormones in the presence of TPA, all of the hormones reversed the TPA-induced reduction in the taurine-transporter activity. Treatment of the cells with CsA led to a significant reduction in taurine-transporter activity, but ionomycin (IM) alone did not affect taurine-transporter activity. However, IM reversed the TPA- and CsA-induced reduction in taurine-transporter activity. These results showed that both IM and the glucocorticoid hormones reversed TPA-induced reductions in taurine-transporter activity but only IM reversed the CsA-induced reduction of transporter activity in the RAW264.7 cell line.

## INTRODUCTION

Taurine (TAU, 2-aminoethanesulfonic acid, $^{+}NH_3CH_2CH_2SO_3^{-}$) is one of the major intracellular free amino acids in mammals. It is involved in a number of important physiological processes, such as neuromodulatory actions[14], cardiac $Ca^{++}$ modulation[20] and anti-oxidant actions[3,8]. Cloning of the cDNA encoding a transporter of taurine has been achieved in the mouse[11], rat[22], canine[25], and human[5]. In the human, the taurine transporter was found to be identical in thyroid[5], placenta[17] and retinal pigment epithelium[12]. The chromosomal localization of the taurine transporter was found to be p24-p26[19]. The homology of amino acid sequences of the transporter is more than 90% among these species. Taurine has been considered to be a neurotransmitter and its transporter has some amino acid homology with other neurotransmitters. However, the taurine transporter is different from the β-alanine transporter[13] or the GABA transporter[16,21].

Recently, there has been increasing interest in the mechanisms of taurine transport in various tissues. There have been many studies about factors affecting taurine-transporter activity. Tyrosine residues are essential for the placental taurine transporter[9]. The activity of the taurine transporter is affected by various extracellular stimuli such as ions, hormones, and drugs. It is also reported that taurine uptake into the macrophage is an important step in protecting against tissue injury. 12-O-Tetradecanoylphorbol 13-acetate (TPA) is a tumor-promoter and can activate protein kinase C (PKC)[1]. TPA is also an inhibitor for monocyte differentiation[4]. Inhibition of RNA transcription by actinomycin D does not alter taurine-transporter activity, but inhibition of protein synthesis by cycloheximide or treatment with TPA prevents taurine-transporter activity[6]. Previously, it has been shown that phorbol ester inhibits the activity of the taurine transporter[7].

Cyclosporine A (CsA) is a cyclic undecapeptide that has potent immunosuppressive activity. It is currently in wide use in organ transplant patients to prevent graft rejection[2]. Recently, it has been demonstrated that CsA has some functional relationship with taurine. For instance, CsA inhibits bile acid-taurine conjugation[26] and tauroursodeoxycholate diminishes cyclosporine-induced cholestasis[15]. Taurine-transporter activity was selectively impaired by CsA in human placental cell[18] and calmodulin might be involved in this inhibition[17]. One of the functions of CsA in cells is an inhibition of the $Ca^{++}$-dependent activation pathway[10]. Therefore from these relationships between taurine and CsA, we attempted in this study to demonstrate a reversal of CsA-induced inhibition of taurine-transporter activity by ionomycin (IM), as well as recovery of taurine-transporter activity with various glucocorticoids in a TPA-treated macrophage cell line.

## MATERIALS AND METHODS

### Materials

The RAW 264.7 cell line, an Abelson leukemia virus-transformed murine macrophage cell line, was obtained from American Type Culture Collection (Rockville, MD, USA). Dulbecco's modified Eagle's medium (DMEM) and [2-$^3$H(N)]-taurine (21.9 Ci/mmol) were bought from Gibco-BRL (Grand Island, NY) and NEN (Boston, MA), respectively. Fetal bovine serum (FBS), penicillin and streptomycin, dexamethasone (DM), cortisone (CS), triamcinolone (TA), hydrocortisone (HCS), prednisone (PSN), prednisolone (PSL), methyl-prednisolone (MPSL), ionomycin (IM), and all other chemicals were obtained from Sigma Chemical (St. Louis, MO). Cyclosporine A (CsA, Sandoz) was kindly provided by Han-mi Pharmaceutical Co., Korea.

### Cell Culture and Taurine Transportation

RAW 264.7 cells were cultured and maintained in 90 mm culture dishes in DMEM medium containing 10% heat-inactivated fetal bovine serum (FBS) and penicillin (100U/ml) /streptomycin (100 μg/ml), and in 6-well plates for taurine-uptake experiments.

Cultures were maintained at 37°C in a 5% $CO_2$ incubator until they reached confluence. When the cells were confluent on the dish, the medium was removed from the monolayer culture and replaced with 2 ml transporter buffer, which contained 25 mM-HEPES/Tris (pH 7.5), 140 mM NaCl, 5.4 mM KCl, 1.8 mM $CaCl_2$, 0.8 mM $MgSO_4$, 5 mM D-glucose. After stabilizing the cells for 30 min at 37°C, 1 μCi of $^3$H-taurine was added to each culture and incubated for 30 min. Then the cells were washed three times with ice-cold $Na^+$ deficient buffer and lysed with 1 ml of 0.2 N NaOH/0.5% SDS. The lysate was mixed with 2 ml cocktail solution and was measured in a liquid scintillation counter.

### Cell Stimulation

For the stimulation of the RAW264.7 cells, confluent monolayer cells in DMEM containing 10% fetal bovine serum were incubated with a steroid for 30 min in the presence of TPA in 6-well plates by using transportation buffer. Then the cells were treated with 1 μCi of [$^3$H]-taurine for another 30 min. The cells were washed three times with cold buffer and then lysed with the lysis buffer. Transported radioactive taurine in the lysate was measured in a liquid scintillation counter. The final concentration of TPA and all steroids were 10 ng/ml and 100 nM, respectively. For other stimulation experiments, similar procedures were followed by using 5 nM CsA and 500 ng/ml IM.

## RESULTS

### Glucocorticoids Reverse TPA-Induced Inhibition of Taurine Transporter Activity

The RAW 264.7 cell was cultured in 6-well culture plates. When 10 ng/ml TPA was applied to this cell line for 30 min, the taurine-transporter activity was reduced to 13.2% of the control value. However, treatment of the cell with various steroid hormones in the presence of TPA partially reversed the loss of transporter activity (Table I). Treatment of the cells with DM or TA in the presence of TPA resulted in the transport of 163,655 ± 7,305 (57.5%) or 137,962 ± 3,045 (48.8%) cpm of [$^3$H]-taurine, respectively. Similarly, treatment with CS or HCS led to the transport of 166,612 ± 7,023 (58.6%) or 132,294 ± 7,374 (46.5%) cpm of [$^3$H]-taurine, respectively. When the cells were treated with PSN, PSL or MPSL in this system taurine transport yielded 156,626 ± 6,001 (55.0%), 165,889 ± 3,596 (58.3%) or 155,576 ± 7,112 (54.7%), respectively.

### Cyclosporine A Inhibits Taurine Transporter Activity

Monolayer RAW264.7 cells in 6-well plates were stabilized with transporter buffer for 30 min and then treated with 5 nM CsA. [$^3$H]-Taurine was then added to the cells for 30 min. The control group transported 421,002 ± 401 cpm of [$^3$H]-taurine. Treatment of the cells with CsA inhibited the activity of the transporter. Treatment of the cells with CsA for 5 min, 30 min, 60 min and 5 hr caused the transport of taurine to decrease to 29,406 ± 443 (7.0% of the control), 180,050 ± 5,446 (42.8%), 172,454 ± 6,839 (41.0%) and 149,672 ± 7,738 (35.6%) cpm of [$^3$H]-taurine, respectively (Table II). Inhibition of taurine-transporter activity was greatest at 5 min.

**Table I**. Effects of various steroid hormones on taurine transport in the murine monocytic cell line, RAW264.7.

| Effector | $[^3H]$-Taurine transportation | |
|---|---|---|
| | cpm | % |
| Control | 284,376 ± 1,780 | 100.0 |
| TPA | 37,674 ± 2,058 | 13.2 |
| TPA + DM | 163,655 ± 7,305 | 57.5 |
| TPA + TA | 137,962 ± 3,045 | 48.5 |
| TPA + CS | 166,612 ± 7,023 | 58.6 |
| TPA + HCS | 132,294 ± 7,374 | 46.5 |
| TPA + PSN | 156,626 ± 6,001 | 55.0 |
| TPA + PSL | 165,889 ± 3,596 | 58.3 |
| TPA + MPSL | 155,576 ± 7,112 | 54.7 |

RAW 264.7 monolayer cells in 6-well plates were stabillized with transporter buffer for 30 min and stimulated with steroid hormones for 30 min. Then $^3$H-taurine was added for 30 min. The final concentrations of TPA and all steroids were 10 ng/ml and 100 nM, respectively. Abbreviations for steroid hormones are as follows: DM, dexamethasone; TA, triamcinolone; CS, cortisone; HCS, hydrocortisone; PSN, prednisone; PSL, prednisolone; MPSL, methylprednisolone.

**Table II.** Effects of CsA on taurine-transporter activity at varying time periods.

| CsA stimulation time | $[^3H]$-Taurine transportation | |
|---|---|---|
| | cpm | % |
| 0 | 421,002 ± 401 | 100.0 |
| 5 min | 29,406 ± 443 | 7.0 |
| 30 min | 180,050 ± 5,446 | 42.8 |
| 60 min | 172,454 ± 6,839 | 41.0 |
| 5 hr | 149,672 ± 7,738 | 35.6 |

RAW 264.7 monolayer cells were stabilized with transporter buffer for 30 min and incubated with CsA for each time period. Then $^3$H-taurine was added for 30 min. Final concentration of CsA was 5 nM.

## Ionomycin Reverses Cyclosporine A-Induced Reduction of Taurine Transporter Activity

RAW 264.7 cells in 6-well culture plates were cultured with CsA and IM for various periods. Treatment of the cells with PBS for 30 min caused the transport of 403,609 ± 2,074 cpm of $[^3H]$-taurine and with CsA for 5 min the transport 125,760 ± 6,121 cpm (31.2% of the control) of $[^3H]$-taurine. When the cells were cultured in the presence of both 5 nM CsA and 500 ng/ml IM for 0.5 min, 5 min and 30 min, 64,863 ± 1,088 (16.1% of the control), 223,005 ± 622 (55.3%) and 294,113 ± 264 (72.9%) cpm of $[^3H]$-taurine were transported, respectively (Table III). Treatment of the RAW264.7 cells with CsA, TPA or IM caused the transport of 26.5%, 41.5% and 97.8% of the control value, respectively. Therefore, IM itself did not affect taurine-transporter activity. However, treatment of the cells with TPA + IM or TPA + CsA + IM caused the transport of 58.3% or 32.0% of the control value, respectively. Therefore, IM seems to reverse the TPA- or TPA + CsA-induced reduction in taurine-transporter activity (Table IV). Furthermore, IM reversed DM + CsA-induced reductions in the transporter activity. When the cells were treated with DM + CsA, they transported 32.5% of the control value. However addition of IM to this system partially restored transporter activity to 71.6% of the control value (Table V).

## DISCUSSION

The transport of taurine into or out of the cell is important for the regulation of various cellular functions. The taurine-transporter responsible for the entry of taurine or β-amino acids into the cell has been partially characterized and its gene has been cloned. However, the presence of the taurine transporter responsible for its secretion from the cell is still unclear. In this study, we investigated cellular signals or factors that regulate taurine-transporter activity by using a blood macrophage cell line. We demonstrated that steroid hormones reversed the TPA-induced reduction of taurine-transporter activity while ionomycin reversed the cyclosporine A-induced inhibition of taurine-transporter activity.

**Table III.** Time course of ionomycin on taurine-transporter activity in CsA-treated cells

| Effector | Time | $[^3H]$-Taurine transportation cpm | % |
|---|---|---|---|
| Control | 30 min | 403,609 ± 2,074 | 100.0 |
| CsA | 5 min | 125,760 ± 6,121 | 31.2 |
| CsA + IM | 0.5 min | 64,863 ± 1,088 | 16.1 |
| CsA + IM | 5 min | 223,005 ± 622 | 55.3 |
| CsA + IM | 30 min | 294,113 ± 264 | 72.9 |

Cells were stimulated with each effector and $[^3H]$-taurine was added for 30 min. Final concentrations of CsA and IM were 5 nM and 500 ng/ml, respectively.

**Table IV.** Effect of IM on taurine-transporter activity in TPA and CsA-treated cells

| Effector | $[^3H]$-Taurine transportation cpm | % |
|---|---|---|
| Control | 428,886 ± 2,239 | 100.0 |
| TPA | 177,738 ± 3,160 | 41.5 |
| CsA | 113,756 ± 1,214 | 26.5 |
| IM | 419,254 ± 1,255 | 97.8 |
| TPA + IM | 250,056 ± 1,249 | 58.3 |
| TPA + CsA + IM | 180,257 ± 8,330 | 42.0 |

Cells were stabilized with transport buffer and incubated with each effector for 5 min. Then 1 μCi of $[^3H]$-taurine was added for 30 min. Radioactivity in the lysate was measured with a liquid scintillation counter. The final concentrations of CsA, TPA and IM were 5 nM, 100 ng/ml and 500 ng/ml, respectively.

**Table V.** Effects of ionomycin on taurine-transporter activity in dexamethasone and CsA-treated cells

| Effector | [$^3$H]-Taurine transportation | |
|---|---|---|
| | cpm | % |
| Control | 465,220 ± 7,696 | 100.0 |
| DM | 382,304 ± 419 | 82.2 |
| CsA | 127,240 ± 9,384 | 27.4 |
| IM | 389,970 ± 2,132 | 83.8 |
| DM + CsA | 151,274 ± 1,328 | 32.5 |
| DM + CsA + IM | 333,029 ± 3,417 | 71.6 |

RAW 264.7 cells were stabilized with transport buffer for 30 min and then incubated with each effector for 0.5 min. One μCi of [$^3$H]-taurine was added for 30 min. The final con-centrations of CsA, DM and IM were 5 nM, 100 nM and 500 ng/ml, respectively.

Most of the taurine transporters are dependent on the presence of $Na^+$ and $Cl^-$ ions[23,24]. In preliminary experiments, we confirmed the presence of the taurine-transporter membrane protein in the RAW264.7 cell line by comparing taurine transport in $Na^+$-containing and $Na^+$-deficient buffers. Furthermore the cell line failed to transport taurine when the cell was saturated with β-alanine (data not shown). In a previous report, we demonstrated that TPA blocked taurine transport in the RAW264.7 cell line and dexamethasone reversed this inhibitory effect[7]. Therefore, in this report we investigated the effects of other steroid hormones on taurine-transporter activity. When RAW264.7 cells were treated with TPA, [$^3$H]-taurine transport was reduced to 13.2% of the control value. However, when steroid hormones, such as DM, CS, TA, HCS, PSN, PSL or MPSL were added to this system, transporter activity recovered approximately 50% (Table I). Unfortunately, no structure-function relationship was evident from the data.

CsA is widely used for the purpose of immunosuppression. Ramamoorthy reported the inhibition of taurine transporter by CsA and that FK-506 did not similarly inhibit transporter activity[18]. In this study, we confirmed both the inhibition of the taurine-transporter by CsA and the reversal of its inhibition by IM, a $Ca^{++}$-transporter. Blocking the activity of the taurine transporter by CsA was maximal at 5 min rather than at 30 or 60 min, indicating that CsA works very rapidly in the cell. Reversal of CsA-induced inhibition of transporter activity by IM was a gradual process. IM led to recovery of transporter activity to 16.1%, 55.3% and 72.9% of the control within 0.5 min, 5 min and 30 min, respectively (Table III). Calmodulin regulates the activity of the taurine transporter, although its function could be blocked by CsA[18]. In our experiments, despite the blocking of the calmodulin effect by CsA, increasing intracellular $Ca^{++}$ by IM restores the calmodulin effect, which leads to the recovery of taurine-transporter activity. IM also reverses the TPA-induced reduction in taurine-transporter activity of RAW264.7 cells (Table IV).

Various factors or intracellular signals affecting the activity of the taurine transporter need to be clarified, in light of the important role of taurine in the mammalian system. Antiinflammatory steroid hormones or immunosuppressant agents could exert their effects directly on the blood macrophage cells. The mechanism of the reversal of the inhibition of taurine-transporter activity by the $Ca^{+2}$-transporter, IM, is in progress to clarify the intracellular signal transduction pathway in the macrophage.

## ACKNOWLEDGMENTS

This research was supported in part by a grant from the Korean Taurine Society and by the Korea Science and Engineering Foundation through the Research Center for New Drug Development at Seoul National University.

## REFERENCES

1. Blumberg, P.M., 1988, Protein kinase C as the receptor for the phorbol ester tumor promoters: sixth Rhoads memorial award lecture, *Cancer Res.*, 48:1.
2. Borel, J.F., Seuer, C., Gubler, H.U., and Stahelin, H., 1976, Biological effects of cyclosporin A: a new antilymphocytic agent, *Agents Actions*, 6: 468-475.
3. Gordon, R.E., Heller, R.F. and Heller, R.F., 1992, Taurine, Nutritional Value and Mechanisms of Action, pp. 319-328, Plenum Press, New York.
4. Hass, R., Brach, M., Kharbanda, S., Giese, G., Traub, P., and Kufe, D., 1991, Inhibition of phorbol ester-induced monocytic differentiation by dexamethasone is associated with down- regulation of c-fos and c-jun (AP-1), *J. Cellular Physiol.*, 149: 125-131.
5. Jhiang, S.M., Fithian, L., Smanik, P., McGill, J., Tong, Q., and Mazzaferri, E.L., 1993, Cloning of the human taurine transporter and characterization of taurine uptake in thyroid cells, *FEBS Letters*, 318:139-144.
6. Jones, D.P., Miller, L.A., Dowling, C., and Chesney, R.W., 1991, Regulation of taurine transporter activity in LLC-PK1 cells: role of protein synthesis and protein kinase C activation, *J. Am. Soc. Nephrol.*, 2: 1021-1029.
7. Kim, H.W., Shim, M.J., Kim, W.B., and Kim, B.K., 1995, Regulation of taurine transporter activity by glucocorticoid hormone, *J. Biochem. Mol. Biol.*, 28: 527.
8. Koyama, I., Nakamura, T., Ogasawara, M., Nemoto, M., and Yoshida,T., 1992, Taurine, Nutritional Value and Mechanisms of Action, pp. 355-359, Plenum Press, New York.
9. Kulanthaivel, P., Cool, D.R., Ramamoorthy, S., Mahesh, V.B., Leibach, F.H., and Ganapathy, V., 1991, Tyrosine residues are essential for the activity of the human placental taurine tran-sporter, *Biochem. J.*, 277: 53-58.
10. Landewe, R.B., Dijkmans, B.A., and Verdonk, M.J., 1996, Persistent CD3-crosslinking down- regulates interleukin-2 responsiveness in interleukin-2-competent cloned T cells: the possible involvement of protein kinase C, *Scand. J. Immunol.*, 44: 45-53.
11. Liu, Q.R., Lopez-Corcuera, B., Nelson, H., Mandiyan, S., and Nelson, N., 1992, Cloning and expression of a cDNA encoding the transporter of taurine and β-alanine in mouse brain, *Proc. Natl Acad. Sci. USA*, 89: 12145-12149.
12. Miyamoto, Y., Liou, G.I., and Sprinkle, T.J., 1996, Isolation of a cDNA encoding a taurine transporter in the human retinal pigment epithelium, *Curr. Eye Res.*, 15: 345-349.
13. Munck, L.K. and Munck, B.G., 1992, Distinction between chloride-dependent transport systems for taurine and beta-alanine in rabbit ileum, *Am. J. Physiol.*, 262: G609-615.
14. Oja, S.S. and Kontro, P., 1990, Taurine: Functional Neurochemistry, Physiology and Cardiology, pp. 69-76, Wiley-Liss, New York.
15. Queneau, P.E., Bertault-Peres, P., Mesdjian, E., Durand, A. and Montet, J.C., 1993, Diminution of an acute cyclosporin-induced cholestasis by tauroursodeoxycholate in the rat, *Transplant-ation*, 56: 530-534.
16. Quinn, M.R. and Miller, C.L., 1992, Taurine allosterically modulates flunitrazepam binding to synaptic membranes, *J. Neurosci. Res.*, 33: 136-141.
17. Ramamoorthy, S., Del Monte, M.A., Leibach, F.H., and Ganapathy, V., 1994, Molecular identity and calmodulin-mediated regulation of the taurine transporter in a human retinal pigment epithelial cell line, *Curr. Eye Res.*, 13: 523-529.
18. Ramamoorthy, S., Leibach, F.H., Mahesh, V.B., and Ganapathy, V., 1992, Selective impair-ment of taurine transport by cyclosporin A in a human placental cell line, *Pediatric Res.*, 32:125-127.

19. Ramamoorthy, S., Leibach, F.H., Mahesh, V.B., Han, H., Yang-Feng, T., Blakely, R.D., and Ganapathy, V., 1994, Functional characterization and chromosomal localization of a cloned taurine transporter from human placenta, *Biochem. J.*, 300: 893-900.
20. Sawamura, A., Azuma, J., Awata, N., Harada, H., and Kishimoto, S., 1990, Taurine: Functional Neurochemistry, Physiology and Cardiology, pp. 207-215, Wiley-Liss, New York.
21. Sivakami, S., Ganapathy, V., Leibach, F.H., and Miyamoto, Y., 1992, The gamma-amino butyric acid transporter and its interaction with taurine in the apical membrane of the bovine retinal pigment epithelium, *Biochem. J.*, 283: 391-397.
22. Smith, K.E., Borden, L.A., Wang, C.D., Hartig, P.R., Branchek, T.A., and Weinshank, R.L., 1992, Cloning and expression of a high affinity taurine transporter from rat brain, *Mol. Pharmacol.*, 42: 563-569.
23. Tiruppathi, C., Bransdisch, M., Miyamoto, Y., Ganapathy, V., and Leibach, F.H., 1992, Constitutive expression of the taurine transporter in a human colon carcinoma cell line, *Am. J. Physiol.*, 263: G625-631.
24. Uchida, S., Kwon, H.M., Preston, A.S., and Handler, J.S., 1991, Expression of Madin-Darby canine kidney cell Na(+)-and Cl(-)-dependent taurine transporter in *Xenopus laevis* oocytes, *J. Biol. Chem.*, 266: 9605-9609.
25. Uchida, S., Kwon, H.M., Yamauchi, A., Preston, A.S., Marumo, F., and Handler, J.S., 1992, Molecular cloning of the cDNA for an MDCK cell Na+- and Cl--dependent taurine transporter that is regulated by hypertonicity, *Proc. Natl. Acad. Sci. USA*, 89: 8230-8234.
26. Vessey, D.A. and Kelley, M., 1995, Inhibition of bile acid conjugation by cyclosporin A, *Biochim. Biophys. Acta*, 1272: 49-52.

# THE ANION-EXCHANGER AE1 IS A DIFFUSION PATHWAY FOR TAURINE TRANSPORT IN RAT ERYTHROCYTES

Rafael Martín del Río and José M. Solís

Servicio de Neurobiología
Departamento de Investigación
Hospital Ramón y Cajal.
Crta Colmenar Km 9
28034 Madrid, Spain

## INTRODUCTION

A putative osmolyte is required to meet two main criteria: to be highly concentrated inside the cell, and to possess a specific pathway across the plasma membrane allowing for its osmotic efflux[12]. In the case of taurine, an admitted osmolyte in several animal species, high intracellular levels are maintained in most mammalian tissues by the action of well characterized sodium/chloride-dependent carriers[14,20,24,25]. The nature of the taurine efflux carrier, however, remains to be assessed, although it has been demonstrated in several cell types[11,17,22] that it consists of a diffusion pathway with pharmacological properties similar to a chloride channel. An alternative pathway for taurine exit has been proposed based on extensive studies with red blood cells (RBCs). This release occurs through a transport system that is sensitive to compounds that are well-known inhibitors of the RBC's anion-exchanger (AE1)[8]. It has been claimed, therefore, that taurine uses this diffusion transport system to exit the erythrocyte[9] (but see[16]). This assumption is supported by the fact that some N-derivative analogs of taurine such as, NBD-taurine, Cl-taurine, NAP-taurine or NIP-taurine, are either substrates and/or competitive inhibitors of this anion-exchanger[1]. In addition, AE1 transports many different organic compounds, including some amino acids such as glycine[15].

It is known that erythrocytes from several fish species contain high levels of intracellular taurine, due to the action of sodium dependent carriers, and also due to a diffusion pathway that allows taurine efflux during the physiological process of regulatory volume decrease (RVD)[5,6]. Intracellular taurine levels of rat red blood cells, however, are very low, in the range of its plasma concentration. This fact indicates the lack of a concentrative transport system in these cells and thus negates an osmolyte role for taurine in rat erythrocytes, but at the same time suggests the existence of an open diffusion pathway between both compartments.

In the present paper a number of taurine uptake experiments have been conducted, trying to ascertain the nature of the taurine diffusion pathway in rat RBCs. The results clearly indicate that this amino acid crosses the red blood cell membrane as a substrate of the AE1 (also called band 3) transport system.

## MATERIALS AND METHODS

Adult Sprague-Dawley rats weighing 150-200 g were decapitated and the blood flowing from the neck wound was poured into 15 ml tubes containing 200 μl of a 5,000 I.U./ml heparin solution. The blood of 2-3 animals was pooled and centrifuged at 2,000 xg for 20 min and the supernatant and the buffy coat (white cells and platelets) were removed by suction. The cell pellet was then washed twice with five volumes of Krebs-Ringer-bicarbonate solution (KRB, in mM: 122 NaCl, 3.0 KCl, 1.2 $CaCl_2$, 1.2 $MgSO_4$, 0.4 $KH_2PO_4$, 25 $NaHCO_3$ and 10 glucose). RBCs, resuspended to original blood volume with KRB, were preincubated for 20 min at room temperature or overnight at 4°C, to ensure equilibrium for water and the ion cellular contents before the start of the experiments. Aliquots of 0.5 ml were then transferred to 1.5 ml Eppendorf tubes and centrifuged for 2 min in a Beckman microfuge (10,000 xg). Supernatants were discarded and 0.3ml of the appropriate solution, as indicated, were added to each fraction. The assays were conducted at 37°C at a final approximate hematocrit of 40%.

In the uptake experiments, after incubation of the RBCs for the indicated time with the amino acid solutions, the transport was stopped by centrifugation (2 min at 10,000 xg). The supernatant, containing the nontransported amino acid, was discarded. The amino acids trapped in the extracellular fluid were eliminated by centrifugation, washing the cells four times with 1 ml of fresh KRB (4x2 min, 10,000 xg). The final pellets were lysed by a twenty-fold dilution with distilled water and the resulting solutions were used to measure the cellular content of both protein and the amino acids[19]. In most taurine uptake experiments a 0.5 ml stopping solution consisting of ice-chilled KRB containing 1 mM flufenamic acid, a fast and potent noncompetitive inhibitor of AE1[3], was added to end the uptake period. The KRB, which used in the ensuing washing procedure, also contained this inhibitor.

The effect of pH was evaluated by resuspending the cells in a modified KRB incorporating TRIS buffer instead of bicarbonate and adjusted to the final pH value with HCl. After a 20 min equilibration, taurine uptake was carried out in the appropriate KRB buffered solution. A 100 mM solution of 2-(3-phenylpropylamino)-5-nitrobenzoic acid (NPPB), a rather specific chloride channel inhibitor, was prepared in dimethyl sulfoxide and stored at -20°C. The final solutions were prepared by dilution with KRB. The AE1 competitive inhibitor pyridoxal-5′-phosphate[2], was present in the incubation medium 20 min before and also during the taurine uptake period. A 2 mM solution of dipyridamole, a noncompetitive AE1 inhibitor[18], was obtained dissolving the powder in an acidic KRB (adjusted to pH, 1.5 with concentrated HCl), from which a convenient concentration for uptake assays (50 μM) in standard KRB was prepared.

## RESULTS AND DISCUSSION

Taurine and GABA, two omega amino acids which have transport systems with very similar functional properties[13] and that are also genetically related[20], were taken up by RBC suspensions in a concentration-dependent manner that was linear under the conditions of the assay (0.5-10 mM, 30 min). In fact, we have observed that taurine uptake is linear up to 40 mM during a 30 min incubation period. When a 10 mM concentration was used, the GABA uptake rate decreased progressively after 30 min of incubation whereas taurine influx remained constant for at least 2 hours. Taurine and GABA transport systems in RBCs are clearly unrelated to the classical amino acid carriers because their uptakes were not affected by a modified KRB without NaCl.

These RBC transporters must actually be two different systems since taurine transport is totally blocked by 10 μM of DIDS whereas GABA uptake remains unchanged even in the presence of 100 μM of this compound[21]. The kinetic behavior of taurine uptake is also very different from the RBC transport of other basic or acidic amino acids such as arginine, aspartic or glutamic acid, when assayed at 10 mM concentration. These amino acids are taken up by rat erythrocytes at very low rates and reach a steady state in a few minutes.

Glycine is also transported by RBCs and it has been published that part of glycine uptake by erythrocytes is due to a DIDS-sensitive cotransport of the glycine anion with a proton via band 3[15]. Figure 1 shows that glycine, at concentrations of 1, 10 or even 20 fold that of taurine in the incubation medium, did not inhibit taurine uptake in RBCs, suggesting the lack of common carriers for both amino acids. However, since the $K_{½}$ for glycine by band 3 in human RBCs has been reported to be higher than 50 mM[15], it is still possible that there is a common binding site for both amino acids but with a much higher affinity for taurine in rat RBCs. We suggest the presence of a sodium and chloride independent taurine transporter in RBCs, not shared with GABA, with a functionally important DIDS- sensitive site which is not inhibited by glycine concentrations that are two orders of magnitude higher than those present in plasma.

Band 3 is a system with an unusually large temperature dependence[7]. We have tested taurine uptake (20 mM, 30 min) by rat erythrocytes at several temperatures ranging from 4 to 37°C. We found a marked dependence of the taurine uptake rate on temperature, with

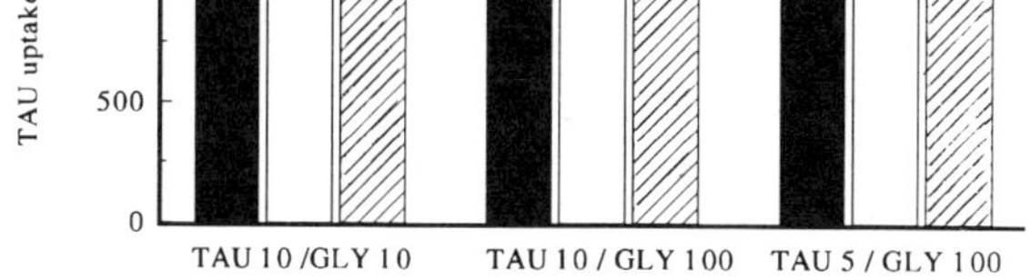


**Figure 1.** RBCs taurine uptake in the presence of high glycine concentrations (hyperosmotic conditions). RBCs were incubated with 5 or 10 mM taurine in the absence (control, solid bars) or in the presence of 10 or 100 mM glycine (striped bars). The empty bars represent the taurine uptake in the presence of sucrose alone, at concentrations matching the osmotic effect of glycine solutions. Values, (pmols/mg prot/60 min) are the mean ± SEM of at least 9 different determinations.

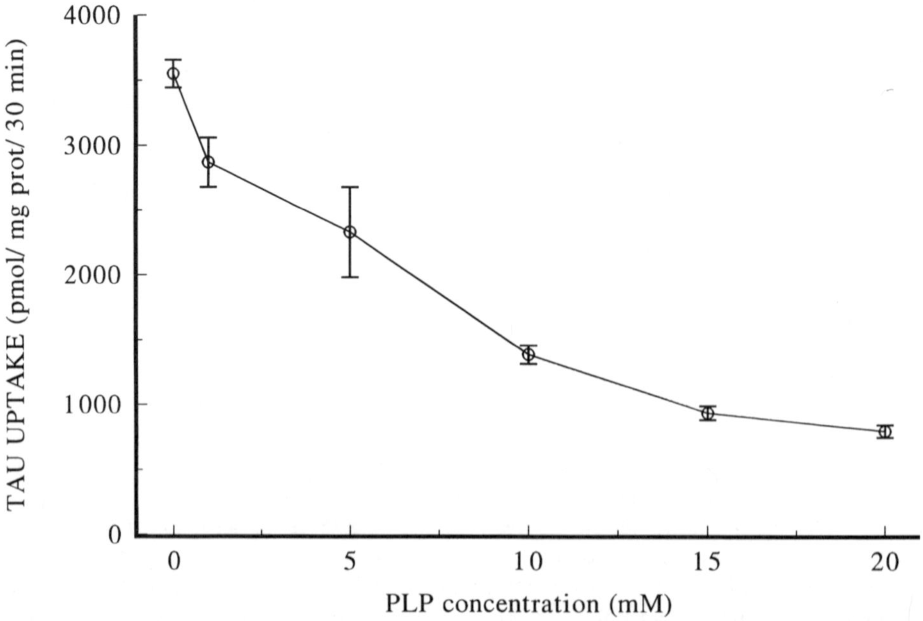


**Figure 2**. Inhibition of taurine uptake by pyridoxal-5′-phosphate (PLP). RBCs were incubated at 37°C with a 20 mM taurine solution for 30 min. Pyridoxal-5′-phosphate at the indicated concentration, was present 20 min before the onset of taurine uptake and also during the influx period. Values, mean ± SEM, represent taurine uptake of three experiments (each in triplicate).

two very different linear components that shift the slope at around 25°C (not shown). It should be remembered that an Arrhenius plot for halide transport by band 3 also shows a clear break around 25°C[7]. These facts suggest that taurine crosses the plasma membrane through a band 3-like transport system rather than by a simple diffusion pathway through the lipid bilayer.

Taurine is a zwitterion at physiological pH, thus the rate of transport for taurine through band 3 should be enhanced when the anion species concentration is favored by increasing the pH of the bath solution. The pH dependence of taurine uptake was evaluated by changing the pH of the incubation medium in the 7.0-9.0 range using TRIS buffer. We observed a marked enhancement of taurine influx with increasing pH (data not shown). It is known that there is no modification of chloride transport by band 3 in this pH range[26]. Thus, if band 3 was a transporter of taurine, the observed increase in amino acid uptake with pH would not be explained by a change in the transporter itself; instead, it should be the consequence of an increment of the anionic species of the amino acid.

Some authors have claimed that the taurine diffusional pathway is a chloride channel but not the chloride-bicarbonate exchanger[16]. We have used several inhibitors for these two permeation systems to distinguish between both possibilities. The general chloride channel blocker, anthracene-9-carboxylate[10], was unable to modify taurine uptake (10 mM) by these cells, even when it was used at a concentration high enough (1.0 mM) to block all tested chloride conductances. NPPB did not have any effect at low micromolar concentrations (0.1-1.0 μM range) at which it is a specific blocker of chloride channels. However, at 10 μM, a concentration in the range at which the inhibitor is a good blocker of band 3[4], NPPB decreased the uptake of 20 mM taurine (Student's *t*-test, taurine uptake in control 4437 ± 269 pmol/mg prot/30min versus 3573 ± 150 pmol/mg prot/30min, in the presence of NPPB, $p<0.05$, $n=12$). These results make very improbable that taurine permeation is through chloride channels in rat RBCs.

We have previously shown that DIDS at 10 μM, a concentration that inhibits AE1 in a specific manner, was able to totally block taurine uptake[21]. We have also used additional criteria to ascertain if band 3 is really a permeation pathway for taurine. Figure 2 shows the concentration inhibition curve for taurine uptake by pyridoxal-5′-phosphate (vitamin $B_6$), a well known competitive inhibitor of $Cl^-/HCO^-_3$ exchange in RBCs[2]. It can be seen that this vitamin inhibits taurine uptake in an apparent competitive manner, at a concentration range in which PLP also inhibits band 3 anion exchange, suggesting the existence of an overlapping site for both compounds at the AE1 transport site.

We next wanted to know if band 3 inhibitors that do not interact with chloride binding sites, were able to block taurine uptake in RBCs. For this purpose, we have used dipyridamole, an agent that is known to inhibit band 3 in a noncompetitive manner[18]. RBC suspensions were incubated with 20 mM taurine for 30 min in the presence and in the absence of 50 μM of dipyridamole. Taurine uptake in control fractions was $91.4 \pm 14.9$ ($n=17$) pmol/mg prot/min whereas in the presence of dipyridamole it decreased to $27.5 \pm 4.5$ ($n=18$) pmol/mg prot/min (Student's *t*-test, $p< 0.001$).

The characteristics of taurine movement through the rat erythrocytes membrane, marked dependence of taurine uptake on temperature, increased influx with increasing pH, and a strong inhibition at pharmacological reasonable concentrations by band 3 inhibitors but not by chloride channel blockers, led us to conclude that the anion-exchanger is a diffusion pathway for taurine in rat erythrocytes.

## ACKNOWLEDGMENTS

This work was supported by the "Ministerio de Educación y Cultura" of Spain (Grant PM95-170). We thank A. Latorre for technical assistance. The presentation of this work was sponsored by "Red Bull España".

## REFERENCES

1. Cabantchik, Z.I. and Greger, R., 1992, Chemical probes for anion transporters of mammalian cell membranes, *Am. J. Physiol.*, 262:C803-827.
2. Cabantchik, Z.I., Balshin, M., Breuer, W., and Rothstein, A., 1975, Pyridoxal phosphate. An anionic probe for protein amino groups exposed on the outer and inner surfaces of intact human red blood cells, *J. Biol. Chem.*, 250:5130-5136.
3. Cousin, J.L. and Motais, R., 1982, Inhibition of anion transport in the red blood cell by anionic amphiphilic compounds. I. Determination of the flufenamate-binding site by proteolytic dissection of the band 3 protein, *Biochim. Biophys. Acta.*, 687:147-155.
4. Eidelman, O., Yanai, P., Englert, H.C., Lang, H.G., Greger, R., and Cabantchik, Z.I., 1991 Macromolecular conjugates of transport inhibitors: new tools probing topography of anion transport proteins, *Am. J. Physiol.*, 260:C1094-C1103.
5. Fugelli, K. and Thoroed, S.M., 1986, Taurine transport associated with cell volume regulation in flounder erythrocytes under unisosmotic conditions, *J. Physiol.*, 374:245-261.
6. Garcia-Romeu, F., Cossins, A.R., and Motais, R., 1991, Cell volume regulation by trout erythrocytes: Characteristics of the transport systems activated by hypotonic swelling, *J.Physiol.*, 440:547-567.
7. Glibowicka, M., Winckler, B., Araníbar, N., Schuster, M., Hanssum, H., Rüterjans, H., and Passow H., 1988, Temeperature dependece of anion transport in the human red blood cell, *Biochim. Biophys. Acta.*, 946:345-358.
8. Goldstein, L., Brill, S.R., and Freund, E.V., 1990, Activation of taurine efflux in hypotonically stressed elasmobranch cells: inhibition by stilbene disulfonates, *J. Exp. Zool.*, 254:114-118.

9. Goldstein, L. and Brill, S.R., 1991, Volume-activated taurine efflux from skate erythrocytes: possible band 3 involvement, *Am. J. Physiol.*, 260:R1014-1020.
10. Greger, R., 1990, Chloride channel blockers, in Methods in Enzymology, Fleischer, B. and Fleischer, S., eds., Academic Press, Orlando, FL., Vol. 191, pp. 793-810.
11. Hall, J.A., Kirk, J., Potts, J.R., Rae, C., and Kirk, K., 1996, Anion channel blockers inhibit swelling activated anion cation, and nonelectrolyte transport in HeLa cells, *Am. J. Physiol.*, 271:C579-C588.
12. Hoffmann, E.K. and Simonsen, L.O., 1989, Membrane mechanisms in volume and pH regulation in vertebrate cells, *Physiol. Rev.*, 69:315-382.
13. Huxtable, R.J., 1989, Taurine in the central nervous system and the mammalian actions of taurine, *Prog. Neurobiol.*, 32:471-533.
14. Jhiang, S.M., Fithian, L., Smanik, P., McGill, J., Tong, Q., and Mazzaferri, E.L., 1993, Cloning of the human taurine transporter and characterization of taurine uptake in thyroid cells, *FEBS Lett.*, 318:139-144.
15. King, A.P. and Gunn, R.B., 1991, Glycine transport by human red blood cells and ghosts: evidence for glycine anion and proton cotransport by band 3, *Am. J. Physiol.*, 261:C814-C821.
16. Kirk, K., Ellory, J.C., and Young, J.D., 1992, Transport of organic substrates via a volume-activated channel, *J. Biol. Chem.*, 267:23475-23478.
17. Lambert, I.H. and Hoffmann, E.K., 1994, Cell swelling activates separate taurine and chloride channels in Ehrlich mouse ascites tumor cells, *J. Membrane Biol.*, 142:289-298.
18. Legrum, B. and Passow, H., 1989, Inhibition of inorganic transport across the human red blood cell membrane by chloride-dependent association of dipyridemole with a stilbene disulfonate binding site on the band 3 protein, *Biochim. Biophys. Acta*, 979:193-207.
19. Lerma, J., Herranz, A.S., Herreras, O., Abraira, V., and Martín el Río, R., 1986, "In vivo" determination of extracellular concentration of amino acids in the rat hippocampus. A method based on brain dialysis and computerized analysis, *Brain Res.*, 384:145-155.
20. Liu, QR., López-Corcuera, B., Nelson, H., Mandiyan, S., and Nelson, N., 1992, Cloning and expression of a cDNA encoding the transporter of taurine and β-alanine in mouse brain, *P. Nat. Acad. Sci.*, 89:12145-12149.
21. Martín del Río, R., Galarreta, M., Menéndez, N., Conejero, C., and Solís J.M., 1996, Taurine is a substrate of the anion-exchanger transport systems, in Taurine 2: Basic and Clinical Aspects, Huxtable, R., Azuma, J., Kuriyama, K., Nakagawa, M., and Baba, A., eds., Plenum Press, New York, pp. 401-407.
22. Pasantes-Morales, H., Chacón, E., Murray, R.A., and Morán, J., 1994, Properties of osmolyte fluxes activated during regulatory volume decrease in cultured cerebellar granule neurons, *J. Neurosci. Res.*, 37:720-727.
23. Passow, H., 1986, Molecular aspects of band 3 protein-mediated anion transport across the red blood cell membrane, *Rev. Physiol. Bioch.*, 103:61-203.
24. Smith, K.E., Borden, L.A., Wang, H.D., Hartig, P.R., Branchek, T.A., and Weinshank, R.L., 1992, Cloning and expression of a high affinity taurine transporter from rat brain, *Mol. Pharmacol.*, 42:563-569.
25. Uchida, S., Moo Kwon, H., Yamauchi, A., Preston, A.S., Marumo, F., and Handler, J., 1993, Molecular cloning of the cDNA for an MDCK cell $Na^+$ and $Cl^-$-dependent taurine transporter that is regulated by hypertonicity, *P. Nat. Acad. Sci.*, 89:8230-8234.
26. Wieth, J.O., Andersen, O.S., Brahm, J., Bjerrum,P.J., and Borders, C.L., 1982, Chloride-Bicarbonate exchange in red blood cells: Physiology of transport and chemical modification of binding sites. *Philos. T. Roy. Soc. London. B*, 299:383-399.

# MOLECULAR CLONING AND FUNCTIONAL EXPRESSION OF AN LLC-PK1 CELL TAURINE TRANSPORTER THAT IS ADAPTIVELY REGULATED BY TAURINE

Xiaobin Han, Andrea M. Budreau, and Russell W. Chesney

Department of Pediatrics
University of Tennessee and the Crippled Children's
Foundation Research Center at Le Bonheur Children's Medical Center
Memphis, TN 38103

## ABSTRACT

Studies have shown that the renal tubular epithelium adapts to alterations in the sulfur amino acid composition of the diet. The renal adaptive response has been described in man, mouse, rat, dog, and pig. The observed phenomenon involves increased or decreased initial rate activity of the NaCl-dependent taurine transporter at the brush border membrane surface of the proximal tubule following dietary manipulation of taurine. A cDNA encoding a taurine transporter has been isolated from LLC-PK1 cells, designated pTAUT, and its functional properties have been examined in *Xenopus laevis* oocytes. The nucleotide sequence of the clone predicts a 621-amino acid protein with about 90% homology to other cloned taurine transporter cDNAs. When expressed in oocytes the transporter displays a $K_m$ of 25 μM and is dependent on the presence of external sodium and chloride, characteristics similar to taurine uptake by LLC-PK1 cells. The abundance of pTAUT mRNA and protein were up-regulated in cells cultured in taurine-free medium as compared with cells cultured in medium containing 500 μM taurine. Activation of PKC by PMA had no effect on adaptive regulation of pTAUT mRNA and protein, indicating that down-regulation of LLC-PK1 cell taurine transport activity by PMA occurs at the post-translational level.

## INTRODUCTION

Taurine (2-aminoethanesulfonic acid) is a small sulfur-containing ß-amino acid present in millimolar quantities in the intracellular space of many tissues and has been demonstrated to serve as an important osmolyte in brain and kidney[1,19]. In mammals, the

*Taurine 3*
Edited by Schaffer *et al.*, Plenum Press, New York, 1998 

taurine total body pool is regulated by the renal proximal tubule, where taurine is transported by a transmembrane protein[2,6]. Studies have shown that the renal tubular epithelium of rats can adapt to alterations in the sulfur amino acid composition of the diet[2,3,5,6]. When dietary intake of taurine or its precursor amino acid is reduced, urinary taurine levels decline due to enhanced renal tubular reabsorption of taurine[4,7]. This adaptive regulation of taurine transport following alterations in external taurine concentration has been described in man, cat, mouse, rat, dog, and pig.

Taurine functions as an osmolyte in MDCK cells[19]. When MDCK cells cultured in isotonic medium were switched to hypertonic medium, the intracellular taurine content doubled. Taurine transport in MDCK cells is dependent on $Na^+$ and $Cl^-$ and is localized primarily in the basolateral plasma membrane. Osmoregulation of net taurine transport is slow, requiring hours to days, and it occurs at the level of mRNA accumulation[18]. In contrast, LLC-PK1 cells respond to hypertonic medium with reduced apical taurine accumulation[13]. Therefore, regulation of taurine transport by hypertonicity and regulation by medium taurine concentration appear to be independent phenomena.

To date, taurine transporter cDNA has been cloned from several species, including man, rat, mouse, and dog. There is as yet no molecular information available about the LLC-PK1 cell taurine transporter, and knowledge about how the renal adaptive response is regulated by dietary manipulation is limited to a description of the phenomenon. This study shows that taurine transporter cDNA of LLC-PK1 cells is highly homologous with two cloned taurine transporters, and that the adaptive response of the transporter to medium taurine concentration is regulated at the level of mRNA accumulation.

## METHODS

A cDNA library was constructed using poly(A)+ RNA isolated from LLC-PK1 cells. In the first strand synthesis reaction, an oligo(dT) primer adapter containing a Sal I site was used. Following second strand synthesis and addition of Not I site adapters, the cDNA was digested with Sal I to generate cDNA with Not I and Sal I termini and ligated to the Not I/Sal I cloning site of a pSPORT vector. Vector-containing cDNA was transferred into *E. coli* JM109 cells. From a total cDNA library of 105 colonies, ~25,000 colonies were plated out at 2,000 colonies/plate and transferred to nitrocellulose membranes. Bacteria on nitrocellulose membranes were lysed, then baked at 80°C for 2 hours. The cDNA probe used for screening was a 900 bp PCR product generated from LLC-PK1 cells by RT-PCR using two primers of rB16a (rat brain taurine transporter). The probe was labelled with [$\alpha$-$^{32}$P] dCTP and hybridization was carried out overnight at 65°C. Post-hybridization washing was done under high stringency conditions. Positive clones were identified and plasmid DNA was isolated, then both strands of the LLC-PK1 cell taurine transporter cDNA were sequenced.

LLC-PK1 cell taurine transporter cDNA was linearized with Sal I for use as a template for *in vitro* transcription, and *in vitro* synthesis of capped transcript with SP6 RNA polymerase was performed. Ovarian lobes were dissected from anesthetized frogs and oocytes were separated by incubation of ovarian fragments for 40 minutes with 2 mg/mL collagenase type II in calcium-free buffer at room temperature. Defolliculated stage V-VI oocytes were selected and incubated overnight at 18°C in media before injection. Thirty nL of water containing 30 ng of capped transcript were injected into each oocyte and injected oocytes were maintained at 18°C in media for 3 days. Oocytes were then transferred to a $Na^+$-containing uptake solution to which was added both radiolabelled and unlabelled taurine. After incubation at room temperature, oocytes were transferred to a 24-well cell culture plate in which they were washed rapidly 5 times with ice-cold

$Na^+$-free uptake solution. Finally, individual oocytes were transferred to miniscintillation vials, solubilized and counted for radioactivity.

LLC-PK1 cells (American Type Culture Collection) were cultured in 5% $CO_2$-95% air with standard medium consisting of Dulbecco's modified Eagle's medium (DMEM) in a 1:1 mixture with Ham's F12 nutrient mixture, plus 10% fetal bovine serum, penicillin (100 U/ml), and streptomycin (100 μg/ml) at 37°C. Cells were subcultured by trypsinization, seeded into 150 x 25 mm dishes, grown to 80% confluence and then cultured in serum-free DMEM/F12 with insulin (5 μg/ml), transferrin (5 μg/ml), prostaglandin $E_1$ ($2.5 \times 10^{-5}$ mg/ml), hydrocortisone ($5 \times 10^{-8}$ M), thyroxine ($5 \times 10^{-12}$ M), with or without taurine (0 or 500 μM) for 1-3 days. For PKC experiments, 100 nM PMA (phorbol 12-myristate 13 acetate, an active phorbol ester known to stimulate PKC activity) was added to the medium 20 min prior to addition of taurine to deplete the cells of PKC. Cells were cultured in medium containing 0 or 500 μM taurine in addition to 100 nM PMA for 24 h, then harvested for quantitation of mRNA.

Thirty μg of total RNA were separated in an agarose gel and transferred to a nylon membrane by overnight capillary blotting in 10x SSC. The Northern blot was hybridized overnight at 42°C with $^{32}P$-labelled pNCT riboprobe. The blot was washed successively in decreasing concentrations of SSC/SDS at 65°C for 30 minutes each and exposed to Kodak film with one intensifying screen at -80° C for 24 h.

A 14 mer C-terminal synthetic peptide (PREPNRWAVEREGAT, 583-596 residues) sequence of predicted taurine transporter (pTAUT) was coupled to BSA using a kit (Imject SuperCarrier EDC System for Peptides, PIERCE). New Zealand White rabbits (Myrtle's Rabbitry, Thompson Station, TN) were immunized with approximately 1000 μg of immunogen and boosted with an equivalent amount of antigen 2 weeks later. Rabbits were bled through the ear vein on days 21 and 36 to measure the antibody response. Antisera were collected on day 49, purified by affinity column chromatography, and stored at -70°C.

Cells were lysed in 50 μl RIPA buffer (1 mM phenylmethylsulfonyl fluoride, 1% deoxycholate, 1% Triton X-100, 0.1% SDS, 150 mM NaCl, 2 mM $MgCl_2$, 10 mM Tris-HCl, pH 7.2) 4 days after injection of pNCT cRNA. The lysates were cleared by centrifugation at 14,000 x *g* for 2 min, and the supernatant was transferred to a new tube. Aliquots of each lysate were separated by electrophoresis on a 12% SDS-polyacrylamide gel and transferred to a nitrocellulose membrane (Millipore) using a semi-dry electrophoretic transfer system (Bio-Rad). Membranes were incubated with either affinity-purified polyclonal antibodies (1:1000 dilution) against the conserved peptide sequence of the taurine transporter, pre-immune IgG, or a mixture of antibody and synthetic peptide. The blots were detected using a chemiluminescence detection kit (ECL, Amersham).

The relative abundance of mRNA is expressed as the ratio of abundance of taurine mRNA bands to abundance of ß-actin mRNA bands on the same membrane. All experiments were performed in triplicate. The data represent the mean ± SEM of three or four experiments. Statistical comparisons were made using one-way ANOVA and Student's t test to determine significant differences in the means.

## RESULTS

The LLC-PK1 cell taurine transporter cDNA is about 2.2 kb long with an open reading frame of 1863 bp, encoding a protein of 621 amino acids (below). The amino acid sequence deduced for the LLC-PK1 cell taurine transporter shows 89% homology with rB16a and about 90% homology with other cloned taurine transporters. Most of the non-homologous residues are located in the third extracellular loop and the C-terminal regions. Some conserved potential sites such as N-linked glycosylation sites, PKC sites, and casein kinase II phosphorylation sites are also present in the pTAUT peptide sequence.

1 MATKEKLSCL KDFHKDILKP SPGKSPGTRP EDEAEGKPPQ REKWPIKMDF
51 VLSVAGGFVG LGNVWRFPYL CYKNGGGAFL IPYFIFLFGG GLPVFFLEVI
101 IGQYTSEGGI TCWEKICPLF AGIGYASIVI VSLLNVYYIV VILAWATYYL
151 FQSFQKDLPW AHCNHSWNTP QCMEDTLRRN ESHWVSLSAA NFTSPVIEFW
201 ERNVLSLSSG IDHPGSLKWD LALCLLLVWL VCFFCIWKGV RSTGKVVYFT
251 ATFPFAMLLV LLVRGLTLPG AGEGIKFYLY NSIKCLEDLK GWNDSGNKIF
301 FSYAICLGAM TSLGSYNKYK YNSYKDCMLL GCLNSGTSFV SGFAIFSILG
351 FMAQEQGVDI ADVAESGPGL AFIPYPKAVT MMPLPTFWSI LFYIMLLLLG
401 LDSHFVELER QITSLVDLYP SFLRKVSPRE IFIAFICCIS YMLGLSMVTE
451 GGMYVCQLFD YYAASGVCLL WVAFFECFAI ARIYGSDNLY DGIENMIGYR
501 PGPCMKYCWA VVTPVLCLGC FIYSLVKYVP LMYNKVYVYP DWAIGLGWSL
551 ALSSILCVPL VMVIRLCQTE GPFLVRLKYL LAPREPNRWA VEREGAMPYS
601 SRSSINSALR KPTHIIVETM M*

pTAUT was functionally expressed in oocytes after injection of pTAUT cRNA. Taurine transport activity of pTAUT was NaCl-dependent and was inhibited by ß-alanine, glycine and GABA. Kinetic studies showed that the affinity of pTAUT expressed in oocytes was 25 μM, which is similar to the $K_m$ found in LLC-PK1 cells (Fig. 1).

As shown in Fig. 2a and b, two transcripts (7.2 kb and 9.6 kb) of pTAUT were detected by a pTAUT riboprobe. The abundance of these two transcripts was increased in cells incubated in taurine-free medium. When cells were incubated in medium containing high taurine (500 μM), the abundance of mRNA was decreased in both transcripts.

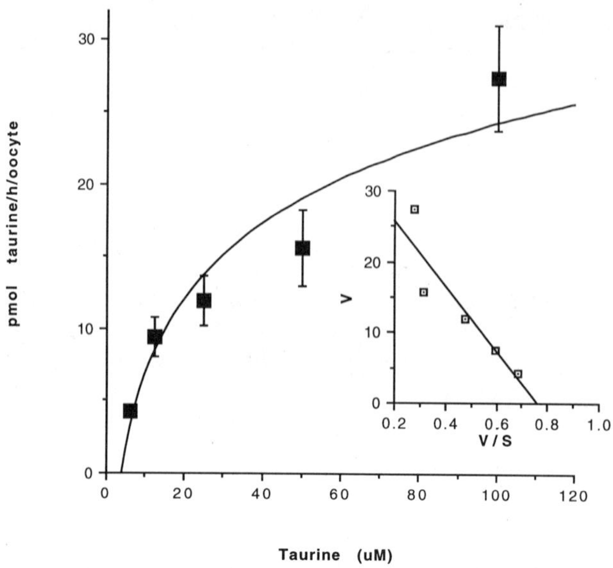


**Figure. 1.** Kinetics of taurine uptake induced in oocytes by injection of 30 ng of pTAUT cRNA. Three days after injection, oocytes were incubated for 1 h at room temperature in $Na^+$-uptake solution containing the indicated concentration of taurine. Results are from 6 oocytes from multiple donors and are expressed as mean ± S.E.

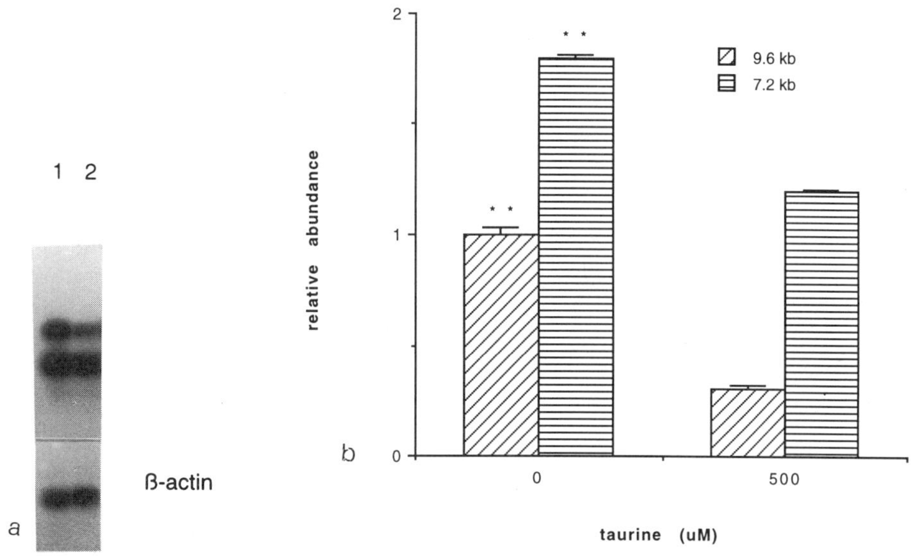


**Figure. 2.** Adaptive regulation of expression of taurine transporter mRNA by taurine. LLC-PK1 cells were cultured in medium containing 0 or 500 μM taurine for 24 h, then cells were harvested for preparation of total cellular RNA. a) Northern blot analysis was carried out as described in Methods. Lane 1, 0 μM taurine; Lane 2, 500 μM taurine. ß-actin was used as an internal control for loading. b) The relative levels of intensity of each band were assayed and quantitated using a BioRad GS-700 densitometer. Values are expressed as means ± SEM of three independent experiments. Data were normalized to the ß-actin densitometric value. $p < 0.05$ vs control.

A polyclonal antibody directed against a 14 mer peptide of the C-terminal of the pTAUT, which is 100% conserved in all cloned taurine transporters, was used to identify taurine transporter protein in both pTAUT-expressing oocytes and LLC-PK1 cells (data not shown). Western blot analysis showed that a single band at 70 kD, presumably the taurine transporter protein, was detected in both pTAUT-expressing oocytes and LLC-PK1 cells. No band was found in uninjected oocytes or when blot analysis was performed using pre-immune IgG as the primary antibody. A combination of the antibody and peptide failed to detect the transporter protein in LLC-PK1 cells, indicating that the antibody is specific to the taurine transporter protein. Further evidence that this antibody detected the taurine transporter protein came from studies in which LLC-PK1 cells were incubated in medium containing 0, 50, or 500 μM taurine. Increased transporter protein was evident in cells incubated with 0 μM taurine, and reduced protein was found in cells cultured in medium containing 500 μM taurine, as compared with cells incubated with 50 μM taurine.

Activation of protein kinase C has been shown to reduce taurine transport into LLC-PK1 cells[15], human placental carcinoma cells[16], and taurine transporter-expressing oocytes[8]. We sought to determine if PKC activation would change the expression of pTAUT mRNA. Addition of 100 μM PMA, a dose known to reduce oocyte taurine uptake by more than 70% (data not shown), had no influence on the abundance of pTAUT mRNA in either the 0 μM taurine (up-regulation) or 500 μM taurine (down-regulation) conditions. Neither the 7.2 kb or 9.6 kb bands were affected in terms of relative abundance by incubation of oocytes with PMA. Incubation of intact cells with PMA for up to 48 h did not alter expression of pTAUT, as analyzed by Northern blot analysis (Fig. 3a, b). Hence it appears that the influence of PKC activation represents a post-translational modification.

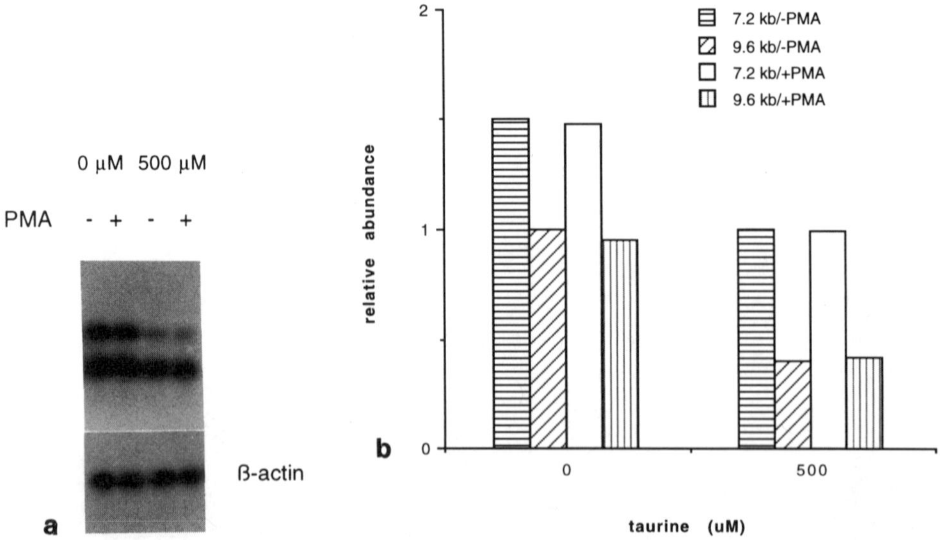


**Figure. 3.** Effect of PKC on adaptive regulation of pNCT mRNA by taurine. PMA (100 nM) was added 20 min before taurine was added. Cells were continuously cultured in medium containing 0 or 500 μM taurine for 24 h. a) Northern blot analysis was done as for Figure 2. Values are means ± SEM of three independent experiments. b) Data were normalized to the ß-actin densitometric value. **$p < 0.01$ vs 500 μM taurine.

## DISCUSSION

This paper describes the cloning of the taurine transporter from LLC-PK1 cells. The cloned protein exhibits 89 - 90% homology with two other taurine transporters found in kidney: pNCT from MDCK cells[18], and rB16a, initially cloned from rat brain but found in rat kidney as well[17]. The specific aim of the paper is to indicate that the clone we developed, the first of proximal tubule origin, indeed does code for the taurine transporter protein, and that it undergoes the renal adaptive response to dietary change that our laboratory has studied for more than 2 decades.

The detection of increased pTAUT mRNA after incubation of cells in taurine-free medium and reduced message after incubation in medium containing 500 μM taurine indicates that expression of the transporter is adaptively regulated by the concentration of taurine in the medium at the level of mRNA accumulation. This phenomenon is also observed when pTAUT is examined in the oocyte expression system. We can speculate that changes in transcription rate may also be found, similar to what we have detected for pNCT[9] and for rB16a[10]. Hence, in the proximal cell of the pig kidney the alteration in taurine transport governed by changes in medium taurine content is on a genomic basis.

In LLC-PK1 cells, the renal adaptive response appears to involve changes in the abundance of pTAUT mRNA as well as in the abundance of a 70 kD protein in the plasma membrane of cells. The abundance of this protein in membranes from oocytes injected with pTAUT cRNA results in increased or decreased taurine uptake, and increased or decreased transepithelial transport (apical to basolateral) of taurine in LLC-PK1 cells grown on filters[14].

The finding that activation of PKC by PMA or other active phorbols will reduce NaCl-dependent taurine transport indicates a role for PKC activation in the regulation of taurine transport. Using a specific antibody, we showed that PKC activation does not modify the abundance of pTAUT protein following incubation under several conditions. Moreover, the amount of the 70 kD protein is not modified by PMA incubation. Incubation of cells with PMA for up to 48 h, a sufficient amount of time to permit genomic action, does not alter protein synthesis or density. Hence, the consistent finding that PKC activation reduces taurine transport in LLC-PK1 cells or mRNA-injected oocytes is most likely due to post-translational modification. Further evidence of this hypothesis is that the antibody to the 4th intracellular loop modifies PKC action[12]. We have preliminary evidence that a single point serine-to-alanine mutation of amino acid 322 will block PKC phosphorylation as well as modification of transport by PMA[11].

The present study allows us to further define the renal adaptive response as one in which a transporter, specific for taurine and inhibitable by ß-alanine and GABA, has been cloned from a stable, commonly employed proximal cell line. This transporter, termed pTAUT, is highly homologous to other cloned taurine transporters. It appears to be presented in 2 forms in LLC-PK1 cells - a 7.2 kb and a 9.6 kb transcript, the abundance of which is modified by cell culture medium taurine content. Alterations in medium taurine content can be viewed as comparable to changes in the dietary intake of taurine or its precursors.

In summary, the relative abundance of taurine transporter protein found in plasma membranes from LLC-PK1 cells and from microinjected oocytes is modified by medium taurine content and can be assessed by a specific antibody. Activation of PKC by PMA has no effect on adaptive regulation of pTAUT mRNA and protein accumulation, indicating that down-regulation of LLC-PK1 cell taurine transporter activity by PMA occurs at the post-translational level.

## REFERENCES


1. Chesney, R.W. 1985. Taurine: Its Biological Role and Clinical Implications. In: "Advances in Pediatrics", L. A. Barnes, ed., Yearbook Medical Publishers, Chicago, pp. 1-42.
2. Chesney, R.W., Gusowski, N., and Dabbagh, S., 1985, Renal cortex taurine content regulates renal adaptive response to altered dietary intake of sulfur amino acids, *J. Clin. Invest.*, 76:2213-2221.
3. Chesney, R.W., Gusowski, N., and Dabbagh, S., 1987, Studies on renal adaptation to altered dietary amino acid intake: Reduced renal cortex taurine content increases the $V_{max}$ of taurine uptake by brush border membrane vesicles, *Pediatr. Nephrol.*, 1:9-15.
4. Chesney, R.W., Gusowski, N., and Friedman, A.L., 1983, Renal adaptation to altered amino acid intake occurs at the luminal brush border membrane, *Kidney Int.*, 24:588-594.
5. Chesney, R.W., Lippincott, S., Gusowski, N., Padilla, M., and Zelikovic, I., 1986, Studies on renal adaptation to altered dietary amino acid intake: Tissue taurine responses in nursing and adult rats, *J. Nutr.*, 116:1965-1976.
6. Friedman, A.L., Albright, P.W., and Chesney, R.W., 1981, Dietary adaptation of taurine transport by rat renal epithelium, *Life Sci.*, 29:2415-2419.
7. Friedman, A.L., Albright, P.W., Gusowski, N., Padilla, M., and Chesney, R.W., 1983, Renal adaptation to alteration in dietary amino acid intake, *Am. J. Physiol.*, 245:F159-F166.
8. Han, X., Budreau, A.M., and Chesney, R.W., 1997, Adaptive regulation of MDCK cell taurine transporter (pNCT) mRNA: transcription of pNCT gene is regulated by external taurine concentration, *Biochim. Biophys. Acta*, 1351:296-304.
9. Han, X., Budreau, A.M., and Chesney, R.W., 1996, Adaptive regulation of MDCK cell taurine transporter (pNCT) mRNA: Transcription of pNCT gene is regulated by external taurine concentration, *Biochim. Biophys. Acta*, in press:
10. Han, X., Budreau, A.M., and Chesney, R.W., 1997, Functional expression of rat renal cortex taurine transporter in *Xenopus laevis* oocytes: Adaptive regulation by dietary manipulation, *Pediatr. Res.*, 41(5):624-631.
11. Han, X., Budreau, A.M., and Chesney, R.W., 1997, Identification of ser-322 as specific site in taurine transporter (pNCT) that is phosphorylated by PKC, *Pediatr. Res.*, 41(4):279A.

12. Han, X., Budreau, A.M., and Chesney, R.W., 1996, Role of conserved peptide in taurine transporter inactivation that is modulated by protein kinase C, *J. Am. Soc. Nephrol.,* 7:2088-2096.
13. Jones, D.P., Miller, L.A., Budreau, A., and Chesney, R.W., 1992, Characteristics of taurine transport in cultured renal epithelial cell lines: Asymmetric polarity of proximal and distal cell lines. In: "Taurine", Lombardini, J. B., Schaffer, S.W., Azuma, J., eds., Plenum Press, New York, pp. 405-411.
14. Jones, D.P., Miller, L.A., and Chesney, R.W., 1993, Polarity of taurine transport in cultured renal epithelial cell lines: LLC-$PK_1$ and MDCK, *Am. J. Physiol.,* 265:F137-F145.
15. Jones, D.P., Miller, L.A., Dowling, C., and Chesney, R.W., 1991, Regulation of taurine transporter activity in LLC-PK cells: Role of protein synthesis and protein kinase C activation, *J. Am. Soc. Nephrol.,* 2:1021-1029.
16. Ramamoorthy, S., Leibach, F.H., Mahesh, V.B., Han, H., Yang, F.T., Blakely, R.D., and Ganapathy, V., 1994, Functional characterization and chromosomal localization of a cloned taurine transporter from human placenta, *Biochem. J.,* 300:893-900.
17. Smith, K.E., Borden, L.A., Wang, C.D., Hartig, P.R., Branchek, T.A., and Weinshank, R.L., 1992, Cloning and expression of a high affinity taurine transporter from rat brain, *Mol. Pharmacol.,* 42:563-569.
18. Uchida, S., Kwon, H.M., Yamauchi, A., Preston, A.S., Marumo, F., and Handler, J.S., 1992, Molecular cloning of the cDNA for an MDCK cell Na(+)- and Cl(-)-dependent taurine transporter that is regulated by hypertonicity [published erratum appears in Proc Natl Acad Sci U S A 1993 Aug 1; 90(15):7424], *Proc. Natl. Acad. Sci. U S A,* 89(17):8230-8234.
19. Uchida, S., Nakanishi, T., Kwon, H.M., Preston, A.S., and Handler, J.S., 1991, Taurine behaves as an osmolyte in Madin-Darby canine kidney cells: Protection by polarized, regulated transport of taurine, *J. Clin. Invest.,* 88:656-662.

# REGULATION OF THE TAURINE CONTENT IN EHRLICH ASCITES TUMOUR CELLS

Ian Henry Lambert
The August Krogh Institute
Biochemical Department
University of Copenhagen
Universitetsparken 13, DK-2100
Copenhagen Ø, Denmark

## INTRODUCTION

Taurine plays quantitatively an important role as an organic osmolyte in anisosmotic cell volume regulation in Ehrlich ascites tumour cells[3]. It is the saturable, pH-sensitive, $Na^+$- and $Cl^-$-dependent β-system that accounts for taurine accumulation[9,10,15], whereas the swelling-activated taurine channel accounts for net loss of cellular taurine following hypotonic exposure[15,16].

When Ehrlich ascites tumour cells are transferred to a hypotonic medium, they initially swell as an almost perfect osmometer, but subsequently regulate their cell volume back towards the initial cell volume. This back regulation is called regulatory volume decrease (RVD) and in the case of the Ehrlich cells involves a net loss of KCl and organic osmolytes, such as taurine[2]. It has been estimated that the initial taurine loss via the swelling-activated taurine efflux pathway is increased four times[15], and that taurine influx via the $Na^+$-dependent β-system is reduced by 70% after transfer of the cell from an isotonic to hypotonic medium with half osmolarity[3]. Taurine accounts for about 10% of the osmolyte loss within the first minutes following cell swelling in a hypotonic medium[16], and the cellular taurine content is reduced from 151 to 27 μmol/g dry wt within 40 min[3] after transfer to the hypotonic medium.

This paper deals with the mechanisms behind the swelling-induced reduction in taurine uptake and the swelling-induced release of taurine in Ehrlich ascites tumour cells.

### Swelling-Induced Reduction in Taurine Uptake Via the β-System

The β-system from mammalian cells has been cloned[5,20,26,30] and it is suggested that the taurine transporter contains 12 membrane spanning regions and putative intracellular phosphorylation sites for both cAMP-activated protein kinase (PKA) and protein kinase C (PKC). The initial rate of taurine influx in Ehrlich cells is significantly stimulated following preincubation with forskolin or dbcAMP ($N^6$, 2′-O-dibutyryl-3′,5′-cyclic adenosine monophosphate) (Fig. 1, left panel). The effect of dbcAMP on the initial rate of taurine influx is totally $Na^+$-dependent, and since forskolin increases the cellular cAMP concentration it has been suggested that cAMP stimulates taurine transport via the β-system in Ehrlich cells[22]. On the other hand, the initial rate of taurine influx in the Ehrlich cell is significantly inhibited following addition of the PKC activator, phorbol 12 myristate 13 acetate[22] and the protein phosphatase inhibitor, calyculin A (Fig. 1, right panel). Furthermore, the effect of forskolin and dbcAMP on the initial rate of taurine influx is suppressed in calyculin A-pretreated Ehrlich cells (Fig. 1, right panel). From these observations it has been proposed by Mollerup and Lambert[22] that the β-system in Ehrlich cells appears in three states depending upon transport activity, *i.e. State I*, *State II*, and *State III* - with low, normal and high taurine transport activity, respectively. A transition from *State II* to *State I* involves PKC mediated phosphorylation, whereas a transition from *State I* to *State II* involves dephosphorylation by calyculin A sensitive phosphatases (probably PP1 or PP2A). Since addition of calyculin A inhibits the initial rate of taurine influx it is suggested that the β-system in unpertubated Ehrlich cells is maintained in *State II* by calyculin A sensitive phosphatases. A transition to *State III* is mediated by cAMP and is only possible when the β-system is in *State II*. $Na^+$ ions increase the affinity of the β-system for taurine in the Ehrlich cells, while they have no effect on total taurine transport capacity. It has been suggested that binding of $Na^+$ ions and a change in the tertiary structure of the β-system is a prerequisite for binding of taurine to the transporter[9]. Treatment with calyculin A reduces the affinity of the β-system for $Na^+$ and taurine in Ehrlich cells and also lowers the coupling ratio between $Na^+$ and taurine[23]. Thus, a transition from *State II* to *State I* primarily affects the binding of $Na^+$ ions to the β-system.

It is evident that dilution of the medium with buffered water reduces the extracellular $Na^+$ and $Cl^-$ concentration and thereby taurine influx via the $Na^+$- and $Cl^-$-dependent β-system in Ehrlich cells. However, cell swelling, without any concomitant reduction in the extracellular $Na^+$ concentration also reduces the initial rate of taurine influx[3]. PKC activity

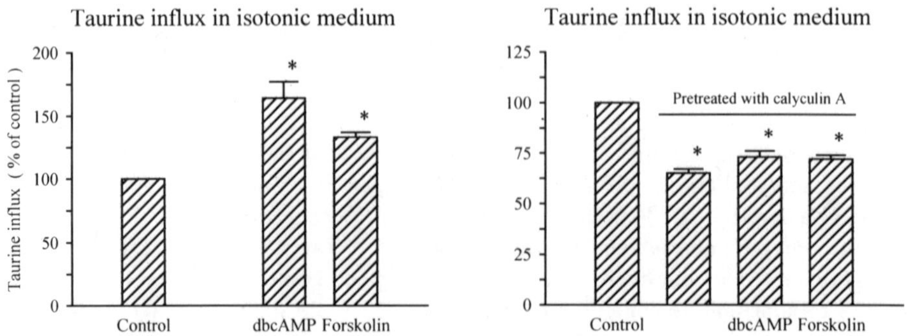


**Figure 1.** Effect of forskolin, dbcAMP, and calyculin A on the initial taurine influx in Ehrlich cells. Ehrlich cells were pre-incubated in standard NaCl medium. $^{14}C$-taurine was added at time zero and 5 samples were drawn within the first 5 min to estimate the increase in cellular $^{14}C$-taurine activity. The initial taurine influx was estimated as the product of the extracellular taurine concentration (1.5 μM) and the rate constant for the initial taurine influx, *i.e.* the slope of the plot of $a^t_c/a^0_m$ *vs.* time, where $a^t_c$ is the cellular $^{14}C$-taurine activity at time t and $a^0_m$ is the activity in the medium at time zero. Forskolin (10 μM), dbcAMP (0.2 mM) and calyculin A (100 nM, right panel) were added 1 min, 1 min and 1.5 min before initiation of influx, respectively. Flux values are presented in percent of initial influx in the control cells ± SEM. The number of paired experiments was 15 (forskolin), 3 (dbcAMP), 5 (calyculin A), 3 (calyculin A plus forskolin) and 4 (calyculin A plus dbcAMP). The figure is redrawn from Mollerup and Lambert[22]. An asterisk indicates in this figure and in the following figures that the value is significantly different from the value in control cells at a 5 % level, as estimated by a Student's t-test.

is unaffected or slightly increased in Ehrlich cells after hypotonic cell swelling[19], a condition which stimulates the transition of the β-system from *State II* to *State I*. On the other hand, if one assumes that dilution of the cellular compartment reduces the activity of calyculin A sensitive phosphatases, then a greater percentage of the β-system will be retained in *State I* following cell swelling. The physiological consequence of this swelling-induced inactivation of the β-system is most probably to avoid re-uptake of taurine during the RVD response.

**Swelling-Induced Taurine Efflux**

It has been demonstrated that the signal transduction cascade, triggered by cell swelling, which leads to the activation of the conductive $K^+$ and $Cl^-$ channels, involves a phospholipase $A_2$ ($PLA_2$) mediated release of arachidonic acid from phospholipids and a 5-lipoxygenase (5-LO) mediated oxidation of the fatty acid to a leukotriene[14]. Leukotriene $D_4$ ($LTD_4$) is thought to be the second messenger that activates the ion transporting pathways[7,11,12,17], whereas $Ca^{2+}$ signalling induced by either $Ca^{2+}$ entry or $Ca^{2+}$ release from intracellular stores, appears to play no role in the activation of the RVD response[6].

Figure 2 (left panel) demonstrates the rate constant for the release of taurine from Ehrlich cells within the first minute following transfer of the cells to hypotonic medium with half osmolarity. The initial taurine efflux rate is obtained by multiplication of the rate constant by the cellular concentration of taurine. Figure 2 shows that inhibition of either $PLA_2$ or 5-LO leads to a reduction in the swelling-induced release of taurine from the Ehrlich cell. Similar results were obtained when $PLA_2$ was directly blocked by addition of RO 31-463 and indirectly by the addition of pimozide, a $Ca^{2+}$/calmodulin inhibitor. Also, inhibition of 5-LO production by addition of ETH 615-139 and NDGA caused a similar degree of inhibition. Figure 2 (right panel) demonstrates that addition of $LTD_4$ (5 and 100 nM), ATP (10 μM) and UTP (10 μM) significantly stimulated the rate constant for the initial rate of taurine efflux from Ehrlich cells suspended in isotonic medium. $LTD_4$ mobilizes $Ca^{2+}$ in Ehrlich cells by increasing $Ca^{2+}$ influx and by releasing $Ca^{2+}$ from intracellular stores ($EC_{50}$ for $LTD_4$-induced $Ca^{2+}$ mobilization ≈ 10 nM[25]). Also, $LTD_4$ causes a depolarization of the plasma membrane in Ehrlich cells[12]. Since taurine release from Ehrlich cells is larger in the absence of extracellular $Ca^{2+}$ and depolarization of the plasma membrane inhibits swelling-induced taurine efflux[15,16], it is proposed that $LTD_4$ directly modulates the taurine releasing system. Exogenous ATP, acting via purinergic receptors, also mobilizes $Ca^{2+}$ in the Ehrlich cell, but in contrast to $LTD_4$, causes a hyper-polarization of the plasma membrane (Pedersen, Pedersen, Droogmans, Prenen and Nilius, unpublished data). Since hyperpolarization of the plasma membrane stimulates swelling-induced taurine efflux in Ehrlich cells[16], the effect of ATP/UTP on taurine efflux (Fig 2, right panel) may involve an ATP-/UTP-induced hyperpolarization of the plasma membrane. It is noted that experiments with purified human 5-LIP have revealed that 5-LIP has an ATP binding site and that ATP is a cofactor for 5-LIP activity[24]. The swelling-induced organic osmolyte pathways in skate hepatocytes and $C_6$ glioma cells require intracellular, nonhydrolytic binding of ATP[27] while swollen HTC rat hepatoma cells require extracellular ATP in order to recover their cell volume[31]. It has been suggested that ATP functions as an autocrine factor in hepatoma cells[31]. Thus, ATP apparently plays an important role in the regulation of the signal transduction cascade that leads to activation of osmolyte channels and in the activation and/or regulation of the channels. Thus, in addition to affecting the activation of $K^+$ and $Cl^-$ transporting channels[14], the hypotonic-induced increase in leukotriene biosynthesis also seems to be responsible for the activation of the taurine releasing system in Ehrlich cells[17]. However, as demonstrated below, the swelling-induced taurine efflux pathway and the small swelling-induced $Cl^-$ channel, although activated by the same second messenger system, represent two distinct types of channels.

## The Swelling-Induced Taurine $Cl^-$ Channel

Several transport systems have been considered candidates for mediating swelling-induced taurine efflux in Ehrlich cells: *i*) the $Na^+$ and $Cl^-$-dependent β-system, *ii*) the anion exchanger, *iii*) the volume sensitive organic osmolyte anion channel, *iv*) the swelling-induced $Cl^-$ channel, *v*) the cAMP-dependent $Cl^-$ channel and *vi*) the taurine channel that is permeable to $Cl^-$.

Taurine efflux from Ehrlich cells suspended in isotonic medium is mediated by a $Na^+$- and $Cl^-$-independent diffusion process, which resembles the β-system with respect to pH and potential sensitivity[9,10,15]. The swelling-induced taurine efflux from Ehrlich cells, on the other hand, has a pH and potential profile which are different from the profiles for steady state taurine efflux[15,16]. Thus, the swelling-induced taurine efflux pathway is different from the taurine transport pathway that dominates under isotonic conditions. The anion exchange system which in this case should exchange cellular taurine for extracellular $Cl^-$ is also not involved in swelling-induced taurine efflux in Ehrlich cells because taurine efflux is unaffected by substitution of gluconate for extracellular $Cl^-$[16].

Swelling-induced taurine efflux from the Ehrlich cell is reduced following depolarization of the plasma membrane, suggesting that taurine might be leaving the Ehrlich cell either as an anion or via an anion channel[16]. However, this notion is inconsistent with the observation that cellular acidification during the RVD response[21] is small compared to the swelling-induced net loss of taurine[3]. Thus, the swelling-induced taurine leak pathway in the Ehrlich cell seems to be a potential-dependent anion channel, coined the taurine channel.

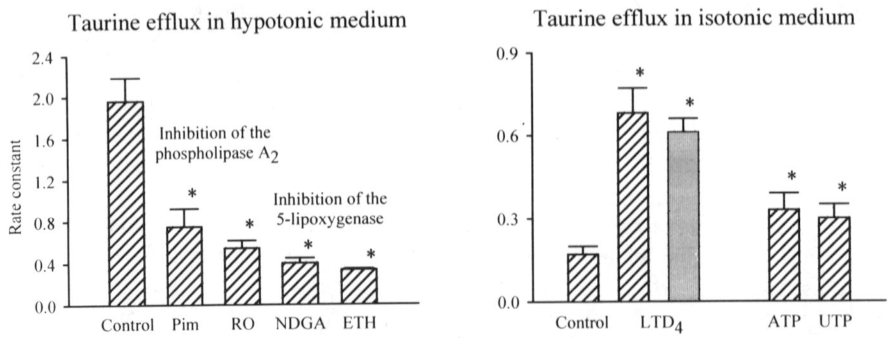


**Figure 2.** Effect of RO 31-4639 (RO), pimozide (Pim), nordihydroguaiaretic acid (NDGA), ETH 615139 (ETH) on the rate constant for the swelling-activated taurine efflux from Ehrlich cells (left panel), and of $LTD_4$, ATP and UTP on the rate constant for taurine efflux from Ehrlich cells suspended in isotonic medium (right panel). Ehrlich cells, equilibrated with $^{14}C$-taurine for 60 min in standard NaCl medium (321 mOsm), were at time zero transferred to either hypotonic $Ca^{2+}$-free NaCl medium (150 mOsm, left panel) or isotonic standard NaCl medium containing 1 mM $Ca^{2+}$ (right panel), and 7 samples were drawn within the first min in order to estimate the release of $^{14}C$-taurine. The rate constant for the initial taurine release ($min^{-1}$ x g cell water/g cell dry wt) was calculated as the product of the cell water content and the slope of the plot $(a^m_t - a^m_0)/a^c_0$ vs time, were $a^m_t$ and $a^m_0$ are the $^{14}C$-taurine activity in the efflux medium at time t and time zero, respectively and $a^c_0$ is the cellular $^{14}C$-taurine activity at time zero. Pimozide (12 μM), RO (2.4 μM), NDGA (50 μM), ETH (10 μM), $LTD_4$ (5 nM, hatched bar; 100 nM, closed bar), ATP (10 μM) and UTP (10 μM) were added at time zero. The rate constants are given as the means ± SEM. The number of experiments in the right panel was 9 (control), 6 (100 nM $LTD_4$), 3 (5 nM $LTD_4$), 6 (ATP) and 4 (UTP). The left panel is redrawn from Lambert and Hoffmann[15].

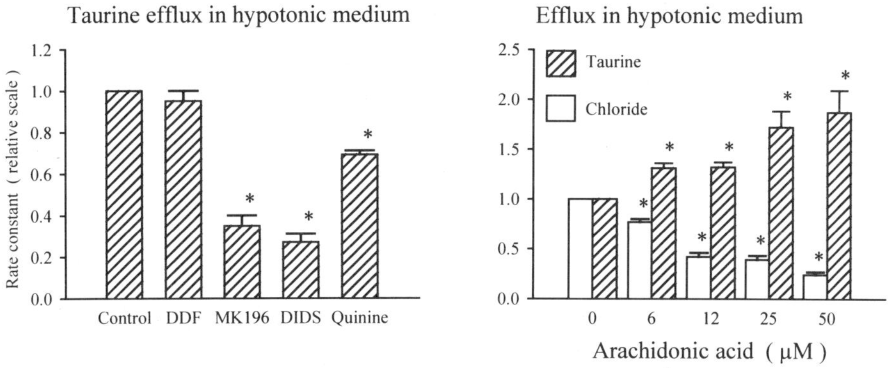


**Figure 3.** Effect of 1,9-dideoxyforskolin (DDF), MK196, DIDS, quinine and arachidonic acid on the rate constant for the swelling-induced taurine efflux, and of arachidonic acid on the rate constant for the swelling-induced $Cl^-$ efflux in Ehrlich cells. Ehrlich cells, equilibrated with $^{14}C$-taurine or $^{36}Cl^-$ in standard NaCl medium (321 mOsm), were at time zero transferred to hypotonic $Cl^-$-free gluconate medium (150 mOsm) and the release of $^{14}C$-taurine and $^{36}Cl^-$ was followed with time. $Cl^-$-free gluconate medium was used in order to avoid $^{36}Cl^-$ exchange via the $Cl^-/HCO_3^-$ exchanger. The rate constants for the initial release (min$^{-1}$ x g cell water/g cell dry wt) were estimated as indicated in the legend to Fig. 2. Cells were preincubated for 1 min with DDF (50 μM). MK196 (1 mM), DIDS (200 μM), quinine (0.5 mM) and arachidonic acid (6 - 50 μM) were added at the time of hypotonic exposure. Values are given relative to control ± SEM. The number of paired experiments was 3 (DDF), 4 (MK196), 3 (DIDS), 3 (quinine) and 3 (arachidonic acid). The data for DIDS, MK196 and arachidonic acid are redrawn from Lambert and Hoffmann[16].

A volume-sensitive organic osmolyte anion channel, designated VSOAC, has been described in different vertebrate cell types[28]. VSOAC is, with a few exceptions, blocked by 1,9-dideoxyforskolin (DDF), extracellular nucleotides, quinine, 4,4′-diisothiocyanostilbene-2,2′-disulfonic acid (DIDS) and polyunsaturated fatty acids[27,28]. From Fig. 3 (left panel) it is seen that swelling-induced taurine efflux from Ehrlich cells is unaffected by the addition of DDF, but is inhibited by the addition of quinine and the anion transport inhibitors DIDS and MK196 (indacrinone). However, addition of quinine significantly depolarizes the cell membrane of the Ehrlich cell[18] and since swelling-induced taurine efflux is inhibited by membrane depolarization[16] it follows that the effect of quinine on efflux in the Ehrlich cell must be secondary to quinine-induced depolarization of the plasma membrane. On the other hand, swelling-induced $Cl^-$ efflux from the Ehrlich cell is inhibited by MK196[16], but unaffected by the addition of DIDS[16], quinine and DDF (the rate constant, given relative to the rate constant for control cells, was in three sets of paired experiments estimated at 1.0 ± 0.2 and 1.3 ± 0.3 following treatment with 0.5 mM quinine and 50 μM DDF, respectively). Furthermore, swelling-induced $Cl^-$ efflux is blocked by the addition of arachidonic acid (Fig. 3, right panel) and other unsaturated fatty acids *e.g.* linoleic acid, linolenic acid, and oleic acid[11], whereas swelling-induced taurine efflux is stimulated by the addition of arachidonic acid (Fig. 3, right panel). Thus, in the Ehrlich cell, the swelling-induced taurine channel and the swelling-induced $Cl^-$ channel represent two different types of channels. Moreover, the taurine channel differs pharmacologically from the VSOAC found in other mammalian cells.

Besides the small (7 pS), swelling-induced $Cl^-$ channel, Ehrlich cells also possess a medium-sized (34 pS) $Cl^-$ channel, which has properties similar to the cAMP activated $Cl^-$ channel in the apical membrane of human airway epithelial cells[1]. Figure 4 reveals that addition of either forskolin or dbcAMP significantly stimulates $Cl^-$ efflux from Ehrlich cells

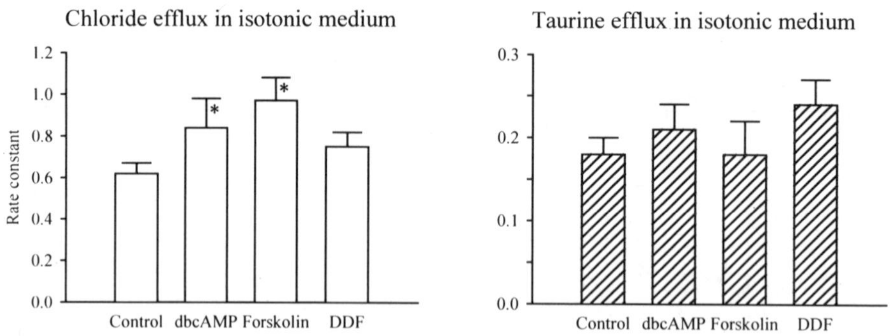


**Figure 4.** Effect of dbcAMP, forskolin and 1,9-dideoxyforskolin (DDF) on the rate constant for the $Cl^-$ efflux and the taurine efflux in Ehrlich cells suspended in isotonic medium. Ehrlich cells, equilibrated with $^{36}Cl^-$ or $^{14}C$-taurine in standard NaCl medium (300 mOsm), were at time zero transferred to isotonic $Cl^-$-free gluconate medium and the release of $^{36}Cl^-$ (left panel) and $^{14}C$-taurine (right panel) was followed for 1 min. $Cl^-$-free gluconate medium was used in order to avoid $^{36}Cl^-$ exchange via the $Cl^-/HCO_3^-$ exchanger. The rate constants for the initial release ($min^{-1}$ x g cell water/g cell dry wt) were estimated as indicated in the legend to Fig. 2. Cells were preincubated for 1 min with forskolin (10 μM), DDF (10 μM), and dbcAMP (0.2 mM). Values are given as mean values ± SEM. The number of experiments for $Cl^-$ efflux was 8 (control), 2 (forskolin), 3 (DDF, dbcAMP) and for taurine efflux 8 (control), 5 (dbcAMP, forskolin, DDF).

suspended in isotonic medium (left panel) but has no significant effect on taurine efflux (right panel). Thus, taurine is not transported via the cAMP sensitive $Cl^-$ channel in Ehrlich cells.

It has previously been shown that in Ehrlich cells suspended in isotonic medium arachidonic acid activates a nonspecific anion transport pathway, which is permeable to $Cl^-$ and propionate[13]. Addition of 25 μM arachidonic acid to Ehrlich cells suspended in isotonic $Ca^{2+}$-free choline Cl medium increases the rate constant for taurine efflux from 0.17 ± 0.01 to 0.41 ± 0.06 ($min^{-1}$ x g cell water/g cell dry wt, 3 paired experiments). Furthermore, the rate constant for taurine efflux after addition of 25 μM arachidonic acid is reduced to 0.17 (min x g cell water/g cell dry wt, 2 experiments) in the presence of 300 μM DIDS. Thus, the DIDS-sensitive channel activated by arachidonic acid is most probably identical to the swelling-induced taurine channel and apparently permeable to inorganic as well as organic anions. It is noted that the DIDS and MK196 sensitive taurine pathway, which accounts for 80% of the initial taurine influx in the ciliate *Tetrahymena pyriformis*, is also stimulated by the addition of arachidonic acid. Moreover, fatty acid-induced taurine influx in *Tetrahymena* is abolished in the presence of MK196[8]. Assuming that $Cl^-$ conductance in hypotonically swollen Ehrlich cells represents $Cl^-$ flux via both the small, swelling-induced $Cl^-$ channel and the swelling-induced taurine channel, it has been estimated that taurine channels contribute less than 5% to total $Cl^-$ conductance[16].

Our current hypothesis is that arachidonic acid released by phospholipase $A_2$ in the Ehrlich cell following osmotic swelling[29] is metabolized to $LTD_4$ via the 5-lipoxygenase system. Arachidonic acid as well as $LTD_4$ both act as activators of the taurine channel. Since taurine loss within the first few minutes following osmotic cell swelling only accounts for about 10% of the total loss of osmolytes in the Ehrlich cell[16], the inhibitory effect of arachidonic acid on the swelling-induced $Cl^-$ channel could represent a protective mechanism against massive stimulation of phospholipase $A_2$. The taurine channel activated by osmotic cell swelling in the Ehrlich cell is a potential dependent anion channel, which is pharmacologically different from the volume-sensitive organic osmolyte anion channel seen in other mammalian cells. The taurine channel also differs from the small, swelling-induced $Cl^-$ channel and the cAMP activated $Cl^-$ channel found in the Ehrlich cell.

## ACKNOWLEDGMENTS

This work was supported by the NOVO foundation and by the Danish Natural Research Council.

## REFERENCES


1. Christensen, O. and Hoffmann, E.K., 1992, Cell swelling activates $K^+$ and $Cl^-$ channels as well as nonselective, stretch-activated cation channels in Ehrlich ascites tumour cells, *J. Membrane Biol.*, 129:13-36.
2. Hoffmann, E.K. and Hendil, K.B., 1976, The role of amino acids and taurine in isosmotic intracellular regulation in Ehrlich ascites mouse tumour, *J. Comp. Physiol.*, 108:279-286.
3. Hoffmann, E.K. and Lambert, I.H., 1983, Amino acid transport and cell volume regulation in Ehrlich ascites tumour cells, *J. Physiol.*, 338:613-625.
4. Hoffmann, E.K., Lambert, I.H., and Simonsen, L.O., 1986, Separate, $Ca^{2+}$-activated $K^+$ and $Cl^-$ transport pathways in Ehrlich ascites tumour cells, *J. Membrane Biol.*, 91:227-24.
5. Jhiang, S.M., Fithian, L., Smanik, P., McGill, J., Tong, Q., and Mazzaferri, E., 1994, Cloning of the human taurine transporter and characterization of taurine uptake in thyroid cells, *FEBS Lett.*, 318:139-144.
6. Jørgensen, N.K., Christensen, S., Harbak, H., Brown, A.M., Lambert, I.H., Hoffmann, E.K., and Simonsen, L.O., 1997, On the role of calcium in the regulatory volume decrease (RVD) response in Ehrlich mouse ascites tumour cells, *J. Membrane Biol.*, 157, 281-299.
7. Jørgensen, N.K., Lambert, I.H., and Hoffmann, E.K., 1996, Role of $LTD_4$ in the regulatory volume decrease in Ehrlich ascites tumour cells, *J. Membrane Biol.*, 151:159-173.
8. Kramhøft, K. and Lambert, I.H., 1997, Taurine transport systems in the ciliate protozoan *Tetrahymena pyriformis*, *Amino Acids*, 12:57-75.
9. Lambert, I.H., 1984, $Na^+$-dependent taurine uptake in Ehrlich ascites tumour cells, *Mol. Physiol.* 6:233-246.
10. Lambert, I.H., 1985, Taurine transport in Ehrlich ascites tumour cells: Specificity and chloride dependence, *Mol. Physiol.*, 7:323-332.
11. Lambert, I.H., 1987, Effect of arachidonic acid, fatty acids, prostaglandins, and leukotrienes on volume regulation in Ehrlich ascites tumour cells, *J. Membrane Biol.*, 98:207-221.
12. Lambert, I.H., 1989, Leukotriene-$D_4$ induced cell shrinkage in Ehrlich ascites tumour cells, *J. Membrane Biol.*, 108:165-176.
13. Lambert, I.H., 1991, Effect of arachidonic acid on conductive $Na^+$, $K^+$ and anion transport pathways in Ehrlich ascites tumour cells under isotonic and hypotonic conditions, *Cell. Physiol. Biochem.*, 1:177-194.
14. Lambert, I.H., 1994, Eicosanoids and cell volume regulation, in: Cellular and Molecular Physiology of Cell Volume Regulation, Ed. K. Strange, CRC Press, Inc: pp. 273-292.
15. Lambert, I.H. and Hoffmann, E.K., 1993, Regulation of taurine transport in Ehrlich ascites tumour cells, *J. Membrane Biol.*, 131:67-79.
16. Lambert, I.H. and Hoffmann, E.K., 1994, Cell swelling activates separate taurine and chloride channels in Ehrlich mouse ascites tumour cells, *J. Membrane Biol.*, 142:289-298.
17. Lambert, I.H., Hoffmann, E.K., and Christensen, P., 1987, Role of prostaglandins and leukotrienes in volume regulation by Ehrlich ascites tumour cells, *J. Membrane Biol.*, 98:247-256.
18. Lambert, I.H., Hoffmann, E.K., and Jørgensen, F., 1989, Membrane potential, anion and cation conductance in Ehrlich ascites tumour cells, *J. Membrane Biol.*, 111:113-132.
19. Larsen, A.K., Jensen, B.S., and Hoffmann, E.K., 1994, Activation of protein kinase C during cell volume regulation in Ehrlich mouse ascites tumour cells, *Biochim. Biophys. Acta*, 1222:477-482.
20. Liu, Q., Lopez-Corcuera, B., Nelson, H., Mandiyan, S., and Nelson, N., 1992, Cloning and expression of cDNA encoding the transporter of taurine and β-alanine in mouse brain, *Proc. Natl. Acad. Sci.*, 89:12145-12149.

21. Livne, A. and Hoffmann, E.K., 1990, Cytoplasmic acidification and activation of $Na^+/H^+$ exchange during regulatory volume decrease in Ehrlich ascites tumor cells, *J. Membrane Biol.*, 114:153-157.
22. Mollerup, J. and Lambert, I.H., 1996, Phosphorylation is involved in the regulation of the taurine influx via the β-system in Ehrlich asites tumor cells, *J. Membrane Biol.*, 150:73-82.
23. Mollerup, J. and Lambert, I.H., 1998, Calyculin A modulates the kinetic constants for the $Na^+$-coupled taurine transport in Ehrlich ascites tumour cells, *Biochim. Biophys. Acta.*, submitted.
24. Noguchi, M., Miyano, M., and Matsumoto, T., 1996, Physicochemical character-ization of ATP binding to human 5-lipoxygenase, *Lipids*, 31:367-371.
25. Pedersen, S., Hoffmann, E.K., Hougaard, C., Jørgensen, N.K., Wybrandt, G.B., and Lambert, I.H., 1997, Leukotriene $D_4$-induced $Ca^{2+}$ mobilization in Ehrlich ascites tumor cells, *J. Membrane Biol.*, 155:61-73.
26. Smith, K.E., Borden, L.A., Wang, C.D., Hartig, P.H., Branchek, T.A., and Weinshank, R.L. 1992, Cloning and expression of a high affinity transporter from rat brain, *Mol. Pharmacol.*, 42:563-569.
27. Strange, K., Emma, F., and Jackson, P.S., 1996, Cellular and molecular physiology of volume-sensitive anion channels, *Am. J. Physiol.*, 270:C711-C730.
28. Strange, K. and Jackson, P.S., 1995, Swelling-activated organic osmolyte efflux: a new role for anion channels, *Kidney Int.*, 48:994-1003.
29. Thoroed, S.M., Lauritzen, L., Lambert, I.H., Hansen, H.S., and Hoffmann, E.K., 1997, Cell swelling activates phopholipase A2 in Ehrlich ascites tumour cells, *J. Membrane Biol.*, 160:47-58.
30. Uchida, S., Kwon, H.M., Yamauchi, A., Preston,A.S., Marumo, F., and Handler, J.S., 1992, Molecular cloning of the cDNA for a MDCK cell $Na^+$- and $Cl^-$-dependent taurine transporter that is regulated by hypertonicity, *Proc. Natl. Acad. Sci.*, 89:8230-8234.
31. Wang, Y., Roman, R., Lidofsky, S.D., and Fitz, J.G., 1996, Autocrine signalling through ATP release represents a novel mechanism for cell volume regulation, *Proc. Natl. Acad. Sci.*, 93:12020-12025.

# FUNCTIONAL CONSEQUENCES OF CALCIUM UPTAKE MODULATION BY TAURINE *IN VIVO* AND *IN VITRO*

E. Trenkner,[1,2] A. El Idrissi,[1,2] R. Dumas,[1] and A. Rabe[1]

[1] New York State Institute for Basic Research and
Developmental Disabilities
1050 Forest Hill Road
Staten Island, New York 10314
[2] CUNY Graduate School and University Center
33 West 42 Street
New York, NY 10036

## INTRODUCTION

The development, function and maintenance of the central nervous system (CNS) tissue are regulated in a timed sequence of inductions by growth factors (GF) and neuroactive amino acids (NAA)[16]. Taurine in particular has been identified as a "balancing factor" in the glutamate system to modulate and stabilize calcium homeostasis or enantiostasis and thus calcium-dependent cellular functions (for reviews[22, 44, 50, 48]).

Taurine's role in enantiostasis[22] is supported by the fact that the highest concentrations of taurine are found in cells that have to respond continuously to depolarizing conditions of their environment in the adult and throughout development. As we have shown in this symposium[16], calcium homeostasis is regulated by taurine in conjunction with growth factors like basic fibroblast growth factor (bFGF) and epidermal growth factor (EGF), but not by brain-derived neurotrophic factor(BDNF), suggesting that alternate mechanisms exist to maintain neuronal integrity and function.

Intracellular calcium levels control fundamental processes of neuronal function[32]. Calcium affects mitochondrial energy metabolism[17], controls the activation and de-activation of calcium-dependent enzymes and second messenger systems through auto -phosphorylation, phosphorylation and de-phosphorylation as well as the regulation of expressing neurotransmitter- and growth factor receptors. Thus taurine maintains a central role in the function of the CNS through modulation of calcium-dependent processes.

Complex processes of learning and memory also appear to be calcium-dependent. Increasing evidence has established that the calcium-permeable glutamate-receptor subtype *N*-methyl-D- aspartate (NMDA)[53] and possibly metabotropic receptors[3, 5], GABA receptor subtypes[41, 7, 24] as well as G-protein linked potassium ($K^+$) channels play an important role

*Taurine 3*
Edited by Schaffer *et al.*, Plenum Press, New York, 1998

during different phases of learning and memory[1,43,38,35]. Pharmacological investigations have implicated the cAMP response element binding protein (CREB) in calcium-dependent signaling pathways[11,27]. Furthermore, calcium-stores in mitochondria may be affected since mitochondrial energy-metabolism is increased with taurine[48]. Therefore, improvement of learning and memory in aged mice, as shown in this study, may reflect the decrease of free radical formation[15, 51] and an increase of the mitochondrial energy metabolism[48, 2].

## METHODS AND RESULTS

### Regulation of Calcium-Uptake by Taurine After Depolarization of Mouse Cerebellar Cells with NMDA and Potassium

The developing cerebellum was used in this study as a model to identify neuronal mechanisms that are affected or modulated by taurine. Primary cerebellar cells have been characterized over time in culture morphologically, functionally and pharmacologically. Cerebellar cell types *in vitro* have been identified to perform in time and space similar to their *in vivo*-counterparts. Moreover, the cerebellum was recently implicated in cognitive functions providing an *in vivo* model to test the extent of manipulated changes on memory[18]. Therefore, this cerebellar cell-culture system appears to be suited to serve as a model to identify various molecular mechanisms in the CNS associated with learning and memory[10, 49, 52].

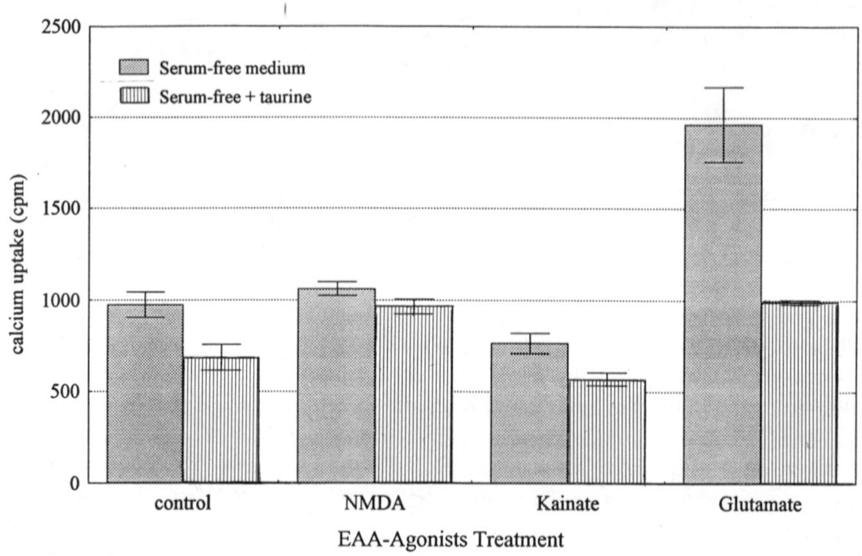


**Figure 1**. Taurine inhibits potassium- and glutamate-induced calcium uptake. P7 cerebellar granule cell cultures were started in horse serum (10%) supplemented BME. After 24h this medium was replaced by serum-free medium (MEM+15% N2 supplement) containing 25 mM KCl. 2μM AraC were added to suppress proliferation of non-neuronal cells. 1mM glutamate, NMDA and kainate, respectively, were added 30 minutes prior to $^{45}Ca^{2+}$ whereas 10 mM taurine was added 24h earlier than the glutamate agonists. Cells were incubated with $^{45}Ca^{2+}$ for 20-30 minutes reaching its maximum uptake after 20 minutes. The assay was performed at room temperature. NMDA-induced calcium-uptake was measured in the presence of 5 μM glycine and 10 μM of DNQX, a non-NMDA antagonist, to enhance the response. In the case of the depolarization with kainate, 10 μM of MK801 were added in order to block the activation of NMDA receptors. Calcium uptake was measured in triplicates and the results of two experiments were presented as the mean.

We have utilized this system to characterize pharmacologically and functionally neuronal cells of the CNS. Early postnatal cerebellar cells (P7) were cultured in serum-free medium as described[16,49]. After 24 h in MEM substituted with 10% horse serum and 25 mM $K^+$, this medium was exchanged to MEM containing 25 mM $K^+$ and 15% N2 supplement[6], respectively. Under these conditions the cerebellar neurons were depolarized throughout the culture period with $K^+$. After two days in serum free medium, neuronal cells were hyperpolarized with $10^{-4}$ M glutamate, NMDA or kainate for 30 minutes in the presence or absence of taurine. When $^{45}Ca^{2+}$ ($2x10^5$ cpm/25 μl assay solution) was added to these hyperpolarized presumptive granule cells, as demonstrated in Fig.1, $^{45}Ca^{2+}$ uptake was significantly higher than when the control cultures were hyperpolarized with glutamate. However, both agonists of NMDA and kainate failed to depolarize granule cells in the presence of $K^+$ above control levels. In cultures pretreated with $10^{-2}$M taurine for 24h, $^{45}Ca^{2+}$ uptake was reduced below control values, suggesting that taurine inhibited or modulated the extent of depolarization/ hyperpolarization and consequently controlled the extent of calcium uptake. On the other hand, values of calcium-uptake into neuronal cells maintained without $K^+$ (non-depolarizing conditions)[16] suggested that NMDA and kainate could depolarize cerebellar neurons in culture. Therefore, we conclude that the receptor-specific glutamate agonists cannot further depolarize neuronal cells in the presence of $K^+$.

**Effect of Taurine on Learning and Memory**

To demonstrate that taurine can enhance learning and memory, mice, known to have deficits as a result of embryonic exposure to alcohol and old age, were chosen for this study[2,12,45]. Dumas & Rabe[14] have shown that mice, ethanol-exposed once as embryos (E9), exhibited a significantly reduced ability to remember learned tasks. This affected neuronal phenotypes[25]. While ethanol irreversibly affected $GABA_A$ -receptors[13,34,42], the functions of $GABA_A$ during aging are not well understood.

Two-year old C57Bl/6J mice were tested for their ability to learn a spatial task in cross shaped water maze. Using the paradigm described by Dumas and Rabe[14], eighteen of the mice had been exposed to a large single oral dose of alcohol as embryos (5.8 g/kg of maternal weight on E9), while 27 other mice were the offspring of control mice similarly treated with isocaloric dextrose. About half of the mice from each early treatment group were exposed to 0.05% taurine in the drinking water, continuously for 6 weeks before and during behavioral testing; the remaining alcohol- and dextrose-treated offspring received no taurine. All mice were given 36 learning trials (6 trials a day for 6 days), and then tested for retention 3 weeks later. The performance was scored positively when a mouse reached the submerged platform in the East arm of the maze, while searches in the three other arms were scored as errors. As shown in Fig. 2, treatment with taurine significantly increased correct responses and significantly reduced the number of errors during the acquisition phase of both ethanol and dextrose-exposed groups of animals. Similarly, taurine significantly preserved retention. Taurine helped improve acquisition and preserved retention in these aged mice to a level comparable to that of young adult mice on the same tasks without taurine.

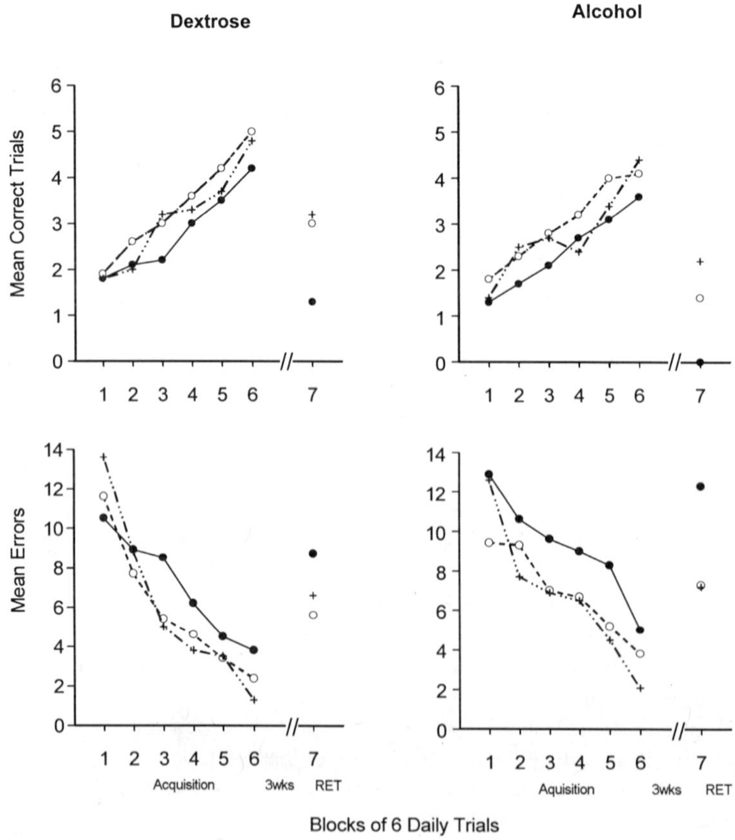


**Figure 2**. Effect of taurine on learning and retention of a spatial task in a cross-shaped water maze. Performance of old C57 Bl/6J mice that had been exposed to alcohol or dextrose as embryos is shown under the influence of taurine (open circles: alcohol n=11, dextrose n=14) and without taurine (solid circles: alcohol n=7, dextrose n=13). Treatment with taurine significantly increased total correct responses ($p<0.001$) and reduced total errors ($p<0.001$) during acquisition for both alcohol- and dextrose-exposed groups. Similarly, taurine significantly improved retention (for all comparisons, $p<0.001$ with the respective day 6 scores as a covariant to control for the level of learning). All groups of mice improved with training (a significant Day effect), but alcohol exposed mice performed more poorly than dextrose-exposed mice with and without taurine (a significant Alcohol effect). There were no significant interactions. Comparison with the performance (crosses) of young adult mice without taurine (dextrose-exposed n=20, alcohol- exposed n=17) taken from another experiment (Dumas and Rabe, 1994) illustrates that taurine improved the performance of the aged mice to a level achieved by young adult mice (3 months). To preserve visual clarity, no measures of variability are included.

## DISCUSSION

A better understanding of how memory works is considered one of the challenges in recent neurobiology. Based on models for learning, including those that demonstrate impairment, the identification of molecules involved and mechanisms proposed, it has become evident that different learning and memory tasks involve different processes that, in turn, require different mechanisms and various sets of molecular compounds[1,43,46]. The mechanism of taurine's action in learning and memory is yet unknown. It had been reported that taurine improves the recovery of learning and memory after induced hypoxia which led to the conclusion that taurine is an anti-hypoxic agent[29,40]. Our results suggest, however, that taurine's involvement in the regulation of intracellular calcium could be directly related. Several mechanisms of learning and memory have been established that require calcium or are related to calcium mediated processes, depending on the type of task to learn. Numerous studies have shown that NMDA-specific glutamate receptors controlled through inhibitory GABA receptors have been linked to long term potentiation (LTP)[4,30,31,43,53]. For example NMDA-receptor specific antagonists like MK801 and AP5 block spatial learning[20,35]. LTP, considered to be a cellular analogue for learning and memory, is achieved through CREB, and a calcium dependent kinase (PKA) that phosphorylates learning-induced proteins[23,27,33]. At the same time, calcium-mediated potassium channels have been shown to control memory. These channels linked to G-proteins induce second messenger systems associated with spatial learning[1,21,39]. A comparison of LTP and G-protein coupled $K^+$ channel recently concluded that both mechanisms fulfill different roles during memory-development[43].

$GABA_A$ and $GABA_B$-receptors play an important role in the NMDA-mediated responses through timed inhibition[26]. The molecular and functional diversity of the $GABA_A$ receptor family has been well documented[41], as has those of the $GABA_B$ receptor-family recently[7,8,9,24]. The $GABA_B$ receptors are coupled to $K^+$-channel linked G-proteins that open $K^+$-channels and close voltage-dependent $Ca^{2+}$ channels and appear to interact with taurine[28]. Both $GABA_A$ and $GABA_B$ receptors have been localized on both pre- and post-synaptic terminals and have been shown pharmacologically to influence learning and memory[13]; e.g. the agonist baclophen interrupted spatial learning while its antagonist attenuated this interruption[37]. Ramirez *et. al.*[38] further demonstrated that $GABA_B$ receptors control the conditioned eye-blink response and therefore, were directly involved in learning. If taurine would react with $GABA_B$ receptors in these paradigms it would act as a $GABA_B$ receptor antagonist.

Taurine appears to affect both calcium- and $K^+$- controlled channel systems as demonstrated in Fig. 1. In early postnatal cerebellar cultures, as well as in hippocampal cultures (unpublished data), taurine controlled calcium homeostasis. In this study, we utilized aging-related learning and memory deficits to demonstrate a possible role of taurine in restoring the ability to learn and retain learned information in animals, functionally compromised by an embryonic exposure to alcohol and old age, or old age alone. These results are consistent with another finding from an unpublished pilot experiment with senescence accelerated mice in which a 6-week treatment with taurine significantly attenuated the passive avoidance learning deficit typical of these mice as they reach middle age.

These results point to a general mechanism of taurine: the modulation and restoration of calcium mediated processes, such as the restoration of age-affected mitochondrial energy metabolism[2,47]. The memory related mechanisms involving CREB, G-protein linked potassium channels and, as proposed by us at this symposium, the dis-inhibition of $GABA_B$ receptors[36], could lead to the opening of $K^+$-channels or the activation and control of NMDA-mediated calcium-channels through GABA release thus restoring alcohol or age related deficiencies of the GABAergic receptor system. Also, calcium-dependent cholinergic- and dopaminergic receptors would be affected by taurine's role in enanti-ostasis[22]. The involvement of taurine in synaptic activities has been emphasized recently by Galaretta *et.al.*[19] strengthening the interpretation of our findings that taurine must act on behavior-controlling neuronal circuitry.

## ACKNOWLEDGMENTS

This study was supported by a grant from NYS Office of Mental Retardation and Developmental Disabilities.

## REFERENCES


1. Alkon, D.L., 1989, Memory storage and neuronal systems, *Scientific American.*
2. Beal, M.F., 1995, Aging, energy, and oxidative stress in neurodegenerative diseases, *Ann. Neuol.*, 18:357-366.
3. Behnisch, T. and Reymann, K.G., 1993, Co-activation of metabotropic glutamate and N-methyl-D-aspartate receptors is involved in mechanisms of long-term potentiation maintenance in rat hippocampal CA1 neurons, *Neuroscience*, 54:37-47.
4. Bliss, T.V.P. and Lomo, 1973, Long-lasting potentiation of synaptic transmission in the dentate area of the anesthesized rabbit following stimulation of the perforant path, *J. Physiol.*, (London), 232:331-356.
5. Bortolotto, Z.A. and Collingridge, G.L., 1993, Characterization of LTP induced by the activation of glutamate metabotropic receptors in area CA1 ofther hippocampus, *Neuropharmacol.*, 32:1-9.
6. Bottenstein, J.E., Skaper, S.D., Varon, S.S., and Sato, G.H., 1980, Selective survival of neurons from chick embryo sensory ganglionic dissociates utilizing serum-free supplemented medium, *Exp. Cell Res.*, 125:183-190.
7. Bowery, N.G., 1993, $GABA_B$ receptor pharmacology, *Annu. Rev. Pharmacol. Toxicol.*, 33:109-147.
8. Bowery, N.G. and Brown, D.A., 1997, The cloning of $GABA_B$ receptors, *Nature*, 386:223-224.
9. Brucato, F.H., Levin, E.D., Mott, D.D., Lewis, D.V., Wilson, W.A., and Swartzwelder, H.S., 1996, Hippocampal long term potentiation and spatial learning in the rat: effects of $GABA_B$ receptor blockade, *Neuroscience*, 74:331-339.
10. Damgaard, I., Trenkner, E., Sturman, J.A., and Schouboe, A., 1996, Effect of $K^+$-and kainate-mediated depolarization on survival and functional maturation of GABAergic and glutamatergic neurons in cultures of dissociated mouse cerebellum, *Neurochem. Res.*, 21:267-275.
11. Dash, P.K., Moore, A.N., Dixon, C.E., 1995, Spatial memory deficits, increased phosphorylation of the transcription factor CREB, and induction of the AP-1 complex following experimental brain injury, *J. Neurosci.*, 15:2030-2039.
12. Desroches, D., White, S.S., and Benno, R.H., 1987, Effects of acute alcohol administration on maternal calcium metabolism and fetal develpment in mice, *Alcohol and Drug Research,* 7:443-452.
13. Devaud, L.L., Fritschy, J.M., Sieghart, W., and Morrow, A.L., 1997, Bidirectional alterations of $GABA_A$ receptor subunit peptide levels in rat cortex during chronic ethanol consumption and withdrawal, *J. Neurochem.*, 69:126-130.
14. Dumas, R.M. and Rabe, A., 1994, Augmented memory loss in aging mice after one embryonic exposure to alcohol, *Neurotoxicol. Teratol.*, 16:605-612.
15. Dykens, J.A., Stern, A., and Trenkner, E., 1987, Kainate-induced death of cerebellar neurons *in vitro* is mediated by xanthine oxidase, *J. Neurochem.*, 49:1222-1228.
16. El Idrissi, A., Harris, C., and Trenkner, E. 1997, Taurine modulates glutamate-and growth factors mediated signaling mechanisms, *This volume*.
17. Ernster, L. and Schatz, G., 1981, Mitochondria: a historical review, *J. Cell Biol.*, 91:227s-255s.
18. Fiez, J.A., 1996, Cerebellar contributions to cognition, *Neuron,* 16:13-15.
19. Galaretta, M., Bustamante, J., Del Rio, R.M., and Solis, J.M., 1996, A new neuromodulatory action of taurine: long-lasting increase of synaptic potentials, in: Taurine 2, Basic and Clinical Aspects, Huxtable et. al., eds., Plenum Press, pp. 463-471.
20. Gorter, J.A. and de Bruin, J.P.C., 1992, Chronic neonatal MK-801 treatment results in an impairment of spatial learning in the adult rat, *Brain Res.*, 580:12-17.
21. Hoffman, D.A., Magee, J.C., Colbert, C.M., and Johnston, D., 1997, $K^+$ channel regulation of signal propagation in dendrites of hippocampal pyramidal neurons, *Nature,* 387:869-875.
22. Huxtable, R.J., 1992, Physiological actions of taurine, *Physiological Rev.*, 72:101-163.
23. Kandel, E.R. and O'Dell, T.J., 1992, Are adult learning mechanisms also used for development?, *Science*, 258:243-245.

24. Kaupmann, K., Huggel, K., Heid, J., Flor, P.J., Bischoff, S., Mickel, S.J., McMaster, G., Angst, C., Bittiger, H., Froestl, W., and Bettler, B., 1997, Expression cloning of $GABA_B$ receptors uncovers similarity to metabotropic glutamate receptors, *Nature*, 386:239-246.
25. Kentroti, S. and Vernadakis, A., 1992, Ethanol administration during early embryo-genesis affects neuronal phenotypes at a time when neuroblasts are pluripotential, *J. Neurosci. Res.*, 33:617-625.
26. Kim, U., Sanchez-Vives, M.V., and McCormick, D.A., 1997, Functional dynamics of GABAergic inhibition in the thalamus, *Science*, 278:130-134.
27. Kogan, J.H., Frankland, P.W., Blendy, J.A., Coblentz, J., Marowitz, Z., Schutz, G., and Silva, A.J., 1996, Spaced training induces normal long-term memory in CREB mutant mice, *Current Biology*, 7:1-11.
28. Kontro, P., Korpi, E.R., and Oja, S.S., 1990, Taurine interacts with $GABA_A$ and $GABA_B$ receptors, in: Taurine: Functional Neurochem. Physiol. and Cardiol., Pasantes -Morales, H., Martin, D.L., Shain, W., Martin del Rio, R., eds., Wiley-Liss, New York, pp. 83-94.
29. Malcangio, M., Bartolini, A., Ghelardini, C., Malmberg-Aiello, P., Franconi, F., and Giotti, A., 1989, Effect of ICV taurine on the impairment of learning, convulsions and death caused by hypoxia, *Psychopharmacol.*, 98:316-320.
30. Malenka, R.C., 1992, The role of post-synaptic calcium in the induction of long term potentiation, *Mol. Neurobiol.*, 5:289-295.
31. Martinez, J.L. and Derrick, B.E., 1996, Long-term potentiation and learning, *Annu. Rev. Psychol.*, 47:173-203.
32. Mattson, M.P., Guthrie, P.B., and Kater, S.B., 1989, A role for $Na^+$-dependent $Ca^{2+}$ extrusion in protection against excitotoxicity, *FASEB J.*, 3:2519-2526.
33. Mayford, M., Abel, T., and Kandel. E.R., 1995, Transgenic approaches to cognition, *Curr. Opin. Neurobiol.*, 5:141-148.
34. Mihic, S.J., Ye, Q., Wick, M.J., and Koltchine, V.V., 1997, Sites of alcohol and volatile anesthetic action on $GABA_A$ and glycine receptors, *Nature*, 389:385-389.
35. Morris, R.G.M., Anderson, E., Lynch, G.S., and Baudry, M., 1986, Selective impairment of learning and blockade of long-term potentiation by an *N*-methyl-D-aspartate receptor antagonist, AP5, *Nature*, 319:774-776.
36. Mott, D.D. and Lewis, D.V., 1991, Facilitation of the induction of long-term potentiation by $GABA_B$ receptors, *Science*, 252:1718-1720.
37. Nakagawa, Y. and Takashima, T., 1997, The $GABA_B$ receptor antagonist CGP36742 attenuates the baclofen- and scopolamine-induce deficit in Morris water maze task in rats, *Brain Res.*, 766:101-106.
38. Ramirez, O.A., Nordholm, A.F., Gellerman, D., Thompson, J.K., and Thomson, R.F., 1997, The conditioned eye-blink response: a role of $GABA_B$ receptors, *Pharmacol, Biochem. and Behavior*, 58:127-132.
39. Sah, P., 1996, $Ca^{2+}$-activated $K^+$ currents in neurons: types, physiological roles and modulation, *Trends Neurosci.*, 19:150-154.
40. Schurr, A., Tseng, M.T., West, C.A., and Rigor, B.M., 1987, Taurine improves the recovery of neuronal function following hypoxia: an *in vitro* study, *Life Sci.*, 40:2059-2066.
41. Seeburg, P.H., Wisden, W., and Verdoorn, T.A., 1990, The $GABA_A$ receptor family: molecular and functional diversity, *Cold Spring Harbor Symposia* (Vol. LV.) pp. 29-40.
42. Shibley, I.A. and Pennington, S.N., 1997, Metabolic and mitotic changes associated with fetal alcohol syndrome, *Alcohol and Alcoholism*, 32:423-434.
43. Shors, T.J. and Matzel, L.D., 1997, Long-term potentiation: What's learning got to do with it? *Brain and Behavioral Sciences*, in press.
44. Sturman, J.A., 1993, Taurine in development, *Physiol. Rev.*, 73:119-147.
45. Swartzwelder, H.S., Wilson, W.A., and Tayyeb, M.I., 1995, Differential sensitivity of NMDA receptor- mediated synaptic potentials to ethanol in immature versus mature hippocampus, *Alcohol Clin. Exp. Res.*, 19:320-323.
46. Thompson, R.F. and Krupa, D.J., 1994, Organization of memory traces in the mammalian brain, *Annu. Rev. Neurosci.*, 17:519-549.
47. Trenkner, E., El Idrissi, A., and Harris, C., 1996, Balanced interaction of growth factors and taurine regulate energy metabolism, neuronal survival, and function of cultured mouse cerebellar cells under depolorizing conditions, in: Taurine 2, Basic and Clinical Aspects, Huxtable, R., et. al., eds., Plenum Press, pp. 507-517.

48. Trenkner, E., Lui, D., Harris, C., and Sturman, J.A., 1994, Regulation of protein kinase C activity by taurine and lalanine during excitotoxicity in cat and mouse cerebellar cultures, in: Taurine in Health and Disease, Huxtable, R. and Michalk, D.V., eds., Plenum Press, Adv. Exp. Med. Biol., NY, pp. 309-316.
49. Trenkner, E., 1991, Cerebellar cells in culture, in: Culturing Nerve Cells, Banker G. and Goslin, K., eds., MIT Press, pp. 283.
50. Trenkner, E., 1990, The role of taurine and glutamate during early postnatal cerebellar development, in: Excitatory amino acids and neuronal plasticity, Ben-Ari, Y., ed., in: Adv. Exp. Med. and Biol., Plenum Press, Vol. 268: pp. 239-244.
51. Trenkner, E. and Dykens, J.A., 1986, Taurine moderates kainate and kynurenine excitotoxicity in vitro, *Soc. Neurosci. Abstr.*, 12:96.
52. Trenkner, E. and Sidman, R., 1977, Histogenesis of mouse cerebellum in microwell cultures: cell reaggregation and migration, fiber and synapse formation, *J. Cell Biol.*, 75:915-940.
53. Wangen, K., Myhrer, T., Moldstad, J.N., Iversen, E.G., and Fonnum, F., 1997, Modulatory treatment of NMDA receptors in neonatal rats affects cognitive behavior, *Dev. Brain Res.*, 99:126-130.

# AMPLIFIED EFFECT OF TAURINE ON PCB-INDUCED HYPERCHOLESTEROLEMIA IN RATS

Hideki Mochizuki,[1] Hiroaki Oda,[2] and Hidehiko Yokogoshi[1]

[1] School of Food and Nutritional Sciences
The University of Shizuoka
52-1 Yada
Shizuoka 422, Japan
[2] Department of Applied Biological Sciences
Nagoya University
Nagoya 464-01, Japan

## INTRODUCTION

The physiological functions of taurine such as membrane-stabilization, antioxidation, detoxification, osmoregulation, neuromodulation and brain-retinal development have been investigated in detail. Currently, we have been examining the hypercholesterolemic action of taurine in several conditions. In a previous study we reported that feeding a normal diet with taurine supplementation did not change the serum cholesterol levels, but in animals fed a high cholesterol diet, in which serum cholesterol significantly increased, taurine supplementation significantly reduced the cholesterol level[11,19]. This hypocholesterolemic action of taurine is explained by an increase in the activity of cholesterol 7 α-hydroxylase and also by the enhancement of fecal steroid excretion. Also, when the serum lipoprotein distribution was monitored by agarose-gel electrophoresis, the heavier very low density lipoprotein (VLDL), namely β-VLDL, significantly increased by feeding a high cholesterol diet. However, taurine repressed this increase and normalized the lipoprotein pattern. Many other investigators also examined the mechanism of hypercholesterolemia induced by feeding a high cholesterol diet, and demonstrated the elevation of both low density lipoprotein (LDL) and VLDL-cholesterol[10,19].

On the other hand, xenobiotics, such as polychlorinated biphenyls (PCB, Aroclor 1248, Mitsubishi Monsant Co., Ltd., Tokyo), are widely distributed in the environment. The administration of xenobiotics to animals causes many metabolic and pathologic changes[17,18,20]: (a) induction of hepatic drug metabolizing enzymes[17], (b) elevation of serum levels of high density lipoprotein (HDL) cholesterol and of apolipoprotein A-1[13,16], (c) accumulation of liver lipids[8,14,20], and (d) an increased concentration of ascorbic acid in urine and tissues[5,9,20]. These metabolic parameters are also influenced by dietary nutrients, such as protein[8] and sulfur-containing amino acids (S-AA). For example, in rats fed a diet containing a low level of soy protein isolate, supplemental methionine and cystine significantly increased these indexes[12,15].

In this study, we investigated whether taurine has cholesterol-lowering action in endogenous hypercholesterolemia induced by PCB administration.

## MATERIALS AND METHODS

Young male rats of the Wistar strain weighing approximately 60 g (SLC Inc., Hamamatsu, Japan) were housed in individual cages and were fed 20% casein diet (basal diet) for 2 days, *ad libitum*, in order to acclimate the rats to our animal facility. Room temperature was kept at 24°C with a 12-hour cycle of light (0700-1900 hours) and dark.

The composition of the basal diet was (in weight percent): casein, 20; mineral mixture (AIN-93), 5.0; corn oil, 5.0; vitamin mixture (AIN-76), 1.0; choline chloride, 0.15; and a mixture of sucrose and corn starch (1:2; w/w) to 100%. The experimental diet was prepared by supplementing with 1 or 5% of taurine with or without 0.02% of PCB to the basal diet at the expense of carbohydrate. Animals were provided these test diets, *ad libitum*, for 10 days (experiment 1) or 14 days (experiment 2). At the end of the experimental period, blood was collected from the cervical wound after decapitation and after 16-hours of fasting, and then centrifuged to obtain serum. Liver lipids were extracted by the method of Folch et al[4]. The concentrations of total cholesterol, HDL-cholesterol, triglycerides, and phospholipids in serum or liver were measured enzymatically by using kits (Wako Pure Chemical, Osaka, Japan). Liver triglycerides were measured by the method of Fletcher[3] (Wako Pure Chemical, Osaka, Japan).

Statistical analyses were performed by using one-way analysis of variance (ANOVA) followed by Duncan's multiple range test[2]. P-values of less than 0.05 are considered to be of statistical significance.

## RESULTS

### Effect of Taurine (1 or 5%) on Cholesterol and Triglyceride in Serum or Liver of Rats Fed 0.02% PCB Diet

The results of experiment 1 are shown in Table 1. There were no significant differences in body weight or food intake with the supplementation of taurine and/or PCB between the four groups. Relative liver weight was heavier in rats fed the PCB-containing diet compared with those rats in the normal group. The addition of 1 or 5% taurine to the PCB diet further increased liver weight over that of the already increased liver weights of the PCB group. However, this taurine-induced change in liver weight was not linearly dependent on taurine. Serum concentrations of total cholesterol, HDL-cholesterol and phospholipids were significantly elevated in rats fed a PCB diet as compared to those fed a normal diet. Supplemental taurine did not influence serum triglyceride levels of rats fed a PCB-containing diet; however, it caused further increases of serum cholesterol, especially HDL-cholesterol. Thus, taurine did not reduce serum cholesterol levels, unlike the observation for taurine in hypercholesterolemia induced by a high cholesterol diet. Moreover, total lipids, cholesterol and triglycerides significantly accumulated in the livers of PCB-fed animals, and taurine enhanced this accumulation. The concentrations of phospholipids in livers of rats fed a PCB-containing diet did not increase as compared with that of rats fed a normal diet, but the simultaneous supplementation of taurine tended to decrease these values. The maximal effect of taurine was observed at 1%.

**Table 1**. Effect of taurine (1 or 5%) on body weight, food intake, liver weight, and lipids in serum and liver of rats fed a 0.02% PCB containing diet for 10 days.

| | Normal | PCB | PCB+1%Tau | PCB+5%Tau |
|---|---|---|---|---|
| Body wt gain (g/10 days) | 55.2 ± 1.2 | 56.8 ± 1.3 | 56.9 ± 3.2 | 53.9 ± 2.1 |
| Final body wt (g) | 142 ± 1.9 | 140 ± 2.6 | 142 ± 2.3 | 137 ± 2.3 |
| Liver wt (g/100 g b.w.) | 3.79 ± 0.2[a] | 4.92 ± 0.3[b] | 7.24 ± 0.2[c] | 6.57 ± 0.6[c] |
| Food intake (g/day) | 14.9 ± 0.6 | 14.2 ± 0.2 | 15.2 ± 1.0 | 15.8 ± 0.5 |
| Serum lipids (mg/dl) | | | | |
| Total cholesterol | 89.5 ± 2.3[a] | 113 ± 6.7[a,b] | 131 ± 7.1[b] | 127 ± 11[b] |
| HDL-cholesterol | 32.2 ± 1.2[a] | 59.2 ± 3.1[b] | 80.2 ± 2.3[c] | 75.2 ± 3.7[c] |
| Triglyceride | 201 ± 27[b] | 134 ± 7.9[a] | 129 ± 13[a] | 142 ± 18[a] |
| Phospholipids | 156 ± 3.1[a] | 233 ± 10[b] | 245 ± 9.2[b] | 228 ± 9.2[b] |
| Liver lipids (mg/g liver) | | | | |
| Cholesterol | 4.08 ± 0.4[a] | 5.49 ± 0.3[b] | 6.46 ± 0.3[c] | 6.68 ± 0.4[c] |
| Triglyceride | 13.8 ± 2.3[a] | 46.7 ± 8.1[b] | 98.6 ± 16[c] | 105 ± 13[c] |
| Phospholipids | 36.3 ± 1.8[b,c] | 44.3 ± 4.8[c] | 28.1 ± 2.5[a,b] | 24.0 ± 2.2[a] |
| Total lipids | 58.2 ± 3.91[a] | 96.5 ± 8.50[b] | 133 ± 16[c] | 135 ± 14[c] |

Values are means ± SEM for 6 rats per group; values with different superscripts are significantly different at $P<0.05$.

**Table 2**. Effect of taurine (3%) on body weight, liver weight, and liver cholesterol of rats fed test diets with or without 0.02% PCB for 14 days.

| | Normal | 3% Tau | 0.02% PCB | PCB+3% Tau |
|---|---|---|---|---|
| Body wt gain (g/14 days) | 67.8 ± 2.3 | 63.3 ± 2.8 | 62.8 ± 5.0 | 57.3 ± 1.3 |
| Final body wt (g) | 143 ± 4.3 | 140 ± 1.3 | 136 ± 3.3 | 137 ± 4.6 |
| Liver wt (g/100 g b.w.) | 3.11 ± 0[a] | 2.93 ± 0.1[a] | 4.45 ± 0.2[b] | 6.72 ± 0.2[c] |
| Serum lipids (mg/dl) | | | | |
| Total cholesterol | 61.2 ± 4.1[a] | 64.0 ± 1.5[a] | 82.6 ± 3.2[b] | 122 ± 3.1[b] |
| HDL-cholesterol | 33.5 ± 0.9[a] | 36.3 ± 2.3[a] | 59.0 ± 2.5[b] | 95.7 ± 2.8[c] |
| Cytochrome P-450 (nmol/g liver) | 7.10 ± 0.7[a] | 6.64 ± 0.4[a] | 45.1 ± 1.5[b] | 61.4 ± 2.6[c] |
| Cytochrome b5 (nmol/g liver) | 6.66 ± 0.8[a] | 6.82 ± 0.5[a] | 13.8 ± 2.2[b] | 11.2 ± 0.5[b] |

Values are means ± SEM for 4 rats per group; values with different superscripts are significantly different at $P<0.05$.

### Effect of Taurine and/or PCB on Serum Cholesterol and Hepatic Drug-Metabolizing Activity by the Indexes of Cytochrome P-450 and b5

The results of experiment 2 are shown in Table 2. Almost the same results for serum total cholesterol and HDL-cholesterol were obtained. That is, PCB induced hypercholesterolemia, and the simultaneous supplementation of taurine further enhanced this hypercholesterolemia.

Serum lipoprotein distribution was also determined (Fig. 1). In the case of PCB-induced hypercholesterolemia, the HDL-fraction increased and taurine reinforced this action.

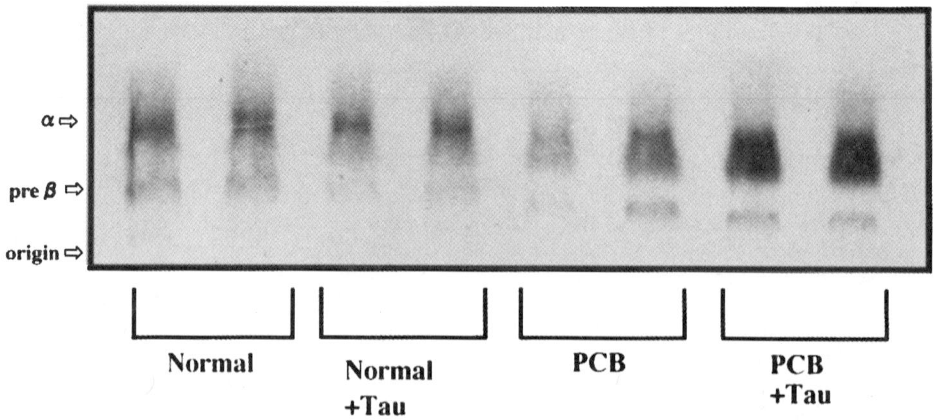


**Figure 1**. Serum lipoprotein distribution of rats fed 0.02% PCB-containing diet with or without 3% taurine by agarose-gel electrophoresis.

In general, it is known that PCB induces drug metabolizing systems. Therefore, the cytochromes p-450 and b5 were determined as an index of drug metabolizing systems in the liver. PCB increased the levels of cytochromes p-450 and b5 in the liver; taurine further enhanced the level of cytochrome p-450. Cytochrome b5 was not altered by the simultaneous supplementation of taurine.

## DISCUSSION

It is well known that dietary supplementation with taurine significantly reduces the hypercholesterolemia induced by feeding a high cholesterol diet. This hypocholesterolemic action of taurine is explained by (1) an increase in the conversion of cholesterol to bile acids due to an enhancement of the activity of hepatic cholesterol 7 α-hydroxylase, the limiting enzyme for bile acid synthesis, and (2) an increase in bile acid conjugation by the enhancement of taurine concentrations in the liver, with the subsequent increase in biliary bile acid secretion[4,19]. Also, in rats fed a high cholesterol diet, serum LDL and VLDL-cholesterol levels were dramatically increased, but the administration of taurine immediately decreased these lipoproteins. Furthermore, when we examined the hypocholesterolemic action of taurine in hypercholesterolemia caused by streptozotocin-induced diabetic rats, taurine decreased the elevation of LDL-cholestrol and VLDL-cholesterol[11]. Therefore, according to our previous results, the mechanism of the hypocholesterolemic action of taurine might be due to the elevation in serum LDL-cholesterol and VLDL-cholesterol.

In the present study, rats fed a PCB-containing diet, exhibited elevated serum HDL-levels, but the elevation in VLDL-cholesterol was only modest[11]. Also, the simultaneous supplementation of the PCB-diet with taurine further accelerated hypercholesterolemia, especially elevating serum HDL-cholesterol and liver lipids. The mechanism underlying the amplified effect of taurine on PCB-induced hypercholesterolemia in rats is not clear, but it is known that the hypercholesterolemia induced by PCB is mainly caused by a stimulation of liver cholesterol synthesis through the activation of 3-hydroxy-3-

methylglutaryl coenzyme A (HMG-CoA) reductase[7]. Since HMG-CoA reductase activity has been found to correlate with the activity of the hepatic drug metabolizing enzymes[1], we determined the concentration of liver microsomal cytochrome P-450 in rats fed PCB containing diets with or without taurine. Taurine significantly increased cytochrome P-450 in liver.

The overall results of these experiments demonstrate that, in endogenous hypercholesterolemia induced by PCB, taurine does not have hypocholesterolemic activity, but rather amplifies the hypercholesterolemia. It is concluded that the action of taurine may be correlated with the activity of the drug metabolizing system.

## REFERENCES


1. Duncan, D.B., 1955, Multiple range and multiple F tests, *Biometrics*, 11:1-42.
2. Cooper, D.Y., Schleyer, H., Levin, S.S., Eisenhardt, R.H., Novack, B.G., and Rosenthal, O., 1979, A reevaluation of the role of cytochrome P-450 as the terminal oxidase in hepatic microsomal mixed function oxidase catalyzed reactions, *Drug Metab. Rev.*, 10:153-185.
3. Fletcher, M.J., 1968, A colorimetric method for estimating serum triglycerides, *Clin. Chim. Acta*, 22:393-397.
4. Folch, J., Lees, M., and Sloane-Stanley, G.H., 1957, A simple method for the isolation and purification of total lipids from animal tissues, *J. Biol. Chem.*, 226:497-509.
5. Horio, F. and Yoshida, A., 1982, Effect of some xenobiotics on ascorbic acid metabolism in rats, *J. Nutr.*, 112:416-425.
6. Kato, N., Tani, T., and Yoshida, A., 1980, Effect of dietary level of protein on liver microsomal drug-metabolizing enzymes, urinary ascorbic acid and lipid metabolism in rats fed PCB-containing diets, *J. Nutr.*, 110:1686-1694.
7. Kato, N. and Yoshida, A., 1980, Effect of dietary PCB on hepatic cholesterogenesis in rats, *Nutr. Rep. Inter.*, 21:107-112.
8. Kato, N. and Yoshida, A., 1981, Effect of various dietary xenobiotics on serum total cholesterol and high density lipoprotein cholesterol in rats, *Nutr. Rep. Inter.*, 23:825-831.
9. Longenecker, H.E., Fricke, H.H., and King, G.C., 1940, The effect of organic compounds upon vitamin C synthesis in the rat, *J. Biol. Chem.*, 135:497-510.
10. Murakami, S., Yamagishi, I., Asami, Y., Ohta, Y., Toda, Y., Nara, Y., and Yamori, Y., 1996, Hypo-lipidemic effect of taurine in stroke-prone spontaneously hypertensive rats, *Pharmacology*, 52:303-313.
11. Nanami, K., Oda, H., and Yokogoshi, H., 1996, Antihypercholesterolemic action of taurine on streptozotocin-diabetic rats or on rats fed a high cholesterol diet, Taurine 2, edited by Huxtable et al., Plenum Press, New York, pp.561-568.
12. Oda, H., Fukui, C., Hitomi, Y., and Yoshida, A., 1991, Alternation of serum lipoprotein metabolism by polychlorinated biphenyls and methionine in rats fed a soybean protein diet, *J. Nutr.*, 121:925-933.
13. Oda, H., Matsushita, N., Hirabayashi, A., and Yoshida, A., 1990, Hyperlipoproteinemia in rats fed polychlorinated biphenyls, *J. Nutr. Sci. Vitaminol.*, 36:117-122.
14. Oda, H., Matsushita, N., Hirabayashi, A., and Yoshida, A., 1994, Cholesterol-rich very low density lipoproteins and fatty liver in rats fed polychlorinated biphenyls, *Biosci. Biotech. Biochem.*, 58:2152-2158.
15. Oda, H., Okumura, Y., Hitomi, Y., Ozaki, K., Nagaoka, S., and Yoshida, A., 1989, Effect of dietary methionine and polychlorinated biphenyls on cholesterol metabolism in rats fed a diet containing soy protein isolate, *J. Nutr. Sci. Vitaminol.*, 35:333-348.
16. Oda, H. and Yoshida, A., 1994, Effect of feeding xenobiotics on serum high density lipoprotein and apolipoprotein A-I in rats, *Biosci. Biotech. Biochem.*, 58:1646-1651.
17. Poland, A. and Knuttson, J.C., 1982, 2,3,7,8-Tetrachlorodibenzo-p-dioxin and related halogenated aromatic hydrocarbons, *Annu. Rev. Pharmacol. Toxicol.*, 22:517-554.
18. Quazi, S., Yokogoshi, H., and Yoshida, A., 1983, Effect of dietary fiber on hypercholesterolemia induced by dietary PCB or cholesterol in rats, *J. Nutr.*, 113:1109-1118.
19. Sugiyama, K., Ohishi, A., Ohnuma, Y., and Muramatsu, K., 1989, Comparison between the plasma cholesterol-lowering effects of glycine and taurine in rats fed on high cholesterol diets, *Agric. Biol. Chem.*, 53:1647-1652.

20. Yokogoshi, H., Mochizuki, S., Takahata, M., Quazi, S., and Yoshida, A., 1983, The hypercholesterolemic effect of caffeine-containing beverages and xanthine-derivatives in rats, *Nutr. Rep. Inter.*, 28:805-814.

# TAURINE CAN AMELIORATE INFLAMMATORY BOWEL DISEASE IN RATS

Miwon Son,[1] Jun Il Ko,[1] Won Bae Kim,[1] Hee Kyoung Kang[2], and Byong Kak Kim[3]

[1] Research Laboratories of Dong-A Pharmaceutical Co., Ltd.
Yongin-shi, Kyunggi-do, 449-900, Korea
[2] Department of Pharmacology
Seoul National University College of Medicine
Chongno-gu, Seoul 110-799, Korea
[3] College of Pharmacy
Seoul National University
Kwanak-gu, Seoul 151-742, Korea

## ABSTRACT

We previously reported that the protective effect of taurine against indomethacin-induced gastric mucosal injury was due to its antioxidant effects which inhibited lipid peroxidation and neutrophil activation. In this sudy, we examined the effect of taurine on reducing the inflammatory parameters of trinitrobenzene sulfonic acid (TNBS)-induced inflammatory bowel disease (IBD) in rats. To induce IBD, rats were given ethanolic TNBS intracolonically. The rats then received 500 mg/kg/day of taurine per orally. The rats were sacrificed one week after IBD induction. Ulceration and inflammation of the distal colon with formation of granuloma in the vehicle-treated IBD rats after two days of administration of TNBS were observed. Treatment with 0.5 g/kg of taurine by the oral route ameliorated colonic damage and decreased the incidence of diarrhea and adhesions. Colon weight (an index of tissue edema) was markedly increased in the IBD rats after adminstration of TNBS, but was significantly lower after taurine treatment. Myeloperoxidase (MPO) activity in the vehicle-treated IBD rats was substantially increased compared with that of the control. The taurine-treated animals showed reduced MPO activity (35% lower) when compared with that of the vehicle-treated animals. Taurine treatment decreased basal and formyl-methionyl leucyl phenylalanine (FMLP) stimulated reactive oxygen generation in colonic tissue of the IBD rat compared with vehicle treatment after one week. These results suggest that administration of taurine reduced the inflammatory parameters in this rat model of IBD by increasing the defenses against oxidative insult.

*Taurine 3*
Edited by Schaffer *et al.*, Plenum Press, New York, 1998

## INTRODUCTION

Taurine, 2-aminoethanesulfonic acid, is one of the most abundant intracellular free amino acids present in various mammalian tissues, including those tissues exposed to high concentrations of oxidants[3,4,25], such as lung, retina, and intestine. Taurine has been shown to possess signficant anti-inflammatory properties in various *in vivo* and *in vitro* models of inflammation[5,8,16,19-20,23]. Although the mechanisms of taurine protection are still unclear, much attention has focused on the ability of taurine to attenuate the oxidative cellular damage through neutrophil modulation and scavenging of reactive oxygen species[6,8,11,19].

It is suggested that the chronically inflamed intestine and/or colon may be subjected to considerable oxidative stress[9,13,26]. Most sources of these oxidants are the phagocytic leukocytes since these cells are known to be present in large numbers in inflamed mucosa and to produce significant amounts of reactive oxygen species in response to certain inflammatory stimuli. Because the colonic mucosa contains relatively small amounts of antioxidant enzymes (e.g. SOD, catalase, GSH peroxidase)[2] it is possible that the gut mucosa may be overwhelmed during times of active inflammation which may result in intestinal injury. This would suggest that potent antioxidants may prove beneficial in the treatment or prophylaxis of IBD[21,27].

In the present study, we investigated the possible anti-inflammatory effect of taurine in the acute colitis rat model.

## MATERIALS AND METHODS

### Animals

Male Sprague-Dawley rats, weighing 200-250 g were used for these experiments. They were housed throughout the study in a room with constant temperature (22°C), humidity (55%), and light (12 h on, 12 h off). Tap water and standard rat pellet food were provided *ad libitum*. For the experiments, rats were randomly selected and divided into 4 groups of 12 animals each.

### TNBS-Induced Rat Colitis Model

Colitis was induced using the technique described by Morris et al[10]. Briefly, in rats lightly anesthetized with ether, a rubber cannula (OD, 2 mm) was inserted into the lumen of the colon via the anus to approximately the splenic flexure (8 cm from the anus). TNBS (30 mg/rat) was dissolved in 50% ethanol (vol/vol) and injected (0.5 ml) into the colon via the rubber cannula.

Two groups of rats were treated, respectively, with taurine (500 mg/kg, once a day) or sulphasalazine (50 mg/kg, once a day) by oral gavage starting 24 h after the induction of colitis. The third group (controls) of rats received an equal volume of a vehicle.

Animals were observed daily for clinical symptoms of colitis, such as weight change and diarrhea (and/or loose stools). All scoring of damage and excision of tissue samples were performed by an observer unaware of the treatment group. The rats in the various treatment groups were randomized before being killed. The rats were weighed and killed by cervical dislocation and the distal 10 cm of colon was removed. The colon was opened by a longitudinal incision, rinsed with tap water, and pinned out on a wax block. Macroscopically visible damage was scored on a 0-10 scale using the scoring system described in Table 1 which takes into consideration the area of involvement and the presence or absence of ulcers. The presence or absence of adhesions between the colon and other

**Table 1**. Criteria for scoring of colonic ulceration and inflammation

| Score | Macroscopic appearance |
|---|---|
| 0 | Normal |
| 1 | Localized hyperemia, no ulcers |
| 2 | Ulceration without hyperemia or bowel wall thickening |
| 3 | Ulceration with inflammation at 1 site |
| 4 | 2 or more sites of ulceration and inflammation |
| 5 | Major sites of damage extending > 1 cm along length of colon |
| 6-10 | When an area of damage extended > 2 cm along length of colon, score was increased by 1 for each additional cm of involvement |

organs was also noted. Tissue samples were excised for subsequent measurement of MPO activity and the ability to synthesize eicosanoids.

### Myeloperoxidase Activity

The MPO activity of colonic tissue was determined by an index of inflammation[7]. Segments of the colon (200-300 mg) were taken from ulcerated and/or inflamed areas and minced finely. Segments of the colon were weighed and suspended in 0.5% hexa-decyl-trimethylammonium bromide (pH 6.5, 50 mg of tissue per ml) and then homo-genized. After freezing and thawing the homogenate three times, the tissue levels of MPO were determined by the technique utilizing 0.0005% hydrogen peroxide as a substrate for the enzyme. One unit of MPO activity is defined as that degrading 1 μmol of peroxide per minute at 25°C and is expressed in units per gram weight (U/g) of wet tissue.

### Luminol-Dependent Chemiluminescence (CL)[12,18]

The mucosa was scraped from the rectum and colon and washed immediately in PBS. The tissue suspension was then divided equally into scintillation test tubes containing 0.5ml of minimal essential medium (MEM; Gibco, USA). After addition of 0.02 mg luminol (Sigma Chemical Co., USA) and 0.14 mg/ml FMLP to the wells of the microplates (LitePlate, Packard), CL was determined using a TopCount (Packard). CL values were expressed as counts/s/mg wet weight of the colon after subtraction of the background.

### Colonic Eicosanoids Formation

Segments of the colon (200-300 mg) were taken from ulcerated and/or inflamed areas, minced finely, and after the addition of Tris buffer (0.5 ml, 50 mM, pH 7.5 at 0°C) centrifuged for 10s at 9000xg as described by Whittle[24]. The supernatant was transferred to microfuge tubes containing indomethacin (final concentration 10 μg/ml) and frozen (dry ice granules, -20°C). Eicosanoids [leukotriene $B_4$ ($LTB_4$) and prostaglandin $E_2$ ($PGE_2$)] were measured by a radioimmunoassay kit (Amersham).

### Statistical Analysis

All data are expressed as means ± SE. Comparisons between groups of nonparametric data, such as the damage scores, were made with the wilcoxon rank-sum test. Comparisons between groups of parametric data (with all statistical analyses) on associated probability (P value) of $\leq 5\%$ was considered significant.

## RESULTS

### TNBS-Induced Colitis

A single intracolonic administration of TNBS to rats produced a syndrome characterized by diarrhea, anorexia, loss of weight, and severe colonic damage. Gain in body weight was reduced by 50% in the TNBS-treated animals whereas weight gain in the saline-treated rats was not reduced. The colon of the TNBS-treated animals showed diffuse hemorrhagic necrosis of the mucosa and thickening of the bowel wall (typically extending 3 to 7 cm along the length). These animals were assigned a median score of 8.0. Severe adhesions of the colon to adjacent organs (intestinal loops, stomach, etc) were noted in 90% of the TNBS-control animals. None of the saline-treated rats presented signs of diarrhea, adhesions, or colonic damage. The mortality rate for the experiment was less than 1%. Animals treated with 500 mg/kg of taurine showed no signs of diarrhea. Oral treatment of colitic rats with 500 mg/kg/day of taurine significantly reversed the decrease in gain in body weight. However, taurine had no significant effect on the incidence of diarrhea (which was approximately 22%). Taurine treatment significantly reduced the number of adhesions (data not shown) and reduced the weight of the colon (an index of edema). The colonic damage score was lower in the group treated with taurine than in the TNBS-control group. This reduction represented a 27% decrease in the damaged area. Sulfasalazine, a drug used in the therapy of inflammatory bowel disease, had a significant effect on all indices of disease and inflammation. However, intracolonic treatment with taurine had no significant effect on all inflammation factors. These differences between oral and intracolonic administration of taurine are presumably due to differences in taurine uptake and should be further investigated.

### Myeloperoxidase Activity

Figure 1 shows MPO activity in the colonic mucosa. In the inflamed tissue of the colitic colon, the MPO activity, indicative of neutrophil infiltration, demonstrated a 3-fold increase 1 week after TNBS treatment compared to noninflamed tissue of the normal colon. These increases of MPO activity were reduced by treatment with taurine or sulfasalazine for 1 week.

### Luminol-Dependent Chemiluminescence

Superoxide levels measured by luminol-dependent chemiluminescence showed that there was approximately a 2.5-fold increase in basal superoxide generation in colon tissue from TNBS-treated colitic rats after 1 week. Treatment with taurine decreased the levels of superoxide anion generation by 20%. When superoxide generation was stimulated with 0.14 mg/ml of FMLP, there was a substantial increase in superoxide generation measured in tissues obtained from TNBS-treated rats after 1 week. Taurine treatment significantly decreased superoxide generation by 50% (Fig. 2).

**Table 2**. Effect of oral taurine treatment on the anti-inflammatory activity in rat TNBS-induced colitis[a]

| Drug | Dose (mg/kg) | Change of weight | | Incidence of diarrhea (%) | Lesion score |
|---|---|---|---|---|---|
| | | body (g/day) | colon (g/10 cm) | | |
| Normal | | 8.5 ± 1.2 | 0.7 ± 0.1 | 0 | 0 |
| Control | | 4.2 ± 1.5 | 2.2 ± 0.3 | 38.1 | 7.0 ± 0.5 |
| Taurine | 500 | 6.7 ± 1.6 | 1.8 ± 0.2 | 29.6 | 5.1 ± 0.4* |
| Sulfasalazine | 50 | 7.5 ± 1.3 | 1.6 ± 0.1 | 14.3 | 3.9 ± 0.7* |

[a] Colitis was induced by intracolonic administration of TNBS/ethanol (30 mg/50% ethanol, 0.5 ml/rat)

[b] Taurine and other drugs were administrated perorally to colitic rats for 1 week.
Data are means ± S.E. (*significantly different from controls at $P<0.05$).

**Table 3.** Effect of intracolonic taurine treatment on the anti-inflammatory activity in rat TNBS -induced colitis[a]

| Drug | Dose (mg/kg) | Change of weight | | Incidence of diarrhea (%) | Lesion score |
|---|---|---|---|---|---|
| | | body (g/day) | colon (g/10 cm) | | |
| Normal | | 6.2 ± 2.4 | 0.6 ± 0.1 | 0 | |
| Control | | 0.8 ± 2.0 | 1.3 ± 0.6 | 51.9 | 8.9 ± 1.9 |
| Taurine | 500 | 1.0 ± 1.9 | 1.2 ± 0.4 | 47.2 | 8.3 ± 1.4 |
| Sulfasalazine | 50 | 3.0 ± 1.8 | 1.2 ± 0.3 | 26.9 | 6.3 ± 2.3 |

[a] Colitis was induced by intracolonic administration of TNBS/ethanol (30 mg/50% ethanol, 0.5 ml/rat)

[b] Taurine and other drugs were intracolonically administrated to colitic rats for 1 week. Data are means ± S.E.

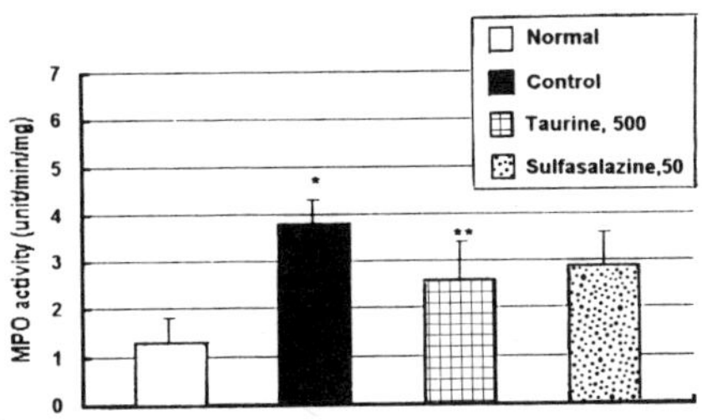


**Figure 1**. Effect of taurine on colonic myeloperoxidase (MPO) in rat TNBS-induced colitis. Taurine was given to colitic rats induced by administration of TNBS for 1 week as described in Materials and Methods. Data are mean ± SE; n=8-12 animals/group (*vs normal, $p<0.05$; **vs control, $p<0.05$).

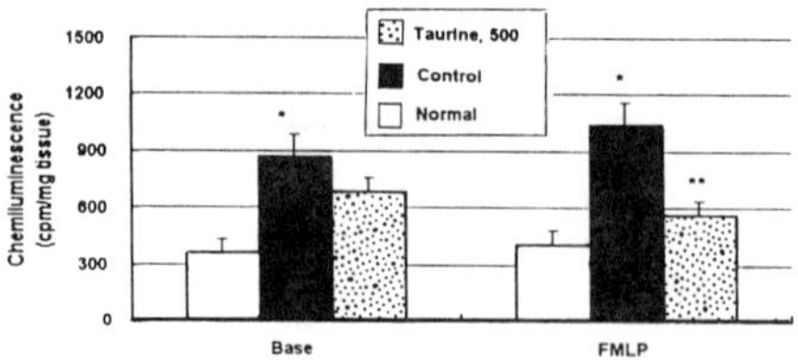


**Figure 2.** Effect of taurine on luminol-dependent chemiluminescence in the mucosa of inflamed colon induced by TNBS. Taurine was given to colitic rats induced by administration of TNBS for 1 week as described in Materials and Methods. Data are means ± SE; n=8-12 animals/group (*vs normal, $p<0.05$; **vs control, $p<0.05$).

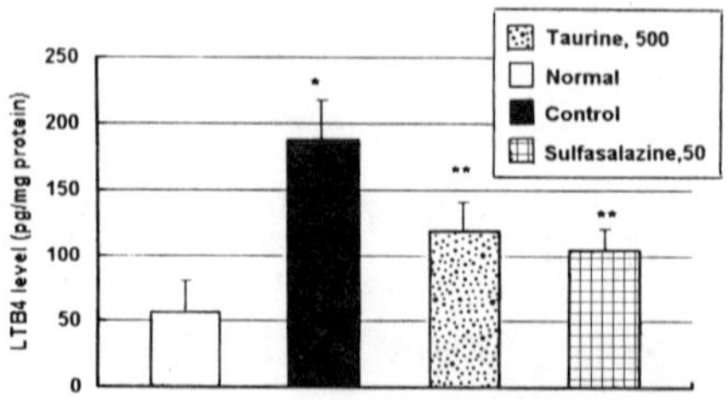


**Figure 3.** Effect of taurine on colonic leukotriene $B_4$ levels in rats with TNBS-induced colitis. Taurine was given to colitic rats induced by intracolonic administration of TNBS for 1 week as described in Materials and Methods. Data are means ± SE; n=8-12 animals/group (*vs normal, $p<0.05$; **vs control, $p<0.05$).

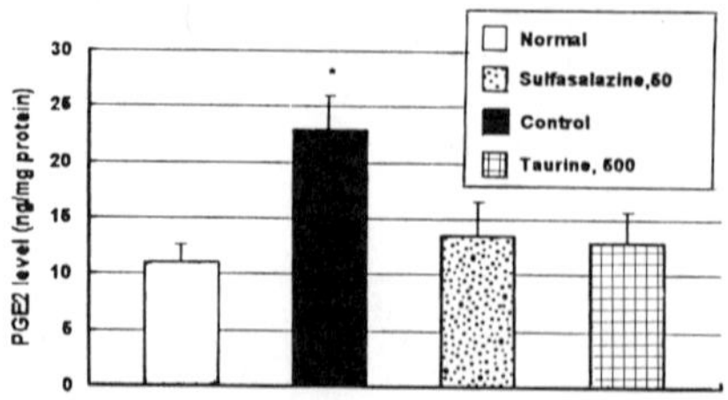


**Figure 4.** Effect of taurine on colonic prostaglandin $E_2$ levels in rats with TNBS-induced colitis. Taurine was given to colitic rat induced by intracolonic administration of TNBS for 1 week as described in Materials and Methods. Data are means ± SE; n=8-12 animals/group (*vs normal, $p<0.05$).

## Colonic Eicosanoids Formation

TNBS-induced colitis resulted in elevated colon levels of leukotriene $B_4$ and prostaglandin $E_2$ when compared to the levels of eicosanoids in the normal group (Figs. 3 and 4). As shown in Figure 3, $LTB_4$ synthesis in inflamed tissue was significantly elevated 1 week after TNBS administration. Treatment with taurine or sulfasalazine significantly reduced colonic $LTB_4$ levels by 38% and 44%, respectively. Levels of $PGE_2$ were increased 1 week after TNBS administration and this increase was inhibited by treatment with taurine or sulfasalazine for 1 week.

## DISCUSSION

In this study, it was observed that there was a reduction in inflammation of taurine-treated animals with inflammatory bowel disease compared with non-treated animals. Thus, measurement of the inflammatory bowel disease index, such as lesion scores, diarrhea, adhesion, and colon and body weight changes, showed reduced responses in taurine-treated animals suggesting that taurine has anti-inflammatory activity.

The mucosal lesion in IBD is characterized by a dense inflammatory cell infiltrate mainly comprised of neutrophils, macrophages and lymphocytes. Although many inflammatory mediators secreted by these cells including the cytokines and leukotrienes together with luminal bacterial products have been implicated in the mucosal injury observed in IBD[15], the molecules that mediate tissue damage remain poorly understood. Recent indirect evidence has, however, implicated reactive oxygen and nitrogen species (RONS) in the pathogenesis of the mucosal lesion[9]. Increased production of RONS has been observed after *in vitro* stimulation of whole colonic mucosa of IBD patients[18]. Furthermore, reactive oxygen scavengers and inhibitors including superoxide dismutase and sulfasalazine have shown anti-inflammatory activities[21,26]. Sulfasalazine, a 5-aminosalicylic acid (5-ASA) derivative, is the first line therapy of IBD. Although its mode of action is still unclear and various activities have been reported, one possibility is that 5-ASA acts as an antioxidant[21].

TNBS treatment caused an increase in MPO activity and superoxide anion generation in animals. Taurine was effective in reducing MPO activity and superoxide generation. Activated granulocytes secrete myeloperoxidase (MPO) into the extracellular medium where it catalyzes the oxidation of $Cl^-$ by $H_2O_2$ to yield HOCl. HOCl reacts rapidly with primary amines to yield chloramines (RNHCl)[26]. Taurine reacts readily with myeloperoxidase-derived HOCl/$OCl^-$ to form long-lived anionic oxidants known as taurochloramine[22]. Therefore, taurine has been proposed to modulate the inflammatory process by scavenging HOCl and preventing feed back inhibition of HOCl on MPO[8].

There is considerable evidence implicating several eicosanoids as key proinflammatory mediators in chronic inflammation including IBD[1,17]. Arachidonic acid metabolism in experimental colitis has been shown to closely resemble its metabolism in human IBD. Administration of taurine to colitic rats reduced colonic $LTB_4$ and $PGE_2$ synthesis. Several studies have shown that taurine reduced $LTB_4$ production by manipulation of the MPO-$H_2O_2$ halide system in human leukocytes[5] and $PGE_2$ production in raw 264.7 cells[14]. Colitic rats induced by TNBS showed an increase in the eicosanoid levels in the colonic mucosa and these increases were inhibited by taurine treatment. Thus, the inhibition of the production of eicosanoids by taurine is likely to be from the formation of taurochloramine in inflammatory cells.

In this study, we demonstrated that taurine has anti-inflammatory effects on inflammatory bowel disease which may be mediated through the inhibition of MPO activity, inhibition of reactive oxygen generation, and modulation of eicosanoids synthesis.

## REFERENCES


1. Boughton-Smith, N.K., Wallace, J.L., Morris, G.P., and Whittle, B.J.R., 1988, The effect of anti-inflammatory drugs on eicosanoid formation in a chronic model of inflammatory bowel disease in the rat, *Br. J. Pharmacol.*, 94:65-72.
2. Buffinton, G.D. and Doe, W.F., 1995, Depleted Mucosal Antioxidant Defences in Inflammatory Bowel Disease, *Free Rad. Biol. Med.*, 19:911-918.
3. Fukuda, K., Hirai, Y., Yoshida, H., Hakajiman, T., and Usii, T., 1982, Free amino acid content of lymphocytes and granulocytes compared, *Clin. Chem.*, 28:1758-1761.

4. Green, T., Fellman, J.H., Eicher, A.L., and Pratt, K.J., 1991, Antioxidant role and subcellular location of hypotaurine and taurine in human neutrophils, *Biochem. Biophys. Acta*, 1073:91-97.
5. Gurujeyalakshmi, G., Iyer, S.N., Hollinger, M.A., and Giri, S.N., 1996, Procollagen Gene Expression is Down-Regulated by Taurine and Niacin at the Transcriptional Level in the Bleomycin Hamster Model of Lung Fibrosis, *J. Pharm. Exp. Ther.*, 277:1152-1157.
6. Kim, C.K., Park, E.K., Quinn, M.R., and Schuller-Levis, G., 1996, The production of superoxide anion and nitric oxide by cultured murine leukocytes and the accumulation of TNF-α in the conditioned media is inhibited by taurine chloramine, *Immunopharm*, 34:89-95.
7. Krawisz, J.E., Sharon, P., and Stenson, W.F., 1984, Quantitative assay for acute intestinal inflammation based on myeloperoxydase activity. Assessment of inflammation in rat and hamster models, *Gastroenterology*, 87:1344-1350.
8. MaLoughlin, D.M., Stapleton, P.P., and Bloomfield, F.J., 1991, Influence of Taurine and a substituted taurine on the respiratory burst pathway in the inflammatory response, *Biochem. Soc. Trans.*, 19:73-78.
9. McKenzie, S.J., Baker, M.S., Buffinton, G.D., and Doe, W.F., 1996, Evidence of Oxidant induced Injury to Epithelial Cells during Inflammatory Bowel Disease, *J. Clin. Invest.*, 98:136-141.
10. Morris, G.P., Beck, P.L., Herridge, W., Szewczuk, M.R., and Wallace, J.L., 1989, Hapten-induced model of chronic inflammation and ulceration in the rat colon, *Gastren-terrolgy*, 96:759-803.
11. Murakami, M., Asgoe, K., Dekigai, H., Kusaka, S., Saita, H., and Kita, T., 1995, Products of nuetrophil metabolism increase ammonia-induced gatric mucosal damage, *Dig. Dis. Sci.*, 40:268-273.
12. Nielsen, O.H., Berild, D., and Ahnfelt-Ronne, I., 1994, In vitro superoxide production by peripheral neutrophils from patients with inflammatory bowel disease, *Mediat. Inflam.*, 3:161-164.
13. Pfeiffer, C.J., and Qiu, B.S., 1995, Effects of Chronic Nitric Oxide Synthase Inhibition on TNB-induced Colitis in Rats, *J. Pharm. Pharmacol*, 47:827-832.
14. Quinn, M.R., Park, E.K., and Schuller-Levis, G., 1996, Taurine chloramine inhibits prostaglandin E2 production in activated raw 264.7 cells by post- transcriptional effects on inducible cyclooxygenase expression, *Immunol Letters*, 50:185-188.
15. Sartor, R.B., 1994, Cytokines in intestinal inflammation: pathophysiological and clinical considerations, *Gastroenterology*, 106:533-539
16. Schuller-Levis, G., Quinn, M.R., Wright, C., and Park, E., 1994, Taurine protects against oxidant-induced lung injury: possible mechanism(s) of action in: Taurine in Health and Disease, R Huxtable, Ed, New York, NY: Pleum Press, pp 31-39.
17. Sharon, P. and Stenson, W.F., 1984, Enhanced synthesis of leukotriene B4 by colonic mucosa in inflammatory bowel disease, *Gastroenterology*, 86:453-460.
18. Simmonds, N.J., Allen, C.R., Stevens, T.R., Van Someren, R.N.M., Blake, D.R., and Rampton, D.S., 1992, Chemiluminescence assay of mucosal reactive oxygen metabolites in inflammatory bowel disease, *Gastroenterology*, 103:186-196.
19. Son, M., Kim, H.K., Kim, W.B., Yang, J., and Kim, B.K., 1996, Protective effect of taurine on indomethacin-induced gastric mucosal injury in: "Taurine 2", Huxtable, R.J., Azuma, J., Kuriyama, K., Nakagawa, M., and Baba, A., eds, Plenum Press, NY. pp: 147-155.
20. Stapleton, P.P., Molloy, A.M., Rogers, S., and Bloomfield, F.J., 1996, Neutrophil taurine in psoriasis, *Ir. J. Med. Sci.*, 165:173-176.
21. Tamai, H., Kachur, J.F., Grisham, M.B., and Gaginella, T.S., 1991, Scavenging Effect of 5-Aminosalicylic Acid on Neutrophil-Derived Oxidants: Possible Contribution to the Mechanism of Action in Inflammatory Bowel Disease, *Biochem. Pharm.*, 41:1001-1006.
22. Test, S.T., Lampert, M.P., Ossanna, P.J., Thoene, J.G., and Weiss, S.J., 1984, Generation of nitrogen-chlorine oxidants by human phagocytes, *J. Clin. Invest.* 74:1341-1349.
23. Wang, Q., Giri, S.N., Hyde, D.M., and Nakashima, J.M., 1989, Effects of Taurine on Bleomycin-Induced Lung Fibrosis in Hamsters, *Proc. Soc. Exp. Biol. Med.* 190:330-338.
24. Whittle, B.J.R., 1981, Temporal relationship between cyclooxygenase inhibition measured as prostacyclin biosynthesis and gastrointestinal damage induced by indomethacin in the rat, *Gastroenterology*, 80:94-98.
25. Wright, C.E., Lin, T.T., Lin, Y.Y., Sturman, J., and Gaull, G., 1985, Taurine scavenges oxidized chlorine in biological systems, *Prog. Clin. Biol. Res.* 179:113-123.
26. Yamada, T. and Grisham, M.B., 1991, Role of neutrophil-derived oxidants in the pathogenesis of intestinal inflammation, *Klin Wochenschift* 69:988-944.
27. Yue, G., Sun, F.F., Dunn, C., Yin, K., and Wong, P.Y.K., 1996, The 21-Aminosteroid Tirilazad Mesylate Can Ameliorate Inflammatory Bowel Disease in Rats, *J. Pharm. Exp. Ther.*, 276:265-270.

# THE EFFECTS OF TAURINE AND BIOGENETICALLY RELATED SULFUR-CONTAINING COMPOUNDS ON THE METABOLISM OF AND HYPOTHERMIA BY ETHANOL IN THE RAT

Vasilios Theofanopoulos and Cesar A. Lau-Cam

St John's University
College of Pharmacy and Allied Health Professions
8000 Utopia Parkway
Jamaica, New York 11439

## INTRODUCTION.

Two main metabolic pathways have been described for the biosynthesis of taurine in mammals: the cysteine sulfinic acid pathway, entailing the formation of cysteine sulfinic acid from cysteine, followed by conversion of cysteine sulfinic acid to hypotaurine; and the cysteamine pathway, where cysteine is sequentially metabolized to phosphopantothenyl cysteine, cysteamine and hypotaurine[7,19]. In turn, hypotaurine is enzymatically oxidized to taurine which, ultimately, is reported to undergo a very limited conversion to isethionic acid in certain organs[3,20]. An alternate minor pathway to taurine may occur through the intermediate formation of cysteic acid from cysteine sulfinic acid[3,13,23].

A body of information is presently available to suggest that acute dosing with taurine precursors such as cysteine[10,11] , cysteine sulfinic acid[8]and cysteic acid[17], but not with the metabolite, isethionic acid[17], can lower the levels of blood ethanol[7,8]; and that taurine and its immediate precursor, hypotaurine, can antagonize some of the physiological biochemical and behavioral properties of ethanol in man and animals[1,2,4,8,9, 12, 16,17]. More recently we have verified that the administration of taurine, in doses $\geq$1.2 mM/kg, to adult rats subsequently treated with a hypothermic dose of ethanol, will not only decrease the blood ethanol concentration (BEC) in a dose-related manner but also attenuate the hypothermic response to an extent proportional to the lowering of the BEC[24]. These effects were subsequently found to stem from the ability of taurine to promote blood ethanol clearance through its activating effect on hepatic alcohol dehydrogenase (ADH) and aldehyde dehydrogenase (ALDH)[24]

The purpose of the present investigation was to determine if the actions manifested by taurine on ethanol metabolism and ethanol-induced hypothermia are shared by those sulfur-containing compounds occurring along its biogenetic pathways, and if such actions are related to any particular structural feature. To our knowledge, these correlations have not been previously established.

## MATERIALS AND METHODS

### Chemicals

Taurine, hypotaurine, cysteamine, cysteine sulfinic acid and isethionic acid sodium salt were obtained from Sigma Chemical Co., St Louis, MO. Cysteic acid was purchased from Acros Organics, Pittsburgh, PA. Absolute ethanol was from United States Industrial and Chemical Co., Louisville, KY. Yeast ADH (365 units/mg of protein) and β-$NAD^+$ were obtained from Sigma Chemical Co.. All other chemicals were of analytical reagent grade and were purchased from Aldrich Chemical Co., Milwaukee, WI, or from Fluka, Ronkonkoma, NY.

### Animals

Male Sprague-Dawley rats, weighing 200-225 g, were purchased from Taconic Farms, Germantown, NY. The animals were housed in plastic cages, 6 per cage, in a temperature regulated (23±1°C), constant humidity, room and on a 12 hr light-12 hr dark cycle. During a 5 day acclimation period, the animals had free access to a pelleted animal diet (Purina Rat Chow, Ralston Purina, St. Louis, MO) and tap water. All experiments were conducted on rats that had been fasted overnight for 12 hr. Control and treatment groups consisted of 6 rats each.

### Treatment Solutions

All treatment solutions were prepared in physiological saline on the day of the experiment. The ethanol solution was 50% w/v and those of the sulfur compounds were 4% w/v.

### Treatment Schedules, Doses and Routes of Administration

All pretreatments were given 30 min prior to an ethanol dose. The doses were: ethanol 2 g/kg; sulfur compounds 1.2 and 2.4 mM/kg. All administrations were by the intraperitoneal (i.p.) route.

### Core Temperature Measurements

Core temperatures were measured with a Model 700C digital thermistor thermometer (Jenco Electronics Ltd., Taiwan, ROC) fitted with a YSI 702A flexible vinyl probe (Yellow Springs Instrument Co., Inc., Yellow Springs, OH). The probe was lightly lubricated with petrolatum and inserted to a depth of 3 cm into the rectal cavity. A reading was recorded after allowing the temperature to stabilize for at least 30 sec. The initial temperature represented the average of two readings taken 30 min before ethanol dosing. Thereafter, readings were taken at 30 min intervals for up to 240 min. Readings were converted to ΔT (°C) values, i.e., the differences from baseline values.

## Assay of Blood and Liver Ethanol

Blood samples for the determination of ethanol were collected at 60, 120 and 240 min after ethanol administration by the orbital sinus technique of Riley[22]. Samples were placed in stoppered polyethylene tubes containing disodium EDTA, stored on ice, and analyzed within 30 min after collection. Rats were sacrificed by decapitation, and their livers were surgically removed using the freeze-clamp technique of Wollenberger *et al.*[27]. A 2 g portion of frozen liver was reduced to a powder with the aid of a mortar and pestle precooled in liquid nitrogen, and next homogenized over an ice bath with a mixture of ice-cold 0.6 M perchloric acid - 40 mM thiourea using a Tissue Tearor electric homogenizer (Biospec Products, Barletsville, MD). The resulting suspension was centrifuged at 4000 x g and 4°C for 15 min. A 0.2 mL aliquot of blood or liver supernatant was analyzed for its ethanol content by the headspace gas chromatographic method of Eriksson *et al.*[6] using n-propanol as an internal standard and helium as a carrier gas.

## Assay of Hepatic ADH and ALDH

For the assay of ADH activity, a portion of frozen liver (about 1 g) was thawed, rinsed twice with physiological saline at 4°C, minced, and homogenized with 9 volumes of 1% Triton X-100 using a Potter-Elvehjem homogenizer. The suspension was centrifuged at 27,000 x g at 4°C for 60 min. Enzyme activity at 22°C was measured in an aliquot of the clear supernatant by following the rate of formation of NADH from $NAD^+$ spectrophotometrically at 340 nm according to Rachamin *et al.*[21]. For the assay of ALDH, the sample was prepared in a similar manner except that the homogenization medium was ice-cold 0.25 M sucrose and the centrifugation step was at 10,000 x g at 4°C for 20 min. The reduction of $NAD^+$ was followed spectrophotometrically at 340 nm and 37°C according to Vasiliou and Marselos[26]. The protein content of the liver supernatants was determined automatically with the Abbott Vision Analyzer using a commercial colorimetric kit (Kit 1430-10, Abbott Laboratories, Abbott Park, IL).

## Assay of Yeast ADH

The activity of yeast ADH was determined spectrophotometrically by the method of Vallee and Hoch[25].

## Statistics

Experimental values were analyzed for statistical significance using unpaired Student's t test and are reported as the means ± SEM. Differences were considered significant when $p \leq 0.05$.

# RESULTS AND DISCUSSION

The effects of taurine and related sulfur-containing compounds on BEC and ethanol-induced hypothermia were evaluated at two doses, 1.2 mM/kg and 2.4 mM/kg. At the lower dose, BEC was decreased by taurine, but was enhanced by cysteamine and cysteine sulfinic acid. It remained about the same in the presence of isethionic acid and hypotaurine (Fig. 1). Doubling the pretreatment dose affected the BEC in a manner dependent on the sulfur-containing compound administered. For example, whereas both isethionic acid and taurine significantly lowered BEC in that order of potency; cysteamine, cysteine sulfinic acid and hypotaurine exerted the opposite effect and to an extent not very different from that noted with the lower doses (Fig. 2). In contrast, both doses of cysteic acid were found to lower BEC only at 60 min post-ethanol (Fig. 1) and not thereafter (Fig. 2). These results agree with those of an earlier study that found cysteic acid (50 mg/ kg) to lower BEC in mice treated 30 min later with ethanol (2.5 g/kg). The reported failure of taurine (30 mg/kg i.p.)[4] and isethionic acid (50 mg/kg i.p.)[8] to decrease BEC in rodents dosed with ethanol (≥2.0 g/kg i.p.) might have been due to underdosing with the sulfur-containing compound.

In mammals, ethanol induces a characteristic drop in body temperature by influencing thermoregulatory mechanisms and the hypothalamic set point for body temperature control. This laboratory has previously determined that the administration of compounds such as fructose[14] or sodium ascorbate[15], will lessen the hypothermic response to ethanol by stimulating blood ethanol clearance through increased hepatic metabolism. The results presented here appear to lend support to this concept. First, taurine, cysteic acid and isethionic acid, the only three sulfur-containing compounds found to attenuate the thermolytic property of ethanol and to hasten the return of the core temperatures to original values, were also the only ones capable of lowering BEC. The effect of cysteic acid differed from that of taurine in that it took place only 60 min post-ethanol regardless of the dose used (Figs. 3 and 4). At present, it is unclear whether the rapid metabolic removal of cysteic acid could account for its rather brief action. In the case of isethionic acid, only the higher dose offered protection against blood ethanol elevation and ethanol-induced hypothermia, in which case the effects were significantly greater than those derived from an equivalent dose of taurine (Fig. 4). In contrast, sulfur-containing compounds not lowering BEC (i.e., cysteine sulfinic acid, cysteamine, hypotaurine), accentuated and prolonged the hypothermic response to ethanol (Figs. 3 and 4). Second, a good correlation existed between the ability of a sulfur-containing compound to influence the clearance of

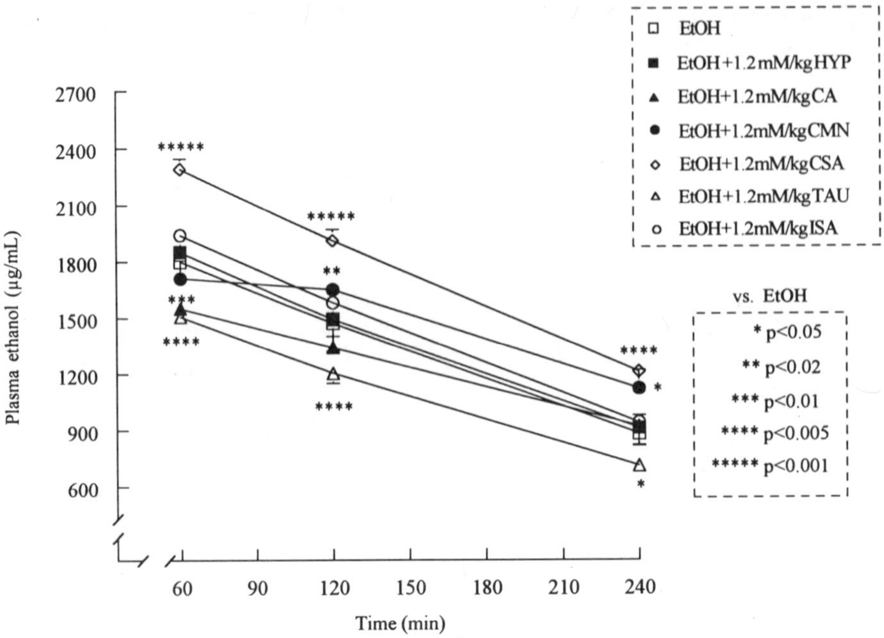


**Figure 1**. Effect of taurine and metabolically-related compounds (1.2 mM/kg i.p.) on the temporal changes in blood ethanol levels of rats receiving 30 min later a 2 g/kg i.p. dose of ethanol. Values are reported as means ± SEM (n = 6). Key: EtOH = ethanol, HYP = hypotaurine, CA = cysteic acid, CMN = cysteamine, CSA = cysteinesulfinic acid, TAU = taurine, and ISA = isethionic acid.

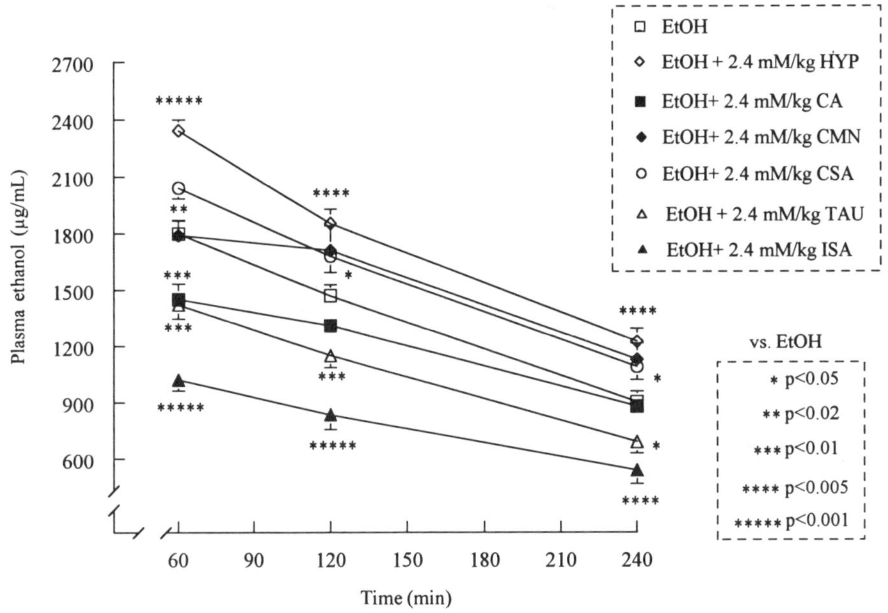


**Figure 2**. Effect of taurine and metabolically-related compounds (2.4 mM/kg i.p.) on the temporal changes in blood ethanol levels of rats receiving 30 min later a 2 g/kg i.p. dose of ethanol. Values are reported as means ± SEM (n = 6). Key: EtOH = ethanol, HYP = hypotaurine, CA = cysteic acid, CMN = cysteamine, CSA = cysteinesulfinic acid, TAU = taurine, and ISA = isethionic acid.

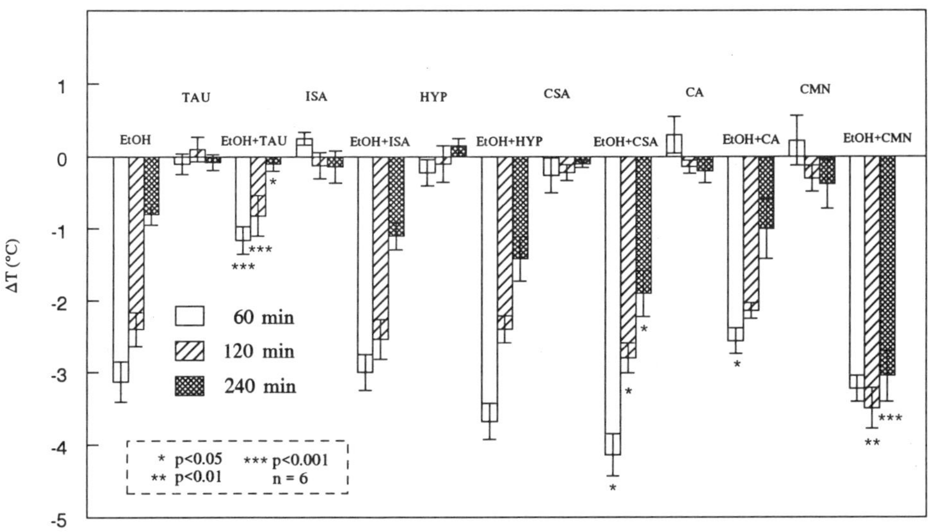


**Figure 3.** Time-course of the effects of taurine and biochemically-related compounds (1.2 mM/kg, i.p. each) on the hypothermic response to a 2 g/kg i.p. dose of ethanol. Values are reported as means ± SEM (n = 6). Comparisons were by unpaired Student's t-test vs. ethanol. Key: EtOH = ethanol, TAU = taurine, ISA = isethionic acid, CSA = cysteinesulfinic acid, CA = cysteic acid, and CMN = cysteamine.

ethanol from the circulation and its action on the hepatic ADH. Thus, in ethanol-treated rats, taurine and the higher dose of isethionic acid were found to significantly stimulate the ADH (Tables 1 and 2). On the other hand, in vitro studies with yeast ADH indicated that while cysteic acid demonstrated a marginal stimulatory action (data not shown), those compounds previously found to enhance BEC demonstrated an inhibitory action (Figs. 5(a)-5(c)). In addition, taurine and isethionic acid were the only test compounds to positively affect hepatic ALDH activity (Tables 3 and 4). In agreement with the results of earlier investigations[5,17,18], neither taurine nor isethionic acid stimulated hepatic ADH and ALDH activities in the absence of ethanol.

The foregoing results suggested the existence of a number of structure-activity correlations. For example, the hepatic metabolism of ethanol was stimulated by those compounds containing a sulfonic acid group in their structure (i.e., isethionic acid, taurine), but was inhibited by those compounds containing either a sulfinic acid (i.e., cysteine sulfinic acid, homotaurine) or a sulfhydryl (i.e., cysteamine) group. Furthermore, two structural modifications of taurine, namely the addition of a β-carboxyl group (like in cysteic acid) and the replacement of the β-amino group by a hydroxyl group (like in isethionic acid) resulted in an attenuation and a potentiation of the actions of taurine on hepatic ADH, respectively.

**Table 1**. Effect of taurine (TAU) on hepatic alcohol dehydrogenase (ADH) activity of rats receiving a fixed dose of ethanol (EtOH).

| | | Liver ADH activity | | |
|---|---|---|---|---|
| Treatment | n | 60 min | 120 min | 240 min |
| Saline | 6 | 511 ± 22 | 511 ± 22 | 472 ± 18 |
| EtOH | 6 | 518 ± 21 | 563 ± 18 | 503 ± 14 |
| TAU, 0.4 mM/kg | 6 | 524 ± 33 | 501 ± 33 | 465 ± 19 |
| TAU, 1.2 mM/kg | 6 | 504 ± 33 | N.D. | N.D. |
| TAU, 2.4 mM/kg | 6 | 510 ± 28 | N.D. | N.D. |
| EtOH + TAU, 0.4 mM/kg | 6 | 456 ± 14* | N.D. | N.D. |
| EtOH + TAU, 1.2 mM/kg | 6 | 635 ± 21* | 632 ± 14* | 568 ± 13* |
| EtOH + TAU, 2.4 mM/kg | 6 | 623 ± 34** | N.D. | N.D. |

The dose of ethanol was 2 g/kg i.p., given 30 min after TAU. Data represent units/g protein/min, means ± SEM. N.D. = not determined. Comparisons were by unpaired Student's t-test vs. EtOH group: *p<0.01, **p<0.005.

**Table 2**. Effect of isethionic acid (ISA) on hepatic alcohol dehydrogenase (ADH) activity of rats receiving a fixed dose of ethanol (EtOH).

| | | Liver ADH activity | | |
|---|---|---|---|---|
| Treatment | n | 60 min | 120 min | 240 min |
| Saline | 6 | 511 ± 22 | 494 ± 22 | 468 ± 22 |
| EtOH | 6 | 518 ± 21 | 510 ± 21 | 486 ± 15 |
| ISA, 2.4 mM/kg | 6 | 524 ± 33 | 502 ± 17 | 458 ± 33 |
| EtOH + ISA, 2.4 mM/kg | 6 | 635 ± 21*** | 559 ± 15* | 557 ± 22** |

The dose of ethanol was 2 g/kg i.p., given 30 min after ISA. Data represent units/g protein/min, means ± SEM. Comparisons were by unpaired Student's t-test vs. EtOH group: *p<0.05, **p<0.02, ***p<0.005.

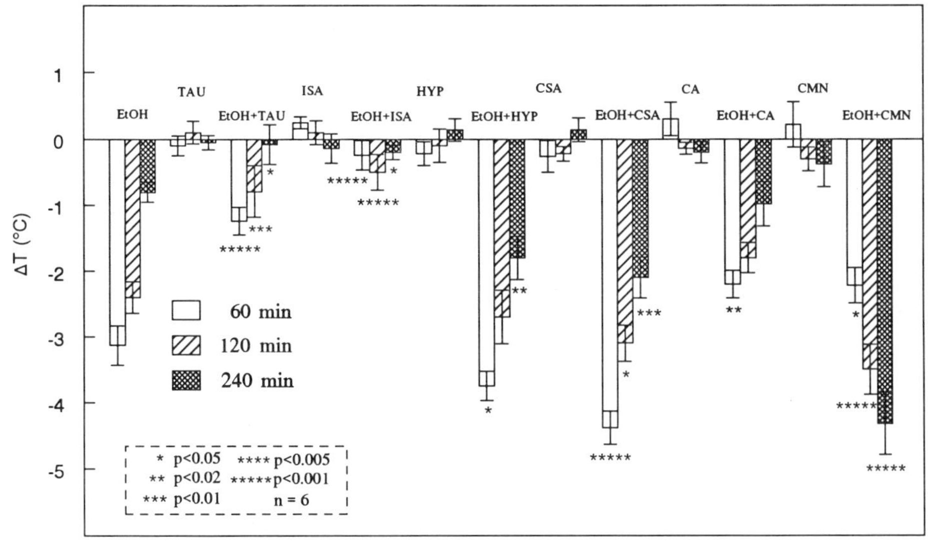


**Figure 4.** Time-course of the effects of taurine and biochemically-related compounds (2.4 mM/kg, i.p. each) on the hypothermic response to a 2 g/kg i.p. dose of ethanol. Values are reported as means ± SEM (n = 6). Comparisons were by unpaired Student's t-test vs. ethanol. Key: EtOH = ethanol, TAU = taurine, ISA = isethionic acid, CSA = cysteinesulfinic acid, CA = cysteic acid, and CMN = cysteamine.

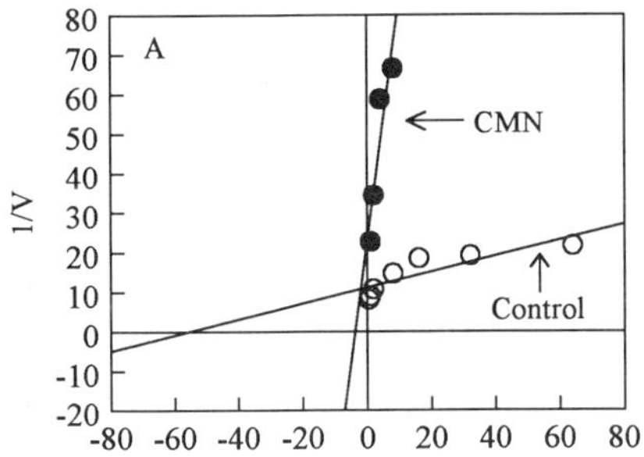


| Treatment | $V_{max}$ ($\Delta A_{340}$) | $K_m$ (M) |
|---|---|---|
| Control | 0.110 | 0.027 |
| CMN | 0.040 | 0.230 |
| CSA | 0.050 | 0.100 |
| HYP | 0.060 | 0.024 |

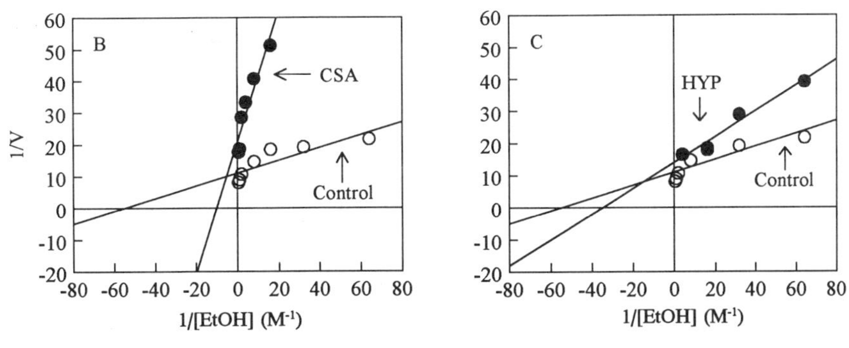


**Figure 5.** Double reciprocal plots of the rate of the enzymatic reaction of yeast ADH as a function of substrate concentration in the absence (Control) and presence of 1 M of a sulfur-containing compound found to increase BEC. Data shown are the mean of duplicate experiments. Key: CMN = cysteamine, CSA = cysteinesulfinic acid, and HYP = hypotaurine.

In summary, the present study has determined that among those compounds lying along the biosynthetic and metabolic pathway of taurine, only those exhibiting a sulfonic acid will positively affect the hepatic metabolism of ethanol and, hence, the clearance.of ethanol from the circulation. These effects appear to be dose related and sensitive to single structural modifications, such as the replacement of the sulfonic acid functionality by either a sulfinic acid or a sulfhydryl acid group. In this regard, taurine is rather unique in that it possesses the structural features that translate into the maximal effects among those sulfur-containing compounds examined in this study.

**Table 3**. Effect of taurine (TAU) on hepatic aldehyde dehydrogenase (ALDH) activity of rats receiving a fixed dose of ethanol (EtOH).

| | | Liver ADH activity | | |
|---|---|---|---|---|
| Treatment | n. | 60 min | 120 min | 240 min |
| Saline | 6 | 1875 ± 52 | 2178 ± 66 | 2145 ± 57 |
| EtOH | 6 | 1949 ± 34 | 2390 ± 34 | 2078 ± 48 |
| TAU, 0.4 mM/kg | 6 | 1795 ± 33 | N.D. | N.D. |
| TAU, 1.2 mM/kg | 6 | 1905 ± 38 | 2298 ± 48 | 2038 ± 79 |
| TAU, 2.4 mM/kg | 6 | 1878 ± 41 | N.D. | N.D. |
| EtOH + TAU, 0.4 mM/kg | 6 | 2081 ± 44 | N.D. | N.D. |
| EtOH + TAU, 1.2 mM/kg | 6 | 2107 ± 54* | 2628 ± 77* | 2376 ± 69*** |
| EtOH + TAU, 2.4 mM/kg | 6 | 2189 ± 51** | N.D. | N.D. |

The dose of ethanol was 2 g/kg i.p., given 30 min after TAU. Data represent units/g protein/min, means ± SEM. N.D. = not determined. Comparisons were by unpaired Student's t-test vs. EtOH group: *p<0.05, **p<0.005, ***p<0.001.

**Table 4**. Effect of isethionic acid (ISA) on hepatic aldehyde dehydrogenase (ALDH) activity of rats receiving a fixed dose of ethanol (EtOH).

| | | Liver ALDH activity | | |
|---|---|---|---|---|
| Treatment | n | 60 min | 120 min | 240 min |
| Saline | 6 | 1962 ± 54 | 1872 ± 42 | 1796 ± 88 |
| EtOH | 6 | 2069 ± 49 | 2078 ± 34 | 2174 ± 48 |
| ISA, 2.4 mM/kg | 6 | 2028 ± 79 | 1895 ± 58 | 1832 ± 94. |
| EtOH + ISA, 2.4 mM/kg | 6 | 2298 ± 52*** | 2335 ± 51* | 2498 ± 84** |

The dose of ethanol was 2 g/kg i.p., given 30 min after ISA. Data represent units/g protein/min, means ± SEM. Comparisons were by unpaired Student's t-test vs. EtOH group: *p<0.05, **p<0.02, ***p<0.01.

## REFERENCES

1. Aragon, C.M.G. and Amit, Z., 1993, Taurine and ethanol-induced conditioned taste aversion, *Pharmacol. Biochem. Behav.*, 44:263-266.
2. Aragon, C.M.G., Trudeau, L.E., and Amit, Z., 1992, Effect of taurine on ethanol-induced changes in open-field locomotor activity, *Psychopharmacology*, 107:337-340.
3. Awapara, J. and Wingo, W.J., 1953, On the mechanism of taurine formation from cysteine in the rat, *J. Biol. Chem.*, 203:189-194.

4. Boggan, W.O., Medberry, C., and Hopkins, D.H., 1978, Effect of taurine on some pharmacological properties of ethanol, *Pharmacol. Biochem. Behav.*, 9:469-472.
5. Dolara, P., Marino, P., and Buffoni, F., 1973, Effect of 2-aminomethanesulphonic acid (taurine) and 2-hydroxyethane sulphonic acid (isethionic acid) on calcium transport by rat liver mitochondria, *Biochem. Pharmacol.*, 22:2085-2094.
6. Eriksson, C.J.P., Sippel, H.W., and Forsander, O.A., 1977, The determination of acetaldehyde in biological samples by headspace gas chromatography, *Anal. Biochem.*, 80:116-124.
7. Federici, G., Ricci, G., Santoro, L., Antonucci, A., and Cavallini, D., 1980, Cysteamine pathway of taurine biosynthesis, in: "Natural Sulfur Compounds", Cavallini, D., Gaull, G.E., and Zappa, V., eds., Plenum Press, New York, pp. 187-193.
8. Ferko, A.P., 1991, Cysteine sulfinic acid can enhance the central depressant effect of ethanol in mice, *Pharmacol. Biochem. Behav.*, 39:653-657.
9. Ferko, A.P., 1987, Ethanol-induced sleep time: interaction with taurine, *Pharmacol. Biochem. Behav.*, 27:235-238.
10. Harada, K., 1982, Effects of L-cysteine on alcohol metabolism - *in vivo* and *in vitro* study, *Jpn. J. Alcoh. Stud. Drug Depend.*, 17:S65-S66.
11. Hirayama, C., Kishimoto, Y., Wakushima, T., and Murawaki, Y., 1983, Mechanism of the protective action of thiol compounds in ethanol-induced liver injury, *Biochem. Pharmacol.*, 32:321-325.
12. Iida, S. and Hikichi, M., 1976, Effect of taurine on ethanol-induced sleeping time in mice, *J. Stud. Alcoh.*, 37:19-26.
13. Jacobsen, J.G. and Smith, L..H., Jr., 1963, Comparison of decarboxylation of cysteine sulphinic acid-1-$^{14}$C and cysteic acid-1-$^{14}$C by human, dog, and rat liver and brain, *Nature*, 200:575-577.
14. Lau-Cam, C.A., Theofanopoulos, V., Thadikonda, K.P., and DeNicola, M., 1992, Effect of fructose on the hypothermic response to ethanol in the rat, *Res. Commun. Subst. Abuse*, 13:203-217.
15. Lau-Cam, C.A., Theofanopoulos, V., and Thadikonda, K.P., 1994, Antagonistic effect of ascorbate on ethanol-induced hypothermia in the rat, *Res. Commun. Subst. Abuse,* 15:33-50.
16. McBroom, M.J., Elkhawad, A.O., and Dlouha, H., 1986, Taurine and ethanol-induced sleeping time in mice: route and time course effects, *Gen. Pharmacol.*, 17:97-100.
17. Messiha, F.S., 1979, Taurine, analogues and ethanol elicited responses, *Brain Res. Bull.*, 4:603-607.
18. Messiha, F.S., 1987, Differential response of NADP-linked hepatic aldehyde dehydrogenase toward taurine: implication for behavioural effects of ethanol, *J. Appl. Toxicol.*, 70:193-196.
19. Pasantes-Morales, H., Chatagner, F., and Mandel, P., 1980, Synthesis of taurine in rat liver and brain *in vivo*, *Neurochem. Res.*, 5:441-451.
20. Peck, E.J., Jr. and Awapara, J., 1967, Formation of taurine and isethionic acid in rat brain, *Biochim., Biophys. Acta*, 141:499-506.
21. Rachamin, G., Macdonald, J.A., Wahid, S., Clapp, J.J., Khanna, J.M., and Israel, Y., 1980, Modulation of alcohol dehydrogenase and ethanol metabolism by sex hormones in the spontaneously hypertensive rat, *Biochem. J.*, 180:483-490.
22. Riley, V., 1960, Adaptation of orbital bleeding technique to rapid serial blood studies, *Proc. Soc. Exp. Biol. Med.*, 104;751-754.
23. Sturman, J.A. and Fellman, J.H., 1983, Methionine metabolism in the rat: accumulation of hypotaurine after partial hepatectomy, in: "Sulfur Amino Acids: Biochemical and Clinical Aspects", Kuriyama, K., Huxtable, R.J., and Iwata, H., eds., Alan R. Liss, New York, pp. 435-447.
24. Theofanopoulos, V. and Lau-Cam, C.A., 1996, Modification of ethanol metabolism by taurine and its impact on ethanol-induced hypothermia in the rat, *Pharm. Res. (Suppl.)* 40:67P.
25. Vallee, B.L. and Hoch, F.L., 1955, Zinc, a component of yeast alcohol dehydrogenase *Proc. Nat. Acad. Sci.*, 41:327-338.
26. Vasiliou, V. and Marselos, M., 1989, Tissue distribution of inducible aldehyde dehydroxygenase activity in the rat after treatment with phenobarbital or methylcholanthrene, *Pharmacol. Toxicol.*, 64:39-42.
27. Wollenberger, A., Ristau, O., and Schoffa, G., 1960. A simple technic for extremely rapid freezing of large pieces of tissue, *Pflueger Arch. Ges. Physiol.*, 270:399-412.

# MODIFICATION BY TAURINE OF THE METABOLISM AND HYPOTHERMIC EFFECT OF ETHANOL IN THE RAT

Vasilios Theofanopoulos and Cesar A. Lau-Cam

St. John's University
College of Pharmacy and Allied Health Professions
8000 Utopia Parkway
Jamaica, New York 11439

## INTRODUCTION

Taurine is a β-aminosulfonic acid which occurs ubiquitously both among animal species and in animal tissues[12]. Although the role of taurine in the conjugation of bile acids is well established, this compound has also demonstrated a myriad of nonmetabolic actions which are exerted throughout the animal body, notably in the heart, brain, reproductive system, muscle and eye[8].

In addition to its antioxidant[7,20,26], hepatoprotective[25], calcium level modulating[6-8], neuromodulating[6,13], membrane stabilizing and osmoregulatory actions[8], evidence has accumulated in recent years to suggest that taurine can also modify certain ethanol-related behavioral[1,2,18,19] and pharmacological[3,9-11,16,20,29,30] effects upon interacting with ethanol *in vivo*. However, despite the considerable amount of work that has been devoted to this subject, the ensuing results are in several instances open to question or even conflict. In laboratory animals, for example, it is reported that taurine will antagonize[3,10-12,17,20], reinstate[6], potentiate[5,16,17], or exert no effect[18,19] on ethanol-induced narcosis. Likewise, the effects of taurine on blood ethanol clearance and hepatic ethanol metabolism have varied from laboratory to laboratory. In one study, for example, the administration of taurine (0.5 g/kg p.o.) to rats dosed 30 min prior to ethanol administration (1.5 g/kg p.o.) was found to decrease blood and liver acetaldehyde, to insignificantly elevate liver ethanol, but to have little effect on blood ethanol[30]. In another study, treating human volunteers with taurine (20 mg/kg p.o.) 1 hour before and 1 hour after a single dose of ethanol (0.5 g/kg p.o.) lowered blood acetaldehyde but not blood ethanol[29]. Furthermore, the blood levels of ethanol-treated mice and rats have been found to remain unchanged[1,2,5,6,16,18,19] or to decrease in a dose-related manner[24] in the presence of taurine. Such a variability of results is probably the result of the great diversity of experimental conditions and test systems used, including the age, species and strain of the animal, the dose of ethanol and taurine, the strength of the ethanol solution, the dosing schedule, and the route of administration.

The aim of the present investigation was to reevaluate the actions of taurine on

*Taurine 3*
Edited by Schaffer *et al.*, Plenum Press, New York, 1998

hepatic metabolism and systemic clearance of ethanol by using a wider range of doses of this compound than previously reported, and to determine if these effects can be modified by certain inhibitors of ethanol metabolism. As an additional index of the effects of taurine on ethanol clearance, the changes in core temperatures elicited by a single hypothermic dose of ethanol were concurrently measured.

## MATERIALS AND METHODS

### Chemicals

Taurine, 4-methylpyrazole and β-alanine were obtained from Sigma Chemical Co. (St. Louis, MO). Absolute ethanol was from United States Industrial and Chemical Co. (Louisville, KY). Yeast ADH (365 units/mg of protein), equine liver ADH (1.2 units/ mg of protein, yeast ALDH (24 units/mg of protein) and β-$NAD^+$ were purchased from Sigma Chemical Co. All other reagents were of analytical reagent grade and were from Aldrich Chemical Co. (Milwaukee, WI) or from Fluka (Ronkonkoma, NY).

### Animals

Male Sprague-Dawley rats rats, 200-225 g in weight, were purchased from Taconic Farms (Germantown, NY). The animals were housed in groups of 6 in plastic cages, in a room maintained at 23°±1°C and constant humidity, on a 12 hr light-12 hr dark cycle. During a 5 day acclimation period the rats had free access to a commercial rat diet (Purina Rat Chow, Ralston Purina, St. Louis, MO) and tap water. All experiments were conducted after the rats had been fasted overnight for 12 hr. Control and treatment groups consisted of 6 rats each.

### Treatment Solutions

All treatment solutions were prepared in physiological saline to contain: ethanol, 50% w/v; taurine, 4% w/v; 4-methylpyrazole, 5% w/v; and β-alanine, 4% w/v.

### Treatment Schedules, Doses and Routes

All pretreatments were administered intraperitoneally (i.p.) or orally (p.o.) 30 min before ethanol excepting 4-methylpyrazole, which was administered 15 min before ethanol. The doses were: ethanol, 2 g/kg i.p. and 4 g/kg p.o.; taurine, 0.2, 0.4, 0.8, 1.2, 2.4 mM/kg i.p. and 0.6, 1.8 mM/kg p.o.; 4-methylpyrazole, 1.2 mM/kg i.p.; and β-alanine, 1.2 mM/ kg, i.p..

### Temperature Measurements

Core temperatures were measured rectally with a Model 700C digital thermistor thermometer (Jenco Electronics Ltd., Taiwan, ROC) fitted with a YSI 702A flexible vinyl probe (Yellow Springs Instrument Co., Inc., Yellow Springs, OH). The probe was lightly lubricated with petrolatum and inserted 3 cm into the rectal cavity. A reading was recorded after allowing the temperature to stabilize for at least 30 sec. The initial temperature was the average of two readings taken 30 min before a dosing with ethanol. Thereafter, readings were taken at 30 min intervals for up to 240 min, unless otherwise indicated. Readings were converted to ΔT (°C) values, i.e., the differences from baseline values.

## Blood and Liver Ethanol

Blood samples for the determination of ethanol were collected at 60, 120 and 240 min into stoppered polyethylene tubes containing disodium EDTA according to the orbital sinus technique of Riley[23]. The samples were stored on ice, and analyzed within 30 min after collection. Rats were sacrificed by decapitation and their livers were surgically removed using the patch-clamp technique of Wollenberger *et al*[31]. A portion of frozen liver (about 2 g) was reduced to a powder with the aid of a mortar and pestle previously cooled with liquid nitrogen, and then homogenized inside an ice bath with a mixture of ice-cold 0.6 M perchloric acid - 40 mM thiourea using an electric homogenizer (Tissue Tearor, Biospec Products, Bartlesville, MD), The resulting suspension was centrifuged at 4000 x g at 4°C for 15 min, and a 0.2 mL portion of the clear supernatant was analyzed for its ethanol content by the head-space gas chromatographic method of Eriksson *et al.*[4] using n-propanol as an internal standard and helium as the carrier gas.

## Hepatic ADH and ALDH Activities

For the assay of ADH activity, a portion of frozen liver (about 1 g) was thawed, rinsed twice with physiological saline at 4°C, minced, and homogenized with 9 volumes of 1% Triton X-100 using a Potter-Elvehjem homogenizer. The suspension was centrifuged at 27,000 x g at 4°C for 60 min, and the enzyme activity was measured at 22°C in an aliquot of the clear supernatant by following the rate of formation of NADH from $NAD^+$ spectrophotometrically at 340 nm according to Rachamin *et al.*[22]. For the assay of ALDH, the sample was prepared in a similar manner except that the homogenization medium was ice-cold 0.25 M sucrose and the centrifugation step was at 10,000 x g at 4°C for 20 min. The reduction of $NAD^+$ at 37°C was followed spectrophotometrically at 340 nm according to Vasiliou and Marselos[28]. The protein content of the liver supernatants was determined automatically with the Abbott Vision Analyzer using a commercial colorimetric kit (Kit 1430-10, Abbott Laboratories, Abbott Park, IL).

## Assay of Yeast ADH

The activity of yeast ADH was determined spectrophotometrically by the method of Vallee and Hoch[25].

## Statistics

Results are reported as mean ± SEM values. Data were analyzed by unpaired Student's t-test. Differences were considered statistically significant when $p<0.05$.

# RESULTS

The intraperitoneal administration of graded doses of taurine (0.2-1.8 mM/kg) prior to a hypothermic (2 g/kg) dose of ethanol resulted in a dose-related effect on the blood ethanol concentration (BEC). As shown in Figure 1, doses of taurine ≥1.2 mM/kg caused a significant decrease in BEC throughout the experimental period, an effect that was also achieved when taurine was given orally (Fig. 2). In contrast, a 0.4 mM/kg dose of taurine elevated BEC.

In rats acutely dosed with ethanol, we have earlier ascertained that the lowering of BEC by an exogenous compound is paralleled by an attenuation of the hypothermic response to an acute dose of ethanol[15,24]. The present study confirms this conclusion. For instance, intraperitoneal doses of taurine ≥ 1.2 mM/kg were found to both attenuate the hypothermia by ethanol and to accelerate the return of the core temperatures to initial values (Fig. 2). However, lower doses of the β-amino acid either accentuated or had no effect on ethanol-induced hypothermia (Figs. 2 and 3). The effects of the route of administration were examined by repeating the same experiments with oral doses of both taurine (0.6 and 1.8 mM/kg) and ethanol (4 g/kg). In this case the results were similar to those attained with an intraperitoneal, 1.2 mM/kg, dose of taurine (Fig. 1 and 4). By itself, taurine was without effect on the core temperatures at all doses tested.

To determine if taurine lowers the BEC upon its entry into the liver, rats were pretreated with β-alanine, a compound that depletes hepatic taurine through competition for the same active cellular uptake mechanism. As shown in Fig. 5, β-alanine attenuated the effect of taurine on the BEC caused a further elevation of BEC in animals not receiving taurine and abolished the effects of taurine on rectal temperature (Fig. 6). Furthermore, the inability of taurine to lower BEC and alter rectal temperature upon pretreatment with the ADH inhibitor, 4-methylpyrazole, (Figs. 7 and 8, respectively)

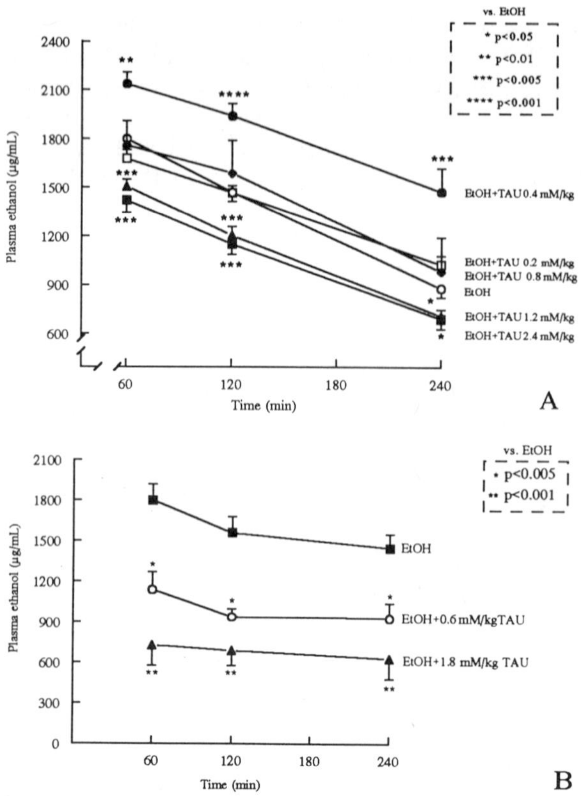


**Figure 1.** Effect of graded doses of taurine (TAU) on blood ethanol levels of rats given a hypothermic dose of ethanol (EtOH) 30 min later. Values represent means ± SEM (n = 6). In (A) both EtOH (2 g/kg) and TAU were administered i.p.; in (B) both EtOH (4 g/kg) and TAU were administered p.o.

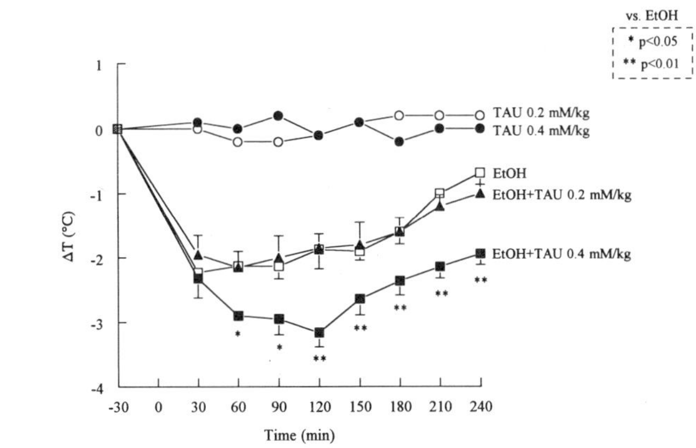


**Figure 2.** Effect of an i.p. treatment with taurine (TAU) on the core temperature of rats given a hypothermic, 2 g/kg, i.p., dose of ethanol (EtOH) 30 min later. Values represent ± SEM (n = 6).

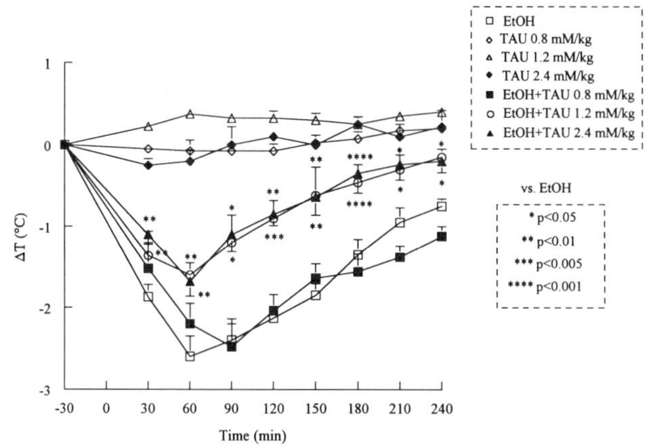


**Figure 3.** Effect of an i.p. treatment with taurine (TAU) on the core temperature of rats given a hypothermic, 2 g/kg, i.p., dose of ethanol (EtOH) given 30 min later. Values represent means ± SEM (n=6).

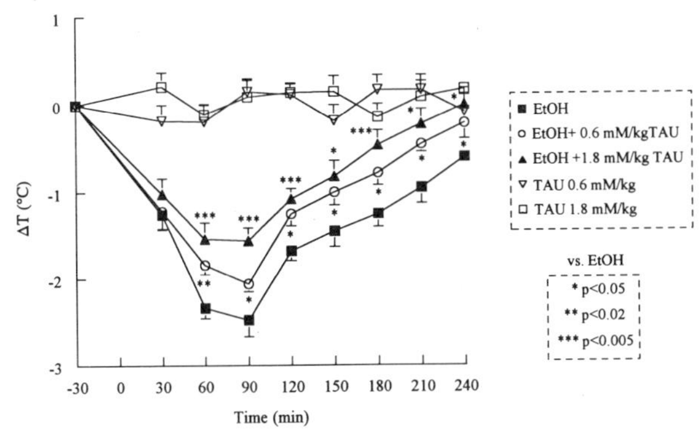


**Figure 4.** Effect of an oral pretreatment with taurine (TAU) on the hypothermic response of rats to an oral, 4 g/kg, dose of ethanol (EtOH). Values are reported as means ± SEM (n=6).

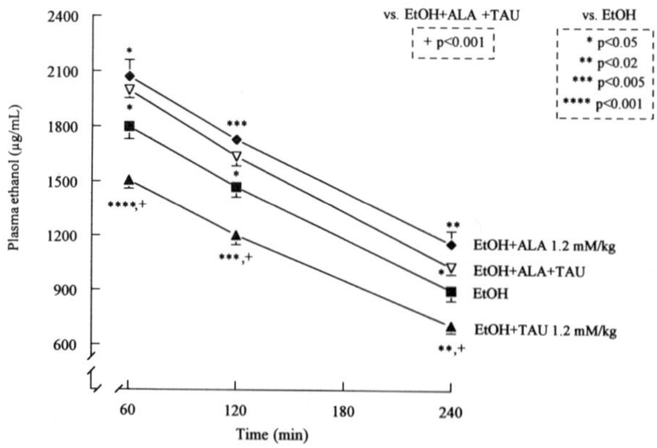


**Figure 5.** Effect of β-alanine (ALA) on the blood ethanol level of rats receiving a 2 g/kg, i.p. dose of ethanol (EtOH) with and without a 30 min pretreatment with taurine (TAU). Values are reported as means ± SEM (n = 6).

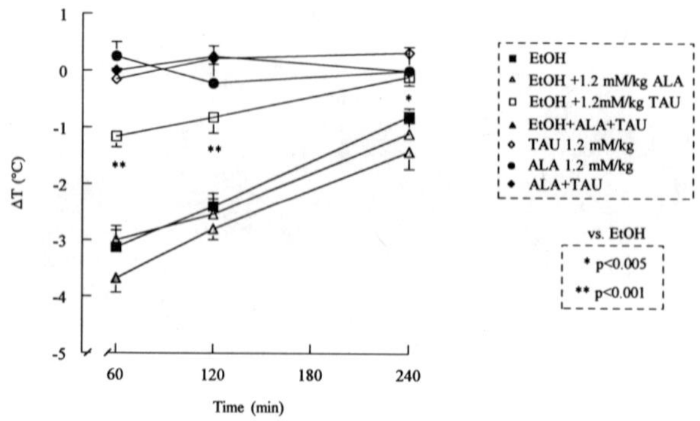


**Figure 6.** Effect of β-alanine (ALA) on the rectal temperature of rats receiving a dose, 2 g/kg, i.p., of ethanol (EtOH) with and without a 30 min pretreatment with taurine (TAU). Values are reported as means ± SEM (n = 6).

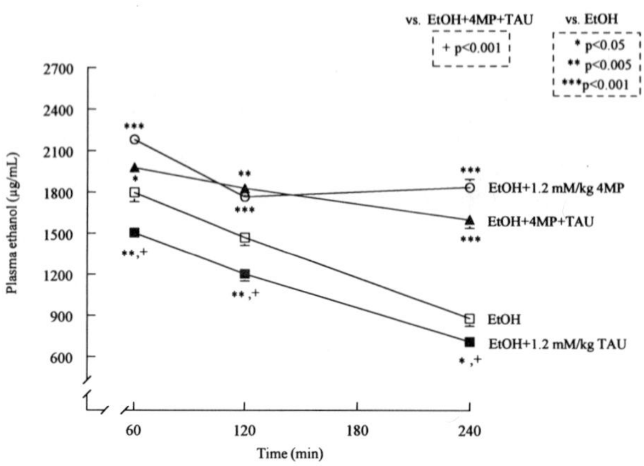


**Figure 7.** Effect of 4-methylpyrazole (4MP) on the blood ethanol levels of rats receiving a 2 g/kg, i.p. dose of ethanol (EtOH) with and without a 30 min pretreatment with taurine (TAU). Values are reported as means ± SEM (n=6).

indicates that taurine is stimulating the clearance of ethanol through a direct effect on hepatic ADH.

This view is supported by the results of three additional experiments, namely the significant and protracted decrease in hepatic ethanol caused by a 1.2 mM/kg dose of taurine (Fig. 9); the parallel increase in hepatic ADH activity at 60 min after ethanol (Fig. 10), an effect that was maintained for at least another 180 min (+13 %, $p<0.01$), but which was abolished by a pretreatment with 4-methylpyrazole (Fig. 10); and the activation of yeast ADH by taurine under *in vitro* conditions ($V_{max}$=0.150, $K_m$=0.011 M with taurine; $V_{max}$=0.110, $K_m$=0.027 M with no taurine). Whereas doses of taurine <1.2 mM/kg were found not to alter the hepatic activity of ADH, at 0.4 mM/kg it acted as an inhibitor (Fig. 10). In addition, the activity of hepatic ALDH was significantly higher in the presence of taurine at 60 min after ethanol (Fig. 11) and remained elevated (+18 %, $p<0.05$) for at least another 180 min.

## DISCUSSION

Since the appearance of reports by Iida and Hisada[10] describing protection by taurine against the depressant (anesthetic) action of ethanol in mice and rabbits, numerous studies have been devoted to confirm this central effect[3,5,6,10-12,16-19,28,29] or to uncover a role for this compound on other physiological actions of ethanol[1-3]. After evaluating the data emanating from these studies, it is apparent that for any experimental attempt to demonstrate an *in vivo* protective effect by taurine, particularly as it relates to toxicity stemming from the metabolism of systemically administered ethanol, it must take into account the level at which taurine is present in the liver. Indeed, the concept of a critical threshold level for liver taurine above which this compound will offer protection against hepatocellular damage in animals and humans has been advanced by Timbrell *et al.*[25].

Earlier attempts to find a correlation between the effects of taurine on ethanol-induced central depression and BEC might have relied on doses of taurine that were far too low to reach the threshold level, or used a route of administration not conductive to such a critical concentration. A case in point is the study by Boggan *et al.*[3] who, after coad-ministering ethanol (3.5 g/kg i.p.) and taurine (45 mg/kg i.p.) to mice to establish a relationship between taurine and ethanol-induced hypothermia and sleeping time, found

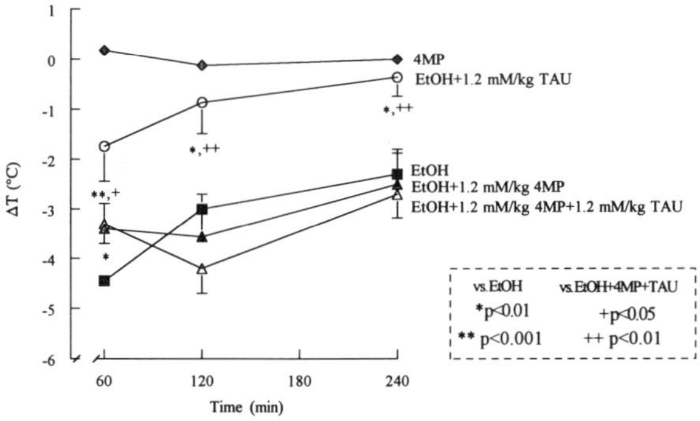


**Figure 8.** Effect of 4-methylpyrazole (4MP) on the rectal temperatures of rats receiving a 2 g/kg, i.p. dose of ethanol (EtOH) with and without a 30 min pretreatment with taurine (TAU). Values are reported as means ± SEM (n = 6).

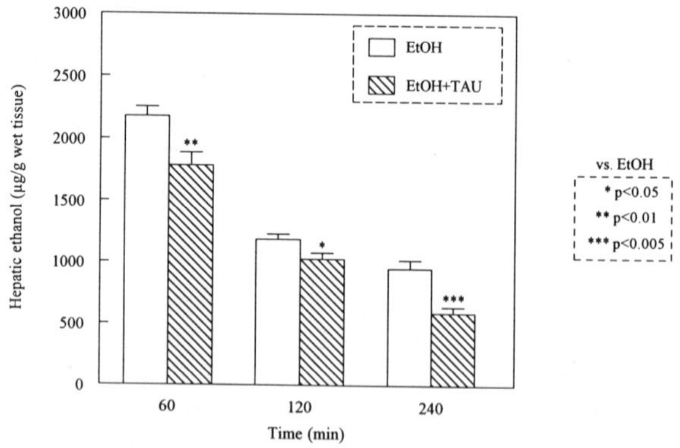


**Figure 9.** Temporal changes in hepatic ethanol levels after the administration of ethanol (EtOH, 2 g/kg, i.p.) with and without a 30 min pretreatment with taurine (TAU, 1.2 mM/kg, i.p.). Values are reported as means ± SEM (n = 6).

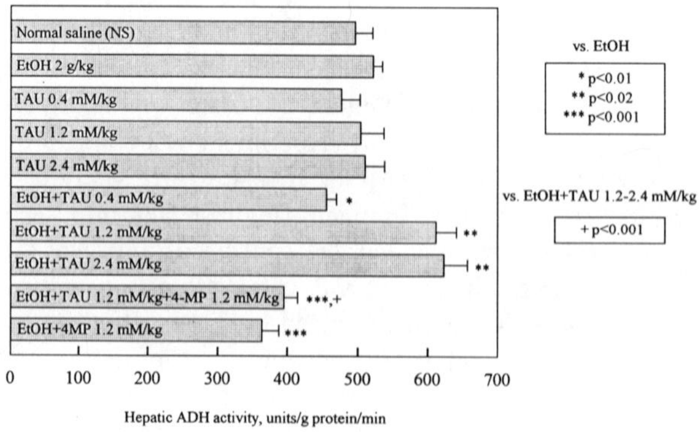


**Figure 10.** Effect of various doses of taurine (TAU) on the 60 min post-ethanol hepatic ADH activity of rats treated with ethanol (EtOH) and ethanol + 4-methylpyrazole (4MP). Values are given as means ± SEM (n = 6).

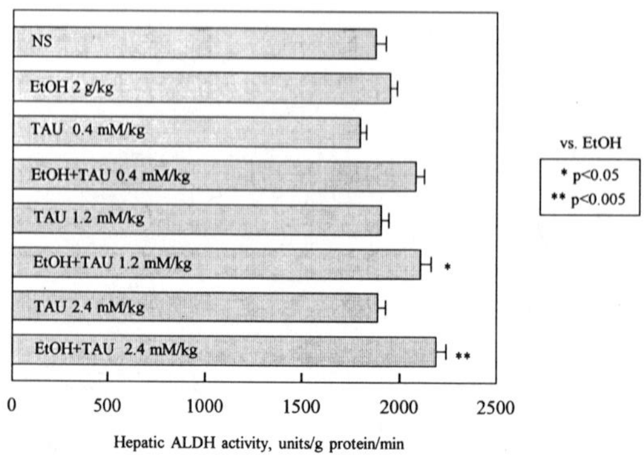


**Figure 11.** Effect of various doses of taurine (TAU) on the 60 min post-ethanol hepatic ALDH activity of rats treated with ethanol (EtOH). Values are given as means ± SEM (n = 6).

taurine to fail to antagonize both the central effects of ethanol and the elevations in BEC. Similarly, Ferko and Bobyock[6] found that an intracerebroventricular dose of taurine (7.5, 15 or 25 μM/kg) given to short sleep mice after ethanol (5.3 g/kg i.p.) was incapable of modifying the onset and duration of the sleep time and BEC at all the doses tested. The same results were obtained in long sleep mice, except that taurine at 15 μM/kg and 25 μmol/kg was able to cause a modest (13-15%) decrease in BEC[6]. To test the action of taurine on open-field locomotor activities due to ethanol, Aragon *et al.*[1] treated mice first with varying doses of taurine (30, 45, or 60 mg/kg i.p.) and next with ethanol (1, 2 or 3 g/kg i.p.). These authors found that while taurine can either lower or increase motor excitation depending on how much ethanol is given, it will not modify BEC. Furthermore, when taurine (45 mg/kg i.p.) was evaluated for its action on ethanol-related conditioned taste aversion (CTA) in rats, it was found that it will either enhance, block or show no effect on CTA depending on the dose of ethanol used, but again without altering BEC[2].

The present study demonstrates that taurine can significantly attenuate the hypothermic response to ethanol. This effect appears to be related to the ability of taurine to promote the clearance of ethanol from the circulation by directly stimulating the hepatic activity of the ADH-ALDH pathway for the metabolic disposal of ethanol. In accordance with the idea of a critical hepatic threshold concentration, doses of taurine ≥1. 2mM/kg are required for this compound to exert an appreciable stimulatory effect and only after undergoing cellular uptake. Moreover, peripherally administered taurine by itself will not alter the activity of the hepatic ADH-ALDH pathway unless ethanol is present. The same results have been previously documented for liver homogenates from mice[18] and rats[19] that had been treated with various doses of taurine and no ethanol. In addition, experiments in rats[29] and human volunteers[30] indicating that taurine will decrease blood acetaldehyde, but not BEC, when administered before an ethanol dose, might have been derived with doses of taurine that were too low to generate optimal hepatic concentrations of this amino acid.

In conclusion, the ability of taurine to promote the clearance of blood ethanol is a reflection of its direct action on the hepatic ADH-ALDH pathway for the metabolism of ethanol. In turn, this ability appears to be closely dependent on the availability of a critical amount of peripheral taurine for uptake into hepatocytes.

## REFERENCES


1. Aragon, C.M.G., Trudeau, L.-E,. and Amit, Z., 1992, Effect of taurine on ethanol-induced changes in open-field locomotor activity, *Psychopharmacology,* 107:337-340.
2. Aragon, C.M.G. and Amit, Z., 1993, Taurine and ethanol-induced conditioned taste aversion, *Pharmacol. Biochem. Behav.*, 44:263-266.
3. Boggan, W.O., Medberry, C., and Hopkins, D.H., 1978, Effect of taurine on some pharmacological properties of ethanol, *Pharmacol. Biochem. Behav.*, 9:469-472.
4. Eriksson, C.J.P., Sippel, H.W., and Forsander, O.A., 1977, The determination of acetaldehyde in biological samples by headspace gas chromatography, *Anal. Biochem.,* 80:116-124.
5. Ferko, A.P., 1987, Ethanol-induced sleep time: interaction with taurine and a taurine antagonist, *Pharmacol. Biochem. Behav.*, 27:235-238.
6. Ferko, A.P. and Bobyock, E., 1989, Effect of taurine on ethanol-induced sleep time in mice genetically bred for differences in ethanol sensitivity, *Pharmacol. Biochem. Behav.,* 31:667- 673.
7. Huxtable, R.J., 1989, Taurine in the central nervous system and the mammalian actions of taurine, *Prog. Neurobiol.*, 32:471-533.
8. Huxtable, R.J., 1992, Physiological actions of taurine, *Physiol. Rev.*, 72:101-163.
9. Huxtable, R.J. and Sebring, L.A., 1986, Towards a unifying theory for the actions of taurine, *Trends Pharmacol. Sci.,*, 481-485.

10. Iida, S. and Hisada, Y., 1974, Effect of taurine and other amino acids on the depressant effects of ethanol in mice, *Jpn. J. Pharmacol. (Suppl.),* 24:91.
11. Iida, H. and Hikichi, M., 1976, Effect of taurine on ethanol-induced sleeping time in mice, *J. Stud. Alcoh.*, 37:19-26.
12. Ikeda, H., 1977, Effects of taurine in alcohol withdrawal, *Lancet,* 2:509.
13. Jacobsen, J.G. and Smith, L.H., Jr., 1968, Biochemistry and physiology of taurine and taurine derivatives, *Pharmacol. Rev.*, 48:424-511.
14. Kuriyama, K., 1990, Taurine as a neuromodulator, *Fed. Proc.,* 39:2680-2684.
15. Lau-Cam, C., Theofanopoulos, V., Thadikonda, K.P., and DeNicola, M., 1992, Effect of fructose on the hypothermic effect to ethanol in the rat, *Res. Commun Subst. Abuse,* 13:203-217.
16. Mattucci-Schiavonne, L. and Ferko, A.P., 1985, Acute effects of taurine and a taurine antagonist on ethanol-induced central nervous system depression, *Eur. J. Pharmacol.,* 113:275-278.
17. McBroom, M.J., Elkhawad, A.O., and Dlouha, H., 1986, Taurine and ethanol-induced sleeping time in mice: route and time course effects, *Gen. Pharmacol.*, 17:97-100.
18. Messiha, F.S., 1979, Taurine, analogues and ethanol elicited responses, *Brain Res. Bull.*, 4:603-607.
19. Messiha, F.S., 1987, Differential response of NADPH-linked hepatic aldehyde dehydrogenase toward taurine: implication for behavioural effects of ethanol, *J. Appl. Toxicol.*, 7:193-196.
20. Nakano, N., 1991, Influence of diazepam and taurine on ethanol-induced sedation in man, *Jpn. J. Physiol. (Suppl.),* 40:67P.
21. Pasantes-Morales, H. and Cruz, C., 1984, Protective effect of taurine and zinc on peroxidation-induced damage in photoreceptor outer segments, *J. Neurosci. Res.,* 11:303-311.
22. Rachamin, G., MacDonald, J.A., Wahid, S., Clapp, J.J., Khanna, J.M., and Israel, Y., 1980, Modulation of alcohol dehydrogenase and ethanol metabolism by sex hormones in the spontaneously hypertensive rat, *Biochem. J.*, 186:483-490.
23. Riley, V., 1960, Adaptation of orbital bleeding technique to rapid serial blood studies, *Proc. Soc. Exp. Biol. Med.,* 104:751-754.
24. Theofanopoulos, V. and Lau-Cam, C., 1996, Modification of ethanol metabolism by taurine and its impact on ethanol-induced hypothermia in the rat, *Pharm. Res. (Suppl.),* 13: S-493.
25. Timbrell, J.A., Seabra, V., and Waterfield, C.J., 1995, The *in vivo* and *in vitro* protective properties of taurine, *Gen. Pharmacol.,* 26:453-462.
26. Trachtman, H., Futterweit, S., and Bienkowski, R.S., 1993, Taurine prevents glucose-induced peroxidation and increased collagen production in cultured rat mesangial cells, *Biochem. Biophys. Res. Commun.*, 191:759-765
27. Vallee, B.L. and Hoch, F.L., 1955, Zinc, a component of yeast alcohol dehydrogenase, *Proc. Nat. Acad. Sci.*, 41:327-338.
28. Vasiliou, V. and Marselos, M., 1989, Tissue distribution of inducible aldehyde dehydrogenase activity in the rat after treatment with phenobarbital or methylcholanthrene, *Pharmacol. Toxicol.,* 64:39-42.
29. Watanabe, A., Hobara, N., and Nagashima, H., 1985, Lowering of liver acetaldehyde but not ethanol concentration by pretreatment with taurine in ethanol-loaded rats, *Experientia*, 41:1421-1422.
30. Watanabe, A., Hobara, N., Kobayashi, M., and Nagashima, H., 1985, Effect of taurine on blood acetaldehyde elevation following alcohol ingestion, *Res. Commun. Subst. Abuse*, 6:247-250.
31. Wollenberger, A., Ristau, O., and Schoffa, G., 1960, A simple technic for extremely rapid freezing of large pieces of tissue, *Pflueger Arch. Ges. Physiol.,* 270:399-412.

# DIETARY TAURINE SUPPLEMENTATION REDUCES PLASMA AND LIVER CHOLESTEROL AND TRIGLYCERIDE LEVELS IN RATS FED A HIGH-CHOLESTEROL OR A CHOLESTEROL-FREE DIET

Taesun Park and Kyungshin Lee
Department of Food and Nutrition
Yonsei University
134 Shinchon-dong, Sudaemun-ku
Seoul, Korea 120-749

## ABSTRACT

The effects of dietary taurine supplementation on plasma and hepatic lipid levels and phospholipid profiles were evaluated in rats fed a high-cholesterol or a cholesterol-free diet. Four groups of male rats were fed one of the following diets for 5 weeks: cholesterol-free diet (CFD); high cholesterol diet (HCD); high cholesterol, high taurine diet (HCHTD); or high taurine diet (HTD). Rats fed a HCHTD had significantly lower plasma levels of total cholesterol (32% reduction), LDL-cholesterol (37% reduction) and triglyceride (43% reduction) than rats fed a HCD alone. Plasma concentrations of total cholesterol, LDL-cholesterol and triglyceride were also significantly reduced in rats fed a HTD compared to rats fed a CFD. Taurine supplementation to the HCD significantly reduced hepatic cholesterol (50% decrease) and triglyceride (30% decrease) levels in rats. Taurine supplementation to the CFD also significantly reduced the hepatic triglyceride concentration (43% decrease) and elevated hepatic free fatty acid levels (77% increase) compared to rats fed only a CFD. These results suggest that dietary taurine supplementation is both hypocholesterolemic and hypotriglyceridemic in rats whose body cholesterol status is high or normal.

## INTRODUCTION

Taurine was initially identified in the bile of an ox in 1827[24]. The most well-documented and the first established physiological function of taurine is bile acid conjugation in the liver[9]. The conjugation of bile acids with taurine or glycine results in the formation of water-soluble bile salts which are essential for micelle formation and fat absorption in the small intestine. The percentage of bile acid conjugated with taurine or glycine appears to depend on the type of species and the relative concentrations of taurine and glycine available in the hepatocyte for conjugation. In species that conjugate bile acids with both taurine and glycine, such as man and rats, taurine conjugation of bile acids has a preference over glycine conjugation. Early studies[4,13] reported that both human and rat liver homogenates exposed to equimolar concentrations of taurine and glycine in the incubation medium conjugated cholic acid predominantly with taurine.

Much attention has been paid to the effects of dietary taurine levels on bile acid metabolism in different species of animals, and the results from the conflicting data indicate that increased taurine conjugation of bile acids profoundly increases bile flow, biliary cholesterol secretion and lipid solubility[2,16,22,23]. In 1959, Herrmann[14] initially reported on the hypocholesterolemic effect of taurine in rats. However, this study was not confirmed for several decades by other investigators. Only recently, a few studies explored the effect of taurine supplementation on serum or liver lipid levels in rats[12,25], guinea pigs[5], rabbits[20] and cats[6] fed different dietary levels of cholesterol. The results were species-dependent.

In the present study we report that long-term dietary taurine supplementation reduces plasma and liver lipid concentrations in rats regardless of their cholesterol dietary intake level.

## MATERIALS AND METHODS

Male Sprague Dawley rats weighing 110-130 g were randomly divided into four groups, and fed one of the following diets for 5 weeks: cholesterol-free diet (CFD), high-cholesterol diet (HCD), high-cholesterol, high-taurine diet (HCHTD) or high-taurine diet (HTD). The CFD contained 650 g carbohydrate (starch : sucrose = 80:20), 180 g casein, 100g corn oil, 40 g mineral mix, 10 g vitamin mix, and 20 g carboxymethyl cellulose sodium salt per kg diet. The compositions of HCD, HCHTD or HTD were identical to CFD except that they included 15 g cholesterol/kg diet, 15 g cholesterol plus 15 g taurine/kg diet, or 15 g taurine/kg diet, respectively. Rats were housed in communal cages in a room with controlled temperature humidity (20-22 and 40-60%, respectively) and under a 12-h light:dark cycle. They were given free access to food and water.

Animals were food-deprived overnight at the end of the feeding period and killed under ether anesthesia between 11:00 and 12:00. Blood was collected from the abdominal vein into a heparinized syringe, and plasma samples were stored at -20°C. The liver was removed immediately after blood collection, quickly freeze-clamped in liquid nitrogen after being weighed, and stored at -80°C until analyses.

Plasma concentrations of total cholesterol, HDL-cholesterol, triglyceride, free fatty acid and total phospholipid were measured enzymatically using commercial kits (Eiken Chemical Co., Japan). The concentration of LDL cholesterol was calculated by the equation described by Friedewald et al[11]. Approximately two grams of liver were homogenized and the lipids were extracted as described by Folch et al[10]. The concentrations of cholesterol and triglyceride in the lipid extracts of liver homogenates were measured colorimetrically by the methods of Zak[26], and Neri and Frings[17], respectively. Total phospholipid content was determined by the inorganic phosphorus assay after ashing of the lipid extracts over a flame[1]. Free fatty acid concentration in the lipid extract was measured enzymatically using a kit (Eiken Chemical Co., Japan). The phospholipid profile in the lipid extract of liver homogenate was evaluated as described by Christie[8] with minor modifications. Major lipid classes were separated by TLC using petroleum ether-diethylether-acetic acid (80:20:2, v/v/v) as the developing solvent system. The band corresponding to phospholipid was scraped off the plate, and dissolved in methanol-benzene (1:4, v/v) for phospholipid extraction. An aliquot of the extract was again loaded on a TLC plate for the separation of phospholipid subclasses using chloroform-methanol-acetic acid-$H_2O$ (100:75:7:4, v/v/v/v) as the developing solvent. Each band was directly analyzed for inorganic phosphorus content as described earlier[1].

Plasma and whole blood samples were deproteinized using sulfosalicylic acid and analyzed for taurine concentration with an automatic amino acid analyzer based on ion-exchange chromatography (Biochrom 20, Pharmacia Biotech, Cambridge, England).

Values in the tables and figures are expressed as mean $\pm$ SEM of eight rats. Effects of dietary treatment (for CFD, HCD and HTHCD only) were analyzed by one-way analysis of variance at $P<0.05$, and differences between means were tested using Duncan's multiple range test. Differences between means for CFD and HTD were tested separately by the Student's t-test.

## RESULTS

Cumulative body weight gain, liver weight and liver to body weight ratio of rats fed the experimental diets are shown in Table 1. Rats fed the experimental diets for 5 weeks gained 150-184 g of body weight; there was no significant difference in the body weight gain among the groups. Liver weight and liver to body weight ratios were significantly increased in rats fed a HCD compared to the values obtained for rats fed a CFD. Taurine supplementation of the HCD significantly reduced these values to a level similar to that of the CFD group. Similarly, feeding a HTD significantly reduced the liver to body weight ratio compared to that for the rats fed a CFD.

Plasma levels of taurine and lipid, along with hepatic lipid concentrations of rats fed the experimental diets, are shown in Table 2. Plasma taurine concentrations were not influenced by the dietary level of cholesterol, but significantly elevated by taurine supplementation. Feeding a HCHTD for 5 weeks resulted in a 560% increase in plasma taurine concentration compared to the value for HCD rats; feeding a HTD resulted in a 497% increase in this value compared to those rats fed a CFD.

Rats fed a HCD exhibited an 81% increase in plasma total cholesterol concentration compared to the value for rats fed a CFD. Plasma total cholesterol levels were significantly reduced (32% reduction) in rats fed a HCHTD compared to those rats fed only a HCD. HDL-cholesterol levels were significantly decreased (46% decrease) in rats

**Table 1.** Liver weight and the liver weight to body weight ratio of rats fed experimental diets for 5 weeks

| | CFD[1] | HCD[2] | HCHTD[3] | HTD[4] |
|---|---|---|---|---|
| Body weight gain (g/5 wks) | 168 ± 8.2 | 169 ± 21 | 150 ± 27 | 184 ± 6.3 |
| Liver wt (g) | 9.4 ± 0.3[b] | 11.1 ± 0.4[a] | 9.6 ± 0.6[b] | 8.8 ± 0.4 |
| Body wt (kg) | 0.32 ± 0.01 | 0.31 ± 0.01 | 0.31 ± 0.01 | 0.33 ± 0.009 |
| Liver wt (g)/Body wt (kg) | 30.2 ± 0.9[b] | 35.5 ± 0.9[a] | 31.3 ± 1.1[b] | 27.0 ± 0.7* |

Values are means ± SEM of 8 rats.

[1] CFD, cholesterol-free diet; [2]HCD, high cholesterol diet; [3]HCHTD, high cholesterol, high taurine diet; [4]HTD, high taurine diet.

Different superscripts in the same row indicate significant differences among dietary groups (P< 0.05 for liver weight ; P<0.01 for liver wt/body wt).

* Significantly different from CFD group by the Student's t-test at P<0.05.

**Table 2.** Plasma and hepatic lipid and taurine levels in rats fed experimental diets.

| | CFD[1] | HCD[2] | HCHTD[3] | HTD[4] |
|---|---|---|---|---|
| PLASMA | | mmol/L | | |
| Taurine (ìmol/L) | 56.5 ± 4.5[b] | 56.0 ± 3.9[b] | 370.3 ± 26.1[a] | 337.1 ± 54.3 |
| Total chol. | 2.2 ± 0.14[b] | 3.9 ± 0.41[a] | 2.7 ± 0.28[b] | 1.33 ± 0.14*** |
| HDL chol. | 0.97 ± 0.06[a] | 0.52 ± 0.05[b] | 0.56 ± 0.03[b] | 0.71 ± 0.06** |
| LDL chol. | 0.63 ± 0.11[c] | 2.85 ± 0.36[a] | 1.75 ± 0.30[b] | 0.38 ± 0.10* |
| Triglyceride | 1.26 ± 0.12[a] | 1.40 ± 0.23[a] | 0.80 ± 0.10[b] | 0.60 ± 0.05*** |
| Free fatty acid | 1.39 ± 0.11 | 1.3 ± 0.11 | 1.3 ± 0.27 | 1.3 ± 0.1 |
| Phospholipid | 1.8 ± 0.15 | 2.0 ± 0.19 | 2.5 ± 0.30 | 2.3 ± 0.4 |
| LIVER | | μmol/g liver | | |
| Cholesterol | 6.2 ± 0.58[c] | 40.1 ± 5.1[a] | 20.3 ± 4.7[b] | 8.0 ± 1.29 |
| Triglyceride | 10.0 ± 1.1[b] | 19.8 ± 2.1[a] | 13.8 ± 2.3[b] | 5.8 ± 0.45** |
| Free fatty acid | 4.7 ± 0.6[b] | 9.7 ± 0.6[a] | 7.5 ± 1.4[a] | 8.3 ± 0.1*** |

Values are mean ± SEM of 8 rats.

[1] CFD, cholesterol-free diet; [2]HCD, high cholesterol diet; [3]HCHTD, high cholesterol, high taurine diet; [4]HTD, high taurine diet.

Different superscripts in the same row indicate significant differences among dietary groups ($P< 0.05$ for triglyceride; $P<0.001$ for the rest of the parameters).

*,**,*** Significantly different from CFD group by the Student's t-test at $P<0.05$, $P<0.01$ or $P<0.001$.

fed a HCD compared to those rats fed a CFD; taurine supplementation of the HCD did not correct this value. A HCD dramatically increased LDL-cholesterol levels in the plasma (284% increase) compared to the values for rats fed a CFD. Also, a HCHTD resulted in a 37% lower LDL-cholesterol concentration than the values obtained for the HCD group. Compared to the values for rats fed a CFD, the HCD increased the percentage of LDL-cholesterol (from 30% to 71%) and reduced the proportion of HDL-cholesterol (from 44% to 13%) to total cholesterol. This change in the proportion of LDL- and HDL-cholesterol, induced by high cholesterol consumption, was reversed by taurine supplementation. A HCHTD, compared to those rats fed a HCD alone, exhibited a reduced percentage of LDL-cholesterol (from 71% to 65%), and an elevation in the proportion of HDL-cholesterol (from 13% to 21%) to total cholesterol. Taurine supplementation of the CFD also significantly decreased the plasma concentration of total cholesterol, HDL cholesterol and LDL-cholesterol compared to values for rats fed the CFD alone. Feeding a HCD did not change plasma triglyceride levels in rats, but taurine supplementation of the HCD significantly lowered plasma triglyceride levels compared to the values for rats fed either a CFD (37% reduction) or a HCD (43% reduction). Rats fed the HTD also exhibited significantly lower plasma triglyceride levels (52% decrease) than those fed the CFD alone. There were no significant differences in plasma free fatty acid and total phospholipid levels among the dietary groups. Feeding the HCD significantly increased hepatic cholesterol (546% increase) and triglyceride (97% increase) levels compared to the rats fed a CFD (Table 2). Elevated hepatic cholesterol and triglyceride concentrations were again significantly reduced by dietary taurine supplementation (50% decrease in cholesterol, and 30% decrease in triglyceride concentrations). Dietary taurine supplementation of the CFD did not influence hepatic cholesterol levels, but resulted in a significant reduction in hepatic triglyceride levels (42% decrease), as well as a significant increase in hepatic free fatty acid levels (72% increase).

**Table 3.** Hepatic phospholipid composition of rats fed experimental diets

| | CD[1] | HCD[2] | HCHTD[3] | HTD[4] |
|---|---|---|---|---|
| | | μmol/g liver (%) | | |
| PC[5] | 6.5 ± 0.21 (32) | 7.0 ± 0.39 (34) | 6.6 ± 0.35 (33) | 5.5 ± 0.52 (30)* |
| PE[6] | 4.5 ± 0.13 (23)[a] | 3.7 ± 0.20 (18)[b] | 4.2 ± 0.23 (21)[a] | 3.5 ± 0.32 (19) ** |
| PS[7] + PI[8] | 3.4 ± 0.22 (17) | 3.8 ± 0.21 (18) | 4.1 ± 0.26 (20) | 4.7 ± 0.23 (26) *** |
| LPC[9] | 3.3 ± 0.13 (16) | 3.6 ± 0.16 (18) | 3.1 ± 0.21 (15) | 2.9 ± 0.17 (11) |
| SM[10] | 2.4 ± 0.17 (12) | 2.4 ± 0.10 (12) | 2.2 ± 0.13 (11) | 2.6 ± 0.09 (12) |
| Total | 20.1 ± 0.40 (100) | 20.5 ± 0.80 (100) | 20.1 ± 0.69 (100) | 19.1 ± 0.8 (100) |
| PC/PE | 1.4 ± 0.05[b] | 1.9 ± 0.08[a] | 1.6 ± 0.12[b] | 1.62 ± 0.08 |

Values are mean ± SEM of 8 rats.
[1] CFD, cholesterol-free diet; [2]HCD, high cholesterol diet; [3]HCHTD, high cholesterol, high taurine diet; [4]HTD, high taurine diet; [5]PC, phosphatidylcholine; [6]PE, phosphatidylethanolamine; [7]PS, phosphatidyl-serine; [8]PI, phosphatidylinositol; [9]LPC, lysophosphatidylcholine; [10]SM, sphingomyelin.
Different superscripts in the same row indicate significant differences among dietary groups (P<0.05 for PE ; P<0.01 for PC/PE).
*,**,*** Significantly different from the CFD diet group by the Student's t-test at P<0.05, P<0.01 or P<0.001.

Table 3 shows the effects of dietary taurine supplementation on the phospholipid composition of rat liver homogenates. Phosphatidylcholine (PC) and phosphatidylethanolamine (PE) constituted 30-34% and 18-23% of the total phospholipid pool in liver homogenates, respectively. Hepatic PE levels in rats fed a HCD were significantly decreased compared to the value for rats fed a CFD; taurine supplementation of the HCD significantly increased this value. However, taurine supplementation of the CFD significantly reduced hepatic PC and PE levels and elevated hepatic PS plus PI levels compared to rats fed only the CFD. Concentrations of other subclasses of phospholipids in rat liver homogenates were not changed by the dietary treatments. Rats fed the HCD had significantly higher PC/PE ratios in the liver homogenates than those rats fed a CFD. Animals fed the HCHTD had significantly lower hepatic PC/PE ratios than rats fed the HCD alone (Table 3).

## DISCUSSION

Dietary taurine supplementation at the level of 15 g/kg diet for 5 weeks markedly suppressed plasma total cholesterol levels in rats fed either the 1.5% cholesterol diet or the cholesterol-free diet. Interestingly, the hypocholesterolemic effects of dietary taurine supplementation in rats fed the HCD was mainly attributed to the dramatic decrease in LDL cholesterol levels; HDL cholesterol levels were unchanged. Previous studies using hypercholesterolemic rats[12,19] and Japanese quail[15] demonstrated that the addition of taurine (1-3%) to the diet significantly reduced plasma total cholesterol levels. However, our results contradict the reports of Petty et al.[20] and Yan et al.[25] showing no significant effect of taurine (administered at 0.2-1% in drinking water) on serum cholesterol levels in rabbits fed a 2% cholesterol diet or rats fed a regular diet. The magnitude of the hypotriglyceridemic effect of taurine supplementation was much larger in our study (43-52% reduction) than the data reported (19% decrease) by Gandhi et al.[12] on rats with hypercholesterolemia. In contrast, Yan et al.[25] demonstrated that serum triglyceride levels were not affected by either taurine supplementation or taurine depletion in rats.

Our results showing a significant depletion (50% reduction) in hepatic triglyceride and cholesterol stores after dietary treatment with taurine were basically in agreement with the previous conclusion that lipid metabolism in rat liver is affected by dietary taurine levels[12,25]. Again these effects of taurine supplementation were more pronounced in our study than previously reported. The effect of dietary taurine on lowering the levels of hepatic lipids appears to be species dependent since studies with guinea pigs[5] and cats[6], species with a limited capacity to synthesize taurine, demonstrated that the taurine supplemented diet increased hepatic triglyceride and cholesterol levels.

Marked accumulation of hepatic triglyceride and cholesterol stores appear to directly contribute to the significant increase in liver weight and the ratio of liver weight/body weight in rats fed the HCD compared to rats fed the CFD. Thus, the decrease in liver weight and the liver to body weight ratio in rats fed taurine supplemented diets is also most likely due to the depletion of hepatic triglyceride and cholesterol stores. These results support the ameliorating effect of taurine on the development of a fatty liver, as described previously in rats[12] and children with simple obesity[18].

The exact mechanisms underlying the cholesterol and triglyceride lowering effects of taurine are not clear at the present time. It has been reported that the decrease in hepatic triglyceride levels was likely due to taurine-mediated inhibition of diacylglycerol:acyl-CoA acyl transferase activity, the key enzyme in hepatic triglyceride synthesis[7]. The most intriguing explanation for the hypocholesterolemic effect of taurine appears to be the enhanced elimination of cholesterol from the body via increased bile acid synthesis. Taurine is known to stimulate bile acid synthesis and secretion of taurine-conjugated primary bile salts, leading to increased utilization of hepatic free cholesterol[12,16,22,25]. Dietary taurine supplementation in hamsters[3] and guinea pigs[16] increases the activity of hepatic cholesterol 7$\alpha$-hydroxylase, the rate-limiting enzyme in bile acid synthesis from cholesterol. *In vitro* studies[22] have shown that preincubation of human hepatoblastoma cells (HepG2) with taurine stimulated high affinity LDL receptor activity, which is known to be regulated in part by the rate of bile acid synthesis[21]. These results should encourage further studies regarding the effect of taurine on the activity and gene expression of important regulatory proteins involved in cholesterol and/or triglyceride metabolism in the liver.

## ACKNOWLEDGMENTS

This work was supported in part by a grant (#971-0603-018-1) from KOSEF (Korea Science and Engineering Foundation).

## REFERENCES

1. Bartlett, G.R., 1959, Phosphorus assay in column chromatography, *J. Biol. Chem.*, 234:466-468.
2. Batta, A. K., Salen, G., Shefer, S., Tint, G.S., and Dayal, B., 1982, The effect of acid and taurine supplementation on biliary bile acid composition, *Hepatology,* 2:811-816.
3. Bellentani, S., Pecorari, M., Cordoma, P., Marchegiano, P., Manenti, F., Bosisio, E., De Fabiani, E., and Galli, G., 1987, Taurine increases bile acid pool size and reduces bile saturation index in the hamster, *J. Lipid Res.*, 28:1021-1027.
4. Bremer, J., 1955, The conjugation of glycine with cholic acid and benzoic acid in rat liver homogenate. Bile acids and steroids, *Acta Chem. Scan.,* 9:268-271.
5. Cantafora, A., Mantovani, A., Masella, R., Mechelli, L., and Alvaro, D., 1986, Effect of taurine administration on liver lipids in guinea pig, *Experientia*, 42:407-408.
6. Cantafora, A., Blotta I., Ross, S.S., Hofmann, A.F., and Sturman, J.A., 1991, Dietary taurine content changes liver lipids in cats, *J. Nutr.,* 121:1522-1528.

7. Cantafora, A., Yan, C.C., Sun, Y., and Masella, R., 1994, Effects of taurine on microsomal enzyme activities involved in liver lipid metabolism of wistar rats, in: "Taurine in Health and Disease", Huxtable, R. and Michalk, D.V. eds., Plenum Press, New York, pp. 99-110.
8. Christie, W.W., 1982, The analysis of complex lipids, in: "Lipid Analysis," Christie, W.W., ed., Pergamon Press, New York, pp. 115-121.
9. Danielsson, H., 1963, Present states of research on catabolism and excretion of cholesterol, *Adv. Lipid Res.*, 1:335-385.
10. Folch, J., 1957, A simple method for the isolation and purification of total lipids from animal tissues, *J. Biol. Chem.*, 226:497-509.
11. Friedewald, W.T., Levy, R.I., and Fredrickson, D.S., 1972, Estimation of low density lipoprotein cholesterol in plasma, without use of the preparative ultracentrifuge, *Clin. Chem.*, 18:499-502.
12. Gandhi, V.M., Cherian, K.M., and Mulky, M.J., 1992, Hypolipidemic action of taurine in rats, Ind. J. Expt. Biol., 30:413-417.
13. Gottfries, A., Scherstén, T., and Ekdahl, P.H., 1966, The capacity of human liver homogantes to synthesize taurocholic and glycocholic acid *in vitro*, *Scand. J. Clin. Lab. Invest.*, 18:643-653.
14. Herrmann, R.G., 1959, Effect of taurine, glycine and â sitosterols on serum and tissue cholesterol in the rat and rabbit, *Circuration Res.*, 7:224-227.
15. Jackson, J.A., and Burns, M.J., 1974, Effects of cystine, niacin and taurine on cholesterol concentration in the Japanese quail with comments on bile acid metabolism, *Comp. Biochem. Physiol.*, 48A:61-68.
16. Kibe, A. , Wake, C. , Kuramoto, T., and Hoshita, T., 1980, Effect of dietary taurine on bile acid metabolism in guinea pigs, *Lipids*, 15:224-229.
17. Neri, B.P., and Frings, C.S., 1973, Improved method for determination of triglycerides in serum. *Clin. Chem.*, 19 (10):1201-1202.
18. Obinata, K., Maruyama, T., Hayashi, M., Watanabe, T., and Nittono, H., 1996, Effect of taurine on the fatty liver of chidren with simple obesity, *Adv. Exp. Med. Biol.*, 403:607-613.
19. Ogawa, H., 1996, Effect of dietary taurine on lipid metabolism in normocholesterolemic and hypercholesterolemic stroke-prone spontaneously hypertensive rats, *Adv. Exp. Med. Biol.*, 403:107-115.
20. Petty, M.A., Kintz, J., and DiFracesco, G.F., 1990, The effects of taurine on atherosclerosis development in cholesterol-fed rabbits, *Eur. J. Pharmacol.*, 180:119-127.
21. Spady, D.K., Stange, E.F., Bilhartz, L.E., and Dietschy, J.M., 1986, Bile acids regulate hepatic low density lipoprotein receptor activity in the hamster by altering cholesterol flux across the liver, *Proc. Natl. Acad. Sci.*, 83:1916-1920.
22. Stephan, Z.F., Armstrong, M.J., and Hayes, K.C., 1981, Bile lipid alterations in taurine-depleted monkeys, *Am. J. Clin. Nutr.*, 34:204-210.
23. Strasberg, S.M., Ilson, R.G., and Paloheimo, J.E., 1983, Bile salt-associated electrolyte secretion and the effect of sodium taurocholate on bile flow, J. Lab. Clin. Med., 101:317-326.
24. Tiedemann, F. and Gmelin, L., 1987, Einige neue bestandtheile der galle des fochsen., *Ann. Physik Chem.*, 9:326-327.
25. Yan, C.C., Bravo, E., and Cantafora, A., 1993, Effect of taurine levels on liver lipid metabolism : an *in vivo* study in the rat, *Proc. Soc. Expt. Biol. Med.*, 202:88-96.
26. Zak, B., 1957, Simple rapid microtechnique for serum total cholesterol, *Am. J. Clin. Pathol.*, 27:583-588.

# Part 7:
# Taurine and the Central Nervous System I: The Brain

# INTERRELATIONSHIP BETWEEN TAURINE AND GABA

Kinya Kuriyama and Tsuneichi Hashimoto[1]

[1] Department of Pharmacology
Kyoto Prefectural University of Medicine
Kawaramachi-Hirokoji
Kamikyo-ku
Kyoto 602, Japan

## INTRODUCTION

β-Aminoethanesulfonic acid (taurine) and γ-aminobutyric acid (GABA) are recognized as major inhibitory neuroactive amino acids distributed in large quantities in various central nervous system (CNS) areas. The role of GABA as an inhibitory neurotransmitter has been well established, whereas that of taurine is still unclear. In the case of GABA, the structure and function of two types of the GABA receptor, $GABA_A$ and $GABA_B$ receptors, have been identified, while the taurine receptor is not yet fully understood. However, some very convincing evidence concerning the functional interrelationships between taurine and GABA in the brain has accumulated.

### CNS Localization

The concentration of GABA in the brain is known to be 200 to 1000 times greater than those of other neurotransmitters, such as dopamine, norepinephrine, acetylcholine and serotonin. The highest concentration of GABA is detected in the substantia nigra, globus pallidus and hypothalamus in the rat, monkey and human brain. Furthermore, the subcellular localization of glutamic acid decarboxylase (GAD), the rate-limiting enzyme in the formation of GABA, is found exclusively in the synaptosomal fraction. On the other hand, high concentrations of taurine are found in the cerebellum, olfactory bulb and striatum in rat brain as well as in retina, pituitary and pineal glands. The activity of cysteinesulfinic acid decarboxylase (CSD) which is believed to be a key enzyme of taurine biosynthesis is found, however, mainly in the cytosolic fraction and only to a lesser extent in the synaptosomal fraction.

Localization of taurine and GABA in the brain at cellular and subcellular levels appear also to be different. For example, in mammalian retina, GABA is localized exclusively in the amacrine, horizontal and bipolar cells, while taurine is found mainly in the rods and cones as well as in Muller's cells[12]. The highest concentration of taurine is known to occur in developing brains, at which time the concentrations of other neuroactive amino acids tend to be low. For example, in developing feline cerebellum, taurine-like immunoreactivity reaches its peak at 4 weeks after birth. Purkinje cells are all taurine-positive in the neonatal cerebellum, whereas these cells in the mature cerebellum appear to contain only GABA and glutamate. In the newborn feline cerebellum, taurine-like immunoreactivity is localized in dendrites, axons of Purkinje and granule cells as well as in glial processes[21]. In female rats, the hypothalamic level of taurine decreases markedly 30 days after birth. In fish retina, GABA is also found in one layer of the horizontal cells and some bipolar cells, whereas taurine is found in both rods and cones. However, other cellular components including glia are devoid of taurine-like immunoreactivity[30]. In the lamprey spinal cord, the most intensive taurine-like immunoreactivity occurs in a population of liquor-contacting cells (tanycytes), located around the central canal, which extends processes to the dorsal, lateral, and ventral margins of the spinal cord. In addition, a proportion of the taurine-immunoreactive cells contains GABA-like immunoreactivity. A moderate level of taurine immunoreactivity is also present in ependymal cells, located around the central canal, as well as in astrocytes throughout all regions of the spinal cord. At the ultrastructural level, taurine-like immunoreactivity shows an even distribution in the cytoplasm of the labeled glial cells. In contrast to the glial labeling, neuronal cell bodies and axons exhibit very low levels of taurine labeling[40].

**Synaptic Release**

It has been well established that synaptic- and depolarization-evoked releases of GABA are suppressed by taurine, possibly by immobilizing intracellular $Ca^{2+}$ and activating the function of the presynaptic GABA autoreceptor. In fact, Kamisaki et al.[13] reported that taurine inhibited KCl (30 mM)-evoked release of GABA from rat cortical synaptosomes in a dose-dependent manner with potencies ($IC_{50}$) of 5 nM. This suppressive effect of taurine on GABA release was counteracted by the addition of phaclofen (a $GABA_B$ receptor antagonist), but not by bicuculline (a $GABA_A$ receptor antagonist). These data suggest that taurine may modify the calcium-related release of GABA. On the other hand, it has been reported that GABA increases extracellular taurine in the hippocampus and this enhancing effect of GABA was accompanied with a decrease in amplitude of the population spikes[16]. Lerma and colleagues[16] suggested that GABA stimulates taurine release, probably by accentuating a counter-transport process for taurine. In the preoptic mediobasal hypothalamic area, it has also been reported that muscimol and baclofen ($GABA_A$ and $GABA_B$ agonists, respectively) increase GABA release and decrease taurine output in immature and adult rats.

Arias et al[2] demonstrated that repeated intracerebroventricular administration of taurine lowered serum luteinizing hormone (LH) as well as hypothalamic LH-releasing hormone (LHRH) levels in peripubertal female rats. In these experimental conditions, taurine treatment doubled hypothalamic GABA levels and caused a delay in the vaginal opening by more than 2 days. Based on these data, Arias and colleagues[2] concluded that supplementation with taurine during the fourth postnatal week reduced LHRH/LH secretions and delayed sexual development by increasing hypothalamic GABAergic activities.

It is also noteworthy that the application of various types of acute stress alters GABA and taurine effluxes in the brain. For example, tail pinch for 3 min induced immediately a pronounced tetrodotoxin-sensitive increase in glutamate and aspartate release as well as increases in taurine and GABA overflow from the locus caeruleus. In contrast, it was reported that immobilized stress for 9 min elevated the release of GABA from the locus caeruleus without affecting taurine release[41]. Rego et al.[37] reported that oxidative stress such as the application of ascorbate/$Fe^{2+}$ also increased the release of glutamate, aspartate, GABA and taurine from cultured retinal cells in the presence of $Ca^{2+}$. Under hypoxic and ischemic conditions, the release of glutamate, aspartate, GABA and taurine are also increased in a $Ca^{2+}$-independent manner. Since hypoxia and/or ischemia-induced release of these amino acids were significantly accentuated in the presence of NMG, which induces the reversal of the $Na^{+}$-dependent transporters, the release was considered to be due to the facilitation of the reversal process of the amino acid transporter. A possible functional significance of the increased release of these inhibitory amino acids such as GABA, glycine and taurine during hypoxia was suggested as protection against neuronal death, caused by excessive release of excitatory amino acids[37].

Guzman et al.[7] reported that nitric oxide (NO) induced a significant release of taurine and GABA. Ohkuma et al[27] also demonstrated that NO and *N*-methyl-*D*-aspartate (NMDA) evoked taurine release from cortical neurons. This elevated release of taurine was mediated by a reversal of the $Na^{+}$-dependent carrier-mediated taurine transport system. On the other hand, it was reported that NO-evoked [$^{3}$H]GABA release from cortical neurons was mediated by two release mechanisms, a $Ca^{2+}$-dependent release system and a reverse process of the $Ca^{2+}$-independent and $Na^{+}$-dependent carrier-mediated GABA uptake system, respectively[28].

## Transport Mechanism

The recent application of molecular biological techniques to the taurine research field led to the isolation and characterization of cDNAs that encode taurine transporters from mouse brain[19], rat brain[42], human thyroid cells[11], human placenta[34,36], human retinal pigment epithelial cells[35] and Madin-Darby canine kidney cells[45]. Taurine transporters all showed extremely high homology to each other, exhibiting >90% similarity in amino acid content. Characteristics of each taurine transporter are summarized in Table 1. These cloned taurine transporters belong to a family of $Na^{+}$- and $Cl^{-}$-dependent neurotransmitter transporters which mediate the uptake of inhibitory amino acids as well as biogenic amine transmitters into presynaptic terminals and surrounding glial cells. The mRNA for the rat taurine transporter was found in the cellular cytoplasm throughout the CNS. *In situ* hybridization for taurine transporter mRNAs indicated that particularly high levels were detected in the retina, olfactory regions, cerebral cortex, hippocampus and cerebellum. On the other hand, four transporters for GABA in the brain were also cloned, termed GAT-1[6], GAT-2[20,48], GAT-3[18], GAT-4[18], respectively. Their characterizations are also summarized in Table 1. They have considerable structural similarity, including 12 putative transmembrane domains, with that of taurine. The homology in GABA transporters was found to be 50 to 70% within each transporter. Jhiang et al.[11] reported that the cloned human taurine transporter from thyroid cells showed about 60% amino acid identity with the dog betaine transporter. Uchida et al.[45] also demonstrated that there was 56% amino acid identity with the MDCK cell betaine/GABA transporter as compared with the taurine transporter in MDCK cells. GAT-2, which was insensitive to nipecotic acid, was thought to be the same as the betaine transporter. GABA uptake by GAT-2, GAT-3 and GAT-4 were inhibited by β-alanine as well as by taurine transporters. Competition between GABA and β-alanine and between taurine and β-alanine was also reported[18,19]. Therefore, it seems likely that the β-alanine transport systems are operated by both GABA and taurine transporters which have a high homology to each other.

**Table 1**. Characteristics of taurine and GABA transporters

a) Taurine transporter

| Tissue sources<br>Molecular structure | mouse brain<br>590aa | rat brain<br>621aa | human thyroid cells<br>619aa | human placenta<br>620aa | MDCK cells<br>655aa |
|---|---|---|---|---|---|
| Km value for Tau<br>for β-alanine | 4.5 μM<br>56 μM | 40 μM | | 6.0 μM | |
| Distribution | brain<br>kidney | brain, lung<br>kidney, heart<br>spleen | muscle, ovary<br>spleen, thyroid,<br>colon, kidney | placenta<br>skeltal muscle<br>heart, brain<br>liver | kidney<br>skeltal muscle<br>ileal mucosa<br>brain, liver |
| Homology with<br>GABA transporter | 52% GAT1<br>62% GAT2 | 51% GAT1<br>62% GAT2<br>62% GAT3 | 80% dog betaine<br>GABA transporter | | 56% MDCK cell<br>betaine/GABA<br>transporter |

b) GABA transporter

| Subtype (mouse) | GAT1 | GAT2 | GAT3 | GAT4 |
|---|---|---|---|---|
| Another name | rGAT1 | bBGT1 | rGAT2 | rGATB, rGAT3 |
| Molecular structure | 598aa | 614aa | 602aa | 627aa |
| Km for GABA | 5.9 μM | 79 μM | 18 μM | 0.6 -1.0 μM |
| Distribution | brain, retina | kidney, brain | all tissue | brain, retina |

## Receptor Function

Taurine increases chloride conductance across neural cell membranes[4,29] by affecting chloride channels that are either bicuculline-sensitive or strychinine-sensitive, and this increase is apparently dependent on the relative number of $GABA_A$ and glycine receptors, respectively[8,10,15]. There are several reports that taurine displaces GABA agonist binding to the $GABA_A$ receptor[5,14,22,31,47], and taurine appears to act as a partial agonist on the benzodiazepine receptor that is coupled to the $GABA_A$ receptor when measured at 37°C[33]. Bureau and Olsen[3] also reported that taurine acts on a subclass of $GABA_A$ receptors and competitively inhibits [$^3$H]muscimol binding to purified $GABA_A$ receptors. They also demonstrated that taurine enhances [$^3$H]flunitrazepam binding to benzodiazepine receptors which are coupled to the $GABA_A$ receptor. Furthermore, their studies indicate that taurine has variable affinities for muscimol binding sites. This evidence is further supported by the fact that the taurine-sensitive sites were enriched in the dentate gyrus, substantia nigra, molecular layer of cerebellum, median hypothalamic nuclei and the hippocampal CA3 regions. It has also been reported that a high concentration of taurine allosterically inhibits the binding of [$^{35}$S]-*t*-butylbicyclo-phosphorothionate (TBPS) to the $GABA_A$ receptor that is coupled to the $Cl^-$ channel in synaptic membranes by interacting with a bicuculline-sensitive site. The potency of taurine inhibition on TBPS binding varies among brain regions in the order of cerebellum > olfactory bulb > cortex[32]. These results suggest that taurine may interact with the $GABA_A$ receptor/benzodiazepine receptor/chloride channel complex as an agonist and potentiate inhibitory GABA neurons in the brain.

## Functional Correlation

Much evidence has accumulated for supporting the presence of functional interactions between GABAergic and taurinergic neurons. Furthermore, Liljequist[17] demonstrated that local applications between GABAergic and taurinergic neurons. Furthermore, Liljequist[17] demonstrated that local applications of muscimol ($GABA_A$ agonist) and baclofen ($GABA_B$ agonist) into the nucleus raphe dorsalis caused a concentration-dependent increase in spontaneous motor activity and food intake. Taurine by itself was found to be ineffective on locomotion or food intake but enhanced the effect of muscimol.

Yokoi et al.[49] reported that guanidinoethanesulfonate (GES: taurocyamine), a taurine uptake inhibitor, induced electrocorticographic spike discharges and these discharges were suppressed by taurine and muscimol. Based on these data, it was suggested that taurine might suppress GES-induced seizure activity, possibly by accentuating the function of the $GABA_A$/benzodiazepine receptor complex. It was also demonstrated that taurine administr-ation inhibited the writhing responses induced by acetic acid administration, and that pretreatment with naloxone and TAG (6-aminoethyl-3-methyl-4H-1,2,4 benzothiazine 1-1-dioxide), a taurine antagonist, antagonized the antinociceptive effect of taurine[39,43]. Furthermore, Serrano et al.[38] reported that the antinociceptive action of taurine was antagonized with bicuculline, a $GABA_A$ antagonist, but not by CGP 35348, a $GABA_B$ antagonist. From these data, these authors suggested that the antinociceptive effect of taurine might be partly mediated by spinal $GABA_A$ receptors.

Osmoregulation is a well known function assigned to taurine. As described in the section on synaptic release, the release of taurine from a variety of cells might be related to the regulation of cellular volume. For example, Haugstad and Langmoen[9] demonstrated that hypo-osmotic stress caused a 10-fold increase of taurine release and over a two-fold increase of GABA from hippocampal slices, and these changes were blocked when extracellular osmolarity was normalized with mannitol. On the other hand, energy deprivations induced by glucose and oxygen removal also caused a 50-fold increase of

GABA release, a 40-fold increase of glutamate release and a two-fold to six-fold increase in taurine, aspartate, glycine, asparagine, serine and alanine release. However, mannitol did not block these increases in amino acid release induced by energy deprivation. They concluded from these results that taurine mainly acted as an osmotic regulator. Nagelhus et al.[26] also demonstrated that the level of taurine in rat cerebellar Purkinje cells was substantially decreased by hypo-osmotic stress induced by intraperitoneal injection of distilled water. The loss of taurine immunoreactivity from Purkinje cells was accompanied by a 70-80% increase in the content of immunoreactive taurine in adjacent glial cells. On the other hand, adjacent sections preincubated with a GABA antiserum did not exhibit any significant changes in taurine in response to the hyposmotic stress.

Under ischemic conditions, Matsumoto et al.[23] determined the evoked release of both excitatory amino acids, glutamate and aspartate, and inhibitory amino acids, GABA and taurine, from the rat hippocampus in a 4-vessel occlusion model using *in vivo* micro-dialysis procedures. Potassium (100 mM)-evoked release of these amino acids was measured on the 5th day after 20 minutes of ischemia. They found that the release of glutamate and aspartate was reduced in the ischemic animals, while the release of GABA and taurine was found to be unchanged and increased, respectively. The immuno-histochemical studies of Torp and his colleagues[44] found a depletion of taurine levels in the pyramidal cell body in the same animal model (4 vessel occlusion model). These data also indicated that ischemia induced the increased release of taurine from cerebral tissues as shown by Matsumoto et al.[23]. The taurine dipeptide, γ-glutamyltaurine, an endogenous constituent of brain tissue, has been considered to be an excitatory amino acid antagonist[46]. γ-glutamyltaurine was found to be elevated 1.6 fold in the CA1 pyramidal cells in the hippocampus during ischemia (4-vessel occlusion model)[1]. These results suggest that γ-glutamyltaurine and taurine may act as protective substances against neuronal toxicity and subsequent neuronal death caused by excitatory amino acids.

The application of taurine for experimental and human epilepsies started over thirty years ago, and it was reported to possess some mild anticonvulsant activities in both humans and experimental animal models[24,25]. Sodium valproate (VPA) is a known drug for the treatment of epilepsy and the mechanism of action of VPA has been thought to be related to the accentuating action of cerebral GABA neurons through the activation of the $GABA_A$ receptor complex as well as the inhibition of GABA metabolism. Although it is true that high doses of VPA cause an increase in whole brain GABA concentration, the mechanisms responsible for this phenomenon have not been fully elucidated. As described in the section on receptor function, taurine has a direct accentuating action on the cerebral $GABA_A$ receptor. Accordingly, the anticonvulsive effect of taurine may be produced through the activation of the $GABA_A$ receptor.

## CONCLUSION

It seems likely that taurine and GABA have a definite interrelationship, both functionally and metabolically, in a variety of brain regions. Especially, recent evidence indicates such interrelationships exist at transporters, receptors and specific neuronal networks. Although a lot of work needs to be accomplished before making clear these interrelationships, it appears that GABA always play a major role as an inhibitory transmitter and taurine acts as a modulator and/or "Waki Yaku" (a side player in the Japanese "Noh" play) in the brain.

## REFERENCES


1. Andiné, P., Orwar, O., Jacobson, I., Sandberg, M., and Hagberg, H., 1991, Extracellular acidic sulfur-containing amino acids and γ-glutamyl peptides in global ischemia: postischemic recovery of neuronal activity is paralleled by a tetrodotoxin-sensitive increase in cysteinesulfinate in the CA1 of the rat hippocampus, *J.Neurochem.*, 57:230-236.
2. Arias P., Feleder, C., Ródriguez, M., Ginzburg, M., Refojo, B., Szwarcfarb, B., and Moguilevsky, J.A., 1995, Repeated intracerebroventricular administration of taurine lowers LH levels and postpones vaginal opening in peripubertal female rats, *Devel. Brain Res.*, 85:137-139.
3. Bureau, M.H. and Olsen, R.W., 1991, Taurine acts on a subclass of $GABA_A$ receptors in mammalian brain *in vitro*, *Eur. J. Pharmac.*, 207:9-16.
4. Curtis, D.R. and Johnson, G.A.R., 1974, Amino acid transmitters in mammalian nervous system, *Ergeb. Physiol.*, 69:97-188.
5. Greenlee, D.V., Van Ness, P.C., and Olsen R.W., 1978, Gamma-aminobutyric acid binding in mammalian brain: receptor-like specificity of sodium-dependent sites, *J. Neurochem.*, 31:933-938.
6. Guastelia, J., Nelson, N., Nelson, H., Czyzyk, L., Keynan, S., Miedel., M.C., Devidson, N., Lester, H.A., and Kanner, B.I., 1992, Cloning and expression of a rat brain GABA transporter, *Science*, 249:1303-1306.
7. Guevara-Guzman, R., Emson, P.C., and Kendrick, K.M., 1994, Modulation of *in vivo* striatal release by nitric oxide and cyclic GMP, *J.Neurochem.*, 62:807-810.
8. Haas, H.L. and Hosli, L., 1973, The despression of brain stem neurons by taurine and its interaction with strychynine and bicuculline, *Brain Res.*, 52:399-402.
9. Haugstad, T.S. and Langmoen, I.A., 1996, Release of brain amino acids during hyposmolar stress and energy deprivation, *J. Neurol. Anesth.*, 8:159-168.
10. Horikoshi, T., Asanuma, A., Yanagisawa, K., Anza, K., and Goto, S., 1988, Taurine and β-alanine acts on both GABA and glycine receptors in Xenopus oocyte injected with mouse brain messenger RNA, *Molec. Brain Res.*, 4:97-105.
11. Jhiang, S.M., Fithian, L., Smanik, P., McGill, J., Tong, Q., and Mazzaferri, E.L., Cloning of the human taurine transporter and characterization of taurine uptake in the thyroid cells, *FEBS Lett.*, 318:139-144.
12. Kalloniatis, M., Marc, R.E., and Murry, R.F., 1996, Amino acid signatures in the primate retina, *J. Neurosci.*, 16:6807-6829.
13. Kamisaki, Y,. Maeda, K., Ishimura, M., Omura, I., and Itoh, T., 1993, Effects of taurine on depolarization-evoked release of amino acids from rat cortical synaptosomes, *Brain Res.*, 627:181-185.
14. Kontro, P. and Oja, S.S., Properties of sodium-independent taurine binding to brain synaptic membranes, in: *Taurine: Biological Actions and Clinical Perspectives* (Oja, S.S., Ahtee, L., Kontro, P., and Passonen, M.K., eds.) Alan R. Liss, New York, pp. 249-259.
15. Kudo, Y., Akiyoshi, E., and Akagi, H., 1988, Identification of two taurine receptor subtypes on the primary afferent terminal of frog spinal cord, *Br. J. Pharmacol.*, 94:1051-1056.
16. Lerma, J., Herranz, A.S., Herreras, O., Munoz, D., Slis, J.M., del Rió, R.M., and Delgado, J.M.R., 1985, γ-Aminobutyric acid greatly increases the *in vivo* extra-cellular taurine in the rat hippocampus, *J. Neurochem.*, 44:983-986.
17. Liljequist, R., 1992, Interaction of taurine and related compounds with GABAergic neurones in the nucleus raphe dorsalis, *Pharmac. Biochem. Behav.*, 44:107-112.
18. Liu, Q-R., Lópes-Corcuera, B., Mandiyan, S., Nelson, H., and Nelson, N., Molecular characterization of four pharmacologically distinct α-aminobutyric acid transporters in mouse brain, *J. Biol. Chem.*, 268:2106-2112.
19. Liu, Q.-R., Lópes-Corcuera, B., Nelson, H., and Mandiyan, S., 1992, Cloning and expression of cDNA encoding the transporter of taurine and β-alanine in mouse brain, *Proc. Natl. Acad. Sci.*, 89:12145-12149.
20. Lópes-Corcuera, B., Liu, Q-R., Mandiyan, S., Nelson, H., and Nelson, N., 1992, Expression of a mouse brain cDNA encoding novel γ-aminobutyric acid transporter, *J. Biol. Chem.*, 267:17491-17493.
21. Lu, P., Imaki, H., Xu, W., and Sturman, J.A., !993, Visualization of taurine, GABA and glutamate in developing feline cerebellum by immunohistochemistry, *Int. J. Neuroscience*, 11:493-505.
22. Malmnein, O. and Kontro, P., 1986, Modulation of GABA-benzodiazepine receptor complex by taurine in rat brain membranes, *Neurochem. Res.*, 11:85-94.
23. Matsumoto, K., Ueda, S., Hashimoto, T., and Kuriyama, K., 1991, Ischemic injury in the rat hippocampus following transient forebrain ischemia: evaluation *in vivo* microdialysis, *Brain Res.*, 543:236-242.

24. Mcgreer, P.L. and Mcgreer, E.G., 1989, Amino acid neurotransmitter, in Basic Neurochemistry, Molecular, Cellular and Medical Aspects, 4th ed., Raven Press, New York.
25. Mutani, R., Monaco, F., Durelli, L., and Delsedime, M., 1975, Level of free amino acids in serum and cerebrospinal fluid after administration of taurine to epileptic and normal subjects, *Epilepsia*, 16:765-769.
26. Nageluhus, E.A., Lehmann, A., and Ottersen, O.P., 1993, Neuronal-glial exchange of taurine during hypo-osmotic stress: A combined immunocytochemical and biochemical analysis in rat cerebellar cortex, *Neurosicence.*, 54:615-631.
27. Ohkuma, S., Katsura, M., Chen, D.-Z., and Kuriyama, K., 1996, Nitric oxide-evoked [$^3$H]taurine release is mediated by reversal of the $Na^+$-dependent carrier-mediated taurine transport system, in: Adv. Exp. Med. Biol. "Taurine 2; Basic and Clinical Aspects", Huxtable, R.J., Azuma, J., Kuriyama, K., Nakagawa, M., and Baba, A., eds., Plenum, Press, New York, pp. 417-425.
28. Ohkuma, S., Katsura, M., Chen, D.-Z., Narihara, H., and Kuriyama, K., 1996, Nitric oxide-evoked [$^3$H]γ-aminobutyric acid release is mediated by two distinct mechanisms, *Mol. Brain Res.*, 36:137-144.
29. Okamoto, K., Kimura, H., and Sasai, I., 1983, Taurine-induced increase of $Cl^-$ conductance of cerebellar purkinje cell dendrites *in vitro, Brain Res.*, 259:319-323.
30. Pow, D.V., 1994, Taurine, amino acid transmitter, and related molecules in the retina of the Australian lungfish *Neoceratodus forsteri*: a light-microscopic immunocyto-chemical and electron-microscopic study, *Cell Tissue Res.*, 278:311-326.
31. Quinn, M.R., 1990, Taurine allosterically modulates binding sites of the $GABA_A$ receptor, in: *Taurine: Functional Neurochemistry, Physiology, and Cardiology* (Pasantes-Morales, H., Martin, D.L., Shain, W., and Martin del Rio R., eds., Wiley-Liss Inc., New York, pp. 121-127.
32. Quinn, M.R. and Harris, C.L., 1995, Taurine allosterically inhibits binding of [$^{35}$S]-*t*-butylbicyclophosphorothinate (TBPS) to rat brain synaptic membranes, *Neuro-pharmacol.*, 34:1607-1613.
33. Quinn, M.R. and Miller, C.L., 1992, Taurine allosterically modulates flunitrazepam binding to synaptic membranes, *Neurosci. Res.*, 33:136-141.
34. Ramamoorthy, S., Ainol, L.S., Leibach F.H., Blalkely, R.D., and Ganapathy, V., 1993, Molecular cloning, expression, chromosomal localization, and regulation of the human placental taurine transporter, *Placenta*, 14:A60.
35. Ramamoorthy, S., Del Monte, M.A., Leibach F.H., and Ganapathy, V., 1994, Molecular identity and calmodulin-mediated regulation of the taurine transporter in a human retinal pigment epithelial cell line, *Curr. Eye Res.*, 523-529.
36. Ramamoorthy, S., Leibach F.H., Mahesh, V.B., Han, H., Yang-Feng, T., Blalkely, R.D., and Ganapathy, V., 1994, Functional characterization and chromosomal local-ization of a cloned taurine transporter from human placenta, *Biochem. J.*, 300:893-900.
37. Rego, C.R., Santos M.S., and Oliveira R., 1996, Oxidative stress, hypoxia, and ischemia-like conditions increase the release of endogenous amino acids by distinct mechanism in cultured retinal cells, *J. Neurochem.*, 66:2506-2516.
38. Serrano, M., Serrano, S.J., Guerrero, M.R., and Fernandes, A., 1994, Role of $GABA_A$ and GABAB receptors and peripheral cholinergic mechanisms in the antinociceptive action of taurine, *Gen. Pharmac.*, 25:1123-1129.
39. Serrano, J.S., Serrano, M.I., Guerrero, R., Ruiz de Valderas, R., and Polo, J., 1990, Antinociceptive effect of taurine and its inhibition by naloxone, *Gen. Pharmac*, 21:333-336.
40. Shupliakov, O., Brodin, L., Srinivasan, M., Grillner, S., Cullheim, S., Storm-Mathisen, J., and Otterson, O.P., 1994, Extrasynaptic localization of taurine-like immunoreactivity in the lamprey spinal cord, *J. Comp. Neurol.*, 347:301-311.
41. Singewald, N., Zhou, G.-Y., and Schneider, C., 1995, Release of excitatory and inhibitory amino acids from the locus coeruleus of conscious rats by cardiovascular stimuli and various forms of acute stress, *Brain Res.*, 704:42-50.
42. Smith, K.E., Borden, L.A., Wang, C.-H. D., Hartig, P.R., Branchek, T.A., and Weinshank, R.L., *J. Pharmacol. Exp. Ther.*, 42:563-569.
43. Smullin, D.H., Schamber, C.D., Skilling, S.E., and Larson, A.A., 1990, A possible role for taurine in analgesia, Taurine: Functional Neurochemistry, Physiology, and Cardiology, Pasantes-Morales, H., Martin, D.L., Shain, W., and Martin del Rio, R., eds., Wiley-Liss Inc., New York, pp. 129-132.
44. Torp, R., Andine, P., Hagberg, H., Karagülle, T., Blackstad, T.W., and Otterson, O.P., 1991, Cellular and subcellular distribution of glutamate-, glutamine- and taurine-like immunoreactivity during forebrain ischemia: a semiquantitative micro-scopic study in rat hippocampus, *Neuroscience*, 41:433-447.

45. Uchida, S., Kwon, H.M., Yamauchi, A., Preston, A.S., Marumo, F., and Handler, J.S., 1992, Molecular cloning of the cDNA for an MDCK cell $Na^+$- and $Cl^-$-dependent taurine transporter that is regulated by hypertonicity, *Proc. Natl. Acad. Sci.*, 89:8230-8234.
46. Varga, V., Janaky, R., Marnela, K.-M., Gulyas, J., Kontro, P., and Oja, S.S., 1989, Displacement of excitatory amino acid receptor ligands by acidic oligo peptides, *Neurochem. Res.*, 14:1223-1227.
47. Williams, M., Risley, E., and Totaro, J.A., 1980, Interaction of taurine and β-alanine with central nervous system neurotransmitter receptors, *Life Sci.*, 26:557-560.
48. Yamauchi, A., Uchida, S., Kwon, H.M., Preston, A.S., Robey, R.B., Garcia-Perez, A., Burg, M.B., and Handler, J.S., 1992, Cloning of a $Na^+$- and $Cl^-$-dependent betaine transporter that is regulated by hypertonicity, *J. Biol. Chem.*, 267:649-652.
49. Yokoi, I., Kabuto, H., and Mori, A., 1993, Effects of GABAmimetics on electro-corticographic spike discharges induced by guanidinoethanesulfonic acid (amido-taurine) in the rat, 1993, *Neurochem. Res.*, 18:533-538.

# REGULATION OF TAURINE BIOSYNTHESIS AND ITS PHYSIOLOGICAL SIGNIFICANCE IN THE BRAIN

Jang-Yen Wu,[1] Xiao Wen Tang,[1] John V. Schloss,[2] and Morris D. Faiman[3]

[1] Department of Physiology and Cell Biology
[2] Department of Medicinal Chemistry
[3] Department of Pharmacology and Toxicology
University of Kansas
Lawrence, KS 66045

## ABSTRACT

Cysteine sulfinic acid decarboxylase (CSAD), the rate-limiting enzyme in taurine biosynthesis, was found to be activated under conditions that favor protein phosphorylation and inactivated under conditions favoring protein dephosphorylation. Direct incorporation of $^{32}P$ into purified CSAD has been demonstrated with $[\gamma^{32}P]$ATP and PKC, but not PKA. In addition, the $^{32}P$ labeling of CSAD was inhibited by PKC inhibitors suggesting that PKC is responsible for phosphorylation of CSAD in the brain. Okadaic acid had no effect on CSAD activity at 10 μM suggesting that protein phosphatase-2C (PrP-2C) might be involved in the dephosphorylation of CSAD. Furthermore, it was found that either glutamate- or high $K^+$-induced depolarization increased CSAD activity as well as $^{32}P$-incorporation into CSAD in neuronal cultures, supporting the notion that the CSAD activity is endogenously regulated by protein phosphorylation in the brain. A model to link neuronal excitation, phosphorylation of CSAD and increase in taurine biosynthesis is proposed.

## INTRODUCTION

Taurine, 2-aminoethanesulfonic acid, is one of the most abundant amino acids in mammals[4]. The physiological role of taurine has received considerable attention since the reports that cats fed a taurine deficient diet developed central retinal degeneration[3] and cardiomyopathy[12]. Now, taurine has been shown to be involved in many important physiological functions e.g., as trophic factor in the development of the CNS[14,15], maintaining the structural integrity of the membrane[9,11], regulating calcium binding and transport[6,8], as an osmolyte[13,19], a neuromodulator[5], a neurotransmitter[7,10,16], and a neuro-protector against L-

*Taurine 3*
Edited by Schaffer *et al.*, Plenum Press, New York, 1998

Glu-induced neurotoxicity[17]. The major pathway for taurine synthesis in the brain is through the oxidation of cysteine to cysteine sulfinic acid (CSA), decarboxylation of CSA to hypo-taurine, and oxidation of hypotaurine[4]. The rate limiting step in this pathway is decarboxylation of CSA by CSA decarboxylase (CSAD)[4].

Although CSAD has been isolated and purified from mammalian brain more than a decade ago[20], little is known with certainty regarding the mechanism of regulation of CSAD activity. Recently, we have reported that the GABA-synthesizing enzyme, L-glutamate decarboxylase (GAD) is activated when it is dephosphorylated presumbly by protein phosphatase (PrP) 2B and is inhibited when it is phosphorylated, presumbly by protein kinase A (PKA)[1,2]. In this communication, evidence is presented to suggest that CSAD is activated when it is phosphorylated presumbly by protein kinase C (PKC) and is inactivated when it is dephosphorylated presumbly by PrP 2C.

## MATERIALS AND METHODS

### CSAD Activity Assay

A radiometric assay was used to determine CSAD activity as previously described[18]. Briefly, 160 µl of CSAD sample, 20 µl of 10 mM CSA (L-[1-$^{14}$C]-CSA, 2.5 µCi/ml) and 20 µl of 100 mM L-glutamate in standard CSAD buffer (50 mM KP, pH 7.2, 1 mM reduced glutathione (GST), 2 mM AET, and 0.4 mM PLP) was incubated in a test tube (15 x 85 mm) sealed with a rubber stopper. A 0.6 ml microfuge tube containing 0.2 ml benzethonium hydroxide was used to trap $^{14}CO_2$ released. The enzymatic reaction was carried out at room temperature for 1 hr with constant shaking. The reaction was then terminated by injecting 0.25 ml of 0.5 N $H2SO_4$ and $^{14}CO_2$ collected was counted.

### Preparation of Synaptosomal Fractions

Fresh porcine brains were homogenized in 0.32 M sucrose (w/v = 15 g: 100 ml) using a glass homogenizer. The homogenate was centrifuged at 1,000 xg for 10 min and the supernatant solution obtained was further centrifuged at 100,000 xg for 30 min. The resulting pellet was the crude synaptosomal preparation. The pellet was resuspended in Kreb-Ringer's phosphate buffer (pH 7.2, containing 123 mM NaCl, 3 mM KCl, 0.4 mM $MgCl_2$, 0.5 mM $NaH_2PO_4$, 0.25 mM $Na_2HPO_4$, and 1 mg/ml glucose) and divided into aliquots for further studies.

### Extraction of CSAD from the Synaptosomal Fraction in the Presence of Kinase Inhibitors

Aliquots of the synaptosomal fraction were centrifuged and the pellets resuspended in standard CSAD buffers contained various kinase inhibitors as indicated. The synaptosomes were then ruptured by sonication (3 x 1 seconds). The suspensions obtained were kept at room temperature for 45 min with constant shaking. CSAD activity was then determined by the CSAD activity assay as described above.

### Phosphorylation and Dephosphorylation of Purified CSAD

Aliquots of the purified CSAD preparations as previously reported[18] were dialyzed in 50 mM Tris/citrate buffer, pH 7.2, containing 1 mM GST, 1 mM AET and 1 µM PLP, overnight. The CSAD samples were treated under the following conditions: (1) PKC buffer alone (containing 1 mM $CaCl_2$, 5 mM $MgCl_2$, 0.3 mg/ml L-α-phosphatidyl-L-serine, 0.06 mg/ml diolein, 0.03% Triton X-100, 0.1 mM ATP, and 100 μCi [$\gamma$-$^{32}$P]ATP); (2) PKC buffer and 200 ng/ml PKC; (3) PKA buffer (containing 5 mM $MgCl_2$, 0.1 mM cAMP, 0.1 mM ATP, and 100 μCi [$\gamma$-$^{32}$P]ATP); plus 150 units PKA catalytic subunit; (4) the same as (2), followed by treatment with calf intestine phosphatase (CIP). The suspensions were incubated at 37°C for 45 min. The reactions were stopped by adding 5 x SDS sample loading buffer and the samples were analyzed by SDS-PAGE/autoradiograhpy as described previously[1,2].

### Phosphorylation of CSAD in Primary Neuronal Cultures

Primary neuronal cultures were prepared as described previously[17] and cultures of 14 days *in vitro* (DIV) were used for the experiments. Briefly, the culture was first washed and then incubated with 1 ml EBSS (116.4 mM NaCl; 5.4 mM KCl; 0.8 mM $MgSO_4$; 1.0 mM $NaH_2PO_4$; 26.2 mM $NaHCO_3$; 1.8 mM $CaCl_2$; 5.6 mM D-glucose; pH 7.4). Eighty-five μCi of [$^{32}$P]-phosphate was added to each dish, and the culture was further incubated for 1 hr at 37°C. The culture was treated for an additional 15 min by adding 1 ml of EBSS containing the following different substances: (1) EBSS alone; (2) EBSS plus 2 mM glutamate; (3) EBSS plus 2 mM glutamate and 25 mM taurine; (4) high $K^+$ EBSS (20.6 mM NaCl; 100.7 mM KCl; 0.8 mM $MgSO_4$; 1.0 mM $NaH_2PO_4$; 26.2 mM $NaHCO_3$; 1.8 mM $CaCl_2$; 5.6 mM D-glucose; pH 7.4). After a 15 min incubation, the media were removed and the dishes washed twice with EBSS buffer. The cells were harvested in CSAD buffer containing phosphatase inhibitor (0.2 mM vanadate), protease inhibitors (1 mM PMSF, 5 mM benzamidine) and 100 μM AOAA. The crude cell extracts were purified by anti-CSAD immunoaffinity column, and then analyzed on SDS-PAGE/autoradiography as described[1,2].

### Effect of Neuronal Stimulation on CSAD Activity in Primary Neuronal Cultures

The conditions were the same as those described above except that [$^{32}$P]-phosphate was omitted. The cell homogenates were assayed for CSAD as described above under CSAD activity assay.

## RESULTS AND DISCUSSION

### Effect of Protein Kinase Inhibitors on CSAD Activity in Synaptosomal Preparations

When CSAD was extracted from porcine brain synaptosomal fractions in the presence of protein kinase inhibitors, CSAD activity decreased (Fig. 1). Among the inhibitors, only those that inhibited PKC activity decreased CSAD activity. PKC inhibitory peptide (100 ng/ml), staurosporine (1 μM), H-8 (10 μM), and chelerythrine (50 μM), decreased CSAD activity to 48 ± 9%, 48 ± 2%, 54 ± 6%, and 57 ± 14% of the control level, respectively. On the other hand, PKA inhibitor peptide had no significant effect on CSAD activity. These results indicate that PKC, but not PKA, is involved in the regulation of CSAD activity. In addition, when 5 mM ATP was included in the extraction, it increased CSAD activity to 172 ± 19% of the control level. In short, CSAD activity increased under conditions that favor protein phosphorylation, while CSAD activity decreased under the conditions that favor dephosphorylation of proteins. Thus, CSAD activity is likely to be regulated by protein phosphorylation.

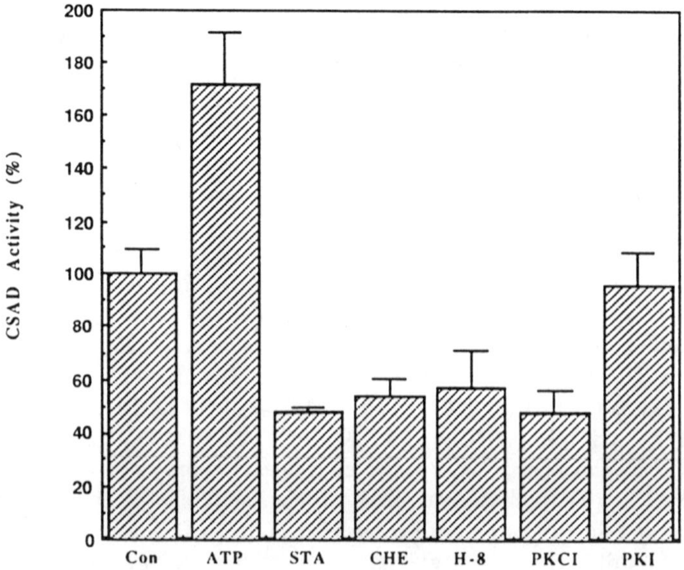


**Figure 1.** Effect of ATP and protein kinase inhibitors on CSAD activity in synaptosomal extracts. CSAD was extracted from synaptosomal preparations in the presence of ATP or protein kinase inhibitors. After incubation for 45 min at room temperature, the extracts were assayed for CSAD activity as described.[18] (See CSAD activity assay) Con: control, ATP: 5 mM ATP, STA: 1 μM staurosporine, CHE: 50 μM chelerythrine, H-8: 10 μM H-8, PKCI: 100 ng/ml PKC inhibitory peptide, and PKI: 100 ng/ml PKA inhibitory peptide.

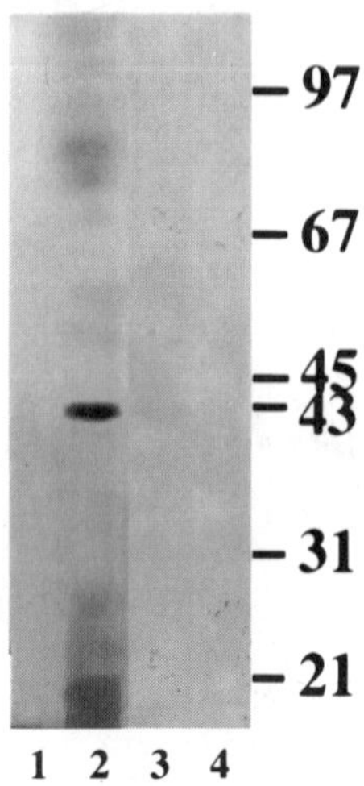


**Figure 2.** Demonstration of $^{32}P$ incorporation into highly purified CSAD and subsequent dephosphorylation by CIP. Purified CSAD was phosphorylated as detailed in Materials and Methods. Lane 1: control in PKC buffer, lane 2: PKC treatment, lane 3: PKA treatment, lane 4: the same as Lane 2 treated with CIP.

### Phosphorylation/Dephosphorylation of Purified CSAD

To demonstrate direct incorporation of [$^{32}$P] into CSAD protein and to identify the kinase involved in this process, a highly purified CSAD preparation and suspected kinases were used in the studies. It was found that purified CSAD was phosphorylated by PKC (Fig. 2, lane 2) but not PKA (lane 3). $^{32}$P-Labeled CSAD was dephosphorylated by CIP treatment (Lane 4).

The activity of CSAD was found to correlate well with its state of phosphorylation as shown in Figure 2. For instance, CSAD activity was increased to 350% by treatment with PKC, but not PKA, similar to the phosphorylation pattern of CSAD. Furthermore, PKC-mediated activation of CSAD activity was abolished by CIP treatment, corresponding to the complete dephosphorylation of CSAD (Fig. 2, Lane 4).

### Effect of Neuronal Stimulation on CSAD Activity in Neuronal Cultures

In order to demonstrate that CSAD activity is regulated *in vivo*, cultured neurons were stimulated either by excitatory neurotransmitter, L-glutamate or by high $K^+$. The results showed that CSAD activity increased when neurons were depolarized by either glutamate or high $K^+$ stimulation (Fig. 3). Furthermore, the increase of CSAD activity by glutamate stimulation was blocked in the presence of 25 mM taurine.

### Effect of Neuronal Stimulation on Incorporation of $^{32}$P into CSAD in Neuronal Cultures

To investigate further whether the activation of CSAD through depolarization of neuron correlated with the degree of phosphorylation of CSAD, $^{32}$P incorporation experiments were also conducted under the same conditions as described in Figure 3.

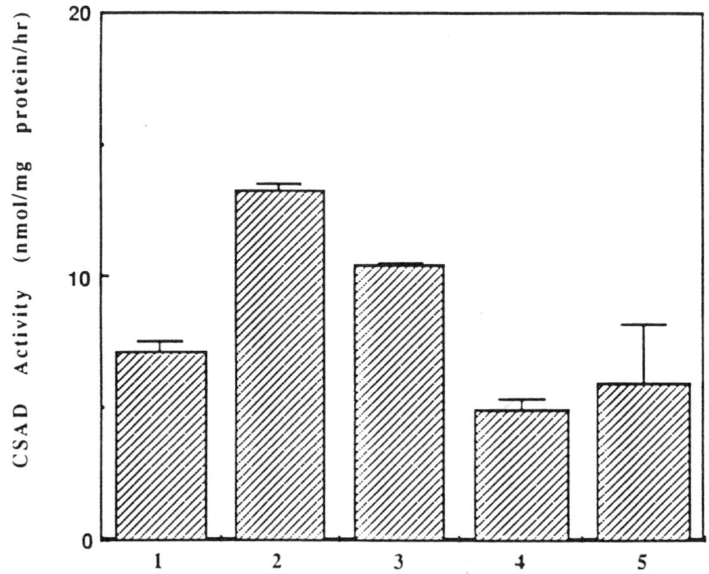


**Figure 3.** Effect of high $K^+$ or glutamate stimulation on CSAD activity in cultured neurons. Cultured neurons (14 DIV) were treated under various conditions as described below. CSAD activity was determined by the CSAD assay and expressed as nmol of $CO_2$ formed/mg of protein/hr (See Materials and Methods for details). (1) Control, (2) High $K^+$, (3) 1 mM glutamate, (4) 1 mM glutamate and 25 mM taurine, and (5) 25 mM taurine alone. Each value is the average of 4 determinations ± SE (N = 4).

In agreement with the activation of CSAD in cultured neurons by neuronal stimulation, the degree of phosphorylation of CSAD was also enhanced by either glutamate- or high $K^+$-stimulation (Fig. 4). Furthermore, the increase of CSAD phosphorylation by neuronal stimulation was also inhibited by 25 mM taurine corresponding to a concomitant inhibition of CSAD activation.

In summary, the following conclusion can be made:

(1) CSAD activity is found to be regulated by protein phosphorylation/dephosphorylation in crude synaptosomal preparations as well as in purified preparations. Phosphorylation activates CSAD while dephosphorylation inactivates it. PKC and PrP-2C have been identified as the possible protein kinase and protein phosphatase, respectively, involved in phosphorylation and dephosphorylation of CSAD.

(2) In neuronal cultures, depolarization induced by either glutamate or high $K^+$ stimulates the incorporation of $^{32}P$ into CSAD and activates CSAD, suggesting that CSAD activity is regulated in vivo through protein phosphorylation.

(3) Taurine inhibits both glutamate- and high $K^+$-induced phosphorylation of CSAD and activation of CSAD in neuronal cultures.

Based on the above observations and our recent findings that taurine at high concentrations, e.g., 25 mM can protect neurons from L-glutamate-induced neuronal injury,[17] the following model is proposed: when neurons are excited or injured by overstimulation, the intracellular free $Ca^{2+}$ level is elevated, resulting in activation of PKC which, in turn, activates CSAD through protein phosphorylation. The activated CSAD then synthesizes more taurine which functions as either an osmolyte or $Ca^{2+}$ regulator to prevent cells from swelling and $Ca^{2+}$ overloading, thus protecting the cells from damage. When enough taurine has accumulated in the cells, its high concentration then inhibits the activation of PKC directly or indirectly (possibly through regulating $Ca^{2+}$ availability), thus shutting down activation of CSAD through inhibition of CSAD phosphorylation by PKC. The activated CSAD soon returns to its inactive state through action of a protein phosphatase, most likely PrP-2C.

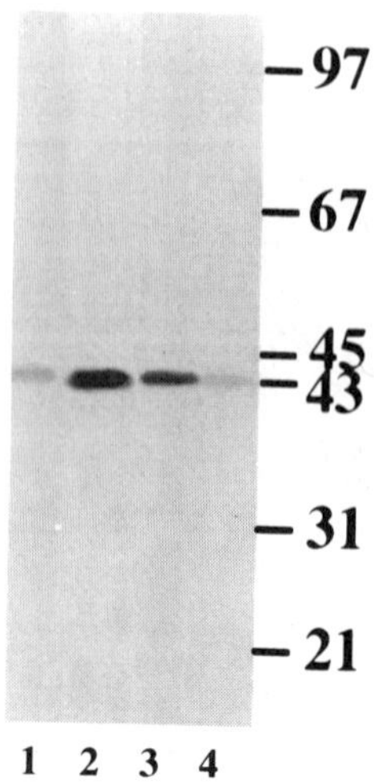


**Figure 4.** Direct phosphorylation of CSAD in cultured neurons. Cultured neurons (14 DIV) were incubated with [$^{32}$P]-orthophosphate for 1 hr, followed by various treatments as indicated. The cells were harvested, sonicated, and analyzed by SDS-PAGE, Western immunoblotting and autoradiography (See Materials and Methods for details). Lane 1, control; Lane 2, high $K^+$ treatment; Lane 3, 1 mM glutamate treatment; Lane 4, 1 mM glutamate and 25 mM taurine.

## ACKNOWLEDGMENTS

The support of this work from National Science Foundation (IBN-9723079), Office of Naval Research (N00014-94-1-0457), the University of Kansas General Research Fund, and NIH (NS20978) is much appreciated. The authors wish to thank Drs. Erik Floor and James Orr for critical review of the manuscript. The expert typing of the manuscript by Sharon Lee Hopkins is greatly appreciated.

## REFERENCES

1. Bao, J., Cheung, W. Y., and Wu, J.-Y., 1995, Brain L-glutamate decarboxylase: Inhibition by phosphorylation and activation by dephosphorylation, *J. Biol. Chem.*, 270:6464-6467.
2. Bao, J., Nathan, B., and Wu, J.-Y., 1994, Role of protein phosphorylation in the regulation of brain L-glutamate decarboxylase activity, *J. Biomed. Sci.*, 1:237-244.
3. Hayes, K. C., Carey, R. E., and Schmidt, S. Y., 1975, Retinal degeneration associated with taurine deficiency in the cat, *Science*, 188:949-951.
4. Jacobsen, J. G. and Smith, L. H., 1968, Biochemistry and physiology of taurine and taurine derivatives, *Physiol. Rev.*, 48:424-511.
5. Kuriyama, K., 1980, Taurine as a neuromodulator, *Fed. Proc.*, 39:2680-2684.
6. Lazarewicz, J. W., Noremberg, K., Lehmann, A., and Hamberger, A., 1985, Effects of taurine on calcium binding and accumulation in rabbit hippocampal and cortical synaptosomes, *Neurochem. Int.*, 7:421-428.
7. Lin, C.-T., Su, Y. Y. T., Song, G.-X., and Wu, J.-Y., 1985, Is taurine a neurotransmitter in rabbit retina? *Brain Res.*, 337:293-298.
8. Lombardini, J. B., 1985, Effects of taurine on calicum ion uptake and protein phosphorylation in rat retinal membrane preparations, *J. Neurochem.*, 45:268-275.
9. Moran, J., Salazar, P., and Pasantes-Morales, H., 1988, Effect of tocopherol and taurine on membrane fluidity of retinal rod outer segments, *Experimental Eye Res.*, 45:769-776.
10. Okamoto, K., Kimura, H., and Sakai, Y., 1983, Evidence for taurine as an inhibitory neurotransmitter in cerebellar stellate interneurons: Selective antagonism by TAG (6-aminomethyl-3-methyl-4H,1,2,4-benzothiadiazine-1,1-dioxide), *Brain Res.*, 265:163-168.
11. Pasantes-Morales, H. and Cruz, C., 1985, Taurine and hypotaurine inhibit light-induced lipid peroxidation and protect rod outer segment structures, *Brain Res.*, 330:154-157.
12. Pion, P. D., Kittleson, M. D., Rogers, Q. R., and Morris, J. G., 1987, Myocardial failure in cats associated with low plasma taurine: A reversible cardiomyopathy, *Science*, 237:764-768.
13. Solis, J. M., Herranz, A. S., Herreras, J., Lerma, J., and DEL Rio, R. M., 1988, Does taurine act as an osmoregulatory substance in the rat brain, *Neurosci. Lett.*, 91:53-58.
14. Sturman, J. A., 1993, Taurine in development, *Physiol. Rev.*, 73:119-147.
15. Sturman, J. A., Moretz, R. C., French, J. H., and Wisniewski, H. M., 1985, Postnatal taurine deficiency in the kitten results in a persistence of the cerebellar external granule cell layer: Correction by taurine feeding, *J. Neuro. Res.*, 13:521-528.
16. Taber, T. C., Line, C.-T., Song, G.-X., Thalman, R. H., and Wu, J.-Y., 1986, Taurine in the rat hippocampus-localization and postsynaptic action, *Brain Res.*, 386:113-121.
17. Tang, X. W., Deupree, D. L., Sun, Y., and Wu, J.-Y., 1996, Biphasic effect of taurine on excitatory amino acid-induced neurotoxicity, in: "Taurine: Basic and Clinical Aspects", Huxtable, R. J., Azuma, J., Nakagawa, M., Kuriyama, A. and Bala, A., eds., Plenum Publishing Co., pp. 499-506.
18. Tang, X. W., Hsu, C.-C., Sun, Y., Wu, E., Yang, C. Y., and Wu, J.-Y., 1996, Multiplicity of brain cysteine sulfinic acid decarboxylase-purification, characterization and subunit structure, *J. Biomed. Sci.*, 3:442-453.
19. Wade, J. V., Olsen, J. P., Samson, F. E., Nelson, S. R., and Pazdernik, T. L., 1988, A possible role for taurine in osmoregulation within the brain, *J. Neurochem.*, 541:740-745.
20. Wu, J.-Y., 1982, Purification and characterization of cysteic/cysteinesulfinic acids decarboxylase and L-glutamate decarboxylase in bovine brain, *Proc. Natl. Acad. Sci. USA*, 79:4270-4274.

# CELL-DAMAGING CONDITIONS RELEASE MORE TAURINE THAN EXCITATORY AMINO ACIDS FROM THE IMMATURE HIPPOCAMPUS

Pirjo Saransaari[1] and Simo S. Oja[1,2]

[1] Tampere Brain Research Center
University of Tampere Medical School
Box 607
FIN-33101 Tampere
[2] Department of Clinical Physiology
Tampere University Hospital
Box 2000
FIN-33521 Tampere
Finland

## INTRODUCTION

High concentrations of excitatory amino acids are neurotoxic and overstimulation of their receptors contributes to neuronal death during cerebral ischemia[35]. Excitatory amino acids are released from neural structures from adult animals during hypoxia and ischemia[1,7,20,23]. Oxygen-derived free radicals formed in hypoxic brain tissue also induce release of excitatory amino acids[22] and thus contribute to ischemia-induced neuronal damage[4]. Of the various brain areas, the hippocampus is particularly vulnerable to ischemic injury[25]. In the newborn, however, the hippocampus is relatively well spared after hypoxia and ischemia[3,5]. The major part of excitatory innervation in the hippocampus is glutamatergic. Taurine also abounds in the hippocampus[11] and taurine-like immunoreactivity has been located in hippocampal interneurons, pyramidal neurons and dentate granule cells[15]. This inhibitory amino acid seems to be essential for the development and survival of neural cells[8,34]. We have now assessed the simultaneous release of endogenous excitatory amino acids, glutamate and aspartate, together with taurine, in the above cell-damaging conditions in hippocampal slices prepared from immature mice.

## MATERIALS AND METHODS

Slices 0.4 mm thick weighing 15-20 mg were prepared from the hippocampi of 7-day-old NMRI mice with a Stadie-Riggs tissue slicer and used immediately in efflux experiments as detailed in Kontro and Oja[9]. At 30 min the medium was in many experiments changed to medium containing 50 mM $K^+$ (potassium stimulation). Glutamate, aspartate and taurine were assayed by high-pressure liquid chromatography as described in Oja and Kontro[18].

Neural cell damage was induced by modifying experimental conditions. In hypoxia the medium was bubbled with $N_2$ gas for 1 h before the experiments and then during efflux. In ischemia the glucose-free medium was bubbled with $N_2$ gas. Free radical production was achieved by exposure to hydrogen peroxide (0.01 %)[6,24]. In $Ca^{2+}$-free medium $Ca^{2+}$ (0.8 mM) was omitted and ethylenediaminetetra-acetate (2.0 mM) added at the beginning of the superfusion experiments[9].

## RESULTS

The concentrations of glutamate, aspartate and taurine were 58.1 ± 2.5, 24.0 ± 0.3 and 204.5 ± 3.8 mmol/kg protein (mean ± SEM, n=4), respectively, in the hippocampus of 7-day-old mice. The basal release of aspartate and glutamate was at the sensitivity limit of our assay and $K^+$ stimulation enhanced release of both amino acids only marginally (Fig. 1). The basal release of taurine was also low, but markedly enhanced by $K^+$ stimulation (Fig. 1). In hypoxic conditions the basal release of taurine was enhanced about 10-fold as compared to normoxic conditions, with the release steadily increasing as a function of time (Fig. 2C). The basal release of glutamate and aspartate was also discernible, though very low when compared to that of taurine (Fig. 2A,B). $K^+$ stimulation was able to evoke the release of the three amino acids markedly, but the responses of aspartate and glutamate were greater than that of taurine. In ischemia the releases of both glutamate and aspartate increased, although glutamate release decreased with time (Fig. 3A,B). Taurine release was strikingly enhanced in ischemia, also gradually decreased with time (Fig. 3C). Both the spontaneous and $K^+$-stimulated releases of glutamate and aspartate were greatly

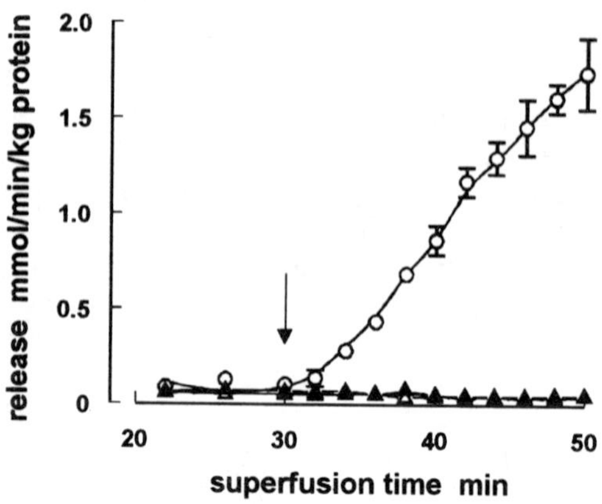


**Figure 1.** $K^+$-Stimulated release of glutamate (-▲-), aspartate (-Δ-) and taurine (-O-) in normoxia from hippocampal slices from 7-day-old mice superfused for 50 min with Krebs-Ringer-Hepes medium. The presence of 50 mM $K^+$ from 30 min onwards is indicated by an arrow. The results are mean values ± SEM of four independent experiments.

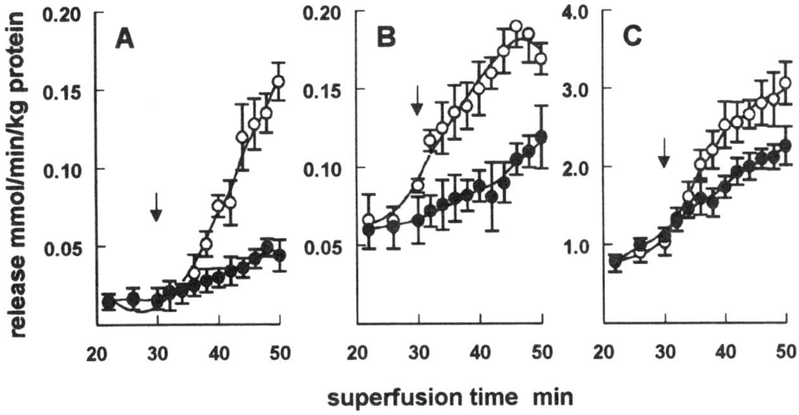


**Figure 2.** Spontaneous (-●-) and $K^+$-stimulated (-O-) release of glutamate (A), aspartate (B) and taurine (C) from hippocampal slices from 7-day-old mice superfused for 50 min with Krebs-Ringer-Hepes medium in hypoxic conditions. The presence of 50 mM $K^+$ from 30 min onwards is indicated by an arrow. The results are mean values ± SEM of four independent experiments.

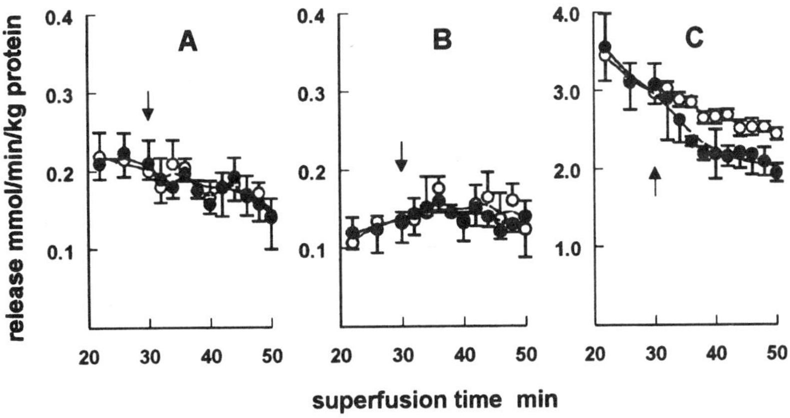


**Figure 3.** Spontaneous (-●-) and $K^+$-stimulated (-O-) release of glutamate, (A) aspartate (B) and taurine (C) from hippocampal slices from 7-day-old mice superfused for 50 min with Krebs-Ringer-Hepes medium in ischemic conditions. The presence of 50 mM $K^+$ from 30 min onwards is indicated by an arrow. The results are mean values ± SEM of four independent experiments.

increased in the pressure of free radicals, with the actual amount released being greater for aspartate than for glutamate (Fig. 4A,B). There was also a massive release of taurine, which steadily increased with time. $K^+$ stimulation had a slight but significant effect (Fig. 4C).

The $Ca^{2+}$-dependence of the release processes was investigated in $Ca^{2+}$-free superfusion media in normal conditions and ischemia. The basal release of taurine was

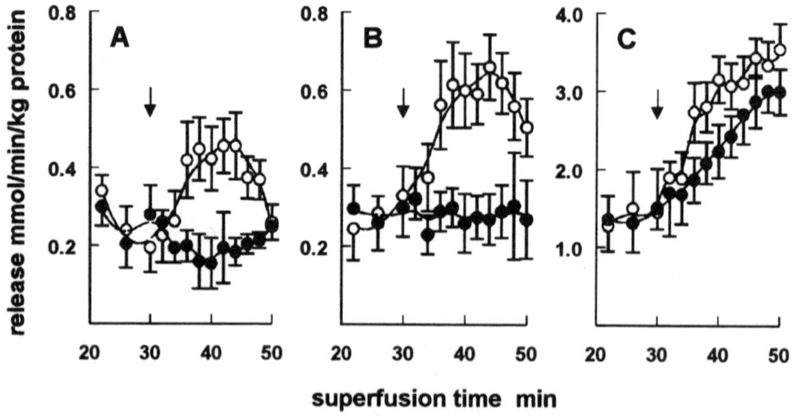


**Figure 4.** Spontaneous (-●-) and $K^+$-stimulated (-○-) release of glutamate (A), aspartate (B) and taurine (C) from hippocampal slices from 7-day-old mice superfused for 50 min in Krebs-Ringer-Hepes medium in free-radical-containing media. The presence of 50 mM $K^+$ from 30 min onwards is indicated by an arrow. The results are mean values ± SEM of four independent experiments.

tripled in $Ca^{2+}$-free media in normoxia, but the relative response to $K^+$ stimulation was reduced (Fig. 5). In ischemia, taurine release was further enhanced in the absence of $Ca^{2+}$, but in this condition there was no significant $K^+$ stimulation. The release of both aspartate and glutamate was not measurable in the $Ca^{2+}$-free condition.

## DISCUSSION

The amino acids studied are not confined to a neurotransmitter pool but also reside in the metabolic pool in neurons and glial cells. The spontaneous release of these amino acids under normal conditions largely reflects their hippocampal concentrations, but cannot be directly equated with the synaptic release. The massive $K^+$-stimulated release of taurine observed in other immature brain tissues[27] was now demonstrated in the hippocampus. Neural cell damage can enhance the hippocampal release of the three amino acids studied by several mechanisms, among them $Ca^{2+}$-dependent exocytosis, $Ca^{2+}$-independent release via reversal of carrier-mediated uptake, indiscriminate opening of ion channels and unspecific leakage through damaged plasma membranes. This latter alternative is not likely, however, since in hippocampal slices the liberation of lactate dehydrogenase (a common marker of plasma membrane damage and nonspecific lysis of neural cells) is not increased by the present type of ischemia[2,23]. We likewise saw no increase in lactate dehydrogenase in superfusion media under ischemic conditions (data not shown). Moreover, no general increase in hippocampal membrane permeability has been found which would allow all amino acids to escape into the extracellular spaces[14]. An activation of volume-sensitive efflux could have contributed, as has been shown in the case of taurine[31]. Intracellular swelling activates stretch-sensitive ion channels and is generally accompanied by a release of both inorganic and organic osmolytes, including taurine[21]. Nonsaturable diffusion of taurine in immature brain tissue is indeed known to be greatly enhanced in ischemic conditions[28].

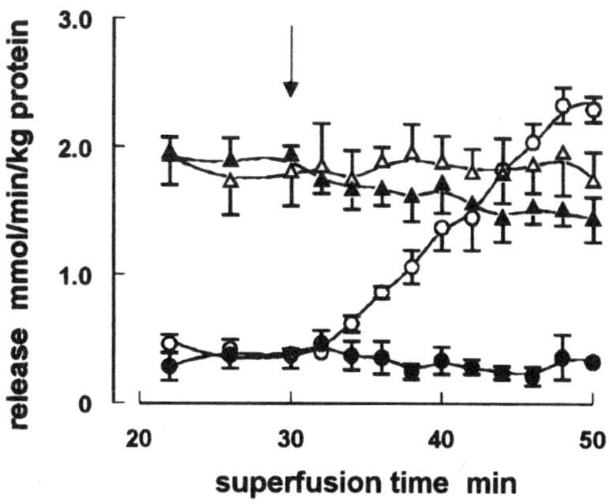


**Figure 5.** Spontaneous (-●-) and $K^+$-stimulated (-O-) release of taurine in normoxia, and spontaneous (-▲-) and $K^+$-stimulated (-Δ-) release of taurine in ischemia from hippocampal slices from 7-day-old mice superfused for 50 min in $Ca^{2+}$-free Krebs-Ringer-Hepes medium. The presence of 50 mM $K^+$ from 30 min onwards is indicated by an arrow. The results are mean values ± SEM of four independent experiments.

The substantial increase in amino acid release in ischemia and other cell-damaging conditions could be due partly to inhibition of reuptake. Glutamate, aspartate and taurine possess $Na^+$-dependent high-affinity uptake systems in the developing brain. However, taurine uptake by cerebral cortical synaptosomes from the 7-day-old mouse is unaffected in most of the present cell-damaging conditions. Only prolonged incubation during ischemia markedly reduce taurine uptake[28]. The enhanced release of excitatory amino acids in hypoxic and ischemic conditions[7,13] may also result from the reversal of $Ca^{2+}$-independent flux by the cell membrane carriers[26,35]. The function of taurine carriers in the cell plasma membrane is similarly reversible[12]. In ischemia, $Ca^{2+}$-dependent release has also been implicated in the case of glutamate[1]. Taurine release under normal conditions is more $Ca^{2+}$-dependent in the immature than the mature hippocampus[29]. Under cell-damaging conditions more taurine is also released from the immature than the mature hippocampus[29]. A part of this release is $Ca^{2+}$-dependent.

The massive amounts of taurine released simultaneously with the excitatory amino acids in the immature hippocampus may constitute an important protective mechanism. Taurine release may limit and prevent excitation from reaching neurotoxic levels. Taurine protects neural cells against excitotoxicity induced by excitatory amino acids[36,37] and prevents harmful metabolic events induced by ischemia or hypoxia[32]. Furthermore, taurine-containing neurons are relatively inert in cerebral ischemia induced by 4-vessel occlusion[39]. Intracellular swelling of neurons and glial cells due to an activation of glutamate receptors is attenuated by extracellularly released taurine[38]. Taurine has also long been known to ameliorate symptoms in epilepsy[17]. Activation of $GABA_A$ receptors increases $Cl^-$ conductance, inducing hyperpolarization and reducing cell excitability[19,33], but only minor amounts of GABA are released in the immature hippocampus[30]. Taurine exhibits the same, albeit less potent, effect on membrane chloride conductance in immature brain tissue[19]. Taurine also attenuates $Ca^{2+}$ influx in the developing brain and antagonizes depolarization-evoked $Ca^{2+}$ efflux[10]. The resistance of the immature hippocampus has been ascribed to lower oxygen consumption and ATP utilization when compared to adults[16]. The elevated extracellular levels of taurine apparently also contribute to the maintenance of homeostasis in the hippocampus upon impending hyperexcitation.

## ACKNOWLEDGMENTS

The skillful assistance of Mrs. Irma Rantamaa and Miss Sari Perttunen and the financial support of the Medical Research Fund of Tampere University Hospital are gratefully acknowledged.

## REFERENCES

1. Benveniste, H., Drejer, J., Schousboe, A., and Diemer, N.H., 1984, Elevation of the extracellular concentrations of glutamate and aspartate in rat hippocampus during transient cerebral ischemia monitored by intracerebral microdialysis, *J. Neurochem.*, 43:1369-1374.
2. Cherici, G., Alesiani, M., Pellegrini-Giampietro, D.E., and Moroni, F., 1991, Ischemia does not induce the release of excitotoxic amino acids from the hippocampus of newborn rats, *Dev. Brain Res.*, 60:235-240.
3. Cherubini, E., Ben-Ari, Y., and Krnjevic, K., 1989, Anoxia produces smaller changes in synaptic transmission, membrane potential and input resistance in immature rat hippocampus, *J. Neurophysiol.*, 62:882-895.
4. Coyle, J.T. and Puttfarcken, P., 1993, Oxidative stress, glutamate and neurodegenerative disorders, *Science*, 262:689-695.
5. Ferriero, D.M., Arcavi, L.J., and Simon, R.P., 1990, Ontogeny of excitotoxic injury to nicotinamide adenine dinucleotide phosphate diaphorase reactive neurons in the neonatal rat striatum, *Brain Res.*, 304:417-424.
6. Gilman, S.C., Bonner, M.J., and Pellmar, T.C., 1994, Free radicals enhance basal release of D-[$^3$H]aspartate from cerebral cortical synaptosomes, *J. Neurochem.*, 62:1757-1763.
7. Globus, M.Y.-T., Busto, R., Dietrich, W.D., Martinez, E., Valdes, I., and Ginsberg, M.D., 1988, Effect of ischemia on the in vivo release of striatal dopamine, glutamate and γ-aminobutyric acid studied by intracerebral microdialysis, *J. Neurochem.*, 51:1455-1464.
8. Huxtable, R.J., 1992, The physiological actions of taurine, *Physiol. Rev.*, 72:101-163.
9. Kontro, P. and Oja, S.S., 1987, Taurine and GABA release from mouse cerebral cortex slices: potassium stimulation releases more taurine than GABA from developing brain, *Dev. Brain Res.*, 37:277-291.
10. Kontro, P. and Oja, S.S., 1988, Effects of taurine on the influx and efflux of calcium in brain slices of adult and developing mice, *Int. J. Neurosci.*, 38:103-109.
11. Kontro, P., Marnela, K.-M., and Oja, S.S., 1980, Free amino acids in the synaptosome and synaptic vesicle fractions of different bovine brain areas, *Brain Res.*, 184:129-141.
12. Korpi, E.R. and Oja, S.S., 1983, Characteristics of taurine release from cerebral cortex slices induced by sodium-deficient media, *Brain Res.*, 289:197-204.
13. Lekieffre, D., Callebert, J., Plotkine, M., and Boulu, R.G. 1992, Concomitant increases in the extracellular concentrations of excitatory and inhibitory amino acids in the rat hippocampus during forebrain ischemia, *Neurosci. Lett.*, 137:78-82.
14. Madl, J.E. and Burgesser, K., 1993, Adenosine triphosphate depletion reverses sodium-dependent neuronal uptake of glutamate in rat hippocampal slices, *J. Neurosci.*, 13:4429-4444.
15. Magnusson, K.R., Clements, J.R., Wu, J.-Y., and Beitz, A.J., 1989, Colocalization of taurine- and cysteine sulfinic acid decarboxylase-like immunoreactivity in the hippocampus of the rat, *Synapse*, 4:55-69.
16. Meibach, R.C., Ross, D.A., Cox, R.D., and Glick, S.D., 1981, The ontogeny of hippocampal energy metabolism, *Brain Res.*, 204:431-435.
17. Oja, S.S. and Kontro, P., 1983, Free amino acids in epilepsy: possible role of taurine, *Acta Neurol. Scand.*, 67, Suppl. 93:5-20.
18. Oja, S.S. and Kontro, P., 1989, Release of endogenous taurine and γ-aminobutyric acid from brain slices from the adult and developing mouse, *J. Neurochem.*, 52:1018-1024.

19. Oja, S.S., Korpi, E.R., and Saransaari, P., 1990, Modification of chloride flux across brain membranes by inhibitory amino acids in developing and adult mice, *Neurochem. Res.*, 15:797-804.
20. O'Regan, M.H., Smith-Barbour, M., Perkins, L.M., and Phillis, J.W., 1995, A possible role for phospholipases in the release of neurotransmitter amino acids from ischemic rat cerebral cortex, *Neurosci. Lett.*, 185:191-194.
21. Pasantes-Morales, H., Alavez, S., Sánchez Olea, R., and Morán, J., 1993, Contribution of organic and inorganic osmolytes to volume regulation in rat brain cells in culture, *Neurochem. Res.*, 18:445-452.
22. Pellegrini-Giampietro, D.E., Cherici, G., Alesiani, M., Carlá, V., and Moroni, F., 1988, Excitatory amino acid release from rat hippocampal slices as a consequence of free-radical formation, *J. Neurochem.*, 51:1960-1963.
23. Pellegrini-Giampietro, D.E., Cherici, G., Alesiani, M., Carlá, V., and Moroni, F., 1990, Excitatory amino acid release and free radical formation may cooperate in the genesis of ischemia-induced neuronal damage, *J. Neurosci.*, 10:1035-1041.
24. Pelmar, T.C., 1995, Use of brain slices in the study of free-radical actions, *J. Neurosci. Meth.*, 59:93-98.
25. Pulsinelli, W.A., Brierley, J.B., and Plum, F., 1982, Temporal profile of neuronal damage in a model of transient forebrain ischemia, *Ann. Neurol.*, 11:491-498.
26. Sánchez-Prieto, J. and González, P., 1988, Occurrence of a large $Ca^{2+}$-independent release of glutamate during anoxia in isolated nerve terminals (synaptosomes), *J. Neurochem.*, 50:1322-1324.
27. Saransaari, P. and Oja, S.S., 1992, Release of GABA and taurine from brain slices, *Prog. Neurobiol.*, 38:455-482.
28. Saransaari, P. and Oja, S.S., 1996, Taurine and neural cell damage: transport of taurine in adult and developing mice, *Adv. Med. Exp. Biol.*, 403:481-490.
29. Saransaari, P. and Oja, S.S., 1997, Enhanced taurine release in cell-damaging conditions in the developing and ageing mouse hippocampus, *Neuroscience*, 79:847-854.
30. Saransaari, P. and Oja, S.S. 1997, Enhanced GABA release in cell-damaging conditions in the adult and developing mouse hippocampus, *Int. J. Dev. Neurosci.*, 15:163-174.
31. Schousboe, A., Morán, J., and Pasantes-Morales, H., 1990, Potassium-stimulated release of taurine from cultured cerebellar granule cells is associated with cell swelling, *J. Neurosci. Res.*, 27:71-77.
32. Schurr, A., Tseng, M.T., West, C.A., and Rigor, B. M., 1987, Taurine improves the recovery of neuronal function following cerebral hypoxia: an in vitro study, *Life Sci.*, 40:2059-2066.
33. Sivilotti, L. and Nistri, A., 1991, GABA receptor mechanisms in the central nervous system, *Prog. Neurobiol.*, 36:35-92.
34. Sturman, J.A., 1993, Taurine in development, *Physiol. Rev.*, 73:119-147.
35. Szatkowski, M. and Attwell, D., 1994, Triggering and execution of neuronal death in brain ischemia: two phases of glutamate release by different mechanisms, *Trends Neurosci.*, 17:359-365.
36. Tang, X.W., Deupree, D.L., Sun, Y., and Wu, J.-Y., 1996, Biphasic effect of taurine on excitatory amino acid-induced neurotoxicity, *Adv. Exp. Med. Biol.*, 403:499-505.
37. Trenkner, E., 1990, The role of taurine and glutamate during early postnatal cerebellar development of normal and weaver mutant mice, *Adv. Exp. Med. Biol.*, 268:239-244.
38. Walz, W. and Allen, A. F., 1987, Evaluation of the osmoregulatory function of taurine in brain cells, *Exp. Brain Res.*, 68:290-298.
39. Wu, J.-Y., Lin, C.-T., Johanssen, F.F., and Liu, J.-M., 1994, Taurine neurons in rat hippocampal formation are relatively inert to cerebral ischemia, *Adv. Exp. Med. Biol.*, 359:289-298.

# EFFECTS OF NO-GENERATING COMPOUNDS ON THE UPTAKE AND RELEASE OF TAURINE IN THE MOUSE BRAIN

Pirjo Saransaari[1] and Simo S. Oja[1,2]

[1] Tampere Brain Research Center
University of Tampere Medical School
Box 607, FIN-33101 Tampere
[2] Department of Clinical Physiology
Tampere University Hospital
Box 2000, FIN-33521 Tampere
Finland

## INTRODUCTION

Nitric oxide (NO) has been ascertained to play an important role in signal transduction in the central nervous system, functioning as a short-lived modulator and/or second messenger[1,32]. There is a functional link between NO and glutamatergic neurotransmission. NO is involved in synaptic plasticity, for example, in long-term potentiation, memory and learning, in the hippocampus[4], which phenomena represent physiological effects of glutamate as a transmitter. Glutamate also participates in the development of ischemic brain injury, and therefore, NO is likely to be associated with mechanisms of neuronal damage[4].

The glutamate-stimulated production of cyclic GMP in brain slices has been shown to be mediated via NO released by an activation of the N-methyl-D-aspartate (NMDA) class of glutamate receptors[6]. NO also affects transport processes of neurotransmitters, e.g., by evoking the release of acetylcholine[20], monoamines[15-17,35], GABA[22,23] and glutamate[33]. On the other hand, inhibitory effects have also been observed in dopamine[2,8], GABA[7,34] and glutamate release[11,12], suggesting that the actions of NO on the release processes are complex. We have recently characterized the release of taurine in hippocampal slices and shown that glutamate receptor agonists, particularly NMDA, potentiate the release in a receptor-mediated manner[28,29]. Moreover, the release is greatly enhanced in ischemic conditions[30,31]. We have now investigated the effects of different NO-generating compounds on taurine uptake and release in the adult mouse brain. The uptake was characterized with cerebral cortical synaptosomal preparations and the release with hippocampal slices using a superfusion system.

## MATERIAL AND METHODS

Synaptosomes prepared from the cerebral cortex of NMRI mice of both sexes aged 3 months were incubated for 10 min with 30 MBq/l of [$^3$H]taurine in Krebs-Ringer-Hepes-glucose medium (pH 7.4) as detailed by Kontro[13]. The kinetics of the uptake were assessed in a concentration range from 1 μM to 10 mM. The kinetic parameters and $IC_{50}$ were estimated by nonlinear regression analysis using the Markwardt algorithm. The protein content of synaptosomes was measured according to Lowry et al[18]. Slices 0.4 mm thick weighing 15-20 mg prepared with a Stadie-Riggs tissue slicer from the hippocampi of mice were preloaded for 30 min with 10 μM (50 MBq/l) [$^3$H]taurine in preoxygenated Krebs-Ringer-Hepes-glucose medium under $O_2$ and superfused with the same medium as in Kontro and Oja[14]. After 30 min the medium was in many cases changed to modified medium. In $Na^+$-free media, NaCl was substituted by equimolar choline chloride and in $Ca^{2+}$-free media equimolar choline chloride, together with 2.0 mM ethylenediaminetetra-acetate, replaced $CaCl_2$. The efflux rate constants of taurine for the time intervals of 20 to 30 min ($k_1$) and 34 to 50 min ($k_2$) were computed as negative slopes for the regression lines of the logarithm of radioactivity remaining in the slices vs. superfusion time[14].

## RESULTS

The uptake of 10 μM [$^3$H]taurine by synaptosomal preparations from mouse cerebral cortex was not affected by the NO-generating compounds 1.0 mM sodium nitroprusside (SNP) and 0.1 mM S-nitroso-N-acetylpenicillamine (SNAP), but 1.0 mM hydroxylamine (HA) reduced it to 77.7 ± 3.5 % of control (mean ± SEM, n = 12). The inhibitors of nitric oxide synthase, nitroarginine (NO-arg) and 7-nitroindazole (7NI) (both 0.1 mM), were also slightly inhibitory. The uptake was also unaffected when the synaptosomes were preincubated for 15 min with 1.0 mM SNP or 0.1 mM SNAP. Preincubation with 1.0 mM HA inhibited taurine uptake to 66.0 ± 1.3 % of control (mean ± SEM, n = 4), this effect being still further accentuated by 0.1 mM 7NI (52.8 ± 4.3 %, n = 4). The effect of HA was concentration-dependent (Fig. 1) with an $IC_{50}$ of 2.16 ± 0.08 mM. The HA (1.0 mM)

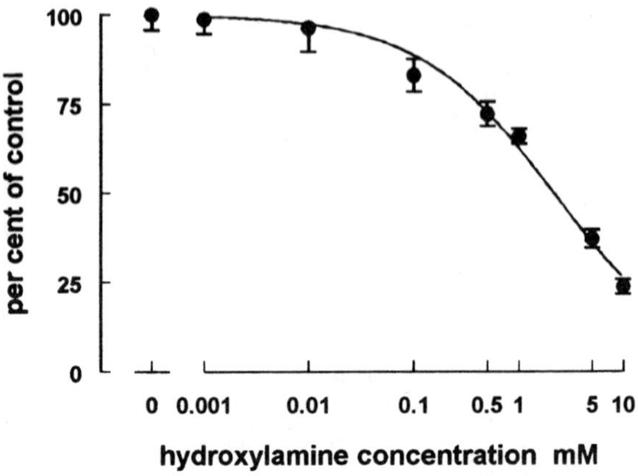


**Figure 1.** The effects of different hydroxylamine concentrations on taurine uptake by synaptosomal preparations from the mouse cerebral cortex. The results (mean ± SEM, n = 4) are percentages of control uptake. The calculated $IC_{50}$ was 2.16 ± 0.08 mM.

inhibition was also dependent on the concentration of taurine. The estimated kinetic parameters for taurine uptake show that the high-affinity transport component was completely absent in the presence of HA (Table 1). $K_m$ for the low-affinity component and nonsaturable uptake were both significantly increased by HA, but the maximal velocity was not changed.

The release of [$^3$H]taurine from mouse hippocampal slices was enhanced when SNAP, SNP and HA were added at the beginning of the superfusion (Fig. 2A), the effect of HA being concentration-dependent (Fig. 2B). The HA-evoked taurine efflux was not affected by $K^+$ stimulation (50 mM), the presence of NMDA (0.1 mM) or NO-arg (0.1 mM) (Table 2). NMDA stimulation was reduced by SNP, SNAP and NO-arg and the $K^+$-stimulated release of taurine by SNAP and NO-arg (Table 2).

**Table 1.** Kinetic parameters of taurine uptake in synaptomal preparations from the mouse cerebral cortex

| Parameter | Control | 1 mM HA |
|---|---|---|
| $K_{m1}$, μM | 18.0 ± 2.0 | - |
| $V_2$, nmol/s/kg | 28.0 ± 3.0 | - |
| $K_{m2}$, mM | 1.14 ± 0.21 | 3.27 ± 0.64* |
| $V_2$, μmol/s/kg | 4.51 ± 0.58 | 4.82 ± 0.37 |
| $K_d$ x $10^{-3}$ $s^{-1}$ | 0.39 ± 0.08 | 1.11 ± 0.12* |

Mean values ± SEM. Significance of differences between control and 1.0 mM hydroxylamine (HA) -containing media: *p<0.01.

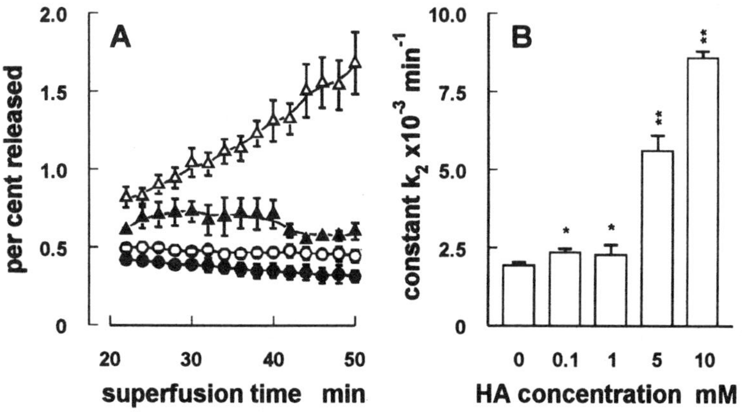


**Figure 2.** (A) Time course of taurine release (-●-) from mouse hippocampal slices in the presence of 0.1 mM S-nitroso-N-acetylpenicillamine (SNAP) (-O-), 1.0 mM sodium nitroprusside (SNP) (-▲-) and 5 mM hydroxylamine (HA) (-Δ-). The slices were superfused for 50 min with these compounds. The results are mean values ± SEM of four independent experiments. (B) Effects of different hydroxylamine concentrations on taurine release from mouse hippocampal slices. The slices were superfused for 50 min with HA. The results are efflux rate constants $k_2$ (mean ± SEM, n = 4-8) for the time interval of 34-50 min. Significant differences from control: *p<0.05, **p<0.01.

**Table 2.** Effects of NO-generating compounds on taurine release from mouse hippocampal slices

| | Efflux rate constants ± SEM ($k_2$ x $10^{-3}$ $min^{-1}$, 34-50 min) | | | |
|---|---|---|---|---|
| | Basal release | $K^+$-Stimulation | NMDA | NO-arg |
| Control | 1.94 ± 0.09 (20) | 3.08 ± 0.14 (6) | 3.38 ± 0.24 (4) | 1.84 ± 0.07 (4) |
| HA | 2.37 ± 0.10* (4) | 2.72 ± 0.14 (8) | 2.82 ± 0.36 (4) | 2.03 ± 0.22 (4) |
| SNP | 2.63 ± 0.20** (4) | 2.85 ± 0.08 (4) | 2.47 ± 0.28* (4) | 2.05 ± 0.15 (4) |
| SNAP | 1.91 ± 0.15 (4) | 2.28 ± 0.11** (4) | 2.53 ± 0.13* (4) | 1.31 ± 0.10** (4) |
| NO-arg | 1.95 ± 0.44 (4) | 2.37 ± 0.04** (4) | 2.31 ± 0.24* (4) | |

The slices were preloaded for 30 min with 10 μM [$^3$H]taurine in Krebs-Ringer-Hepes medium (pH 7.4) and superfused with 0.1 mM hydroxylamine (HA), sodium nitroprusside (SNP), S-nitroso-N-acetylpenicillamine (SNAP) or nitroarginine (NO-arg) for 50 min. $K^+$ ions (50 mM), 0.1 mM NMDA or 0.1 mM NO-arg were added at 30 min. Results are the rate constants $k_2$ (± SEM) for the period of 34-50 min. Number of independent experiments in parenthesis. Significant differences from the control: *p<0.05, **p<0.01.

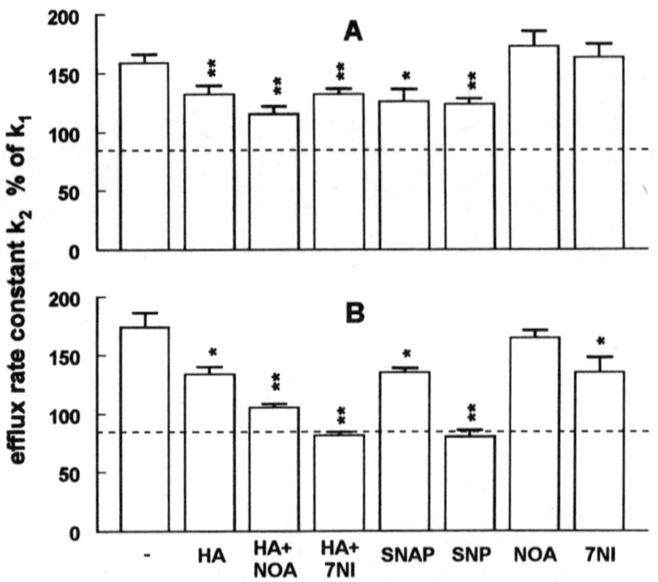


**Figure 3.** Effects of NO-generating compounds on $K^+$ (50 mM)-stimulated (A) and NMDA (0.1 mM)-evoked (B) taurine release from mouse hippocampal slices. The tested compounds were added to the superfusion medium at 30 min. The results are efflux rate constants $k_2$ (mean ± SEM, n = 4-8) for the time interval of 34-50 min as percentages of control. NOA = nitroarginine. Significant differences from the corresponding control: *p<0.05, **p<0.01.

The results were somewhat different when the NO-generating compounds were added to the superfusion media at 30 min. The $K^+$-stimulated release of taurine was significantly inhibited by HA, SNP and SNAP, but not by NO-arg and 7NI (Fig. 3A). The HA effect was not altered by 7NI but was enhanced by NO-arg. The NMDA-evoked release was also reduced by HA, SNAP and 7NI. The SNP inhibition was very pronounced, but NO-arg was without effect (Fig. 3B). The effect of HA was potentiated by both NO-arg and 7NI.

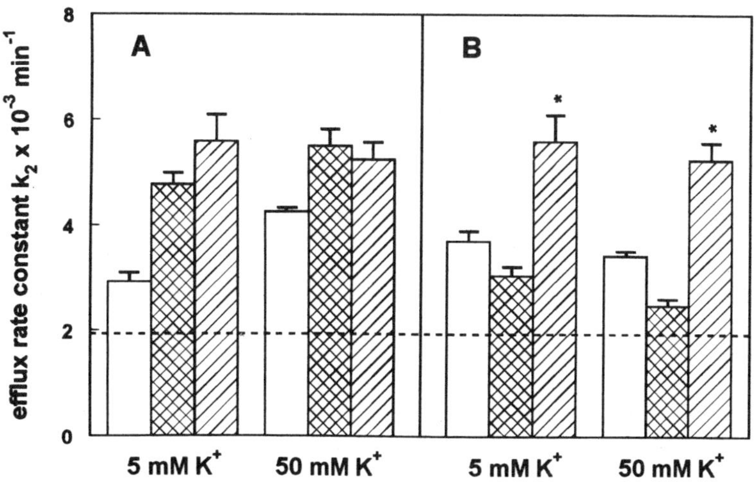


**Figure 4.** Effects of $Ca^{2+}$-free (A) and $Na^{+}$-free (B) media on HA-evoked release of taurine from mouse hippocampal slices. The slices were superfused for 50 min in the above conditions. The bars show the release in $Ca^{2+}$-free or $Na^{+}$-free media (open bars), in $Ca^{2+}$- or $Na^{+}$-free and 5 mM HA-containing media (cross-hatched bars) and in $Ca^{2+}$-, $Na^{+}$-, and 5 mM HA-containing media (hatched bars). The results are efflux rate constants $k_2$ (mean ± SEM, n = 4) for the time interval of 34-50 min. Significant differences when compared to the HA-evoked release: *$p<0.01$.

When $Ca^{2+}$ was omitted from the beginning of the superfusion, the HA-evoked release, both at normal-$K^{+}$ and high-$K^{+}$, remained the same (Fig. 4A). When $Na^{+}$ was absent from the beginning both release processes were significantly reduced (Fig. 4B). One mM hypotaurine or β-alanine had no effect on the HA-evoked release, though they potentiated taurine release by 90 ± 9 and 113 ± 17 per cent (mean ± SEM, n=4), respectively, when applied alone.

## DISCUSSION

The NO-releasing agents produce NO by different mechanisms. SNP forms NO extracellularly by spontaneous decay. SNAP, being presumably membrane-impermeant, also releases NO in the extracellular space[32]. SNP and SNAP were without any effect on taurine uptake, though they have previously been shown to inhibit striatal dopamine uptake by lowering $V_{max}$[25] and hippocampal glutamate uptake by affecting both $K_m$ and $V_{max}$[26]. In contrast to SNP and SNAP, HA can be regarded as an intracellular generator of NO, since it penetrates easily into cells and is broken down by a catalase-dependent reaction[5]. Its effects might be expected to mimic the increase in intracellular NO production. HA inhibited taurine uptake concentration-dependently. Intracellular NO production thus reduces taurine uptake, affecting the transporters in the membrane directly. The physical properties of membranes may also be altered by NO, since nonsaturable diffusion of taurine was markedly increased, in addition to changes in carrier-mediated transport. This kind of mechanism constitutes a novel type of trans-synaptic retrograde regulation of transport. It is particularly noteworthy that an increase in NO production abolished high-affinity taurine uptake, which is thought to remove taurine from the synaptic cleft after its neuromodulatory actions. Moreover, an inhibition of taurine reuptake by NO may provide neuroprotection against excitotoxicity.

Hippocampal taurine release was modified by the NO donors, both inhibitory and excitatory effects being discernible. The basal release of taurine was increased by HA and SNP, indicating that NO production evokes taurine release directly without the involvement of glutamate receptors. SNP and SNAP also enhance taurine release from cortical neurons[3] and from the striatum in vivo[8], but the concentation-dependent potentiation of HA was shown for the first time here. HA also stimulates the basal release of transmitters, e.g., that of noradrenaline, dopamine, acetylcholine, GABA and glutamate[16,17]. Thus, NO is capable of acting directly upon presynaptic terminals to enhance as a retrograde messenger the release of both transmitters and taurine.

While the NO-producing compounds increased the basal release of taurine, the $K^+$- and NMDA-evoked releases were generally reduced. The activation of NMDA receptors is known to induce NO synthesis in the hippocampus[6], whereas the endogenous production of NO is responsible for a substained blockade of NMDA receptor activity[19]. The NO-generating agents, particularly SNP, diminish NMDA-mediated changes in intracellular $Ca^{2+}$ in neurons, independent of the NMDA-receptor redox state and cGMP production[9,24], and block glutamate accumulation[21]. Accordingly, the NMDA-evoked release of taurine was now strongly reduced by the NO donors, especially by SNP. The inhibition of NO synthesis by NO-arg also inhibited taurine release when NO-arg was added at the beginning of superfusions. The location and timing of NO synthesis are apparently important for the effects of NO to manifest themselves, as suggested by Izumi et al[10]. Moreover, different doses of NO donors can increase or decrease neurotransmitter release, as in the case of SNP-induced striatal dopamine release[36]. A biphasic modulation of GABA release by SNAP and nitric oxide synthase inhibitors has also been observed in the hippocampus in vivo[7,33]. It appears that the overall degree of neuronal excitability may depend on the concentration of NO and subsequent increases or decreases in the release of inhibitory (taurine and GABA) and excitatory amino acids.

The release of taurine has been shown to be mediated by at least two different mechanisms, $Ca^{2+}$-dependent exocytotic release and reversal of $Na^+$-dependent carrier-mediated uptake[27]. The NO-evoked release of neurotransmitters seems to operate by means of the same mechanisms[23]. Moreover, the HA-stimulated striatal release of dopamine occurs by a $Ca^{2+}$-dependent process[17], while the NO-evoked release is independent of $Ca^{2+}$ [35]. In the present study, the HA-evoked release of taurine was independent of $Ca^{2+}$. This is in keeping with the SNAP-induced release of taurine from cortical neurons[3], but not with the $Ca^{2+}$-dependent release in the striatum in vivo[8]. On the other hand, the HA-induced release of taurine was strongly reduced in $Na^+$-free media, indicating a reversed operation of carrier-mediated transport, as in cortical neurons[3]. However, there was no trans-stimulation by the uptake inhibitors, hypotaurine and β-alanine, in the presence of HA. The absence of additivity of hypotaurine/β-alanine and HA effects may indicate that they are mediated by the same process which is maximally stimulated by each effector.

In conclusion, the results show that taurine uptake by cerebral cortical synaptosomal preparations is markedly reduced by the NO donor HA. HA and other NO-generating agents, SNP and SNAP, are able to modify the basal, $K^+$- and NMDA-evoked release of taurine in hippocampal slices. In general, the basal release was increased but the evoked releases reduced. The NO-induced transport of taurine is in principle similar to those of neurotransmitters.

## ACKNOWLEDGMENTS

The skillful technical assistance of Mrs. Oili Pääkkönen and Mrs. Irma Rantamaa and the financial support of the Medical Research Fund of Tampere University Hospital are gratefully acknowledged.

## REFERENCES

1. Bruhwyler, J., Chleide, E., Liégeois, J.F., and Carreer, F., 1993, Nitric oxide: a new messenger in the brain, *Neurosci. Biobehav. Rev.*, 17:373-384.
2. Bugnon, O., Schaad, N.C., and Schorderet, M., 1994, Nitric oxide modulates endogenous dopamine release in bovine retina, *Neuroreport*, 5:401-404.
3. Chen, D.-Z., Ohkuma, S., and Kuriyama, K., 1996, Characterics of nitric oxide-evoked [$^3$H]taurine release from cerebral cortical neurons, *Neurochem. Int.*, 28:601-607.
4. Dawson, V.L. and Dawson, T.M., 1996, Nitric oxide actions in neurochemistry, *Neurochem. Int.*, 29:97-110.
5. DeMaster, E.G., Raij, L., Archer, S.L., and Weir, E.K., 1989, Hydroxylamine is a vasorelaxant and a possible intermediate in the oxidative conversion of L-arginine to nitric oxide, *Biochem. Biophys. Res. Commun.*, 163:527-533.
6. Garthwaite, J., 1991, Glutamate, nitric oxide and cell-cell signalling in the nervous system, *Trends Neurosci.*, 14:60-67.
7. Getting, S.J., Segieth, J., Ahmad, S., Biggs, C.S., and Whitton, P.S., 1996, Biphasic modulation of GABA release by nitric oxide in the hippocampus of freely moving rats in vivo, *Brain Res.*, 717:196-199.
8. Guevara-Guzman, R., Emson, P.C., and Kendrick, K.M., 1994, Modulation of in vivo striatal transmitter release by nitric oxide and cyclic GMP, *J. Neurochem.*, 62:807-810.
9. Hoyt, K.R., Tang, L.-H., Aizenman, E., and Reynolds, I.J., 1992, Nitric oxide modulates NMDA-induced increases in intracellular $Ca^{2+}$ in cultured rat forebrain neurons, *Brain Res.*, 592:310-316.
10. Izumi, Y., Clifford, D.B., and Zorumski, C.F., 1992, Inhibition of long-term potentiation by NMDA-mediated nitric oxide release, *Science*, 257:1273-1276.
11. Kamisaki, Y., Maeda, K., Ishimura, M., Omura, H., Moriwaki, Y., and Itoh, T., 1994, NO enhancement by nitric oxide of glutamate release from P2 and P3 synaptosomes of rat hippocampus, *Brain Res.*, 644:128-134.
12. Kamisaki, Y., Wada, K., Nakamoto, K., and Itoh, T., 1995, Nitric oxide inhibition of the depolarization-evoked glutamate release from synaptosomes of rat cerebellum, *Neurosci. Lett.*, 194:5-8.
13. Kontro, P., 1984, Comparison of taurine, hypotaurine, and ß-alanine uptake in brain synaptosomal preparations from developing and adult mouse, *Int. J. Dev. Neurosci.*, 2:465-470.
14. Kontro, P. and Oja, S.S., 1987, Taurine and GABA release from mouse cerebral cortex slices: potassium stimulation releases more taurine than GABA from developing brain, *Dev. Brain Res.*, 37:277-291.
15. Lonart, G. and Johnson, K.M., 1995, Characterization of nitric oxide generator-induced hippocampal [$^3$H]norepinephrine release. I. The role of glutamate, *J. Pharmacol. Exp. Ther.*, 275:7-13.
16. Lonart, G., Wang, J., and Johnson, K.M., 1992, Nitric oxide induces neurotransmitter release from hippocampal slices, *Eur. J. Pharmacol.*, 220:271-272.
17. Lonart, G., Cassels, K.L., and Johnson, K.M., 1993, Nitric oxide induces calcium-dependent [$^3$H]dopamine release from striatal slices, *J. Neurosci. Res.*, 35:192-198.
18. Lowry, O.H., Rosebrough, N.J., Farr, A.L., and Randall, R.J., 1951, Protein measurement with the Folin phenol reagent, *J. Biol. Chem.*, 193:265-275.
19. Manzoni, O., Prezeau, L., Marin, P., Deshager, S., Bockaert, J., and Fagni, L., 1992, Nitric oxide-induced blockade of NMDA receptors, *Neuron*, 8:653-662.
20. Montague, P.R., Gancayco, C.D., Winn, M.J., Marchase, R.B., and Friedlander, M.J., 1994, Role of NO production in NMDA receptor-mediated neurotransmitter release in cerebral cortex, *Science*, 263:973-977.
21. Oh, S. and McCaslin, P.P., 1995, The iron component of sodium nitroprusside blocks NMDA-induced glutamate accumulation and intracellular $Ca^{2+}$ elevation, *Neurochem. Res.*, 20:779-784.

22. Ohkuma, S., Narihara, H., Katsura, M., Hasegawa, T., and Kuriyama, K., 1995, Nitric oxide induced [$^3$H]GABA release from cerebral cortical neurons is mediated by peroxynitrite, *J. Neurochem.*, 65:1109-1114.
23. Ohkuma, S., Katsura, M., Chen, D.-Z., Narihara, H., and Kuriyama, K., 1996, Nitric oxide-evoked [$^3$H]γ-aminobutyric acid release is mediated by two distinct release mechanisms, *Mol. Brain Res.*, 36:137-144.
24. Omerovic, A., Leonard, J.P., and Kelso, S.R., 1994, Effects of nitroprusside and redox reagents on NMDA receptors expressed in Xenopus oocytes, *Mol. Brain Res.*, 22:89-96.
25. Pogun, S., Baumann, M.H., and Kuhar, M.J., 1994, Nitric oxide inhibits [$^3$H]dopamine uptake, *Brain Res.*, 641:83-91.
26. Pogun, S., Dawson, V., and Kuhar, M.J., 1994, Nitric oxide inhibits [$^3$H]dopamine transport in synaptosomes, *Synapse*, 18:21-26.
27. Saransaari, P. and Oja, S.S., 1992, Release of GABA and taurine from brain slices, *Prog. Neurobiol.*, 38:455-482.
28. Saransaari, P. and Oja, S.S., 1994, Taurine release from mouse hippocampal slices: effects of glutamatergic substances and hypoxia, *Adv. Med. Exp. Biol.*, 359:279-287.
29. Saransaari, P. and Oja, S.S., 1997, Taurine release from the developing and ageing hippocampus: stimulation by agonists of ionotropic glutamate receptors, *Mech. Ageing Dev.*, in press.
30. Saransaari, P. and Oja, S.S., 1997, Enhanced taurine release in cell-damaging conditions in the developing and ageing mouse hippocampus, *Neuroscience*, 79:847-854.
31. Saransaari, P. and Oja, S.S., 1997, Glutamate-agonist-evoked taurine release from the adult and developing mouse hippocampus in cell-damaging conditions, *Amino Acids*, in press.
32. Schuman, E.R. and Madison, D.V., 1994, Nitric oxide and synaptic function, *Annu. Rev. Neurosci.*, 17:153-183.
33. Segieth, J., Getting, S.J., Biggs, C.S., and Whitton, P.S., 1995, Nitric oxide regulates excitatory amino acid release in a biphasic manner in freely moving rats, *Neurosci. Lett.*, 200:101-104.
34. Semba, J., Sakai, M., Miyoshi, R., and Kito, S., 1995, $N^G$-monomethyl-L-arginine, an inhibitor of nitric oxide synthase, increases extracellular GABA in the striatum of the freely moving rat, *Neuroreport*, 6:1298-1300.
35. Stewart, T.L., Michel, A.D., Black, M.D., and Humphrey, P.P.A., 1996, Evidence that nitric oxide causes calcium-independent release of [$^3$H]dopamine from rat striatum in vitro, *J. Neurochem.*, 66:131-137.
36. Zhu, X.-Z. and Luo, L.-G., 1992, Effect of nitroprusside (nitric oxide) on endogenous dopamine release from rat striatal slices, *J. Neurochem.*, 59:932-935.

# THE *IN VIVO* RELEASE OF TAURINE IN THE STRIATONIGRAL PATHWAY

Maria Alessandra Colivicchi,[1] Loria Bianchi,[1] John P. Bolam,[3] Francesca Galeffi,[1] Maria Frosini,[2] Mitri Palmi,[2] Gianpietro Sgaragli,[2] and Laura Della Corte[1]

[1] Dipartimento di Farmacologia Preclinica e Clinica "M. Aiazzi Mancini"
Università degli Studi di Firenze
Vle G.B. Morgagni 65
50134 Firenze, Italia
[2] MRC Anatomical Neuropharmacology Unit
Mansfield Road
Oxford, OX1 3TH
UK
[3] Istituto di Scienze Farmacologiche
Università degli Studi di Siena
Via E.S. Piccolomini 170
53100 Siena, Italia

## INTRODUCTION

Cortical information is processed in the neostriatum (STR) and then transmitted to the output nuclei of the basal ganglia, the entopeduncular nucleus (EP) and the substantia nigra pars reticulata (SNr). In the mammalian neostriatum the majority of neurones are the medium-size spiny neurons, sending their axons to the globus pallidus (GP) or the EP/SNr, by either the "direct" or the "indirect" pathways[26]. The "direct" pathway consists of a direct projection from the neostriatum to the EP/SNr, while the "indirect" pathways consist of a striatal projection to the GP and thence to the output nuclei directly or via the subthalamic nucleus (STN). Striatonigral projection neurons, neurons of the GP, SNr and EP are GABAergic, whereas those of the STN use glutamate. Activation of the "direct" pathway thus leads to inhibition in the EP/SNr whereas activation of the "indirect" pathways produces the opposite effect i.e. disinhibition; the imbalance between the two is thought to underlie disorders of movement[1,6,12,15].

The inhibitory amino acid taurine (2-aminoethanesulphonic acid) has also been proposed to subserve a neurotransmitter or neuromodulatory role in the direct striatonigral pathway[9,11]. Uptake and release studies of exogenous radiolabelled taurine suggest that neurons identified as medium-size, densely spiny striatonigral neurons contain taurine and release it at their terminals in the SNr.

Recent studies using dual probe microdialysis have demonstrated that it is possible to selectively stimulate the release of endogenous GABA and possibly taurine from the local axon collaterals of striatonigral neurons by intrastriatal application of the excitatory amino acid non-NMDA receptor agonist, kainic acid (KA), and detect at the same time the release of GABA and taurine from their axon terminals in the SNr[3,4]

The first aim of the present study was to examine the KA-evoked release of endogenous taurine in comparison to that of GABA from the indirect pathway at the level of the GP, i.e. pallidonigral neurons using dual probe microdialysis.

The SNr contains a population of GABAergic neurons, which possess local axon collaterals[13] and receive excitatory (glutamatergic) inputs derived from the subthalamic nucleus[23], the cortex[22] and possibly the mesopontine[17]. *In vivo* microdialysis studies in the freely moving rats have demonstrated that stimulation of excitatory amino acid receptors, which are presumably associated with the excitatory inputs, by perfusion with kainic acid (KA) induces a DNQX-sensitive release of GABA[4]. At least some of this released GABA may be derived from the terminals of striatonigral neurons. Since taurine has also been proposed to be present in these neurons and has been shown to be released in the SNr in response to stimulation of striatonigral neurons[3,11], the second aim of the present study was to determine whether endogenous taurine is released in response to excitatory amino acid receptor stimulation in a manner similar to that of GABA.

## MATERIALS AND METHODS

The experiments were performed on male Sprague-Dawley rats. Vertical single cannula microdialysis probes were implanted stereotassically into the right GP (2 mm exposed tip) and the ipsilateral SNr (1 mm exposed tip) or in the SNr alone, as previously described[3]. Twenty four hours later, the probes, of the now freely moving rats, were perfused with artificial cerebral spinal fluid (CSF) at a rate of 1.2 μl/min. After a perfusion period of 2 h the GP and the SNr or the SNr alone were exposed to 100 μM KA in CSF, alone or in the presence of 10 or 100 μM DNQX (6,7-dinitroquinoxaline-2,3-dione) or 3 or 10 μM TTX (tetrodotoxin) throughout the experiment. Fractions of the perfused fluid (20 min) were collected for the 2 h period before the KA perfusion and for up to 3 h afterwards. GABA and taurine in the perfusate fractions were derivatised, separated by HPLC and the orthophtalaldehyde derivative was detected fluorimetrically[10] and expressed either as pmol of amino acid/μl of perfusate or as the area under the concentration-time curve after stimulation as percentage of the basal area prior to stimulation. Data were evaluated by analysis of variance of the original values as described previously[4].

## RESULTS

### Effect of the intrastriatal application of KA on the release of GABA and taurine from the pallidonigral pathway

The intrapallidal application of the selective non-NMDA agonist, KA (100 μM, 20 min) significantly enhanced the local release of both GABA (+365%, stimulated area as % of basal area) and taurine (+161%), whereas only the release of GABA (+368%) was significantly enhanced distally at the nigral site (Table 1). When KA was applied to the

GP in the presence of the selective non-NMDA receptor antagonist DNQX (100μM) or the sodium channel blocker TTX (10 μM), the GABA output was blocked both locally and distally in the SNr. Similarly to GABA, the taurine output in the GP was also blocked by the presence of TTX or DNQX.

**Table 1.** Effect of DNQX and TTX on GABA and taurine output induced by intrapallidal KA application, simultaneously detected in GP and SNr, and by intranigral application, detected locally in the SNr.

| | | Output (stimulated area % of basal area) | | | |
|---|---|---|---|---|---|
| Intrapallidal Treatment | n | Pallidal Probe | | Nigral Probe | |
| | | GABA | Taurine | GABA | Taurine |
| KA 100 μM | 5 | 465 ± 135* | 261 ± 54* | 468 ± 81* | 131 ± 33 |
| KA 100 μM + DNQX 100 μM | 4 | 126 ± 35 | 106 ± 10 | 193 ± 86 | 103 ± 24 |
| KA 100 μM + TTX 10 μM | 3 | 108 ± 18 | 102 ± 15 | 110 ± 25 | 105 ± 26 |
| Intranigral Treatment | n | | | Nigral Probe | |
| | | | | GABA | Taurine |
| KA 100 μM | 5 | | | 257 ± 15* | 167 ± 11* |
| KA 100 μM + DNQX 10 μM | 4 | | | 89 ± 24 | 104 ± 19 |
| KA 100 μM + TTX 3 μM | 3 | | | 128 ± 27 | 97 ± 15 |

*$p<0.05$, MANOVA on original area values. Figures represent mean ± approx. s.e.m.

### Effect of intranigral application of KA on the release of GABA and taurine

The intranigral application of KA (100 μM, 20 min) significantly enhanced the local release of both GABA (+157%) and taurine (+ 67%) (Table 1). However, the release of GABA and taurine was abolished when KA was applied in the presence of DNQX (100 μM) or TTX (10 μM).

## DISCUSSION

The expected consequences of activation of the "direct" and "indirect" pathways, in terms of neurotransmitter release, are summarised in Table 2.

Stimulation of the direct striatonigral neurons should lead to an increase of GABA release from their axon collaterals and from their terminals in the SNr, whereas a reduction of GABA release and a facilitation of the release of glutamate should be expected from stimulation of the indirect pathway. Activation of the indirect one at the pallidal level, however, should have the opposite effect of increasing GABA release and decreasing that of glutamate in SNr.

**Table 2.** Expected consequences of activation of the "direct" and "indirect" pathways, in terms of neurotransmitter release

| | | |
|---|---|---|
| **STIMULATION OF NEOSTRIATUM RESULTS IN:** | | |
| • **GABA release in substantia nigra (SNr)** | ⇑ | **- direct (*striato-nigral*) pathway** |
| • **globus pallidus (GP) activity** | ⇓ | **- indirect (*striato-pallidal*) pathway** |
| • **GABA release in SNr** | ⇓ | **- indirect (*pallido-nigral*) pathway** |
| • **glutamate/aspartate release in SNr** | ⇑ | **- indirect (*pallido-STN-nigral*) pathway** |
| **STIMULATION OF GLOBUS PALLIDUS RESULTS IN:** | | |
| • **GABA release in SNr** | ⇑ | **- indirect (*pallido-nigral*) pathway** |
| • **glutamate/aspartate release in SNr** | ⇓ | **- indirect (*pallido-STN-nigral*) pathway** |

Previous *in vivo* studies[4], using dual probe microdialysis, have demonstrated that indeed it is possible to selectively stimulate the release of endogenous GABA from the local axon collaterals of striatonigral neurons, by intrastriatal application of the excitatory non-NMDA agonist KA and detect, at the same time, the release of GABA from their axon terminals in the SNr. Furthermore, in agreement with the suggested presence of taurine in a subpopulation of striatonigral neurons[9,11], taurine was also released, both locally in the STR and distally in the SNr[3]. Like GABA, the release of taurine from the distal probe was DNQX- and TTX-sensitive, indicating its dependence on neostriatal non-NMDA receptor activation and propagation of impulse along the striatonigral pathway. The release detected locally, however, differed significantly in that it was far greater and insensitive to DNQX or TTX[3].

In the present set of experiments KA stimulation of the GP induced a DNQX- and TTX-sensitive release of GABA both at the local site of stimulation and distally in the SNr. Taurine behaved like GABA showing an increased local release that was blocked by the presence of DNQX and TTX. Distally in the SNr, however, the release of taurine was unchanged. These observations are consistent with the known chemical anatomy of the basal ganglia. They demonstrate that the release of GABA in the SNr is dependent upon the activation of excitatory non-NMDA receptors present in the GP and the propagation of the action potential along the pallidonigral pathway. Thus, they are consistent with the findings indicating that there is a GABAergic projection from the GP to the SNr[27]. Furthermore, they support the presence of taurine in striatonigral but not in pallidonigral neurons, consistent with previous observations where the terminals that accumulated exogenous taurine in the SNr were morphologically characterised as those originating in the STR rather than in the GP[11].

The other finding of the present study is that endogenous GABA and taurine are also released from the SNr in response to local perfusion with KA. This release is mainly due to the stimulation of excitatory amino acid receptors of the non-NMDA type, as the enhanced release of both taurine and GABA was blocked by DNQX. These data are consistent with the presence of excitatory amino acids in some of the afferents of the SNr (see Introduction for references) and the presence of non-NMDA excitatory amino acid receptors[2,19]. The stimulated release of both of the amino acids was also abolished in the presence of TTX indicating a dependence on fast sodium channels and thus on axon potential propagation. This suggests that the enhanced release in response to KA is not an effect directly on GABA- or taurine-containing axon terminals but more likely to be an effect at the level of neuronal cell bodies or dendrites. Thus the release is likely to be derived from the local axon collaterals of GABA neurons in the nigra[13] or due to an indirect action of KA on GABA-containing terminals mediated by a class of neurons containing a different transmitter e.g. the dopaminergic neurons. Since there is no evidence for taurine-containing neurons in the SNr then the latter mechanism is more applicable to the release of taurine. If we thus make the conclusion that the DNQX- and the TTX-dependence of the taurine release indicates a neuronal origin of the stimulated release, then it is likely to have been derived from afferent axon terminals. These data thus provide additional support for the possible association of taurine with the striatonigral pathway[9,11]. The data, however, do not preclude the notion that at least part of the released taurine and GABA is derived from non-neuronal sources.

If we now compare the behaviour of the release of taurine induced by local application of KA in the different areas of the basal ganglia, it is interesting to observe that, while GABA is released in a quantitatively similar DNQX- and TTX-sensitive manner in the 3 examined areas, STR[3], GP and SNr (see table 1), the KA-induced release of taurine is characterized by some striking differences (Figure 1). The behaviour of taurine in the STR is completely different from that of GABA, i.e. much higher in magnitude and insensitive to both DNQX and TTX, whereas the behaviour of the KA-induced release of taurine in GP and SNr is similar, in terms of both magnitude and sensitivity to DNQX and TTX, to that of GABA. The lack of effect of both DNQX and TTX on the local release of taurine induced by KA in the STR suggests the involvement of mechanism(s) different from that responsible either for the local release of GABA induced by KA in this area or for the local release of taurine induced by KA in GP and SNr. Such mechanisms are independent of DNQX-sensitive receptors and do not involve the propagation of action potentials. As discussed earlier[3], one possibility is that the large increase of taurine output observed in the neostriatum may be an excitotoxic effect of the locally applied KA. Excitatory amino acid receptor agonists, however, have been reported to cause toxic actions in different regions of the mammalian brain through activation of their specific receptors[8,14]. Cellular swelling is an early component of the toxicity produced by these agonists, due to $Na^+$ entry through receptor-operated channels and subsequent entry of $Cl^-$ plus water[7,24]. Since taurine release has been observed following several stimuli inducing osmotic stress in the rat brain[25,28-29], taurine released by excitatory amino acids may be involved in correcting osmotic imbalances, thus behaving as an osmolite. It is interesting to note that part of the KA-induced release of taurine from rat hippocampus, either in *in vitro*[18] or in *in vivo*[20] conditions has been suggested to be related to excitotoxicity/osmotic stress.

The involvement of taurine in cell volume control to maintain structural glial-neuronal integrity does not preclude its ability to simultaneously or subsequently act as an inhibitory modulator. Recent immunohistochemical and biochemical studies have suggested that this indeed may be the case in the supraoptic nucleus[21]. In this brain area taurine released from pituicytes, the astrocytes of the posterior pituitary, under basal and hypo-osmotic conditions may act to suppress axon terminal depolarisation and thereby depress release of neuro-hypophysial peptides. Mechanisms involving osmolality-related release of taurine from glial cells could contribute, at least in part, to the massive DNQX- and TTX-insensitive taurine release observed in the STR following KA stimulation. Part of the released taurine, however, could still originate from the axon collaterals of the striatonigral neurons to exert its inhibitory effects. Those non-selective non-neuronal mechanisms, however, do not seem to be operative, when the same concentration of KA is applied to the GP or SNr. In these areas the KA-induced release of taurine behaves similarly to that of GABA, suggesting a neuronal origin for both of these amino acids.

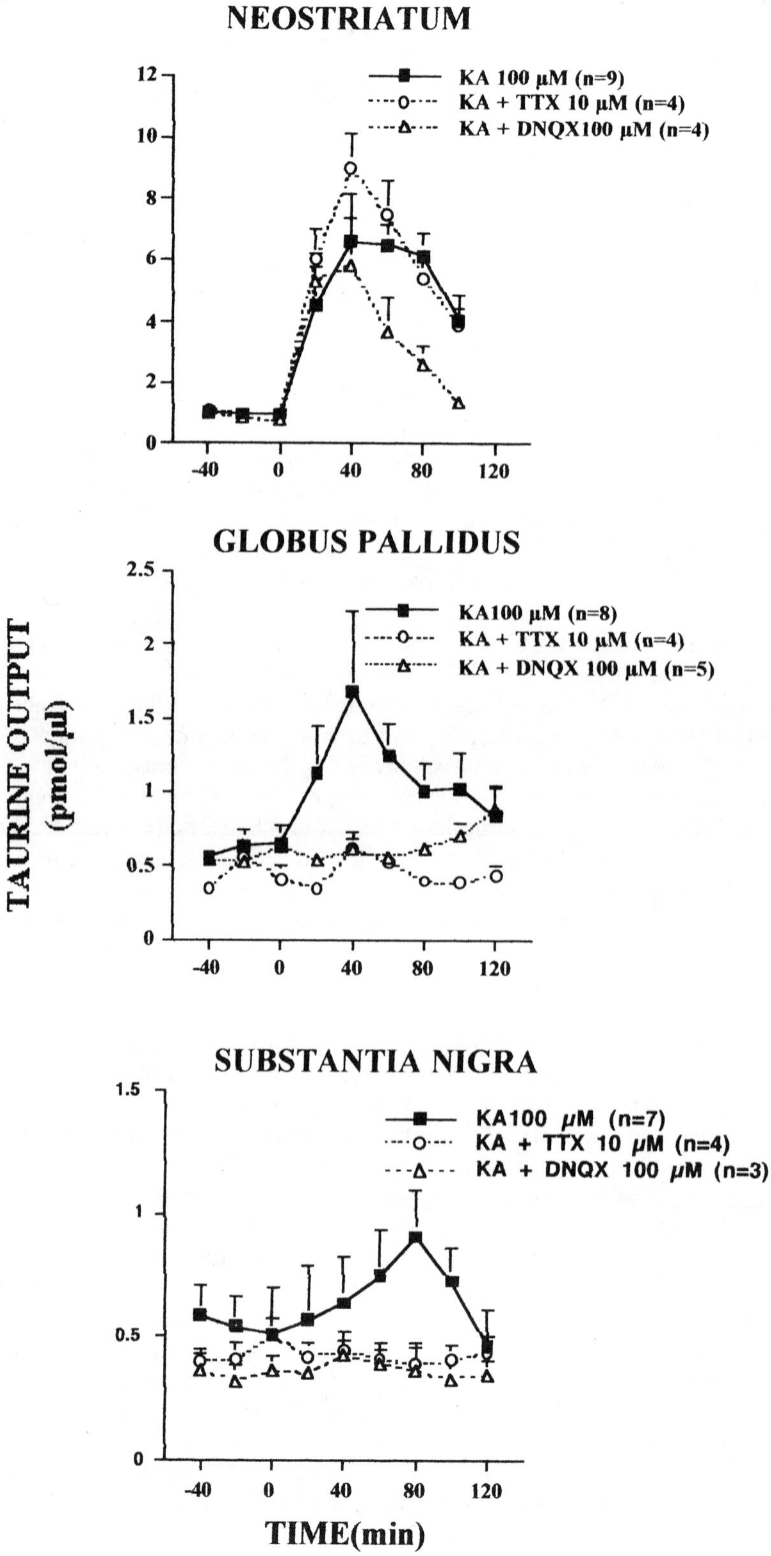


**Figure 1.** Time course of the release of taurine induced by locally applied KA (at time 0 for 20 min) to the STR (upper panel), GP (center panel) and SNr (lower panel) and the effect of TTX 10 μM and DNQX 100 μM (present throughout the experiment). Figures represent mean ± S.E.M.

In terms of the possible final targets of taurine released from the striatonigral neurons, taurine has been shown to behave like GABA and glycine in inhibiting the firing rate of nigral dopaminergic neurones by activation of specific glycine-[16,30] and GABA-linked[30] $Cl^-$ conductances. Furthermore, taurine has been shown to bind to a $GABA_A$ receptor subclass, which is particularly enriched in the substantia nigra[5]. The possibility of the existence of specific taurine recognition sites and the physiological significance of the selective neuronal release of taurine in the brain, however, remain to be established.


## ACKNOWLEDGMENTS

This work was supported by the European Community (BMH1 CT94-1402), CNR-Bilateral Project (contract n. 96.00086.04), Ministero dell'Universita' e della Ricerca Scientifica e Tecnologica, Roma, Italy and the Medical Research Council, UK.


## REFERENCES


1. Albin, R.L., Young, A.B., and Penney, J.B., 1989, The functional anatomy of basal ganglia disorders, *TINS*, 12:366-375.
2. Bernard, V., Gardiol, A., Faucheux, B., Bloch, B., Agid, Y., and Hirsch, E.C., 1996, Expression of glutamate receptors in the human and rat basal ganglia. Effect of dopaminergic denervation on AMPA receptor gene expression in the striatopallidal complex in Parkinson's disease and rat with 6-OHDA lesion, *J. Comp. Neurol.*, 368:553-568.
3. Bianchi, L., Bolam, J.P., Galeffi, F., Frosini, M., Palmi, M., Sgaragli, G.P., and Della Corte, L., 1996, *In vivo* release of taurine from rat neostriatum and substantia nigra, in Adv. Exp. Med. Biol. "Taurine 2: Basic and Clinical Aspects", Huxtable, R.J., Azuma, J., Kuriyama, K., Nakagawa, M., and Baba, A., eds., Plenum Press, New York, Vol. 403, pp 427-433.
4. Bianchi, L., Sharp, T., Bolam, J.P., and Della Corte, L., 1994, The effect of kainic acid on the release of GABA in rat neostriatum and substantia nigra, *NeuroReport*, 5:1233-1236.
5. Bureau, M.H. and Olsen, R.W., 1991, Taurine acts on a subclass of $GABA_A$ receptors in mammalian brain in vitro, *Eur. J. Pharmacol.*, 207:9-16.
6. Chevalier, G. and Deniau, J.M., 1990, Disinhibition as a basic process in the expression of striatal functions, *TINS*, 13:277-280.
7. Choi, D.W., 1987, Ionic dependence of glutamate neurotoxicity, *J. Neurosci.*, 7:369-379.
8. Choi, D.W., Koh, J., and Peters, S., 1988, Pharmacology of glutamate neurotoxicity in cortical cell culture: attenuation by NMDA antagonists, *J. Neurosci.*, 8:185-196.
9. Clarke, D.J., Smith, A.D., and Bolam, J.P., 1983, Uptake of [$^3$H]taurine into medium-size neurones and into identified stratonigral neurones in the rat neostriatum, *Brain Res.*, 289:342-348.
10. Della Corte L., Bianchi L., Federico S., and Michelassi S., 1993, In vivo HPLC estimation of extra-cellular aspartate, glutamate, taurine and GABA in rat striatum: improved methodology, *Eur. J. Neurosci.*, Suppl., 6:261.
11. Della Corte, L., Bolam, J.P., Clarke, D.J., Parry, D.M., and Smith, A.D., 1990, Sites of [$^3$H]taurine uptake in the rat substantia nigra in relation to the release of taurine from the striatonigral pathway, *Eur. J. Neurosci.*, 2:50-61.
12. De Long, M.R., 1990, Primate models of movement disorders of basal ganglia origin, *TINS*, 13:281-285.
13. Deniau, J.M., Kitai S.T., Donoghue J.P., and Grofova, I., 1982, Neuronal interactions in the substantia nigra pars reticulata through axon collaterals of the projection neurons, *Exp. Brain Res.*, 47:105-113.
14. Frandsen, A., Drejer, J., and Schousboe, A., 1989, Direct evidence that excitotoxicity in cultured neurones is mediated via N-methyl-D-aspartate (NMDA) as well as non-NMDA receptors, *J. Neurochem.*, 53:297-299.
15. Gerfen, C.R., 1992, The neostriatal mosaic: Multiple levels of compartmental organisation, *TINS*, 15:133-139.
16. Häusser, M.A., Yung, W.H., and Lacey, M.G., 1992. Taurine and glycine activate the same $Cl^-$ conductance in substantia nigra dopamine neurones, *Brain Res.*, 571:103-108.

17. Lavoie, B. and Parent, A., 1994, Pedunculopontine nucleus in the squirrel monkey: cholinergic and glutamatergic projections to the substantia nigra, *J. Comp. Neurol.*, 344:232-241.
18. Magnusson, K.R., Koerner, J.F., Larson, A.A., Smullin, D.H., Skilling, S.R., and Beitz, A.J., 1991, NMDA-, kainate- and quisqualate-stimulated release of taurine from electrophysiologically monitored rat hippocampal slices, *Brain Res.*, 549:1-8.
19. Martin, L.J., Blackstone, C.D., Levey, A.J., Huganir, R.L., and Price, D.L., 1993, AMPA glutamate receptor subunits are differentially distributed in rat brain, *Neuroscience*, 53:327-358.
20. Menendez, N., Solìs, J.M., Herreras, O., Herranz, A.S., and Martìn del Rìo, R., 1990, Role of endogenous taurine on the glutamate analogue-induced neurotoxicity in the rat hippocampus in vivo, *J. Neurochem.*, 55:714-717.
21. Miyata, S., Matsushima, O., and Hatton, G.I., 1997, Taurine in rat posterior pituitary: localization in astrocytes and selective release by hypoosmotic stimulation, *J. Comp. Neurol.*, 381:513-523.
22. Naito, A. and Kita, H., 1994, The cortico-nigral projection in the rat: an anterograde tracing study with biotinylated dextran amine, *Brain Res.*, 637:317-322.
23. Rinvik, E. and Ottersen, O.P., 1993, Terminals of subthalamonigral fibres are enriched with gluatamte-like immunoreactivity: an electron microscopic, immunogold analysis in the cat, *J. Chem. Neuroanat.*, 6:19-30.
24. Rothman, S.M., 1985, The neurotoxicity of excitatory amino acids is produced by passive chloride influx, *J. Neurosci.*, 5:1483-1489.
25. Pasantes-Morales, H. and Schousboe, A., 1988, Volume regulation in astrocytes: a role for taurine as an osmoeffector, *J. Neurosci.*, 20:505-509.
26. Smith, A.D. and Bolam, J.P., 1990, The neural network of the basal ganglia as revealed by the study of synaptic connections of identified neurones, *TINS*, 13:259-285.
27. Smith, A.D. and Bolam, J.P., 1991, Convergence of synaptic inputs from the striatum and the globus pallidus onto identified nigrocollicular cells in the rat: a double anterograde labelling study, *Neuroscience*, 44:45-73.
28. Solis, J.M., Herranz, A.S., Herreras, O., Lerma J., and Martin del Rio, R., 1988, Does taurine act as an osmoregulatory substance in the rat brain? *Neurosci. Lett.*, 91:53-58.
29. Wade, J.V., Olson, J.P., Samson, F.E., Nelson, S.R., and Pazdernik, T.L., 1988, A possible role for taurine in osmoregulation within the brain, *J. Neurochem.*, 51:740-745.
30. Ye, G.-l., Tse, A.C.O., and Yung, W.-h., 1997, Taurine inhibits rat substantia nigra pars reticulata neurons by activation of GABA- and glycine-linked chloride conductance, *Brain Res.*, 749:175-179.

# TAURINE-INDUCED POTENTIATION IS PARTIALLY REVERSED BY LOW-FREQUENCY SYNAPTIC STIMLATION

Nuria Del Olmo,[1] Julián Bustamante,[2] Rafael Martín del Río,[1] and José M. Solís[1]

[1] Servicio de Neurobiología
Depto. de Investigación
Hospital Ramón y Cajal
28034-Madrid
[2] Depto. de Fisiología
Facultad de Medicina
Universidad Complutense
28040-Madrid, Spain

## INTRODUCTION

Neuroinhibitory actions of taurine in the brain are well known to be mediated by the activation of $GABA_A$ receptors[11]. We have recently reported a new neuromodulatory taurine action in the hippocampus independent of this GABA receptor type[8,9]. The new taurine action consists of a long-lasting potentiation of synaptic transmission, that is induced when taurine is applied at a concentration of 5-10 mM over a 10-30 min period. Excitatory postsynaptic potentials (EPSP) increase during taurine application, remaining elevated long after taurine withdrawal (at least three hours). This new taurine action could have important relevance to brain function, since persistent changes in the synaptic strength are envisaged as the cellular substrate of learning and memory[4]. Taurine-induced potentiation, unlike long-term potentiation (LTP) induced by high frequency synaptic stimulation of afferent fibers, is independent of NMDA receptor activation[8]. Nevertheless, taurine-induced potentiation has some points in common with LTP, primarily because its induction requires a rise in intracellular calcium, and both potentiation phenomena occlude mutually (Del Olmo et al., in preparation).

With the aim of assessing similarities between taurine-induced potentiation and LTP, we have studied whether the persistent enhancement of synaptic transmission evoked by taurine was reversable by low-frequency synaptic stimulation (LFS), a procedure known to reverse potentiation in those synapses that previously developed LTP[2,3,6,15].

## MATERIALS AND METHODS

The experiments were carried out on transverse hippocampal slices (400 μm thick), obtained from female Sprague-Dawley rats (200-250 g), using standard procedures. Briefly, after the rat was decapitated, its brain was rapidly removed and placed in ice-cold standard medium (composition in mM: NaCl 119, $NaHCO_3$ 26.2, KCl 2.5, $KH_2PO_4$ 1, $MgSO_4$ 1.3, $CaCl_2$ 2.5 and glucose 11) pregassed with 95% $O_2$ and 5% $CO_2$. Hippocampi were then dissected free of surrounding tissue, and transversal slices were cut by a manual chopper. Prior to recording, the slices were maintained in an interface holding chamber[14] at room temperature (21-25°). After at least 1 hr, a slice was transferred to a submersion-type recording chamber[14], and continuously perfused (flow rate 1.5-2 ml/min) with the standard medium equilibrated with 95% $O_2$ and 5% $CO_2$. Experiments were performed at 30-32°C.

Field excitatory postsynaptic potentials (fEPSP) were evoked in the CA1 area by stimulating Schaffer collateral-commissural fibers with electrical pulses (20-50 μA, 40 μsec, every 15 sec) applied through a bipolar microelectrode located in the stratum radiatum, and were recorded with low resistance micropipettes filled with 2 M NaCl (2-3 MΩ) connected to an Axopatch-1D amplifier. Evoked responses were digitized at 25 KHz (TL-1 interface, Axon Instruments) and stored and analyzed on a 386 IBM compatible computer using pCLAMP software (Axon Instruments). Synaptic strength was calculated using the slope of the initial rising phase of the fEPSP. Values were normalized with respect to the mean values of the responses at the control period, which was before the application of taurine. We used a program developed by J. Bustamante for these calculations.

To avoid taurine activation of $GABA_A$ receptors during the experiments, the standard perfusion solution contained 100 μM picrotoxin, an antagonist of these receptors. To prevent epileptiform discharges caused by picrotoxin, a surgical cut was made between CA1 and CA3, and the concentrations of both calcium and magnesium in the standard medium were increased to 4 mM.

Data are expressed as means ± SEM. Statistical differences were assessed by two-way ANOVA and two-tailed Student´s t-test.

## RESULTS AND DISCUSSION

In a series of experiments we studied whether the application of a LFS protocol (1 Hz during 7 min) was able to reverse the synaptic potentiation induced by the perfusion of 10 mM taurine during a 30 min time period. We interspersed control experiments in which 1 Hz low-frequency stimulation was not delivered with experiments in which LFS was applied 20 min after taurine withdrawal. As it is shown in Figure 1, taurine perfusion under control conditions induced fEPSP potentiation that lasted at least one hour after taurine washout (144.3 ± 4.1% at 55-60 min of washout). This level of potentiation was partially reduced in those slices where LFS was applied (123.7 ± 7.8% at 55-60 min of washout; $p=0.024$). This result is qualitatively similar to those reported by other authors studying the attenuating effect of LFS on LTP, but differs in the magnitude of the reversal. The taurine-induced potentiation is the result of two long-lasting phenomena, an increase in the synaptic efficacy and an increase in the axon excitability. Since the last phenomenon is not affected by 1 Hz LFS[12], it is logical to propose that LFS only reverses the component of taurine-induced potentiation due to the enhancement of synaptic efficacy. In fact, the component of taurine-induced potentiation remaining after LFS quantitively matches the percentage of potentiation due to the increase in axon excitability[8]. Never-

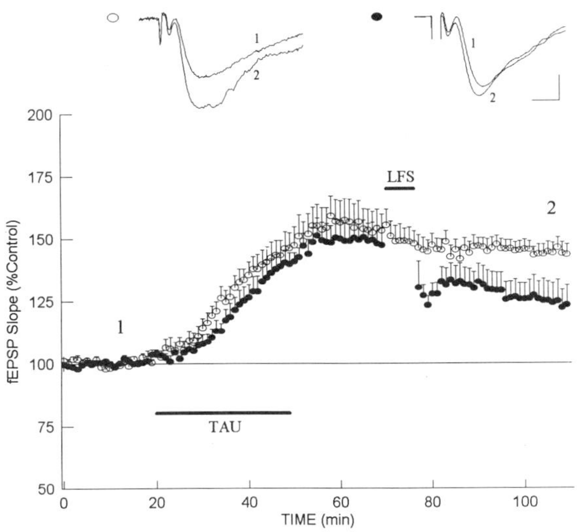


**Figure 1.** Taurine-induced potentiation is partially reversed by LFS. The plots contain summary data that were normalized to the mean value of the 20 min control period and expressed as the mean ± SEM/min. Black horizontal bars indicate either the taurine perfusion period (10 mM, 30 min) or LFS application (1 Hz, 7 min). Open circles represent control experiments (n=9) without LFS, while closed circles denote experiments (n=5) in which LFS was delivered 20 min after taurine withdrawal. The upper traces show representative averaged fEPSP tracing (n=9) recorded at the indicated times on the graph. Calibration: 0.5 mV, 2 msec.

theless, we cannot discount the possibility that the LFS protocol used here was not the most suitable for inducing a complete reversal of taurine effect, since the magnitude of the reversal attained with LFS depends on the frequency and duration of the stimulation protocol, as well as on the time-lag between the induction of LTP and the application of LFS[6,15].

It has been reported[5] that LFS application under basal conditions evokes a long-term depression of synaptic potentials in slices from young rats but not in those from adult rats. Since adult rats were used in this study the LFS protocol did not induce a long-term depression when applied during the control perfusion period (see Fig. 2). However, this lack of LFS effect under control conditions yielded an unexpected result when taurine (10 mM during 30 min) was subsequently added. During the taurine perfusion period, the fEPSP increased in a manner analogous to that seen in control experiments in which LFS was not applied. However, after taurine withdrawal, the fEPSP potentiation decayed faster than in the control experiments (118.5 ± 3.5% and 143.1 ± 5.4% of basal levels one hour after taurine withdrawal, respectively, $p=0.003$). The suppressive effect of a prior LFS application was not inhibited when it was applied in the presence of 50 μM D,L-amino-5-phosphonovalerate, a NMDA receptor antagonist, indicating that this type of glutamate receptor is not involved in the modulatory action of LFS on the subsequent taurine-induced potentiation (data not shown). Other authors[7,10] have also described the suppressive effect of a previous synaptic stimulation, which did not alter basal synaptic transmission, but did impair the subsequent induction of LTP. Yet, these effect was found to depend on NMDA receptor activation. These intriguing phenomena have been termed metaplasticity (the plasticity of synaptic plasticity) by Abraham and Bear[1].

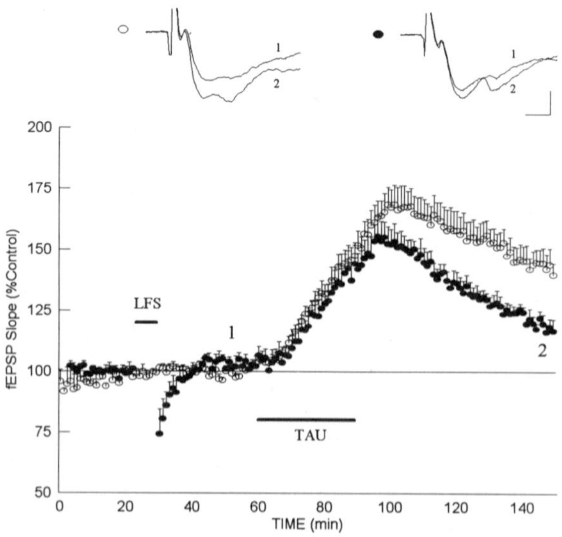


**Figure 2.** LFS applied prior taurine exposure affects subsequent potentiation. The characteristics of the graph and traces are as in Figure 1. Open circles represent control experiments (n=7) in which a baseline of 1 hour was recorded before taurine perfusion, and closed circles represent those experiments (n=6) in which a LFS protocol was applied 30 min before taurine perfusion.

In conclusion, the long-lasting synaptic potentiation induced by taurine resembles LTP with respect to the modulation of their induction mechanisms by some patterns of synaptic stimulation, indicating that taurine is affecting directly or indirectly some of the enzymatic cascades involved in the processes of synaptic plasticity. In this context, it is promising, to search for molecular targets underlying the taurine-induced potentiation, such as the reported actions of taurine on the phosphorylation state of specific proteins[13].

## ACKNOWLEDGMENTS

This work was supported by the "Ministerio de Educación y Cultura" (Grant PM95-170). N. Del Olmo is recipient of a "Ministerio de Educación y Cultura" fellowship. We thanks M.J. Asensio for technical assistance. The presentation of this work was sponsored by "Red Bull España".

## REFERENCES

1. Abraham, W.C. and Bear M.F., 1996, Metaplasticity: the plasticity of synaptic plasticity, *Trends Neurosci.*, 19:126-130.
2 Barrionuevo, G., Schottler, F., and Lynch, G., 1980, The effects of repetitive low frequency stimulation on control and "potentiated" synaptic responses in the hippocampus, *Life Sci.*, 27:2385-2391.
3. Bashir, Z.I. and Collingridge, G.L., 1994, An investigation of depotentiation of long-term potentiation in the CA1 region of the hippocampus. *Exp. Brain Res.*, 100:437-443.
4. Bliss, T.V.P. and Collingridge, G.L., 1993, A synaptic model of memory: long-term potentiation in the hippocampus, *Nature*, 361:31-39.
5. Dudek, S.M. and Bear, M.F., 1993, Bidirectional long-term modification of synaptic effectiveness in the adult and immature hippocampus, *J. Neurosci.*, 13:2910-2918.

6. Fujii, S., Saito, K, Miyakawa, I., and Kato, H., 1991, Reversal of long-term potentiation (depotentiation) induced by tetanus stimulation of the input to CA1 neurons of guinea pig hippocampal slices, *Brain Res.*, 555:112-122.
7. Fujii, S., Kuroda Y., Miura, M., Furuse, H., Sasaki, H., Kaneko, K., Ito, K., Chen, Z., and Kato, H., 1996, The long-term suppressive effect of prior activation of synaptic inputs by low-frequency stimulation on induction of long-term potentiation in CA1 neurons of guinea pig hippocampal slices, *Exp. Brain Res.*, 111:305-312.
8. Galarreta, M., Bustamante, J., Martín del Río, R., and Solís, J.M., 1996, Taurine induces a long-lasting increase of synaptic efficacy and axon excitability in the hippocampus, *J. Neurosci.*, 16:92-102.
9. Galarreta, M., Bustamante, J., Martín del Río, R., and Solís, J.M., 1996, A new neuromodulatory action of taurine: long-lasting increase of synaptic potentials, in Adv. Exp. Med. Biol. "Taurine 2: Basic and Clinical Aspects", Huxtable, R.J., Azuma, J., Nakagawa, M., Kuriyama, K., and Baba, A., eds., Plenum Press, New York, Vol. 403, pp. 463-471.
10. Huang, Y.-Y., Colino, A., Selig, D.K., and Malenka, R.C., 1992, The influence of prior synaptic activity on the induction of long-term potentiation, *Science*, 255:730-733.
11. Huxtable, R.J., 1989, Taurine in the central nervous system and the mammalian actions of taurine, *Prog. Neurobiol.*, 32:471-533.
12. Kobayashi, K., Manabe, T., and Takahashi, T., 1996, Presynaptic long-term depression at the hippocampal mossy fiber-CA3 synapse, *Science*, 273:648-650.
13. Lombardini, J.B., 1994, The inhibitory effects of taurine on protein phosphorylation: Comparison of various characteristics of the taurine-affected phosphoproteins present in rat retina, brain and heart, in Adv. Exp. Med. Biol. "Taurine in Heath and Disease", Huxtable, R.J. and Michalk, D.V., eds., Plenum Press, New York, Vol. 359, pp. 9-17.
14. Nicoll, R.A. and Alger, B.E., 1981, A simple chamber for recording from submerged brain slices, *J. Neurosci. Methods*, 4:153-156.
15. Stäubli, U. and Chun, D., 1996, Factors regulating the reversibility of long-term potentiation, *J. Neurosci.*, 16:853-860.

# TREATMENT OF RAT BRAIN MEMBRANES WITH TAURINE INCREASES RADIOLIGAND BINDING

Pattie Green,[1] Ralph Dawson, Jr.,[1] David R. Wallace,[2] and Jeannie Owens[2]

[1] Department of Pharmacodynamics
University of Florida
Gainesville, Florida 32610
[2] Department of Pharmacology and Physiology
Oklahoma State University
Tulsa, OK 74107

## INTRODUCTION

Taurine is known to have numerous actions to modulate membrane stability and function[9]. Taurine has also been proposed to have antioxidant functions and several studies have demonstrated that taurine can attenuate lipid peroxidation caused by various toxic insults[22]. Our laboratory has studied the protective effects of taurine against oxidative stress caused by metal-stimulated catecholamine oxidation[5]. Taurine was shown to protect proteins from oxidative damage in both *in vitro* and *in vivo* experiments[5]. We previously observed in pilot studies that taurine could prevent the loss of dopamine receptors caused by exposing tissue homogenates to ferric chloride and dopamine[6]. We hypothesized that taurine might function to protect receptor proteins from oxidative damage. We also noted in these pilot studies that taurine alone in the absence of any oxidative damage could enhance radioligand binding.

Previous studies have also shown that taurine increases β-adrenergic binding sites[21], flunitrazepam binding to GABA receptors[13,14] and TCP binding to the phencyclidine (PCP) binding site on the N-methyl-D-aspartate (NMDA) receptor[17]. Taurine has also been shown to inhibit binding to cardiac $\alpha_1$ -adrenergic receptors[7]. The ability of taurine to alter ligand binding could be due to it acting directly at sites on receptors to mimic the actions of glycine or GABA[2,8,16,18]. This would be an unlikely mechanism for taurine's actions on dopamine or α-adrenergic receptors.

We performed a series of experiments to determine if taurine modulated $\alpha_2$-adrenergic binding sites in brain homogenates using the conditions we had previously established in assays that were used to determine taurine's inhibition of oxidative damage to dopamine receptor binding sites[6]. We also examined in more detail the interaction of taurine with

D3 dopamine receptors. Finally, we studied taurine's actions to modulate ligand binding to the PCP site in the NMDA channel, a binding site not directly coupled to a G-protein. Our initial hypothesis was that taurine may increase receptor binding by preventing oxidative damage to receptor proteins that normally occurs due to tissue processing. We present data that suggest taurine may stabilize membranes and improve recovery of specific ligand binding sites identified using traditional radioligand binding methods.

## MATERIALS AND METHODS

Cerebral cortex was obtained from adult female Sprague-Dawley (SD) rats and immediately homogenized in 50 volumes (w/v) ice cold 50 mM Tris buffer (pH=7.4) containing 1 mM $Na_2EDTA$ and 0.1 mM phenylmethanesulphonyl fluoride. The homogenate was centrifuged at 39,000 x g for 10 min at 4°C. The pellet was resuspended in buffer and centrifuged again. The pellet was resuspended in buffer and stored frozen at -80°C until used in receptor binding assays. The homogenates from several rats were pooled and aliquoted for use in the assays for adrenergic and NMDA receptors. Some preliminary studies were performed in whole brain homogenates from adult male SD rats prepared as described above. The homogenates were thawed and centrifuged as described above and were resuspended in 50 volumes (w/v) 100 mM sodium phosphate buffer (pH=7.4). These homogenates were incubated for 30 min at room temperature in a shaking water bath (20-22°C) in the presence or absence of taurine or other amino acids. The homogenates were centrifuged and the pellet was surface washed. The pellet was resuspended in ligand binding buffer.

Tris buffer (50 mM, pH=7.4) was used to measure 2 nM $^3$H-p-aminoclonidine (PAC) binding as previously described[4]. The assay was conducted in a 1.0 ml volume and the assay tubes contained 1 mM $MgCl_2$ and 2 nM $^3$H-PAC with 10 μM yohimbine used to define nonspecific binding. The samples were incubated at room temperature for 30 min prior to termination of the assay by filtration on GF/B filters. MK801 binding was measured in 5 mM Tris buffer that contained 10 nM $^3$H-MK801, 30 μM glycine, and 100 μM glutamate. Nonspecific binding was defined using 100 μM MK801. The assay was terminated by filtration using GF/C filters after a 30 min incubation at room temperature.

Striatal and nucleus accumbens homogenates from adult male SD rats were prepared in a similar fashion as described above and were used to measure dopamine (D3) binding sites. Briefly, the tissue was homogenized in 50 mM Tris buffer containing 1 mM $Na_2$ EDTA (pH=7.4) and then centrifuged at 27,000 x g for 10 minutes at 4°C. The pellets were subsequently washed twice and recentrifuged. The tissue homogenate was incubated in phosphate buffer for 30 minutes at room temperature in the presence or absence of taurine. The final pellet was resuspended in approximately 250 volumes (w/v) of 100 mM sodium phosphate buffer (pH=7.4). D3 dopamine binding sites were assayed as previously described[19,20]. The homogenate was incubated 90 minutes at room temperature with 4 nM (+)-7-hydroxy-N,N-di-n-[3H]propylaminotetralin ($^3$H-DPAT). Saturation analysis was performed using 8 concentrations of $^3$H-DPAT (0.1-8.92 nM). Dopamine (10 μM) was used to define nonspecific binding. The binding assay was terminated by filtration onto GF/C filters.

A Brandel tissue harvester was used to filter and wash the samples in all assays. Bound radioactivity was determined using scintillation spectrophotometry. Protein was assayed in the homogenates by the method of Bradford[1].

## RESULTS AND DISCUSSION

Taurine increased the binding of $^3$H-PAC to cortical membranes in a concentration dependent manner (Fig. 1). Amino acids with similar structure (β-alanine, hypotaurine) also stimulated ligand binding while other sulfur containing compounds (cysteine and glutathione) did not significantly increase binding sites (Fig. 2). The fact that glutathione did not increase $^3$H-PAC binding argues against an antioxidant mechanism of action for taurine in preventing receptor degradation under this set of conditions. We have previously found that glutathione at equimolar concentrations was much better at inhibiting protein oxidation than taurine[5]. GABA was also without effect on $^3$H-PAC binding (Fig. 3). The failure of GABA to significantly increase binding sites argues against the actions of taurine being mediated via GABA receptor sites.

In preliminary experiments, taurine increased the binding of two different $\alpha_2$-adrenergic antagonists (rauwolscine & idazoxan). Whereas agonists may bind preferentially to high affinity states of the receptor, antagonists bind to both high and low

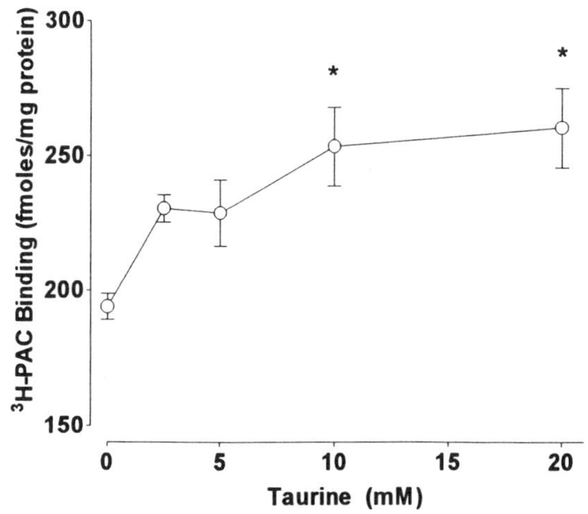


**Figure 1.** The effects of increasing concentrations of taurine to increase $^3$H-PAC binding to rat cortical membranes. *$p<0.05$ versus no taurine condition. N=6 replications.

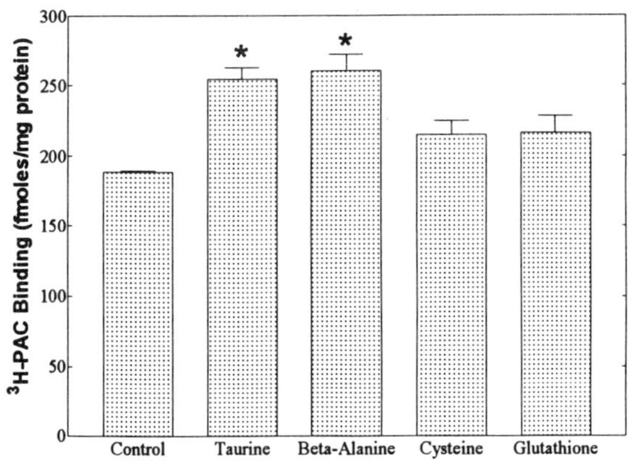


**Figure 2.** The effects of taurine and related amino acids (10 mM) on PAC binding in cortical homogenates. Taurine and β-alanine significantly (*$p<0.05$) increased the number of binding sites. N=4 replications.

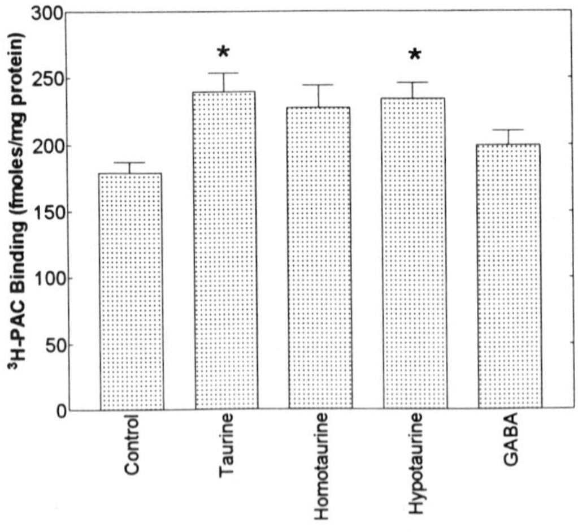


**Figure 3.** The effects of taurine and other related amino acids (10 mM) on PAC binding to rat cortical membranes. Taurine and hypotaurine significantly (*$p<0.05$) increased binding sites. N=4 replications.

affinity states of adrenergic receptors[3]. The increase in $^3$H-PAC binding caused by taurine was observed in the presence of 1 mM $MgCl_2$ which shifts the receptor to the agonist preferring state[3]. This suggests that taurine was not acting to alter the receptor affinity state by simply shifting $\alpha_2$-adrenergic binding sites to the agonist preferring state.

Taurine (20 mM) significantly increased striatal D3 dopamine binding sites, but higher concentrations did not significantly alter binding (Fig. 4). The increase in D3 binding sites was due to an increase in $B_{max}$. A saturation analysis was performed with 10 mM taurine present in the preincubation period which resulted in an approximately 40% increase in $B_{max}$ versus membranes not treated with taurine (Table 1). The affinity (Kd) was also slightly decreased by taurine when compared to the control (Table 1). These data suggest that taurine's main action in these experiments was to increase the number of binding sites. Comparable results were observed in experiments using homogenates from the nucleus accumbens (Table 1). When taurine was present in the preincubation period, but the homogenates were not centrifuged (NC) prior to the binding assay, taurine had no effect to increase radioligand binding. Thus, taurine's major action appeared to be to pre-

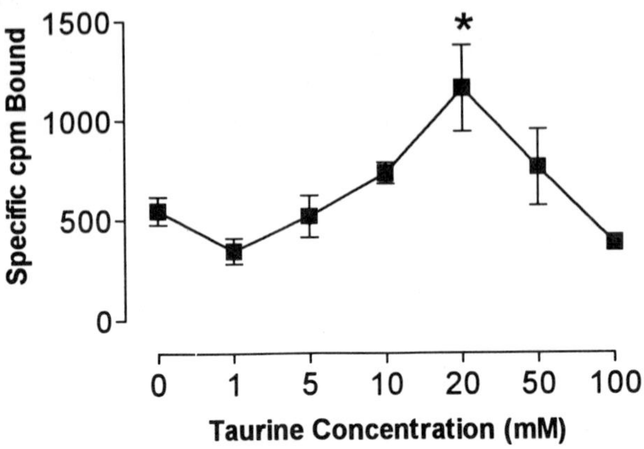


**Figure 4.** The effects of taurine on $^3$H-DPAT binding to striatal membranes. Taurine at 20 mM significantly (*$p<0.05$) increased binding of $^3$H-DPAT to D3 dopamine binding sites. N=6.

**Table 1.** The Effect of Taurine on D3 Dopamine Receptor Affinity and Number

| Experimental Condition | Kd (nM) | $B_{max}$ (fmoles/mg wet weight) |
|---|---|---|
| Striatum/No taurine | 4.3 ± 0.7 | 6.3 ± 0.4 |
| Striatum/10 mM taurine | 6.7 ± 1.9 | 8.8 ± 1.3 |
| Striatum/No taurine/NC | 5.2 ± 1.5 | 7.5 ± 1.0 |
| Striatum/10 mM taurine/NC | 6.3 ± 1.8 | 6.0 ± 0.9 |
| N. Accumbens/No taurine | 1.2 ± 0.3 | 2.5 ± 0.5 |
| N. Accumbens/10 mM taurine | 2.4 ± 0.5 | 4.2 ± 0.4 |

The data are the means of 3 separate replications. NC=tissue not centrifuged prior to the ligand binding assay.

vent the loss of binding sites that was associated with the centrifugation step. Similar results were also obtained when a single concentration $^3$H-PAC was used to measure $\alpha_2$-adrenergic binding sites in homogenates not subjected to centrifugation prior to the binding assay (data not shown). Taurine placed directly in the receptor binding assay also had no effect on PAC binding. The action of taurine appeared to be to prevent the loss of binding sites rather than actually increasing available binding sites. These results support a role for taurine at physiological intracellular concentrations (10-20 mM) to stabilize receptor proteins in the membrane.

The effect of taurine to stimulate radioligand binding was not limited to G-protein coupled receptors ($\alpha_2$-adrenergic or D3 dopamine binding sites) since MK801 binding was also increased by 20 mM taurine (Fig. 5). The increased MK801 binding occurred in the presence of 30 μM glycine so an allosteric interaction to facilitate the binding to the channel (PCP) site on the NMDA receptor would appear unlikely. Other preliminary experiments suggested that the increase in radioligand binding mediated by taurine treatment was due to membrane stabilization and a reduction in the loss of soluble receptor protein from the tissue homogenates. The total protein content of the taurine treated membranes was 94 ± 4% of the control preparations for all the experiments conducted using cerebral cortex, so an effect of taurine to enhance total protein recovery was not evident. Taurine had no effect on ligand binding when the preincubation of the homogenates was carried out in Tris buffer. Phosphate buffer was used in the preincuba-

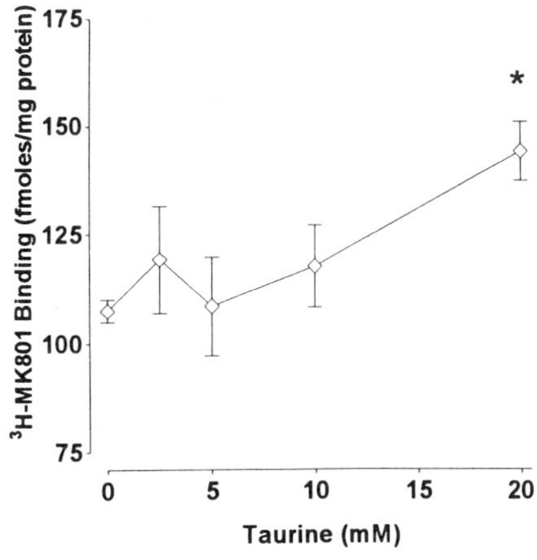


**Figure 5.** The effects of taurine on MK801 binding to cortical homogenates. Taurine (20mM) significantly (*p<0.05) increased MK801 binding. N=4 replications.

tions because previously we had found that Tris buffer attenuated taurine's ability to inhibit metal-stimulated catecholamine oxidation. This effect is likely due to the metal chelating and free radical scavenging properties of Tris buffer. Tris buffer may also have membrane stabilizing actions. We hypothesize that taurine may prevent receptor protein loss by a membrane stabilizing action similar to that previously reported[10].

Taurine has been shown to interact with a number of receptor systems[2,7,8,11-13,15-18,21]. Taurine appears to act as an agonists at both glycine and GABA receptors[2,8,16,18]. Taurine at micromolar concentrations can stimulate flunitrazepam binding to the GABA receptor at 37°C, but inhibit binding a 4°C[13,15]. The stimulatory effects on the GABA receptor appear due to a specific interaction at the GABA recognition site on the $GABA_A$ receptor protein[15,16]. The inhibitory actions are speculated to be a general membrane fluidity interaction via taurine's reported interactions with membrane phospholipids[15]. Saransarri and Oja[17] have shown that 10 μM taurine can stimulate binding to the PCP binding site in NMDA receptor channel. This effect could be blocked by glycine antagonists and appears to be mediated by taurine's action at the glycine site on the NMDA receptor[12,17]. We found that taurine at high concentrations (20 mM) could stimulate MK801 binding to the PCP site in the NMDA receptor channel in the presence of both glycine and glutamate. GABA had no effect on $\alpha_2$-adrenergic binding, so we believe that the stimulatory effects observed at $\alpha_2$-adrenergic, D3 dopamine and PCP binding sites were not mediated via either GABA or glycine recognition sites. Franconi et al.[7] reported that taurine (1-20 mM) could inhibit $\alpha_1$-adrenergic antagonist binding to ventricular membranes that had been preincubated (30 min at 25°C) with taurine. This effect was reflected as a change in Bmax and could not be produced by β-alanine. We found that β-alanine could increase $\alpha_2$-adrenergic binding sites but sulfur containing amino acids (Cys, GSH) were ineffective. Taurine has been shown to decrease membrane phospholipid mobility and increase membrane fluidity in the presence of calcium[14]. Huxtable and Bresssler[10] demonstrated that taurine could improve the preservation of the membrane integrity and reduce membrane damage as indexed by functional measures of membrane calcium binding and uptake. Furthermore, the protective effect of taurine was not mimicked by cysteine. Taurine may decrease the loss of ligand recognition sites in membranes by stabilizing the membrane and preventing solubilization or degradation of the receptor. Taurine may act to inhibit processes unrelated to oxidative damage to prevent proteolysis or protein degradation. Taurine may also exhibit antioxidant properties in the local membrane environment that cannot be mediated by glutathione or cysteine. Our results suggest caution in interpreting studies of receptor function in tissue preparations isolated in the presence of high concentrations of taurine. As we have demonstrated, taurine may have nonspecific membrane actions to increase the number of ligand binding sites for multiple neurotransmitters.


## ACKNOWLEDGMENT

The authors wish to acknowledge the technical assistance of Joel Cherdack.


## REFERENCES


1. Bradford, M.M., 1976, A rapid and sensitive method for the quantitation of microgram quantities of protein utilizing the principle of protein-dye binding, *Anal. Biochem.*, 72:248-254.
2. Bureau, M.H. and Olsen, R.W., 1991, Taurine acts on a subclass of $GABA_A$ receptors in mammalian brain in vitro, *Eur. J. Pharmacol.*, 207:9-16.

3. Bylund, D.B. and U'Prichard, D.C., 1983, Characterization of α1- and α2-adrenergic receptors, *Intl. Rev. Neurobiol.*, 24:343-431.
4. Dawson, R. and Oparil, S., 1987, Renal catecholamines and $\alpha_2$-adrenergic receptors in salt-related and genetic hypertension, *Pharmacology*, 34:131-142.
5. Dawson, R., Tang, E., Shih, D., Hern, H., Hu, M., Baker, D., and Eppler, B., in press, Taurine inhibition of iron-stimulated catecholamine oxidation, in Adv. Exp. Med. Biol. "Taurine....",. Huxtable, R., Lombardini, J.B., and Schaffer, S., eds., Plenum Press, ....
6. Eppler, B., Wallace, D.R., Pelleymounter, M.A., Booze, R.M., Liu, S., and Dawson, R., 1996, Taurine-dopamine interactions in the striatum, *Soc. Neurosci. Abst.* 22:1495.
7. Franconi, F., Bennardini, F., Matucci, R., Stendardi, I., Failli, P., Manzini, S., and Giotti, A., 1986, Functional and binding evidence of taurine inhibition of α-adrenoceptor effects on guinea-pig ventricle, *J. Mol. Cell. Cardiol.*, 18:461-468.
8. Horikoshi, T., Asanuma, A., Yanagisawa, K., Anzai, K., and Goto, S., 1988, Taurine and β-alanine act on both GABA and glycine receptors in *Xenopus* oocyte injected with mouse brain messenger RNA, *Mol. Brain Res.*, 4:97-105.
9. Huxtable, R., 1989, Taurine in the central nervous system and the mammalian actions of taurine, *Prog. Neurobiol.*, 32:471-533.
10. Huxtable, R. and Bressler, R., 1973, Effect of taurine on a muscle intracellular membrane, *Biochim. Biophys. Acta*, 323:573-583.
11. Kontro, P., and Oja, S.S., 1990, Interactions of taurine with $GABA_B$ binding sites in mouse brain, *Neuropharmacology*, 29:243-247.
12. Lopez-Colome and Fragoso, G., 1995, Glycine stimulation of glutamate binding to chick retinal pigment epithelium, *Neurochem. Res.*, 20:887-894.
13. Medina, J.H. and De Robertis, E., 1984, Taurine modulation of the benzodiapzepine-γ-aminobutyric acid receptor complex in brain membranes, *J. Neurochem.*, 42:1212-1217.
14. Nakashima, T., Shima, T., Sakai, M., Yama, H., Mitsuyoshi, H., Inaba, K., Matsumoto, N., Sakamoto, Y., Kashima, K., and Nishikawa, H., 1996, Evidence of a direct action of taurine and calcium on biological membranes, *Biochem. Pharmacol.*, 52:173-176.
15. Quinn, M.R. and Miller, C.L., Taurine allosterically modulates flunitrazepam binding to synaptic membranes, *J. Neurosci. Res.*, 33:136-141.
16. Quinn, M.R. and Harris, C.L., 1995, Taurine allosterically inhibits binding of [$^{35}$S]-t-butylbicyclo-phosphorothionate (TBPS) to rat brain synaptic membranes, *Neuropharmacology*, 34:1607-1613.
17. Saransaari, P. and Oja, S.S., 1993, Phencyclidine-binding sites in mouse cerebral cortex during development and ageing: effects of inhibitory amino acids, *Mech. Ageing Dev.*, 68:125-136.
18. Schmieden, V., Kuhse, J., and Betz, H., 1992, Agonist pharmacology of neonatal and adult glycine receptor α subunits: identification of amino acid residues involved in taurine activation, *EMBO J.*, 11:2025-2032.
19. Wallace, D.R. and Booze, R.M., 1995, The dopamine D3-selective radioligand, [$^3$H]-7-OH-DPAT, identifies sigma receptors in rat basal forebrain, *J. Neurochem.*, 64:700-710.
20. Wallace, D.R. and Booze, R.M., 1997, Upregulation of (+)-7-hydroxy-N,N-di-n-[3H]propy-laminotetralin binding following intracerebroventricular administration of nitric oxide generator, *Neurochem. Res.*, 22:163-170.
21. Williams, M., Risley, E.A., and Totaro, J.A., 1980, Interaction of taurine and β-alanine with central nervous system neurotransmitter receptors, *Life Sci.*, 26:557-560.
22. Wright, C.E., Tallan, H.H., and Lin, Y.Y., 1986, Taurine: biological update, *Ann. Rev. Biochem.*, 55:427-453.

# TAURINE MODULATES GLUTAMATE- AND GROWTH FACTORS-MEDIATED SIGNALING MECHANISMS

A. El Idrissi,[1,2] C. Harris,[3] and E.Trenkner[1,2]

[1] CUNY Graduate School and University Center
33 West 42 Street
New York, NY 10036
[2] New York State Institute for Basic Research and
Developmental Disabilities
1050 Forest Hill Road
Staten Island, New York 10314
[3] CSI / IBR Center for Developmental Neuroscience and
Developmental Disabilities
2800 Victory Blvd.
Staten Island, NY 10314

## INTRODUCTION

In the central nervous system, the formation and maintenance of neuronal connections are regulated, in part, through a balanced interaction between intracellular and extracellular signals. Under physiological conditions, these signals regulate neuronal maturation, survival and function. Slight changes of this balance however, result in significant functional and structural changes, leading to pathological conditions, loss of function and subsequently to cell death. Prominent among these signals are growth factors and neuro-active amino acids (NAAs). The purpose of these studies was to examine the combined effects of growth factors (Gfs) and neuroactive amino acids (NAAs) on mouse cerebellar granule cells (CGC) survival, energy-metabolism, calcium homeostasis, and protein phosphorylation. We have demonstrated that taurine at physiological concentrations had a neurotrophic effect and protected neurons against glutamate excitotoxicity. These effects were partially mediated through the modulation of intracellular calcium homeostasis[34,17]. Here we report that also cellular energy metabolism was affected by taurine. Furthermore, as a consequence of its calcium modulatory role, taurine regulated protein kinase C (PKC) activity during glutamate depolarization. Finally, taurine down-regulated the glutamate-induced phosphorylation of a specific set of proteins. We further demonstrated that these various effects of taurine were selectively modulated by brain-derived neurotropic factor (BDNF) and basic fibroblast growth factor (bFGF), suggesting that NAAs, the mitochondrial energy-metabolism and growth factors together regulate neuronal survival and function. It is, therefore, of considerable importance to identify the different environmental signals that interact to regulate the development and maintenance of the integrity of neuronal functions, in order to better understand mechanisms that could lead to abnormal development.

## MATERIAL AND METHODS

### Cell Preparation

Cerebellar granule cells were prepared from 7- to 8-day-old (P7/8) mice (C57 Bl/6J) as previously described[31,32]. For serum-free cultures, 24 h after plating in BME + 10 % horse serum, culture medium was replaced by serum-free medium consisting of MEM supplemented with 15% N-2 supplement[3].

### Determination of Mitochondrial Electrochemical Gradient (mt ECG)

Quantitative determination of rhodamine 123 uptake was performed as described by Chen[8]. Rhodamine uptake was normalized to total protein.

### $^{45}Ca$ Uptake

$^{45}Ca^{2+}$ uptake was determined as described by Eimerl and Shramm[15]. Cells were depolarized with glutamate for 20 min at room temperature, washed twice with Locke's solution and dissolved in 0.5 ml NaOH. $^{45}Ca^{2+}$ associated with cells was counted in a Beckman-scintillation-counter.

### Protein Kinase C Assay

PKC was assayed by measuring the incorporation of $^{32}P$ from [$\gamma$-$^{32}P$]ATP into a synthetic peptide from myelin basic protein (MBP) as a specific substrate for PKC according to the supplier's instruction. PKC-specific activity was expressed in pmol/min and normalized to total protein.

### Electrophoresis and Immunobloting

Proteins were separated by 2D electrophoresis on a 7.5 % SDS- polyacrilamide gel and transferred to a nitrocellulose membrane. Blots were blocked with 0.2 % NFDM in D-PBS and incubated overnight at +4$^0$C with a mouse monoclonal anti-phosphotyrosine antibody at 0.4 Fg/ml (Santa Cruz). After two washes in blocking buffer, the membranes were incubated for 1h at room temperature with a rabbit anti-mouse immunoglobulin antibody (1:2,000) diluted in blocking buffer.

## RESULTS

### 1. Taurine down-regulates $^{45}Ca^{2+}$ uptake during glutamate depolarization:

We have previously demonstrated the excitoprotective effects of taurine[17,34]. Here we report that calcium modulation may be one of the mechanisms by which taurine exerts its neurotrophic and neuroprotective effects. As shown in Fig. 1, glutamate induced a significant increase in $^{45}Ca^{2+}$ uptake into cerebellar granule cells grown in serum-free medium. This glutamate-induced $^{45}Ca^{2+}$ uptake was significantly reduced when the cells were pretreated for 24h with 10 mM taurine. Taurine alone however, had no effect on $^{45}Ca^{2+}$ uptake as compared to baseline.

## 2. Effects of BDNF, bFGF and taurine on calcium uptake

BDNF and bFGF, similar to taurine, modulated calcium influx in response to glutamate. As shown in Fig. 2, bFGF at 1 and 10 ng/ml significantly reduced glutamate-induced $^{45}Ca^{2+}$ uptake to almost control levels, suggesting that both taurine and bFGF, act

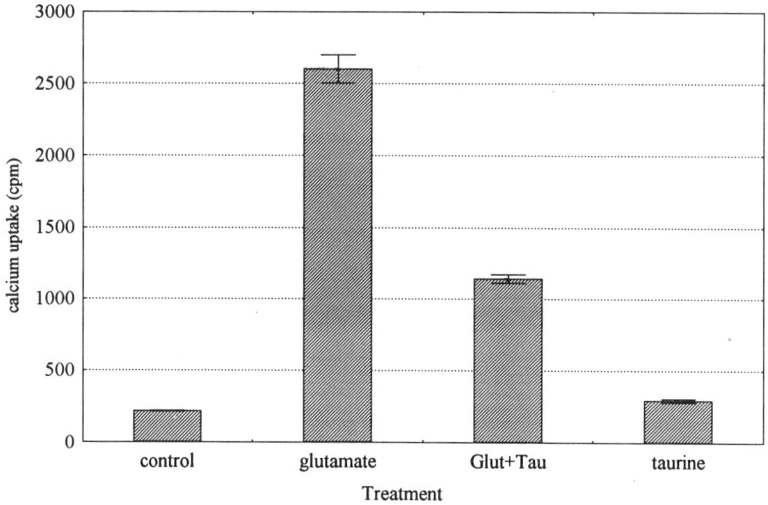


**Figure 1**. Taurine down-regulates glutamate-induced calcium uptake. The role of taurine ($10^{-2}$M) was analyzed after glutamate depolarization (30 min) in serum-free medium. Data represent the mean ± SEM of three different experiments.

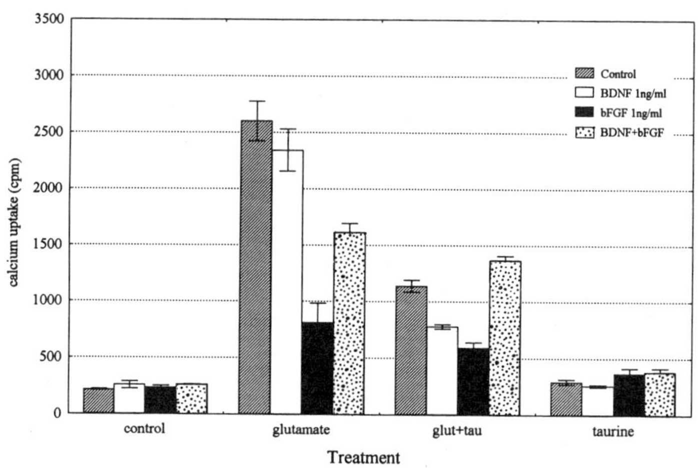


**Figure 2**. Regulation of glutamate-induced calcium uptake by taurine and growth factors. Cultures were treated with taurine (10mM), BDNF (1ng/ml) and bFGF (1ng/ml) 24h prior to glutamate depolarization. Data represent the mean ± SEM of three different experiments.

through the regulation of calcium influx to protect neurons from glutamate excitotoxicity. On the other hand, BDNF at 1 ng/ml had an insignificant effect on glutamate-induced $^{45}Ca^{2+}$ uptake (Fig. 2), suggesting that BDNF-mediated neuroprotection is mediated through other mechanisms. BDNF and bFGF (1 ng/ml) together significantly reduced the glutamate-induced $^{45}Ca^{2+}$ uptake to levels ranging between those of BDNF and bFGF (Fig. 2), indicating that BDNF can affect bFGF activity.

The effects of BDNF, bFGF and taurine on cell survival and calcium uptake require long-term treatment (24 h) of cultures with these factors to achieve maximum effects. In order to characterize the role of taurine in the presence of BDNF and bFGF, glutamate-induced $^{45}Ca^{2+}$ uptake was measured in cultures pretreated with these factors. As expected, a significant reduction in glutamate-induced $^{45}Ca^{2+}$ uptake was observed when cultures were pretreated with taurine (Fig. 2), these levels were comparable to those obtained with bFGF. Taurine was more efficient in reducing glutamate-induced calcium uptake when cultures were pretreated with either BDNF or bFGF (1 ng/ml) alone. Furthermore, when cultures were pretreated with all three factors, glutamate-induced $^{45}Ca^{2+}$ uptake exceeded the levels seen with taurine alone or in combination with either growth factor. In the absence of glutamate however, $^{45}Ca^{2+}$ uptake in the presence of taurine, BDNF and bFGF, was similar to the baseline of untreated controls.

### 3. Increased vulnerability to FCCP in serum-free culture conditions

We first determined whether serum affects the MtECG and influences the degree of vulnerability, in cultured CGCs with or without the addition of taurine and growth factors. In Fig. 3 removal of the serum from the culture medium resulted in a significant decrease in baseline of the MtECG as compared to cultures grown in serum-containing medium. Addition of BDNF (10 ng/ml) or BDNF and taurine (10 mM) to serum-free cultures increased MtECG over baseline. 1mM Glutamate significantly reduced energy levels with or without serum, however the reduction was more pronounced in serum-free medium. After uncoupling the MtECG with carbonyl cyanide p-(trifluoromethoxy) phenyl hydrazone (FCCP) at concentrations that reduce only slightly the mtECG (0.1 μM) a slight but not significant decrease of rhodamine 123-accumulation was observed. Serum-free cultures, on the other hand, were more affected by FCCP than in the presence of serum, indicating a possible cause for the increased vulnerability. Combination of subtoxic concentrations of FCCP (0.1 μM) and glutamate (0.1 mM) caused a significant decrease in the energy levels comparable to those obtained with 10 times the amount of glutamate (1 mM) alone. When added, these data suggest that the degree of susceptibility to glutamate depends on the function of the mitochondria energy metabolism. The reduced mtECG that leads to reduction of energy levels under serum-free medium may well explain the increased vulnerability of granule cells to glutamate.

### 4. Taurine, BDNF and bFGF down-regulate glutamate-induced PKC activity

PKC has been shown as one of the calcium-activated enzymes linking glutamate receptors activation to excitotoxicity[12]. In this study, we investigated the effects of taurine and growth factors on the activity of this enzyme. As has been reported earlier[14,33], glutamate caused a significant increase in PKC activity. Pretreatment of the cultures with taurine (10 mM), however, prevented the glutamate-induced increase in PKC activity, and resulted in activity levels significantly lower than untreated controls (Fig. 4). Similarly, addition of BDNF or bFGF (10 ng/ml) prevented the increase in PKC activity in response to glutamate (Fig. 4). Pretreatment with taurine and bFGF caused a significant decrease in PKC from baseline levels, while taurine and BDNF did not. These data suggest that taurine, BDNF and bFGF modulate PKC activity through regulation of intracellular calcium, as these factors were found to affect intracellular calcium concentrations.

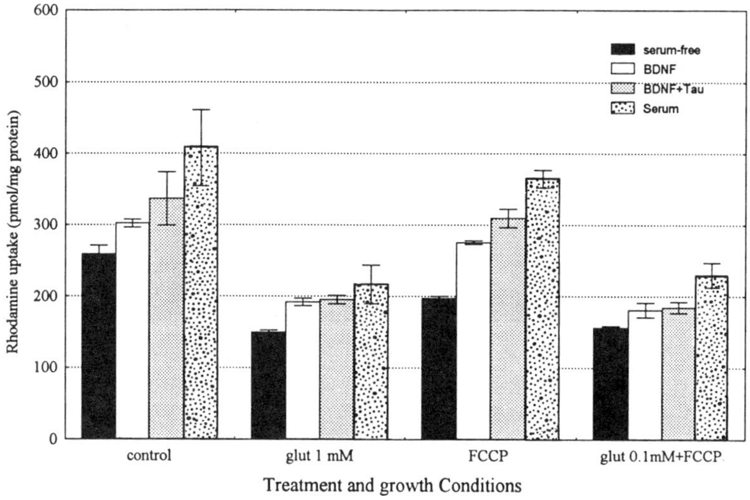


**Figure 3**. Taurine and BDNF restore mitochondrial energy levels. Serum removal resulted in a significant decrease in rhodamine uptake. Addition BDNF (10 ng/ml) or BDNF and taurine (10 mM) resulted in an increase in the MtECG. Glutamate ($10^{-3}$M) resulted in a significant decrease in rhodamine uptake under all culture conditions. When cultures where treated with FCCP (0.1FM), they became more susceptible to glutamate, as ($10^{-4}$M) glutamate caused a similar decrease in the MtECG as compared to $10^{-3}$M glutamate. Data represent the mean ± SEM of three different experiments.

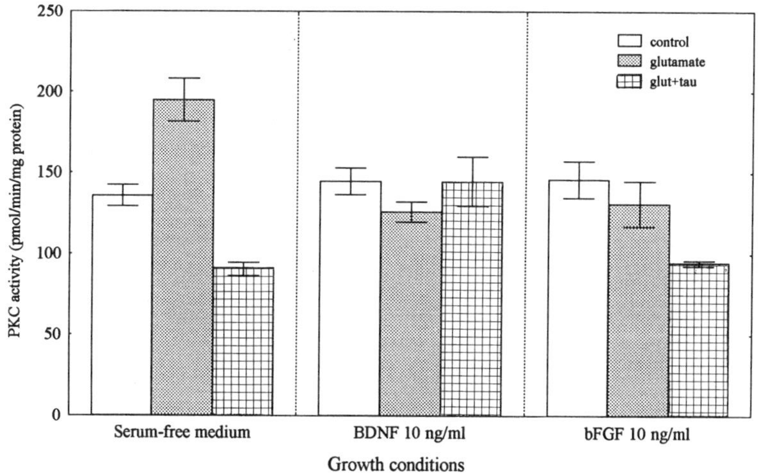


**Figure 4**. Regulation of PKC by taurine, BDNF and bFGF. Glutamate ($10^{-3}$M) caused a significant increase in PKC activity, which was down-regulated by taurine ($10^{-2}$M). In the presence of BDNF and bFGF (10 ng/ml) glutamate failed to activate PKC above control levels. When taurine was combined with bFGF, PKC activity was significantly reduced. Data represent the mean ± SEM of three separate experiments.

## 5. Taurine down-regulate glutamate- and BDNF-induced phosphorylation

Addition of 1mM glutamate to CGCs rapidly stimulated protein phosphorylation. Fig. 5 shows a typical Western blot of two-dimensional gel electrophoretic analysis of phosphoproteins (PP) identified with anti-phosphotyrosine monoclonal antibodies. In the presence 1 mM glutamate, a rapid protein phosphorylation was observed mainly in three sets of polypeptides of apparent molecular weight of 95 (PP95), 55 (PP55) and 20 KDa (PP20). Each set of the same molecular weight PP consisted of at least three labeled spots with different apparent isoelectric points. Pretreatment of CGCs with 10 mM taurine resulted in complete inhibition of the phosphorylation of the PP95, PP55 and a significant decrease in the phosphorylation of the PP20 as shown by quantitative analysis of the phosphorylated proteins (Fig. 6). In untreated or taurine-treated cells, there were no detectable tyrosine-phosphorylated proteins.

BDNF(10 ng/ml) induced different phosphorylation pattern in CGCs, mainly a strong phosphorylation of PP95 and PP55 (Fig. 7). Stimulation with glutamate significantly increased the phosphorylation of the PP55 but not the PP95. Quantitative analysis of the phosphorylated proteins revealed that the phosphorylation levels of the PP95 were not significantly affected by glutamate and taurine. Whereas the phosphorylation of the PP55 was significantly increased by glutamate, and brought back to baseline levels by pretreatment with taurine (Fig. 8), these results clearly indicate that stimulation of CGCs with excitotoxic doses of glutamate for 10-15 min or BDNF 24 h stimulates tyrosine kinase activity. Taurine on the other hand has a modulatory role on tyrosine phosphorylation through reducing glutamate-stimulated phosphorylation.

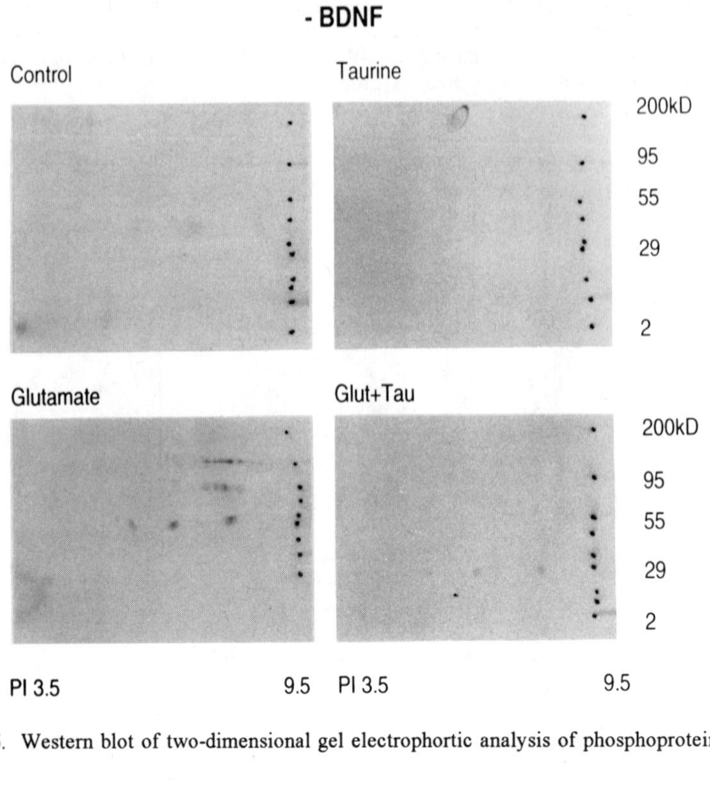


**Figure 5.** Western blot of two-dimensional gel electrophortic analysis of phosphoproteins (pp.)

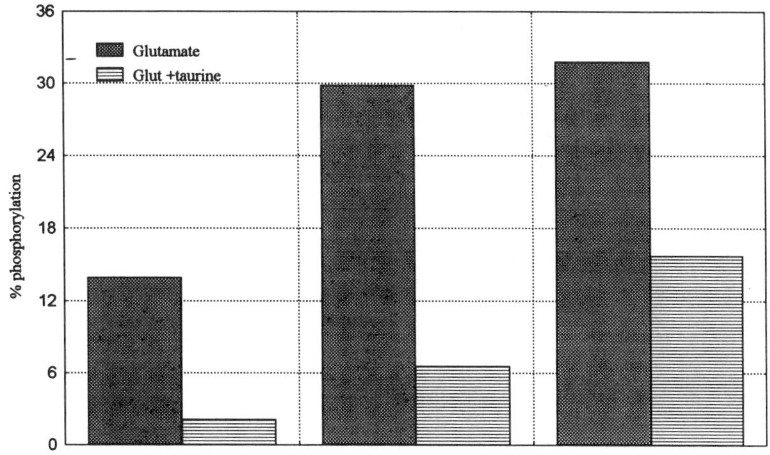


**20 KD Phosphoproteins (isoelectric focusing points between pH 6.5 - 9.5)**

**Figure 6**. Taurine down-regulates glutamate-induced phosphorylation. Quantitative analysis of phosphorylated proteins shows that addition of glutamate ($10^{-3}$M) resulted in a phosphorylation of a set of 20 KD proteins, with three different isoelectric focusing points. The presence of taurine significantly reduced glutamate-induced tyrosine phosphorylation.

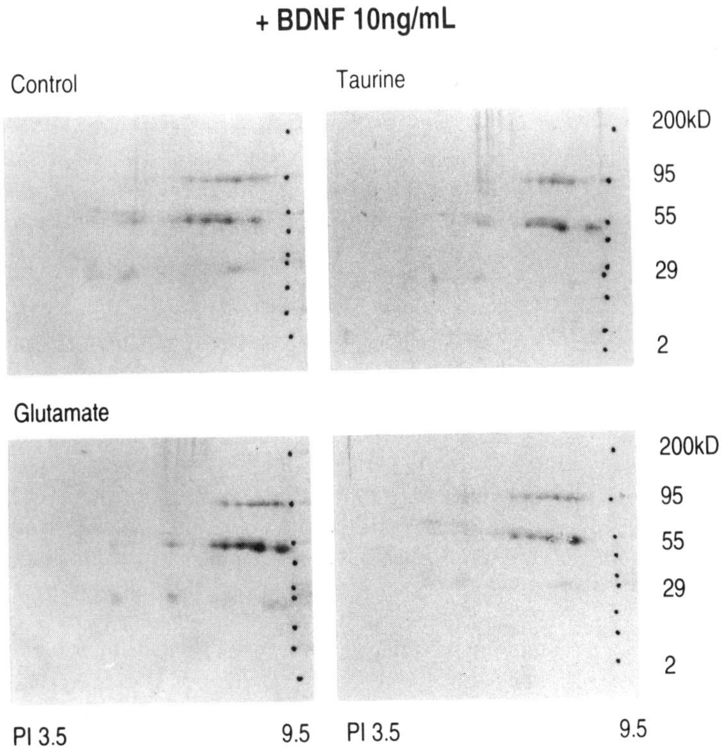


**Figure 7**. Brain-derived neurotropic factor (BDNF) induced phosphorylation patterns in mouse cerebellar granule cells.

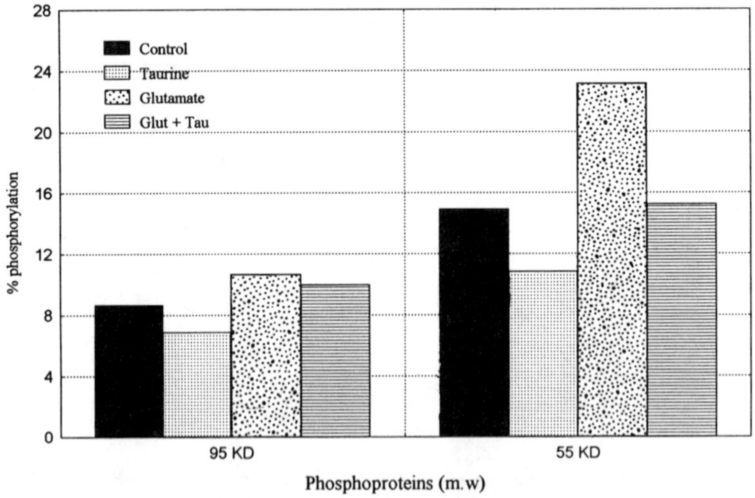


**Figure 8**. Taurine differentially down-regulates protein phosphorylation. When cultures were treated with BDNF (10 ng/ml), different proteins were tyrosine-phosphorylated, mainly a set of 95 KD PP and 55 KD PP. Taurine and glutamate affected only the phosphorylation 55 KD.

## DISCUSSION

Biological cell-systems develop, mature and function in a complex environment that changes depending on developmental stages and location in the organism. Neuroactive amino acids and growth factors are known to regulate crucial steps in the developing nervous system[1,21,24,30]. They are pivotal to development, to maintenance as well as to the function of neuronal cell populations through activation and regulation of second messenger system. Moreover, depending on the timed sequence of events, and the concentration of these factors, they control whether a given neuron will survive or die[11,24].

In these studies we tried to determine whether and how these factors interact during crucial steps in cerebellar development. One of the essential mechanisms in neuronal development is the regulation and maintenance of calcium homeostasis. Calcium ions are ubiquitous intracellular second messengers, acting as key regulators of numerous cellular functions. Because of this, neurons must tightly regulate the free cytosolic calcium ion concentration to allow efficient $Ca^{2+}$-dependent signaling to occur. Furthermore, loss of intracellular calcium homeostasis has been shown to be at the origin of numerous brain pathologies, including glutamate excitotoxicity[10,11,16,25,26]. We have chosen representatives from distinct groups of these signaling molecules that are known to regulate calcium-homeostasis under depolarizing conditions: taurine and glutamate as representatives of neuroactive amino acids, and bFGF and BDNF as examples of two distinct growth factor families. Since all these molecules are present at a given time in the intracellular and the extracellular environment, we have tried for the first time to characterize their interactions and ask the questions whether these molecules induce the same cellular regulatory systems and could substitute for each other, whether the regulation of calcium influx is synergistic or additive and whether these mechanisms are complementary depending on concentration and availability?

One common denominator in the function between bFGF, BDNF and taurine appears to be, among others, their ability to regulate calcium homeostasis[9,25,26]. Our earlier studies, as well as this study, have shown that the neuroactive amino acid taurine prevents glutamate excitotoxicity in mouse cerebellar cells through modulation of calcium uptake[17,34], and down-regulates the degree of calcium uptake under depolarizing conditions to control levels, thus offering alternative pathways and interactive reactions that can regulate calcium homeostasis. This regulatory role of taurine depends on growth conditions, and mitochondrial energy metabolism, as well as the presence of growth factors. When compared to our previous experiments[17,34], serum-free medium provided the least trophic environment for cerebellar granule cells. In serum-free medium, comparable to the composition of CSF, glutamate-induced calcium uptake into cerebellar cells was significantly higher than in the presence of serum[17,34]. However, when incubated for 24 h prior to glutamate-depolarization in the presence of bFGF and/or taurine, calcium uptake was reduced almost to control values (Fig. 2). Although less pronounced in the presence of serum, a similar trend was observed. Under serum-free conditions, mitochondrial energy metabolism was low compared to serum-containing media (Fig. 3) while calcium influx was significantly higher. Both results suggest a possible mechanism for the observed increase in glutamate vulnerability (Fig. 3). Our earlier studies have shown that neuronal cells survived glutamate excitotoxicity, a calcium mediated toxicity[10,11,19,22], with the addition of taurine and/or bFGF[34]. This study demonstrates that the neuroprotective effects of taurine and bFGF are mediated through the regulation of intracellular calcium (Figs. 1 & 2). Furthermore, taurine and bFGF interacted additively (Fig. 3), suggesting compensatory pathways for the regulation of calcium homeostasis.

In contrast to taurine and bFGF, BDNF did not directly affect glutamate-induced calcium uptake, though it protected cerebellar cells against glutamate excitotoxicity. Although the regulation of intracellular calcium homeostasis offers a plausible explanation for the neuroprotective effect of taurine and bFGF, the results of this study inferred a different mechanism for BDNF. Based on the increase in rhodamine uptake in the presence of BDNF (Fig. 4), one of the possible mechanisms should involve the mitochondrial electrochemical gradient.

Energy metabolism is recognized as one of the fundamental factors that control the required balance between maintenance of neuronal structures and their function during development and throughout adult life[2,26]. Also neuronal development and maintenance of function depends on cellular energy[4,6,7]. We have as others recently demonstrated that depleted cellular energy-levels caused by malfunction of the mitochondrial electrochemical gradient (MtECG) increased the vulnerability towards excitotoxins and neurotoxins leading to neuronal malformation and subsequently to cell death[5,617,34]. The decrease in energy levels may well explain the increased vulnerability to glutamate, as demonstrated with the removal of serum from culture medium (Fig. 4). Taurine and BDNF synergistically induced an increase in rhodamine uptake, as indicator for increased mitochondrial function.

This study also examined other calcium-dependent reactions like PKC activity and protein kinase-dependent phosphorylation. PKC represents a family of structurally related isozymes[17,28] that are involved in many aspects of neuronal function, such as neurotransmitter release and modulation of membrane excitability[13,18,28]. Glutamate depolarization resulted in a significant increase in PKC activity (Fig. 4). This PKC induction was prevented by taurine and bFGF. When added together in different ratios, taurine and bFGF interacted additively to regulate PKC activity (Fig. 4) as well as calcium influx, suggesting that the calcium-modulatory effects of taurine and bFGF may underlie their PKC regulation. However, this mechanism may not apply for BDNF's down-regulation of PKC activity, since BDNF did not directly affect calcium influx.

Preliminary results indicated that protein phosphorylation is highly regulated by NAAs and Gfs. Western blot analysis of phosphoproteins identified by anti-phosphotyrosine antibodies revealed specific sets of BDNF- and glutamate-induced phosphoprotein. Consistent with previous findings[23], taurine's regulatory role has been observed again in the pattern of glutamate-induced phosphoproteins, in the presence of BDNF. On the other

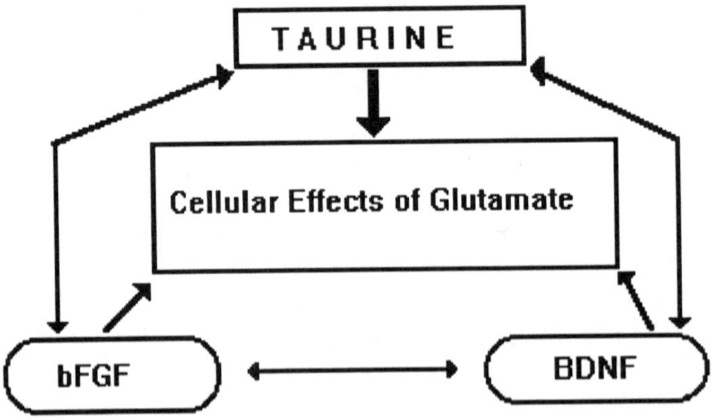


**Figure 9**. Proposed model for the interaction between growth factors and neuroactive amino acids. Glutamate's effects can be excitatory or excitotoxic depending on its extracellular concentration. In the presence of other factors, such as taurine, BDNF and bFGF, the effects of glutamate can be significantly modulated. Other factors may also be present to affect cells in their *in vivo* environment. The net effect of these factors is also regulated by the cellular energy level which was shown to be significantly affected by external and internal stimuli.

hand, in the presence of BDNF, glutamate-depolarization resulted in a phosphorylation of a different set of proteins, one of which was down-regulated by taurine (Fig. 8). These findings suggest that distinct tyrosine kinases induced by BDNF and glutamate may be involved, since the interaction of BDNF with its receptor trkB[29] triggers a cascade of phosphorylation events resulting in the observed phosphorylation pattern. On the other hand glutamate presumably activates different calcium-dependent, tyrosine kinases. A mechanism by which taurine modulates glutamate-induced phosphorylation is suggested by its calcium-modulatory role. Although the exact mechanisms of taurine's action are not fully understood, taurine plays a significant role in the regulation of glutamate-induced signaling events. In particular taurine has opposing effects to those of glutamate.

In summary, this study clearly demonstrates that interactive mechanisms between growth factors and NAAs are operational. Furthermore, some of these factors induce the same regulatory mechanism, e.g. taurine and bFGF reduced calcium uptake; taurine and BDNF increased rhodamine uptake; taurine, bFGF and BDNF all decreased cell death and cell vulnerability. These findings suggest that cells utilize alternative pathways to regulate and maintain their function. These pathways are accessible to cells depending on their developmental stages and the availability of these factors. Here we propose a model that describes the interaction between these factors (Fig 9). Although additional factors may be involved, we conclude that under conditions where more than one factor is present, these factors regulate each other's function to promote an optimal environment for cell survival and function and taurine plays a pivotal role in this regulated interaction.

## ACKNOWLEDGMENTS

This study was supported by a grant from NYS Office of Mental Retardation and Developmental Disabilities.

## REFERENCES


1. Barde, Y-A., 1989, Trophic factors and neuronal survival, *Neuron,* 2: 1525-1534.
2. Beal, M.F., 1995, Aging, energy, and oxidative stress in neurodegenerative diseases, *Ann. Neuol.,* 18:357-366.
3. Bottenstein, J.E., Skaper, S.D., Varon, S.S., and Sato, G.H., 1980, Selective survival of neurons from chick embryo sensory ganglionic dissociates utilizing serum-free supplemented medium, *Exp. cell Res.,* 125:183-190.
4. Budd, S.L. and Nicholls, D.G., 1995, Protein kinase C-mediated suppression of the presynaptic adenosine A1 receptor by a faciliatory metabotropic glutamate receptor, *J. Neurochem.,* 65:615-621.
5. Budd, S.L. and Nicholls, D.G., 1995, A reevaluation of the role of mitochondria in neuronal $Ca^{2+}$ homeostasis, *J. Neurochem.,* 66:403-411.
6. Budd, S.L. and Nicholls, D.G., 1996, Mitochondria, calcium regulation, and acute glutamate excitotoxicity in cultured cerebellar granule cells, *J. Neurochem.,* 67:2282-2291.
7. Budd, S.L., Castilho, R.F., and Nicholls, D.G., 1997, Mitochondrial membrane potential and hydroethidine-monitored superoxide generation in cultured cerebellar granule cells, *FEBS Lett.,* 415:21-24.
8. Chen, L.B., 1989, Fluorescent labeling of mitochondria, *Methods in Cell Biology,* 29:103-123.
9. Cheng, B. and Mattson, M.P., 1991, NGF and bFGF protect rat hippocampal and human cortical neurons against hypoglycemic damage by stabilizing calcium homeostasis, *Neuron,* 7:1031-1041.
10. Choi, D.W., 1987, Ionic dependence of glutamate neurotoxicity, *J. Neurosci.,* 7:369-379.
11. Choi, D.W., 1988, Glutamate neurotoxicity and diseases of the nervous system, *Neuron,* 1: 623-634.
12. Choi, D.W., 1990, The role of glutamate neurotoxicity in hypoxic-ischemic neuronal death. *Ann. Rev. Neurosci.,* 13:171-182.
13. Coffey, E.T., Sihra, T.S., and Nicholls, D.G., 1993, Protein kinase C and the regulation of glutamate exocytosis from cerebrocortical synaptosomes, *J. Biol. Chem.,*268:21060-21065.
14. Eboli, M.L., Mercanti, D., Ciotti, M.T., Aquino, A., and Castellani, L., 1994, Glutamate-induced protein phosphorylation in cerebellar granule cells: role of protein kinase C, *Neurochem, Res.,* 19:1257-1264.
15. Eimerl, S. and Schramm, M., 1993, Resuscitation of brain neurons in the presence of $Ca^{2+}$ after toxic NMDA receptor activity, *J. Neurochem.,* 61:518-525.
16. Eimerl, S. and Schramm, M., 1995, Resuscitation of brain neurons in the presence of $Ca^{2+}$after toxic NMDA-receptor activity, *J. Neurochem.,* 65:739-734.
17. El Idrissi, A., Harris, A., and Trenkner, E. 1996, Neurotrophins, neuro-active amino acids and the mitochondrial respiratory chain together maintain neuronal survival and function, *Abstr. Soci. Neurosci.,* 22:770.11
18. Farago, A. and Nishizuka, Y., 1990, Protein kinase C in transmembrane signaling, *FEBS Lett.,* 268:350-354.
19. Foster, A.C., Gill, R. and Woodruff, G.N., 1988, Neuroprotective effects of MK801 in vivo: selectivity and evidence for delayed degeneration mediated by NMDA receptor activation, *J. Neurosci.,* 8:4745-4754.
20. Hartley, D.M., Kurth, M.C., Bjerkness, L., Weiss, J.H., and Choi, D.W., 1993, Glutamate receptor-induced $^{45}Ca^{2+}$ accumulation in cortical cell culture correlates with subsequent neuronal degeneration, *J. Neurosci.,* 13:1993-2000.
21. Huxtable, R.J., 1992, The physiological actions of taurine, *Physiol. Rev.,* 72:101-163.
22. Gill, R., Foster, A.C., and Woodruff, G.N., 1987, Systemic administration of MK-801 protects against ischemia-induced hippocampal neurodegeneration in the gerbil, *J. Neurosci.,* 7:3343-3349.
23. Lombardini, J.B., 1994, The inhibitory effects of taurine on protein phosphorylation: comparisons of various characteristics of the taurine-affected phosphoproteins present in rat retina, brain and heart, in: Adv. Exp. Med. Biol. "Health and Disease", Michalk, D.V. and Huxtable, R.J., eds., Plenum Press, New York, Vol. 359, pp 9-17.
24. Mattson, M.P., 1988, Neurotransmitters in the regulation of cytoarchitecture, *Brain Res. Rev.,* 13:179-212.
25. Mattson, M.P. and Cheng, B., 1993, Growth factors protect neurons against excitotoxic/ ischemic damage by stabilizing calcium homeostasis, *Stroke,* 24:1136-1140.
26. Mattson, M.P., Zhang, Y., and Bose, S., 1993, Growth factors prevent mitochondrial dysfunction, loss of calcium homeostasis, and cell injury, but not ATP depletion in hippocampal neurons deprived of glucose, *Exp. Neurol.,* 121:1-13.
27. Nishizuka, Y., 1988, The molecular heterogeneity of protein kinase C and its implications for cellular regulation, *Nature.* 334:661-665.

28. Nishizuka, Y., 1988, The heterogeneity and differential expression of multiple species of the protein kinase C family, *Biofactors*, 1:17-20.
29. Squinto, S.P., Stitt, T.N., Aldrich, T.H., Davis, S., Bianco, S.M., Radziejewski, C., Glass, D.J., Masiakowski, P., Furth, M.E., and Valenzuela, D.M., 1991, trkB encodes a functional receptor for brain-derived neurotrophic factor and neurotrophin-3 but not nerve growth factor, *Cell*, 65:885-893.
30. Sturman, J.A., 1993, Taurine in development, *Physiol. Rev.*, 73:119-147.
31. Trenkner, E. and Sidman, R.L., 1977, Histogenesis of mouse cerebellum in microwell cultures: cell reaggregation and migration, fiber and synapse formation, *J. Cell. Biol.*, 75:915-940.
32. Trenkner, E., 1991, Cerebellar cells in culture, in: Culturing Nerve Cells, Banker, G. and Goslin, K. eds., MIT Press, pp. 283-307.
33. Trenkner, E., Liu, D.J., Harris, C., and Sturman, J.A., 1994, Regulation of protein kinase C activity by taurine and β-alanine during excitotoxicity in cat and mouse cerebellar cultures, in: Adv. Exp. Med. Biol. "Health and Disease", Michalk, D.V. and Huxtable, R.J., eds., Plenum Press, New York, Vol. 359, pp. 309-316.
34. Trenkner, E., El Idrissi, A., and Harris, C., 1996, Balanced interaction of growth factors and taurine regulate energy metabolism, neuronal survival and function of mouse cerebellar granule cells under depolarizing conditions, in: Adv. Exp. Med. Biol. "Taurine 2", Huxtable, R.J., Azuma, J., Kuriyama, K., Nakagawa, M., and Baba, A., eds., Plenum Press, New York, Vol. 403:507-517.

# EFFECT OF TAURINE ON HUMAN FETAL NEURON CELLS: PROLIFERATION AND DIFFERENTIATION

Xue-Cun Chen,[1] Zhi-Ling Pan,[1] Dong-Sheng Liu,[1] and Xiaobin Han[2]

[1] Institute of Nutrition and Food Hygiene
CAPM
29 Nan Wei Road
Beijing 100050, China

[2] Department of Pediatrics
University of Tennessee
Memphis, Tennessee 38103

## ABSTRACT

The purpose of this study was to investigate the effect of taurine on human fetal brain neuron cell proliferation and differentiation using a glial-free, pure cerebral neuronal culture grown in a serum-free environment. We found that taurine was necessary for neuronal survival and neurite extension. Taurine, on the other hand, has a trophic effect on the human fetal brain cell, promoting both proliferation and differentiation. Results showed that DNA synthesis of the neurons was increased in a dose-dependent manner when neurons were cultured in the medium containing taurine (100-6400 μM). The protein content of neuronal cells was also significantly increased in the neurons treated with taurine as compared to the control. At day 15, the expression of neuron-specific enolase (NSE) was only detected in the neurons cultured in the medium containing taurine. These results establish taurine as a putative human fetal brain neurontrophic factor in the process of human brain development.

## INTRODUCTION

The role of taurine in the developing mammalian nervous system has recently been the subject of considerable recent research[9,11]. It has been known for some time that the concentration of taurine in the newborn mammalian brain is several-fold greater than in the mature brain of the same species, and that the decrease in levels occurs appproximately at the time of weaning[10]. The level of taurine in human milk is about 10 times higher than that in cow's milk[2]. Studies have shown that taurine must be provided to the infant from mother's milk[2]. Infants fed various formulas have urine and plasma taurine concentrations

*Taurine 3*
Edited by Schaffer *et al.*, Plenum Press, New York, 1998

lower than infants fed pooled human milk[5,6,8]. Other studies have demonstrated that taurine-depleted kittens have a persistence of cells in the external granule cell layer with reduced levels of taurine. These observations suggest that taurine may play an important role in the development of the newborn infant, particularly in brain development. The establishment of a pure neuronal culture, and identification of new growth factors using these cultures, have represented milestones on the study of neuronal development in the peripheral nervous system. A pure neuronal population grown in a defined chemical medium is a necessary condition for the molecular investigation of brain neuronal development because of the obscure and uncontrollable environment surrounding neurons in the intact central nervous system. This system allows the investigation of growth factors essential for the development of cerebral neurons in a serum-free environment.

In these studies, the effect of taurine on brain development was investigated *in vitro*. Study results showed that taurine may function as a neurotrophic factor in the process of human brain development. When neurons were incubated in medium containing 200 μM taurine, proliferation of the neurons was significantly increased as compared to the control.

## METHODS AND MATERIALS

### Cell Culture

Fragments of human fetal brain between 12-14 weeks of gestation (estimated by menstrual history and ultrasonic examination) were obtained from legal abortions. The cerebral hemispheres were replaced in HBSS medium to prepare cell suspensions with the method of mechanical cell dissociation. Subsequently the cells were seeded on plastic culture dishes (35 mm) precoated with poly-D-lysine (Sigma) by an overnight incubation followed by 3 washes with triple-distilled water. For the initial 24 h incubation the cells were maintained in medium containing 10% fetal calf serum (FCS). Thereafter, the cultures were kept in a chemically defined medium (1:1 MEM/F-12, transferrin 10 mg/ml, putrescine 100 μM, insulin 5 mg/ml, sodium selenite 30 nM) that was changed every third day. Cultures were maintained in a humidified atmosphere of 95% air/5% $CO_2$ at 37°C. All results reported were obtained from cells taken from a total of 10 fetuses.

### Cell Counts

Cell viability in suspensions of dissociated cells was determined by the ability of viable neurons to exclude the dye trypan blue. Aliquots of cell suspension were mixed 1:1 with saline containing 0.8% trypan blue. Ten minutes later stained and unstained cells were counted with the aid of a microscope (Nikon Diaphot). The viability of cultured neurons was determined by morphological criteria as previously described[7]. The results represent four separate cultures.

### [$^3$H]-Thymidine Incorporation and Protein Assay

Human fetal brain neuron cells were cultured in 24-well plates at 1 x $10^5$ cells per well in medium containing 0 or 25 μg/ml taurine for 3-12 days. They were then incubated with [$^3$H]-thymidine for another 4 h. The amount of [$^3$H]-thymidine incorporated into DNA was determined by liquid scintillation counting after the cells were collected on glass-fiber filters. The protein content of the cells was assayed by the method of Bradford.

### Measurement of Neuron-Specific Enolase (NSE)

Human fetal brain neuron cells were cultured in 24-well plates at 1 x $10^5$ cells per well in medium containing 0 or 25 μg/ml taurine for 15-25 days before expression of NSE was determined as described previously [1].

### Statistics

Data were analyzed by one-way analysis of variance (ANOVA). The Student's t-test was used to compare significant differences between the two experimental groups. The values represent means ± SEM three independent experiments.

## RESULTS

### Induction of Neurite Outgrowth Induced by Taurine

After 2 h of plating, small neutites were observed. Taurine induced neurite outgrowth at concentrations of as low as 2.5-25 μg/ml. The neurites network was formed after 48 h of treatment of taurine (25 μg/ml), which was faster than that of the control cells.

### Effects of Taurine on Primary Cerebral Neuronal Cell Survival

Neuronal survival of cultures of cerebral hemisphere cells maintained in the presence or absence of taurine was determined by morphological criteria on day 5. Taurine significantly increased the number of surviving neuronal cells in a dose-dependent manner (12.5 - 200 μg/ml), as compared to control ($p < 0.01$) (Fig.1). In control cells, there was a progressive loss of cells over the course of the experiment, with a marked reduction in cell number by day 20. The cells survived for 4-5 weeks only in chemically defined medium lacking taurine. However, cell survival was significantly enhanced in the presence of taurine from weeks 5 to 7.

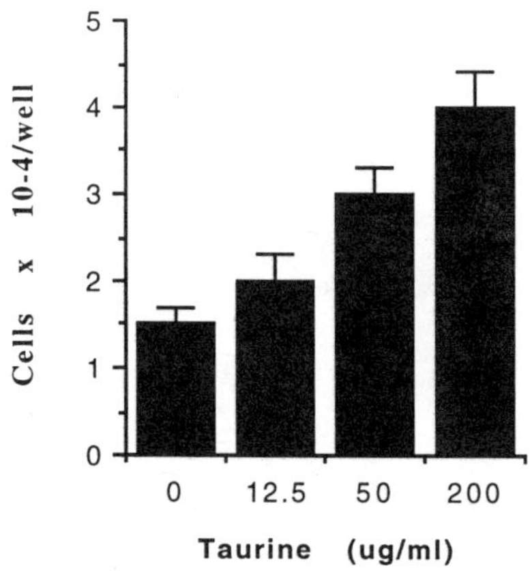


**Figure 1.** Effect of taurine on survival of neuronal cells. Neurons were cultured in medium containing taurine as indicated for 3 days and then the viability of the cells was determined.

## Effect of Taurine on Neuronal Cell Proliferation

The effect of taurine on neuronal cell proliferation was assessed by [$^3$H]-thymidine incorporation into DNA. The rate of DNA synthesis of taurine-treated cells was significantly increased in a dose-dependent manner (Fig. 2). Proliferation of human neuronal cells began as early as 2 h after seeding, as observed by phase contrast microscopy (data not shown). Six days after plating, proliferation of the cells reached a maximum (Fig. 3). Proliferation of the neuronal cells was significantly increased at all time points examined in the neuronal cells treated with taurine (25 μg/ml) ($p < 0.05$). In contrast to cell proliferation, protein content of neurons showed two peaks (at day 10 and day 20) that were observed in both groups of cells (Fig. 4). The protein content was significantly higher in the cells treated with taurine compared to control ($p < 0.001$).

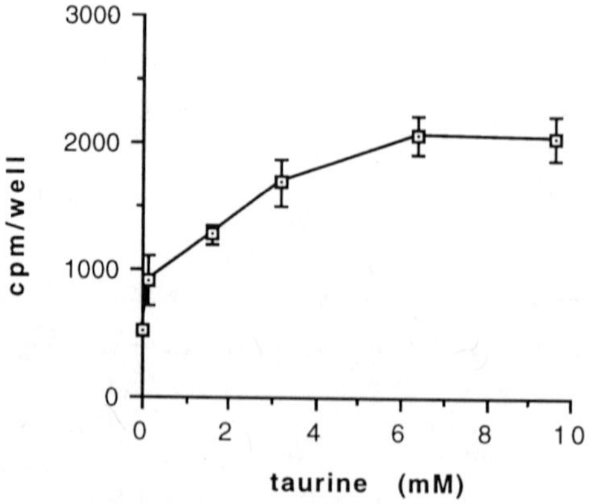


**Figure 2.** Dose-dependent effect of taurine on DNA synthesis of neuronal cells. Neurons were treated with various levels of taurine, in addition to [$^3$H]-thymidine for 3 days, and then the amount of [$^3$H]-thymidine incorporated into DNA was counted. Each value represents the mean ± SEM of three independent experiments.

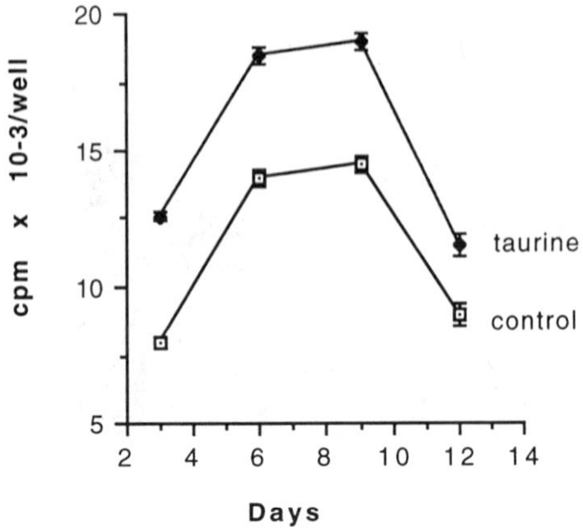


**Figure 3.** Effect of taurine on the proliferation of human neurons. Neurons were cultured in medium containing 0 μM or 200 μM taurine for the indicated period of time before DNA synthesis of the neurons was measured. Each value represents the mean ± SEM of three independent experiments. P<0.01 taurine vs control for in all time points.

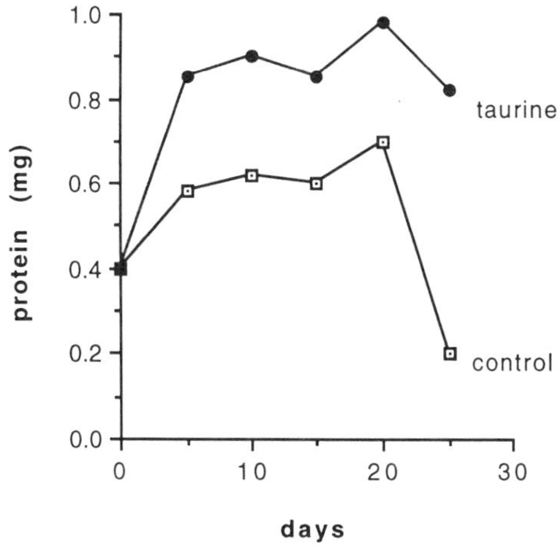


**Figure 4.** Effect of taurine on protein content of human neurons. Neurons were cultured in medium containing 0 μM or 200 μM taurine for 25 days. Protein content was measured at the indicated day. Each value represents averages from three independent experiments.

## Effect of Taurine on the Expression of NSE of Human Neuron Cells

NSE has been used as a marker to the extent of neuron cell differentiation. In this study, we found that the expression of NSE was only detected in neurons cultured in medium containing taurine after 15 days in culture. Expression of NSE was up-regulated in the cells treated with taurine compared to the control (Fig. 5).

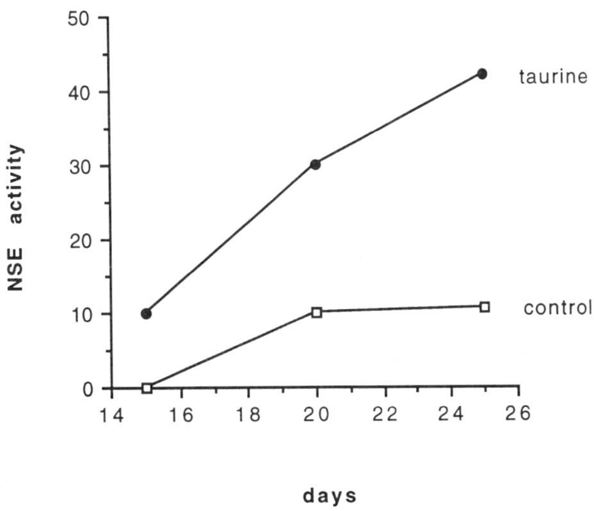


**Figure 5.** Effect of taurine on expression of NSE in human neurons. Neurons were cultured in medium containing 0 μM or 200 μM taurine for 25 days before expression of NSE was measured. Each value represents averages from three independent experiments.

## DISCUSSION

Primary cultures of dissociated nervous tissue have been utilized as model systems in the study of the central nervous system. These cell cultures are potentially important tools for investigating morphological and biochemical aspects of nerve and glial cell diferentiation, intracellular regulatory mechanisms, cellular interactions, and drug effects.

Previously, it has been shown that taurine is essential for growth and survival of mammalian cells. In fact, lymphoblastoid cells do not grow in the absence of taurine[3], which myocytes exhibit poor viability in the absence of taurine[4]. Taurine depletion causes degenerative changes in retinal cells. However, sufficient knowledge regarding the effect of taurine on human neuron cells development is lacking. In the present study, the effect of taurine on human brain cell proliferation and differentiation was studied using cultured human fetal cerebral hemisphere neurons. Our study showed that neurites start to outgrow 2 h after seeding. By day 2, neurons begin to divide, indicating the cells are undergoing proliferation. The proliferation reaches a peak by day 9. By day 15, the cells begin the process of differentiation and by days 20-25, the neurons start the process of apoptosis. Interestingly, we found that neurite outgrowth is induced by treatment of taurine. Taurine promotes the proliferation of neurons without affecting the proliferation pattern. cAMP may be involved in the neurotrophic effect of taurine on human brain neurons, since the level of cAMP is elevated in the neurons treated with taurine, as compared to control neurons (data not shown). We also found that taurine induces the differentiation of the neurons as determined by NSE expression, a marker for mature neuronal tissue[1]. When the neurons are cultured in the medium containing taurine (200 $\mu$M), the expression of NSE is up-regulated. Most importantly, the differentiating neurons do not undergo apoptosis as observed in control neurons (at day 25) until 50 days after seeding, suggesting that taurine protects against programmed neuron death. The mechanism by which taurine functions as a neurotrophic factor needs to be further determined.

## CONCLUSION

In this study, human fetal brain neuron cells were used as a model. We found that taurine was necessary for human fetal brain neuronal survival and neurite extension. Taurine, on the other hand, has a trophic effect on human fetal brain cells proliferation and differentiation. These results establish taurine as a putative human fetal brain neuron-trophic factor in the process of human brain development.

## ACKNOWLEDGMENTS

This work was supported in part by grants from the National Natural Science Foundation of China to Drs. X. Chen (39070719) and X. Han (39070720).

## REFERENCES

1. Bock, E., 1978, Nervous system specific proteins, *J. Neurochem.*, 30:7-14.
2. Gaull, G.E., 1982, Taurine in the nutrition of human infant, *Acta Paediatr. Scand. Suppl.*, 296:38-47.
3. Gaull, G.E., Wright, C.E., and Tallan, H.H. 1983. Taurine in human lymphoblastoid cells: uptake and role in proliferation, Alan R Liss, Inc., New York, pp. 297-304.
4. Hunter, G.E., 1986, Adult ventricular myocytes isolated from CHF 146 and CHF 147 cardiomyopathic hamster, *Can. J. Physiol. Pharm.*, 46:1503-1506.

5. Jarvenpaa, A.L., Rassin, D.K., Raiha, N.C., and Gaull, G.E., 1982, Milk protein quantity and quality in the term infant, II. Effects on acidic and neutral amino acids, *Pediatrics,* 70:221-230.
6. Martensson, J. and Finnstrom, O., 1985, Metabolic effects of a human milk adapted formula on sulfur amino acid degradation in full-term infants., *Early Hum. Dev,,* 11:333-339.
7. Mattson, M.P., Dou, P., and Kater, S.B., 1988, Outgrowth-regulating actions of glutamate in isolated hippocampal pyramidal neurons, *J. Neurosci,* 8:2087-2100.
8. Rassin, D.K., Gaull, G.E., Jarvenpaa, A.L., and Raiha, N.C.R., 1983, Feeding the low-birth-weigth infant. II Effects of taurine and cholesterol supplementation on amino acids and cholesterol, *Pediatrics,* 71:179-186.
9. Sturman, J.A., 1983, Taurine in nutrition research, *Prog. Clin. Biol Res.,* 125:281-295.
10. Sturman, J.A. and Gaull, G.E., 1975, Taurine in the brain and liver of the developing human and monkey, *J. Neruochem,* 25:831-835.
11. Sturman, J.A. and Hayes, K.C., 1980, The biology of taurine in nutrition and development, *Adv. Nutr. Res.,* 3:231-299.

# Part 8:
# Taurine and the Central Nervous System II: The Retina

# THE ROLE OF TAURINE IN OSMOTIC, MECHANICAL, AND CHEMICAL PROTECTION OF THE RETINAL ROD OUTER SEGMENTS

Andranik M. Petrosian and Jasmine E. Haroutounian

Buniatian Institute of Biochemistry
National Academy of Sciences of Armenia
Sevag St. 5/1
Yerevan 14, Armenia

## SUMMARY

The ability of the photoreceptor cell to resist osmotic stress was examined by incubating isolated frog retina in medium of varying osmolality. An electron microscopic analysis of the rod outer segment following a severe hypoosmotic insult revealed connections between adjacent disks and between disk rims and the plasma membrane, which presumably provide mechanical stability to the rod outer segment. One surprising result was the extent of the damage incurred by the electrical signaling pathway of the photoreceptor cells subjected to a 50 mOsm insult; only the distal PIII component of the ERG remained unaffected. Thus, the rod outer segment is particularly resistant to osmotic-induced injury, presumably because of the effective osmoregulatory actions of taurine. Incubation of retina with tauret, retinylidentaurine, uncovered rose-like hexagonal structures on the surface of the rod outer segment. These structures purportedly consist of connections between disk rims and the plasma membrane of the rod outer segments. Based on the influence of tauret, it is likely that the calcium dependence of these channels is selective for retinoids. These data are discussed relative to taurine's role in the process of rhodopsin regeneration and in the protection of the rod outer segments against osmotic, mechanical and light induced damage.

## INTRODUCTION

Taurine is required for proper maintenance of retinal function, not only in terms of cytoprotection but also relative to physiological function[7,9,12-14,18,20,25,28]. Although the basis underlying these effects has not been definitively established, numerous actions of taurine have been described. These include modulation of $Ca^{2+}$ transport[18,21], regulation of protein phosphorylation[18], promotion of cell proliferation and regeneration[1,5,17], enhancement of phagocytosis[3], stimulation of rod development[1] and osmoregulation[6,10,15,16,21].

Recent studies suggest that one of the most important functions of taurine in the retina is the maintenance of osmotic homeostasis[19,21]. Since the distribution of taurine within the retina varies[29], however, one would predict that certain layers of the retina may be more susceptible to osmotic stress-induced injury. To examine this idea, frog retina were subjected to hypoosmotic stress before being examined microscopically. This procedure was not only useful in assessing cell damage, but also in analyzing underlying cytoskeletal structures that are exposed by the osmotic insult. The studies indicate that major differences exist in the ability of the various retinal cells to resist a hypoosmotic insult.

## MATERIALS AND METHODS

### Animals

Frogs (Rana ridibunda, Rana temporaria and Rana pipiens) were used throughout the study. The animals were maintained at 10-20°C in transparent jars made of glass and subjected to standard light cycles. The frogs were dark adapted for at least 3 hr before onset of the experiment. After dark adaptation at 16-20°C, all tissue procedures were performed in dim red light.

### Transmission and Scanning Electron Microscopy

After decapitation, the eyes of Rana temporaria were removed, and then retina were carefully dissected free of pigment epithelium in isoosmotic medium (8-12°C). Initially, the isolated retina were incubated in isoosmotic medium (225 mOsm) containing the following: NaCl-85mM, $NaHCO_3$-10mM, KCl-2.5mM, $CaCl_2$-1mM, $MgSO_4$-O.2mM, Glucose-10mM, and a sodium phosphate buffer (pH 7.5). When required, medium of lower tonicity (100, 50 and 25 mOsm) were prepared from isotonic medium by dilution. After a 10 - 15 min incubation in medium containing the chosen tonicity, retina were treated with solutions of the same tonicity containing 1% $OsO_4$. After dehydration, samples were embedded in epon.

Rana ridibunda were used in several experiments. Dark adapted retina from Rana ridibunda were detached in medium containing the following: NaCl-85mM, $NaHCO_3$-10mM, KCl-2.5mM, $CaCl_2$-1mM, $MgSO_4$-O.2mM, Glucose-10mM, Tris HCl-10 mM (pH 7.4) and either 0 or 5 mM trans tauret. After a 30 min incubation, the retina were transiently rinsed in tauret free medium, placed on a copper grid and quickly frozen in liquid nitrogen. To prepare the samples for freeze fracture, glycerol was used as a cryoprotectant. The freeze-fracture procedure was a modification of the method described by Popov et al.[27]. Fracturing was performed using a cooled knife (-150 C°). Platinum-carbon replication was carried out at -150°C and $3.10^{-4}$ Pa using electron beam evaporators. All samples were embedded in epon-araldite and examined using a JEM-100B or a JEM-7 electron microscope.

In some experiments, high resolution scanning electron microscopy was used to investigate the structure of the rod outer segments of retina obtained from the frog, Rana pipiens. Dark adapted retina were incubated for 30 min in standard isoosmotic medium containing either 0 or 10 mM trans tauret. After the incubation period, the samples were rinsed in tauret free saline and then treated first with 2.5% glutaraldehyde followed by 1% $OsO_4$. The critical point drying method was used for sample drying. After chromium coating the samples according to the method of Apkarian[2], the samples were examined using the SEM DS-130F.

## RESULTS AND DISCUSSION

Previous studies have shown that useful structural information can be obtained by monitoring the response of the rod outer segment to changes in medium osmolality[3,8,11,24,26]. The procedure also provides useful data on the consequences of osmotic stress. Figure 1A shows the normal appearance of rod outer segments in isoosmotic medium (230 mOsm). Upon lowering the osmolarity of the incubation medium from 230 mOsm to 100 mOsm, no major change in morphology of the rod outer segment was observed. The most apparent effects were a widening of the intradisk spaces and a curving of the disk membrane surface (Fig. 1B). A further reduction in the medium osmolality to 50 mOsm resulted in substantial disk swelling, breaches in the regular stacking arrangement of the disks and further widening of the intradisk space (Fig. 1C). Nonetheless, the appearance of the rod outer segments in 50 mOsm medium largely resembled those incubated in isoosmotic medium, in part because the plasma membrane remained completely intact and well preserved. As expected, the most dramatic changes in the rod outer segment occurred following exposure to 22.5 mOsm medium. The dramatic drop in osmolality triggered an "osmotic explosion", severely damaging the plasma membrane. Swelling became so acute in the 22.5 mOsm medium that one could visualize connections between the disks and the plasma membrane, as well as between adjacent disks (Fig. 1D). Surprisingly, in the 50 mOsm medium, most cells quickly lost their electrical activity, although the PIII of the

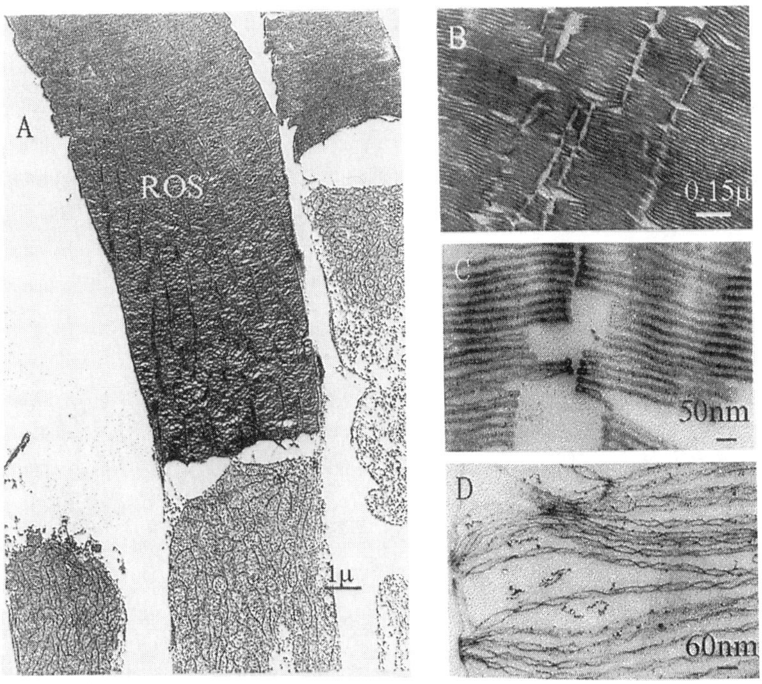


**Figure 1**. A - Transmission electron micrograph of dark adapted Rana ridibunda retinal photoreceptor cells at normal osmolarity - 230 mOsm. ROS- rod outer segment. Changes in the structure of the rod outer segment following incubation in medium containing the following osmolality: B - 100 mOsm, C - 50 mOsm and D - 22.5 mOsm.

ERG could be generated upon light stimulation. These data indicate that the rod outer segment is very resistant to osmotic stress; only after dramatic reductions in the osmolality of the incubation medium did significant alterations in the cell's morphology take place. This contrasts with other retinal cells, such as the cone cells, which are highly susceptible to osmotic stress. This differential response of various retinal cell types to osmotic stress could have one practical application, i.e., a hypoosmotic step could be introduced into the preparation of enriched rod outer segments.

Two factors presumably contribute to the unusual stability of the rod outer segments, the peculiar cytoskeletal structure and the existence of an effective chemical osmoregulatory system. Figure 2 reveals some important features of the cytoskeletal structure, including the connection between the disks and the cell membrane. The surface of the rod outer segment resembles a segmented cylinder (Fig. 2A and 2B). Upon freeze fracturing the retina, rose-like structures are uncovered, which appear to be outward gated connectors between the disks and the plasma membrane (Fig. 2C). These structures become more apparent in the presence of tauret, suggesting that they may be related to retinoid transport. Within the central regions of the hexagonal structure are dark areas, which presumably correspond to pores that exist in either an open or closed state. These structures remain

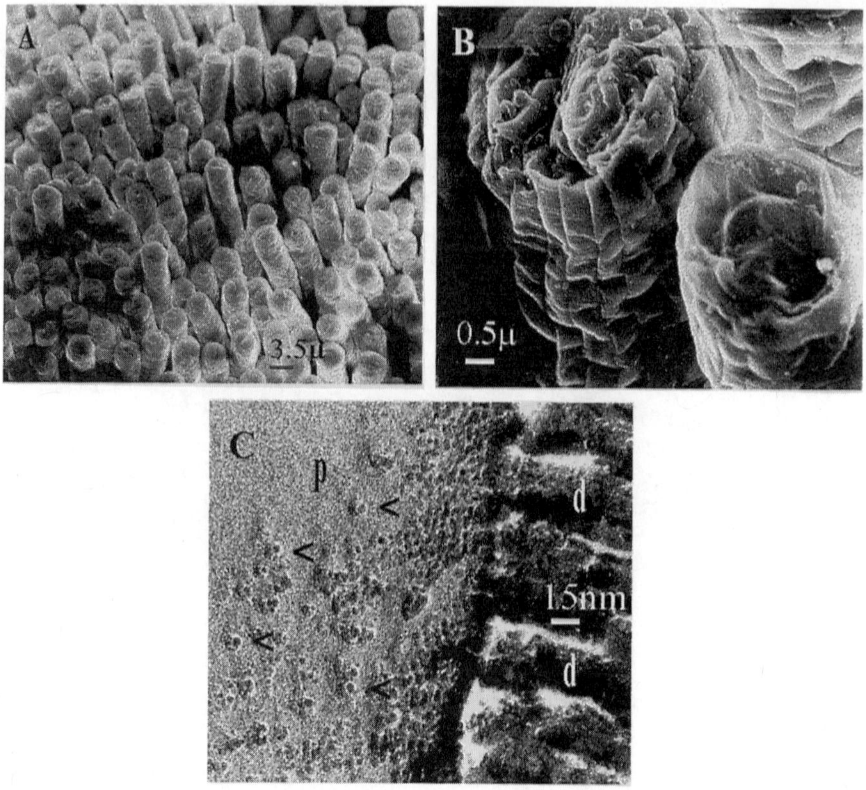


**Figure 2**. A - High resolution scanning electron micrograph of rod outer segments obtained from the retina of Rana pipiens. Both A and B show views of the retain taken from the brain side. C - A fragment of freeze-fractured Rana ridibunda retinal rod outer segments after dark incubation in medium containing 10 mM all trans tauret. p - outer surface of the ROS plasma membrane, d - disk membrane. Arrows reveal rose-like outward gates connections between the disks and the plasma membrane.

intact despite the large osmotic pressure gradient that develops across the rod outer segment during hypo-osmotic shock.

The osmoregulatory system of the rod outer segment is designed to prevent cellular damage by stimulating the uptake and release of organic and inorganic osmolytes. One of the important organic osmolytes of this system is taurine. Part of its effectiveness as an osmolyte is the existence of multiple transporting systems that utilize taurine as a substrate, including a transporter localized to the inner segment[13], a chloride type channel situated on the plasma membrane[15,16], a chloride channel found on the disk membrane and a molecular tauret-related channel[23].

Besides functioning as an effective osmolyte in rod outer segments, we have recently shown that taurine also regulates retinoid transport between the pigment epithelium and the photoreceptor of the retina[22,23,25]. This process assumes an important cytoprotective role in the eye during exposure to intense light. As a consequence of light exposure, rhodopsin is bleached, causing a rapid release of all trans retinal. Since high levels of free all trans retinal sensitize the cells to light-induced damage, the retinoid must be rapidly removed from the rod outer segment to prevent severe cell damage. The elimination process initially involves a reaction of taurine with retinal to form a Schiffs' base product, tauret, which can then readily leave the rod outer segment. Another endogenous defense mechanism against light-induced damage are the antioxidants, of which hypotaurine is an important member[20]. Hypotaurine has attracted attention because it not only is able to scavenge damaging free radicals generated during photo-injury, but also undergoes oxidation to taurine, which can then serve to eliminate all trans retinal from the cell.

Previously, we have shown that tauret may also contribute to regeneration of damaged photoreceptor cells[25]. It is well known that rhodopsin is incapable of spontaneously recovering from light-induced bleaching. However, incubation of rhodopsin-bleached retina with 11-cis tauret leads to a restoration of rhodopsin function in 84-90% of the damaged cells[25]. The discovery of tauret has reinforced the concept that taurine serves as a very important cytoprotective agent in the retina. At the molecular level, taurine's important actions include osmoregulation, maintenance of $Ca^{2+}$ homeostasis, modulation of protein phosphorylation and promotion of retinoid transport. Because of the importance of these processes, it is not surprising that taurine is required for normal function of the retina.

## ACKNOWLEDGMENTS

We thank Dr. Victor I. Popov (Moscow, Pushchino) and Dr. Robert P. Apkarian (Atlanta) for assisting in the freeze-fracture and the high resolution scanning microscopy studies, respectively. We also wish to express our deepest gratitude to Professor Ryan J. Huxtable for making our participation in the Taurine Symposium'97 possible.

## REFERENCES

1. Altshuler, D., Lo-Turco, J.J., Rush, J., and Cepko, A., 1993, Taurine promotes the differentiation of a vertebrate retinal cell type in vitro, *Development*, 119:1317-1328.
2. Apkarian, R.A., 1994, Analysis of high quality monoatomic chromium films used in biological high resolution scanning electron microscopy, *Scanning Microscopy*, 8:289-301.
3. Cohen, A.J., 1971, Electron microscopic observations on form changes in photoreceptor outer segments and their saccules in response to osmotic stress, *J. Cell Biol.*, 48:547-565.
4. Dontsov, A.E., Sakina, N.L., and Boulton, M.A., 1991, Taurine stimulates phagocytosis in retinal pigment epithelium, *Bio. Membranes (Russ.)*, 8:1197-1198.

5. Gabrielian, K., Wang, H.M., Ogden, T.E., and Ryan, S.J., 1992, *In vitro* stimulation of retinal pigment epithelium proliferation by taurine, *Curr. Eye Res.*, 11:481-487.
6. Goldstein, L., Davis, E.M., 1994, Taurine, betaine, and inositol share a volume-sensitive transporter in skate erythrocyte cell membrane, *Am. J. Physiol.*, 267:426-431.
7. Hayes, K., Garey, R., and Schmidt, S., 1975, Retinal degeneration associated with taurine deficiency in the cat, *Science*, 188:945-951.
8. Heller, Jr., Ostwald, T., and Bok, D., 1971, The osmotic behavior of rod photoreceptor outer segment disks, *J. Cell Biol.*, 48:633-649.
9. Imaki, H., Jacobson, S.G., Kemp, C.M., Knighton, R.W., Neuringer, M., and Sturman, J., 1993, Retinal morphology and visual pigment levels in 6- and 12- month-old rhesus monkeys fed a taurine-free human infant formula, *J. Neurosci. Res.*, 36:290-304.
10. Kanli, H., Fugelli, K., and Terreros, D., 1995, Taurine efflux is a part of the osmoregulatory response in proximal renal tubules of the teleost Carassius auratus, in: "Abstracts of International Taurine Symposium 1995", Osaka, Japan, pp. 100.
11. Korenbrot, J.I. and Cone, R.A., 1972, Dark ionic flux and the effect of light in isolated rod outer segments, *J. Gen. Physiol.*, 60:20-45.
12. Lake, N., 1989, The role of taurine in retinal function, in: "Extracellular and Intracellular Messengers in the Vertebrate Retina", Alan R.Liss, Inc., pp. 61-86.
13. Lake, N. and Orlowski, J., 1995, Cellular studies of the taurine transporter, in: "Abstracts of International Taurine Symposium 1995", Osaka, Japan. pp. 43.
14. Lake, N. and Verdone-Smith, C., 1989, Immunocytochemical localization of taurine in the mammalian retina, *Curr. Eye Res.*, 8:163-173.
15. Lambert, I.H. and Hoffmann, E.K., 1994, Cell swelling activates separate taurine and chloride channels in Ehrlich mouse ascites tumor cells, *J. Memb. Biol.*, 142:289-298.
16. Lehmann, A. and Ottersen, O.P., 1993, Neuronal-glial exchange of taurine during hypo-osmotic stress: a combined immunocytochemical and biochemical analysis in rat cerebellar cortex, *Neuroscience*, 540:615-631.
17. Lima, L., Drujan, B., and Matus, P., 1990, Spatial distribution of taurine in the teleost retina and its role in retinal tissue regeneration, in: "Taurine: Functional Neurochemistry, Physiology and Cardiology", Pasantes-Morales, H., Martin, D.L., Shain, W., and Martin del Rio, R., eds., Wiley-Liss, Inc., New York, Vol. 351, pp. 103-112.
18. Lombardini, J.B., 1991, Taurine: retinal function, *Brain Res. Rev.*, 16:151-169.
19. Pasantes-Morales, H., 1986, Current concepts on the role of taurine in the retina, in: "Progress in Retina Research", Osborne, N. and Chader, J., eds., Vol. 5, pp. 207.
20. Pasantes-Morales, H. and Cruz, C., 1985, Taurine and hypotaurine inhibit light-induced lipid peroxidation and protect rod outer segment structure, *Brain Res.*, 330:154-157.
21. Pasantes-Morales, H., Lopez-Escalera, R., and Macedo, M.D., 1989, Taurine and ionic fluxes in photoreceptors, in: "Extracellular and Intracellular Messengers in the Vertebrate Retina", Alan R. Liss, Inc., pp. 87-104.
22. Petrosian, A.M. and Haroutounian, J.E., 1988, On the possibility of retinal transportation in rods by tauret through special channels, Abstr. 13th All-Union Conf. on Electron Microscopy (Russ.), Zvenigorod, Moscow.
23. Petrosian, A.M. and Haroutounian, J.E., 1990, Tauret: further studies of the role of taurine in retina, in: "Taurine: Functional Neurochemistry, Physiology and Cardiology", Pasantes-Morales, H., Martin, D. L., Shain, W., and Martin del Rio, R., eds., Wiley-Liss, Inc., New York, Vol. 351, pp. 471-475.
24. Petrosian, A.M. and Haroutounian, J.E., 1994, On the nature of linkage between retinal rod outer segment disk and cellular membrane, in: Proc. 52nd Ann. Meeting of the MSA, Bayley, G.W. and Garran-Reed, A.J., eds., San Francisco Press Inc., pp. 28-29.
25. Petrosian, A.M., Haroutounian, J.E., Zueva, L.V., 1996, A taurine-related endogenous substance in the retina and its role in vision, in: "Taurine 2", Huxtable R.J., et al. eds., Plenum Press, New York, Vol. 403, pp. 333-342.
26. Petrossian, A.M., Mchitarian, R.G., and Stepanian, S.M, 1979, Investigation of retinal rod intradisk spaces by X-ray and electronmicroscopic histochemistry, Proceed. 7th All-Union Conf., "Local X-ray and spectral investigations and their apply", Chernogolovka, Moscow, pp. 166-168.
27. Popov, V.I, Matorin, D.N., Gostimsky, S.V., Tageeva, S.V., and Alakhverdov, B.L., 1981, Ultrastructural organization of chloroplast membranes in mutants of Pisum sativum L. with impaired activity in the photosystems, *Planta*, 151:512-524.
28. Schmidt, S.Y., Berson, E.L., Watson, G., and Huang, C., 1977, Retinal degeneration in cats fed casein. III, Taurine deficiency and ERG amplitudes, *Invest. Ophthal. Vis. Sci.*, 16:673-678.

29. Voaden, M.J., Lake, N., Marshall, J., and Morjaria, B., 1977, Studies on the distribution of taurine and other neuroactive amino acids in the retina, *Exp. Eye Res.*, 25:249-257.

# EFFECTS OF TAURINE AND LIGHT ON RETINAL GABA CONTENT AND THE EFFLUX OF $^{14}C$-GABA AND $^{14}C$-ASPARTATE FROM FROG RETINA

Jasmine E. Haroutounian and Andranik M. Petrosian

Buniatian Institute of Biochemistry
National Academy of Sciences of Armenia
Sevag ST. 5/1
Yerevan 14, Armenia

## SUMMARY

GABA content of isolated, dark adapted frog retina was found to be 3.15 ± 0.28 mM. After 30 minutes of exposure to intense light (200 1x), retinal GABA levels increased about 70%. Interestingly, incubation of dark adapted retina for 30 minutes with medium containing 0.4 mM taurine also led to a 70% increase in GABA levels. Since the light-induced elevation in GABA content was reduced over 50% by a simultaneous injection of 0.02 mM strychinine, it is likely that the light-induced GABA change is partly mediated by the release of taurine from the retina seen after light exposure. However, incubation of isolated retina with medium containing increasing concentrations of taurine (1, 2 and 20 mM), caused a progressive rise in $^{14}C$-GABA efflux from retina that was preloaded with 2.2 μM GABA and exposed to dim light (0.05 1x). It was also shown that taurine (1 and 5 mM) dramatically reduced $^{14}C$-aspartate efflux from retina preloaded with radioactive aspartate and exposed to dim light conditions. By comparison, intense light stimulation (40 1x) reduced basal $^{14}C$-aspartate efflux while dark exposure increased $^{14}C$-aspartate loss from the isolated retina. We found that taurine depressed the b-wave signal of frog retina, with the maximum effect occurring at a concentration of 1 mM. Addition of strychnine (0.4 mM) reversed the taurine effect on the b-wave, indicating that taurine receptors must be present in the inner retina. By contrast, taurine (0.1 - 20 mM) had no effect on the P111 component of the ERG initiated by either aspartate or cobalt. However, taurine exerted a modest depressant activity on P111 initiated by glutamate. The significance of these data relative to the putative neurotransmitter function of taurine in the inner retina is discussed.

## INTRODUCTION

Taurine is found in high concentrations in retina, with the highest content found in the photoreceptor cells[8,9,10,19]. These elevated levels are maintained by a ß-amino acid transporter situated on the inner segment near the photoreceptor cell nucleus[7]. It has been proposed that taurine functions either as a neuromodulator or a neurotransmitter in the retina. In support of this idea, it has been shown that light triggers the release of taurine from isolated retina and from isolated rod outer segments[5,13]. Simultaneously, the light causes an elevation in retinal GABA content, an effect that may be caused by taurine release[3,15]. The fact that an intravitreal injection of taurine reduces the amplitude of the ERG b-wave[1,11,12] while strychnine reverses this effect[2], further suggests that taurine might regulate the actions of at least one of the inhibitory neurotransmitters. These data suggest that an interaction may exist between GABA and taurine. Thus, the aim of the present study was to evaluate the interaction between taurine and both GABA and aspartate. Of particular interest was the degree of taurine involvement in light-mediated increases in retinal GABA levels. Also of interest was the effect of taurine on both GABA and aspartate efflux from frog retina. The data support the theory that taurine may be involved in chemical transmission in the retina.

## MATERIAL AND METHODS

### Animals

Frogs (Rana ridibunda) were used throughout the study. Experiments were performed predominantly in the periods of October-December or February-March. The animals were maintained at 10-20°C in transparent jars made of glass and subjected to standard light cycles. The frogs were dark adapted at least 3 hours before onset of the experiment.

### Isotope Measurements

Following the initial dark adaptation at 16-20°C, all procedures were performed in dim red light. Eyes of Rana ridibunda were removed following decapitation, placed in standard medium (6-8°C) and the retina were then carefully dissected free of the pigment epithelium. The isolated retina were then incubated in standard medium containing the following: NaCl-85 mM, $NaHCO_3$-10 mM, KCl-2.5 mM, $CaCl_2$-1 mM, $MgSO_4$-0.2 mM, glucose-10 mM and 10 mM Tris HCl buffer (pH 7.5). In some experiments, a Tris HCl 10 mM HEPES buffer was substituted for the standard buffer. The standard and HEPES buffers were equilibrated with a 95% $O_2$, 5% $CO_2$ mix. After a 10 min preincubation period, the retina were transferred to incubation tubes containing standard medium supplemented with either 2.2 µM $^{14}C$-GABA (Amersham, England) or 2.5 µM $^{14}C$-aspartate (Amersham, England). After a 15 min incubation in low intensity light, the retina were quickly rinsed in medium lacking the isotope, filtered and then placed in a 2 ml perfusion chamber. The retina were perfused with medium at a rate of 1 ml/min, which allowed the introduction of taurine, as well as exposing the retina to the appropriate degree of light. Perfusate samples were collected every 2 min. The radioactivity of all samples was measured using a scintillation counter (SL-4221, Intertechnique, France) which exhibited an efficiency for $^{14}C$ of 95%.

To determine retinal GABA levels, detached retina were weighed and homogenized in perchloric acid. The resulting extract was assayed according to the fluorometric method described by Udenfriend[18].

## Electrophysiological Measurements

Dark-adapted retina were detached in medium containing the following: NaCl-85 mM, $NaHCO_3$-10 mM, KCl-2.5 mM, $CaCl_2$-1 mM, $MgSO_4$-0.2 mM, glucose-10 mM, Tris HCl-10 mM (pH 7.5). The medium was saturated with 95% $O_2$-5% $CO_2$. Isolated retina were placed on white filter paper located in a transparent chamber covered with a lucid block containing a hole with a diameter of 4 mm. A connection between the retina chamber and the buffer reservoir permitted alterations in the composition of the medium. For the electroretinogram recordings, two Ag-AgCl light-shielded electrodes were used to reduce the input resistance to less than 1 mOhm. Light of a given intensity and frequency was adjusted to produce flashes with a duration of 50 msec. All responses were amplified with a four channel a.c. low noise amplifier containing a band 0.1 - 100 Hz.

## RESULTS AND DISCUSSION

The GABA content of dark adapted frog retina incubated for 30 min in Krebs-bicarbonate buffer at 20°C is 3.15 mM, with the highest levels found in the horizontal and amacrine cells of the inner nuclear layer. After a 30 min illumination with intense white (200 1x) light, GABA content in isolated frog retina rose to 5.37 mM (Fig. 1). Interestingly, following a 30 min incubation of dark adapted retina with medium containing 0.2 mM taurine, GABA content also increased about 70% to a value of 5.42 mM, indicating that taurine exerts a similar effect on retinal GABA content as illumination. To determine if light-induced elevations in GABA content might be mediated by taurine, isolated frog retina were incubated in Krebs-bicarbonate buffer containing 20 μM strychnine prior to illumination with intense white light (200 1x). Strychnine reduced the light-induced increase in retinal GABA content by half, resulting in a final concentration of 4.31 mM (Fig. 1).

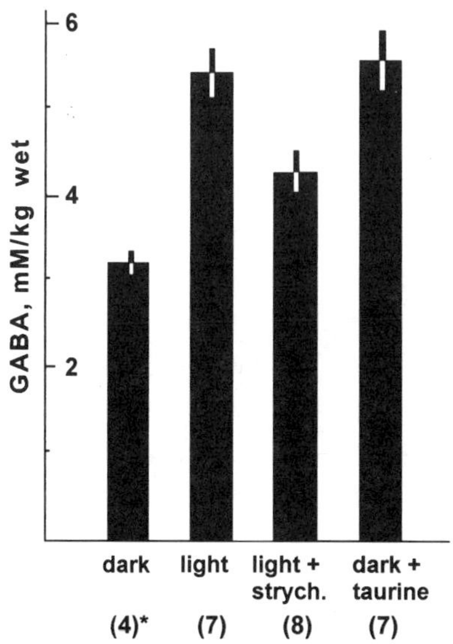


**Figure 1.** Effect of light (200 lx, 30 min), taurine (0,4 mM) and strychnine( 0.02 mM) on GABA content of frog retina ( )* - number of measurements. Confidence interval 95%.

Previously, it has been shown that dark stimulation of frog retina after low intensity light (0.05 lx) causes a marked release of $^{14}$C-GABA, while stimulation with intense light leads to a transient increase followed by a significant decrease in $^{14}$C-GABA efflux from the retina[4]. Therefore, we anticipated that exposure of the retina to medium containing taurine would reduce GABA efflux during low intensity light exposure. However, Figures 2 and 3 show that in most experiments, taurine exerted a dose-dependent (1 - 20 mM) increase in $^{14}$C-GABA efflux from isolated retina preloaded with 2.2 μM radioactive GABA. Although this effect of taurine was transient, we felt that taurine was not affecting either $Cl^-$ channel flux or GABA reuptake. Rather, taurine's actions appeared to resemble the response seen in amacrine or bipolar cells of the inner retina. In these cells it has been shown that taurine potentiated veratridine, ouabain and $K^+$-induced release of $^3$H-GABA, implying that the amino acid exhibits neuromodulatory activity. In a related study, Kamisaki et al.[6] provided data supporting a role of taurine in the activation of presynaptic $GABA_B$ autoreceptors in the cerebral cortex. This view contrasts with the conclusion of Sivakami et al.[15], who observed a competition between taurine and GABA for uptake, suggesting that the two effectors share a common transporter.

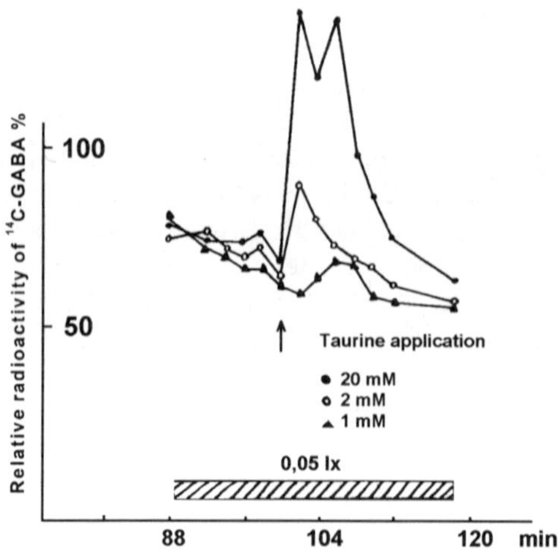


**Figure 2.** Effect of taurine (20, 2 and 1 mM) on $^{14}$C- GABA efflux from the isolated frog retina at background illumination 0.05 lx.

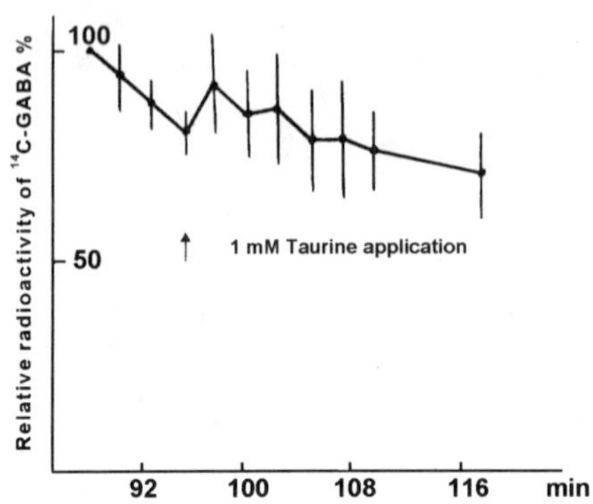


**Figure 3.** Effect of taurine (1 mM) on $^{14}$C - GABA efflux from the isolated frog retina at background illumination 0.05 lx. Number of experiments - 6, confidence interval 95%.

To assess the idea that taurine may modulate the actions of aspartate in the retina, the effect of taurine administration on $^{14}$C-aspartate efflux from frog retina was examined. Taurine (1 and 5 mM) depressed $^{14}$C-aspartate efflux from retina exposed to dim light (0.05 lx) (Fig. 4). In fact, the actions of taurine on $^{14}$C-aspartate efflux resembled the effect of intense illumination (40 lx) on the retina. By comparison, dark stimulation led to an increase in $^{14}$C-aspartate efflux from the retina (Fig. 5). We believe that the light-dependent outflow of $^{14}$C-asparatate from the retina is related to transmitter release from the synaptic endings of the photoreceptor cell, which depolarize in the dark and hyperpolarize following light stimulation.

We also examined the effect of taurine on the b-wave of the ERG, as well as on aspartate, glutamate and cobalt isolated P111 component. In agreement with earlier studies[1,11,12], we found that taurine slightly depressed the b-wave signal at a concentration of 0.1 mM, but significantly reduced the b-wave signal at 1 mM (Fig. 6A). Strychnine (0.4 mM) partially reversed the taurine-mediated depression of the b-wave, supporting the

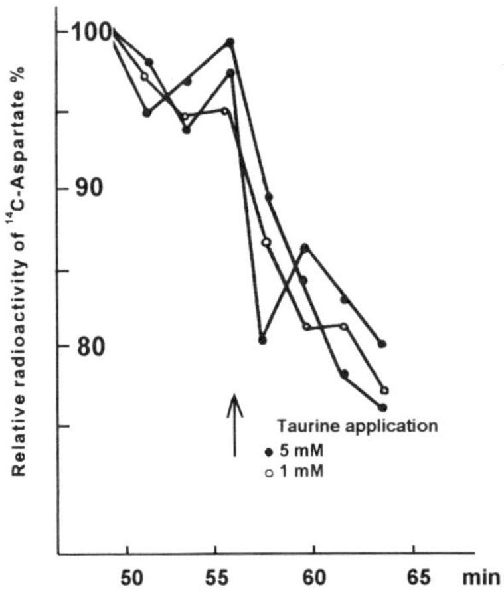


**Figure 4.** Effect of taurine (1 and 5 mM) on $^{14}$C-aspartate efflux from the isolated frog retina at background illumination 0.05 lx.

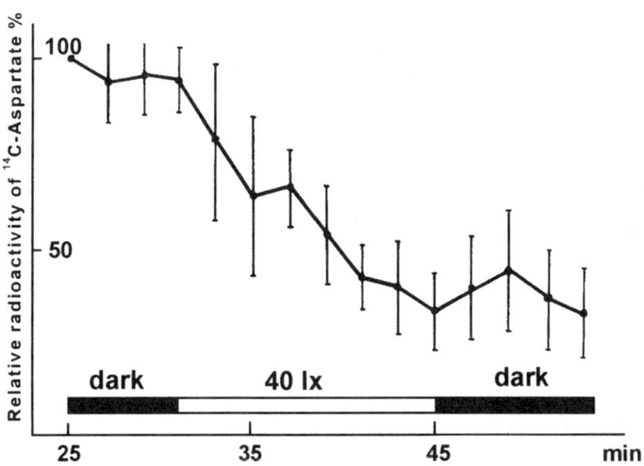


**Figure 5.** Effect of light (40 lx) and dark stimulus on $^{14}$C - aspartate efflux from the isolated frog retina. Number of experiments - 6, confidence interval 95%.

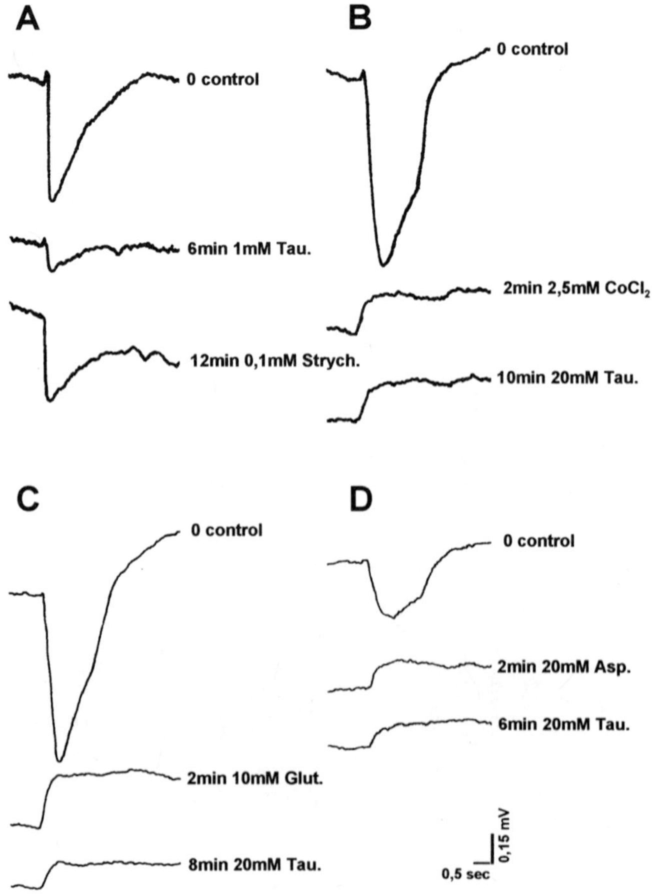


**Figure 6**. Effects of taurine on the ERG and PIII of the frog retina. A - 1 mM taurine and 0.1 mM strychnine action on the ERG. 20 mM taurine action on the: B - 2,5 mM $Co^{2+}$; C - 10 mM glutamate; D - 10 mM aspartate isolated PIII component of the ERG in the frog retina. Calibration: 0.15 mV, 0.5 sec.

notion that taurine receptors exist in the inner retina. However, taurine had no detectable effect on either aspartate or cobalt isolated P111 component (Figs. 6B and 6D). In the case of cobalt, but not of aspartate, this result was anticipated since the metal inhibits taurine binding to its receptor[20]. In contrast to the cobalt and aspartate results, taurine significantly attenuated the glutamate response (Fig. 6C).

Although these studies support the neuromodulatory theory of taurine action, this theory is not universally accepted. One of the most significant findings in favor of a neuromodulatory or neurotransmitter function for taurine is the identification of taurine receptors that appear to be coupled to a G-protein mediated second messenger system[16,17,20]. Also, the existence of an endogenous substance capable of modulating "taurinergic" transmission supports the theory[19]. To clarify the debate, further studies are required to ascertain the role of taurine in the retina.

## ACKNOWLEDGMENTS

We wish to thank Professor Felix G. Gribakin (Sant Petrebourgh) for his assistance in completion of the electrophysiological studies.

## REFERENCES

1. Bonaventure, N. and Wioland, N., 1976, GABA and Taurine: Implication in Organization of receptive Fields of Ganglon Cells in the Frog Retina, *Docum. Ophthal. Proc. Series,* 15:251-255.
2. Bonaventure, N., Wioland, N., and Mandel, P., 1974, Antagonists of the putative inhibitory transmitter effects of taurine and GABA in the retina, *Brain Res.*, 80:281-289.
3. Graham, L. T., Baxter, C. F., and Lolley, R. N., 1970, In vivo influence of light or darkness on the GABA system in the retina of the frog (Rana pipiens), *Brain Res.*, 20:379-388.
4. Haroutounian, J.E., Gevorgian, G.A., and Petrosssian, A.M., 1981, The effect of illumination level on $^{14}$C-GABA efflux from isolated frog retina, *J.Evolutional Biochem. and Physiol.(Russ.)*, 17:542-546.
5. Haroutounian, J.E. and Petrossian, A.M., 1982, The action of illumination level on the efflux of $^{3}$H-taurine from the isolated frog retina, *Armenian Biol.J (Russ),* 35:259-264.
6. Kamisaki, Y., Maeda, K., Ishimura, M., Ornura, H., and Itch, T., 1993, Effects of taurine on depolarization-evoked release of amino acids from rat cortical synaptosomes, *Brain Res.*, 627:181-185.
7. Lake, N. and Orlowski, J., 1995, Cellular studies of the taurine transporter. Abstracts of International Taurine Symposium 1995, Taurine: Basic and Clinical Aspects, Osaka, Japan, pp. 43.
8. Lake, N. and Verdone-Smith, C., 1989, Immunocytochemical localization of taurine in the mammalian retina, *Current Eye Research*, 8:163-173.
9. Lima, L., Drujan, B., and Matus, P., 1990, Spatial distribution of taurine in the teleost retina and its role in retinal tissue regeneration, in: Prog. Clin. Biol. Res. "Taurine: Functional Neurochemistry, Physiology and Cardiology", Pasantes-Morales, H., Martin, D.L., Shain, W., and Martin del Rio, R., eds., Wiley-Liss, Inc. New York, Vol. 351, pp. 103-112.
10. Orr, H.T., Cohen, A.I., and Lowry, O.H., 1976, The distribution of taurine in the vertebrate retina, *J. Neurochem.*, 26:609-611.
11. Pasantes-Morales, H., Bonaventure, N., Wioland, N., and Mandel, P., 1973, Effects of intravitreal injections of taurine and GABA on chicken electroretinogram, *Int. J. Neurosci.*, 5:235-241.
12. Roberts, E. and Kutyama, K., 1968, Some correlations in studies of the GABA system, *Brain Res.*, 8:1-35.
13. Salceda, R., Lopez-Colome, A.M., and Pasantes-Morales, H., 1977, Light-stimulated release of 35-S-taurine from frog retinal outer segment, *Brain Res.*, 135:186-191.
14. Sivakami, S., Ganapathy, V., Leibach, F.H., and Miyamoto, Y., 1992, The gamma aminobutyric acid transporter and its interaction with taurine in the apical membrane of the bovine retinal pigment epithelium, *Biochem. J.*, 283:391-397.
15. Starr, M.S., 1975, Effect of light stimulation on the synthesis and release of GABA in rat and frog retinae, *Brain Res.*, 100:343-353.
16. Tang, X.W., Tsai, W.H., and Wu, J.Y., 1992, Taurine receptor: kinetic analysis and pharmacological studies, *Adv. Exp. Med. Biol.*, 315:263-268.
17. Tang, X.W., Yarom, M., Carlson, R.G., Vander-Velde, D., Huang, P.Y., Lee, Y.H., Seah, E.C., Deupree, D., and Wu, J.Y., 1993, Isolation of endogenous modulators for the $GABA_A$ and taurine receptors, *Neurochem. Int.,* 23: 485-493.
18. Udenfriend, S., 1962, in: Fluorscence assay in biology and medicine, pp. 185, Acad. Press, NY.
19. Voaden, M.J., Lake, N., Marshall, J., and Morjaria, B., 1977, Studies on the distribution of taurine and other neuroactive amino acids in the retina, *Exp. Eye Res.,* 25:249-257.
20. Wu, J.Y., Tang, X.W., and Tsai, W.H., 1992, Taurine receptor: kinetic analysis and phrmacological studies, *Adv. Exp. Med. Biol.*, 315:263-268.

# TAURINE-STIMULATED OUTGROWTH FROM THE RETINA IS IMPAIRED BY PROTEIN KINASE C ACTIVATORS AND PHOSPHATASE INHIBITORS

Lucimey Lima and Suzana Cubillos
Laboratorio de Neuroquímica
Centro de Biofísica y Bioquímica
Instituto Venezolano de Investigaciones Científicas
Apdo. 21827
Caracas 1020-A, Venezuela

## ABSTRACT

Taurine increases neurite elongation of post-crush goldfish retinal explants, as well as the number of outgrowing isolated cells from goldfish and rat retina in culture. The trophic effect of taurine is related to an elevation in calcium flux rather than increased cell proliferation. Since taurine regulates phosphorylation in rat retina, we investigated if this process could be involved in the mechanism of taurine action on outgrowth. Control and taurine-supplemented post-crush goldfish retinal explants were cultured in the presence of protein kinase C activators or phosphatase inhibitors, and the length of neurites was measured after five days in culture. In some cases, there was an inhibition of the stimulatory effect of taurine without a modification in basal outgrowth. In others, outgrowth of control explants was also reduced. A certain level of protein phosphorylation seems to be critical for the trophic effect of taurine in the retina.

## INTRODUCTION

Mechanisms modulating phosphorylation of cellular proteins are important in the regulation of outgrowth and differentiation. In support of this idea, it has been shown that protein kinase (PKC)-immunoreactive somata shows a peak at postnatal day 11 and a decrease up to adulthood in rabbit retina[6]. Several studies indicate the presence of multiple subtypes of PKC with differential localization and function in the retina of various vertebrates, such as rabbit[26], rat[1,9], monkey[8], human[14], newt [29], and teleosts[7,15,27]. A number of these isozymes have been implicated in neuronal regeneration and growth. For instance, PKC isoform beta, present in the nerve cell bodies of adult mouse sciatic nerve in culture and upregulated after injury, appears to contribute to regeneration although its activity is not indispensable for the process[36]. Other PKC isozymes may also play a role in regeneration since the inhibition of PKC isozymes results in the regeneration of a filamentous network in transgenic mice overexpressing light medium chain neurofilaments. Thus, the impairment in protein phosphorylation may lead to the accumulation of neurofilaments, as seen in neurodegenerative diseases[5]. Alpha-PKC activity increases during wound-healing of the cornea, which also leads to an increase in kinase substrate growth associated protein-43[19]. Moreover, it has recently been reported that a develop-mentally regulated phosphorylated isoform of microtubule-associated protein 1B is expressed in regenerating axons of the sciatic nerve of the rat[3], which plays a role in axonal extension[35]. PKC is involved in these processes, since the nonspecific PKC inhibitor, KT5926, prevents nerve growth factor-induced outgrowth and regeneration of neurites in PC12 and PC12-C41 cells[34]. On the other hand, the phosphatase inhibitor okadaic acid completely inhibits *in vitro* regeneration of frog sciatic nerve[33].

Taurine-like immunoreactivity is higher in certain immature cell types than in mature brain. This reactivity increases during regeneration[23], which is in accordance with the putative link between taurine and nerve injury. It has been reported that the severity of the diabetic neuropathy is aggravated by taurine depletion[32]. In addition, taurine plasma concentration, but not essential nutritional amino acids, is decreased in the plasma of patients with cuban optic neuropathy[10]. Taurine-stimulated outgrowth from goldfish retinal explants does not involve cell proliferation but only cell lengthening[24]. It is partially mediated through an increase in intracellular calcium[18], an effect perhaps related to protein phosphorylation. In this regard, it is significant that taurine can modulate the phosphorylation of specific proteins in the retina of the rat[20]. A 20 kD protein, whose phosphorylation is reduced by taurine, might correspond to histone H2B[21]. Therefore, to establish the mechanism of taurine action as a trophic agent, we have used protein kinase activators and protein phosphatase inhibitors in order to study a possible modification of taurine's effect on neurite length of post-crush retinal explants of the goldfish.

## MATERIALS AND METHODS

Goldfish (*Carassius auratus*), 4-6 cm in length were adapted to darkness for 30 min and anesthetized in 0.05% tricaine (Sigma) prior to dissection of the retina. The lesion of the optic nerve was performed 10-12 days before explantation, by pulling the eye ahead of the orbit and crushing the nerve with fine forceps[17]. The retina was dissected, placed in a small volume of culture medium, and chopped in a McIlwain tissue chopper into 500 μm squares. Explants were placed (10-15 per dish) in poly-L-lysine (Sigma) pre-coated culture flasks. Nutrient medium (2 ml per dish) contained Leibovitz (L-15, Gibco), 20 mM HEPES, 10% fetal calf serum, and 0.1% w/v gentamicin (Sigma). Cytosine arabinose (Sigma, 2 mM) was added 24 h after plating. The post-crush explants were cultured in the absence or in the presence of 4 mM taurine (Sigma), and the PKC activators: phorbol 12,13-dibutyrate (PDBu, 0.05-50 μM); N-n-heptyl-5-chloro-1-naphthalenesulfonamide (SC-10, 0.05-10 μM); (-)-7-octylindolactam (OCL, 1-50 nM), and the negative control 4-α-phorbol 12-myristate 13-acetate (4-α-PMA, 0.01-2 μM); or the phosphatases inhibitors, okadaic acid (OKA, 0.01-50 nM) and calyculin A (CAL, 0.1-2 nM). All reagents were purchased from Research Biochemicals International, except OCL, which was obtained from Calbiochem. The drugs were dissolved in dimethylsulfoxide to a final maximal concentration of 0.07%. Control explants in the absence or in the presence of taurine were included in each group of experiments in order to make the adequate comparison. This was necessary because the basal length of the neurites and the percentage of stimulation with taurine varied between groups of animals.

The outgrowth of post-crush retinal explants was determined after 5 days in culture.

The length of the neurites (20-30 per explant or less, if the number of neurites was lower) was measured in µm with an ocular micrometer. The neurite to be measured was evaluated from the border of the explant to its maximal length, which might include the fusion of fibers emitted by the same explant. Statistical differences were determined by analysis of variance.

## RESULTS AND DISCUSSION

### Effect of Protein Kinase C Activators on Outgrowth from Goldfish Retinal Explants

The addition of the PKC activator, PDBu, at the low concentration (0.05 µM) did not affect the outgrowth of goldfish retinal explants in the absence or in the presence of taurine. However, at concentrations from 0.1 to 10 µM, the phorbol ester reduced the length of neurites in control and taurine supplemented cultures (Table 1). Interestingly, the non-phorbol activator of PKC, SC-10, did not modify the extension of the neurites from control explants, but decreased the trophic effect of taurine (Table 1). OCL, another non-phorbol activator of PKC, did not decrease the outgrowth from control cultures at a concentration of 1 nM, but did impair the effect of taurine. Yet, at higher concentrations, OCL decreased outgrowth in the presence and absence of taurine (Table 2). As expected, at a low concentration, the inactive phorbol, 4-α-PMA, did not modify outgrowth of either the control or taurine-supplemented explants. The fact that basal outgrowth was normally maintained in the presence of several PKC activators, while stimulated outgrowth was blocked, indicates a role for PKC in the stimulated condition. Moreover, the results suggest that taurine may affect the same pathway of outgrowth as PKC, explaining the ability of both factors to accelerate neurite elongation.

In accordance with previous studies, it is likely that taurine reduces the phosphorylation of selective proteins and that the activation of PKC might block the effect of the amino acid without completely blocking outgrowth. It is important to recall that one of the molecular mechanisms underlying signal transduction is the phosphorylation of key

**Table 1.** Length of neurites of goldfish retinal explants in the presence of 4 mM taurine and of protein kinase C activators.

| Concentration (µM) | | | Concentration (µM) | | |
|---|---|---|---|---|---|
| PDBu | Control (µm) | Taurine (µm) | SC-10 | Control (µm) | Taurine (µm) |
| 0 | 54.83 ± 4.66 | 65.11 ± 3.92* | 0 | 49.51 ± 2.43 | 58.33 ± 3.17* |
| 0.05 | 61.15 ± 2.51 | 61.85 ± 3.92 | 0.05 | 43.15 ± 3.50 | 40.33 ± 3.91** |
| 0.1 | 37.83 ± 3.09* | 49.90 ± 2.31** | 0.1 | 45.58 ± 2.34 | 47.47 ± 3.32** |
| 5 | 38.79 ± 2.99* | 45.88 ± 2.09** | 5 | 49.31 ± 3.28 | 32.61 ± 4.54** |
| 10 | 37.89 ± 3.20* | 44.71 ± 3.04** | 10 | 42.52 ± 3.24 | 38.02 ± 2.23** |

Each value is the mean ± S.E.M., n = 35-84. ANOVA: PDBu Control $F_{(4,265)}=13.35$, $p < 0.001$; Taurine $F_{(4,271)}=8.51$, $p < 0.001$; Both $F_{(9,532)}=10.93$, $p < 0.001$. SC-10 Control $F_{(4,342)}=1.24$, $p = 0.29$; Taurine $F_{(4,341)}=8.13$, $p < 0.001$; Both $F_{(9,683)}=4.82$, $p < 0.001$. * $p < 0.05$ denotes significant difference from Control 0, ** $p < 0.05$ denotes significant difference from Taurine 0.

proteins[13]. Vertebrate retina contain eight subspecies of PKC[37], which control dopamine release, modulate glutamate receptor function, phosphorylate $GABA_C$ receptors, control cytoskeletal interactions, have been implicated in the control of spinule formation, feedback control inositol phosphate signalling, and play a role in nerve regeneration and cone motility[4,16,36,37]. In addition PKC and protein kinase A (PKA) are involved in light-independent elongation of photoreceptors[27].

In the present study, we found that the stimulation of phosphorylation through PKC activators impairs the outgrowth and affects the stimulatory effect of taurine. The lack of complete blockade of outgrowth could be due to the concentration of the activator, but may also indicate that other mechanisms may be involved in outgrowth or that a critical level of phosphorylation is required to maintain a certain degree of neurite elongation. It is interesting to note that the 20 kD protein, whose phosphorylation is decreased by taurine, appears to be histone H2B, a protein involved in nuclear function[21]. However, histone phosphorylation may not be directly involved in the regulation of outgrowth by PKC because PKC is upregulated after injury in adult mouse sciatic nerves[36], indicating that the required substrates must be phosphorylated during regeneration. This observation is in accordance with the high level of phosphorylation observed in lamprey after injury, which contributes to the extension of the neurites[28]. Although PKC appears to play a central role in outgrowth, the presence of multiple isozymes with differing functions has led to conflicting data. According to our results the antiproliferative effect of transforming growth factor beta 2 and that of metabotropic glutamate agonists in the adult retina involves activation of PKC[14]. Similarly, intracranial or intraocular injections of a general kinase inhibitor or a PKC inhibitor prevent the retinotopic sharpening in the goldfish[30]. On the other hand, Jian et al.[16] found that calmodulin-dependent kinase, and not PKC or PKA, is involved in neurite elongation of goldfish retina. This apparent discrepancy may be related to differences in the concentration or the time of exposure.

It has been demonstrated that the taurine transporter in various cell lines is down-regulated by PKC-mediated protein phosphorylation[12,22,25]. The first evidence of PKC-induced regulation of taurine uptake into fresh retinal cells was recently shown[11]. Interestingly, the trophic effect of taurine in goldfish retinal explants is blocked by the simultaneous addition of the uptake blocker, guanidinoethane sulfonate[24]. Thus, it is possible that the impairment of the taurine effect by PKC activation could be mediated by a reduction in taurine uptake, a conclusion consistent with the notion that the taurine effect is dependent upon the size of the intracellular taurine pool.

**Table 2.** Length of neurites of goldfish retinal explants in the presence of 4 mM taurine and of the protein kinase C activator (-)-7-octylindolactam.

| Concentration (nM) OCL | Control (μm) | Taurine |
|---|---|---|
| 0 | 79.75 ± 2.75 | 96.67 ± 2.76 |
| 1 | 72.10 ± 3.14 | 68.71 ± 2.77** |
| 5 | 51.79 ± 3.40* | 49.90 ± 2.31** |
| 10 | 51.58 ± 3.40* | 25.11 ± 2.69** |

Each value is the mean ± S.E.M., n = 68-80. ANOVA: Control $F_{(3,286)}=21.01$, $p < 0.001$; Taurine $F_{(3,270)}=138.22$, $p < 0.001$; Both $F_{(7,556)}=55.80$, $p < 0.001$ * $p < 0.05$ denotes a significant difference from Control 0, ** $p < 0.05$ denotes a significant difference from Taurine 0.

**Effect of Phosphatase Inhibitors on Outgrowth from Goldfish Retinal Explants.**

Since the process of protein phosphorylation consists of i) protein substrates, which are often hard to identify; ii) protein kinases; and iii) protein phosphatases, we decided to test the effect of inhibitors of protein phosphatases, such as okadaic acid (OKA) and calyculin A (CAL), on outgrowth. As shown in Fig. 1A, OKA did not inhibit outgrowth at low concentrations, although there was a small decrease in neurite elongation at 1 nM. However, the stimulatory effect of taurine was completely blocked by low concentrations of OKA, i.e., at 1 nM the length of the control and the taurine supplemented cultures were nearly identical. By contrast, the phosphatase inhibitor CAL inhibited outgrowth from the control and taurine supplemented cultures at all concentrations tested (Fig. 1B).

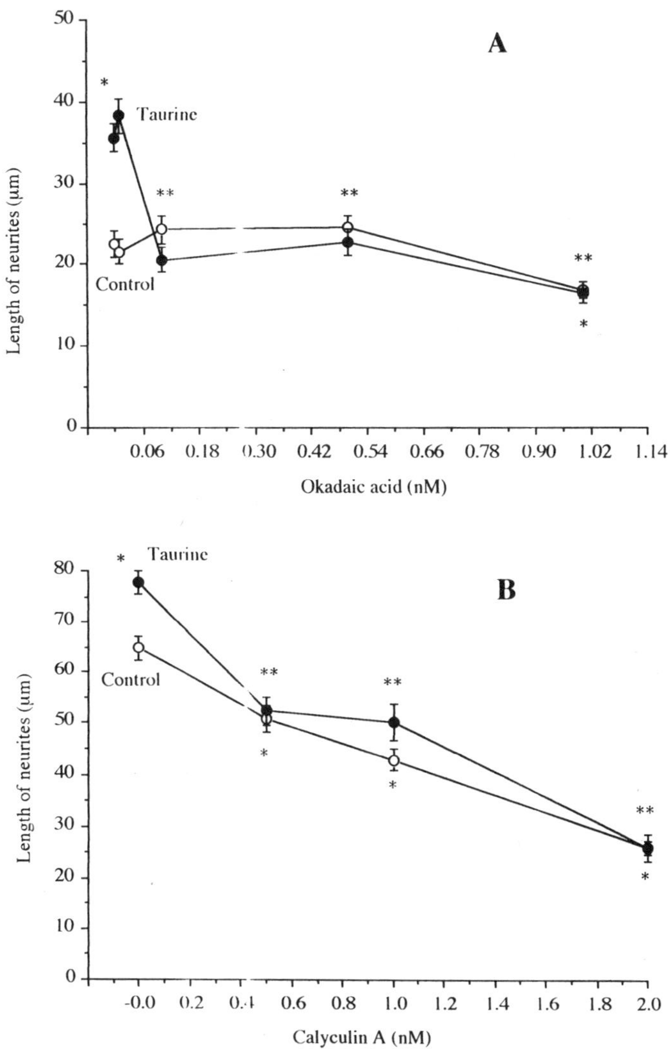


**Figure 1.** Each value is the mean ± S.E.M., n = 20-63. ANOVA: A) OKA Control $F_{(4,391)}=8.06$, $p < 0.001$; Taurine $F_{(4,390)}=32.06$, $p < 0.001$; Both $F_{(9,781)}=21.92$, $p < 0.001$. B) CAL Control $F_{(3,283)}=42.00$, $p < 0.001$; Taurine $F_{(3,225)}=89.30$, $p < 0.001$; Both $F_{(7,508)}=54.50$, $p < 0.001$. * $p < 0.05$ denotes a significant difference from Control 0, ** $p < 0.05$ denotes a significant difference between Taurine 0 and other Taurine values.

Developing chick brain and retinal Müller cells express various types of phosphatases[31], which are involved in the elongation of rod photoreceptors in teleosts[27], and also constitute part of the mechanism for the inhibition of exocytotic and transporter-dependent dopamine release in bovine retina[2]. As shown in the present report, an increase in the net phosphorylation level of goldfish explants in culture by the use of phosphatase inhibitors produces a similar effect as that of PKC activators, supporting the notion that PKC is involved in the regeneration of the retina.

Interestingly, the taurine-mediated increase in neurite elongation and density of the fibers occurs in retinal explants. In isolated cells, either from goldfish or rat, taurine only increases the number of outgrowing cells, but not the length of the axon[24]. This suggests that an interaction between cell populations is required for the trophic effect of taurine. The modulation of protein phosphorylation by taurine could take place directly in ganglion cells or could be the result of an interaction between specific elements of the explant, such as the Müller and the ganglion cells.

A proper balance between the phosphorylating and the dephosphorylating reactions is critical for the regeneration of peripheral nerves[33], as well as those of the central nervous system, such as the optic nerve. The factors involved in regeneration after crushing and following taurine exposure need to be investigated. At the present time, an attractive hypothesis, based on the importance of protein phosphorylation in the regulation of taurine transport, is that inhibition of taurine uptake by the cells might be the cause of the impairment in outgrowth observed with PKC activators or phosphatase inhibitors in the retina.

## ACKNOWLEDGMENT

This work was supported by a Grant S1-723 from Consejo Venezolano de Investigaciones Científicas y Tecnológicas (CONICIT). We appreciate the secretarial assistance of Mrs. Isabel Otaegui.

## REFERENCES

1. Barnett, N.L. and Osborne, N.N., 1995, Prolonged bilateral carotid artery occlusion induces electrophysiological and immunochemical changes to the rat retina without causing histological damage, *Exp. Eye Res.*, 61:83-90.
2. Bugnon, O., Ofori S., and Schorderet, M., 1995, Okadaic acid modulates exocytotic and transporter-dependent release of dopamine in bovine retina *in vitro*, *Naunyn-Schmiedebergs Arch. Pharmacol.*, 351:53-59.
3. Busch, M.S., Tonge, D.A., Woolf, C., and Gordon-Weeks, P.R., 1996, Expression of a developmentally regulated, phosphorylated isoform of microtubule-associated protein 1B in regenerating axons of the sciatic nerve, *Neuroscience*, 73:553-563.
4. Campenot, R.B., Draker D.D., and Senger D.L., 1994, Evidence that protein kinase C activities involved in regulating neurite growth are localized to distal neurites, *J. Neurochem.*, 63:868-878.
5. Carter, J.E., Gallo, J.M., Anderson, V.E., Anderton, B.H., and Robertson, J., 1996, Aggregation of neurofilaments in NF-L transfected neuronal cells: regeneration of the filamentous network by a protein kinase inhibitor, *J. Neurochem.*, 67:1197-2004.
6. Casini, G., Grassi, A., Trasarti, L., and Bagnoli, P., 1996, Developmental expression of protein kinase C immunoreactivity in rod bipolar cells of the rabbit, *Vis. Neurosci.*, 13:817-831.
7. Fernández, E., Cuenca, N., García, M., and De Juan, J., 1995, Two types of mitocondria are evidenced by protein kinase C immunoreactivity in the Müller cells of the carp retina, *Neurosci. Lett.*, 183:202-205.

8. Fukuda, K., Saito, N., Yamamoto, M., and Tanaka, C., 1994, Immunocytochemical locali-zation of the alpha-, beta I-, beta II- and gamma-subspecies of protein kinase C in the monkey visual pathway, *Brain Res.*, 658:155-162.
9. Ghalayini, A.J., Koutz, C.A., Wetsel, W.C., Hannun, Y.A., and Anderson, R.E., 1994, Immuno-localization of PKC zeta in rat photoreceptor inner segments, *Curr. Eye Res.*, 13:145-150.
10. González-Quevedo, A., Obregón, F., Santiesteban, R., Fernández, R., and Lima L., 1997, Plasma taurine concentration is decreased in patients with cuban optic neuropathy, in International Taurine Satellite Symposium, 16th Biennal Meeting International Society for Neurochemistry, pp. 60.
11. Guerra, A. and Lima, L., 1997, Taurine transport in isolated cells from the rat retina during ontogeny: possible modulation by phosphorylation, in International Taurine Satellite Symposium, 16th Biennal Meeting International Society for Neurochemistry, pp. 60.
12. Han, X., Budreau, A.M., and Chesney R.W., 1996, Role of conserved peptide in taurine transporter inactivation modulated by protein kinase C, *J. Am. Soc. Nephrol.*, 7:2088-2096.
13. Hemmings, H.C., Nairnm A.C., McGuinness, T.L., Huganir, R.L., and Greengard, P., 1989, Role of protein phosphorylation in neuronal signal transduction, *FASEB J.*, 3:1583-1592.
14. Ikeda, T. and Puro, D.G., Regulation of retinal glial cell proliferation by antiproliferative molecules, *Exp. Eye Res.*, 60:435-443.
15. Janssen-Bienhold, U., Buschmann-Gebhardt, B., and Weiler, R., Phorbol ester binding sites in the fish retina: correlation with stimulation of endogenous phosphorylation and protein kinase C activation, *J. Neurochem.*, 65:744-753.
16. Jian, X., Hidaka, H., and Schmidt, J.T., 1994, Kinase requirement for retinal growth cone motility, *J. Neurobiol.*, 25:1310-1328.
17. Lima, L., Matus, P., and Drujan, B., 1988, Taurine effect on neuritic growth from goldfish retinal explants, *Int. J. Devl. Neurosci.*, 6:417-424.
18. Lima, L., Matus, P., and Drujan, B., 1992, The trophic role of taurine in the retina. A possible mechanism of action, in Taurine, Lombardini, J.B., Schaffer, S.W. and Azuma, J., eds., Plenum Press, New York, pp. 287-294.
19. Lin, N. and Bazán, H.E., 1995, Protein kinase C substrates in corneal epithelium during wound healing: the phosphorylation of growth associated protein-43 (GAP-43), *Exp. Eye Res.*, 61:451-459.
20. Lombardini, J.B., 1994, The inhibitory effects of taurine on protein phosphorylation: comparison of various characteristics of the taurine-affected phosphoproteins present in rat retina, brain and heart, in Taurine in Health and Disease, Huxtable, R.J. and Michalk, D., eds., Plenum Press, New York, pp. 9-17.
21. Lombardini, J.B., 1997, Increased phosphorylation of specific rat cardiac and retinal proteins in taurine-depleted animals: isolation and identification of the phosphoproteins, in International Taurine Satellite Symposium, 16th Biennal Meeting International Society for Neurochemistry, p. 47.
22. Loo, D.D., Hirsch, J.R., Sarkar, H.K., and Wright, E.M., 1996, Regulation of the mouse retinal taurine transporter (TAUT) by protein kinases in Xenopus oocytes, *FEBS Lett.*, 392:250-254.
23. Magnusson, K.R., 1994, Changes in the localization of taurine-like immunoreactivity during development and regeneration in the rat brain, *Adv. Exp. Med. Biol.*, 359:235-243.
24. Matus, P. and Lima, L., (1997) Differential effect of taurine and serotonin on the outgrowth from explants or isolated cells of the retina, *Int. J. Devl. Neurosci.* (in press).
25. Mollerup, J. and Lambert, I.H., 1996, Phosphorylation is involved in the regulation of the taurine influx via the beta-system in Erlich ascites tumor cells, *J. Membr. Biol.* 150:73-82.
26. Osborne, N.N., Wood, J., and Müller, A., 1995, The inluence of experimental ischaemia on protein kinase C and the GABAergic system in the rabbit retina, *Neuropharmacology*, 34:1279-1288.
27. Pagh-Roehl, K., Lin, D., Su, L., and Burnside, B., 1995, Phosducin and PP33 are in vivo targets of PKA and type 1 or 2A phosphatases, regulators of cell elongation in teleost rod inner-outer segments, *J. Neurosci.*, 15:6475-6488.
28. Pijak, D.S., Hall, G.F., Tenicki, P.J., Boulos, A.S., Lurie, D.I., and Selzer, M.E., 1996, Neurofilament spacing, phosphprylation, and axon diameter in regenerating and uninjured lamprey axons, *J. Comp. Neurol.*, 368:569-581.
29. Saito, T., Kaneko, Y., Maruo, F. Niino, M., and Sakaki, Y., 1994, Study of the regenerating newt retina by electrophysiology and immunohistochemistry (bipolar- and cone-specific antigen localization), *J. Exp. Zool.*, 270:491-500.
30. Schmidt, J.T., 1994, C-kinase manipulations disrupt activity-driven retinotopic sharpening in regenerating goldfish retinotectal projection, *J. Neurobiol.*, 25:555-570.
31. Shock, L.P., Bare, D.J., Klinz, S.G., and Maness, P.F., 1995, Protein tyrosine phosphatases expressed in developing brain and retinal Müller glia, *Brain Res. Mol. Brain Res.* 28:110-116.

32. Stevens, M.J., Lattimer, S.A., Kamijo, M., Van-Huysen, C., Sima, A.A., and Greene, D.A., 1993, Osmotically-induced nerve taurine depletion and the compatible osmolyte hypothesis in experimental diabetic neuropathy in the rat, *Diabetologia*, 36:608-614.
33. Svensson, B., Ekström, P.A., and Edström, A., 1995, Okadaic acid and cultured frog sciatic nerves: potent inhibition of axonal regeneration in spite of unaffected Schwann cell prolife-ration and ganglionic protein synthesis, *J. Neurochem.*, 64:1000-1007.
34. Teng, K.K. and Greene, L.A., 1994, KT5926 selectively inhibits nerve growth factor-dependent neurite elongation, *J. Neurosci.*, 14(Pt 1):2625-2635.
35. Tonge, D.A., Golding, J.P., and Gordon-Weeks, P.R., 1996, Expression of a developmentally regulated, phosphorylated isoform of microtubule-associated protein 1B in sprouting and regenerating axons in vitro, *Neuroscience*, 73:541-551.
36. Wiklund, P., Ekdtröm, P.A.R., Edbadh, M., Tonge, D., and Edström, A., 1996, Protein kinase C and mouse sciatic nerve regeneration, *Brain Res.*, 715:145-154.
37. Wood, J.P., McCord, R.J., and Osborne, N.N., 1997, Retinal protein kinase C, *Neurochem. Int.*, 30:119-136.

# UNIQUE PHARMACOLOGICAL INTERACTIONS OF TAURINE AND CHELERYTHRINE IN THE RETINA

Julius D. Militante[1] and John B. Lombardini[1,2]

[1] Departments of Pharmacology
[2] Ophthalmology & Visual Sciences
Texas Tech University Health Sciences Center
Lubbock, TX 79430

## ABSTRACT

The effects of taurine and chelerythrine (CHT) on ATP-dependent calcium uptake and the phosphorylation of the ~20 kDa phosphoprotein in the retina were compared. In the absence of the CHT, taurine stimulated ATP-dependent calcium uptake and attenuated the phosphorylation of the ~20 kDa phosphoprotein. On the other hand, CHT produced the opposite results in the absence of taurine. When the two agents were used simultaneously, it was found that CHT non-competitively inhibited the action of taurine to stimulate calcium uptake, while taurine non-competitively inhibited the action of CHT to stimulate the phosphorylation of the ~20 kDa phosphoprotein. The data present an unusual pharmacological mechanism for controlling the signal transduction pathway involving the two distinct cellular processes being studied. Given the unique data, a control system is proposed in which the function of the ~20 kDa phosphoprotein is linked to the stimulation of ATP-dependent calcium uptake.

## INTRODUCTION

The cells in the photoreceptor layer of the retina are responsible for processing the light stimulus that the eye receives into information that the brain can understand. The transduction of this signal requires the delicate regulation of calcium flux in the rod outer segments[1,4]. Taurine, a free amino acid found in high concentrations in the retina, has been consistently shown to potentiate ATP-dependent calcium uptake in whole retina and rod outer segment preparations[8]. Nutritional deficits and artificial depletion of taurine have been found to be responsible for visual dysfunction. Thus, taurine has always been regarded as essential in normal visual function.

Taurine has also been shown to modulate other cellular functions in the rat retina aside from calcium flux. It has been shown to decrease the *in vitro* phosphorylation of a specific ~20 kDa retinal protein[10]. The protein is found throughout the retina but appears to be concentrated in the mitochondrial fraction, though it probably is not an integral part of the mitochondrial membrane. The phosphorylation reaction predominantly involves serine residues but also involves threonine residues. Taurine depletion in the live rat produces the opposite result, that is, the phosphorylation of the ~20 kDa protein is increased[14]. It can be assumed that the function of the protein is altered by taurine through phosphorylation modulation. Preliminary results from amino acid sequence analyses of a section of PAGE gel containing the ~20 kDa protein indicate that the preparation was not homogeneous and contained two phosphoproteins in an abundance ratio of approximately 4 to 1 (unpublished data). One phosphoprotein was identified as histone H2B which is probably not applicable to these studies on ATP-dependent calcium uptake while the second protein is unknown and potentially pertinent to these studies.

The use of a benzophenanthridine alkaloid called chelerythrine (CHT) to modulate both ATP-dependent calcium flux[16] and the phosphorylation of the ~20 kDa protein[12], as well as interact with taurine, have allowed the study of the relationship between calcium uptake and protein phosphorylation. CHT is known as a very potent inhibitor of protein kinase C (PKC)[3], though it has also been found to inhibit other cellular processes such as the $Na^+$, $K^+$-ATPase[2]. Through a series of experiments, the effects of chelerythrine were studied relative to the effects of taurine on ATP-dependent calcium uptake and the phosphorylation of the ~20 kDa protein. It was found that CHT exhibited the exact opposite effects as taurine: CHT increases the phosphorylation of the ~20kDa protein[12] , and decreases ATP-dependent calcium uptake[16]. Taurine and CHT were also found to antagonize each other's effects in a non-competitive manner. The pattern suggests that the phosphorylation of the ~20kDa protein and ATP-dependent calcium uptake are linked events. Thus, it is possible that both taurine and CHT are acting on one of the processes in order to produce their effects on the other.

## MATERIALS AND METHODS

### Materials

Chelerythrine chloride was purchased from LC Laboratories, Woburn, MA, and [γ-$^{32}$P]ATP and $^{45}CaCl_2$ were obtained from New England Nuclear, Boston, MA. BCA protein assay reagent was obtained from Pierce Chemicals, Rockford, IL. Aminoethylhydrogen sulfate (AEH) was obtained from Aldrich Chemical Co. Isethionic acid (ISA) was obtained from Nutritional Biochemicals Corp. Guanidinoethanesulfonic acid (GES) was synthesized according to the procedure of Morrison et al.[17] β-Alanine was obtained from Sigma Chemical Co.

### Preparation of Membrane Homogenate and Mitochondrial Subcellular Fraction

For the membrane homogenate, adult Wistar rats were euthanized with $CO_2$ and the eyes were immediately removed from the animal. The eyes were then stored at -80°C until used. The eyes were thawed and retinal tissue was teased out of the eye cup in 0.32 M sucrose while on ice. All subsequent procedures were done on ice to maintain a 2°C temperature. The tissue was centrifuged for 15 minutes at 16,000 *g*, washed in 20 mM bicarbonate, recentrifuged as before and then washed in sodium-bicarbonate buffer [$NaHCO_3$, 50 mM, NaCl, 50 mM; KCl, 50 mM; $KH_2PO_4$, 1.2 mM; $MgCl_2$, 2 mM (Kuo and Miki, 1980[5])] with $CaCl_2$ added to a final concentration of 10 μM. The tissue was recentrifuged, resuspended in sodium bicarbonate-$CaCl_2$ buffer and gently homogenized.

For the mitochondrial subcellular fraction, 0.3 M mannitol was used instead of 0.32 M sucrose. Retinal tissue was dissected out as before and the rod outer segments (ROS) were removed by vortex-mixing for 6s, allowing the tissue to settle, and then decanting (and discarding) the supernatant which contained the ROS. The remaining tissue components were gently hand-homogenized with 10 up-and-down strokes in a Potter-Elvehjem homogenizer and centrifuged at 150 x *g* for 15 minutes to remove cell debris. The supernatant was centrifuged at 12,500 x *g* for 15 minutes and the pellet was suspended in the mannitol solution. The tissue preparation (minus the ROS) was then layered on a discontinuous Ficoll gradient (8%, 16%, and 20% in 0.3 M mannitol) and centrifuged at 63,000 x *g* for 1 h in a swinging bucket rotor. The pellet at the bottom of the gradient contained mitochondria and unidentified cellular components. This fraction referred to as the mitochondrial fraction was then suspended in sodium-bicarbonate-$CaCl_2$ buffer.

**ATP-Dependent Calcium Uptake Assay**

The incubation medium contained sodium-bicarbonate buffer, which was kept in ice until the start of the reaction. Reagents such as ATP, taurine and chelerythrine, were added in the appropriate concentrations, including identical amounts of $^{45}CaCl_2$ (400,000-500,000 dpm) in a final concentration of 10 μM $CaCl_2$. The reaction tubes were then preincubated in a shaking water bath set at 37°C for 2 minutes. Retinal homogenate (100-300 μg) was added to start the reaction, making a final volume of 250 μl, and the mixture was incubated for an additional 2 minutes. The reaction was terminated by adding 3 ml of ice-cold sodium-bicarbonate buffer and immediately filtering on a Millipore glass fiber filter (Whatman GF/B filter). The filter was washed three times with 3 ml of the above sodium-bicarbonate buffer and then counted for radioactivity with Aquasol scintillation fluid. The amount of $^{45}$calcium taken up by the retinal tissue was determined by subtracting the counts retained on the filter after a zero-time incubation with retinal preparation.

**Phosphorylation Assay and Polyacrylamide Gel Electrophoresis (PAGE)**

The incubation mixture containing sodium bicarbonate buffer, mitochondrial fraction (~0.1 mg protein), and additives such as chelerythrine chloride and taurine was preincubated for 2 minutes in a shaking water bath at 37°C. The phosphorylation reaction was started by the addition of [$\gamma$-$^{32}$P]ATP (20 μCi, 10 μM). The system was incubated for 6 minutes which was determined to be in the linear time range for optimal phosphorylation. The reaction was then stopped by adding 0.3 ml of gel electrophoresis sample buffer [60 mM Tris-HCl (pH 6.8), 2% sodium dodecyl sulfate, 10% glycerol, 2 mM mercaptoethanol, and 0.00125% Bromophenol blue] and immediately boiling for 5 minutes. Aliquots of the incubation mixture were subjected to one-dimensional polyacrylamide gel electrophoresis on 12% gels according to the method of Laemmli[6]. The gels were dried and exposed to X-ray film to visualize the incorporation of radioactive phosphate into the various proteins. The amount of radioactive phosphate in the ~20 kDa phosphoprotein was quantified by densitometry (BIORAD scanning densitometer, Model 1650).

**ATPase Assay**

The ATPase assay was adapted from commonly used procedures[19] and the sodium-bicarbonate medium used for the calcium uptake assay. Briefly, all the appropriate reagents including 10 μM $CaCl_2$ were added as with the calcium uptake assay except for the $^{45}CaCl_2$. Instead, equal amounts of [$\gamma$-$^{32}$P]ATP (400,000 - 500,000 dpm) were added to each incubation mixture and the reaction tubes were preincubated in a 37°C water bath for 2 minutes. Retinal homogenate was then added as before and the tubes were incubated for an additional 2 minutes. The reaction was terminated by adding 250 μl of 5% perchloric acid, and the radioactivity was extracted into *n*-butanol/benzene (1:1) from which an aliquot was taken and counted in a scintillation counter. Control tubes were incubated in the same length of time and radioactivity was counted the same way, but the perchloric acid was added before preincubation and the addition of the retinal homogenate. ATPase activity was determined by subtracting experimental values from control counts.

### Phosphatase Assay

Retinal membrane preparations were phosphorylated as described above. After a 5-minute incubation, non-radioactive ATP (1.2 mM) was added to the system plus 20 mM taurine or taurine analogue. The incubation reaction was then incubated for an additional 10 minutes at 37°C. The reaction was stopped by the addition of 1.0 ml of gel electrophoresis sample buffer and boiled as above. A control for these experiments was also performed as described for the phosphorylation system. Four milliliters of trichloroacetic acid (10%) were added to an aliquot (0.05 ml) of the incubation mixture to precipitate the retinal membrane proteins. The precipitate was washed four times with 4 ml of 10% TCA, collected on 0.65-μm Millipore filter, and then counted for radioactivity. Radioactivity retained on the Millipore filter from a control experiment ([$\gamma$-$^{32}$P]ATP added after the incubation mixture was boiled) was subtracted from the experimental value.

### Protein Assay

Protein concentrations were determined by the method of Lowry et al.[15] with bovine serum albumin utilized as the standard. For some experiments, the amount of protein used was assayed using the bicinchoninic acid (BCA) method (Pierce Chemical Co.)

## RESULTS AND DISCUSSION

Table 1 summarizes the different effects of taurine and CHT on both ATP-dependent calcium uptake and phosphorylation of the ~20 kDa protein. Taurine stimulates ATP-dependent calcium uptake while CHT is inhibitory[7,16]. Both taurine and CHT were found to have no effect on calcium uptake in the absence of ATP[16]. Vice-versa, taurine was found to inhibit the phosphorylation of the ~20 kDa protein while CHT is stimulatory [9,12]. Taurine causes half-maximal stimulation of calcium uptake at around 31.6 ± 6.1 mM, a concentration that is not significantly different from the concentration calculated to cause half-maximal inhibition of the phosphorylation of the ~20 kDa protein (34.2 ± 2.1 mM). Similarly, CHT causes half-maximal inhibition of calcium uptake and half-maximal stimulation of protein phosphorylation at concentrations not significantly different from each other (37.3 ± 4.6 and 31.5 ± 9.6 μM, respectively).

Combination treatment was next performed with both taurine and CHT being used simultaneously in the calcium uptake and phosphorylation assays. CHT was used to inhibit the action of taurine in ATP-dependent calcium uptake, and taurine was used to inhibit the action of CHT on the phosphorylation of the ~20 kDa protein. Figure 1A shows the increase in ATP- dependent calcium uptake as percent of control levels (0 mM taurine) at different concentrations of taurine[16]. ATP-dependent calcium uptake was inhibited by 100 μM CHT, and this inhibition was not overcome by increasing the concentration of taurine, indicating a non-competitive type of inhibition. Figure 1B shows a similar experiment, but in this instance, taurine at 30 mM concentration was used to inhibit the action of CHT,

**Table 1.** Effect of taurine and CHT on ATP-dependent calcium uptake and ~20 kDa protein phosphorylation

| | ATP-dependent calcium uptake | ~ 20 kDa protein phosphorylation |
|---|---|---|
| Stimulatory factor ($SC_{50}$) | 31.6 ± 6.1 mM taurine[13] (n = 5) | 37.3 ± 4.6 μM CHT[10] (n = 6) |
| Inhibitory factor ($IC_{50}$) | 31.5 ± 9.6 μM CHT (n = 4) | 34.2 ± 2.1 mM taurine[9] (n = 4) |

which increases the phosphorylation of the ~20 kDa protein[12]. As with the ATP-dependent calcium uptake experiments, the competition was found to be non-competitive in that increased CHT concentrations did not overcome the effects of taurine.

To study the possible mechanisms responsible for the effect of taurine on phosphorylation, phosphatase activity was measured. Figure 2 shows the effects of taurine and other taurine analogues on retinal phosphatase activity[7]. It was found that taurine at 20 mM concentration, along with AEH (aminoethylhydrogen sulfate), ISA (isethionic acid) and GES (guanidinoethanesulfonic acid) at the same concentration, do not have any effect on phosphatase activity. Thus, the decrease in phosphorylation of the ~20 kDa protein cannot be attibuted to phosphatase-catalyzed dephosphorylation. In the same way, inhibition of phosphatase activity does not seem to be linked to the phosphorylation of the ~20 kDa protein. With okadaic acid, which is a phosphatase inhibitor, no significant change in the phosphorylation of the ~20 kDa protein was observed (Table 2).

Because CHT is a known PKC inhibitor, PKC modulators (Table 2) were used to study the possible modulation of ATP-dependent calcium uptake and the phosphorylation of the ~20 kDa protein by PKC. It was found that phorbol esters, which activate PKC, exhibited no effect on the phosphorylaiton of the ~20 kDa protein[11]. Similarly, the PKC inhibitor, K252b, exhibited no effect on calcium uptake[16].

ATPase activity could theoretically modulate ATP-dependent calcium uptake and may be affected by phosphorylation. Therefore, the effect of taurine and CHT on ATPase

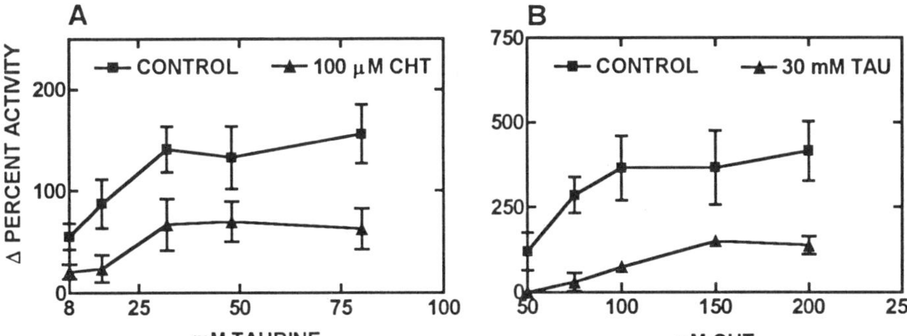


**Figure 1.** A. Concentration-response of the effect of taurine on ATP-dependent calcium uptake in the presence and absence of 100 μM CHT. Values are in percent change over control wherein control is at 0 mM taurine. B. Concentration-response of CHT for the phosphorylation of the ~20 kDa protein in the presence and absence of 100 mM taurine. Values are in percent change over control wherein control is at 0 μM CHT.

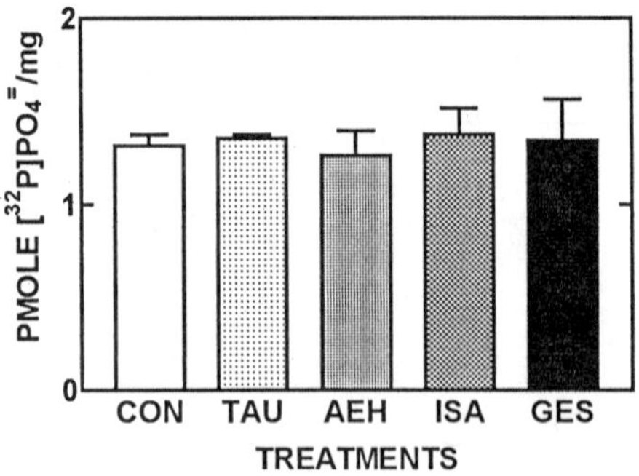


**Figure 2.** The effect of equimolar concentrations (30 mM) of taurine (TAU) and taurine analogues on phosphatase activity. AEH = aminoethylhydrogen sulfate, ISA = isethionic acid, GES = guanidinoethanesulfonic acid.

activity was studied in the retina. ATPase ativity was found to be unaffected by taurine and other taurine analogues (Fig. 3). In contrast, CHT was found to potently inhibit ATPase activity (Fig. 4).

Inasmuch as the effects of taurine and CHT on ATP-dependent calcium uptake and and ~20 kDa protein phosphorylation occur in the absence of the other effector, the interaction between the two effectors does not represent classical pharmacological inhibition since inhibitors by definition exhibit affinity but no efficacy on their own. It is possible that taurine and CHT act through totally different signalling pathways to produce their effects on the same system, whether it be ATP-dependent calcium uptake or protein phosphorylation.

CHT is a potent PKC inhibitor but also exerts other biological effects[2,3]. The stimulation and the inhibition of PKC have been shown not to affect the phosphorylation of the ~20 kDa protein and retinal ATP-dependent calcium uptake, respectively (Table 2). This suggests that CHT and taurine do not cause their effects through PKC modulation. It was also found that taurine does not alter phosphatase activity (Fig. 3) and that okadaic acid inhibition of the retinal phosphatases does not affect the phosphorylation of the ~20 kDa protein (Table 2). Though the negative evidence does not completely rule out a role for a phosphatase in the effects of CHT, it suggests strongly that both taurine and CHT do not regulate phosphatase activity, at least in terms of their effects on phosphorylation of the ~20 kDa protein.

Taurine has been shown not to have a significant effect on ATPase activity in the

**Table 2.** Various treatments that produced no significant effects on ATP-dependent calcium uptake and phosphorylation of the ~20 kDa protein in the retina.

| | AGENT | MECHANISM OF ACTION |
|---|---|---|
| Phosphorylation of the ~20 kDa protein | phorbol ester<br>okadaic acid | PKC activator[9]<br>phosphatase inhibitor[11] |
| ATP-dependent calcium uptake | K252b | PKC inhibitor[15] |

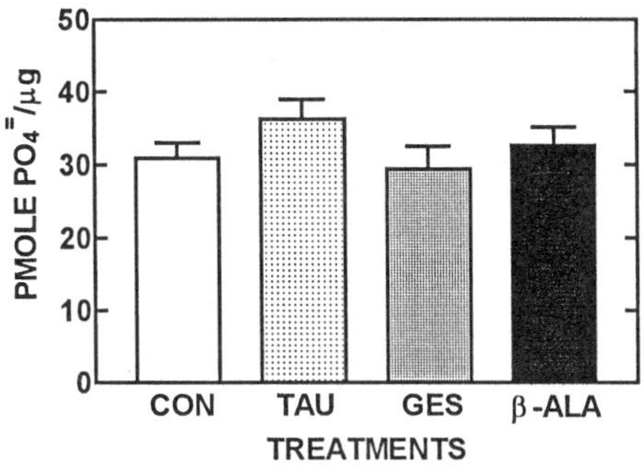


**Figure 3.** The effect of equimolar concentrations (32 mM) of taurine and taurine analogues on ATPase activity. GES = guanidinoethanesulfonic acid, β-ALA = β-alanine.

retina[7]. This is important to note, given that both taurine stimulation and CHT inhibition of calcium uptake are dependent on ATP. CHT inhibits ATPase activity in the retina (Fig. 4)[16], therefore, effects on calcium uptake in the absence of taurine may occur through this mechanism. But it is clear that the action of CHT to antagonize the effects of taurine on calcium uptake does not involve ATPase modulation[16]. Taurine may be involved in the activation of an ATPase-independent calcium uptake mechanism, such as calcium channel opening, and not with active processes like calcium transporters.

It is interesting that taurine exhibits two contrasting effects, one being the stimulation of ATP-dependent calcium uptake and the other inhibition of protein phosphorylation, both of which could be antagonized effectively by the same agent, which in this case is CHT (Table 1). It is highly unlikely that taurine and CHT produce their effects in these two systems through a mechanisms in which both ATP-dependent calcium uptake and the ~20 kDa phosphoprotein are affected in exactly the opposite manner. It is even more unlikely that taurine and CHT produce their effects in a manner totally exclusive for each system. Rather, it makes sense to assume that an interaction exists between calcium flux and the function of the ~20 kDa phosphoprotein, such that taurine and CHT act through different pathways to modulate one of these systems which in turn produces its effect on the other system. The non-competitive nature of the observed inhibition of taurine on CHT stimulation of the phosphorylation of the ~20 kDa phosphoprotein and of CHT on taurine-stimulated ATP-dependent calcium uptake supports this notion (Fig. 1 and 2).

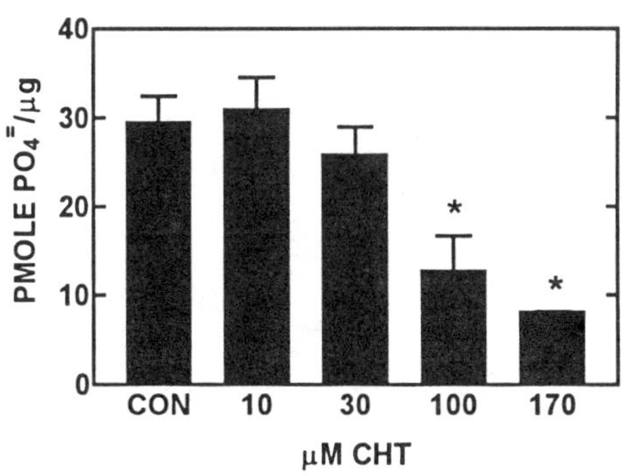


**Figure 4.** The concentration-response of CHT on ATPase activity. (* $p < 0.05$, compared to control, Student's t-test.)

This reasoning leads to the hypothesis that the function of the ~20 kDa phosphoprotein is linked to ATP-dependent uptake of calcium in the retina. What could be the nature of the dependence between the two events? The *in vitro* phosphorylation of the ~20 kDa protein is accomplished in conditions of very low ATP concentration (~20 μM). Since the effects of taurine and CHT on ATP-dependent calcium uptake occur only at much higher ATP concentrations (~1.2 mM)[16], the effects of taurine and CHT on the phosphorylation of the ~20 kDa protein must be independent of their effects on calcium uptake. In fact, it has been shown that this phosphorylation is not affected by calcium ions[11]. On the other hand, it is likely that under conditions in which taurine stimulates calcium uptake, i.e., in the presence of 1.2 mM ATP and 10 μM calcium, taurine would simultaneously inhibit the phosphorylation of the ~20 kDa phosphoprotein. The same idea can be applied to the effects od CHT. Although the nature of the calcium flux and the identity of the ~20 kDa phosphoprotein are unknown, it is clear that if there is a link between the activity of the two, it is the calcium flux that must be dependent on the function of the ~20 kDa phosphoprotein.

Figure 5 is a schematic illustration of the pathways that might be involved in this relationship, including the various conditions that might characterize these pathways. Taurine and CHT modulate the phosphorylation of the ~20 kDa protein in a phosphatase- and PKC-independent manner. The ~20 kDa protein, in turn, modulates some form of ATP-dependent calcium utpake system, probably the opening of a calcium channel. The dual regulation of the ~20 kDa phosphoprotein is non-competitive, suggesting modulation through different phosphorylation sites on the molecule. Theoretically, taurine and CHT each affect a different set of kinases and phosphorylation sites. The distinct sites, $P_x$ for taurine and $P_y$ for CHT, are shown. By inhibiting phosphorylation at its own distinct sites, taurine causes activation of the protein and its activity causes an increase in calcium uptake. CHT stimulates phosphorylation of the ~20 kDa protein at different sites, a modification which inactivates the protein and in turn decreases calcium uptake.

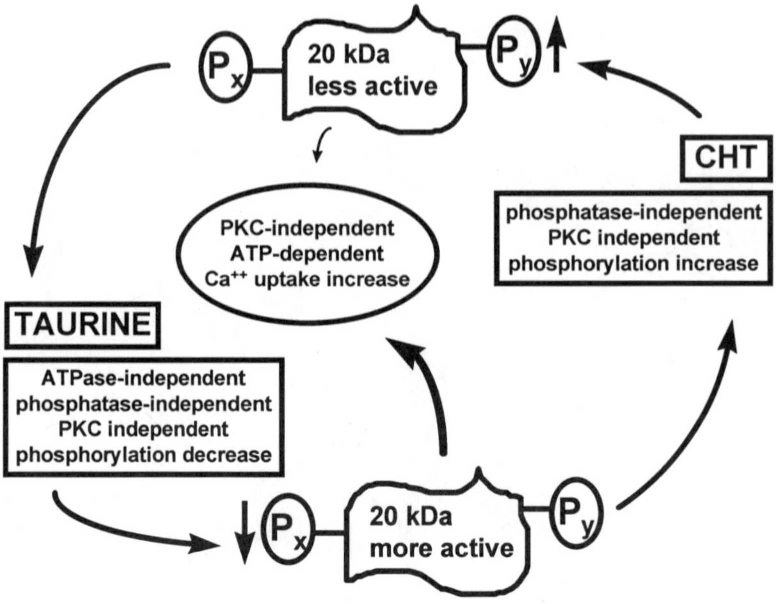


**Figure 5**. Schematic diagram of the possible link between ATP-dependent calcium uptake and the phosphorylation of the ~20 kDa protein, and of the pharmacological modulation of the two systems by taurine and by CHT.

## ACKNOWLEDGMENTS

This work was funded by grants from the RGK Foundation of Austin, Texas, and by the Taisho Pharmaceutical Co., LTD. of Tokyo, Japan.

## REFERENCES


1. Baylor, D., 1996, How photons start vision, *Proc. Natl. Acad. Sci.*, 93:560-565.
2. Cohen, H.G., Seifen, E.E., Straub, K.D., Tiefenback, C., and Stermitz, F.R., 1978, Structural specificity of the NaK-ATPase inhibition by sanguinarine, an isoquinoline benzophenanthridine alkaloid, *Biochem. Pharmacol.*, 27:2555-2558.
3. Herbert, J.M., Augereau, J.M., Gleye, J., and Maffrand, J.P., 1990, Chelerythrine is a potent and specific inhibitor of protein kinase C, *Biochem. Biophys. Res. Comm.*, 172:993-999.
4. Koch K.-W., 1995, Control of photoreceptor proteins by $Ca^{2+}$, *Cell Calcium,* 18:314-321.
5. Kuo, C.-H. and Miki, N., 1980, Stimulatory effect of taurine on Ca-uptake by disk membranes, *Biochem. Biophys. Res. Comm.*, 94:646-651.
6. Laemmli, U.K., 1970, Cleavage of structural proteins during the assembly of the head of bacteriophage T4, Nature, 227:680-685.
7. Lombardini, J.B., 1985, Effects of taurine on calcium ion uptake and protein phosphorylation in rat retinal membrane preparations, *J. Neurochem.,* 45:268-275.
8. Lombardini, J.B., 1991, Taurine: retinal function, *Brain Res. Rev.*, 16:151-169.
9. Lombardini, J.B., 1992, Effects of taurine on the phosphorylation of specific proteins in subcellular fractions of the rat retina, *Neurochem. Res.*, 17(8):821-824.
10. Lombardini, J.B., 1993, Partial characterization of an ~20K $M_r$ retinal protein whose phosphorylation is inhibited by taurine, *Biochem. Pharmacol.,* 46:1445-1451.
11. Lombardini, J.B., 1993, The inhibitory effects of taurine on protein phosphorylation: comparison of various characteristics of the taurine-affected phosphoproteins present in rat retina, brain and heart, *Adv. Exp. Med. Bio.*, 359:9-17.
12. Lombardini, J.B., 1995, Paradoxical stimulatory effect of the kinase inhibitor chelerythrine on the phosphorylation of a ~20K M protein present in the mitochondrial fraction of rat retina, *Brain Res.*, 673:194-198.
13. Lombardini, J.B., Liebowitz, S.M., and Chou, T.-C., 1989, Analogues of taurine as stimulators and inhibitors of ATP-dependent calcium ion uptake in rat retina: combination kinetics, *Mol. Pharmacol.*, 36:256-264.
14. Lombardini, J.B., Young, R.S.L., and Props, C.L, 1996, Taurine depletion increases phosphorylation of a specific protein in the rat retina, *Amino Acids*, 10:153-165.
15. Lowry, O.H., Rosebrough, N.J., Farr, A.L., and Randall, R.J., 1951, Protein measurement with the Folin phenol reagent, *J. Biol. Chem.*, 193:265-275.
16. Militante, J.D. and Lombardini, J.B., 1997, The effect of taurine on chelerythrine inhibition of calcium uptake and ATPase activity in the rat retina, *Biochem. Pharmacol.*, submitted for publication.
17. Morrison, J.F., Ennor, A.H., and Griffiths, D.E., 1958, The preparation of barium monophosphotaurocyamine, *Biochem. J.*, 68:447-452.
18. Nakanishi, S., Matsuda, Y., Iwahashi, K., and Kase, H., 1986, K-252b, c and d, potent inhibitors of protein kinase C from microbial origin, *J. Antibiotics,* 39:1066-1071.
19. Ottlecz, A., Garcia, C.A., Eichberg, J., and Fox, D.A., 1993, Alterations in retinal $Na^+$, $K^+$-ATPase in diabetes: streptozotocin-induced and Zucker diabetic fatty rats, *Curr. Eye Res.* 12:1111-1121.

# INCREASED PHOSPHORYLATION OF SPECIFIC RAT CARDIAC AND RETINAL PROTEINS IN TAURINE-DEPLETED ANIMALS: ISOLATION AND IDENTIFICATION OF THE PHOSPHOPROTEINS

John B. Lombardini

Departments of Pharmacology and Ophthalmology
and Visual Sciences
School of Medicine
Texas Tech University Health Sciences Center
Lubbock, TX 79430

## ABSTRACT

Partial depletion of the tissue levels of taurine in the alive animal stimulates *in vitro* phosphorylation of both an ~20 kDa (94%) and an ~44 kDa (85%) protein present in subcellular fractions of the rat retina and heart. Tissue levels of taurine were reduced by feeding the animals 1.5% guanidinoethanesulfonic acid (GES) in their drinking water for 6 weeks. The increase in the *in vitro* phosphorylation of both the ~20 kDa and ~44 kDa proteins was reversed when the animals were subsequently given 1.5% taurine in their drinking water for an additional 6 weeks. Isolation, purification, tryptic digestion, and peptide sequence analysis of the retinal ~20 kDa phosphoprotein suggest that it is histone H2B. However, the results are ambiguous due to an impurity (<25%) in the retinal preparation. Sequence analysis of the ~44 kDa phosphoprotein indicates that it is pyruvate dehydrogenase.

## INTRODUCTION

The inhibitory effects of taurine on the phosphorylation of specific proteins present in mitochondrial fractions of rat heart and retina have been a prime focus of our laboratory for approximately the last 13 years. During this time span a number of interesting points have been learned about these proteins and the intricacies of their phosphorylation. In earlier reviews various characteristics of the taurine-affected phosphoproteins were compared with respect to a variety of parameters such as: the effects of taurine analogues, kinase activators and kinase inhibitors, the determination of the isoelectric points, and the

*Taurine 3*
Edited by Schaffer *et al.*, Plenum Press, New York, 1998

specific amino acid(s) that accepts the phosphate group[5,8,9,11,13]. The proteins in question have molecular masses of ~20 kDa (retina) and ~44 kDa (heart) and are different from each other not only in the obvious, i.e, the molecular mass, but also are unique proteins as demonstrated by their different characteristics. In this chapter I wish to report on the manipulation of the phosphorylation of these proteins in the taurine-depleted animal and also to report on the tentative identification of both proteins.

## MATERIALS AND METHODS

Guanidinoethanesulfonic acid (GES) was synthesized according to the procedure of Morrison et al.,[17]. [$\gamma$-$^{32}$P]ATP (30 Ci/mmole) was purchased from New England Nuclear Corporation. Experimental animals (Wistar rats) were maintained on drinking water which contained 1.5% GES or 1.5% taurine for the times indicated in the figure legends. Control animals were mantained on untreated water.

The mitchondrial fraction of the rat hearts was prepared by differential centrifugation techniques as previously described[12]. The retinal preparation was prepared on a Ficoll mannitol discontinuous gradient and contained mitochondria and unidentified subcellular components. The incubation mixture for the phosphorylation assay contained bicarbonate buffer and the subcellular (~0.1 mg protein) fraction[12,6]. The mixture was preincubated for 2 min in a shaking water bath set at 37°C and the phosphorylation reaction was started by the addition of [$\gamma$-$^{32}$P]ATP (20 μCi, 10 μM). The complete assay system was incubated for 6 min and then stopped by adding 0.3 ml of gel electrophoresis sample buffer and immediately boiling for 5 min. Aliquots of the incubation mixture were subjected to PAGE on 12% gels according to the method of Laemmli[3]. When the electrophoretic run was finished the gels were stained with Coomassie Brilliant Blue R, destained, and dried on a slab gel drier (Hoefer Pharmacia BioTech). The dried gels were then exposed to X-ray film (overnight, -80°C) to visualize the location of the phosphoproteins.

Isolation of the phosphoproteins for amino acid sequence analysis was performed as follows: the retinal or heart preparations were phosphorylated and then subjected to electrophoresis as described above except that when the electrophoretic run was finished the gels were not stained (but washed with water) and not dried. X-Ray film was used to visualize the location of the ~20 and ~44 kDa phosphoproteins. The portions of the gel (separate gels) corresponding to the ~20 or ~44 kDa proteins were excised and the gel fragments were placed in the well of a tricine-buffered gel (13%T/6%T) and re-electrophoresed for 2 hrs at 30 volts (constant voltage) and then for either 24 (heart proteins) or 48 hrs (retinal proteins) at 95 volts[23]. At the end of the run the gel was stained with Coomassie Brilliant Blue R, destained, dried, and exposed to X-ray film for 1-2 days (-80°C). The radioactive band was then cut out of the gels and sent to either the Protein & Nucleic Acid Facility at Stanford University, Stanford, CA., or the Biomedical Mass Spectrometry Laboratory at the University of Virginia Health Sciences Center, Charlottesville, VA, for digestion, HPLC separation of the peptides, and sequencing of the peptides.

Electroretinogram (ERG) recordings were performed as previously described[15]. Briefly, the rats were dark-adapted for 16 hours prior to testing and then anesthetized with sodium pentobarbital (30 mg/kg in 10% alcohol). Mydriacyl (1%) was used topically to dilate the pupils. The eyelid was retracted and a cotton wick electrode attached to a chlorided silver wire was placed on the cornea. A second chlorided wire placed in a small cut in the cheek of the rat served as the indifferent electrode. A needle electrode was inserted in the rat's tail to provide the electrical ground connection. The stimulus was a Grass Xenon Flash Lamp placed approximately 6 inches from the eye. The resulting ERG was composed almost exclusively of a b-wave which was measured as the maximum positive excursion from the baseline. The recordings consist of a single ERG response, typically the first response obtained. The ERG was amplified with a Grass amplifier.

## RESULTS AND DISCUSSION

### Rat Retina: ~20 kDa Phosphoprotein[15]

Partial depletion of the taurine content in the rat retina was accomplished by introduction of 1.5% GES in the drinking water. Taurine levels decreased by 50% after 1 week of GES treatment and by 80% after 16 weeks (Fig. 1). Treatment with GES up to 22 weeks did not further increase the taurine depletion. The amplitude of the electroretinogram (ERG) b-wave did not change after 2 weeks of GES treatment but then dropped precipitously at 3 weeks, reaching a maximum at 16 weeks (Fig. 1). This reduced level, a 60% decrease in the amplitude, was maintained for up to 22 weeks.

After 6 weeks of GES treatment when the retinal taurine content was reduced by 70% and the amplitude of the b-wave was reduced by 50%, *in vitro* phosphorylation of a specific protein present in the mitochondrial fraction of the rat retina with an approximate molecular weight of ~20 kDa was increased by 94% (Fig. 2, left columns). The increased phosphorylation of this ~20 kDa protein observed after GES treatment was reversed when the animals were treated with taurine (1.5%) in the drinking water for an additional 6 weeks (Fig. 2, right columns). Treatment of the rats for 6 weeks with taurine (1.5%) alone had no effect on the phosphorylation of the ~20 kDa protein (Fig. 2, middle columns).

Results from these studies show that a substantive increase in phosphorylation of an ~20 kDa protein present in the mitochondrial fraction of the rat retina occurs when the taurine content of the retina was decreased and the amplitude of the ERG b-wave was reduced. It is not entirely clear how or whether these observations are related, but if they are, it may be that taurine regulates the phosphorylation of certain, specific proteins that are essential to the retinal light response but not to the structural integrity of the neurons.

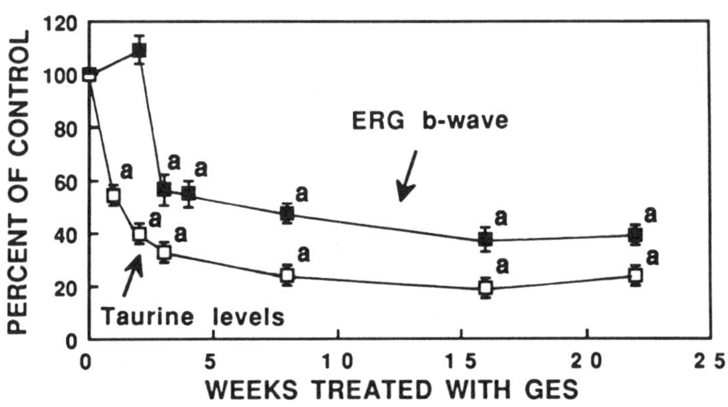


**Figure 1.** ERG b-waves and taurine levels in the retinas of rats treated with GES (11/2%) in the drinking water for 22 weeks. ERG measurements and determination of taurine content are described in the Materials and Methods section. Data are expressed as the percent of control values (means ± SEM). Each mean for ERG measurements contain 5-8 animals. Each mean for taurine content contains 6-7 animals. Statistical differences between control values and the values obtained form animals maintained on GES for up to 22 weeks were determined by analysis of variance and the Duncan's multiple-range test ($^{a}P < 0.05$). Taurine levels (□), ERG b-wave (■). (Reprinted from ref. 15 with permission of Springer-Verlag.)

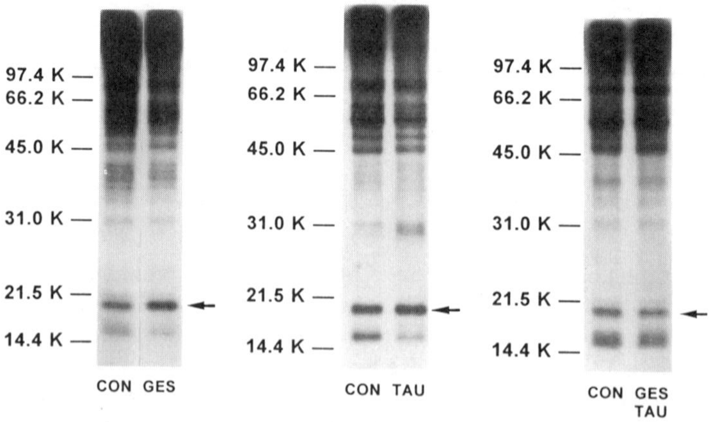


**Figure 2.** Autoradiograph from a 1-dimensional (PAGE) gel of the effects of taurine depletion, taurine treatment, and reversal of taurine depletion on the phosphorylation of an ~20 K $M_r$ protein present in a mitochondrial fraction of the rat retina (representative experiment). Left column: Animals were treated for 6 weeks with GES (1.5%) in the drinking water. Middle column: Animals were treated for 6 weeks with taurine (1.5%) in the drinking water. Right column: Animals were treated for 6 weeks with GES (1.5%) followed by 6 weeks of treatment with taurine (1.5%) in the drinking water. Description of the phosphorylation assay, PAGE, autoradiography, and densitometry are described in the Materials and methods section. Marker proteins with molecular weights ranging from 14.4 to 97.4 K are indicated. CON control; GES guanidinoethanesulfonic acid: TAU taurine. The arrow indicates the location of a protein with an approximate molecular weight of ~20 K. (Reprinted from ref. 15 with permission of Springer-Verlag.)

Isolation of the ~20 kDa phosphoprotein on a 1-dimensional polyacrylamide gel using traditional glycine buffers followed by re-electrophoresing the cut out portion of the gel which corresponds to the ~20 kDa protein on a tricine-buffered gel resulted in sufficient pure protein for digestion and sequence analysis. The sequence analysis results indicate that the ~20 kDa phosphoprotein is histone H2B. It has been previously reported that histone H2B from various tissue sources can be phosphorylated on the serine-32 residue by a cyclic nucleotide-independent histone kinase[21]. Our previous data also indicate that the rat retinal ~20 kDa protein is phosphorylated on a serine residue[10], and cAMP and cGMP have no effect on its phosphorylation[11].

However, other non-histone peptides (not present in protein databases) were observed at a presence of < 25% of the abundance of the histone peptides and thus suggest that the sample contained more than one protein. These results that the ~20 kDa phosphoprotein preparation is not homogenous are thus ambiguous and a definitive conclusion that the phosphoprotein is histone H2B cannot be made at this time.

**Rat Heart: ~44 kDa Phosphoprotein[14,16]**

After 6 weeks of GES (1.5%) treatment the taurine levels of the rat heart were reduced to approximately 25% of control values. *In vitro* phosphorylation of an ~44 kDa protein present in the mitochondrial fraction of the rat heart was increased by 85% (Fig. 3, left columns). The increase in the phosphorylation of this specific ~44 kDa protein after

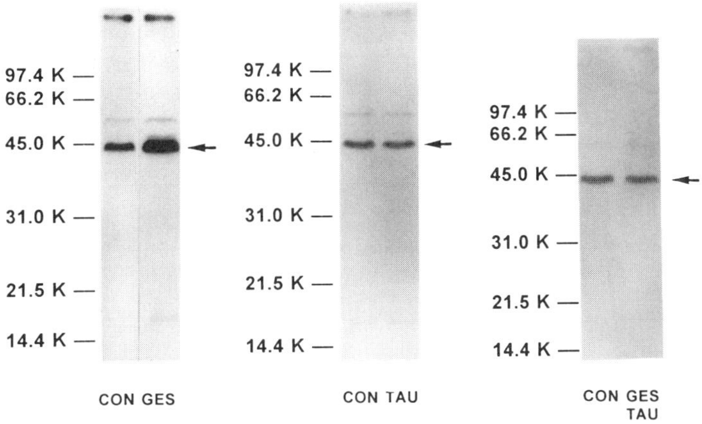


**Figure 3.** Autoradiography from a 1-dimensional (PAGE) gel of the effects of taurine depletion, taurine treatment, and reversal of taurine depletion on the phosphorylation of an ~44 kDa protein present in a mitochondrial fraction of the rat heart (representative experiment). Left column: animals were treated for a minimum of 6 weeks with GES (1.5%) in the drinking water. Middle column: animals were treated for 6 weeks with taurine (1.5%) in the drinking water. Right column: animals were treated for 6 weeks with GES (1.5%) followed by 6 weeks of treatment with taurine (1.5%) in the drinking water. Description of the phosphorylation assay, PAGE, autoradiography, and densitometry measurements are described in the Materials and Methods section. Marker proteins with molecular weights ranging from 14.4 to 97.4 K are indicated. CON=control: GES=guanidinoethanesulfonic acid: TAU=taurine. The arrow indicates the location of a protein with an approximate molecular weight of ~44 kDa. (Reprinted from ref. 14 with permission of Academic Press Limited.)

GES treatment was reversed when the animals were subsequently given 1.5% taurine in their drinking water for an additional 6 weeks (Fig. 3, right columns). Taurine (1.5%) alone for a period of 6 weeks had no effect on the phosphorylation of the ~44 kDa protein.

The ~44 kDa phosphoprotein was isolated (similar techniques as described above for the retinal protein) and digested. Sequence analysis indicated that this phosphoprotein is pyruvate dehydrogenase, an important enzyme for aerobic energy metabolism in the cell. The ~44 kDa phosphoprotein isolated from the mitochondrial fraction of the rat heart and pyruvate dehydrogenase have similar properties: 1) Molecular weight of pyruvate dehydrogenase is 43.2 kDa in the mouse liver[1] and 41 kDa in the bovine kidney[19] and bovine heart[20]. Commercial pig heart pyruvate dehydrogenase migrates on 1-dimensional polyacrylamide gels in the same region as the rat heart radioactive ~44 kDa phosphoprotein. 2) Pyruvate dehydrogenase is a known phosphoprotein[18]. 3) The ~44 kDa protein is phosphorylated only in a serine residue(s) which is consistent with the reported serine phosphorylation sites for the mouse liver, mouse testis, human somatic, and human testis pyruvate dehydrogenases[1]. 4) It has been reported that cAMP has no effect on the phosphorylation of pyruvate dehydrogenase[24] which is consistent with our data that cAMP also has no effect on the phosphorylation of the ~44 kDa protein[11].

Taurine has been reported to have significant effects on energy metabolism in cardiac tissue. It has been demonstrated that taurine affects myocardial metabolism due to stimulation of glycolysis and aerobic metabolism[22] and decreases the levels of pyruvate and the ratio of ATP/ADP in perfused rat hearts exposed to taurine[4]. Taurine-depleted cardiac tissue demonstrates altered myocardial high energy phosphate metabolism due to elevated long chain acylcarnitine content and reduced long chain fatty acyl CoA levels[2]. These results presented herein that taurine affects the phosphorylation of pyruvate dehydrogenase are additional data that suggest taurine has an effect on energy metabolism in the cell.

## ACKNOWLEDGMENTS

Ms. Josie L. Aleman and Ms. Tonia Borrego are acknowledged for their secretarial assistance. This work was supported in part by grants from the RGK Foundation of Austin, TX, and the Taisho Pharmaceutical Co. Ltd. of Tokyo, Japan.

## REFERENCES

1. Fitzgerald, J., Hutchison, W.M., and Dahl, H-H.M., 1992, Isolation and characterization of the mouse pyruvate dehydrogenase E1a genes, *Biochim. Biophys. Acta.*, 1131:83-90.
2. Harada, H., Allo, S., Viyuoh, N., Azuma, J., Takahashi, K., and Schaffer, S.W., 1988, Regulation of calcium transport in drug-induced taurine-depleted hearts, *Biochim. Biophys. Acta.*, 944:273-278.
3. Laemmli, U.K., 1970, Cleavage of structural proteins during the assembly of the head of bacterioophage $T_4$, *Nature*, 227:447-452.
4. Lampson, W.G., Kramer, J.H., and Schaffer, S.W., 1983, Potentiation of the actions of insulin by taurine, *Can. J. Physiol. Pharmacol.*, 61:457-463.
5. Lombardini, J.B., 1985, Inhibition by taurine of the phosphorylation of rat retinal membranes, in: "Progress in Clinical and Biological Research, Taurine: Biological Actions and Clinical Perspectives" Oja, S.S., Ahtee, L., Kontro, P. and Paasonen, M.K. eds., Alan R. Liss Press, New York, 179: 383-396.
6. Lombardini, J.B., 1985, Effects of taurine on calcium ion uptake and protein phosphorylation in rat retinal membrane preparations, *J. Neurochem.*, 45:268-275.
7. Lombardini, J.B., 1988, Effects of taurine and mitochondrial metabolic inhibitors on ATP-Dependent calcium ion uptake in synaptosomal and mitochondrial subcellular fractions of rat retina, *J. Neurochem.*, 51:200-205.
8. Lombardini, J.B. and Liebowitz, S.M., 1989, Taurine modifies calcium ion uptake and protein phosphorylation in rat heart, in: "Taurine and the Heart" Iwata, H., Lombardini, J.B. and Segawa, T. eds., Kluwer Academic Publishers, Massachusetts, pp. 117-137.
9. Lombardini, J.B., 1992, Effects of taurine on protein phosphorylation in mammalian tissues, in: "Advances in Experimental Medicine and Biology, Taurine: Nutritional Value and Mechanisms of Action" Lombardini, J.B., Schaffer, S.W. and Azuma, J. eds., Plenum Press, New York, Vol. 315, pp. 309-318.
10. Lombardini, J.B., 1993, Partial Characterization of a ~ 20 K $M_r$ Retinal protein whose phosphorylation is inhibited by taurine, *Biochem. Pharmacol.* 46:1445-1451.
11. Lombardini, J.B., 1994, The inhibitory effects of taurine on protein phosphorylation: Comparison of various characteristics of the taurine-affected phosphoproteins present in rat retina, brain and heart, in: "Advances in Experimental Medicine and Biology, Taurine In Health And Disease" Michalk, D.V. and Huxtable, R.J. eds., Plenum Press, New York, Vol. 359, pp. 9-17.
12. Lombardini, J.B., 1994, Inhibitory effects of taurine and taurine analogues on the phosphorylation of a ~44 K $M_r$ phosphoprotein, *J. Mol. Cell. Cardioy* 26: 1675-1689.
13. Lombardini, J.B. and Props, C., 1996, Effect of kinase inhibitors and taurine analogues on the phosphorylation of specific proteins in mitochondrial fractions of rat heart and retina, in: "Advances in Experimental Medicine and Biology, Taurine 2: Basic and Clinical Aspects" Huxtable, R., Azuma, J., Kuriyama, K., Nakagawa, M. and Baba, A. eds., Plenum Press, New York, Vol. 403, pp. 343-350.
14. Lombardini, J.B., 1996, Taurine depletion in the intact animal stimulates *in vitro* phosphorylation of an ~44 kDa protein present in the mitochondrial fraction of the rat heart, *J. Mol. Cell. Cardiol.*, 28:1957-1961.
15. Lombardini, J.B., Young, R.S.L., and Props, C., 1996, Taurine depletion increases phosphorylation of a specific protein in the rat retina, *Amino Acids*, 10:153-165.
16. Lombardini, J.B., 1997, Identification of a specific protein in the mitochondrial fractions of the rat heart whose phosphorylation is inhibited by taurine, *Amino Acids*, 12:139-144.

17. Morrison, J.F., Ennor, A.H., and Griffiths, D.E., 1958, The preparation of barium monophosphotaurocyamine, *Biochem. J.,* 68:447-452.
18. Pettit, F.H., Pelley, J.W., and Reed, L.J., 1975, Regulation of pyruvate dehydrogenase kinase and phosphatase by acetyl-CoA/CoA and NADH/NAD ratios, *Biochem. Biophys. Res. Commun.*, 65:575-582.
19. Pettit, F.H. and Reed, L.J., 1982, Pyruvate dehydrogenase complex from bovine kidney and heart, *Meth. Enzymol.,* 89:376-391.
20. Reed, L.J. and Yeaman, S.J., 1987, Pyruvate dehydrogenase, *The Enzymes*, 18:77-95.
21. Romhányi, T., Seprödi, J., Mészáros, Gy., Antoni, F., and Faragó, A., 1982, The intrinsic substrate specificity of a cyclic nucleotide-independent protein (histone) kinase, *FEBS Letters,* 144:223-225.
22. Schaffer, S.W., Kramer, J.H., Lampson, W.G., Kulakowski, E., and Sakane, Y., 1983, Effect of taurine on myocardial metabolism: role of calmodulin, in: "Sulfur amino acids: biochemical and clinical aspects". Kuriyama, K., Huxtable, R.J., Iwata, H. eds., Alan R. Liss, New York, *Prog. Clin. Biol. Res.,* 125:39-50.
23. Schägger, H. and Von Jagow, G., 1987, Tricine-sodium dodecyl sulfate - polyacrylamide gel electrophoresis for the separation of proteins in the range from 1 59 100 kDa, *Anal. Biochem.,* 166: 368-379.
24. Technikova-Dobrova, Z., Sardanelli, A.M., and Papa, S., 1993, Phosphorylation of mitochondrial proteins in bovine heart: Characterization of kinases and substrates, *FEBS Letters*, 322:51-55.

# EXTENSIVE TAURINE DEPLETION AND RETINAL DEGENERATION IN CATS TREATED WITH ß-ALANINE FOR 40 WEEKS

Humi Imaki, Jeffrey Messing, and John A. Sturman

Department of Developmental Biochemistry
Institute for Basic Research in Developmental Disabilities
1050 Forest Hill Road
Staten Island, NY 10314

## INTRODUCTION

It has now been established that cats are dependent on dietary sources of taurine to maintain their body reserve, and when deprived, they develop a variety of disorders including retinal degeneration, myocardial failure, depressed immune responsiveness and extensive reproductive deficits[37,38,39]. Feline central retinal degeneration, a photoreceptor degenerative condition originally described as bilaterally symmetrical generalized focal lesions of unknown etiology[3,4,26], was among the first to be clearly linked to taurine deficiency[16]. The small incipient lesions visible by fundus ophthalmoscope as hyper-reflective round or oval foci in the area centralis, extend nasally and temporally in advanced stages of taurine deprivation accompanied by significant reductions in rod and cone ERG amplitudes[5,6,34]. The progression of retinal degeneration, however inevitable, is a relatively slow process and takes at least 6 to 12 months for cats fed a completely taurine-free diet to show the first signs. Moreover, tissue taurine levels are lowered only to the point that equilibrium with biosynthesized taurine is reached, with considerable individual variation, no matter how long cats are maintained on such a diet. In an attempt to reduce tissue taurine levels further, we provided adult female cats with ß-alanine which competes with taurine for the same transport system, thereby limiting taurine uptake[21,25]. Previously, we reported that tissue taurine concentrations were greatly reduced while large amounts of ß-alanine accumulated in visceral tissues and neural tissues of acutely taurine-depleted cats after 20 weeks of ß-alanine treatment[42]. The morphological changes associated with photoreceptor degeneration in the retinas of these cats were considerably more extensive and widespread than any described in cats rendered taurine-deficient by means of taurine-free diet alone[17,40]. In the present report we describe the consequence of extreme taurine depletion attained by extending the period of ß-alanine treatment to 40 weeks.

## MATERIALS AND METHODS

Female domesticated cats raised in the Animal Colony of the Institute for Basic Research were fed a completely defined, taurine-free, purified diet (BioServe, Frenchtown, New Jersey, USA), or the same diet supplemented with 0.05% taurine and some maintained *ad libitum* and plain tap water for at least 2 years. They were then given water containing 5% ß-alanine as the sole drinking source for 40 weeks while being kept on their original diets. Other cats, fed the taurine-supplemented diet and plain water throughout, served as controls. Upon completion of the feeding period, all cats were anesthetized with sodium pentobarbital (Nembutal, Abbott Laboratories, North Chicago, IL, USA), and right eyes enucleated were perfused by injecting cold 2.5% glutaraldehyde and 1% paraform-aldehyde in 0.1 M Sorensen's phosphate buffer (pH 7.4), and stored in the same fixative at 4°C. The posterior eye cups were dissected radially around area centralis, postfixed in 2% $OsO_4$ for 1-2 hrs., dehydrated and embedded in EM-bed 812. Transverse sections of retina were collected from each sector 1-2 mm off the area centralis. Semithin (0.75 μm) sections were stained with toluidine blue and photographed in a Zeiss Axiophot using Kodak TMAX-100 film. Thin sections were stained with uranyl acetate and lead citrate and examined in a Philips EM300 operated at 80 kv. Tissue from the left eye and other body parts were rapidly dissected and processed for determination of taurine and ß-alanine concentration by reverse-phase HPLC as described previously[41].

## RESULTS

The concentrations of taurine and ß-alanine in various ocular tissue from cats used in the current and previous studies are shown in Table 1. After 40 weeks of ß-alanine consumption, the taurine concentration in all ocular tissue was greatly reduced, approximately 75% in cats fed the taurine-supplemented diet and 90% in cats fed taurine-free diet relative to the control cats. These values were roughly half of the concentration found after 20 weeks of ß-alanine treatment for each tissue except for tapetum and lens which exhibited steeper reductions in the first 20 weeks. The concentration of ß-alanine, on the other hand, was markedly increased in all tissues examined, but changes were more pronounced in cats fed the taurine-free than -supplemented diet; displaying an inverse relationship to the taurine concentration. Thus, the increase in the ß-alanine level was especially large in tapetum and retina, which showed drastic reductions in taurine concentration upon ß-alanine treatment.

Light microscopic examination of retinas from these cats revealed signs of degeneration in practically every area studied. These abnormalities were similar in pattern, but definitely more magnified than those observed in the retinas of cats after 20 weeks of ß-alanine treatment. The severity and extent of the retinal lesions were closely correlated with the degree of reduction in retinal taurine concentration, which in general, reflected taurine content of the diet. Instead of the normal 12-13 rows, the outer nuclear layer of the superior retina was comprised of 8-9 rows of photoreceptor nuclei in a cat fed the taurine-supplemented diet and ß-alanine, and its retinal taurine reduced to 25% of the control level (Figs. 1a & 1b). In another cat from the same dietary group, which registered much lower retinal taurine concentration, the outer nuclear layer from the same region exhibited less than a complete row of nuclei, many of which appeared pyknotic. Moreover, neither inner nor outer segments were recognizable in this cat (figure not shown), a pattern indistinguishable from that observed in the corresponding area of retina from a cat fed a taurine-free diet and contain-ing an equally low (1/7 of control) retinal taurine (Fig. 1c). Photoreceptors were completely absent from the same area of another

**Table 1.** Concentration of taurine and ß-alanine in ocular tissues of cats fed diet with (+T) or without (-T) 0.05% taurine, and with or without 5% ß-alanine for 20 or 40 weeks

| Diet | ß-Ala | Retina | Tapetum | Lens | Optic Nerve |
|---|---|---|---|---|---|
| | | Taurine (μmoles/g wet weight) | | | |
| +T[1)] | 0 | 42.4 ± 4.4 | 11.1 ± 3.6 | 6.64 ± 2.3 | 3.29 ± 1.08 |
| +T[2)] | 20 wks | 21.8 ± 8.9 | 3.43 ± 1.57 | 1.32 ± 0.56 | 1.00 ± 0.57 |
| +T[3)] | 40 wks | 9.29 ± 8.87 | 3.39 ± 1.95 | 1.54 ± 0.91 | 0.75 ± 0.31 |
| -T[1)] | 0 | 15.9 ± 5.5 | 5.78 ± 2.94 | 1.68 ± 0.88 | 1.37 ± 0.58 |
| -T[2)] | 20 wks | 6.48 ± 3.42 | 1.20 ± 0.23 | 0.69 ± 0.36 | 0.63 ± 0.42 |
| -T[3)] | 40 wks | 3.01 ± 0.34 | 0.95 ± 0.34 | 0.20 ± 0.09 | 0.34 ± 0.17 |
| | | ß-Alanine (μmoles/g wet weight) | | | |
| +T[2)] | 20 wks | 5.52 ± 1.42 | 3.42 ± 1.50 | 3.66 ± 1.95 | 0.21 ± 0.10 |
| +T[3)] | 40 wks | 10.73 ± 6.99 | 12.17 ± 3.85 | 4.71 ± 1.46 | 1.79 ± 1.22 |
| -T[2)] | 20 wks | 9.44 ± 3.03 | 8.71 ± 2.67 | 3.03 ± 0.53 | 2.32 ± 1.23 |
| -T[3)] | 40 wks | 17.00 ± 10.68 | 34.86 ± 20.23 | 5.36 ± 2.23 | 3.98 ± 2.94 |

1 Sturman and Messing (1991). Mean ± S.D. of 10 to 30 cats.
2 Imaki and Sturman (1995). Mean ± S.D. of 5 cats each.
3 Mean ± S.D. of 4 cats each.

cat fed the taurine-free diet and recorded the lowest (approx. 1/30 of control) retinal taurine concentration (Fig. 1d).

Similar, but more pronounced reductions in the thickness of the outer nuclear layer and the number of photoreceptor cells were found in the temporal and nasal retina in each cat (Figs. 2 and 3). The inferior retina, on the other hand, revealed less dramatic changes, although there was a similar tendency toward diminishing photoreceptor cell number, which was proportional to the degree of taurine depletion (Fig. 4). Thus, as noted in cats given ß-alanine for 20 weeks, the degree of photoreceptor degeneration seemed to be dependent on the retinal region, in increasing order of inferior < superior < temporal < nasal sectors (c.f. Figs. 1-4). Other changes readily recognized in the most severely affected retina included greatly increased proportion of Müller cells, with their thick processes traversing the inner plexiform through the nerve fiber layers (Figs. 1c, 2d & 3d), some swollen cells and processes in the inner nuclear and plexiform layers (Fig. 4d), and reductions in the number and size of certain cells and blood vessels in the outer plexiform through the inner retina. The retinal epithelium also appeared greatly attenuated in areas where photoreceptors were missing, but the tapetum retained a normal structure which was indistinguishable from that of control cats.

Under the electron microscope, the earliest lesions most frequently seen in the retinas with the least taurine depletion included photoreceptors whose inner and outer segments were shorter and wider, often twisted or bent, in contrast to those in the normal control retina, which were distinctively uniform and cylindrical, aligned in parallel with one another (Figs. 5a). At slightly more advanced stages of degeneration, the inner and outer segments were further shortened and bulged, or convoluted and interposed by abrupt, irregular subretinal spaces, indicating substantial decreases in their mass (Fig. 5b). In such photoreceptors, mitochondria in the inner segments also were greatly shortened, or rounded up, and membrane discs in the outer segments formed smaller, disarranged stacks

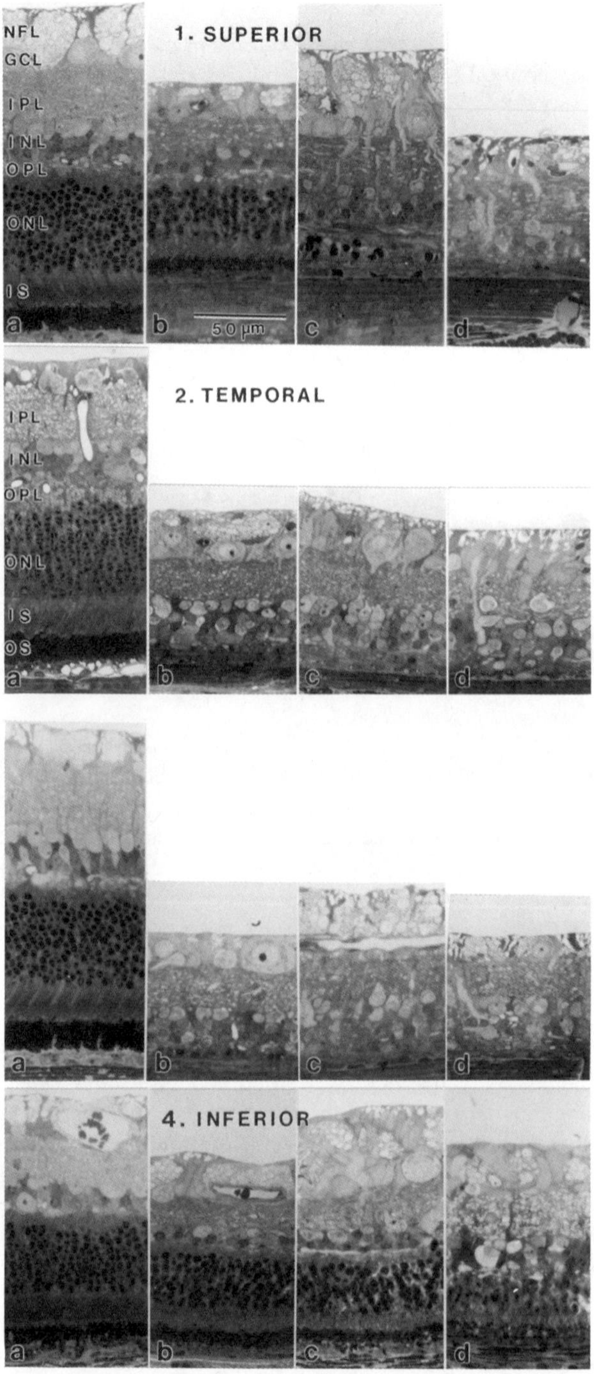


**Figures 1 to 4.** Light micrographs (LMs) of 1) superior, 2) temporal, 3) nasal and 4) inferior retina from a) a control cat, b) a taurine-supplemented, ß-alanine-treated (+T) cat, and, c) and d) taurine-deprived, ß-alanine-treated (-T) cats. Toluidine blue. GCL: ganglion cell layer; INL: inner nuclear layer; IPL: inner plexiform layer; IS: inner segments; NFL: nerve fiber layer; ONL: outer nuclear layer; OPL: outer plexiform layer; OS: outer segments. The scale bar in 1b) applies to all LMs.

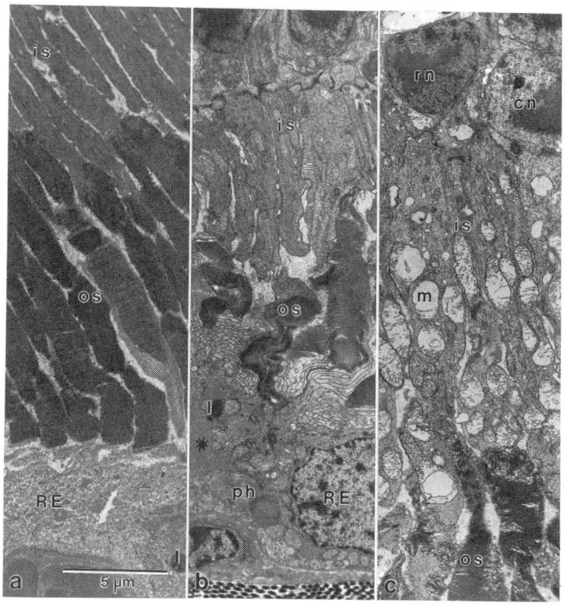


**Figure 5.** Electron micrographs (EMs) of nasal retinas showing that the inner and outer segments (is, os) of photoreceptors are conspicuously shortened and convoluted, or vesiculated and frayed in -T cats (b and c) in comparison to uniformly elongate and straight counterparts in a control cat (a). The retinal epithelial cells (RE) in b) show an increased number of phagosomes (ph) and areas engorged by dense granular material (*). The inner segments and perinuclear cytoplasm of photoreceptors in c) are filled with swollen mitochondria (m) and vacuolar profiles. The scale bar in 5a) applies to all EMs. cn: cone nuclei; l: lipofuscin granules; rn: rod nuclei.

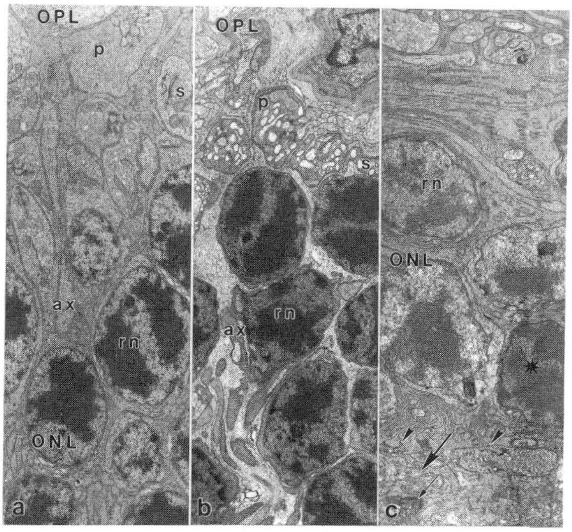


**Figure 6.** EMs showing portions of outer nuclear and outer plexiform layers (ONL, OPL) of retina from a) a control cat, b) a -T cat, and, c) a +T cat. The rod nuclei (rn), axons (ax), pedicles (p) and spherules (s) in b) are more angular in profile and electron-dense than those in the control cat retina. The ONL in c) is comprised only of 2 or 3 rows of photoreceptor nuclei, some of which are distended or flattened while others are condensed (*). Small remnants of inner and outer segments (large and small arrows) are found in the attenuated subretinal space demarcated by indistinct outer limiting membrane (arrowheads) and irregular or obscure apical surface of retinal epithelial cells.

(Figs. 6a & 6b). The retinal epithelial cells appeared slightly compressed or uneven in height, and showed increased numbers of phagosomes and lipofuscin granules and irregularly oriented microvilli, especially in regions where the photoreceptor outer segments were disorganized or missing (Figs 5a & 5b). In addition, many of these condensed epithelial cells revealed large pockets, which were occupied exclusively by dense fine granular material. Photoreceptors in some regions exhibited generalized swelling, as manifested by greatly expanded mitochondria in the inner segments, large vacuolar profiles in the perinuclear and axonal cytoplasm and loose disrupted plasma membranes associated with ruffled or vesiculated disc membranes in the outer segments (Fig. 5c).

In the more severely affected areas, the outer nuclear layer consisted of only one or two rows of photoreceptor nuclei, with a few scattered remnants of inner and outer segments barely discernible in contracted subretinal space partially filled with flocculant or amorphous material (Fig. 6c). Photoreceptor nuclei displayed various shapes and densities, ranging from relatively normal oval to flattened, from pale and large to mottled and crenated, compact and very dense pyknotic figures. The nuclei were surrounded by cytoplasm often containing irregular vacuoles and membranous debris. When photoreceptors were absent, the retinal epithelial cells abutted directly on the cells of the inner nuclear layer, or in apposition to intervening phagocytic macrophages or microglia (Figs. 7a & 7b). The interface between them was either very complex, where sheet-like processes of Müller cells folded with one another and entwined with apical microvilli of epithelial cells, or relatively simple, where the outer limiting membrane was focally interrupted and underlying epithelial cells bore a small number of microvilli. The epithelial cells in such areas usually were further reduced in thickness, but contained even larger numbers of phagosomes scattered throughout the cytoplasm, particulary around the mass of fine dense granular material which occupied an increasingly larger portion of the cell. The cells comprising the inner nuclear layer appeared relatively intact, although the nuclei of the bipolar and horizontal cells were more flattened, and the cytoplasm contained increased numbers of vacuolar or membranous profiles, as did that of the amacrine cells (Figs. 7a & 7b). Most Müller cells were enlarged, their bulbous nuclei surrounded by abundant cytoplasm filled with fine fibrils, and their stout processes stretched and spread over the epithelial cells. This is in stark contrast to the pattern seen in the normal cat retina in which Müller cell perikarya were rather inconspicuous, with their small, angular nuclei and meager cytoplasm appearing to fill space among a more robust population of neurons and blood vessels. Some Müller cells exhibited distinctively lighter cytoplasmic matrices and an increased number of rough ER and elongated mitochondria. Other Müller cells were engorged by fine fibrillar material, often accompanied by thicker filaments or tubules in the peripheral zone (Figs. 7b & 7c). Also observed were Müller cells, whose nuclei were juxtaposed with the epithelial cells (Fig. 7c), a phenomenon never observed in the normal retina in which Müller cells are situated in the mid-inner level of the inner nuclear layer. Proximally, thick processes of Müller cells filled with profuse fibrils and filaments were found to course around neuronal somata and processes and occasional blood vessels in the inner plexiform layer, ganglion cell and nerve fiber layers. In some retinas, these processes were so massive that they occupied most of the inner plexiform layer, in contrast to the delicate and scanty processes spread among the mesh of neuropils in the normal retina (Figs. 8a & 8b). The fine structure of other cells encountered in the inner retina, i.e. ganglion and endothelial cells, astrocytes and microglia, did not appear to be affected substantially, although there seemed to be varying degrees of the decline of both the number and size of ganglion cells and blood vessels, and augmentation of vacuolar intercellular spaces, which often were infiltrated by microglia.

## DISCUSSION

The results of biochemical and morphological studies presented here confirm the earlier report that addition of ß-alanine in drinking water elicits prompt taurine depletion and profound morphological changes in the retina of mature cats, the species known to develop characteristic retinal degeneration by deprivation of dietary taurine. The current study demonstrated that extending the period of ß-alanine treatment from 20 to 40 weeks reduces the tissue taurine concentration in these animals to unprecedented levels, resulting in complete photoreceptor loss in substantial areas of the retina. For example, after 40 weeks of ß-alanine feeding, the taurine concentration of ocular tissues from cats fed the taurine-free diet dropped to less than 10% of the normal level or roughly half that found in cats treated for only 20 weeks. The concentration of ß-alanine in the tissue of these cats increased proportionally to the reduction in taurine levels, again confirming the trend observed in cats treated with ß-alanine for 20 weeks. However, there were considerable individual variations in tissue taurine and ß-alanine concentrations, especially among cats fed the taurine-supplemented diet. Two out of 4 of these cats had lower retinal taurine concentration than one cat fed the taurine-free diet, which in turn had more than twice as much retinal taurine as the other 3 cats in the same group, each having approximately ½ the normal level. The source of such large individual variation is not clear, but it is more likely due to genetic and physiological differences in the ability of individuals to metabolize and retain body taurine than to other factors, such as the total amount of dietary taurine or ß-alanine consumed. There was no correlation between tissue taurine concentration and body weight gain or loss or other physical conditions. The cats showed no obvious signs of ailment other than blindness. Barnett and Burger[2,8] have reported substantial individual variations in taurine concentration in plasma among cats raised on taurine-free or low taurine diets.

The variation in the severity of retinal degeneration appears to be a mere reflection of the variation in the retinal taurine concentration, which in turn, was usually dependent on the taurine content of the diet. Individual variations in the rate of progression and extent of retinal lesions, as detected by ophthalmoscopy, have been noted among cats raised on semipurified diets[2,8] and among cats fed a taurine-deficient vegetarian diet[23]. By applying fundus photography and retinal wholemount histology, Leon et al.[23] confirmed an earlier observation that the oval hyperreflective lesion, which corresponded the area of the retina with a marked reduction in thickness and a loss in the photoreceptor layer, started at the area centralis and spread as a horizontal band across the fundus as the period of taurine deprivation were prolonged. They reported further that in the most advanced cases, the band merged with peripheral annular lesions. Paradoxically, one of eight cats did not develop any lesions even after 1.6 years while other cats acquired stabilized lesions which did not increase in size and resembled the oval dark-bordered lesions observed in the retina of cats fed diets replenished with taurine after developing the typical depletion-mediated lesions. Our current observation of widespread morphological changes in the nasal and temporal regions of the β-alanine treated retina which were more extensive than other regions of the retina, are in agreement with the characteristic lesion topography described in the retina of cats deprived of dietary taurine. However, β-alanine treatment differs with dietary taurine deficiency relative to the involvement of the superior retina, and to a lesser degree the inferior retina. This appears to be attributable to the more pronounced state of taurine deprivation in these cats compared with all the others studied to date. Since the extreme peripheral retina was trimmed before tissue processing in the current study, it is not known if any of our cats developed the peripheral annular lesion described by Leon et al.[23]

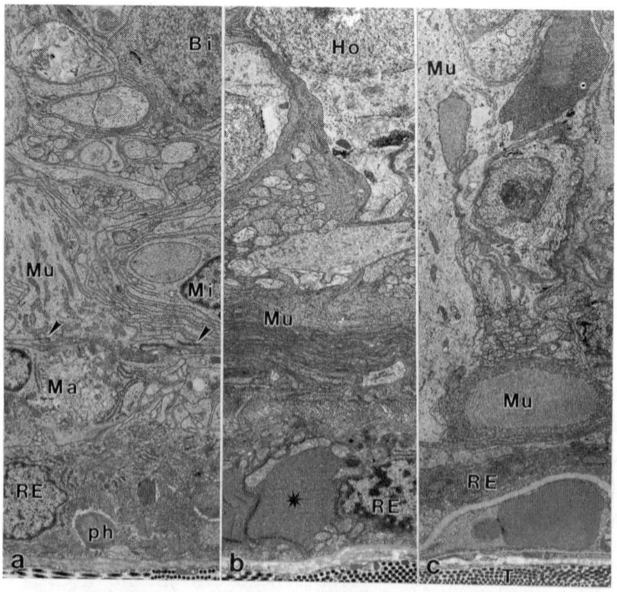


**Figure 7.** EMs of retinas from -T cats showing parts of retinal epithelial cells (RE) lining the cells of the inner nuclear layer in the absence of intervening photoreceptor cells. In a), a microglia (Mi) and a macrophage (Ma) are situated beside the outer limiting membrane (arrowheads). In b) and c), thin sheet-like or blunt endfeet-like processes of Müller cells (Mu) are in direct and intimate contact with the apicalurface of RE which are compressed and engorged by mass of dense granular material (*) and/or spherical phagosomes (ph) of varying size. Note extremely variable cytoplasmic matrix density and organelle population of different Müller cells. Bi: bipolar cells; Ho: horizontal cells, T: tapetum.

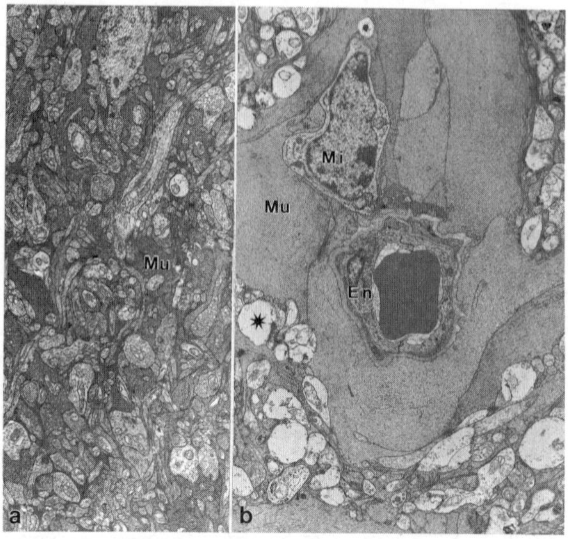


**Figure 8.** EMs showing parts of the inner plexiform layer in the retina of a) a control cat, and b) a -T cat. In contrast to fine and intricate synaptic organization with neuronal processes and endings predominating and forming tight meshwork with equally fine processes of Müller cells (Mu), in the normal retina as seen in a), the inner plexiform layer of the affected retina in b) is engrossed by thick processes of Müller cells (Mu) filled with evenly distributed fibrillar material, which invade clusters of neuronal processes, many of which are grossly swollen or coalesced (*). En: endothelial cells; Mi: microglia.

The unique distribution of photoreceptor degeneration induced by taurine deficiency has elicited several theories on the mechanisms involved. The early and consistent ophthalmoscopic and histological findings that the degenerative process begins in the area centralis, where cone photoreceptors have the highest density[35] and where more cones than rods appear affected in early stages[16], pointed toward the involvement of cones in the initial event and in the progression of the degenerative lesions[2]. This view is supported by electrophysiological data that in the early stages, both a- and b- wave implicit times of only the cones, but not the rods, were delayed, despite a reduction in ERG amplitudes in both the rods and cones[28]. In cats with moderately advanced retinal degeneration, however, rod implicit times also were prolonged, with pronounced reductions in amplitude indicating extensive involvement of rods[16,28,34]. These observations are consistent with the widespread reduction and disappearance of photoreceptor cells, especially of rods from the retina of cats deprived of taurine for prolonged periods. However, these data do not explain the unique developmental pattern of lesions in the taurine deficient retina. Applying imaging fundus reflectometry, which can be used to obtain relative levels of visual pigment, Jacobson et al.[18] demonstrated that rhodopsin is preferentially lost from the horizontal streak encompassing the area centralis and the midperipheral nasal focus in retina of taurine-deprived cats. Interestingly, these areas, which correspond to the horizontal hyper-reflective band of lesion, display the highest rhodopsin levels in normal cat retina, indicating that they not only represent areas with the highest population of ganglion cells[36] but also of rod photoreceptors[20]. These findings suggest that while cones may play an important role in the formation of the initial focal lesion in the area centralis, the subsequent development follows the region with a high rod population, which compete for available taurine[18,20]. They also explain the peculiar pattern of lesions since the region of the retina with the highest content of rhodopsin absorbs more light and therefore is more susceptible to potential damage from light-activated free radicals. Moreover, there is evidence that taurine might retard the formation of lipid peroxidation products and protects membranes of photoreceptor outer segments which have been exposed to light[27]. In support of this hypothesis, Cocker & Lake[10] demonstrated that dark maintenance doubled rhodopsin levels and reduced the ERG deficit, the onset of which paralleled the loss of retinal taurine observed in the GES-treated (Sprague Dawley) rats exposed to low-intensity cyclic light. However, based on electroretinographic recordings, Rapp et al.[30] found that taurine-depleted pigmented rats were not more susceptible to light-induced retinal damage, although the retinal taurine concentration was reduced by 63% and rod outer segments showed some discernible shortening and disintegration. On the other hand, Hageman and Schmidt[15] demonstrated that reductions in the b-wave were significantly greater in the albino than in the pigmented rat, indicating that pigmentation appears to afford some protection against light in moderately taurine-depleted (50% in retina of both groups) rats. To determine if light intensity per se, rather than light deprivation or ocular pigmentation, could affect retinal changes due to taurine depletion, Rapp et al.[29] used GES-treated albino rats raised in either dim or relatively bright light and concluded that taurine depletion and environmental lighting act synergistically to cause photoreceptor degeneration. These rats exhibited the greatest cell loss in the central superior region of the retina, the region most susceptible to acute light stress[31]. This latter observation may imply that the central-superior retina is inherently more sensitive to light damage than the inferior region or simply that the former area receives more light in natural and experimental lighting conditions, and thus more photoreceptor damage. The distinctive triangular sheet of tapetum lucidum, which covers a large part of the central-superior retina in cats, significantly increases the intensity of reflected light, thereby exposing nearby photo-receptors to additional photons and possibly contributing to the development of light-induced lesions, i.e., if indeed ambient light plays a synergistic role in taurine deficiency retinopathy. It is of interest to note that unilateral tarsorrhaphy applied to taurine-depleted cats by Leon et. al.[23] did not affect the ophthalmoscopically detectable lesion topography, leading the authors to conclude that light does not have a major influence in determining the pattern of lesion. Instead, they suggested that the action of "equivalent light" (abnormal continuous activation of the phototransduction cascade), which is postulated to play a role in certain types of pathological retinal degeneration[13], might lead to a scenario in which photoreceptors with the highest content of rhodopsin and cascade molecules, would be most adversely affected resulting in the formation of the horizontal band. Whatever the mechanism involved in the pathological progression, the similarity in both regional variations in the rate of injury and the extent of photoreceptor degeneration suggest that common factors are involved in β-alanine induced and dietary taurine deficient retinopathy in cats. In both conditions, it appears that photoreceptors with a high cell density and elevated rhodopsin content, such as those located in the central and superior retina and seem to play more active and important roles, are more vulnerable to taurine depletion than those with shorter and less tightly packed outer segments, such as those situated peripherally and inferiorly.

The electron microscopic examination confirmed that the fine structural features of the retinal lesions in ß-alanine-treated cats are indistinguishable from those observed in cats fed the taurine-free diet, except for the accelerated rate of development and severity attained at the end of the experimental period. The degeneration of the photoreceptor is first observed in the outer segments, which become bent, twisted, or frayed as the disc membranes dilate, vesiculate, and split, resulting in the loss of their orderly arrangement of tightly-packed, precise stacks. Meanwhile, the inner segments shorten and detract, as their major organelle, the mitochondria, contract, diverge or distend. Many nuclei assume a crenated contour and an increased density, with some acquiring a crumpled chromatin and a pyknotic appearance. The outer and inner segments continue to disintegrate and diminish in size and number before they become rudimentary and eventually disappear. At the same time, nuclei are gradually eliminated so that inevitably the vestigial outer nuclear layer contains only a single incomplete row of flattened dense nuclei. Eventually these nuclei vanish, bringing the inner retinal cells in direct apposition to the retinal epithelium. Such an extreme consequence of retinal degeneration has been reported in some cats fed diets containing no or very low taurine for extended periods, but definitive lesions appear to be confined to small, limited areas, and rarely are accompanied by apparent changes in retinal cells other than the photoreceptors[16,23,26,28]. Often the retinas of ß-alanine-treated cats exhibiting extensive photoreceptor degeneration associated with discrete changes in certain neighboring cells appear more comparable to certain inherited retinal degeneration conditions, than the typical taurine deficiency retinopathy. For example, the dramatic increase and expansion of the Müller cell processes seen in the retina of ß-alanine-treated cats have been observed in rd mice and Irish setters with rod cone dysplasia[1,9,33], although a notable proliferation of Müller cells was noted once in the retina of a cat fed a casein-based, taurine-deficient diet[16]. These morphological changes of Müller cells are interpreted as an adaptation to an altered microenvironment, reactive hypertrophy or gliosis, similar to that reported for other hereditary retinal degenerations and clinical conditions[12,14,24,43]. Müller cells are known to adapt their shape and function to both the structure and functional state of their neuronal neighbors[32]. Similarly, although a marked reduction in arteriole number was noted in some retina of cats fed a synthetic taurine-deficient diet[26], reduction in the number of retinal vessels, especially in the outer plexiform layer, observed in the retina of ß-alanine-treated cats resembles phenomena reported in kittens with rod-cone dysplasia[22,24] and rd mice[7,9]. The rationale for such vascular regression is not clear, but it has been attributed to an abnormal interaction between the dysplastic photoreceptor cells and the growing retinal capillary buds in the retina of Rdy-affected kittens[22]. In the case of the ß-alanine-induced retinopathy, it presumbly is a simple response to the reduced thickness of the cell layers following disappearance of photoreceptor cells. The irregular thickness, electron density and numerical density of the organelles of the retinal epithelial cells are frequently observed in ß-alanine-treated cats, as they are in Rdy cats[11,22] and rd and rds mice[9,19]. These characteristics appear to be a consequence of changes in the photoreceptor cells.

## REFERENCES


1. Aguirre, G.D. and Rubin, L.F., 1975, Rod-cone dysplasia (progressive retinal atrophy) in Irish Setters, *J. Am. Vet. Med. Assoc.*, 166:157-164.
2. Barnett, K.C. and Burger, I.H., 1980, Taurine deficiency retinopathy in the cat, *J. Small Anim. Pract.*, 21:521-534.
3. Bellhorn, R.W., Aguirre, G.D., and Bellhorn, M.B., 1974, Feline central retinal degeneration, *Invest. Ophthalmol.*, 13:608-616.
4. Bellhorn, R.W. and Fischer, C.A., 1970, Feline central retinal degeneration, *J. Amer. Vet. Med. Assoc.*, 157:842-849.
5. Berson, E.L., Hayes, K.C., Rabin, A.R., Schmidt, S.Y., and Watson, G., 1976, Retinal degeneration in cats fed casein. II. Supplementation with methionine, cysteine, or taurine, *Invest. Ophthalmol.*, 15:52-58.
6. Berson, E.L., Watson, G., Grasse, K.L., and Szamier, R.B., 1981, Retinal degeneration in cats fed casein. IV. The early receptor potential, *Invest. Ophthalmol. Vis. Sci.* 21:345-350.
7. Blanks, J.C. and Johnson, L.V., 1986, Vascular atrophy in the retinal degenerative rd mouse, *J. Comp. Neurol.*, 254:543-553.
8. Burger, H.H. and Barnett, K.C., 1982, The taurine requirement of the adult cat, *J. Small Anim. Pract.*, 23:533-537.
9. Caley, D.W., Johnson, C., and Liebelt, R.A., 1972, The postnatal development of the retina in the normal and rodless CBA mouse: a light and electron microscopic study, *Am. J. Anat.*, 133:179-211.
10. Cocker, S.E. and Lake, N., 1989, Effects of dark maintenance on retinal biochemistry and function during taurine depletion in the adult rat, *Visual Neurosci.*, 3:33-38.
11. Curtis, R., Barnett, K.C., and Leon, A., 1987, An early-onset retinal dystrophy with dominant inheritance in the Abyssinian cat, *Invest. Ophthalmol. Vis. Sci.*, 28:131-139.
12. Ekström, P., Sanyal, S., Narfström, K., Chader, G.J., and VanVeen, T., 1988, Accumulation of glial fibrillary acidic protein in Müller radial glia during retinal degeneration, *Invest. Ophthalmol. Vis. Sci.*, 29:1363-1371.
13. Fain, G.L. and Lisman, J.E., 1993, Photoreceptor degeneration in vitamin A deprivation and retinitis pigmentosa: the equivalent light hypothesis, *Exp. Eye Res.*, 57:335-340.
14. Fan, W., Lin, N., Sheedlo, H.J., and Turner, J.E., 1996, Müller and RPE cell response to photoreceptor degeneration in aging Fischer rats, *Exp. Eye Res.*, 63:9-18.
15. Hageman, G.S., and Schmidt, S.Y., 1987, Taurine-deficient pigmented and albino rats: early retinal abnormalities and differential rates of photoreceptor degeneration, *Prog. Clin. Biol. Res.*, 247:497-515.
16. Hayes, K.C., Rabin, A.R., Berson, E.L., 1975, An ultrastructural study of nutritionally induced and reversed retinal degeneration in cats, *Am. J. Pathol.*, 78:505-524.
17. Imaki, H., Sturman, J.A., 1995, Taurine depletion and retinal degeneration resulting from dietary ß-alanine in adult cats, *Biomed. Lett.*, 51:111-125.
18. Jacobson, S.G., Kemp, C.M., Barruat, F.-X., Chaitin, M.H., and Faulkner, D.J., 1987, Rhodopsin topography and rod-mediated function in cats with the retinal degeneration of taurine deficiency, *Exp. Eye Res.*, 45:481-490.
19. Jansen, H.G. and Sanyal, S., 1984, Development and degeneration of retina in rds mutant mice: electron microscopy, *J. Comp. Neurol.*, 224:71-84.
20. Kemp, C.M., Faulkner, D.J., and Jacobson, S.G., 1988, The distribution and kinetics of visual pigments in the cat retina, Invest. *Ophthalmol. Vis. Sci.*, 29:1056-1065.
21. Larsson, O.M., Griffiths, R., Allen, I.C., Schousboe, A., 1986, Mutual inhibition kinetic analysis of γ-amino butyric acid, taurine, and ß-alanine high-affinity transport into neurons and astrocytes: evidence for similarity between the taurine and ß-alanine carriers in both cell types, *J. Neurochem.*, 47:426-432.

22. Leon, A. and Curtis, R., 1990, Autosomal dominant rod-cone dysplasia in the Rdy cat. 1. Light and electron microscopic findings, *Exp. Eye Res.*, 51:361-381.
23. Leon, A., Levick, W.R., and Sarossy, M.G., 1995, Lesion topography and new histological features in feline taurine deficiency retinopathy, *Exp. Eye Res.*, 61:731-741.
24. Lewis, G.P., Matsumoto, B., Fisher, S.K., 1995, Changes in the organization and expression of cytoskeletal proteins during retinal degeneration induced by retinal detachment, *Invest. Ophthalmol. Vis. Sci.*, 36:2404-2416.
25. Martin, D.L., Shain, W., 1979, High affinity transport of taurine and ß-alanine and low affinity transport of γ-aminobutyric acid by a single transport system in cultured glioma cells, *J. Biol. Chem.*, 254:7076-7084.
26. Morris, M.L., 1965, Feline degenerative retinopathy, *Cornell Vet.*, 55:295-308.
27. Pasantes-Morales, H. and Cruz, C., 1985, Taurine and hypotaurine inhibit light-induced lipid peroxidation and protect rod outer segment structure, *Brain Res.* 330:154-157.
28. Rabin, A.R., Hayes, K.C., and Berson, E.L., 1973, Cone and rod responses in nutritionally induced retinal degeneration in the cat, *Invest. Ophthalmol.*, 12:694-704.
29. Rapp, L.M., Thum, L.A., and Anderson, R.E., 1988, Synergism between environmental lighting and taurine depletion in causing photoreceptor cell degeneration, *Exp. Eye Res.*, 46:229-238.
30. Rapp, L.M., Thum, L.A., Tarver, A.P., and Wiegand, R.D., 1987, Retinal changes associated with taurine depletion in pigmented rats, *Prog. Clin. Biol. Res.*, 247:485-495.
31. Rapp, L.M. and Williams, T.P., 1980, A parametric study of retinal light damage in albino and pigmented rats, in: "The Effects of Constant Light on Visual Processes", Williams, T.P. and Baker, B.N., eds., Plenum Press, New York, pp. 135-159.
32. Reichenbach, A., Schneider, A., Leibnitz, L., Reichelt, W., Schaaf, P., and Schümann, R., 1989, The structure of rabbit retinal Müller (glial) cells is adapted to the surrounding retinal layers, *Anat. Embryol.*, 180:71-79.
33. Sanyal, S. and Bal, A.K., 1973, Comparative light and electron microscopic study of retinal histogenesis in normal and rd mutant mice, Z. Anat. Entwickl., Gesch, 142:219-238.
34. Schmidt, S.Y., Berson, E.L., Watson, G., and Huang, G., 1977, Retinal degeneration in cats fed casein. III. Taurine deficiency and ERG amplitudes, *Invest. Ophthalmol. Vis. Sci.*, 16:673-678.
35. Steinberg, R.H., Reid, M., and Lacey, P.L., 1973, The distribution of rods and cones in the retina of the cat. (Felis domesticus), *J. Comp. Neurol.* 148:229-248.
36. Stone, J., 1965, A quantitative analysis of the distribution of ganglion cells in the cat's retina, *J. Comp. Neurol.*, 124:337-352.
37. Sturman, J.A., 1986, Nutritional taurine and central nervous system development, Ann. N.Y. Acad. Sci., 477:196-213.
38. Sturman, J.A., 1992, Review: taurine deficiency and the cat, in Adv. Med. Exp. Med. Biol. "Taurine: Nutritional Value and Mechanisms of Action", Lombardini, J.B., Schaffer, S.W., Azuma, J., eds., Plenum Press, New York, Vol. 135, pp. 1-5.
39. Sturman, J.A., 1993, Taurine in development, *Physiol. Rev.*, 73:119-147.
40. Sturman, J.A., Lu, P., Messing, J.M., and Imaki, H., 1996, Depletion of feline taurine levels by ß-alanine and dietary taurine restriction, in: "Taurine 2. Basic and Clinical Aspects, Huxtable, R.J., Azuma, J., Kuriyama, K., Nakagawa, M., Baba, A., eds, (*Adv. Exp. Med. Biol.* 403) Plenum Press, New York, pp. 19-36.
41. Sturman, J.A., Messing, J.M., 1991, Dietary taurine content and feline reproduction and outcome, *J. Nutr.*, 121:1195-1203.
42. Sturman, J.A., Messing, J.M., 1996, Depletion of feline taurine levels by ß-alanine and dietary taurine restriction, *Nutr. Res.*, 16:789-795.
43. Tyler, N.K. and Burns, M.S., 1991, Alterations in glial cell morphology and glial fibrillary acidic protein expression in urethan-induced retinopathy, *Invest. Ophthalmol. Vis. Sci.*, 32:246-256.
44. West-Hyde, L. and Buyukmihci, N., 1982, Photoreceptor degeneration in a family of cats, *J. Am. Vet. Med. Assoc.*, 181:243-247.

# Part 9:
# Taurine and Human Nutrition

# THE ROLE OF TAURINE IN INFANT NUTRITION

Russell W. Chesney,[1] Richard A. Helms,[2] Michael Christensen,[2]
Andrea M. Budreau,[1] Xiaobin Han,[1] and John A. Sturman[3]

[1] The University of Tennessee College of Medicine
[2] Departments of Pediatrics and Clinical Pharmacy
The Pediatric Pharmacology Research Unit and Crippled Children's Foundation Research Center at Le Bonheur Children's Medical Center
The University of Tennessee College of Pharmacy
Memphis, Tennessee
[3] Department of Developmental Biochemistry
Institute for Basic Research in Developmental Disabilities
Staten Island, New York

## INTRODUCTION

*"Know thyself'; Do not upon God presume to scan".*
*The proper study of mankind is man".*

Alexander Pope
1688-1744

Beginning in 1981 and continuing throughout the 1990's, the organic compound taurine has been added to infant formulas and total parenteral nutrition (TPN) solutions in countries around the world. This process of taurine supplementation was based upon the findings of studies first begun in Scandinavia in the late 1970's[23,61,64,70] as well as on studies in cats[36] and primates[72]. Although by 1984 the US Food and Drug Administration was convinced to permit the addition of taurine to purified infant formulas, many studies since have supported this decision. This review will examine the evidence that supports the addition of taurine to formulas as well as to TPN solutions, much of it fairly recent and, clearly, in the post-marketing phase of the FDA decision. In accordance with the admonitions of Alexander Pope's famous couplet from his Essay on Man quoted above, this review will focus upon studies performed in humans. While animal studies are always critical in terms of a deeper understanding of the role of taurine, the final value of any biomedical research should be understood in terms of improvements in human health.

*Taurine 3*
Edited by Schaffer *et al.*, Plenum Press, New York, 1998

## REVIEW

Taurine (2-aminoethanesulfonic acid) is a small sulfur-containing β-amino acid present in millimolar quantities in the intracellular space of many tissues that has been demonstrated to serve as an important organic osmolyte in brain and kidney. As an osmolyte it contributes to the process of cell volume regulation, which is especially critical following hypo- or hyperosmolar stresses[9,80]. In mammals, taurine and inorganic sulfate are the major end products of sulfur amino acid metabolism[34]. Taurine is the predominant intracellular free amino acid of most tissues in mammals but is involved in few biochemical reactions. Conjugation with bile acids to form water soluble bile salts, essential for fat absorption, is the most well documented chemical reaction in which taurine participates[19,33]. Taurine was first isolated from ox bile in 1827 and, hence, was named for the ox, *Bos taurus*[74]. It is, accordingly, the only amino acid with its own zodiac sign.

Taurine is found in high concentrations in the brain, which led neuroscientists to speculate that it should have (a) specific function(s)[67]. Many early studies demonstrated that especially elevated concentrations of taurine were found in newborn and neonatal brain, often several-fold higher than in adult brain of the same species[68]. In 1975, it became apparent that the decline in brain taurine concentration between birth and maturity was occurring during the period of natural weaning for the species of interest[69]. This observation focused attention on the taurine content of milk. A systematic study indicated that taurine was detectable in the milk of many mammalian species, including humans, and it was frequently found in high concentrations[62,63]. This fact led to the hypothesis that there might be a nutritional role for taurine in the developing infant. When radiolabelled taurine was injected into the peritoneum of a lactating rat, it passed into the milk and accumulated in the brain of the suckling pups, suggesting that this route might be an important source for CNS taurine[71].

Still lacking was conclusive evidence that taurine served any essential role. However, the finding of Hayes and his groups that cats fed taurine-free diets suffered retinal degeneration, accompanied by decreased retinal and plasma taurine concentrations[35,70], opened the possibility that taurine was essential for retinal development. In follow-up studies, it was shown that taurine depletion occurred in all feline tissues whenever dietary taurine was removed[45]. All felines are obligatory omnivores because they have low biosynthetic activity of a rate limiting enzyme, cysteine-sulfinic decarboxylase (CSD), involved in taurine synthesis[55]. The offspring of taurine-deficient female cats have a large number of neurologic defects including tapetal and retinal degeneration, delayed cerebellar granule cell division and migration, and abnormal cortical development[66]. Thus, taurine is an essential amino acid in the cat.

Because of reduced cystathionase activity in pre-term human infants and because of the finding that human milk contained substantial taurine (45 to 50 mg/dL), the notion of taurine as a semi-essential or conditionally essential amino acid arose[60]. An understanding of the special biochemical needs of the developing neonate has resulted in the concept that certain amino acids are developmentally essential for growth (histidine, cysteine), either because of the ontogeny of enzymes that catalyze synthesis of certain amino acids (cysteine, taurine and tyrosine)[70], and/or due to renal immaturity of amino acid reabsorption mechanisms (taurine, possibly glycine)[5]. This concept has now reached full acceptance in nutrition science circles. Although the semi-essential nature of cysteine and glycine is still not conclusively established, taurine is considered conditionally essential.

A number of features contribute to the conditionally essential role of taurine. 1) The activity of the enzyme cystathionase, which catalyzes the formation of cysteine from methionine, is low in man and is especially low or even absent in term and pre-term infant liver[70]. 2) While there is no significant maturational change in CSD activity, which catalyzes taurine synthesis from cysteine, the activity in humans is low as compared to other species (including the rat), although it is higher than in cats[67]. 3) The concentration of taurine in cow's milk, the major protein source for most of the world's infant formula, is quite low[66]. 4) Renal immaturity, particularly in the pre-term infant, does not permit the complete reabsorption of taurine from the glomerular ultrafiltrate; hence renal losses are quite large despite reduced plasma and (presumably) tissue taurine levels[12]. 5) The ingestion of adequate nutrients is a constant source of concern in the pre-term infant, because the intestine of the pre-term infant cannot match the placenta for transfer of nutrients[54]. The entire digestive tract is immature, including swallowing mechanisms, peristalsis, and digestive enzyme activity, and is ill-equipped to meet the very high metabolic demand of the rapidly growing infant[66]. A substantial number of term infants as well as pre-term infants may require partial or total parenteral nutrition solutions in an attempt to meet caloric needs[54]. Since neither cow's milk nor soy protein-based formulas contain substantial taurine concentrations, and because taurine is not present in protein hydrolysate TPN solutions, it must be supplemented.

The absence of taurine from soy protein-based milk and its low content in cow's milk led one group of investigators to compare plasma and urine taurine content in cow's milk and human milk-fed groups[23]. These studies compared protein quantity and quality in purified formulas with pooled human milk. Formula-fed infants had higher free amino acid levels in plasma and urine than human milk-fed infants. Taurine was the only exception, and both plasma and urine values fell throughout the duration of these studies. Cow's milk formula had scant concentrations of taurine due to removal of casein during processing. Further studies in term infants showed the same decline in plasma and urine taurine content following cow's-milk-based formula feeding[22] and added evidence to that discussed in a review of prior literature[66]. These results raised the spectre that purified infant formula might result in harmful consequences related to taurine depletion. They also resulted in a debate as to whether taurine supplementation was required for infant formula despite the lack of any obvious clinical evidence of retinal or growth abnormalities in several decades of formula use[8,66].

Two important observations led to an intensifying of this debate. First, infant rhesus monkeys fed from birth with a human protein hydrolysate formula, Nutramigen (Meade Johnson Nutrition, Evansville, IN), were divided into taurine-free and taurine-supplemented groups. In contrast to taurine-supplemented animals, monkeys fed formula alone showed significant retinal abnormalities on postmortem examination at age 26 months[72]. Disorientation, disintegration and vesiculation of the membrane arrays and outer segment disruption of cone photoreceptors was found. These changes were not evident in monkeys who were taurine-supplemented. Second, a series of reports indicated that humans (infants, children and adults) whose calories arose from TPN solutions (based upon protein hydrolysates and devoid of taurine), had reduced plasma taurine values[26]. Of great biologic importance was the finding that children receiving more than 95% of their calories from TPN for extended periods of time had, in addition to low plasma taurine concentrations, retinal abnormalities including visible changes upon ophthalmoscopic examination and electroretinographic abnormalities. Both low plasma values and retinal changes were essentially corrected by the addition of taurine to TPN solutions[2, 81].

Further support for the addition of taurine has come from other nutritional trials. In one controlled and randomized trial, pre-term infants were fed either taurine-free or supplemented formula[58, 79]. The investigators in this study had originally planned to evaluate 50 infants to achieve sufficient statistical power, but the trial was stopped after the study of 37 infants for ethical considerations. The results indicated that infants receiving taurine-supplemented formula had more mature brain stem auditory-evoked responses and a significant decline in the interval between stimulus and response. No differences in ERG responses were evident between the two groups.

A large multicenter British trial examined the consequences of an early feeding regimen for pre-term infants on short-term and long-term clinical developmental outcomes[47-52]. This decade-long study evaluated the long-term consequences of a nutrient-enriched pre-term formula as compared with a standard term formula. The enriched pre-term formula, used alone or as a supplement to human milk, consisted of 2.0 gm protein/dL vs 1.45 gm/dL in standard formula, and additional quantities of sodium, calcium, phosphate, copper, zinc, the fat soluble vitamins D, E and K, water soluble vitamins, carnitine and taurine (40.8 mmol/dL versus 0 in standard formula). This feeding trial continued for a median of 4 weeks after birth. Blind evaluation of infants at age 18 months demonstrated a developmental advantage for those infants receiving the pre-term formula, particularly in terms of motor function, which suggests involvement of the cerebellum. This part of the brain has a low taurine content in pre-term infants as compared with term infants[39] and much of its development occurs postnatally. The infants fed the pre-term formula also had a slight improvement in social development quotient. Taurine may not be responsible for all these benefits, since other nutrients were present. Among the data derived from this study was the finding that pre-term infants fed maternal milk showed a significant intelligence quotient advantage over infants who never received maternal milk when assessed at both 18 months and 7.5 to 8 years of age. However, since pre-term infants have a greater prevalence of clinical and subclinical defects and behavior disorders during adolescence than do normal birth-weight infants[59], any behavioral advantage may be biologically advantageous.

The manufacturers of infant formula in several European nations began, in 1981, to supplement cow's milk-based infant formula with taurine. The US FDA approved this formulation change in 1984. Currently, taurine is added to cow's milk and soy protein-based formulas virtually world wide[27].

The development of TPN was not unlike the development of infant formula. Throughout the 1970's a number of advances were made in which additions to the ultimate formulation occurred as nutritional deficiencies or inadequacies became apparent. Among the substances added were appropriate amounts of divalent cations, vitamins, micronutrients and lipids. Since original formulations contained protein hydrolysates as their amino acid source, taurine was not included. In a now classical series of papers, a group of investigators from UCLA were able to describe findings of taurine-deficiency in patients with bowel disease, either inflammatory bowel disease or short gut syndrome, who required prolonged TPN therapy. This group had noticed that animals fed diets lacking in taurine had reduced plasma and tissue values for taurine and later developed retinal dysfunction. Since parenteral nutrition was deficient in taurine, they sought evidence of taurine deficiency in 21 children and 23 adults receiving long term home TPN for an average of $27 \pm 13$ μm/L (SD) months. Fasting plasma taurine values were reduced in children to 26 $\pm$ 13 μM/L in comparison to $57 \pm 16$ μM/L in controls, $p < .001$. In adults who consumed less than 25 percent of calories by the enteral route, plasma taurine concentrations were significantly decreased and inversely correlated with the duration of TPN. Abnormalities were found in the ERG of 8 children. Addition of taurine to TPN solutions restored plasma taurine to normal with normalization of ERG's in 3 of 4 children examined. Taurine values again fell one year after taurine was discontinued. Having fulfilled Koch's postulates, they concluded that children and possibly adults receiving long-term TPN have a nutritional requirement for taurine[24-26,82,83].

Another problem identified by early students of TPN was the relative insolubility of cysteine, a taurine precursor[60,66]. Vinton et al showed that patients given taurine- and cysteine-free TPN solutions had significantly reduced plasma, platelet, and urine taurine concentrations: 54%, 48%, and 16%, respectively, of the values from controls (normal, healthy children of comparable ages)[83]. The greatest reductions in plasma taurine values were found in children who were receiving less than 5% of calories by the enteral route. Lymphocyte taurine concentrations were not different in TPN patients versus healthy controls. Plasma methionine and cysteine values were unchanged by TPN. There existed a direct correlation between plasma and platelet taurine values and between plasma and urine taurine values. Both plasma and platelet taurine levels tended to be directly correlated with age and, in children with prolonged TPN of more than one year, the length of total parenteral nutrition.

The impact of prolonged use of taurine-free TPN solutions on plasma taurine concentrations appears to be consistent. Collectively, these studies showed that plasma taurine concentrations fell in patients on taurine-free solutions and were restored or maintained at normal levels in taurine-supplemented solutions. This also appears to be true in beagle pups,[53] or in rats receiving TPN[85]. In general, the amino acid profiles of other amino acids are not affected by TPN. Taurine values in plasma are not influenced by illness or respiratory status[4].

Urinary taurine values fall in infants or children receiving taurine-deficient TPN. In one prospective controlled study plasma and urinary taurine concentrations were measured weekly between postnatal weeks 3 and 18 in seven infants (gestational age < 28 weeks, birth weight 600 - 1000 g) who received taurine-free TPN for 32 - 49 days, and eight sick infants who received formula or human milk from day 3 or 4 of life. Ten healthy full-term neonates served as controls. VLBW pre-term infants were found to have the lowest plasma values, in the range of 2 µmoles/dL, yet high urinary taurine values which fell with age. A group of larger pre-term infants provided with taurine-free TPN solutions were able to conserve taurine by increasing taurine reabsorption. Zelikovic *et al* speculated that the absence of taurine in TPN solutions administered to very low birth weight infants and the limited capacity of the immature kidney to adapt to low taurine intake by "up-regulation" of tubular transporter activity may result in depleted taurine body pools during the first weeks of life when brain growth, especially cerebellar, is so rapid[86]. This inability of the immature nephron to adequately conserve taurine could potentially have a deleterious influence on the developing brain and retina, which indicates a need for taurine supplementation.

An extensive evaluation of the influence of long-term TPN on eye function was performed[82]. Some patients demonstrated granularity of retinal pigmented epithelium or ophthalmoscopy. Approximately half of children and one-third of adults showed at least one and usually two ERG abnormalities. Evaluation of other micronutrient and lipid values relevant to vision were assessed: zinc and vitamin E were normal; vitamin A was increased; selenium, linoleic and linolenic acid values were reduced.

One study examined the effect of taurine supplements on the synthesis of neutral and acidic sterols and on fat absorption in pre-term and full-term infants[84]. They found that taurine supplementation resulted in lower cholesterol synthesis with higher bile acid excretion and fatty acid absorption in pre-term infants with a gestational age under 33 weeks, but not in older pre-term or term infants. This could potentially be related to an expanded hepatic and total body taurine pool size. Other investigators have demonstrated improved fat absorption and reduced steatorrhea in children suffering from cystic fibrosis after taurine supplements[16, 20].

A recent study involving 41 healthy human milk-fed pre-term infants followed over the first 60 days of life has shown that duodenal bile acids were low during the first 2 weeks of life, but that serum bile acids rose over the first month of life. The taurine:glycine ratio was high in duodenal bile acids aspirated from the duodenum, but declined with age[3]. These findings are similar to those that taurine-supplemented formula feeding results in a bile acid:taurine/glycine ratios much more like human milk than unsupplemented cow's milk or soy protein formula[56].

The taurine content of human milk of nonvegetarians and lacto-ovovegetarians falls during the course of lactation[44]. The taurine content of human milk from non-vegetarians is higher than in milk from lacto-ovovegetarians. This study indicated that the taurine content of human milk was different during the course of lactation for nonvegetarians and lacto-ovovegetarians, and that for infants of lacto-ovovegetarians, intake fell by approximately one-third. A recent examination of the influence of vegetarian diets on growing children includes taurine along with iron, vitamin $B_{12}$, and vitamin D as nutrients which need to be considered in children being raised as macrobiotic vegetarians[65].

The influence of taurine supplements to the development of hearing has been examined in pre-term and term infants. Taurine can be localized to the organ of Corti in the inner ear[32]. Tyson et al[78] suggested that taurine supplements could positively affect auditory brainstem-evoked responses (ABR), although this finding is somewhat in dispute[57]. The supplementation of taurine (4.5 mg/dL) to human milk-fed term infants already ingesting $3.5 \pm 0.3$ mg/dL in mother's milk appeared to lengthen the ABR[21]. The findings of this study were, in fact, the opposite of Tyson's study in pre-term infants, where ABR latency was prolonged in the absence of taurine. However, using another hearing test, Dhillon found that transient evoked otoacoustic emissions were significantly higher in breast-fed than in formula-fed infants, whose taurine intake was negligible.

Taurine also appears to optimize the growth and proliferation of human fetal brain cells obtained from the cerebrum, free of glial cells and in a serum-free system. Taurine addition to the medium improved neuron survival and neurite extension. Taurine availability also increased the incorporation of thymidine into brain cell DNA[31].

The finding of reduced plasma taurine concentrations in pre-term or term infants receiving taurine-deficient formula or in infants and children receiving taurine-free TPN solutions is clearly the result of a combination of factors. Among these are reduced cystathionase activity[60,66], the relatively low CSD activity of man in general[60,67] and reduced dietary intake of taurine or other sulfur-containing amino acids[60,66]. However, perhaps just as important is the unique inability of the renal epithelium of the immature kidney to conserve amino acids. The extent of taurinuria is even greater in pre-term infants than in term infants[8].

One of the factors that appears to maintain total body taurine pool size is the renal reabsorptive mechanism of the proximal tubule[6,14]. Taurine is virtually unique among amino acids in that its body pool size is effectively regulated by the kidney. Unlike most amino acids, which are completely reabsorbed by the proximal tubule and appear in the urine in scant amounts, taurine is not completely reabsorbed. During periods of adequate dietary intake of taurine or other sulfur amino acids, relatively large amounts (10 to 60% of the filtered load) are excreted into the urine. However, as taurine availability is reduced, both animals and human are capable of reducing taurine excretion. This adaptive response to dietary taurine intake has been localized to the renal proximal tubule brush border membrane, where the Na-dependent taurine transporter activity is shared with other β-amino acids and γ- aminobutyric acid (GABA)[11]. A greater than two-fold increase in the initial rate of taurine accumulation by renal brush border membrane vesicles (BBMV) has been found in rats or cats receiving low taurine diets as compared with controls[11,14]. Kinetic analysis of this uptake process reveals that the affinity of the taurine transporter (or $K_m$) is unaffected by diet, whereas a significant increase exists in the initial taurine transport rate ($V_{max}$) in response to the low taurine diet. The physiologic effect of increased taurine transporter rate is enhanced reabsorption of taurine from the glomerular ultrafiltrate with resultant decreased urinary taurine excretion and maintenance of taurine tissue content[13]. Diminished transporter activity following a high taurine diet results in diminished taurine reabsorption, greater urinary taurine excretion and maintenance of plasma and tissue taurine levels, particularly in six separate areas within the brain where taurine serves as an important osmolyte[11,13,14]. This adaptive regulation is equally relevant in human nutrition, where dietary taurine restriction results in reduced urinary taurine excretion[7].

The physiologic and intracellular signal for the renal conservation or hyperexcretion of taurine has recently been elucidated. Although plasma taurine concentrations demonstrate a significant correlation with the fractional excretion of taurine[10], an even stronger correlation exists between renal cortex taurine concentration and both plasma and urinary values for taurine[12]. Notably, increased taurine transporter activity was detected in renal cortex slices depleted of taurine by β-alanine feeding or fasting despite no changes in plasma taurine levels[10]. Studies performed in two continuous renal cell lines grown in culture, where intracellular taurine content can be readily manipulated, mimic the physiologic responses observed in the rat renal brush-border transporter[41]. These findings support a role for intracellular taurine as the factor responsible for the renal adaptive response rather than the extracellular, or plasma, concentration. The mechanisms by which intracellular taurine elicits changes in transporter activity are unclear, but molecular biologic studies provide useful insights.

The taurine transporter of Madin-Darby canine kidney (MDCK) cells (derived from a Cocker spaniel) has been cloned and sequenced[80]. The transcription rate of the MDCK cell taurine transporter gene is influenced by changes in media taurine concentration[29]. The same is true for the taurine transporter of the LLC-$PK_1$ (porcine kidney) cell line[30]. Using these cDNA probes the abundance of mRNA can be ascertained in cells grown in high, normal, or low taurine content (500, 50, or 0 μM). Northern blots revealed that the abundance of specific mRNA from taurine transporter cDNA was increased in cells grown in taurine-free conditions as compared with normal cell culture media. Down-regulation of message was evident in cells incubated in 500 μM taurine after 12 hours of exposure, and mRNA abundance was appreciably reduced following 72 hours exposure. A polyclonal antibody was raised to a synthetic peptide based on the amino acid sequence of the 4th intracellular loop between membrane spanning domains VI and VII. Western blots using this antibody showed increased or decreased transporter protein concentration in the membrane[30]. The abundance of this 70 kd protein was directly related to mRNA abundance. Thus, expression of the taurine transporter was up-or down-regulated by medium taurine at the level of mRNA expression and protein synthesis. The signal for regulation of renal taurine transporter activity seems to be the intracellular content of taurine, and the action is genomic. The renal adaptive response to dietary taurine intake, and resultant regulation of total taurine body pool size, implies a gene-nutrient interaction[29,80].

Urinary taurine concentration is higher in relation to a given plasma and tissue taurine concentration in neonatal rats versus mature animals[12]. Increased excretion of taurine is secondary to diminished reabsorption. The immaturity of renal taurine transport is manifest as failure of the renal adaptive response to emerge until a rat is at least 2 weeks old, or until the pre-term human infant is several weeks old[86]. There are several characteristics of the immature kidney which could influence insertion of the taurine transporter protein, such as enhanced permeability to sodium (the driving force for taurine uptake), reduced activity of $Na^+K^+$ATPase (which would lead to failure to pump $Na^+$ out of the cell), or changes in phospholipid content and fluidity of the apical membrane, all of which could account for physiological aminoaciduria of immaturity. Other potential factors include enhanced efflux of taurine across renal cell membranes into the urinary space, or reduced transcription or translation rates.

A nutritional consequence of renal immaturity in terms of amino acid reabsorption is a significant reduction in plasma taurine concentrations, which in turn may influence development of the retina and brain of the very-low-birthweight (VLBW) pre-term infant[12,86]. The urinary taurine concentration of these VLBW neonates is elevated, with a fractional excretion of filtered load of 38 - 60%, in contrast to less than 10% in term infants. VLBW infants receiving taurine-free formulas or TPN solutions still have high urinary taurine excretion. In contrast, the term neonate deprived of taurine by diet excretes less than 1% of filtered load and maintains normal plasma taurine levels. The failure of the immature kidney to demonstrate the renal adaptive response to variations in dietary taurine intake can put the pre-term infant at risk for taurine deficiency[86].

Two fairly recent studies have examined the capacity of taurine-containing TPN solutions to restore sulfur amino acid plasma and urinary levels to normal; Helms, *et al*,[38]; used four different formula mixtures: (A) low methionine, cysteine and taurine; (B) acute high methionine; (C) steady state high methionine; (D) high methionine and cysteine. Methionine values were elevated compared to a formula-fed reference group after employing all formulations. Cysteine was elevated only in groups A and D, and was lowest in group B. Taurine values were within the normal range of 1-16 μmol/dL. However, infants receiving taurine-free solutions had lower plasma taurine values despite higher intakes of methionine and/or cysteine (Fig. 1). All infants in this study were in the 2.3 to 3.0 kg weight range.

In a follow-up study, 18 infants weighing 2.1 ± 0.5 kg and with a mean age of 34 weeks, received a modified TrophAmine® formulation. Infants received 6.5 ± 0.6 mg/kg/day of taurine, excreted 3.7 ± 7.8 mg/kg/day, had a fractional excretion of 17 ± 15% and a retention of 43 ± 116% of intake under steady state conditions (Fig. 2). The fractional excretion of taurine correlated with body weight, indicating that it was a function of renal maturity.

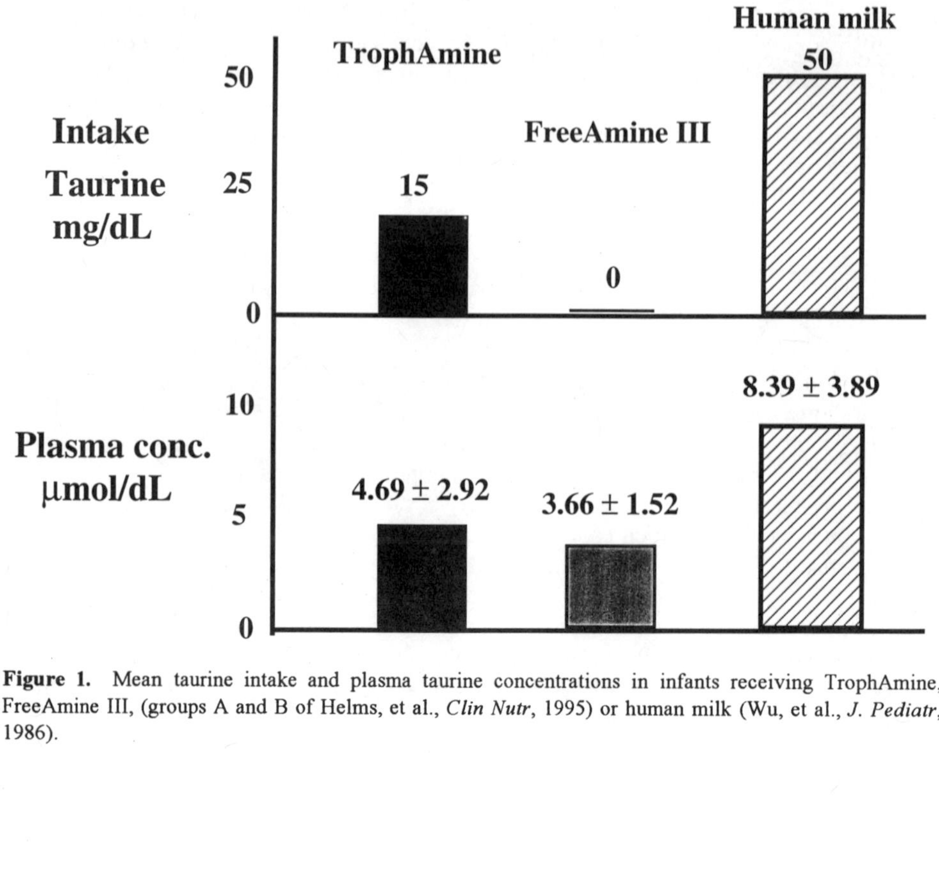


**Figure 1.** Mean taurine intake and plasma taurine concentrations in infants receiving TrophAmine, FreeAmine III, (groups A and B of Helms, et al., *Clin Nutr*, 1995) or human milk (Wu, et al., *J. Pediatr*, 1986).

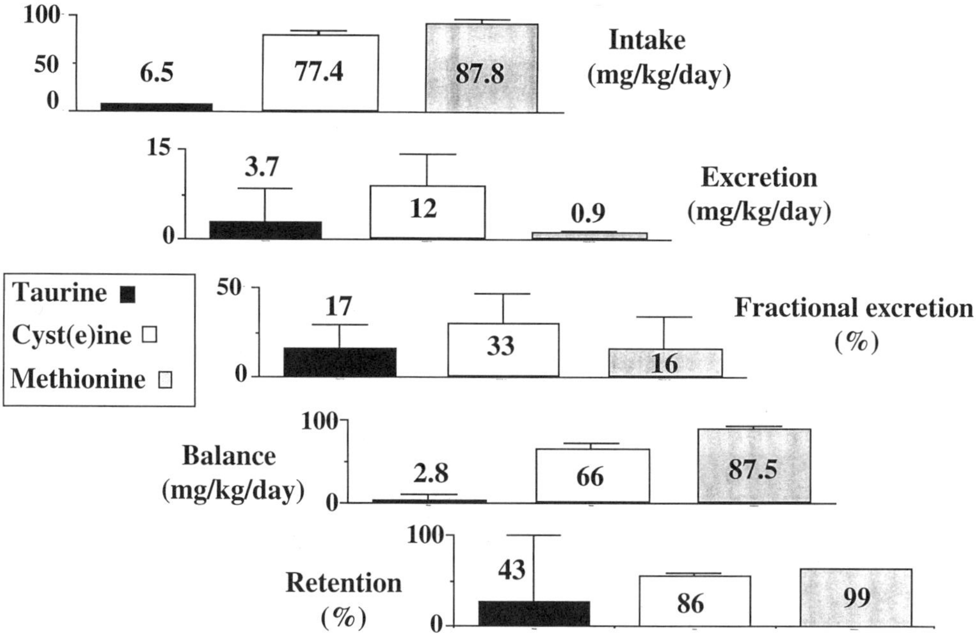


**Figure 2.** Intake, excretion, balance, and retention of taurine, cyst(e)ine, and methionine

Early TPN solutions were associated with a very high incidence of cholestasis. Taurine has been considered as a factor that reduces the impact of cholestatic problems in infants on long-term TPN[1,60,73]. Short-term taurine deficiency does not appear to influence hepatocellular function in children under 10 days of age[17], but in children receiving TPN over months can result in hyperbilirubinemia, increased serum transaminase values, hypoproteinemia, and abnormal coagulation profiles. Liver histology demonstrated fibrosis, fatty replacement, and coarse cholestasis[18]. In contrast, taurine-supplemented solutions resulted in reduced hepatocellular damage in 40 of 41 infants who received TPN for 5 to 21 days, admittedly a shorter time period[37]. In a study using isolated rat liver perfusion, addition of taurine to the TPN solution affected hepatosteatosis and altered lidocaine metabolism; lidocaine clearance was impaired by taurine treatment[85]. While the pathogenesis of TPN solution-induced hepatocellular disease is not fully established, several groups have noted that the addition of taurine, along with other maneuvers, improves liver function. A reduction in the formation and secretion of bile has been well documented[28]. One theory is that the amino acid mixture either has or lacks some factor which is important in formulation/secretion of bile which is modified by taurine. In one study in guinea pigs, bile flow was improved by taurine supplementation and bile acid composition showed an increase in the secondary bile acid, 7-ketolithocholate, and a concomitant reduction in chemodeoxycholate[28]. Comparable findings were found after formula feeding in term infants when taurine was added[56]. Most studies demonstrated that taurine significantly alters the taurine/glycine conjugation ratio[28,37,53,56].

## SUMMARY

The importance of taurine in the diet of pre-term and term infants has not always been clearly understood and is a topic of interest to students of infant nutrition. Recent evidence indicates that it should be considered one of the "conditionally essential" amino acids in infant nutrition[43,60,66]. Plasma values for taurine will fall if infants are fed a taurine-free formula or do not have taurine provided in the TPN solution. Urine taurine values also fall, which is indicative of an attempt by the kidney to conserve taurine[66,87]. The very-low-birth-weight infant, for a variety of reasons involving the maturation of tubular transport function, cannot maximally conserve taurine by enhancing renal reabsorption and, hence, is potentially at greater risk for taurine depletion than larger pre-term or term infants, and certainly more than older children who have taurine in their diet.

Taurine has an important role in fat absorption in pre-term and possibly term infants[56,60,66] and in children with cystic fibrosis[16,20]. Because taurine-conjugated bile acids are better emulsifiers of fat than glycine-conjugated bile acids, the dietary (or TPN) intake has a direct influence on absorption of lipids. Taurine supplementation of formulas or TPN solutions could potentially serve to minimize the brain phospholipid fatty acid composition differences between formula-fed and human milk-fed infants[15].

Taurine appears to have a role in infants, children, and even adults receiving most (>75%) of their calories from TPN solutions in the prevention of granulation of the retina and electroencephalographic changes[3,24,82]. Taurine has also been reported to improve maturation of auditory-evoked responses in pre-term infants, although this point is not fully established[21].

Clearly, taurine is an important osmolyte in the brain and the renal medulla. At these locations, it is a primary factor in the cell volume regulatory process, in which brain or renal cells swell or shrink in response to osmolar changes, but return to their previous volume according to the uptake or release of taurine[46]. While there is a dearth of clinical studies in man concerning this volume regulatory response, studies in cats, rats, and dog kidney cells indicate the protective role of taurine in hyperosmolar stress[40,42,75-77]. The infant depleted of taurine may not be able to respond to hyper- or hyponatremic stress without massive changes in neuronal volume, which has obvious clinical significance. The fact that the brain content of taurine is very high at birth and falls with maturation may be a protective feature, or compensation for renal immaturity.

Defining an amino acid as "conditionally essential" requires that deficiency result in a clinical consequence or consequences which can be reversed by supplementation[60,66]. In pre-term and term infants, taurine insufficiency results in impaired fat absorption, bile acid secretion, retinal function, and hepatic function, all of which can be reversed by taurine supplementation. Therefore, this small β-amino acid, taurine, is indeed conditionally essential.

## ACKNOWLEDGMENTS

This work was supported by a LeBonheur Professorship in Pediatrics; National Institutes of Health Grants DK 37223 and U10-HD-31326; State of Tennessee Center of Excellence in Pediatric Pharmacokinetics and Therapeutics; Crippled Children's Foundation Research Center and Le Bonheur Small Grants (RWC) and by the New York State Office of Mental Retardation and Developmental Disabilities; National Institute of Child Health and Human Diseases Grants HD-11129, HD-16634, and HD-18678; L'Association Canadienne de l'Ataxie de Freidreich; The Sanford Foundation; and Mr. And Mrs. Christopher T.G. Fish (JAS).

## REFERENCES


1. Adamkin, D.D., Radmacher, P., and Rosen, P., 1995, Comparison of a neonatal versus general-purpose amino acid formulation in preterm neonates, *J. Perinatol.*, 15(2):108-113.
2. Ament, M.E., Geggel, H.S., and Heckenlively, J.R., 1986, Taurine supplementation in infants receiving long-term total parenteral nutrition, *J. Am. Coll. Nutr.*, 5:127.
3. Boehm, G., Braun, W., Moro, G., and Minoli, I., 1997, Bile acid concentrations in serum and duodenal aspirates of healthy preterm infants: Effects of gestational and postnatal age, *Biol. Neonate*, 71(4):207-214.
4. Burger, U., Wolf, H., Fritsch, U., and Bauer, M., 1983, Parenteral nutrition in preterm infants: Influence of respiratory treatment and effect of different amino acid compositions, *J. Pediatr. Gastroenterol. Nutr.*, 2(4):644-652.
5. Chesney, R., Gusowski, N., and Theissen, M., 1984, Developmental aspects of renal β-amino acid transport. IV. Brush border membrane response to altered intake of sulfur amino acids, *Pediatr. Res.*, 18(1984):611.
6. Chesney, R.W., 1987, New functions for an old molecule, *Pediatr. Res.*, 22:755-759.
7. Chesney, R.W. 1989. Nutritional consequences of aminoacidurias. *In* Nitrution and the Origin of Disease, Halstead C.H., ed., Academic Press, New York, pp. 161.
8. Chesney, R.W., 1988, Taurine: Is it required for infant nutrition?, *J. Nutr.*, 118:6.
9. Chesney, R.W. 1985. Taurine: Its Biological Role and Clinical Implications, in: Advances in Pediatrics, Barnes L.A., ed., Yearbook Medical Publishers, Chicago, pp. 1-42.
10. Chesney, R.W., Gusowski, N., and Dabbagh, S., 1985, Renal cortex taurine content regulates renal adaptive response to altered dietary intake of sulfur amino acids, *J. Clin. Invest.*, 76:2213-2221.
11. Chesney, R.W., Gusowski, N., and Friedman, A.L., 1983, Renal adaptation to altered amino acid intake occurs at the luminal brush border membrane, *Kidney Int.*, 24:588-594.
12. Chesney, R.W., Gusowski, N., Zelikovic, I., and Padilla, M., 1986, Developmental aspects of renal β-amino acid transport. V. Brush border membrane transport in nursing animals - effect of age and diet, *Pediatr. Res.*, 20:890-895.
13. Chesney, R.W., Lippincott, S., Gusowski, N., Padilla, M., and Zelikovic, I., 1986, Studies on renal adaptation to altered dietary amino acid intake: Tissue taurine responses in nursing and adult rats, *J. Nutr.*, 116:1965-1976.
14. Chesney, R.W., Zelikovic, I., Jones, D.P., Budreau, A., and Jolly, K., 1990, The renal transport of taurine and regulation of renal sodium-chloride-dependent transporter activity, *Pediatr. Nephrol.*, 4:399-407.
15. Cockburn, F., 1994, Neonatal brain and dietary lipids, *Arch. Dis. Child,* 70:F1.
16. Colombo, C., Arlati, S., Curcio, L., et al., 1988, Effect of taurine supplementation on fat and bile acid absorpton in patients with cystic fibrosis, *Scand. J. Gastroenterol.*, 23(supple)(143):151.
17. Cooke, R.J., Whitington, P.F., and Kelts, D., 1984, Effect of taurine supplementation on hepatic function during short-term parenteral nutrition in the premature infant, *J. Pediatr. Gastroenterol. Nutr.*, 3(2):234-238.
18. Cooper, A., Bets, J.M., Pereira, G.R., and Ziegler, M.M., 1984, Taurine deficiency in the severe hepatic dysfunction complicating total parenteral nutrition, *J. Pediatr. Surg.*, 19(4):462-466.
19. Danielson, H., 1963, Present states of research on catabolism and excretion of cholesterol, *Adv. Lipid Res.*, 1:335.
20. Darling, B.P., Lepage, G., Leroy, C., et al., 1985, Effect of taurine supplements on fat absorption in cystic fibrosis, *Pediatr. Res.*, 19:578.
21. Dhillon, S.K., Davies, W.E., Hopkins, P.C., and Rose, S.J., in press, Effects of dietary taurine on auditory function in full term infants, *International Taurine Symposium Proceedings*.
22. Gaull, G.E. and Rassin, D.K. 1980. Taurine in development and nutrition, in: Proceedings of the 72nd Ciba Foundation Symposium on Sulfur in Biology, *Excerpta Med.*, Amsterdam, pp. 271.
23. Gaull, G.E., Rassin, D.K., Räihä, N.C.R., and Sturman, J.A., 1977, Milk protein quantity and quality in low-birthweight infants. III. Effects on sulfur amino acids in plasma and urine, *J. Pediatr.*, 90:348-355.
24. Geggel, H., Ament, M., and Heckenlively, J., 1985, Nutritional requirement for taurine in patients receiving long-term, parenteral nutrition., *N. Engl. J. Med.*, 312:142.
25. Geggel, H., Ament, M.E., Heckenlively, J.R., et al., 1982, Evidence that taurine is an essential amino acid in children receiving total parenteral nutrition (abstract), *Clin. Res.*, 30:486A.
26. Geggel, H.S., Heckenlively, J.R., and Martin, D.A., 1982, Human retinal dysfunction and taurine deficiency, *Doc. Ophthamol. Proc. Ser.*, 31:199.

27. Georgieva, O.V., Kruglik, V.I., Nedospasova, L.V., and Tsyriapkin, V.A., 1995, Taurine in the composition of a new dry milk product, Femilak-2, intended for improving nutrition of nursing mothers, *Vopr. Pitan,* 2:33-36.
28. Guertin, F., Roy, C.C., Lepage, G., Perea, A., Giguere, R., Yousef, I., and Tuchweber, B., 1991, Effect of taurine on total parenteral nutrition-associated cholestasis, *Parenter. Enteral Nutr.,* 15(3):247-251.
29. Han, X., Budreau, A.M., and Chesney, R.W., 1997, Adaptive regulation of MDCK cell taurine transporter (pNCT) mRNA: transcription of pNCT gene is regulated by external taurine concentration, *Biochim. Biophys. Acta,* 1351:296-304.
30. Han, X., Budreau, A.M., and Chesney, R.W., 1996, Molecular cloning and expression of an LLC-PK1 cell taurine transporter that is regulated by taurine, *J. Am. Soc. Nephrol.,* 7(9):1299.
31. Han, X., Chen, X.C., and Zhiling, P., 1992, The effect of taurine on human fetal brain cell proliferation in tissue culture, *Nutr. Res.,* 12:179-185.
32. Harding, N.J. and Davies, W.E., 1993, Cellular localisation of taurine in the organ of Corti, *Hearing Res.,* 65:211-215.
33. Haslewood, G.A.D., 1964, The biological significance of chemical differences in bile salts, *Biol. Res.,* 39:291-296.
34. Hayes K., 1988, In: Modern Nutrition in Health and Disease, M. Shils M., and Young V., eds., Lea and Fehniger, Philadelphia, pp.
35. Hayes, K., Rabin, A., and Berson, E., 1975, An ultrastructural study of nutritionally induced and reversed retinal degeneration in cats., *Am. J. Pathol.,* 78:505.
36. Hayes KC and Carey RE, S.S., 1975, Retinal degneration associated with taurine deficiency in the cat, *Science,* 188:949-51.
37. Heird, W.C., Dell, R.B., Helms, R.A., Greene, H.L., Ament, M.E., Karna, P., and Storm, M.C., 1989, Amino acid mixture designed to maintain normal plasma amino acid patterns in infants and shildren requiring parenteral nutrition, *Pediatrics,* 83:433-442.
38. Helms, R.A., Christensen, M.L., Storm, M.C., and Chesney, R.W., 1995, Adequacy of sulfur amino acid intake in infants receiving parenteral nutrition, *J. Nutr. Biochem.,* 6:462-466.
39. Huppi, P.S., Posse, S., Lazeyras, F., Burri, R., Bossi, E., and Herschkowitz, N., 1991, Magnetic resonance in preterm and term newborns: 1H-spectroscopy in developing human brain, *Pediatr. Res.,* 30(6):574-578.
40. Jones, D.P., Miller, L.A., Budreau, A., and Chesney, R.W. 1992. Characteristics of taurine transport in cultured renal epithelial cell lines: Asymmetric polarity of proximal and distal cell lines. *In* Taurine, Lombardini, J.B., Schaffer, S.W., and Azuma, J., eds., Plenum Press, New York, pp. 405-411.
41. Jones, D.P., Miller, L.A., and Chesney, R.W., 1990, Adaptive regulation of taurine transport in two continuous renal epithelial cell lines, *Kidney Int.,* 38:219-226.
42. Jones, D.P., Miller, L.A., and Chesney, R.W., 1993, Polarity of taurine transport in cultured renal epithelial cell lines: LLC-$PK_1$ and MDCK, *Am. J. Physiol.,* 265:F137-F145.
43. Karan, S., 1991, Taurine in infant nutrition, *Indian J. Pediatr.,* 58(3):311-316.
44. Kim, E.S., Cho, K.H., Park, M.A., Lee, K.H., Moon, J., Lee, Y.N., and Ro, H.K., 1996, Taurine intake of Korean breast-fed infants during lactation, *Adv. Exp. Med. Biol.,* 403:571-577.
45. Knopf, K., Sturman, J.A., and Armstrong, M., 1978, Taurine: An essential nutrient for the cat, *J. Nutr.,* 108:773.
46. Law, R.O., 1991, Amino acids as volume-regulatory osmolytes in mammalian cells, *Comp. Biochem. Physiol.,* 99A(3):263-277.
47. Lucas, A., Bishop, N.J., King, F.J., and Cole, T.J., 1992, Randomized trial of nutrition for preterm infants after discharge, *Arch. Dis. Child,* 67(3):324-327.
48. Lucas, A., Gore, S.M., Cole, T.J., et al, 1984, Multicentre trial on feeding low birthweight infants: Effects of diet on early growth, *Arch. Dis. Child,* 59:722.
49. Lucas, A., Morley, R., Cole, T.J., Lister, G., and Leeson, P.C., 1992, Breast milk and subsequent intelligence quotient in children born preterm, *Lancet,* 339(8788):261-264.
50. Lucas, A., Morley, R., Cole, T.J., Gore, S.M., Lucas, P.J., Crowle, P., Pearse, R., Boon, A.J., and Powell, R., 1990, Early diet in preterm babies and development status at 18 months, *Lancet,* 335(8704):1477-1481.
51. Lucas, A., Morley, R., Cole, T.J., Gore, S.M., Davis, J.A., Bamford, M.F., and Dossetor, J.F., 1989, Early diet in preterm babies and developmental status in infancy, *Arch. Dis. Child,* 64(11):1570-1578.
52. Lucas, A., Morley, R., Cole, J.T., and Gore, S.M., 1994, A randomised multicentre study of human milk versus formula and later development in preterm infants, *Arch. Dis. Child,* 70(2):F141-F145.
53. Malloy, M.H., Rassin, D.K., Gaull, G.E., and Heird, W.C., 1981, Development of taurine metabolism in beagle pups: Effects of taurine-free parenteral nutrition, *Biol. Neonate,* 40:1-8.

54. Morley, R., Cole, T., Powell, R., and Lucas, A., 1988, Mother's choice to provide breast milk and developmental outcome, *Arch. Dis. Child,* 63(11):1382-1385.
55. Morris, J.G., Rogers, Q., and Pacioretty, L.M., 1990, Taurine: An essential nutrient for cats, *J. Small Animal Prac.*, 31:502.
56. Okamoto, E., Rassin, D.K., Zucker, C.L., Salem, G.S., and Heird, W.C., 1984, Role of taurine in feeding the low-birth-weight infant, *J. Pediatr.*, 104:936-940.
57. Perlman, M., 1989, Taurine and auditory system maturation, *Pediatrics,* 83:796-798.
58. Perlman, M., 1989, Taurine and auditory system maturation (letter)., *Pediatrics,* 83:796.
59. Pharoah, P.O., Stevenson, C.J., Cooke, R.W., and Stevenson, R.C., 1994, Prevalence of behaviour disorders in low birthweight infants, *Arch. Dis. Child,* 70(4):271-274.
60. Rassin, D.K., 1994, Essential and non-essential amino acids in neonatal nutrition, in: Protein Metabolism During Infancy. Nestlé Nutrition Workshop Series, Vol. 33, Niels, C.R., and Räihä, N.C.R., eds., Nestec Ltd., Vevey/Raven Press, Ltd., New York, pp. 183-195.
61. Rassin, D.K., Gaull, G.E., and Räihä, N.C.R., 1977, Milk protein quantity and quality in low-birth-weight infants, *J. Pediatr.*, 90:356.
62. Rassin, D.K., Räihä, N.C.R., and Gaull, G.E., 1981, Protein and taurine nutrition in infants. in: Textbook of Gastroenterology and Nutrition in Infancy, Lebenthal, E., ed., Raven Press, Ltd., New York. 391.
63. Rassin, D.K., Sturman, J.A., and Gaull, G.E., 1977, Taurine in milk: Species variation., *Pediatr. Res.*, 11:463A.
64. Rassin, D.K., Sturman, J.A., and Gaull, G.E., 1978, Taurine and other free amino acids in milk of man and other mammals, *Early Hum. Dev.*, 2:1-13.
65. Sander, T.A., 1995, Vegetarian diets and children, *Pediatr. Clin. North AM,* 42(4):955-965.
66. Sturman, J. and Chesney, R., 1995, Taurine in pediatric nutrition, *Pediatr. Clin. N. Amer.*, 42(4):879-897.
67. Sturman, J.A., 1993, Taurine in development, *Physiol. Rev.*, 73:119.
68. Sturman, J.A., 1988, Taurine in development, *J. Nutr.*, 118:1169.
69. Sturman, J.A. and E., G.G., 1975, Taurine in the brain and liver of the developing human and monkey, *J. Neurochem.*, 25:831-835.
70. Sturman, J.A., Gaull, G.E., and Räihä, N.C.R., 1970, Absence of cystathionase in human fetal liver. Is cystine essential?, *Science,* 169:74-76.
71. Sturman, J.A., Rassin, D.K., and Gaull, G.E., 1977, Taurine in developing rat brain: Transfer of [$^{35}$S]taurine to pups via the milk, *Pediatr. Res.*, 11:28.
72. Sturman, J.A., Wen, G.Y., Wisniewski, H.M., and Neuringer, M.D., 1984, Retinal degeneration in primates raised on a synthetic human infant formula., *Int. J. Dev. Neurosci.*, 2:121-130.
73. Thornton, L. and Griffin, E., 1991, Evaluation of a taurine-containing amino acid solution in parenteral nutrition, *Arch. Dis. Child,* 66(1 Spec No):21-25.
74. Tiedmann, F. and Gmelin, L., 1827, Binige nede bestandtheile der galle des ochsen, *Annalen der Physik und Chemie,* 9:326.
75. Trachtman, H., Barbour, R., Sturman, H.A., and Finberg, L., 1988, Taurine and osmoregulation: taurine is a cerebral osmoprotective molecule in chronic hypernatremic dehydration, *Pediatr. Res.*, 23(1):35-39.
76. Trachtman, H., Del Pizzo, R., and Sturman, J.A., 1990, Taurine and osmoregulation III: Taurine deficiency protects against cerebral edema during acute hyponatremia, *Pediatr. Res.*, 27:85.
77. Trachtman, H., Futterweit, S., and Bienkowski, R.S., 1993, Taurine prevents glucose-induced lipid peroxidation and increased collagen production in cultured rat mesangial cells, *Biochem. Biophys. Res. Comm.*, 191:759.
78. Tyson, J.E., Lasky, R., Flood, D., Mize, C., Picone, T., and Paule, C.L., 1989, Randomized trial of taurine supplementation for infants ≤1,300-gram birth weight: Effect on auditory brainstem-evoked responses, *Pediatrics,* 83:406-415.
79. Tyson, J.E., Mize, C., and Lasky, R., 1989, More on taurine, *Pediatrics,* 83:1072-1073.
80. Uchida, S., Kwon, H.M., Yamauchi, A., Preston, A.S., Marumo, F., and Handler, J.S., 1992, Molecular cloning of the cDNA for an MDCK cell Na(+)- and Cl(-)-dependent taurine transporter that is regulated by hypertonicity [published erratum appears in Proc. Natl. Acad. Sci., 1993 Aug 1; 90(15):7424], *Proc. Natl. Acad. Sci.*, 89(17):8230-8234.
81. Vandeqoude, M.F.J. and Delleeuw, I.H., 1985, Taurine requirement with parenteral nutrition, *N. Engl. J. Med.*, 313:120.

82. Vinton, N.E., Heckenlivley, J.R., Laidlaw, S.A., Martin, D.A., Foxman, S.R., Ament, M.E., and Kopple, J.D., 1990, Visual function in patients undergoing long-term total parenteral nutrition, *Am. J. Clin. Nutr.*, 52(5):895-902.
83. Vinton, N.E., Laidlaw, S.A., Ament, M.E., and Kopple, J.D., 1987, Taurine concentrations in plasma, blood cells, and urine of children undergoing long-term total parenteral nutrition, *Pediatr. Res.*, 21(4):399-403.
84. World Health Organization Twenty-Seventh World Health Assembly, 1972, Infant Nutrition and Breastfeeding. In: Official Records of the World Health Organization, Vol. 217, pp. 20.
85. Zaman, N., Tam, Y.K., Jewell, L.D., and Coutts, R.T., 1996, Effects of taurine supplementation in parenteral nutrition-associated hepatosteatosis and lidocaine metabolism. A study using isolated rat liver perfusion, *Drug Metab. Dispos.*, 24(5):534-541.
86. Zelikovic, I., Chesney, R.W., Ahlfors, C.E., Pityer, R., and Friedman, A.L., 1988, Very low birth weight (VLBW) infants receiving prolonged total parenteral nutrition (TPN) are taurine depleted because of renal immaturity, *Pediatr. Res.*, 23:495A.
87. Zelikovic, I., Chesney, R.W., Friedman, A.L., and Ahlfors, C.E., 1990, Taurine depletion in very low birth weight infants receiving prolonged parenteral nutrition: Role of renal immaturity, *J. Pediatr.*, 116:301-306.

# QUANTITATION OF TAURINE AND SELENIUM LEVELS IN HUMAN MILK AND ESTIMATED INTAKE OF TAURINE BY BREAST-FED INFANTS DURING THE EARLY PERIODS OF LACTATION

Eul-Sang Kim,[1] Jung-Seup Kim, Kum-Ho Cho, Kyu-Han Lee, and Yuzo Tamari[2]

[1] Department of Food Science and Nutrition
Dankook University
Seoul 140-714, Korea
[2] Department of Chemistry
Konan University
Kobe 658, Japan

## INTRODUCTION

Nutrient requirements for term infants are often based on the composition of human milk and the average volume of milk consumed within a 24-h period[25]. The role of taurine in human nutrition and physiology has been intensively reviewed[1,9,10,12,36,40]. Taurine is one of the most abundant free amino acids in human milk, being exceeded only by glutamic acid[27,29]. The importance of taurine as a milk component was suggested by studies showing reduced taurine concentrations in plasma and urine of human infants fed synthetic formulas with little or no taurine as compared to those of infants fed human milk[4]. Primates, including humans, may also depend on exogenous sources to maintain normal taurine levels in milk, as their ability to synthesize taurine from endogenous sources is also restricted[13]. Gaull et al.[5] have suggested that human milk supplies the newborn infant with a supply of taurine to compensate for a low capacity for taurine biosynthesis.

Also, selenium is of particular interest in infant nutrition because human milk serves as a single food source of nutrients during the most rapid period of growth. Selenium is an essential component of glutathione peroxidase, an enzyme that protects cell membranes from oxidative damage by hydrogen peroxide and organic hydroperoxides[42]. In blood and many tissues, the activity of glutathione peroxidase is directly related to selenium levels which is dependent on dietary selenium intake[43].

We have previously shown that the amount of taurine intake from human milk is different between infants of non-vegetarian and lacto-ovo-vegetarian mothers[17]. The purpose of this study is to assess the changes on taurine and selenium levels of human milk and the daily intake of taurine and selenium in breast-fed infants during the early period of lactation.

## SUBJECTS AND METHODS

Thirty-five lactating women at the Red Cross Hospital in Seoul were recruited. Milk samples of about 2 to 10 ml were collected manually every day during the first 5 days of milk secretion, and at the 15th and 30th day postpartum. Taurine content was determined by HPLC equipped with a RF-detector after derivatization with 4-dimethylaminoazobenzene-4'-sulfonylchloride (DABS-Cl)[35].

Selenium levels were determined by atomic absorption spectrophotometry with a hydride generator. Prior to analysis, the milk was treated with a mixture of nitric and perchloric acids to digest the organic substances[18].

The intake of taurine and selenium by infants was estimated from the mean of infant milk intakes[16,31], obtained during a survey performed in our laboratory in 1992; total intake of infant milk could not be measured in this study.

For the statistical analyses, ANOVA, Duncan's multiple range test, and regression analysis were performed using SAS.

## RESULTS

Table 1 shows taurine levels in human milk during the course of lactation. B indicates days after the start of milk secretion and P indicates days postpartum. Since the onset of milk secretion in each individual was not the same, B and P are slightly different.

The taurine levels of colostrum were not significantly different among the women during the first 5 days of lactation. The maximum taurine level was found on B4th day, however, this value was not significantly higher compared to the other time periods. At the 15th day postpartum, the taurine levels were slightly but not significantly decreased. Also, at the 30th day, the taurine level was slightly lower compared with that at the 15th day. The present data indicate that taurine levels in human milk tend to decrease during the early lactation period.

**Table 1.** The levels of taurine and selenium in human milk.

| Days | (n) | Taurine | Selenium |
|---|---|---|---|
| | | nmol/ml | ng/ml |
| B 1 | (18) | 419.2 ± 156.2$^{a}$ | 41.9 ± 34.9$^{a}$ |
| B 2 | (22) | 354.7 ± 171.3$^{ab}$ | 30.9 ± 17.0$^{a}$ |
| B 3 | (24) | 398.6 ± 169.5$^{a}$ | 25.5 ± 8.3$^{ab}$ |
| B 4 | (21) | 446.3 ± 186.3$^{a}$ | 21.9 ± 7.5$^{bc}$ |
| B 5 | (19) | 416.9 ± 167.3$^{a}$ | 20.1 ± 7.9$^{c}$ |
| P 15 | (23) | 358.9 ± 125.3$^{ab}$ | 15.0 ± 6.4$^{d}$ |
| P 30 | (18) | 304.4 ± 93.8$^{b}$ | 15.2 ± 5.1$^{d}$ |

Mean ± SD. B: days after start of milk secretion. P: days postpartum.
Values in the same column not sharing a common superscript are significantly different ($P<0.05$).

Figure 1 shows the change on taurine levels of human milk during the course of lactation. As can be observed, the taurine level significantly decreased with the course of the lactation period ($P<0.05$).

Selenium levels in human milk are shown in Table 1. During the B3rd day, the selenium level was slightly decreased, however, the decrease was not significant. At the B5th day, the selenium level was significantly lower than those values of the first B3 days. The highest value was found on the B1st day. At 15 and 30 days postpartum, milk samples contained significantly lower levels of selenium compared with those during the first B5 days. These data show a similar trend to that of the taurine levels which also decreased progressively during the early lactation period.

Table 2 shows the comparison of taurine levels between colostrum and mature human milk. Colostrum samples were collected from the B1st to the B5th day, and mature milk samples at 15 and 30 days postpartum. As expected, colostrum contained significantly higher levels of taurine compared with mature milk.

Figure 2 shows the changes in the selenium levels of human milk during the lactation period. As can be observed, the selenium levels significantly decreased during the course of lactation ($P<0.001$).

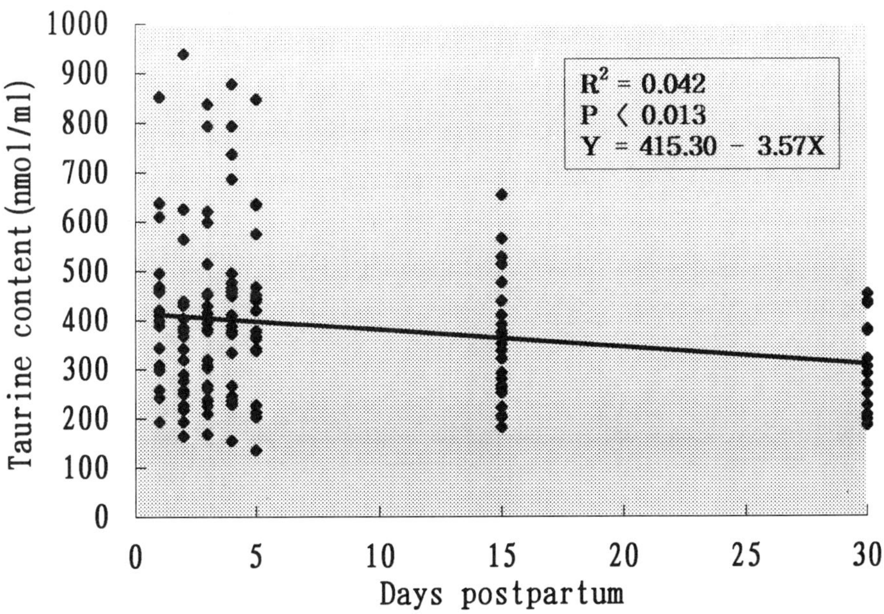


**Figure 1.** Changes in the taurine levels of human milk during the course of lactation.

**Table 2.** Levels of taurine and selenium in human milk.

| Human milk | (n) | Taurine | Selenium |
|---|---|---|---|
| | | nmol/ml | ng/ml |
| Colostrum | (104) | 405.8 ± 174.4[a] | 28.6 ± 19.6[a] |
| Mature | (41) | 335.0 ± 114.5 | 15.1 ± 8.3 |

Mean ± SD. Colstrum: the first 5 days (from B1 to B5) after start of milk secretion.
Mature human milk: 15th and 30th days postpartum.
Values are significantly different between colostrum and mature human milk ($P<0.05$).

The comparison of selenium levels between colostrum and mature human milk is shown in Table 2. The selenium levels of colostrum were significantly greater than that of mature human milk.

Figure 3 shows the relationship between the selenium and taurine levels in human milk. An attempt was made to determine whether the levels of taurine were correlated with the levels of selenium in human milk. Figure 3 shows the relationship between the two nutrients. However, there was no correlation.

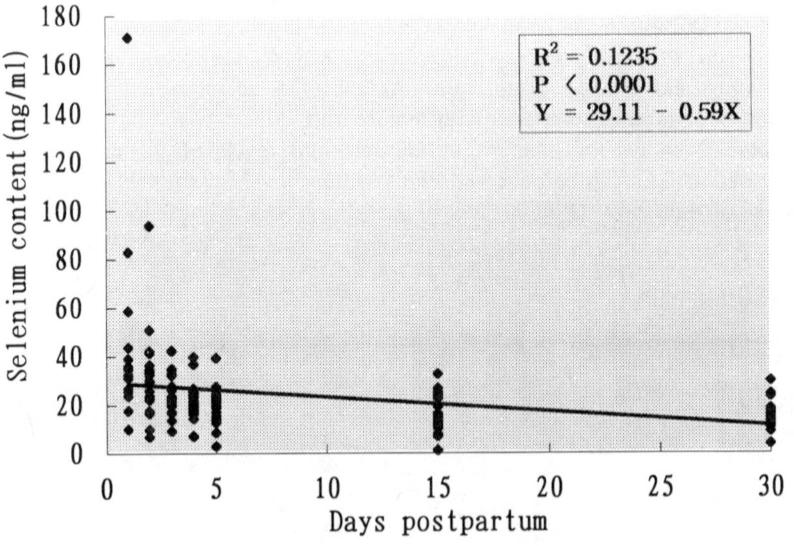


**Figure 2.** Changes in the selenium levels of human milk during the course of lactation.

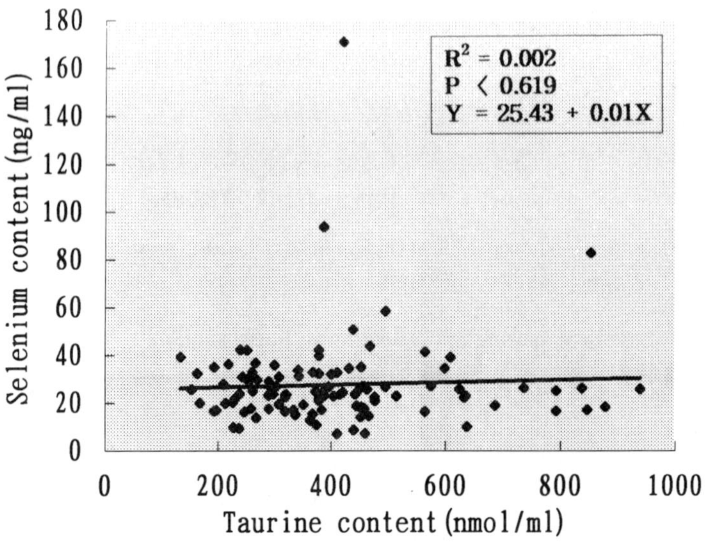


**Figure 3.** Relationships between the selenium and taurine levels of human milk during the course of lactation.

Table 3 shows the average milk intake of Korean infants measured during the first 5 days, and at 15 and 30 days postpartum. This result was obtained from a survey performed in our laboratory in 1992[16,31]. Milk intake of infants was dramatically increased during the early period of lactation, especially during the initial 5 day period. These data were used for estimatiing taurine and selenium intake in this study.

Based on previous results[16,31], we also estimated taurine intake of infants during the early period of lactation (Table 4). The estimation of taurine intake was calculated as the taurine content of milk multiplied by the milk intake of infants. The estimated intake steadily increased during the lactation period. However, as presented earlier, taurine levels progressively decreased during the course of lactation. Therefore, the elevation on taurine intake is a result of the dramatic increase in milk intake. We expect that our results offer useful information on the utility of adding taurine to milk formula in Korea. In most infant formula in Korea, the content of taurine, according to the manufacturing company, is 24 to 27 mg per 100 g of formula.

Selenium intake of infants was estimated by the same procedure (Table 5). The estimated intake of selenium also increased during the course of lactation showing a trend similar to that observed for taurine.

**Table 3.** Intake of human milk in infants.

| Days postpartum | 1 | 2 | 3 | 4 | 5 | 15 | 30 |
|---|---|---|---|---|---|---|---|
| g/day (n=25) | $57.6^{g}$ ± 32.1 | $176.1^{f}$ ± 75.1 | $294.4^{e}$ ± 63.0 | $342.6^{d}$ ± 65.0 | $412.0^{c}$ ± 85.2 | $532.0^{b}$ ± 109.0 | $695.0^{a}$ ± 129.0 |

Mean ± SD. These data were quoted from Kim, S.B. and Seol, M.Y., et al.[16,31]. Values in the same row not sharing a common superscript are significantly different (P<0.05).

**Table 4.** Estimated intake of taurine during the early periods of lactation.

| Days postpartum | 1 | 2 | 3 | 4 | 5 | 15 | 30 |
|---|---|---|---|---|---|---|---|
| n | 18 | 22 | 24 | 21 | 19 | 23 | 18 |
| μ mol/day | 24.3 ± 13.4 | 62.4 ± 26.6 | 117.2 ± 25.1 | 153.1 ± 29.0 | 171.8 ± 35.4 | 190.9 ± 39.1 | 211.6 ± 39.3 |
| mg/day | $3.0^{f}$ ± 1.7 | $7.8^{e}$ ± 3.3 | $14.7^{d}$ ± 3.1 | $19.2^{c}$ ± 3.6 | $21.5^{bc}$ ± 4.4 | $23.9^{ab}$ ± 8.9 | $26.5^{a}$ ± 4.9 |

Mean ± SD. Values in the same row not sharing a common superscript are significantly different (P<0.05).

## DISCUSSION

In the present study, the results are very similar to our previous observations[17] and to those of others[2,29,33,34] that taurine levels of human milk significantly decrease during the course of lactation, but taurine levels of colostrum are significantly higher than mature milk. Kim et al.[17] reported that the taurine level of human milk was 434 nmol/ml in colostrum 3~5 days postpartum and thereafter gradually decreased until 30 days postpartum.

**Table 5.** Estimated intake of selenium during the early period of lactation.

| Days postpartum | 1 | 2 | 3 | 4 | 5 | 15 | 30 |
|---|---|---|---|---|---|---|---|
| n | 18 | 22 | 24 | 21 | 19 | 23 | 18 |
| μ mol/day | 30.8 ± 17.0 | 68.9 ± 29.4 | 90.0 ± 20.4 | 95.1 ± 18.0 | 104.9 ± 21.7 | 101.1 ± 20.8 | 133.7 ± 24.8 |
| μg/day | 2.4[d] ± 1.3 | 5.4[c] ± 2.3 | 7.5[b] ± 1.6 | 7.5[b] ± 1.4 | 8.3[b] ± 1.7 | 8.0[b] ± 1.6 | 10.6[a] ± 2.0 |

Mean ± SD. Values in the same row not sharing a common superscript are significantly different ($P<0.05$).

Rassin et al.[29] reported that, before 3 days postpartum, the taurine level of human milk was 410 nmol/ml, but that by the 5th day postpartum it decreased to 340 nmol/ml. Shubat et al.[33] observed in their study that the taurine level of colostrum decreased gradually after attaining 400 nmol/ml. Furthermore, Erbersdobler et al.[2] reported that after 3-6 days postpartum, the taurine level of human milk in German women was 400 nmol/ml. Stipanuk et al.[34] reported that the concentration of taurine in milk (mol/100 ml defated milk) decreased between days 1 and 21 of lactation and that the levels on day 14 and 21 were significantly less than those on day 1 of lactation. Stipanuk and colleagues[34] suggested that a source of taurine for secretion into the early milk may originate from the enlarged pool of hepatic taurine. These results are consistent with our findings in this study.

Other findings[7,26,28,30,37] are not in agreement with our observations. These differences might be due to the limited sample size and the length of the study period. For instance, the reported taurine intake is very similar to our previous results for 15 and 30 days postpartum. Little previous research has been done to determine the taurine intake of infants. More research using large samples is required to answer this important question. However, results from our studies demonstrate that the taurine intake of infants is 3.0~21.5 mg/day in colostrum and 23.9~26.5 mg/day in mature milk.

The daily requirement of taurine for humans has not been determined. However, high quantities of taurine are normally excreted in urine from individuals regularly consuming meat[34], indicating that the high taurine intake from diets including meat largely exceeds the endogenous requirement of the amino acid. Data on taurine excretion in infants are also not available and unfortunately the taurine content in urine was not measured in the present study.

It has been suggested that taurine plays an important role in the development of the brain[9,36] and as a growth modulator[3,8]. It has also been reported that taurine deficiency is associated with abnormalities in the electroretinograms of patients maintained on long-term parenteral nutrition[6]. Because of the lack of the enzyme for taurine biosynthesis during the neonatal period[36,40,41], precursor sulfur-containing amino acids can not serve to remedy any dietary deficiency in taurine. Therefore, special concern should be given to the composition of formula for premature infants.

The selenium content of milk in the present study gradually decreased from 42 to 15 ng/ml during the course of lactation. The selenium content of colostrum also rapidly decreased during the course of lactation[11,38]. This high secretion of selenium immediately after delivery seems to be related to the maintenance of glutathione peroxidase in newborns, because the enzyme activity of glutathione peroxidase in erythrocytes and plasma, as well as selenium concentrations in serum, are lower in newborns than in their mothers[20,21].

Trafikowaska et al.[39] reported that supplementation of lactating mothers with selenium-enriched yeast rapidly increases the selenium concentration and glutathione peroxidase activity in maternal blood components, and that the selenium concentration of human milk and the selenium supply in infants can be effectively increased by supplementation of lactating mothers with Se-enriched yeast.

In the present study, the selenium content is 29 ng/ml in colostrum and 15 ng/ml in mature milk. Kim et al.[18] showed that the selenium levels were 13.1, 11.5 and 9.8 μg/L at 1, 2 and 3 months postpartum, respectively.

Recently, Trafikowaska et al.[39] reported that the selenium concentration in human milk in Poland was below 10 ng/ml in mature milk and the selenium intake by breast-fed infants was about 6 μg/day. Jochum et al.[14] reported that in German women the selenium content of mature human milk was also 10 ng/ml. Therefore, the selenium content of mature milk is estimated to be in the range of 10 to 20 μg/L in most countries of the world[19], with the exception being low and high selenium areas.

The present results may be used to indicate a causal relationship between dietary selenium intake and the levels of selenium in human milk. Furthermore, comprehensive data[32] on the level of selenium in the breast milk of women living in 17 different states in the United States demonstrate significantly higher concentrations of selenium in high selenium areas such as South Dakota (28 ± 3 μg/L) compared to lower selenium areas such as Ohio (13 ± 1 μg/L). These data were further supported by the results from studies on selenium supplementation performed on lactating mothers thereby demonstrating a strong correlation between dietary selenium intake and selenium in their milk[23,24].

Reported data for the selenium content of colostrum is minimal. However, our data were similar to a report in Japan[38]. The estimated intake of selenium was 2.4~8.3 μg/day in colostrum and 8.0~10.6 μg/day in mature milk. There have also been only a few reports on the investigation of selenium intake in newborns. The US RDA on selenium for infants between birth and 6 months of age is 10 μg/day[25]. The lower intake in colostrum is due to the low intake volume of human milk.

Kim et al.[18] reported that the mean dietary selenium intake of infants was almost constant at 8.38, 8.35 and 6.98 μg/day at 1, 2 and 3 months postpartum, respectively, and the mean daily intake on a body weight basis was gradually decreased during the course of lactation. All of the infants studied, however, were healthy, gaining weight and height according to the norms of typical Korean infants[15].

Recently, the selenium intake by breast fed infants in Poland was about 6 μg/day in mature milk[39]. Kumpulanien et al.[22] showed that the dietary selenium intake of Finnish infants was 8.0 ± 1.8 μg/day at 1 month and 4.7 ± 1.1 μg/day at 3 months postpartum.

Further studies are necessary to determine the factors influencing the selenium concentration of human milk and the transport mechanisms of selenium from maternal blood to breast milk. In our studies we attempted to determine if the taurine and selenium levels in human milk are correlated. However, no apparent correlation was observed.

## SUMMARY


With these results we report the following:

1. Taurine levels in human milk decreased slightly during the early lactation period. The concentration of taurine (406 ± 174 nmol/ml) in colostrum was significantly higher than that (335 ± 115 nmol/ml) in mature milk.
2. Selenium content of human milk also decreased slightly during the early lactation period. The content of selenium (28.6 ± 19.6 ng/ml) in colostrum was significantly higher than that (15.1 ± 5.9 ng/ml) in mature milk.

3. A correlation was not found between the taurine and selenium content of human milk.
4. The intake of taurine and selenium by breast-fed infants progressively increased with days postpartum. These results were due to the significant increase in milk intake by infants.
5. It is suggested that the taurine and selenium levels in colostrum are more concentrated than those in mature milk. However, the absolute intake of taurine and selinium by infants are higher in mature milk.

## ACKNOWLEDGMENTS

This work was supported by grants from Dong-A Pharmaceutical Co. Ltd. of Seoul, Korea and Konan University, Kobe, Japan and Dankook University, Seoul, Korea.

## REFERENCES


1. Chesney, R., 1988, Taurine: Is it required for infant nutrition? *J. Nutr.*, 118: 6-10.
2. Erbersdobler, H.F., Trautwein. E., and Greulich, H.G.,1994, Determination of taurine in milk and infant formula diets, *Eur. J. Ped.,* 142:133-134.
3. Gaull, G.E., 1983, Taurine in human milk: growth modulator or conditionally essential amino acid? *Pediatr. Gastroenterol. Nutr.,* 2:S266-S271.
4. Gaull, G.E., Rassin, D.K., Raiha, N.C.R., and Heionen, K., 1977, Milk protein quantity and quality in low birth-weight infants. III. Effects on sulfurcontaining amino acids in plasma and urine, *J. Pediatr.*, 90:348-355.
5. Gaull, G.E., Sturman, J.A., Wen, G.Y., and Winiewski, H.M., 1982, In membranes in growth and development, progress in clinical and biological research series, 91:349-355, Hoffman, J.F. and Giebisch, G.H., eds., New York: Liss.
6. Geggel, H.S., Ament, M., Heckenlively, J.R., Martin, D.A., and Kopple, J.D., 1985, Nutritional requirement for taurine in patients receiving long-term parenteral nutrition, *N. Engl. J. Med.*, 312:142-146.
7. Harzer, G., Franzke, V., and Birdels, J.G., 1984, Human milk nonprotein nitrogen components: changing patterns of free amino acids and urea in course of early lactation, *Am. J. Clin. Nutr.*, 40:303-309.
8. Hayes, K.C., Stephan, Z.F., and Sturman, J.A., 1980, Growth depression in taurine-depleted infant monkeys, *J. Nutr.,* 110:119-125.
9. Hayes, K.C. and Sturman J.A., 1981, Taurine in metabolism. *Ann. Rev. Nutr.,* 1:401-425.
10. Hayes, K.C. and Trautwein, E.A., 1994, Taurine, in: "Modern Nutrition in Health and Disease", 8th edition, Shils, M.E., Olson, J.A., and Shike, M., eds., Lea and Febiger, Philadelphia, pp. 477-485.
11. Higashi, A., Tamari, Y., Kuroki, Y., and Matsuda, I., 1983, Longitudinal changes in selenium content of breast milk, *Acta. Paediatr. Scand.*, 72:433-436.
12. Huxtable, R.J., 1992, Physiological actions of taurine, *Physiol Rev.*, 72:101-163.
13. Jacobsen, J.G. and Smith, L.H., 1968, Biochemistry and physiology of taurine and taurine derivatives, *Physiol Rev.*, 48:424-511.
14. Jochum, F., Fuchs, A., Menzel, H., and Lombeck, I, 1995, Selenium in German infants fed breast milk or different formulas, *Acta. Paediatr.*, 84:859-862.
15. Keum, H.K., 1994, A longitudinal study on volume and contents of selenium and zinc of human milk, Ph. D. thesis in DanKook University.
16. Kim, S.B., 1992, A study on milk volume and contents of thiamine and riboflavin in the human milk of lactating women with Caesarian section for 7 days after delivery: Master thesis, Dankook University.
17. Kim, E.S., Cho, K.H., Park, M.A., Lee, K.H., Moon, J., Lee, Y.N., and Ro, H.K., 1996, Taurine intake of Korean breast-fed infants during lactation, *Adv. Exp. Med. Biol.,* 403:571-577.
18. Kim, E.S., Keum, H.K., and Tamari, Y., 1996, Selenium intake in breast-fed infants during course of lactation, *J Food Sci. Nutr.,* 1:230-233.

19. Kumpulainen, J., 1989, Selenium: Requirement and supplementation, *Acta. Paediatr. Scand. Suppl.*, 351:114-117.
20. Kumpulainen, J., Salmenpera, L., Siimes, M.A., Koivistoinen, P., Lehto, J., and Perheentupa, J., 1987, Formula feeding results in lower selenium status than breast-feeding : a longitudinal study, *Am. J. Clin. Nutr.*, 45:49-53.
21. Kumpulainen, J., Salmenpera, L., Siimes, M.A., Koivistoinen, P., and Perheentupa, J., 1985, Selenium status of exclusively breast-fed infants as influenced by maternal organic or inorganic selenium supplementation, *Am. J. Clin. Nutr.*, 42:829-835.
22. Kumpulainen, J, Vuori, E., Kuitunen, P., Makinen, S., and Kara, R., 1983, Longitudinal study on the dietary selenium in Finland, Internat. *J. Vit. Nutr. Res.*, 53:420-426.
23. McGuire, M.K., Burgert, S.L., Milner, J.A., Glass, L., Kummer, R., Deering, R., Boucek, R., and Picciano, M.F., 1993, Selenium status of infants is influenced by supplementation of formula or maternal diets, *Am. J. Clin. Nutr.*, 58:643-648.
24. Moser, P.B., Reynolds, R.D., Acharya, S., Howard, M.P., Andon, M.B., and Lewis, S.A., 1988, Copper, iron, zinc and selenium dietary intake and status of Nepalese lactating women and their breast-fed infants, *Am. J. Clin. Nutr.*, 47:729-734.
25. National Research Council, 1989, Recommended dietary allowances. 10th ed. Washington, D.C. : National Academy Press.
26. Pamblanco, M., Portoles, M., Paredes, C., Ten, A., and Comin, J., 1989, Free amino acids in preterm milk from mothers delivering appropriate or small for gestational age infants, *Am. J. Clin. Nutr.*, 50:17-28.
27. Pasantes-Morales, H., Lopez, I., and Ysunza, A., 1995, Taurine content in breast milk of Mexican women from urban and rural areas, *Achives of Med. Res.*, 26:47-52.
28. Rana, S.K. and Sanders, T.A.B., 1986, Taurine concentration in the diet, plasma, urine and breast milk of vegans compared with omnivores, *Br. J. Nutr.*, 56:17-28.
29. Rassin, D.K., Sturman, J.A., and Gaull, G.E., 1978, Taurine and other free amino acids in milk of man and other mammals, *Early Human Dev.*, 2:1-13.
30. Sato, H., Tada, H., Yanagida, M., and Yamaguchi, K., 1983, The postnatal changes of trace nutrients in human breast milk, *Sulfur Amino Acids,* 6:297-300.
31. Seol, M.Y., Kim, E.S., and Keum, H.K., 1993, A longitudinal study on human milk intake in exclusively breast-fed infants, *Korean J. Nutr.*, 26:414-422.
32. Shearer, T.R. and Hajimarkos, D.M., 1975, Geographic distribution of selenium in human milk, *Arch. Environ. Health,* 30:230-233.
33. Shubat, P.J., Parker, H., and Huxtable, R.J., 1989, Effect of suckling and diurnal influences on the concentrations of taurine and other free amino acids in milk, Eur. J. Clin. Nutr., 43:675-680.
34. Stipanuk, M.H., Kuo, S.M., and Hirschberger, L.L., 1984, Changes in maternal taurine in response to pregnancy and lactation, *Life Science*, 35:1149-1155.
35. Stochi, V., Palma, F., Piccoli, G., Biagiarelli, B., Cucciarimi, L., and Magnani, M., 1994, HPLC analysis of taurine in human plasma sample using the DABS-Cl reagent with sensitivity at picomole level, *J. Liq. Chromatogr.*, 17(2):347-357.
36. Sturman, J.A. and Hayes, K.C., 1980, The biology of taurine in nutrition and development, in: "Advances in Nutritional Research, Volume III", Draper, H.H. ed., Plenum Press, New York, pp 231-299.
37. Svanberg, U.F., Gebre-Medhin, M., Ljungqvist, B., and Olsson, M., 1977, Breast milk composition in Ethiopian and Swedish mothers. III. Amino acids and other nitrogenous substances, *Am. J. Clin. Nutr.,* 30:499-507.
38. Tamari, Y., Chayama, K., and Tsuji, H., 1995, Longitudinal study on selenium content in human milk particularly during early lactation compared to that in infant formulas and cow's milk in Japan, *J. Trace Elements Med. Biol.,* 9:34-39.
39. Trafikowska, U., Zachara, B.A., Wiacek, M., Sobkowiak, E., and Czerwionka-Szaflarska, M., 1996, Selenium supply and glutathione peroxidase activity in breastfed Polish infants, *Acta Paediatr.*, 85:1143-1145.
40. Wright, C.E., Tallan, H.H., Lin, Y.Y., and Gaull, G.E., 1986, Taurine biological update, *Ann. Rev Biochem.,* 55:427-453.
41. Yamaguchi, K., 1990, Nutrition and metabolism of sulfur amino acid, in: "Nutrition: Proteins and amino acids", Yoshida, A., Naito, H., Niiyama, Y., and Suguki, T., eds., Japan Sci Soc Press, Tokyo, pp 165-183.
42. Zachara, B.A., 1992, Mammalian selenoproteins, *J. Trace Elem. Electrolytes Health Dis.*, 6:137-51.

43. Zachara, B.A., Mikotajczak, J., and Trafikowska, U., 1993, Effect of various selenium (Se) intakes on tissue Se levels and glutathione peroxidase activities in lambs, *J. Vet. Med. A.*, 40:310-318.

# PLASMA CONCENTRATION OF TAURINE IS HIGHER IN MALNOURISHED THAN CONTROL CHILDREN: DIFFERENCES BETWEEN KWASHIORKOR AND MARASMUS

Lucimey Lima and Erica Jaffé

Laboratorio de Neuroquímica
Centro de Biofísica y Bioquímica
Instituto Venezolano de Investigaciones Científicas
Apdo. 21827, 1020-A
In collaboration with Instituto Nacional de Nutrición
Caracas, Venezuela

## SUMMARY

Plasma free amino acids were determined in the plasma of severely malnourished children under two years of age. A total of thirty-one patients and eleven controls were evaluated: seventeen cases of kwashiorkor, eight cases of marasmus, and six cases of marasmic-kwashiorkor. Fasting plasma samples were taken in the morning on the day of admission. Fasting plasma samples were also taken from nine patients at discharge after two months in the hospital where they only received a balanced diet as treatment. A partial reversal of the signs of malnutrition was observed at discharge. In the whole group of patients at admission, lower concentrations of tyrosine, methionine, tryptophan, and leucine and higher concentrations of aspartate, glutamate, and taurine were observed compared to controls. Taurine continued to be elevated in the malnourished group at the time of discharge. Marasmic children, as compared to controls, had high aspartate and low tryptophan levels, but taurine levels were not significantly different from controls. Kwashiorkor patients had low tyrosine, methionine, tryptophan, and lysine, and significantly higher taurine plasma levels. The elevated concentration of taurine might be the result of a redistribution of this amino acid to provide specific tissues with the required amount for development.

## INTRODUCTION

Studies concerning peripheral markers in malnutrition have been done in different populations, however, there is not a complete evaluation in malrnourished children of Latin America. Moreover, only lately, have biochemical differences between marasmus and kwashiorkor been established. For instance, retinol levels are reduced in marasmic children and the tocopherol/lipids-ratio is lowered in kwashiorkor[4]. However, these authors found high plasma levels of ubiquinone. In general, it seems that there is a depletion of antioxidants in malnourished children, although the increase of the antioxidant vitamin ubiquinone in kwahiorkor is not well understood. In addition, malnourished children show a reduction in highly polyunsaturated fatty acids, cholesterol esters, phospholipids, and phosphatidylcholine in erythrocytes, but children with kwashiorkor have lower levels of linoleic acid metabolites and docosahexaenoic acid than marasmic patients, suggesting a probable increase in lipid peroxidation in the kwashiorkor syndrome[22]. Antioxidant status in children with kwashiorkor, as determined by the concentration of red blood cell antioxidant enzymes, such as catalase, superoxide dismutase and glutathione peroxidase, differs with respect to normally nourished and marasmic children[34]. It is known that total plasma proteins and albumin are lower in malnourished children, but hypoproteinaemia is mainly found in kwashiorkor[19]. Decreased plasma carnitine levels have also been demonstrated in malnutrition, being significantly lower in kwashiorkor[41].

Since taurine was demonstrated to be present in a high concentration in the milk of various animal species, including human, potential nutritional value was given to this amino acid[37,38]. Plasma levels of amino acids are the result of different steps of regulation, such as dietary intake, protein synthesis, protein degradation, synthesis, and catabolism[7]. Taurine is not a constituent of proteins, and thus the balance of plasma concentration is not directly related to protein turnover, but rather to interorgan distribution, and to regulation of the taurine transporter in the kidney by adaptive response or by phosphorylation[13,14].

In the present study we report the values of plasma amino acids in a well-nourished population under two years of age, and we compare them with plasma amino acid values obtained from patients with severe malnutrition. The results were then separated into groups of patients with kwashiorkor and/or marasmus. Plasma amino acid samples obtained at discharge were paired compared with those taken at admission.

## MATERIALS AND METHODS

### Population

The patient population of the study consisted of thirty-one children under two years of age (6 to 20 month), of both sexes, who were admitted to the Unity Menca de Leoni of Instituto Nacional de Nutrición (INN) in Caracas, Venezuela, with severe malnutrition according to the parameters established by the Panamerican Health Organization[21]. The patients exhibited signs of marasmus (mainly emaciation), kwashiorkor (mainly edema) or both. For the analysis, the children were considered as a whole group or in subgroups of kwashiokor (17 patients) and marasmus (8 patients), excluding the children with mixed malnutrition. A sample of venous blood was taken at admission (7:30 to 8:00 a.m., after fasting) in heparinized tubes. After centrifugation, a solution of 20% sulfosalicylic acid was added to the supernatant, giving a final concentration of 5%. At the end of a second centrifugation, the supernatant was kept at -20° until analysis. All patients remained in the hospital for a mean time of two months, receiving a balanced diet, according to the nutritional requirements established by the INN. Patients left the hospital unit with improvement in mental performance, and disappearance of signs, such as emaciation and edema. Blood samples were taken from nine of the patients leaving the hospital. A control group consisted of children from the same socioeconomical level, IV (work-men families) according to Graffar and modified by Méndez Castellanos[26], i.e., equal range of ages, both sexes, and without apparent disease checked by a regular clinical follow up.

### Determination of amino acids

The amino acids were determined by high performance liquid chromatography with fluorescent detection by a modified method[9,31]. The system consisted of a Model 2150 solvent delivery pump, a Model 2152 LC controller, a Model 2157 auto-sampler injector (LKB, Broma, Sweden), and a Spectroflow 980 fluorescence detector (Kratos Analytical). Output from the detector was recorded with a Model 2220 recording integrator (LKB, Broma, Sweden). A LC-18 column was used (4.6 x 100 mm, 5 mm) (Supelco). Potassium borate buffer (50 ml, 0.4 M pH 7.4) was added to 50 ml of plasma, previously centrifuged at 35,000 g for 20 min. Derivatization was carried out by addition of 100 ml of the plasma-borate buffer mixture to 25 mg *o*-phthalaldehyde (Sigma), 500 ml methanol, 25 ml mercaptoethanol (1 g/ml), and 4.5 ml 0.4 potassium borate buffer pH 10.4, followed by incubation at 65°C for 10 min. The reaction was stopped by placing the tubes on ice. After centrifugation at 30,000 g for 2 min, aliquots of the supernatant were injected into the chromatographic system. The mobile phases were: Buffer A, 50 mM sodium phosphate pH 6.3, 0.3% tetra-hydrofurane and 5% acetonitrile, and Buffer B, acetonitrile. The main gradient in percentage of Buffer B was: 0-25 min, 15%; 25-45 min, 20%; 45-60 min, 30%; 60-70 min, 50%; 70-75 min, 90%; 75-78 min, 95%; 78-80 min, 0%. Amino acids were quantified with the use of an external standard and expressed in nmol/ml. Stock solutions of amino acids (Sigma, 1 mg/ml) were prepared in 0.5 N HCl and stored at -20°C. Each value is the mean ± S.E.M. Comparisons were done by Students' $t$ test; significance was considered if $p < 0.05$.

## RESULTS AND DISCUSSION

The children were discharged from the hospital when they had a significant reduction in specific clinical signs[20]. Plasma concentrations of the amino acids were similar to those reported for populations close in age with the present control group[1,10,48]. However, there is no report of plasma amino acid levels for children up to two years old. The concentrations of tyrosine, methionine, tryptophan and leucine were significantly lower in the patients upon admission to the hospital compared to controls (Table 1). Reduced protein intake could be one of the reasons for the lower levels of these amino acids. Some hepatic enzymes, such as tryptophan pyrrolase and tyrosine aminotransferase, are induced during fasting and starvation[33,35,46] and could contribute to modifications in the plasma amino acids. The decrease in the plasma concentration of leucine is an indication of the low intake of protein[8]. Also, the oxidative metabolism of this amino acid increases in muscles during starvation[5].

Glutamate and aspartate were increased in the plasma of patients upon admission compared to the control group (Table 1). Plasma ammonia concentration is elevated in children with marasmus, but significantly decreases after four weeks of nutritional rehabilitation[30]. Hyperammonemia could be the cause of the increase in plasma aspartate and glutamate in the circulation, and might be related to the activities of alanine or aspartate aminotransferases and glutamate dehydrogenase which favor ammonia detoxification. In a reported study[30], circulating transaminases, urea, or urea cycle metabolites were not altered by the malnutrition. However, urea has been reported to be low in children with marasmus or kwashiorkor and has been found to rise after treatment for kwashiorkor[40]. It is interesting to note that recovery of several anthropometric and bio-

**Table 1.** Plasma free amino acid concentrations in control and malnourished children upon hospital admission.

| Amino acid | Control (11) | All patients (31) | Marasmus (8) | Kwashiorkor (17) |
|---|---|---|---|---|
| Asp | 6.45 ± 0.63 | 10.24 ± 1.04[a] | 11.16 ± 1.75[a] | 10.66 ± 1.68 |
| Glu | 54.38 ± 5.71 | 77.33 ± 6.44[a] | 81.56 ± 16.91 | 76.23 ± 9.01 |
| Gln | 320.91 ± 42.9 | 305.20 ± 19.10 | 308.83 ± 29.06 | 328.28 ± 22.93 |
| Ala | 199.01 ± 35.13 | 130.47 ± 32.06 | 229.53 ± 126.72 | 98.96 ± 5.74 |
| Asn | 31.20 ± 3.21 | 50.74 ± 12.95 | 65.93 ± 35.99 | 35.32 ± 3.49 |
| Ser | 422.29 ± 28.42 | 438.68 ± 41.08 | 569.53 ± 45.75 | 404.72 ± 58.07 |
| His | 267.64 ± 46.56 | 439.92 ± 138.92 | 344.81 ± 68.64 | 310.42 ± 61.69 |
| Gly | 267.20 ± 62.16 | 220.33 ± 34.37 | 306.81 ± 97.73 | 214.43 ± 42.89 |
| Thr | 172.89 ± 23.79 | 130.23 ± 34.37 | 153.87 ± 40.06 | 125.45 ± 15.55 |
| Arg | 66.24 ± 8.57 | 56.81 ± 10.15 | 97.14 ± 32.66 | 48.99 ± 8.80 |
| Tau | 33.91 ± 3.64 | 68.01 ± 13.54[a] | 84.36 ± 41.33 | 77.35 ± 15.99[a] |
| Tyr | 44.12 ± 8.44 | 18.60 ± 2.08[a] | 25.54 ± 3.79 | 18.48 ± 2.95[a] |
| Val | 346.16 ± 46.95 | 289.92 ± 21.20 | 351.54 ± 35.23 | 284.53 ± 28.61 |
| Met | 15.01 ± 4.03 | 7.33 ± 0.90[a] | 9.98 ± 1.10 | 7.04 ± 1.45[a] |
| Trp | 36.98 ± 7.23 | 11.92 ± 1.66[a] | 17.67 ± 3.52[a] | 11.58 ± 2.25[a] |
| Ile | 65.71 ± 15.63 | 58.22 ± 8.49 | 72.28 ± 14.45 | 57.34 ± 12.67 |
| Phe | 177.80 ± 23.97 | 156.44 ± 14.62 | 175.41 ± 25.04 | 156.76 ± 22.68 |
| Leu | 231.03 ± 28.23 | 147.82 ± 14.62[a] | 200.14 ± 24.99 | 142.44 ± 17.89[a] |
| Lys | 122.31 ± 15.14 | 115.21 ± 20.88 | 205.22 ± 73.76 | 95.57 ± 9.54[a,b] |

Each value is the mean ± S.E.M., nmol/ml ([a]$P < 0.05$ with respect to control, [b]$P < 0.05$ with respect to marasmus).

chemical parameters, as well as psychomental behaviour, are obtained by only providing a balanced diet for relative short periods of time, such as 18 days[40] or 4 weeks[30], or 8 weeks as noted in the present study.

A low plasma taurine concentration was expected for the malnourished group of children, however, the levels of this sulfur amino acid were significantly increased in patients upon admission with respect to the control group (Table 1). Taurine plays an important role during development and regeneration in humans and in animals[24,26,36,37,38]. A deficiency in taurine levels in children receiving parenteral nutrition has been associated with changes in the normal electric pattern of the retina[2,32]. In addition, it has been observed that preterm infants administered taurine and carnitine enriched formula have advantages in motor and mental function with respect to those receiving a standard formula[25,29]. In this context, it is well known that the human cerebellum, which is the most vulnerable structure with regard to early malnutrition, is inmature at the time of birth[16]. The results of a study using formula alone or formula supplemented with taurine to feed preterm infants led to the conclusion that the group receiving taurine in the formula had more mature brain stem auditory evoked responses[43]. On the other hand, a preliminary study indicates that there are no changes in the audiological assessment of the hearing ability between term infants receiving low or high levels of taurine in their diet[6].

Several studies indicate that a dietary source of taurine is very important for the cat. For instance, a decrease in taurine intake of pregnant mothers results in a variety of abnormalities in the retina and in the brain of the kittens[37]. Severe taurine depletion in the adult cat produces retinal degeneration[39]. In addition, rat glial cell function and neuronal migration are affected by taurine deficiency[28]. The elevation of taurine in malnourished children could be related to an internal movement of this amino acid from tissues with great capacity for synthesis and storage, such as the liver and muscle, to other compartments like the central nervous system. For instance, taurine concentration in human liver is relatively high with respect to other amino acids, with only aspartate being greater[3]. Hepatic taurine might not only be for the production of bile salts, but also as a source of taurine for the circulation. The kidney could play an important role in the maintenance of taurine levels. Thus, increased renal reuptake of taurine due to a reduction in intake probably occurs in this group of patients. The kidney reabsorbs taurine according to the dietary composition, and adaptive regulation of transport activity is induced by either increasing or decreasing sodium- or chloride- dependent transport of taurine[13].

Peripheral modifications of some amino acids linked to a central dysfunction, such as convulsive disorders[12,17,20,44], do not necessarily mean that these changes occur in the brain. Rather, they are indexes of disturbances in metabolism, parallel to possible central alterations, or could also be the result of compartmental distribution of molecules that can cross the blood-brain barrier. Plasma taurine can be taken up by the beta-amino acid transport system in the central nervous system[18], and an elevation of circulating concentrations of taurine might influence central levels. In addition to the considerable number of reports concerning the importance of taurine during development and regeneration in several species, there is also evidence of the relevance of critical levels of this amino acid, i.e., the concentration dependency of taurine as a trophic factor in goldfish retina is bell-shaped[23], and embryonic exposure to high levels of taurine in the chick can be toxic[45]. Another interesting point is that taurine is a potent antioxidant[50], and there is an alteration in the levels of circulating antioxidants in malnutrition, among them an increase in ubiquinone seen in kwashiorkor[4]. However, other antioxidant compounds or peripheral enzymes are decreased in malnourished children[34]. It is possible that the increases in ubiquinone and taurine are related to a protective response to the decrease in the antioxidative state.

Protein-calorie malnutrition is important in some regions of the world, but prevalence and accurate quantification of data are difficult to obtain. The term protein-calorie malnutrition is relatively recent and includes a wide variety of different degrees of clinical situations. Ten years after the description of kwashiorkor[49] this syndrome was accepted as such, and in the fifties it was established that the main cause of it was a deficiency in protein intake[42]. However, marasmus also includes energy deficits. The main sign of kwashiorkor is edema, and that of marasmus is emaciation. Concerning the etiology of the two diseases, it is simple to designate one as due to a decrease in protein intake, and the other to a low intake in calories. It has been proposed for kwashiorkor that there is an intoxication by aflatoxins[15] or an excess of free radicals[11]. These two forms of malnutrition have been classified[47] in terms of the presence or absence of edema. Besides the overlap in the two clinical conditions, there are clear biochemical differences between both diseases. We have found interesting differences in amino acid levels between the two groups of patients, despite the relative low number of patients. Probably most easy to explain are the amino acid data, such as the levels of tyrosine, methionine, tryptophan, leucine and lysine, which are shown as lower in the whole group of patients and in kwashiorkor compared to the control, but with the exception of tryptophan are not reduced in marasmus (Table 1). The increase in aspartate levels observed in marasmus, as we already mentioned, could be due to the hyperammonemia described in this entity[30]. The elevation in taurine, significant in patients with kwashiorkor related to controls, remained significantly higher at discharge (Table 2). It could be explained as a compensatory rise in antioxidative activity.

Taurine supplementation is probably not indicated in cases of severe malnutrition, since several mechanism of regulation could be activated to solve the deficiency. Rather, determining the peripheral levels of this amino acid might help. Amelioration of the clinical signs after receiving a balanced diet has been reported for many, however, normal

**Table 2.** Plasma free amino acid concentrations in nine control and malnourished children at admission and at discharge.

| Amino acid | Patients at admission (9) | Patients at discharge (9) |
|---|---|---|
| Asp | 8.40 ± 1.73 | 8.78 ± 0.94 |
| Glu | 49.17 ± 6.89 | 64.32 ± 6.70 |
| Gln | 314.12 ± 31.41 | 433.75 ± 84.10 |
| Ala | 203.42 ± 114.13 | 80.96 ± 10.02 |
| Asn | 95.36 ± 43.70 | 72.23 ± 35.13 |
| Ser | 333.76 ± 76.30 | 430.94 ± 62.41 |
| His | 169.48 ± 49.88 | 188.28 ± 27.77 |
| Gly | 384.27 ± 100.70 | 335.77 ± 95.57 |
| Thru | 92.58 ± 15.18 | 172.74 ± 29.99 |
| Arg | 47.02 ± 9.04 | 83.38 ± 29.94 |
| Tau | 95.41 ± 35.41 | 62.70 ± 10.84 |
| Tyr | 18.83 ± 2.90 | 36.30 ± 5.76[a] |
| Val | 287.32 ± 40.00 | 397.60 ± 29.14 |
| Met | 7.64 ± 1.36 | 19.22 ± 3.13[a] |
| Trp | 7.56 ± 1.79 | 34.12 ± 4.19[a] |
| Ile | 84.90 ± 16.90 | 122.20 ± 21.65 |
| Phe | 164.32 ± 17.35 | 161.92 ± 23.24 |
| Leu | 132.60 ± 23.81 | 232.66 ± 32.43[a] |
| Lys | 145.11 ± 71.53 | 137.59 ± 15.74 |

Each value is mean ± S.E.M., nmoles/ml ([a]$P < 0.05$ with respect to patients at admission).

parameters were not reached in all cases. In contrast, even if all the amino acids altered at the time of admission are corrected by the time of discharge, the levels of taurine continue to be elevated, and might therefore be an indication of the borderline situation of these children even if they partially recover from the disease.

## ACKNOWLEDGMENTS

This work was partially supported by Grant S1-723 from Consejo Venezolano de Investigaciones Científicas y Tecnológicas (CONICIT). We thank Bc. F. Obregón for technical assistance and Mrs. I. Otaegui for secretarial help. This work was done thanks to Dr. C. Rebrig from the Instituto Nacional de Nutrición in Caracas. We appreciate the comments of Dr. W. Jaffé.

## REFERENCES

1. Adamkin, D.D., Radmacher, P., and Rosen, P., 1995, Comparison of a neonatal versus general-purpose amino acid formulation in preterm neonates, *J. Perinatol,*. 15:108-113.
2. Ament, M.E., Geggel, H.S., Heckenlively, J.R., Martin, D.A., and Kopple, J., 1986, Taurine supplementation in infants receiving long-term total parenteral nutrition., *J. Am. Coll. Nutr.,* 5:127-135.
3. Barle, H., Ahlman, B., Nyberg, B., Andersson, K., Essen, P., and Wernmman, J., 1996, The concentrations of free amino acids in human liver tissue obtained during laparoscopic surgery, *Clin. Physiol.,* 16:217-227.
4. Becker, K., Botticher, D., and Leichsenring, M., 1994, Antioxidant vitamins in malnourished Nigerian children, *Int. J. Vitam. Nutr. Res.,* 64:306-310.

5. Buse, M.G., Biggers, J.F., Crier, C., and Buse, J.F., 1973, The effect of epinephrine, glucagon, and nutritional state on the oxidation of branched amino acids and pyruvate by isolated hearts and diaphragms of the rat, *J. Biol. Chem.*, 248:697-706.
6. Davies, W.E., Hopkins, P.C., Rose, S.J., and Dhillon, S.K., 1996, The influence of different taurine diets on hearing development in normal babies, *Adv. Exp. Med. Biol.*, 403:631-637.
7. Felig, P., 1975, Amino acid metabolism in man, *Ann. Rev. Biochem.*, 44:933-955.
8. Fernstrom, J.D., Wurtman R.J., Hammarstrom-Wiklund B., Rand, W.M., Munro, H.N. and Davidson, C.S., 1979, Diurnal variations in plasma concentrations of tryptophan, tyrosine and other amino acids: Effect of dietary protein intake, *Am. J. Clin. Nutr.*, 32:1912-1922.
9. Fernstrom, M.H. and Fernstrom J.D., 1980, Rapid measurement of free amino acids in serum and CSF using high-performance liquid chromatography, *Life Sci.*, 29:2119-2130.
10. Giovanini, M., Gostoni, C., Fiocchi, A., Bellú, R., Trojan, S. and Riva, E., 1994, Antigen-reduced infant formulas versus human milk: growth and metbolic parameters in the first 6 months of live, *J. Am. Coll. Nutr.*, 13:357-363.
11. Golden, H.M.N. and Ramdanth, D., 1987, Free radicals in the pathogenesis of kwashiorkor, *Proc. Nutr. Soc.*, 46:53-68.
12. Haines J.L., Rich, S.S., Tsai, M.Y., and Anderson, V.E., 1985, Altered amino acid levels in multiply affected sibships with seizures, *Epilepsia* 26:642-648.
13. Han, X., Chesney, R.W., Budreau, A.M., and Jones, D.P., 1996, Regulation of expression of taurine transport in two continuous renal epithelial cell lines and inhibition of taurine transporter by site-directed antibody, *Adv. Exp. Med. Biol.*, 403:173-191.
14. Han, X., Budreau, A.M., and Chesney, R.W., 1996, Role of conserved peptide in taurine transporter inactivation by protein kinase C, *J. Am. Soc. Nephrol.*, 7:2088-2096.
15. Hendrickse R.G., 1984, The influence of aflatoxins on child health in the tropics with particular reference to kwashiorkor, *Trans. R. Soc. Trop. Med.*, 78:436-441.
16. Huppi, P.S., Posse, S., Lazeyras, F., et al. (1991) Magnetic resonance in preterm and term newborns. 1H-spectroscopy in developing human brain, *Pediatr. Res.*, 30, 574-578.
17. Huxtable, R.J., Laerd, H., Lippincott, S.E., and Walson, P., 1983, Epilepsy and the concentrations of plasma amino acids in humans, *Neurochem. Int.*, 5:125-135.
18. Huxtable, R.J., 1989, Taurine in the central nervous system and the mammalian actions of taurine, *Prog. Neurobiol.*, 32:471-533.
19. Ibrahim, S.A., Eltom, A.M., Abdul-Rahman, A.M., and Saeed, B.O., 1994, Correlation of some biochemical parameters with clinical features of protein energy malnutrition, *East. Afr. Med. J.*, 71:77-83.
20. Ko, F.-J., Chiang, C.-H., Liu, W.-J., and Chiang, W., 1993, Alteration of amino acid in plasma and cerebrospinal fluid of children with seizure disorders, *Kaohsiung J. Med. Sci.*, 9:131-142.
21. Latham, M.C., 1991, Malnutrición proteico-energética. In (Organización Pan-americana de la Salud, OPS, and Instituto Internacional de Ciencias de la Vida, ILSI, No. 532, Sixth Ed.), *Conocimientos actuales sobre nutrición* (North America), pp. 45-55.
22. Leichsenring, M., Sutterlin, N., Less, S., Baumann, K., Anninos, , and Becker, K., 1995, Polyunsaturatd fatty acids in erythrocyte and plasma lipids of children with severe protein-energy malnutrition, *Acta Paediatr.*, 84:516-520.
23. Lima, L., Matus, P., and Drujan, B., 1988, Taurine effect on neuritic growth from goldfish retinal explants, *Int. J. Devl. Neurosci.*, 6:417-424.
24. Lima, L., Matus, P., and Drujan, B., 1992, The trophic role of taurine in the retina. A possible mechanism of action. In: Lombardini, J.B., Schaffer, S. W., and Azuma J., eds., Taurine. Nutritional Value and Mechanisms of Action, New York: Plenum Press, pp. 287-294.
25. Lucas, A., Morley, R. Cole, T.J., and Gore, S.M., 1994, A randomised multicentre study of human milk, *Arch. Dis. Child,* 70:F141-F146.
26. Matus, P., Cubillos, S., and Lima, L., 1997, Differential effect of taurine and serotonin on the outgrowth from explants or isolated cells of the retina, *Int. J. Devl. Neurosci.*, 16:785-793.
27. Méndez Castellanos, H., 1986, Investigación sobre la consistencia de las variables utilizadas en el método de Graffar modificado, *Arch. Vlnos. Puericul. Pediat.*, 49:93-104.
28. Morán, J., Maar, T., Gegelashvili, G., Bock, E., Schousboe, A., and Pasantes-Morales, H., 1996, Taurine deficiency and neuronal migration, *Adv. Exp. Med. Biol.*, 403:519-526.
29. Morley, R., Cole, T.J., Powell, R., and Lucas, A., 1988, Mother's choice to provide breast milk and developmental outcome, *Arch. Dis. Child.*, 63:1382-1385.
30. Morsy, M.R., Madina, H., Sharaf, S.A., Soliman, A.T., Elzalabany, M.M., and Ramadan, M.A.F., 1994, Hyperammonemia in marasmic children, *J. Trop. Ped.*, 40:97-99.

31. Qureshi, G.A. and Qureshi, A. R., 1989, Determination of free amino acids in biological samples: problems of quantitation, *J. Chromatogr.*, 491:281-289.
32. Rassin, D.K. and Malloy, M.H., 1985, Taurine requirement with parenteral nutrition (Letter), *N. Engl. J. Med.*, 313:120-121.
33. Salter, M., Satanley, J.C., Fisher, M.J., and Pogson, C.I., 1984, The influence of starvation and tryptophan administration on the metabolism of phenylalanine, tyrosine and tryptophan in isolated rat liver cells, *Biochem. J.*, 221:431-438.
34. Sive, A.A., Subotzky, E.F., Malan, H., Dempster, W.S., and Heese, H.D., 1993, Red blood cell antioxidant enzyme concentrations in kwashiorkor and marasmus, *Ann. Trop. Paediatr.*, 13:33-38.
35. Smith, S.R., Pozefsky, T., and Chetri, M.K., 1974, Nitrogen and amino acid metabolism in adults with protein-calorie malnutrition, *Metabolism*, 23:603-618.
36. Sturman, J.A., 1986, Nutritional taurine and central nervous system development, *Ann. N.Y. Acad. Sci.*, 477:196-213.
37. Sturman, J.A., 1993, Taurine in development, *Physiol. Rev.*, 73:119-147.
38. Sturman, J.A. and Chesney, R.W., 1995, Taurine in pediatric nutrition, *Pediatr. Nutr.*, 42:879-897.
39. Sturman J.A., Lu, P. Messing, J.M., and Imaki, H., 1996, Depletion of feline taurine levels by ß-alanine and dietary taurine restriction, *Adv. Exp. Med. Biol.*, 403:19-36.
40. Taiwo, O.O. and Thomas, K.D., 1992, Plasma biochemical parameters in Nigerian children with protein energy malnutrition, *East. Afr. Med. J.*, 69:428-32.
41. Tanzer, F., Uzunsel, S., and Atalay, A., 1994, Plasma free carnitine levels in children and malnutrition, *Turk. J. Pediatr.*, 36:133-137.
42. Trowell H.C., Davies, J.N.P., and Dean, R.F.A., 1954, Kwashiorkor, Edward Arnold, London.
43. Tyson, J.E., Lasky, R., Flood, D., Mize, C., Picone T., and Paule C.L., 1989, Randomized trial of taurine supplementation for infants <1,300-gram birth weight: effect on auditory brainstem-evoked responses, *Pedriatics*, 83:406-415.
44. van Gelder, N.M., Janjua, N.A., Metrakos, K., MacGibbon, B., and Metrakos, J.D., 1980, Plasma amino acid in 3/sec spike wave epilepsy, *Neurochem. Res.*, 5:659-671.
45. van Gelder, N.M. and Bélanger F., 1988, Embryonic exposure to high taurine: a possible nutritional contribution to Friedreich's ataxia, *J. Neurosci. Res.*, 20:383-389.
46. Welch, A.N. and Badawy, A.A.-B., 1980, Tryptophan pyrrolase in heme regulation. Experiments with administered hematin and the relationship between the heme saturation of tryptophan pyrrolase and the activity of 5-aminolaevulinate synthase in rat liver, *Biochem. J.*, 192:403-410.
47. Wellcome Trust Working Party, 1970, Classification of infantile malnutrition, *Lancet*, 2:302-303.
48. Widhalm, K., Wayne Stargel, W., Burns, T.S.,and Tschanz, C., 1994, Evaluation of clinical and biochemical parameters in children after comsuption of microparticulated protein fat substitute (Simplesse), *J. Am. Coll. Nutr.*, 13:392-396.
49. Williams C.D., 1933, A nutritional disease of chilhood associated with a maize diet, *Arch. Dis. Child.*, 8:423-433.
50. Wright, C.E., Tallan, H.H., and Lin, Y.Y., 1986, Taurine: biological update, *Ann. Rev. Biochem.*, 55:427-453.

# THE EFFECT OF DIETARY SULFUR-CONTAINING AMINO ACIDS ON CALCIUM EXCRETION

Xi Bing Wang and Xi He Zhao

Institute of Nutrition and Food Hygiene,
Chinese Academy of Preventive Medicine
Beijing 100050, P. R. China

## ABSTRACT

The relationships between dietary protein and sulfur amino acid (methionine and cystine or taurine) intakes and urinary calcium excretion were examined both in animals and in young men. Thirty-two adult Wistar rats were divided into 4 groups, i.e., basal diet (group I), supplemented with albumin (II), methionine and cystine (III), or taurine (IV). During the 5-week feeding period, food consumption was recorded and 48 h urine samples were collected 4 times for each rat. Urinary calcium, creatinine and sulfate were measured. The results showed that the calcium and sulfate excretion in rats in group II and III were significantly higher than rats in the basal diet group. In contrast, supplementing a basal diet with taurine did not increase sulfate excretion and failed to induce hypercalciuria.

The same result was also observed in the study carried out in Chinese young men. An increase in protein intake from 67 g to 107 g caused an increase in urinary calcium and sulfate. Supplementation with methionine and cystine in an amount to simulate those in the high protein diet had a similar effect. Adding taurine to the diet had no effect on urinary calcium and sulfate excretion. About 60 percent of the supplemented taurine in the diet was detected in the urine.

## INTRODUCTION

Previous studies showed that an increase in the intake of protein, methionine or cystine would result in increased excretion of calcium[3,8]. This effect has been attributed to a decrease in the renal reabsorption of calcium caused by increased sulfate production from the oxidation of excess sulfur amino acids. Anthony[1] reported that administration of taurine resulted in an increased excretion of inorganic sulfates in the urine. Whether the dietary addition of taurine will also cause an increase in urinary calcium loss has not yet been reported. This study investigated the effects of supplementing protein, methionine and cystine or taurine on the urinary excretion of calcium and sulfate in rats and humans.

## MATERIALS AND METHODS

### Animal Study

Thirty-two female Wistar rats, weighing 186-208 g, were randomly divided into four groups each containing 8 rats. Animals in the first group were fed a basal diet which contained protein 12% (provided as casein), calcium 0.6%, and phosphorus 0.3%. For the 2nd group, ovo-albumin was added to the basal diet which provided a protein content of 28%. For the 3rd group, the basal diet was supplemented with methionine (5.6 g/kg) and cystine (5.2 g/kg) which made the content of methionine and cystine equal to those in the high protein diet. For the 4th group, taurine (7.4 g/kg) was added to the basal diet which made the molar concentration of the sulfur amino acids equal to that in the diet of the 2nd and 3rd groups.

The rats were housed in metabolic cages and fed their respective diets and deionized water *ad libitum* for 5 weeks. Body weight was recorded weekly. Urine samples were collected 4 times: before giving the experimental diet and on the 2-5th, 17-20th, and 30-33th days during the experimental diet period. For each time period, 48 h urine samples were collected. Levels of creatinine, calcium, and sulfate were measured.

### Human Study

Seven healthy young men volunteered for a 4-stage experiment. These men were successively provided with the following 4 kinds of diets (each for 9 days): 1) basal diet, 2) basal diet supplemented with extra protein, 3) basal diet supplemented with taurine, and 4) basal diet supplemented with methionine and cystine. Distilled water was supplied as their drinking water.

The basal diet provided protein 67 g, calcium 550 mg (including the 180 mg calcium provided as calcium tablets) and phosphorus 1520 mg each day. The high protein diet provided an extra 40 g of protein as egg white which was given with lunch (20 g) and dinner (20 g). The basal diet supplemented with methionine and cystine provided methionine 1.9 g and cystine 1.6 g every day in order to simulate the sulfur amino acids in the high protein diet. The taurine diet supplemented taurine 2.4 g which made the molar concentration of sulfur amino acids equivalent to that in the diet supplemented with methionine and cystine. Sugar was used to adjust the caloric intake to a relatively constant level throughout the experiment.

In each of the 9-day experimental periods, the first 5 days were spent in accommodating to the new diet after which the calcium metabolic studies were carried out. Seventy-two hour feces and urine samples were collected. Levels of fecal calcium and urinary calcium, creatinine, sulfate, and taurine were determined. The contents of protein, calcium, and phosphorus in the diets were also determined during the periods of the metabolic studies.

## RESULTS

### Animal Study

No significant differences in food consumption and in increases in body weight were observed among rats of the four groups. The level of urinary creatinine gradually increased with an increase in body weight. This result was found to be most prominent in animals on the high protein diet. These animals also had sigificantly higher creatinine excretion levels than those in the other three groups (Table 1).

Rats in the high protein and methionine plus cystine supplemented group had significantly higher urinary calcium and sulfate levels than the rats in the basal diet group (Tables 2 and 3). There was no difference in urinary calcium and urinary sulfate between the taurine supplemented group and the basal diet group.

**Human Studies**

In human studies involving seven young men, an increase in protein intake from 67 g to 107 g while maintaining calcium and phosphorus intake constant caused an increase in urinary calcium and sulfate levels and a decrease in calcium balance (Tables 4 and 5). Supplementing with methionine and cystine in an amount equal to that in the high protein diet had a similar effect. Adding taurine to the diet had no effect on urinary calcium, urinary sulfate levels and on calcium balance. Calcium absorption was not affected by increasing protein intake or supplementation with the sulfur-containing amino acids.

**Table 1.** Urinary creatinine levels of rats in different groups (mg/24h)

| Metabolic Period | Basal Diet | High Protein | Met + Cys | Taurine |
|---|---|---|---|---|
| I | $5.17 \pm 0.33^a$ | $5.92 \pm 0.19^a$ | $5.49 \pm 0.31^a$ | $5.32 \pm 0.20^a$ |
| II | $5.82 \pm 0.23^a$ | $6.91 \pm 0.18^b$ | $5.66 \pm 0.21^a$ | $5.43 \pm 0.28^a$ |
| III | $6.91 \pm 0.20^a$ | $8.45 \pm 0.18^b$ | $7.15 \pm 0.45^a$ | $6.87 \pm 0.36^a$ |
| IV | $7.46 \pm 0.29^a$ | $8.45 \pm 0.27^a$ | $7.65 \pm 0.37^a$ | $7.46 \pm 0.36^a$ |

Values in a row not sharing a common superscript are significantly different ($p < 0.05$).

**Table 2.** Urinary calcium excretion levels of rats (mg/24h)

| Metabolic Period | Basal Diet | High Protein | Met + Cys | Taurine |
|---|---|---|---|---|
| I | $0.75 \pm 0.23^a$ | $0.88 \pm 0.17^a$ | $0.83 \pm 0.16^a$ | $1.12 \pm 0.25^a$ |
| II | $0.84 \pm 0.19^a$ | $3.00 \pm 0.30^c$ | $2.05 \pm 0.31^b$ | $1.31 \pm 0.23^a$ |
| III | $0.83 \pm 0.04^a$ | $2.78 \pm 0.37^c$ | $1.44 \pm 0.23^b$ | $0.82 \pm 0.09^a$ |
| IV | $0.91 \pm 0.08^a$ | $2.81 \pm 0.38^c$ | $1.74 \pm 0.28^b$ | $1.25 \pm 0.23^{ab}$ |

Values in a row not sharing a common superscript are significantly different ($p < 0.05$).

**Table 3.** Urinary sulfate levels of rats in different groups (mmol/24 h)

| Metabolic Period | Basal Diet | High Protein | Met + Cys | Taurine |
|---|---|---|---|---|
| I | $0.22 \pm 0.02^a$ | $0.24 \pm 0.04^a$ | $0.21 \pm 0.03^a$ | $0.21 \pm 0.04^a$ |
| II | $0.23 \pm 0.02^a$ | $1.95 \pm 0.13^c$ | $1.51 \pm 0.09^b$ | $0.27 \pm 0.05^a$ |
| III | $0.31 \pm 0.04^a$ | $2.18 \pm 0.11^c$ | $1.78 \pm 0.11^b$ | $0.37 \pm 0.03^a$ |
| IV | $0.30 \pm 0.02^a$ | $1.75 \pm 0.05^c$ | $1.43 \pm 0.08^b$ | $0.37 \pm 0.02^a$ |

Values in a row not sharing a common superscript are significantly different ($p < 0.05$).

**Table 4.** Calcium metabolism in humans during different dietary periods (Mean ± SD)

| Diet | Ca intake mg/d | Urinary Ca mg/d | Fecal Ca mg/d | Ca balance mg/d | Ca absorption mg/d | % |
|---|---|---|---|---|---|---|
| BD | 543 ± 1.5 | 183 ± 17.4$^{a}$ | 313 ± 12.2$^{a}$ | 49.0 ± 12.7$^{a}$ | 230 ± 13.1$^{a}$ | 42.3 ± 2.4$^{a}$ |
| BD+Pr | 556 ± 1.9 | 219 ± 14.5$^{b}$ | 325 ± 23.9$^{a}$ | 11.7 ± 8.9$^{a}$ | 231 ± 23.1$^{a}$ | 41.7 ± 4.2$^{a}$ |
| BD+T | 542 ± 2.5 | 185 ± 21.9$^{a}$ | 307 ± 17.8$^{a}$ | 50.0 ± 22.3$^{a}$ | 235 ± 1.63$^{a}$ | 43.4 ± 3.1$^{a}$ |
| BD+M+C | 542 ± 2.2 | 216 ± 21.4$^{ab}$ | 325 ± 21.6$^{a}$ | 1.1 ± 16.5$^{b}$ | 217 ± 2.20$^{a}$ | 40.1 ± 4.0$^{a}$ |

1. BD = basal diet, BD+Pr = high protein diet, BD+T = basal diet supplemented with taurine, BD+M+C = basal diet supplemented with methionine and cystine.
2. Values in a column not sharing a common superscript are significantly different (P<0.05) .

**Table 5.** Urinary creatinine and sulfate levels in humans during different dietary periods (Mean ± SD)

| Diet | Urinary creatinine mmmol/24 h | Urinary sulfate mmol/24 h | Urinary taurine mmol/day |
|---|---|---|---|
| BD | 12.8 ± 0.44$^{a}$ | 38.2 ± 2.25$^{a}$ | 0.88 ± 0.45$^{a}$ |
| BD+Pr | 13.1 ± 0.35$^{a}$ | 82.0 ± 2.20$^{b}$ | 1.07 ± 0.36$^{ac}$ |
| BD+T | 12.8 ± 0.62$^{a}$ | 41.7 ± 2.17$^{a}$ | 13.38 ± 1.59$^{b}$ |
| BD+M+C | 12.8 ± 0.35$^{a}$ | 88.1 ± 5.05$^{b}$ | 1.64 ± 0.70$^{b}$ |

Values in a column not sharing a common superscript are significantly different (P<0.05).

The urinary excretion of taurine in different dietary periods is shown in Table 5. There was no significant differences between the high protein diet and the basal diet periods. Urinary excretion of taurine was higher in the period with methionine plus cystine than in the periods with either the basal diet or a supplemented high protein diet. The highest taurine excretion occurred in the period with the taurine supplemented diet. The amount of taurine in the urine was approximately 65% of the taurine added to the diet.

## DISCUSSION

The results of this study indicate that a high protein diet results in increased urinary calcium loss in both rats and humans. In the human study, an average increase of urinary calcium of 36 mg/d and a shift of calcium balance in a negative direction were observed when the protein intake was increased from 67 g/d to 107 g/d. Linkswiler and his colleagues[4] also reported negative calcium balance and doubled urinary calcium excretion when protein intake was increased from 50 g/d to 140 g/d. However, using meat as the source of protein, Spencer[5,6] did not observe significant changes in urinary calcium excretion or in calcium balance when protein intake was raised from 1 g/kg.bw. to 2 g/kg.bw. It is believed that the discrepancy between the results of these experiments is caused principally by the difference in phosphorus content in the experimental diets[5]. Linkswiler and collegues[4] used purified protein with phosphorus intake kept constant at about 1000 mg/d while Spencer[5,6] used meat as the source of protein. Increased intake of meat will also result in an increase of phosphorus intake, and it is well-known that an increase of phosphorus has the effect of reducing urinary calcium excretion. In our experiments, the phosphorus intake was kept at a constant level. These results are consistent with those of Linkswiler and colleagues[4].

Both the animal and human experiments in our studies showed that supplementing a diet with methionine and cystine induces increased urinary excretion of calcium and sulfate. The urinary excretion of calcium correlated positively with the urinary excretion of sulfate. Urinary sulfate and even urinary calcium was not increased by supplementing the diet with taurine. These results indicate that the hypercalciuria effect caused by a high protein diet is dependent upon the sulfate generated from oxidation of sulfur amino acids contained in the protein. Whiting and Draper[7] also observed that the calciuretic effect of different dietary proteins in rats was in the same order as their sulfur amino acid content.

Anthony[3] reported that taurine has the effect of promoting excretion of sulfate in zinc deficient rats. However, in this study sulfate excretion and calcium excretion was not promoted by taurine. Taurine is now being used widely to fortify infant foods, dairy products and other foods in China. The observation in this human study that supplementing diets with taurine in the amount 2.4 g/d did not cause hypercalciuria indicates that the amount of taurine currently permitted to be used in China (0.3-0.5g/kg) for food fortification is unlikely to cause calcium loss.

We have made a survey of 700 women from 5 regions of China[2] on their dietary intake and urinary excretion of calcium and acids. The results indicate that the amount of urinary calcium is positively correlated with the pH of the urine. Intake of foods of animal origin other than dairy products positively correlated with urinary calcium excretion. In contrast, the intake of foods of plant origin is negatively correlated with urinary calcium excretion. Sulfur amino acids are a group of agents which contribute to the acidity of urine, together with other factors affecting the acid-base status of the diet. It is thus concluded, that sulfur amino acids have a profound effect on the urinary excretion of calcium.

## REFERENCES


1. Anthony, W.L., Woosley, R.L., and Hsu, J.M., 1971, Urinary excretion of radiosulfur following taurine-35S injection in zinc deficient rats, *Proc. Soc. Exp. Biol. Med.*, 138:989-992.
2. Hu, J., Zhao, X., and Parpia, B., 1993, Dietary intakes and urinary excretion of calcium and acids, *Am. J. Clin. Nutr.*, 58:398-406.
3. Kerstetter, J.E. and Allen, L.H., 1990, Dietary protein increases urinary calcium, J. Nutr., 120:134-136.
4. Shuette, S.A., Zemel, M.B., and Linkswiler, H.M., 1980, Studies on the mechanism of protein-induced hypercalcuiuria in older men and women, *J. Nutr.*, 110:305-315.
5. Spencer, H., 1983, Further studies of the effect of a high protein diet as meat on calcium metabolism, *Am. J. Clin. Nutr.*, 37:924-929.
6. Spencer, H., 1988, Do protein and phosphorus cause calcium loss? *J. Nutr.*, 118:657.
7. Whiting, S.J., and Draper, H.H., 1980. The role of sulfate in the calciuria of high protein diets in rats, *J. Nutr.*, 110:212-222.
8. Yuen, D.E., 1984, Effect of dietary protein on calcium metabolism in man, *Nutr. Abs. and Rev.*, 54:447-459.

# TAURINE CONTENT IN CHINESE FOOD AND DAILY TAURINE INTAKE OF CHINESE MEN

Xihe Zhao, Jianbin Jia, and Ying Lin

Institute of Nutrition and Food Hygiene
Chinese Academy of Preventive Medicine
Beijing 100050, P.R. China

## ABSTRACT

The taurine content in Chinese food, including seafood, fresh water fish, meats and some plants, was examined in this study. Seafood was freshly collected from 4 coastal areas in China. Meat and plant food samples were obtained from food markets. The highest concentration of taurine was found in crustaceans and molluses (300-800 mg per 100 g edible portion). The amount of taurine in fish was variable. Beef, pork and lamb contained taurine in concentrations ranging from 30-160 mg per 100 g. No taurine was detected in hen eggs and plants.

The daily taurine intake of representative Chinese men (18-45 years old, 60 kg body weight, light physical activity) was also studied in 1990 as a part of the Total Diet Study. Representative food samples were collected from 12 provinces in 4 areas of China. Samples were then combined and cooked according to food categories (meat, seafood, vegetables, etc). The combined meat and seafood samples were analyzed for taurine. The daily taurine intake of a standard man in the 4 areas was calculated based on the amount of food intake obtained from the dietary survey and the taurine concentration in the analyzed food samples. The result showed that the daily taurine intake of a standard Chinese man in the 4 different test areas ranged from 34 to 80 mg per day.

## INTRODUCTION

Although the physiological role and nutritional function of taurine have been intensely investigated[2,6], the source of taurine in various foods has been relatively neglected. There is little information concerning the taurine content of foods. Food composition tables do not include data on taurine. Therefore, it is not possible to calculate the taurine content in most meals from available published references.

In the present study, taurine content in aquatic products, meat, internal organs of stock and chicken and some plants was measured. The daily taurine intake of representative Chinese men was also studied as a part of a Chinese total diet study in 1990[1].

## METHODS

### Food Sampling and Preparation

Various seafood was freshly collected from 4 coastal areas in China including Dalian, Qingdao, Zhoushan, and Tanggu. Meat, egg, beans were purchased from food markets in Beijing.

All samples were analyzed uncooked. Edible parts of the seafood and lean meat were homogenized in a Waring blender. Beans and nuts were ground into fine powder by a mortar before taurine analysis.

### Storage of Samples

Samples, after processing, were either analyzed for taurine immediately or put into tightly sealed containers and stored in a -30°C freezer. Taurine content after storage for 6 or 9 months (Table 3) was compared with results obtained right after the sample collections.

### Determination of Taurine Content

1) Processed samples were further homogenized with 0.1 mol/L hydrochloric acid. 2) The homogenate was placed in a volumetric flask and brought to volume with deionized water. 3) An aliquot of the supernatant was transferred to a test tube and sulfosalicylic acid solution was added to remove the precipitated protein. 4) Taurine content was analyzed with a Beckman AA Analyzer model 6300 using a lithium citrate elution system.

### Determination of Taurine Consumption

Representative food samples were collected from 4 areas of China, with each region comprised of the various 3 provinces. Samples were combined and cooked according to food categories (meat, seafood, vegetables, etc.). The combined meat and seafood samples were analyzed for taurine. The daily taurine intake of the representative man (18-45 years old, 60 kg body weight, light physical activity) in the 4 areas was calculated based on the amount of food intake obtained from the dietary survey and the taurine concentration in the food samples[1].

## RESULTS AND DISCUSSION

### Taurine Content of Aquatic Products

Table 1 lists the taurine content of a wide variety of aquatic organisms that are used as a food source in China.

**Table 1.** Taurine content of aquatic products (mg/100g edible portion)

| Food | Moisture % | Taurine |
|---|---|---|
| Spiral shell (Rapana venosa) | 79.0 | 520 |
| Conch (Neverita didyma) | 71.6 | 851 |
| Scallop (Argopecten irradians lamark) | 84.4 | 332 |
| Blood clam (Anadara subvrenata) | 85.7 | (439,795)* |
| Clam (Venerupis variegata, Amygdala) | 87.9 | 496 |
| Mussel (Mytilus edulis) | 82.5 | 349 |
| Prawn (Penaeus orientalis, natural) | 77.3 | 143 |
| Prawn (Penaeus orientalis) | 80.4 | 110 |
| Zhou's prawn (Palaemon carinicaude, Exopalaemon) | 78.4 | 181 |
| Swimming crab (Portunus trituberculatus) | 80.5 | 279 |
| Cuttlefish (Sepia esculenta) | 81.2 | 673 |
| Octopus (Octopus vulgaris) | 87.0 | 380 |
| Ray (Raja porosa gunther) | 79.7 | 280 |
| Conger pike (Muraenesox yamaguchiensis) | 74.6 | 193 |
| Flatfish (Pleuronichthys cornutus) | 81.1 | 314 |
| Tonguefish (Cynoglossus) | 79.4 | 256 |
| White Chinese croaker (Argyrosomus argentatus) | 77.6 | 187 |
| Yellow croaker (Pseudosiaena polyactis Bleeker) | 85.6 | 90 |
| Spotted maigre (Nibea albiflora) | 79.0 | 225 |
| Baby croaker (Collichthys niveatusJordan et Starks) | 79.9 | 64 |
| Silver pomfret (Pampus argenteus) | 72.5 | 41 |
| Hairtail (Trichiurus haumela) | 72.3 | (38,75)* |
| Yellow crucian carp (Setipinna taty, Valenciennes) | 74.9 | 240 |
| Black scraper (Navodon septentrionalis) | 79.0 | (196,309)* |
| Kelp (Laminaria japonica Aresch) | 87.3 | ND |
| Grass crap (Ctenopharyngodon idellus) | 81.1 | 185 |
| Silver carp (Hypophthalmichthys molitrix) | 78.4 | 90 |
| Crucian carp (Carássius auratus Linnaeus) | 80.2 | 205 |
| Variegated carp (Aristichthys nobilis, Richardson) | 83.3 | 94 |

The mean of 2 or 3 samples; * The difference between 2 samples is large; ND not detected

Conches, blood clams, clams and cuttlefish had taurine levels as high as 500-900 mg in 100 g of the fresh edible portion. Fish had varied taurine content, which were relatively high in rays and flatfish and relatively low in silver pomfret, yellow croaker and baby croaker. Natural and cultivated prawns had similar taurine content (100 mg in 100 g edible parts).

Pasantes-Morales reported[4] that the taurine content in every 100 g of clams and octopus flesh was 520 mg and 390 mg respectively, which was similar to the results of this study. Laidlaw and his colleagues[3] reported that the taurine content (per 100 g edible parts) in mussel and scallops was 655 mg and 827 mg, respectively which was greater than the results of this study.

In this study we carried out 2 to 3 samplings in the same region or in different regions. A few samples from different sampling had significantly different taurine content, indicating that the taurine in aquatic foods was also affected by the ecological environment and the stage of development of the animal, in addition to the species of the animal.

**Taurine Contents in Skeletal and Visceral Flesh of Livestock and Poultry**

The taurine content in meat (Table 2) appeared to be related to the site of sampling. It is interesting to note that the taurine content in the chicken leg was more than 10 times that in the chicken breast.

Laidlaw and his associates[3] have also reported that the taurine content in red meat of chickens and turkeys is 10-fold more than that in white meat of the same species.

In this study, we have not been able to detect taurine in kelp, eggs, black rice, red beans and peanuts. However, Pasantes-Morales showed that some species of beans and nuts contain small amounts of taurine (<0.5 mg/100 g), which is below the detection limit of the methodology used in this study.

Table 3 shows that the taurine content of various foods was basically unchanged after 6 or 9 months storage at -30°C.

Table 4 shows the average daily taurine intake by representative Chinese men in 4 different areas of China. Since taurine was not detected in grains, eggs and other foods of plant origin and since the taurine content of milk is also very low, the daily taurine intake of these men was calculated as taurine consumed from seafood plus taurine from meat.

**Table 2.** Taurine content of meat and internal organs of stock and chicken (mg/100 g edible portion)

| Food | Moisture % | Taurine |
|---|---|---|
| Foreham, pork | 71.5 | 116 |
| Tenderloin, pork | 76.0 | 122 |
| Liver, pig | 72.0 | 42 |
| Heart, pig | 79.0 | 201 |
| Kidney, pig | 87.0 | 120 |
| Tenderloin, beef | 75.0 | 28 |
| Hindham, beef | 76.4 | 64 |
| Foreleg, mutton | 78.4 | 166,259 |
| Breast, chicken | 71.7 | 26 |
| Leg, chicken | 74.3 | 379 |

**Table 3.** Taurine content before and after storage (mg/100 g edible portion)

| Samples | Duration (month) | Before | After |
|---|---|---|---|
| Spiral shell (Rapana venosa) | 6 | 580 | 531 |
| Conch (Neverita didyma) | 6 | 785 | 737 |
| Prawn (Penaeus orientalis) | 6 | 112 | 109 |
| Black scraper (Navodon septentrionalis) | 6 | 309 | 326 |
| Right eyed flounder (Cleisthenes herzensteini) | 6 | 244 | 246 |
| Shellfish (Argopecten irradians lamark) | 9 | 276 | 250 |
| Swimming crab (Portunus trituberculatus) | 9 | 321 | 313 |
| Silver pomfret (Pampus argenteus) | 9 | 41 | 44 |
| Yellow crucian carp (Setipinna taty, valenciennes) | 9 | 274 | 288 |
| Prawn (Penaeus orientalis, natural | 9 | 130 | 117 |

**Table 4.** Daily taurine intake of representative Chinese men

| Area | Taurine from seafood | Taurine from meat | Daily taurine intake |
|---|---|---|---|
| Northern 1 | 39.4 | 18.0 | 57.4 |
| Northern 2 | 11.6 | 21.9 | 33.5 |
| Southern 1 | 52.6 | 27.1 | 79.7 |
| Southern 2 | 13.6 | 26.9 | 40.5 |

The men in Northern area #1 had the lowest taurine intake (33.5 mg/d) while taurine intake of men in Southern area #2 reached 79.7 mg/d. Rana and colleagues[5] reported that the taurine intake by British male omnivores averaged 76 mg/d, which is close to the amount of taurine consumed by the representative Chinese men in Southern area #2.

From the results of this study, it can be reported that the taurine content in some fresh seafood is as high as 0.5-1%. If this seafood is produced as a dry powder, then only 1 g of such powder is necessary to fortify 100 g of infant food. The amount of taurine permitted to be added to food in China is 0.3-0.5 g per kilogram.

## REFERENCES


1. Chen, J. and Gao, J, 1993, The Chinese totall diet study in 1990. Part II. Nutrients, *J. AOAC International*, 76:1206-1208.
2. Huxtable, R.J., 1992, Physiological action, *Physiological Review*, 72:1, 101-163.
3. Laidlaw, S.A, Grosenor, M., and Kopple, J.D., 1990, The taurine content of common foodstuff, *J. Parent. Ent. Nutr.*, 14:183-188.
4. Pasantes-Morales, H., Quesada, O, Alcocer, L., and Olea, R.S. 1989, Taurine content in foods, *Nutr. Rep. Int.*, 40:793-801.
5. Rana, S.K and Sanders T.A., 1986, Taurine concentration in the diet, plasma, urine and breast milk of vegans compared with omnivores, *British J. Nutr.*, 56:17-27.
6. Rudman, D. and Feller, A. 1986, Evidence for deficiencies of conditionally essential nutrients during total parenteral nutrition, *J. Am. Coll. Nutr.*, 5:101-106.

# EFFECTS OF DIETARY TAURINE ON AUDITORY FUNCTION IN FULL TERM INFANTS

Suki K. Dhillon,[1] W. Ewart Davies,[2] Peter C. Hopkins,[2] and Stephen J. Rose[3]

[1] Hearing Services Centre
City Hospital Birmingham
[2] Department of Pharmacology
Medical School
University of Birmingham
[3] Department of Paediatrics
Heartlands Hospital
Birmingham, England

## ABSTRACT

Three groups of neonates fed taurine supplemented infant formula, non-supplemented infant formula or breast milk, respectively, were studied from birth to 12 weeks of age. In addition to the measurement of whole blood taurine content, auditory function was monitored using auditory brainstem responses (ABRs) and transient evoked otoacoustic emissions (TEOAEs). The results showed a significant reduction in whole blood taurine concentration in the non-supplemented formula group. In addition, there was a significant drop in whole blood taurine levels in all 3 groups over the first four weeks of life. ABR wave latencies were significantly shorter in the non-supplemented group, with wave V showing the greatest reductions. Falling taurine levels after full term birth may aid synaptic maturation/efficiency within the auditory system. TEOAE responses were significantly larger over the low to mid frequencies in the breast fed group suggesting improved middle ear function.

## INTRODUCTION

The taurine content of human milk is vastly higher than that of mature bovine milk[12]. Various effects of taurine-free formula feeding have been documented[2,7]. The main finding in both term and pre-term infants was a reduction in urine and plasma taurine concentrations. Low plasma taurine values have been associated with electrophysiologically detected retinal abnormalities in children receiving long-term parenteral nutrition[3], although no such abnormalities were found in pre-term infants given taurine-free enteral feeding[15]. Retinal degeneration has also been documented in cats and monkeys fed taurine deficient diets[4,6].

Taurine plays an important role in pregnancy and development of the cat[13]. Kittens fed by taurine deficient mothers had delayed auditory brainstem responses (ABRs), although this change was only significant in the first two weeks of life[17]. In contrast, mature cats fed a taurine-free diet exhibited shortened ABR latencies when compared with controls[18]. Pre-term infants fed non-supplemented infant formula had delayed ABR latencies accompanied by lowered plasma concentrations[15]. Taurine levels in fetal and neonatal mammalian brain are high, with the amount per unit weight decreasing during maturation suggesting a role in early development[14].

In addition, taurine has been shown to regulate $Ca^{2+}$ fluxes in nerve terminals[10] and to reduce transmitter release[8]. Although it is unlikely to be a neurotransmitter, it may play a role in membrane stabilisation or neuromodulation. High concentrations of taurine have also been detected in the mammalian cochlea, with the majority of it being located in the supporting cells of the organ of Corti[16] suggesting a possible osmoregulatory role.

In this study the effect of dietary taurine manipulation on hearing development was investigated. Three groups of 20 neonates were studied from birth to 12 weeks of age. One group received a normal casein/whey infant formula with taurine supplementation (4.5mg/100ml), another received the identical formula with no added taurine (0.4mg/100ml), and a third group was entirely breast fed ($3.51 \pm 0.26$mg/100ml). Auditory system maturation was monitored using ABRs and transient evoked otoacoustic emissions (TEOAEs) were used to monitor cochlear function. Whole blood taurine content was also measured.

## MATERIALS AND METHODS

Only normal full term infants, who met the selection criteria as listed in Table 1, were included in the study. Infants were allocated to the breast fed or bottle fed group according to maternal choice. The bottle fed infants were then blindly and randomly assigned to either the taurine-supplemented or non-supplemented infant formula. The gestational age, birth weight, birth length, head circumference and Apgar scores were documented for each infant.

In order to reduce subject generated noise and record clearly defined responses, the auditory measurements were undertaken while the child was asleep. Where this was not possible the assessments were made with the child awake but still.

**Table 1.** Selection criteria

| |
|---|
| No family history of congenital deafness |
| Born 38-42 weeks gestational age |
| No observed abnormalities |
| Birth weight $\geq$ 2.5kg |
| Apgar 1 min $\geq$ 7 |
| Apgar 5 min $\geq$ 8 |
| Bilirubin < 10mg/dL |
| Mother - Rubella immune |
| Negative VDRL |

### ABR Measurement

After preparing the skin, Ag/AgCl electrodes were attached to the high forehead and each mastoid, with a low forehead earth. All electrode impedances were below 10kΩ. Simultaneous two channel recording of ipsilateral and contralateral ABR waveforms was undertaken using 100μs click stimulation presented via TDH 39 earphones. The clicks were presented to each ear in turn at a repetition rate of 19.1/s, and at intensities of 75, 55, 35 and 25dBnHL. At least one replication was recorded at each intensity. A minimum of 1200 averaged stimuli were recorded for each response, with a filter band width of 100Hz to 3000Hz. Consistently replicated responses were analysed off-line. Waves I, III and V were identified and the interpeak intervals, I-III, III-V and I-V calculated.

### TEOAE Measurement

Responses were recorded using the commercially available Otodynamics ILO88 system. The neonatal probe was sealed into the ear canal and used to present click stimuli at suprathreshold levels. Stimulus intensities recorded within the individual ear canal ranged from 75 to 85 dB peak SPL. A minimum of 50 and a maximum of 260 averaged stimuli were used to record the response. The noise rejection level was set at between 48 and 55dBSPL. Probe position was monitored throughout the test and each response was replicated. Consistent responses were analysed off-line to determine the intensity of the TEOAEs in 1/3 octave frequency bands.

**Table 2.** Birth parameters for each dietary group

| Birth parameter | Dietary group | n | mean |
|---|---|---|---|
| Gestational age (weeks) | High taurine | 20 | 40.2 ± 0.29 |
| | Low taurine | 20 | 40.0 ± 0.31 |
| | Breast | 20 | 40.0 ± 0.21 |
| 1 min. Apgar | High taurine | 20 | 8.25 ± 0.18 |
| | Low taurine | 20 | 8.20 ± 0.20 |
| | Breast | 20 | 8.30 ± 0.16 |
| 5 min. Apgar | High taurine | 20 | 9.30 ± 0.13 |
| | Low taurine | 20 | 9.20 ± 0.14 |
| | Breast | 20 | 9.20 ± 0.14 |
| Birth weight (kg) | High taurine | 20 | 3.49 ± 0.11 |
| | Low taurine | 20 | 3.34 ± 0.11 |
| | Breast | 20 | 3.54 ± 0.11 |
| Birth length (cm) | High taurine | 20 | 53.00 ± 0.55 |
| | Low taurine | 20 | 52.85 ± 0.67 |
| | Breast | 20 | 54.40 ± 0.85 |
| Head circumference (cm) | High taurine | 20 | 34.75 ± 0.29 |
| | Low taurine | 20 | 34.55 ± 0.32 |
| | Breast | 20 | 35.05 ± 0.35 |

**Whole Blood Taurine Measurement**

Regular blood samples were taken using the heel-prick technique. About 100μl of blood was collected in a heparinised container and refrigerated prior to analysis. The taurine content was then measured using high pressure liquid chromatography[5].

Each of the assessments was undertaken at birth and then repeated at 2,4,8 and 12 weeks of age. The initial test session was performed 1-4 days after birth to allow the ears to clear of fluid and debris.

## RESULTS

The birth parameters for each of the dietary groups were not significantly different for any of the three groups, as shown in Table 2.

The change in whole blood taurine content for each of the dietary groups with age is shown in Figure 1. Two way analysis of variance by age and feed, showed that the whole blood taurine content was significantly lower in the non-supplemented low taurine group, than in the other two groups ($P=0.013$). In addition, all three groups showed a significant reduction in the whole blood taurine content from birth to 4 weeks of age ($P<0.0001$).

An increase in ABR wave latency with reducing stimulus intensity can be clearly seen in Figure 2. Maturational changes in the ABR are demonstrated in Figure 3, with all wave latencies reducing with increasing age. The effects of diet and age on the latency of wave I, III and V can be seen in Figure 4. The greatest fall in wave I latency occurred over the first two weeks of life, with little change after the 4 week test, whereas the latency of wave III and V continued to fall throughout the period of the study.

The variables of age, stimulus intensity and diet were analysed using a three way analysis of variance. In agreement with the literature, the effects of age and stimulus intensity on wave latency were highly significant. In addition, differences between the dietary groups were also highly significant. Wave V latency was significantly shorter in

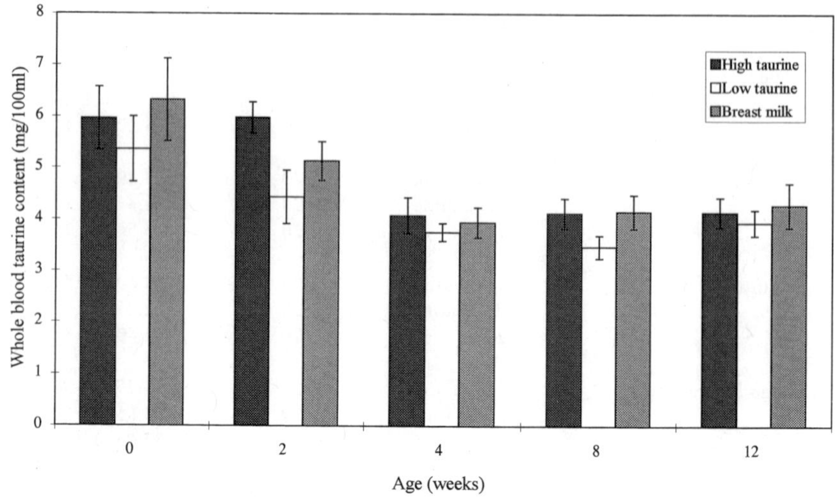


**Figure 1.** Whole blood taurine content for each dietary group from birth to 12 weeks of age (mean ± s.e.m.).

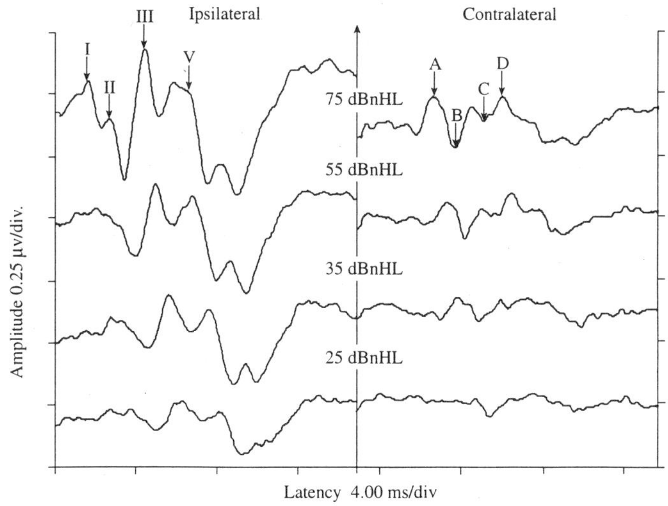


**Figure 2.** Typical ABR showing the reduction in wave latency with increasing stimulus intensity for both the ipsilateral and contralateral response.

the low taurine group when compared with that of the high taurine group (P<0.0001, mean difference = 0.21ms) and breast milk (P<0.0001, mean difference = 0.17ms). Similarly, wave III latency was significantly shorter in the low taurine group when compared with that of the high taurine group (P<0.0001, mean difference = 0.13ms) and breast milk (P<0.0001, mean difference = 0.11ms). A significant shortening of Wave I latency was also observed in the low taurine group, although the mean differences were not as great. Statistical analysis of the inter-peak intervals showed no significant changes in the I-III and I-V interval, although the III-V interpeak interval was again significantly shorter in the low taurine group.

On comparing TEOAE results across the 1/3 octave frequency bands, the responses over the low to mid frequencies were significantly larger in the breast fed group, 1260Hz (P=0.015), 1587Hz (P<0.0001) and 2000Hz (P=0.036). The results for the TEOAE responses recorded over the 1/3 octave band centered at 1260Hz are shown in Figure 5. The breast fed group can be seen to have consistently larger responses, particularly at the birth test.

## DISCUSSION

As a result of the dietary restrictions imposed during the study, the whole blood taurine content of the low taurine group was significantly reduced when compared to the other two groups. This indicates that whole blood taurine concentrations are considerable dependent on dietary taurine intake. Differences in the whole blood taurine content between the groups were observed even at the first test session undertaken within four days of birth. This suggests that the neonate loses taurine rapidly after birth and in the absence of high levels of dietary intake, the concentrations in blood are quickly depleted. Sharp falls in whole blood taurine content were seen across all three dietary groups in the first four weeks of life. The concentration of taurine in neonatal whole blood was higher than that found in adult whole blood even in the low taurine group, suggesting that taurine is conserved by the developing fetus. The fall in taurine levels soon after birth appears to represent a washing out of taurine from the blood and probably other tissues. Such a rapid depletion in taurine levels would be unlikely to occur unless it conferred a positive advantage on the neonate.

TEOAEs are either significantly reduced or absent in the presence of middle ear disorder and the larger responses recorded over the low to mid frequencies in the breast fed group may confirm the reduced incidence of middle ear disorder in this group[1].

Shortened ABR latencies in the low taurine group were unexpected. This could indicate some degree of middle ear disorder in the high taurine and breast fed group. However, this is unlikely as the two types of formula milk were randomly allocated and also, in view of the TEOAE results, the breast fed group showed improved middle ear function. Additionally, any middle ear effects would have had the greatest impact on wave I latency, whereas it was wave V which showed the largest changes.

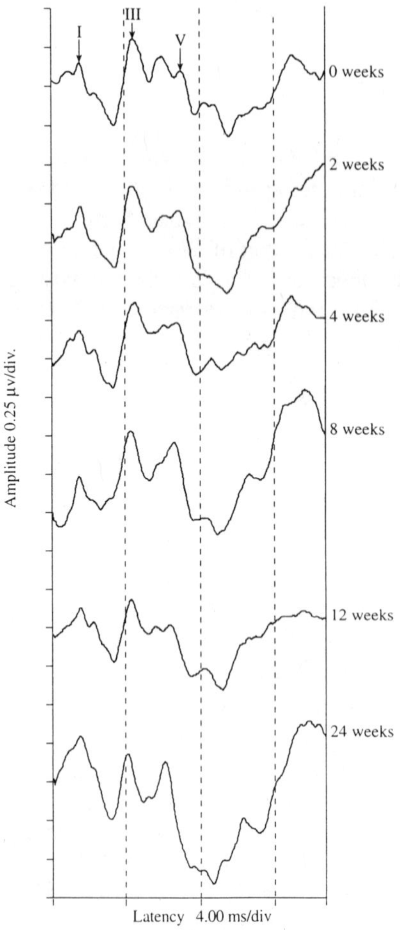


**Figure 3.** Typical ABR showing the maturational changes in wave latency with increasing age (stimulus intensity = 75dBnHL).

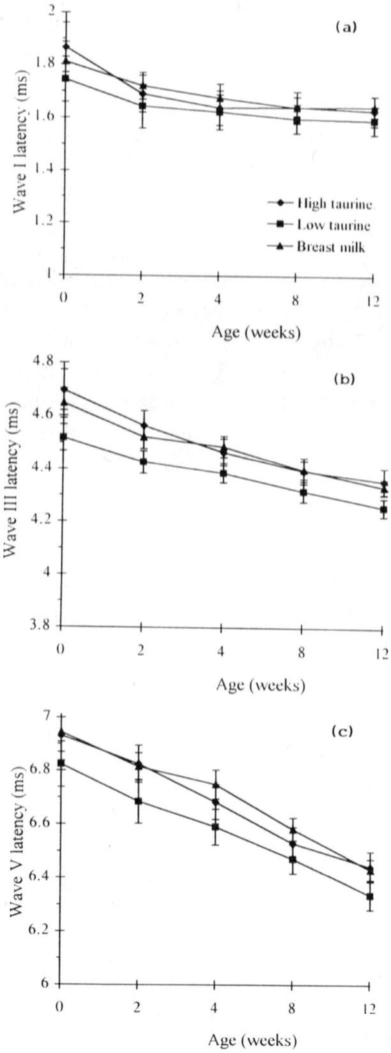


**Figure 4.** Changes in ABR latency with increasing age for each dietary group (mean ± s.e.m.), (a) wave I, (b) wave III, (c) wave V. Stimulus intensity = 75dBnHL

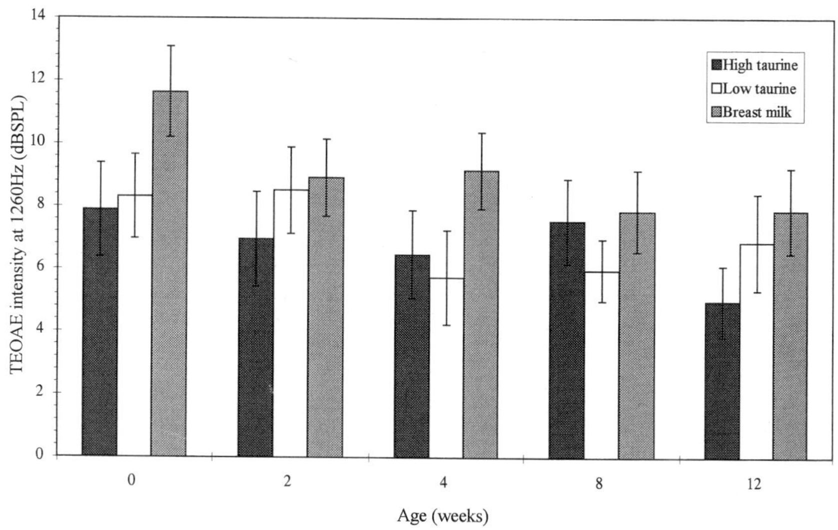


**Figure 5.** TEOAE intensity in 1/3 octave band centered at 1260Hz. Data displayed for each dietary group and age (mean ± s.e.m.).

An earlier study reported the opposite finding, with a low taurine diet producing a delay in ABR latency[15]. However, this earlier work was done on pre-term infants and it may indicate that the role of taurine in the developing fetus is different from that in the full term infant. It has been suggested that ABR maturational changes in the premature infant may be dominated by myelination effects, whereas those in the term infant may relate to synaptic maturation which follows a longer time course[11]. ABR wave delay in kittens feeding on taurine deprived mothers was also only observed during auditory immaturity[17], whereas in mature cats taurine-free feeding produced a shortening of ABR wave latencies[18].

The latencies of the ABR component waves are dependent on neuronal firing rate, synchrony of firing and synaptic efficiency. Taurine has been shown to depress neuronal firing[9] and to reduce neurotransmitter release at nerve endings[8]. High taurine concentrations in the fetus may be related to myelination/axonal development, whereas at term falling taurine levels could aid synaptic maturation. Wave V is generated higher up the auditory pathway and therefore it's latency is dependent on the efficiency of a greater number of synapses than the earlier more peripheral waves. Any improvement in synaptic/axonal efficiency would therefore cause greater reductions in wave V latency, as observed in this study.

## ACKNOWLEDGMENTS

We would like to thank Cow and Gate Limited for providing the infant formula and Action Research for their financial support. In addition, we are very grateful to Sheila Cross for typing the manuscript.

## REFERENCES

1. Duncan, B., Ey, J., Holberg, C. J., Wright, A. L., Martinez, F. D., and Taussig, L. M., 1993, Exclusive breast feeding for at least 4 months protects against otitis media, *Pediatrics*, 91:867-872.

2. Gaull, G. E., Rassin, D. K., Raiha, N. C. R., and Heinonen, K., 1977, Milk protein quantity and quality in low birth weight infants III, Effects on sulfur amino acids in plasma and urine, J. Pediatr. 90:348-355.
3. Geggel, H. S., Heckenlively, J. R., Martin, D. A., Ament, M. E., and Kopple, J. D., 1982, Human retinal dysfunction and taurine deficiency, *Doc. Ophthal. Proc. Ser.*, 31:199-207.
4. Hayes, K. C., Carey, R. E., and Schmidt, S. Y., 1975, Retinal degeneration associated with taurine deficiency in the cat, *Science*, 188:949-951.
5. Hopkins, P. C., Kay, I. S., and Davies, W. E., 1989, A rapid method for the determination of taurine in biological tissue, *Neurochem. Int*, 15:429-432.
6. Imaki, H., Moretz, R., Wisniewski, H., Neuringer, M., and Sturman, J., 1987, Retinal degeneration in 3 month old Rhesus monkey infants fed a taurine free human infant formula, *J. Neurosci. Res.*, 18:602-614.
7. Jarvenpaa, A-L., Rassin, D. K., Raiha, N. C. R., and Gaull, G. E., 1982, Milk protein quantity and quality in the term infant II, Effects on acidic and neutral amino acids, *Pediatrics*, 70:221-230.
8. Kamisaki, Y., Wada, K., Nakamoto, K., and Itoh, T., 1996, Effects of taurine on GABA release from synaptosomes of rat olfactory bulb, *Amino Acids*, 10:49-57.
9. Kontro, P. and Oja, S. S., 1983, Taurine and synaptic transmission, *Med.Bio*l., 61:79-82.
10. Pasantes - Morales, H. Arzate, N. E., and Cruz, C., 1982, The role of taurine in nervous tissue, its effects on ionic fluxes, in: "Taurine in Nutrition and Neurology" Huxtable, R. J. and Pasantes - Morales, H., eds., Plenum Press, New York, pp. 273-292.
11. Ponton, C. W. and Eggermont, J. J., 1996, Differential development of human ABR III-IV and IV-V intervals: implications for myelination and synaptic maturation, *Audiology*, 3:109.
12. Rassin, D. K., Sturman, J. A., and Gaull, G. E., 1978, Taurine and other free amino acids in milk of man and other mammals, *Early Hum. Dev*., 2:1-13.
13. Sturman, J. A. and Messing, J. M., 1991, Dietary taurine content and feline reproduction and outcome, *J. Nutr.*, 121:1195-1203.
14. Sturman, J. A., Rassin, D. K., and Gaull, G. E., 1978, Taurine in the development of the central nervous system, in: "Taurine and Neurological Disorders", Barbeau, A. and Huxtable, R. J., eds., Raven Press, New York, pp. 49-71.
15. Tyson, J. E., Laskey, R., Flood, D. Mize, C., Picone, T., and Paule, C. L., 1989, Randomized trial of taurine supplementation for infants ≤ 1,300 gram birth weight: effect on auditory brainstem evoked responses, *Pediatrics*, 83:406-415.
16. Usami, S. and Ottersen, O.P, 1995, The localization of taurine-like immunoreactivity in the organ of Corti: a semiquantitative, post-embedding, immunoelectron microscopic analysis in the rat with some observations in the guinea pig, *Brain Res.*, 676:277-284.
17. Vallecalle-Sandoval, M. H., Heaney, G., Sersen, E., and Sturman, J. A., 1991, Comparison of the developmental changes of the brainstem auditory evoked response (BAER) in taurine supplemented and taurine deficient kittens, *Int. J. Dev. Neurol,*. 9:571-579.
18. Vallecalle-Sandoval, M. H., Heaney, G., Sersen, E., and Sturman, J. A., 1993, Comparison of the brainstem auditory evoked response (BAER) in taurine deficient and taurine supplemented cats, *Nutr. Res.*, 13:475-487.

# INDEX